Principles and Applications of Therapeutic Ultrasound in Healthcare

Principles and Applications of Therapeutic Ultrasound in Healthcare

Yufeng Zhou

CRC Press
Taylor & Francis Group
Boca Raton London New York

CRC Press is an imprint of the
Taylor & Francis Group, an **informa** business

CRC Press
Taylor & Francis Group
6000 Broken Sound Parkway NW, Suite 300
Boca Raton, FL 33487-2742

First issued in paperback 2020

© 2016 by Taylor & Francis Group, LLC
CRC Press is an imprint of Taylor & Francis Group, an Informa business

No claim to original U.S. Government works

ISBN-13: 978-1-4665-1027-2 (hbk)
ISBN-13: 978-0-367-65866-3 (pbk)

Library of Congress Cataloging-in-Publication Data

Zhou, Yufeng, 1974- , author.
   Principles and applications of therapeutic ultrasound in healthcare / Yufeng Zhou.
   p. ; cm.
   Includes bibliographical references and index.
   ISBN 978-1-4665-1027-2 (hardcover : alk. paper)
   I. Title.
   [DNLM: 1. Ultrasonic Therapy--methods. 2. High-Energy Shock Waves--therapeutic use. WB 515]

RM862.7
615.8'323--dc23   2015018525

Visit the Taylor & Francis Web site at
http://www.taylorandfrancis.com

and the CRC Press Web site at
http://www.crcpress.com

Contents

Preface

A sound wave is a form of mechanical energy that propagates in a medium. Particles of a material oscillate about a fixed point rather than move with the wave itself during sonication. Acoustics, the science of sound, including its production, transmission, and effects (i.e., biological and psychological effects), is an interdisciplinary and ancient science, dating back to the sixth century BC work on the musical sounds by Pythagoras. Major advances in acoustics occurred in the eighth century because of the application of calculus to elaborate the theories of sound wave propagation. Lord Rayleigh in England combined previous knowledge with his own copious contributions to the field in his monumental work *The Theory of Sound* in 1877. The frequency range of normal human hearing is from 16 Hz to 15–20 kHz (in children and young adults). The mechanical vibration beyond this limit is known as ultrasound. The sound produced by way of mechanical vibration is usually audible. Not until the discovery of the piezoelectric effect by Jacques and Pierre Curie in 1880 did the investigation and application of ultrasound become feasible and popular. The first technological application of ultrasound was an attempt by Paul Langevin to detect submarines during World War I in 1917. Ultrasound imaging for human beings (sonography) was developed after World War II using a principle and a structure similar to sonar systems, but with higher frequency for better resolution. Now, sonography has become a popular and important diagnostic modality in modern health care and medicine.

The first large-scale application of ultrasound was around World War II, using sonar systems to navigate submarines. It was soon realized that the high-intensity ultrasound waves were heating and killing fish, which led to research in ultrasound-induced bioeffects. Thus, ultrasound became a medical modality that has been used by physical and occupational therapists for therapeutic effects since the 1940s. Although it was practiced earlier than diagnostic ultrasound, therapeutic ultrasound has attracted more attention and acceptance only since the 1990s. Ultrasound therapy is a simple and noninvasive procedure administered in a clinical setting by a trained professional or even at home by the patient themselves with appropriate instruction and training. It can be performed one or more times per day depending on the purpose of therapy. Therapeutic ultrasound has a little lower frequency range (of about 40 kHz–5.0 MHz) than diagnostic, but higher intensity (from about 0.5 W/cm^2 to >10,000 W/cm^2). Current successful applications of therapeutic ultrasound include physiotherapy for ligament sprains, muscle strains, tendonitis, joint inflammation, plantar fasciitis, metatarsalgia, facet irritation, impingement syndrome, bursitis, rheumatoid arthritis, osteoarthritis, and scar tissue adhesion, high-intensity focused ultrasound (HIFU) for tumor ablation, shock wave lithotripsy (SWL) for kidney stone fragmentation, targeted delivery of encapsulated drug, ultrasound hemostasis for bleeding, cancer therapy, effective and irreversible disruption of the blood–brain barrier (BBB), and ultrasound-assisted thrombolysis

(sonothrombolysis) for ischemic stroke patients using either focused ultrasound (FUS) or unfocused ultrasound. However, the effectiveness of therapeutic ultrasound for some diseases, such as pain relief, musculoskeletal injuries, and soft tissue lesions, still remains questionable in clinics.

Despite the practice of and research in ultrasound therapy, the bioeffects it induces are still not very clear, but can be classified as thermal and nonthermal effects. An increase in the molecular vibration in the tissue can result in heat generation, and ultrasound can be used to produce thermal changes in the tissues, such as enhanced metabolic activities of cells and stimulated circulation. In the thermal mode, ultrasound is most effective in heating the periosteum, collagenous tissues (i.e., ligament, tendon, and fascia) and fibrotic muscle and will require a relatively high intensity, preferably in continuous mode. The effect of hyperaemia at a temperature of 40°C–45°C is therapeutic and helpful in initiating the resolution of chronic inflammatory states. In addition, acoustic waves also generate nonthermal effects in nature, such as cavitation, microstreaming, and acoustic streaming, which may attribute a greater importance to the outcome. Cavitation relates to the formation of gas-filled voids within tissues and body fluids. The presence of acoustically induced cavities enhances the associated acoustic streaming phenomena, a small-scale eddying of fluids near a vibrating gas bubble. Transient cavitation bubbles collapse quickly, releasing a large amount of energy that is detrimental to tissue viability. Sodium ion permeability is altered, resulting in changes in the cell membrane potential and in the enzyme control mechanisms of various metabolic processes, especially concerning protein synthesis and cellular secretions. Enhancement in tissue relaxation, local blood flow, inflammatory response, and scar tissue breakdown are found after exposure to ultrasound. The effect of the increase in local blood flow can be used to reduce local swelling and chronic inflammation and promote bone fracture healing. The operating parameters of therapeutic ultrasound (i.e., frequency, emitting area, focal length, pulse duration, pulse repetition frequency, exposure duration, acoustic intensity) can be adjusted to achieve the desired outcome. In situations where a heating effect is not desirable, such as a fresh injury with acute inflammation, ultrasound can be delivered in a pulsed rather than a continuous wave form.

Although ultrasound therapy has been emerging as an effective modality in medical practice and a large number of studies have been carried out extensively in this field, there are few books discussing it systematically. Assembling papers from different journals may be time consuming and ineffective in the study. This book presents the physical and engineering principles that underlie the use of therapeutic ultrasound and illustrates their successful application in clinics. The text is intended to be helpful to senior undergraduate and graduate students in biomedical engineering, those majoring in bioacoustics, engineers of ultrasound therapy instruments, and clinical operators and physicians for its application. Chapters 2 through 5 present the fundamental knowledge of bioacoustics, starting from the derivation of wave equation, mathematical solution of the wave propagation, phenomena of reflection, refraction, and transmission in the acoustic field from different acoustic sources, the radiation pattern of ultrasound transducers, and the acoustical properties of biological tissues. Only linear acoustics are discussed in this part. Because of the high intensity in ultrasound therapy, nonlinear acoustics are considered and demonstrated in Chapter 6. Cavitation, an important phenomenon that is associated with ultrasound-induced bioeffects, the design of ultrasound transducers, and characterization method of the produced acoustic field are discussed in Chapters 7 through 9. Then, the important applications of ultrasound therapy are discussed in Chapters 10 through 14, including highlights of the working principle, research progress, system structure, summary of clinical outcomes, and critical commentary on the challenges in this field. The directions for future development and investigation in this novel technology are discussed in Chapter 15.

Hopefully, with the publication of this book, more people can be involved in the research and application of therapeutic ultrasound in order to develop this novel technology and increase its acceptance by both patients and physicians.

Author

Yufeng Zhou is an assistant professor at the School of Mechanical and Aerospace Engineering at Nanyang Technological University, Singapore. He earned his BS and MS degrees in electrical engineering from Nanjing University, China, in 1996 and 1999, respectively, and a PhD in mechanical engineering from Duke University, Durham, North Carolina, in 2003. Dr. Zhou's research interests are focused on biomedical ultrasound, including ultrasound diagnosis (i.e., elastography) and therapy (i.e., shock wave lithotripsy to break calculus, high-intensity focused ultrasound for cancer ablation, and ultrasound-mediated drug/gene delivery). He is a senior member of the Institute of Electrical and Electronics Engineers and an advisory editorial board member of the journal *Ultrasound in Medicine and Biology*.

1 Introduction

Ultrasound is oscillating sound pressure with frequency greater than the hearing threshold of humans (Skudrzyk 1971, Temkin 1981, Frank and Walker 1998). It has thus similar physical properties as audible sound except that humans cannot hear it (Graff 1981, Kinsler et al. 1982, Kuttruff 1991). The hearing threshold for a healthy young adult is approximately 20,000 Hz, although it varies from person to person (Figure 1.1). A well-known example of ultrasound is the hunting activities of bats (Popper and Fay 1995) that have been using a variety of ultrasonic ranging (echolocation) techniques for millions of years (Figure 1.2). Bats emit ultrasonic waves, compare the outgoing pulse with the returning echoes, and then form detailed images of their surroundings in their brain and auditory nervous system, which allows them to detect, localize, and even classify their prey in complete darkness. The emitted acoustic intensity by a bat is as high as 130 dB. The frequency of the echoes produced by bats is beyond 100 kHz, possibly up to 200 kHz (Popper and Fay 1995). Similar echo-location techniques are used by dolphins (Thomas et al. 2003), and sonar (**SO**und **N**avigation **A**nd **R**anging) system is used to detect underwater targets (Hill 1962, Edgetron 1986).

Sonar uses acoustic propagation to navigate and detect vessels and submarines. The first human use of underwater sound was by Leonardo Da Vinci in 1490, who used to listen to the sound in a tube that was inserted into the water to detect vessels (Edgetron 1986, Frank and Walker 1998). During World War I, the need to detect submarines prompted extensive research and application of underwater acoustics generated by quartz, and later by piezoelectric and magnetostrictive transducers (Donald et al. 1958, Hill 1962, Frederick 1965, Ristic 1983).

Medical ultrasound is usually classified into diagnosis and therapy, differing in the operation parameters, such as power, intensity, energy (dose), pulse repetition frequency, frequency, pulse duration, and the total ultrasound exposure time (Brown 1968, Morse and Ingard 1968). In sonography, a short acoustic burst emitted from a piezoelectric transducer is reflected from tissue structures within the body, received by the transducer, and is converted into an electrical signal for further processing. The properties (i.e., acoustic impedance) and location of the biological structure are determined by the amplitude of the reflected pulse and the time delay between the transmitted and received bursts, respectively. A two-dimensional gray-scale image, known as B-mode (brightness modulation), can be displayed by sweeping the ultrasonic beam over the region of interest (Wells 1969) (Figure 1.3). In addition, blood circulation can also be observed using the Doppler effect, where the frequency shift depends on the speed and direction of flow, as in cardiology and oncology. The most popular application of sonography is to diagnose fetuses during routine and emergency prenatal care because of its nonionizing

nature (Donald et al. 1958). According to a Cochrane review, routine obstetric ultrasound in early pregnancy (<24 weeks) appears to confirm fetal viability; determine the location of the fetus (intrauterine vs. ectopic); check the location of the placenta in relation to the cervix, the number of fetuses (multiple pregnancy), and major physical abnormalities; assess fetal growth, fetal movement, and heartbeat; and determine the sex of the baby. However, the U.S. Food and Drug Administration (FDA) discourages, but not prohibits, its use as records of fetal videos and photos (Figure 1.4). Furthermore, sonography is also being used to diagnose soft tissues, including cardiac, renal, liver and gallbladder (hepatic), musculoskeletal, ligaments and tendons, ophthalmic, superficial (i.e., testicle), thyroid, salivary glands, and lymph nodes for their size, structure, and pathological changes in real time by radiologists in clinics for at least 50 years and has become one of the most popular tools in modern healthcare (Erikson et al. 1974, Devey and Wells 1978, Edler 1991, Levi 2011). Because of its real-time nature, sonography is often used to guide interventional procedures, such as fine-needle aspiration (FNA) or biopsy of masses for cytology or histology testing in the breast, thyroid, liver, kidney, lymph nodes, muscles, and joints. Recently, increasing use of sonography is found in trauma and first aid cases in most emergency centers. The U.K. Department of Health has reported that during the period of 2005–2006, non-obstetric ultrasound examinations constituted more than 65% of the total number of sonography procedures. This technology is quite cheap and portable in comparison to other diagnostic modalities such as magnetic resonance imaging (MRI), computed tomography (CT), and single-photon emission computed tomography (SPECT). Remote ultrasound diagnosis in teleconsultation, such as scientific experiments in space or mobile sports team diagnosis, is possible now. However, because of the great attenuation and significant difference of acoustic impedance in comparison to that of soft tissue, both the bone and gas (e.g., the lungs) are not good candidates for sonography using the most common frequency range of 1–20 MHz.

If properly performed in clinics, sonography poses no significant risks to the patient, such as chromosome breakage and cancer development, because it does not use mutagenic ionizing radiation. Thus, it is generally described as safe. However, ultrasonic energy at higher intensity levels has two potential physiological effects: enhancing inflammatory response and heating soft tissue. Molecular friction caused by acoustic wave propagation is considered the major reason for the heating. Such a thermal effect becomes less significant *in vivo*, especially in the hypervascular tissues (i.e., liver), as tissue perfusion dissipates some of the heat. As a mechanical pressure wave, ultrasound may cause microscopic bubbles in living tissues to expand and collapse (a phenomenon called

FIGURE 1.1 Classification of sound waves in terms of the frequency range.

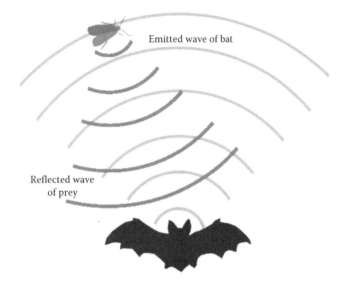

FIGURE 1.2 Insect-hunting activities of bats using ultrasonic echolocation techniques.

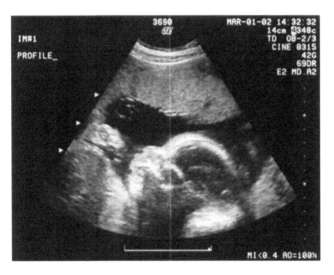

FIGURE 1.3 Medical B-mode ultrasound images.

acoustic cavitation) and distortion of the cell membrane, influencing ion fluxes and intracellular activity. In 2008, the American Institute of Ultrasound in Medicine (AIUM) published a 130-page report titled "American Institute of Ultrasound in Medicine Consensus Report on Potential Bioeffects of Diagnostic Ultrasound," stating that there are indeed some potential risks in medical ultrasound applications, including "postnatal thermal effects, fetal thermal effects, postnatal mechanical effects, fetal mechanical effects, and bioeffects considerations for ultrasound contrast agents" (Bioeffects Committee of the American Institute of Ultrasound in Medicine 2008).

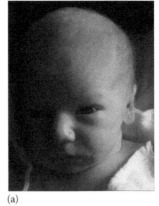
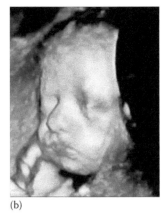

(a)   (b)

FIGURE 1.4 Comparison of the newborn baby (a) and the three-dimensional ultrasound images taken before the birth (b).

Several studies have already demonstrated the harmful effects of sonography on mammalian fetuses (Nyborg 2001). A Yale study in 2006 suggested that a 30 min ultrasonic exposure affects fetal brain development in mice, with no migration and scattered distribution of rodent brain cells, which may be linked to "mental retardation and childhood epilepsy to developmental dyslexia, autism spectrum disorders and schizophrenia" (Callen 2011). However, there have been no findings to show that these side effects could occur in humans, which needs to be investigated more in the future. Therefore, in order to minimize the possible risks, most doctors would advocate the As Low As Reasonably Achievable (ALARA) principle, keeping the scanning time and power settings as low as possible but consistent with sonographic examination.

Therapeutic ultrasound falls into two categories: inducing thermal or mechanical effects to stimulate or accelerate normal physiological response, and producing controlled, well-selective tissue destruction (Wells 1977, ter Haar 2001, Vaezy et al. 2001). Unfocused ultrasound has been employed in physical therapy since the 1930s and focused ultrasound since the 1950s (Fry 1958). Both structural and functional changes produced in exposures of the central nervous system to focused ultrasound have been described: "by appropriate control of the dosage conditions, it is possible to produce either reversible or selected irreversible changes" (Fry 1958). In the past two decades, therapeutic ultrasound has developed rapidly (Brenner 2001, Robertson and Baker 2001). Focused ultrasound ablation was considered by *TIME* magazine as one of the 50 most inspired ideas, innovations, and revolutions of 2011. In the article, it is stated: "Magnetic-resonance-imaging (MRI) and ultrasound technologies are each remarkable in their own right, but combine them and you get something life-changing. A technique called focused ultrasound uses MRI pictures to guide multiple beams of acoustic energy into a concentrated hot spot deep inside the body to heat and melt away tumors or other growths like uterine fibroids. A version of the device is being tested to tweak brain regions to relieve pain and even the tremors associated with Parkinson's disease"

(Brock-Abraham et al. 2011). With well-controlled dosage, ultrasound has the following therapeutic applications (Robertson and Baker 2001):

- Shock wave lithotripsy: breaking calculi, such as kidney stones and gallstones, into fragments small enough to be discharged from the body without undue difficulty, by special acoustic pulses (shock waves)
- Noninvasive tissue ablation by high-intensity focused ultrasound (HIFU) or focused ultrasound surgery (FUS) at a lower frequency (250–2000 kHz) than that of medical diagnostic ultrasound but at intensity several orders higher (Bailey et al. 2003, Kennedy 2005). If guided by MRI, the treatment modality is referred to as magnetic resonance–guided focused ultrasound (MRgFUS)
- Acoustic targeted drug delivery: delivering various drugs to specific tissues at the acoustic intensity ($0–20$ W/cm^2)
- Cleaning teeth in dental hygiene
- Use of focused ultrasound sources for cataract treatment by phacoemulsification
- Stimulation of bone growth and disruption of the blood–brain barrier by low-intensity ultrasound for drug delivery
- Lipectomy and liposuction carried out by the means of ultrasound
- Sonothrombolysis: transcranial Doppler (TCD) ultrasound used together with tissue plasminogen activator (tPA) to enhance the shrinkage or fragmentation of blood clot and discannalization in ischemic stroke patients
- Low-intensity pulsed ultrasound used for stem cell growth (i.e., therapeutic tooth and bone regeneration) (Matheson 1971)
- Killing of bacteria by the synergistic action of ultrasound and antibiotics
- Promotion of nutrient supply to eukaryotic cell tissue cultures by ultrasound

REFERENCES

Bailey MR, Khokhlova VA, Sapozhnikov OA, Kargl SG, Crum LA. Physical mechanisms of the therapeutic effect of ultrasound (a review). *Acoustical Physics* 2003;49:437–464.

Bioeffects Committee of the American Institute of Ultrasound in Medicine. American Institute of Ultrasound in Medicine Consensus report on potential bioeffects of diagnostic ultrasound: Executive summary. *Journal of Ultrasound in Medicine* 2008;27:503–515.

Brenner AE. More on Moore's law. *Physics Today* 2001;54:84.

Brock-Abraham C, Carbone N, Dodds E, Kluger J, Park A, Rawlings N, Suddath C, Sun FF, Thompson M, Walsh B, Webley K. The 50 best inventions. *Time* 2011;178:55–82.

Brown TG. Design of medical ultrasonic equipment. *Ultrasonics* 1968;6:107–111.

Callen PW. *Ultrasonography in Obstetrics and Gynecology.* Philadelphia, PA: Elsevier Health Science, 2011.

Devey GB, Wells PNT. Ultrasound in medical diagnosis. *Scientific American* 1978;238:98–106.

Donald I, MacVicar J, Brown TG. Investigation of abdominal masses by pulsed ultrasound. *Lancet* 1958;1:1188–1195.

Edgetron HG. *Sonar Images.* Englewood Cliffs, NJ: Prentice Hall, 1986.

Edler I. Early echocardiography. *Ultrasound in Medicine and Biology* 1991;17:425–431.

Erikson KR, Fry FJ, Jones JP. Ultrasound in medicine: A review. *IEEE Transactions on Sonics and Ultrasonics* 1974;SU-21:144–170.

Fay RR, Grinnell AD. Hearing by bats. In: Popper A (ed.) *Springer Handbook of Auditory Research.* New York: Springer-Verlag, 1995, pp. 1–36.

Frank F, Walker J. *Fundamentals of Noise and Vibration.* London, U.K.: E & FN Spon, 1998.

Frederick JR. *Ultrasonic Engineering.* New York: Wiley, 1965.

Fry WJ. Intense ultrasound in investigations of the central nervous system. In: Tobias CA, Lawrence JH, eds., *Advances in Biological and Medical Physics.* New York: Academic Press, 1958, pp. 281–348.

Graff KE. A history of ultrasonics. In: Mason WP, Thurston RN, eds., *Physical Acoustics.* New York: Academic Press, 1981, pp. 2–97.

Hill MN. *Physical Oceanography.* Boston, MA: Harvard University Press, 1962.

Kennedy JE. High-intensity focused ultrasound in the treatment of solid tumours. *Nature Reviews Cancer* 2005;5:321–327.

Kinsler LE, Frey AR, Coppens AB, Sanders JV. *Fundamentals of Acoustics.* New York: Wiley, 1982.

Kuttruff H. *Ultrasonics: Fundamentals and Applications.* New York: Elsevier, 1991.

Matheson AJ. *Molecular Acoustics.* New York: Wiley, 1971.

Morse PM, Ingard KU. *Theoretical Acoustics.* New York: McGraw-Hill, 1968.

Nyborg WL. Biological effects of ultrasound: Development of safety guidelines. Part II: General review. *Ultrasound in Medicine and Biology* 2001;27:301–333.

Ristic VM. *Principles of Acoustic Devices.* New York: Wiley, 1983.

Robertson VJ, Baker KG. A review of therapeutic ultrasound: Effectiveness studies. *Physical Therapy* 2001;81:1339–1350.

Skudrzyk E. *The Foundations of Acoustics.* New York: Springer-Verlag, 1971.

Temkin S. *Elements of Acoustics.* New York: Wiley, 1981.

ter Haar GR. Acoustic surgery. *Physics Today* 2001;54:29.

Thomas J, Moss CF, Vater M. Advances in the study of echo-location in bats and dolphin. Chicago, IL: Chicago University Press, 2003.

Vaezy S, Andrew M, Kaczkowski P, Crum L. Image-guided acoustic therapy. *Annual Review of Biomedical Engineering* 2001;3:375–390.

Wells PNT. *Physical Principles of Ultrasonic Diagnosis.* London, U.K.: Academic Press, 1969.

Wells PNT. *Biomedical Ultrasonics.* London, U.K.: Academic Press, 1977.

2 Wave Equation

Acoustic waves are the organized vibrations of molecules or atoms of a medium that is able to support the propagation of these waves. When the frequency of the vibration is above the audible range, the waves are known as ultrasonic waves. As their frequency increases, the wavelength of these waves gets progressively smaller, and this small size accounts for some of the unique resolution capabilities of ultrasound when compared with ordinary sound waves. The characteristics of the acoustic field are expressed by the acoustic pressure p, particle velocity of the medium v, and change of density ρ'. For example, in acoustic wave propagation, at a certain time point, the acoustic pressure at different locations has varying values; meanwhile, the acoustic pressure is a time-varying function at every point in space. In other words, the acoustic pressure is a temporally and spatially varying function. The relationship between acoustic pressure and its space and time variation is called the acoustic wave equation. The theory is inherently approximate, and there are many different versions that differ slightly or greatly from each other depending on what idealizations are made at the outset.

2.1 FUNDAMENTAL EQUATIONS FOR AN IDEAL FLUID

It has already been shown that, during an acoustic disturbance, the acoustic pressure p, the velocity v of the particle in the medium, and density change ρ' are all related. As a macroscopic phenomenon, acoustic vibration must satisfy fundamental physical laws, such as Newton's second law and mass conservation law. For simplicity, the following assumptions are made to derive the general expression of the acoustic wave equation:

1. The medium is an ideal fluid without viscosity, through which there is no energy loss during the acoustic wave propagation.
2. The medium is still (i.e., no initial velocity) and homogeneous in the equilibrium state, so the static pressure P_0 and the static density ρ_0 are both constants.
3. The compression and expansion of the medium are an adiabatic process during acoustic wave propagation: that is, there is no thermal exchange between neighboring media due to the temperature difference.
4. Acoustic vibration is a small-amplitude one, so $p \ll P_0$, $v \ll c_0$, $\xi \ll \lambda$, and $\rho' \ll \rho_0$ or $s_\rho = (\rho'/\rho_0) \ll 1$, where c_0 is the speed of sound, ξ is the medium displacement, λ is the wavelength, and s_ρ is the ratio of the density changes.

2.1.1 MOTION EQUATION

An incremental volume element of the supporting material, as shown in Figure 2.1, is considered to develop the acoustic wave equation. Assume that the size of this volume element is much smaller than the wavelength, so the variations in quantities throughout the volume are negligible. However, the volume element is sufficiently large to contain many molecules or atoms so that the whole material can be considered to be made up of continuous "particles." The length of this tiny element along the longitudinal axis is dx, the area of the faces perpendicular to the longitudinal axis is S, and the volume of this element is Sdx. The mass of this volume is $m = \rho Sdx$, with the density of the material ρ. Because the volume element is small compared to the wavelength, ρ undergoes only a small change throughout the volume, which may be approximated by a single quantity ρ'. There is pressure variation in the longitudinal direction of the material so that the pressures on the left and right sides of the incremental volume element are $P_0 + p$ and $P_0 + p + dp$, respectively. The force applied on the left side of the incremental volume element is $F_1 = (P_0 + p)S$. Because there is no shear force in an ideal fluid and the internal force is always perpendicular to the surface, F_1 is along the x_+ direction. The corresponding force applied on the right side of the incremental volume element is $F_2 = (P_0 + p + dp)S$ along the x_- direction. Therefore, the resultant force applied to the element along the x direction is $F = F_1 - F_2 = -S\frac{\partial p}{\partial x}dx$. Applying Newton's second law yields

$$\rho Sdx\frac{dv}{dt} = -\frac{\partial p}{\partial x}Sdx \quad \text{or} \quad \rho\frac{dv}{dt} = -\frac{\partial p}{\partial x} \quad (2.1)$$

which describes the relationship between the acoustic pressure p and particle velocity v of the medium.

2.1.2 CONTINUITY EQUATION

Continuity equation is actually the conservation of mass, which states that the difference of mass flowing in and out of the volume element in unit time is equal to the addition and subtraction of mass in it.

Assume the volume of the element, as shown in Figure 2.1b, is Sdx, and the particle velocity and density of the medium on the left side of the volume element are $(v)_x$ and $(\rho)_x$, respectively. So, the mass flowing into and out of the volume element from the left and right sides in unit time are $(\rho v)_x S$ and $(\rho v)_{x+dx} S$, respectively. We take the first-order approximation of the Taylor expansion of $(\rho v)_{x+dx}S$ as $\left[(\rho v)_x + \frac{\partial(\rho v)_x}{\partial x}dx \right]S$.

Therefore, the net mass flowing into the volume element in

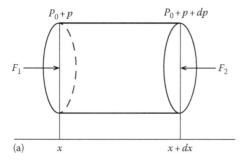

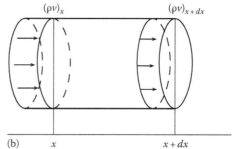

FIGURE 2.1 Illustration of a unit cylindrical element along the x-axis (a) with different acoustic pressure applied to the entry and exit surface to derive the motion equation, and (b) with the mass flowing into it to derive the continuity equation.

unit time is $-\dfrac{\partial(\rho v)}{\partial x}Sdx$. The addition of mass to the volume element means increase in density. If the increase in density in unit time is $\partial\rho/\partial t$, the corresponding mass increase is $\dfrac{\partial\rho}{\partial t}Sdx$. Thus, it follows that

$$-\frac{\partial(\rho v)}{\partial x}Sdx = \frac{\partial\rho}{\partial t}Sdx \quad \text{or} \quad -\frac{\partial}{\partial x}(\rho v) = \frac{\partial\rho}{\partial t} \quad (2.2)$$

which describes the relationship between the particle velocity v of the medium and its density ρ.

2.1.3 Thermodynamic Principle

For a certain volume element, the pressure P_0, density ρ_0, and temperature T_0 are used to describe its equilibrium status. However, in response to the pressure due to the acoustic disturbance, the density and temperature do not vary independently, but synergistically, which is described by the thermodynamic equation. Even at low frequency, the acoustic wave propagation is quite fast. The expansion and compression of the volume element take a much shorter time than thermal conduction. So, the acoustic propagation is considered as an adiabatic process because the medium does not carry out thermal exchange with its neighbors in such time. Therefore, the pressure P is considered to be only a function of the density ρ, that is,

$$P = P(\rho) \quad (2.3)$$

So, the small variations of pressure and density induced by the acoustic disturbance satisfy

$$dP = \left(\frac{dP}{d\rho}\right)_s d\rho \quad (2.4)$$

where the subscript "s" stands for the adiabatic process. When the medium is compressed, the pressure and density increase, $dP > 0$, $d\rho > 0$; meanwhile, in the expansion phase, the pressure and density decrease, $dP < 0$, $d\rho < 0$. So, the coefficient $(dP/d\rho)_s$ is always positive, represented by c^2

$$dP = c^2 d\rho \quad (2.5)$$

It is well known that the adiabatic thermodynamic equation for an ideal gas is

$$PV^\gamma = \text{const} \quad (2.6)$$

where γ is the adiabatic index, $\gamma = \dfrac{C_P}{C_V} = \dfrac{f+2}{2}$, C_P is the specific heat at constant pressure, C_V is the specific heat at constant volume, and f is the number of degrees of freedom (= 3 for a monatomic gas and 5 for a diatomic gas and collinear molecules). For a certain mass of an ideal gas, Equation 2.6 becomes

$$\frac{P}{\rho^\gamma} = \text{const} \quad (2.7)$$

So, c is calculated as

$$c^2 = \frac{\gamma P}{\rho} \quad (2.8)$$

For an ideal fluid, the relationship between pressure and density is more complicated than that for an ideal gas, Equation 2.8. But from the compressibility or the volume elastic coefficient of the medium, one can obtain

$$c^2 = \left(\frac{dP}{d\rho}\right)_s = \frac{dP}{(d\rho/\rho)_s\rho} \quad (2.9)$$

Considering that the mass of the medium is fixed as $\rho dV + V d\rho = 0$, we have

$$\left(\frac{d\rho}{\rho}\right)_s = -\left(\frac{dV}{V}\right)_s \quad (2.10)$$

Substituting this equation into Equation 2.9 yields

$$c^2 = \frac{dP}{(d\rho/\rho)_s\rho} = \frac{dP}{-(dV/V)_s\rho} = \frac{1}{\beta_s\rho} = \frac{K_s}{\rho} \quad (2.11)$$

where
dV/V is the relative increase of volume
$\beta_s = -\dfrac{(dV/V)}{dP}$ is the adiabatic volume compression coefficient
$K_s = 1/\beta_s = \dfrac{dP}{-(dV/V)}$ is the adiabatic volume elastic coefficient

In the adiabatic process, the negative sign shows that the pressure and volume change in opposite directions. It can be seen from Equation 2.11 that for a liquid c^2 is a function of ρ.

2.2 SMALL-AMPLITUDE 1D ACOUSTIC WAVE EQUATION

The fundamental equations in an ideal fluid described in the previous section illustrate that the relationships between all acoustic variables are nonlinear in general. It is not possible to use a single variable to represent the acoustic wave equation. If the amplitude of the acoustic wave is very small, the temporal and spatial changes of the acoustic variables (p, v, ρ') are small. So their higher orders can be neglected. The three fundamental equations can be simplified as follows.

With good justification for the power levels normally encountered in ultrasound diagnosis, it is assumed that the variation in density ρ of the medium due to the action of the waves is a very small percentage of the static density. That is, if the density is written as the sum of two parts

$$\rho = \rho_0 + \rho' \tag{2.12}$$

where
ρ_0 is the static medium density
ρ' is the time-varying change in density, then it may be assumed that $\rho' \ll \rho_0$

In partial differential form using Eulerian coordinates, the acceleration of a particle in the medium dv/dt consists of two parts: the local acceleration $\partial v/\partial t$, and the varying acceleration $\dfrac{\partial v}{\partial x}\dfrac{dx}{dt} = v\dfrac{\partial v}{\partial x}$ due to the motion of particle in the space. Therefore, Equation 2.1 can be rewritten as

$$(\rho_0 + \rho')\left(\frac{\partial v}{\partial t} + v\frac{\partial v}{\partial x}\right) = -\frac{\partial p}{\partial x} \tag{2.13}$$

Ignoring high-order variations, the preceding equation becomes

$$\rho_0\frac{\partial v}{\partial t} = -\frac{\partial p}{\partial x} \tag{2.14}$$

Similarly, Equation 2.2 can be simplified as

$$-\rho_0\frac{\partial v}{\partial x} = \frac{\partial \rho'}{\partial t} \tag{2.15}$$

For small-amplitude acoustic waves, $(dP/d\rho)_s$ in the vicinity of equilibrium status (P_0, ρ_0) can be expanded as

$$\left(\frac{dP}{d\rho}\right)_s = \left(\frac{dP}{d\rho}\right)_{s,0} + \left(\frac{d^2P}{d\rho^2}\right)_{s,0}(\rho - \rho_0) + \cdots \tag{2.16}$$

The subscript "0" stands for the value at equilibrium. Since $\rho - \rho_0 = \rho'$ is very small, the high-order variation in Equation 2.16 can be ignored, that is,

$$\left(\frac{dP}{d\rho}\right)_s \approx \left(\frac{dP}{d\rho}\right)_{s,0} \tag{2.17}$$

If c_0 is defined as

$$c_0^2 = \left(\frac{dP}{d\rho}\right)_{s,0} \tag{2.18}$$

it is clear that c_0^2 is approximately a constant for small-amplitude acoustic waves. For an ideal gas, Equation 2.18 becomes

$$c_0^2 \approx \left(\frac{dP}{d\rho}\right)_{s,0} = \frac{\gamma P_0}{\rho_0} \tag{2.19}$$

Meanwhile, for a fluid in general, Equation 2.18 becomes

$$c_0^2 \approx \left(\frac{dP}{d\rho}\right)_{s,0} = \frac{1}{\beta_s\rho_0} \tag{2.20}$$

After the approximations described earlier for small-amplitude acoustic waves, Equation 2.5 can be simplified as

$$p = c_0^2\rho' \tag{2.21}$$

Overall, after ignoring higher-order variation of small-amplitude acoustic waves, the three fundamental equations, that is, Equations 2.1, 2.2, and 2.5, have been simplified as linear equations, that is, Equations 2.14, 2.15, and 2.21, respectively. From these three equations, arbitrarily two variables among p, v, and ρ' can be eliminated. For example, differentiating Equation 2.21 with respect to t and then substituting it into Equation 2.15 give

$$\rho_0 c_0^2 \frac{\partial v}{\partial x} = -\frac{\partial p}{\partial t} \tag{2.22}$$

Differentiating Equation 2.22 with respect to t gives

$$\rho_0 c_0^2 \frac{\partial^2 v}{\partial t\partial x} = -\frac{\partial^2 p}{\partial t^2} \tag{2.23}$$

Substituting Equation 2.14 into Equation 2.23 gives

$$\frac{\partial^2 p}{\partial x^2} = \frac{1}{c_0^2}\frac{\partial^2 p}{\partial t^2} \tag{2.24}$$

which is the acoustic wave equation at small amplitude in a homogeneous, ideal fluid. Similarly, from Equations 2.14, 2.15, and 2.21, by eliminating p, ρ' or p, v, one can derive the wave equation in the form of v or ρ'. Moreover, it is noted that since high-order variations in Equation 2.24 are ignored, this expression is called the linear acoustic wave equation.

2.3 THREE-DIMENSIONAL WAVE EQUATION

In the previous section, the 1D wave equation was derived based on the assumption that the acoustic field in the y and z directions was uniform. However, in the general case, such an assumption is not valid (the acoustic field in the x, y, and z directions are nonuniform). The derivation of three fundamental equations of the medium and the consequent wave equation are similar to those of 1D case, but both pressure p and velocity v are now vectors.

The 3D versions of motion and continuity equations are

$$\rho_0 \frac{d\vec{v}}{dt} = -\text{grad} \, \vec{p} \qquad (2.25)$$

$$-\text{div}(\rho_0 \vec{v}) = \frac{\partial \rho'}{\partial t} \qquad (2.26)$$

where

$\text{grad} = \frac{\partial}{\partial x}\hat{i} + \frac{\partial}{\partial y}\hat{j} + \frac{\partial}{\partial z}\hat{k}$ is the gradient operator

div is the divergence operator, $\text{div}(\rho\vec{v}) = \frac{\partial(\rho v_x)}{\partial x} + \frac{\partial(\rho v_y)}{\partial y}$

$+ \frac{\partial(\rho v_z)}{\partial z}$, and v_x, v_y, v_z are the components of velocity along three coordinates, respectively

The expression of thermodynamics principle is the same as Equation 2.5.

Eliminating v and ρ' using a similar method as in the 1D case, one obtains the wave equation for the 3D small-amplitude acoustic wave as

$$\Delta p = \frac{1}{c_0^2} \frac{\partial^2 p}{\partial t^2} \qquad (2.27)$$

where $\Delta = \nabla^2 = \frac{\partial^2}{\partial x^2} + \frac{\partial^2}{\partial y^2} + \frac{\partial^2}{\partial z^2}$ is the Laplacian operator in Cartesian coordinates. If the medium is incompressible, then its density in the flow should remain constant. Consequently, the conservation of mass requires $\nabla \cdot v = 0$. Although much of the analysis of fluid mechanics is based on this idealization, it is ordinarily inappropriate for acoustics because the compressibility of the medium plays a major role in the propagation of sound.

2.4 VELOCITY POTENTIAL

The particle velocity $\vec{v}$ can be obtained from the acoustic pressure p in the wave equation by applying the motion equation (2.25):

$$\begin{cases} v_x = -\frac{1}{\rho_0} \int \frac{\partial p}{\partial x} dt \\ v_y = -\frac{1}{\rho_0} \int \frac{\partial p}{\partial y} dt \\ v_z = -\frac{1}{\rho_0} \int \frac{\partial p}{\partial z} dt \end{cases} \qquad (2.28)$$

It can be easily seen that

$$\frac{\partial v_x}{\partial y} - \frac{\partial v_y}{\partial x} = 0$$

$$\frac{\partial v_x}{\partial z} - \frac{\partial v_z}{\partial x} = 0 \quad \text{or} \quad \text{rot} \, \vec{v} = 0 \qquad (2.29)$$

$$\frac{\partial v_y}{\partial z} - \frac{\partial v_z}{\partial y} = 0$$

where rot is the rotation operator. This equation shows that the small-amplitude acoustic field in the ideal fluid medium is nonrotational motion. On the other hand, it is known that if the rotation of a vector is equal to zero, this vector should be the gradient of a scalar function. Equation 2.28 can be rewritten as

$$v_x = -\frac{1}{\rho_0} \int \frac{\partial p}{\partial x} dt = -\frac{\partial}{\partial x} \int \frac{p}{\rho_0} dt$$

$$v_y = -\frac{1}{\rho_0} \int \frac{\partial p}{\partial y} dt = -\frac{\partial}{\partial y} \int \frac{p}{\rho_0} dt \qquad (2.30)$$

$$v_z = -\frac{1}{\rho_0} \int \frac{\partial p}{\partial z} dt = -\frac{\partial}{\partial z} \int \frac{p}{\rho_0} dt$$

A new scalar quantity Φ is defined as

$$\Phi = \int \frac{p}{\rho_0} dt \qquad (2.31)$$

Then, Equation 2.30 becomes

$$v_x = -\frac{\partial \Phi}{\partial x}$$

$$v_y = -\frac{\partial \Phi}{\partial y} \quad \text{or} \quad \vec{v} = -\text{grad} \, \Phi \qquad (2.32)$$

$$v_z = -\frac{\partial \Phi}{\partial z}$$

Thus, the particle velocity $\vec{v}$ can be expressed as a gradient of Φ called the velocity potential, which represents the impulse momentum of a unit mass of the medium due to the acoustic disturbance. Furthermore, the wave equation can also be expressed in the form of the velocity potential Φ. Equation 2.31 is rewritten as

$$p = \rho_0 \frac{\partial \Phi}{\partial t} \qquad (2.33)$$

Differentiating with respect to t on both sides of Equation 2.21 yields

$$\frac{\partial p}{\partial t} = c_0^2 \frac{\partial \rho'}{\partial t} \qquad (2.34)$$

Substituting Equation 2.28 into Equation 2.34 gives

$$\frac{\partial p}{\partial t} = -c_0^2 \text{div}(\rho_0 \vec{v}) \qquad (2.35)$$

Differentiating Equation 2.33 with respect to t and then substituting it into Equation 2.35 give

$$\rho_0 \frac{\partial^2 \Phi}{\partial t^2} = -c_0^2 \text{div}(\rho_0 \vec{v}) \quad (2.36)$$

Substituting Equation 2.32 into it gives

$$\Delta\Phi = \frac{1}{c_0^2} \frac{\partial^2 \Phi}{\partial t^2} \quad (2.37)$$

It is clear that, similar to the acoustic pressure, the velocity potential Φ is a scalar quantity. It is convenient to solve Φ from the wave equation (2.37) and then to derive the particle velocity $\vec{v}$ and acoustic pressure p using Equations 2.32 and 2.33, respectively.

2.5 SPEED OF SOUND

The coefficient c_0 in Equation 2.18 is the speed of sound in the medium, showing the compressibility of medium in response to the acoustic disturbance. If the compressibility of the medium is large (e.g., as for a gas), the variation in density induced by the acoustic pressure will also be large. According to the definition, c_0 is rather small, or the change of the volume element needs a long time to transfer to the surrounding. In contrast, if the compressibility of the medium is small, the variation in density induced by the acoustic pressure will also be small. Correspondingly, c_0 is large and the change of the volume element can be transferred to the neighboring elements quickly. For an ideal rigid body (i.e., noncompressible), c_0 will be infinite and the variation of a volume element can be transferred to all the other elements right away. Altogether, it is shown that the compressibility of the medium is related to the speed of sound.

The speed of a small-amplitude acoustic wave in the ideal gas can be calculated using Equation 2.19. For air ($\gamma = 1.402$, $P_0 = 1.013$ N/m^2, $\rho_0 = 1.293$ kg/m^3 at $T = 0°$C), c_0 ($0°$C) = 331.6 m/s. The sound speed c_0 is dependent on the equilibrium parameters of the medium (i.e., pressure P, volume V, and temperature T). The equation of state for an ideal gas is as follows

$$PV = \frac{m}{M}RT \quad (2.38)$$

where
m is the mass of gas in the sample
M is the molar mass of the gas
$R = 8.314$ J/mol · K is the universal gas constant

Then, Equation 2.19 can be rewritten as

$$c_0 = \sqrt{\frac{\gamma P_0}{\rho_0}} = \sqrt{\frac{\gamma R}{M} T_0} \quad (2.39)$$

If the temperature is expressed in degree Celsius, $t°$C, as $T_0 = 273 + t$, the sound speed is

$$c_0 = \sqrt{\frac{\gamma R}{M}(273+t)} \approx c_0(0°\text{C}) + \frac{c_0(0°\text{C})}{273 \cdot 2}t \approx 331.6 + 0.6t \quad (2.40)$$

For example, at $20°$C, the sound speed in air is $c_0 = 344$ m/s.

It should be noted that the wave equation derived applies equally well to all types of waves: sound as well as electromagnetic waves, such as light, microwaves, and radio waves. The phase velocity of these waves is

$$c_0 = \frac{1}{\sqrt{\rho_0 K}} \quad (2.41)$$

which is a function of the average density ρ_0 and compressibility K of the medium in which the waves are propagating. The values of these constants may be readily obtained from handbooks for most materials. For water ($\rho_0 = 998$ kg/m^3, $K = 45.8 \times 10^{-11}$ m^2/N at $20°$C), the speed of sound is calculated as $c_0 = 1480$ m/s. Because the relationship between pressure and water density is quite complicated, theoretical derivation of the expression of sound speed at different temperatures is difficult. However, according to experimental measurements, the sound speed increases by about 4.5 m/s for every temperature increment of $1°$C. It will be seen that the phase velocity of ultrasound in most soft biological tissue is within $\pm5\%$ of its value in water. Thus, at a given frequency, the velocity and wavelength in tissue will be similar to those in water. This small wavelength is the basic reason why ultrasound has good resolution in imaging instruments. An electromagnetic wave at the same frequency would have a much longer wavelength because of its much higher phase velocity in tissue.

It should be noted that the sound speed c_0 is the acoustic energy propagation speed in the medium, which is different from the particle vibration velocity v of the medium itself. For example, if $p_A = 0.1$ Pa (approximately the sound pressure of human speech), it can be shown that $v_A = \dfrac{p_A}{\rho_0 c_0} \approx 2.5 \times 10^{-4}$ m/s. It is clear that $v_A \ll c_0$ as the assumption validity of small-amplitude wave.

2.6 ACOUSTIC IMPEDANCE

The acoustic impedance of the medium in an acoustic field is defined as the ratio of the acoustic pressure to the particle velocity, that is,

$$z = \frac{p}{v} \quad (2.42)$$

Applying Equations 2.32 and 2.33, it can be seen that the acoustic impedances of a forward (along positive x) and a reflected (along negative x) planar waves are

$$z = \rho_0 c_0 \quad (2.43)$$

$$z = -\rho_0 c_0 \tag{2.44}$$

The acoustic impedance is a real number in an ideal medium, which shows that there is no energy storage at every location of the planar acoustic field and the acoustic energy at the previous location can be transferred to the next one without loss. The role of acoustic impedance in acoustics is similar to the concept of electrical impedance in electronics, describing the characteristics of energy propagation. The unit of acoustic impedance is Rayleigh: 1 Rayleigh = 1 N·s/m^3.

2.7   SOLUTION OF WAVE EQUATION

For the solution of the 1D wave equation (2.24), the static acoustic field of a harmonic source is of interest because most acoustic sources vibrate harmonically and the response of an arbitrary time-varying vibration can be obtained by the addition of individual responses to harmonic functions at different frequencies using the principles of linear superposition and Fourier analysis. It is assumed that the 1D wave equation (2.24) has a solution of the form

$$p(t,x) = p(x)e^{j\omega t} \tag{2.45}$$

where ω is the angular frequency of the harmonic vibration of the acoustic source. So, the differential equation for $p(x)$ is

$$\frac{d^2 p(x)}{dx^2} + k^2 p(x) = 0 \tag{2.46}$$

where $k = \dfrac{\omega}{c_0}$ is the wave number. Substituting Equation 2.41 into the definition of wave number gives

$$k^2 = \rho_0 K \omega^2 \tag{2.47}$$

which is known as the dispersion equation, or Brillouin equation, for a specific medium because of the appearance of the constants ρ_0 and K. The general solution of the ordinary differential equation (ODE) (2.46) has a combination of sine and cosine functions, or complex numbers. Usually for an acoustic wave propagating toward infinite space, the solution in complex numbers is more appropriate:

$$p(x) = Ae^{-jkx} + Be^{jkx} \tag{2.48}$$

where A and B are two arbitrary constants determined by the boundary conditions. Substituting Equation 2.48 into Equation 2.46 gives

$$p(t,x) = Ae^{j(\omega t - kx)} + Be^{j(\omega t + kx)} \tag{2.49}$$

The first item represents the wave propagating along positive x direction, and the second one for the wave along the negative x direction. For planar wave propagation in an infinite medium, if no reflector is present in the propagation path, there is no reflected wave or $B = 0$. So, Equation 2.49 can be simplified as

$$p(t,x) = Ae^{j(\omega t - kx)} \tag{2.50}$$

If there is a sound source at $x = 0$ with the acoustic pressure $p_A e^{j\omega t}$, $A = p_A$ and the acoustic pressure in the field is

$$p(t,x) = p_A e^{j(\omega t - kx)} \tag{2.51}$$

and the particle velocity is

$$v(t,x) = v_A e^{j(\omega t - kx)} \tag{2.52}$$

where $v_A = p_A / \rho_0 c_0$.

Therefore, the characteristics of acoustic field expressed by Equations 2.51 and 2.52 are summarized as follows:

1. Assume that the wave at an arbitrary location $x = x_0$ at time $t = t_0$ moves to the location of $x_0 + \Delta x$ after time Δt without energy dissipation, so

$$p(t_0, x_0) = p(t_0 + \Delta t, x_0 + \Delta x) \tag{2.53}$$

Substituting it into Equation 2.51 gives

$$e^{j(\omega \Delta t - k \Delta x)} = 1 \quad \text{or} \quad \Delta x = c_0 \Delta t \tag{2.54}$$

which illustrates that, using the complex number in Equation 2.49, the forward and reflected waves can be separated from the solution of the wave equation.

2. At an arbitrary time t_0, the track of the particles with the same phase ϕ_0 is

$$\omega t_0 - kx = \phi_0 \tag{2.55}$$

$$x = \frac{\omega t_0 - \phi_0}{k} = \text{const} \tag{2.56}$$

which shows that the particles with the same phase consist of a plane for this kind of wave propagation; so, it is usually called a plane wave.

3. From Equation 2.54, one obtains

$$c_0 = \frac{\Delta x}{\Delta t} \tag{2.57}$$

where c_0 is the wavefront propagation distance in unit time, or the sound propagation speed.

Overall, Equations 2.51 and 2.52 describe the acoustic field produced by a plane wave propagating along positive x direction at velocity c_0. The amplitudes of pressure p_A and particle velocity v_A of plane wave are distance-invariant in a homogeneous ideal medium, or there is no attenuation in the acoustic wave propagation, because no viscosity is considered for energy loss and the area of the plane wavefront is constant. In addition, at any location of the planar acoustic field, the pressure and particle velocity have the same phase. The particle displacement can be calculated from Equation 2.52 as

$$\xi = \int v\,dt = \frac{v_A}{j\omega} e^{j(\omega t - kx)} \tag{2.58}$$

At any arbitrary location x_0, the particle displacement can be written as

$$\xi = \frac{v_A}{\omega} e^{-j\left(kx_0 + \frac{\pi}{2}\right)} e^{j\omega t} = \xi_A e^{j(\omega t - \alpha)} \tag{2.59}$$

where ξ_A and α are both constants. It is clear that the particle at x_0 vibrates around the equilibrium position, not moving away. In fact, such a particle vibration around the equilibrium position transfers the energy of the sound source to the neighboring particles and consequently toward the farther ones.

This choice of an exponential or sinusoidal/cosine waveform is not as restrictive as it seems for two reasons: (1) many sources emit ultrasound waves that are approximately sinusoidal, such as narrowband or continuous-wave transducers, and (2) even an arbitrary wave shape, like a narrow time pulse, can be decomposed into its sinusoidal Fourier components, each of which will have a form with a unique amplitude and phase. This possibility of breaking down a complex wave shape into its more basic sinusoidal components is assured because superposition is allowed in the solution of a linear equation and the 1D wave equation is linear.

The ray associated with a plane wave is simply a vector pointing in the direction of propagation of the wave and is perpendicular to the phase fronts. If an observer is stationary in space at some position, the wave will travel past with a sinusoidal time variation.

2.8 SPHERICAL WAVES

Since the acoustic field in the x, y, and z directions is inhomogeneous, in general the 3D wave equation (2.27) should be solved. However, in certain situations that keep the shape of the wavefront exactly the same in a constant propagation direction, such as the spherical surface of the wavefront and propagation direction at the radial direction r for a homogeneous spherical wave, the wave equation can be simplified to a specific format in order to get a simple solution.

Assume that there is an acoustic wave with arbitrary wavefront propagating in the space and that the normal direction of the wavefront is the propagation direction of the acoustic wave along r. Select a small volume element between two wavefronts with distance dr dissected by a small solid angle, as shown in Figure 2.2. Because the propagation is only in the direction of r with a small amplitude, the motion equation is linearized as

$$\rho_0 \frac{\partial v}{\partial t} = -\frac{\partial p}{\partial r} \tag{2.60}$$

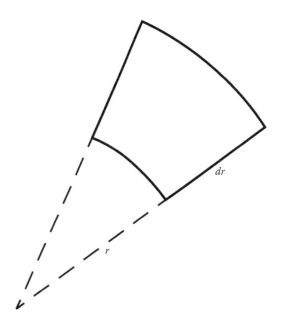

FIGURE 2.2 Cross section of a small volume element for an acoustic wave propagating along the direction r with an arbitrary wavefront that is perpendicular to the wave propagation direction. The volume element is formed by a small angle between two neighboring wavefronts with the interval distance of dr.

The thermodynamic equation is irrelevant to the shape of the volume element. However, the continuity equation of a spherical wave should be rewritten since the area of the wavefront varies with r. Assuming that the area of the wavefront at r is S, particle velocity is $\vec{v}$, and density is ρ, the mass flowing into this volume element in unit time is $\rho v S$. At $r + dr$, the mass flowing out the volume element is $\rho v S + \frac{\partial(\rho v S)}{\partial r} dr$. So, the net mass flowing into this element is $-\frac{\partial(\rho v S)}{\partial r} dr$. Since this volume element is very small, its mass is approximately $\rho S dr$ and the change in this mass at unit time is $\frac{\partial(\rho S dr)}{\partial t}$. Using the conservation of mass, we can write

$$-\frac{\partial(\rho v S)}{\partial r} dr = \frac{\partial(\rho S dr)}{\partial t} \tag{2.61}$$

Since $\rho = \rho_0 + \rho'$ and ρ_0 is time-invariant for a small-amplitude acoustic wave, the preceding equation can be simplified as

$$-\rho_0 \frac{\partial(vS)}{\partial r} dr = S \frac{\partial \rho'}{\partial t} \tag{2.62}$$

which, when multiplied by c_0^2/S on both sides and then expanded, gives

$$-\rho_0 \frac{c_0^2}{S} \left(S \frac{\partial v}{\partial r} + v \frac{\partial S}{\partial r} \right) = c_0^2 \frac{\partial \rho'}{\partial t} \tag{2.63}$$

Differentiating Equation 2.21 and substituting it into the equation above give

$$\frac{\partial p}{\partial t} = -\rho_0 c_0^2 \left[\frac{\partial v}{\partial r} + v \frac{\partial(\ln S)}{\partial r} \right] \quad (2.64)$$

Combining Equations 2.60 and 2.64 and eliminating v give

$$\left[\frac{\partial^2 p}{\partial r^2} + \frac{\partial p}{\partial r} \frac{\partial(\ln S)}{\partial r} \right] = \frac{1}{c_0^2} \frac{\partial^2 p}{\partial t^2} \quad (2.65)$$

It shows that when the wavefront shape $S(r)$ is known, preceding equation can be used to solve the acoustic pressure p.

2.9 CYLINDRICAL WAVES

For cylindrically symmetric waves, there is no dependence on the azimuthal angle or on the axial coordinate, so the Laplacian in cylindrical coordinates reduces to

$$\nabla^2 = \frac{1}{r} \frac{\partial}{\partial r} \left(r \frac{\partial}{\partial r} \right) \quad (2.66)$$

where r is the radial distance from the symmetry axis. Consequently, the wave equation takes the form

$$\frac{\partial^2(\sqrt{r}\,p)}{\partial r^2} - \frac{1}{c_0^2} \frac{\partial^2(\sqrt{r}\,p)}{\partial t^2} + \frac{\sqrt{r}\,p}{4r^2} = 0 \quad (2.67)$$

Then, the wave equation (2.24) can be written in either of the following forms:

$$\frac{d^2 p}{dr^2} + \frac{1}{r} \frac{dp}{dr} + \frac{1}{c_0^2} \frac{d^2 p}{dt^2} = 0 \quad (2.68)$$

$$\frac{d^2(\sqrt{r}\,p)}{dr^2} + \left[k^2 + \frac{1}{4r^2} \right] \sqrt{r}\,p = 0 \quad (2.69)$$

The solution of the latter, which corresponds to an outgoing wave, is

$$p = A H_0^{(1)}(kr) \quad (2.70)$$

where $H_0^{(1)}(*)$ is the Hankel function of the first kind, which asymptotically approaches the limit

$$\lim_{kr \to \infty} H_0^{(1)}(kr) = \left(\frac{2}{\pi kr} \right)^{1/2} e^{-i\pi/4} e^{ikr} \quad (2.71)$$

For cylindrical waves that are not of constant frequency, an outgoing solution can be taken as

$$p = \int_{-\infty}^{\infty} \frac{F(t - R/c_0)}{R \, dz_0} \quad (2.72)$$

where

$$R = \sqrt{r^2 + z_0^2}$$

F is an arbitrary function

Waveforms of outward propagating cylindrical waves tend to get distorted with increasing propagation distance, especially at small r. However, at larger values of r, it is often a good approximation to neglect the last term in Equation 2.67, resulting in an approximate solution of the generic form

$$p(r, t) \approx \frac{f(r - ct)}{\sqrt{r}} \quad (2.73)$$

which is similar to the expression for an outgoing spherical wave; only here the amplitude drops off with r as $1/\sqrt{r}$. The latter approximate expression is consistent with the constant-frequency solution given by Equation 2.70 when the Hankel function is replaced by its asymptotic limit Equation 2.71.

The fluid velocity induced by outgoing cylindrical waves is not as simply related to the corresponding sound pressure as that induced by a plane wave, although symmetry dictates that the velocity must be in the appropriate radial direction when the propagation is cylindrically symmetric. In the constant-frequency case, an expression may be obtained from the radial component such that the velocity is expressed in terms of the Hankel function of first order. However, a simple and approximate result emerges in the limit of a large radial distance r, this being the plane wave relation, Equation 2.46. Here, a large r implies that it is large compared to a characteristic wavelength, or that it is large compared to c divided by a characteristic angular frequency.

2.10 WAVE IN SOLID

When a force is applied to a solid, a distortion will be produced correspondingly, which is described by the strain. The internal forces acting within a deformable body are measured as stress in continuum mechanics. Assume that there is a point A with coordinate (x, y, z) inside a solid and that its displacements in the x, y, and z directions are ξ, η, and ζ, respectively. The corresponding displacements at a neighboring location C $(x + dx, y + dy, z + dz)$ are $\xi + d\xi$, $\eta + d\eta$, and $\zeta + d\zeta$. The discrepancies of displacement between A and C are obtained by using the Taylor series expansion:

$$\begin{cases} d\xi = \dfrac{\partial \xi}{\partial x} dx + \dfrac{\partial \xi}{\partial y} dy + \dfrac{\partial \xi}{\partial z} dz \\[2mm] d\eta = \dfrac{\partial \eta}{\partial x} dx + \dfrac{\partial \eta}{\partial y} dy + \dfrac{\partial \eta}{\partial z} dz \\[2mm] d\zeta = \dfrac{\partial \zeta}{\partial x} dx + \dfrac{\partial \zeta}{\partial y} dy + \dfrac{\partial \zeta}{\partial z} dz \end{cases} \quad (2.74)$$

So, the distortion in the solid can be described by nine strain tensor components:

$$\begin{vmatrix} \dfrac{\partial \xi}{\partial x} & \dfrac{\partial \xi}{\partial y} & \dfrac{\partial \xi}{\partial z} \\[2mm] \dfrac{\partial \eta}{\partial x} & \dfrac{\partial \eta}{\partial y} & \dfrac{\partial \eta}{\partial z} \\[2mm] \dfrac{\partial \zeta}{\partial x} & \dfrac{\partial \zeta}{\partial y} & \dfrac{\partial \zeta}{\partial z} \end{vmatrix} \tag{2.75}$$

$$\begin{cases} \dfrac{\partial \xi}{\partial x} = \varepsilon_{xx}, \quad \dfrac{\partial \eta}{\partial y} = \varepsilon_{yy}, \quad \dfrac{\partial \zeta}{\partial z} = \varepsilon_{zz} \\[2mm] \dfrac{\partial \eta}{\partial x} + \dfrac{\partial \xi}{\partial y} = \varepsilon_{xy} = \varepsilon_{yx} \\[2mm] \dfrac{\partial \xi}{\partial z} + \dfrac{\partial \zeta}{\partial x} = \varepsilon_{xz} = \varepsilon_{zx} \\[2mm] \dfrac{\partial \zeta}{\partial x} + \dfrac{\partial \xi}{\partial y} = \varepsilon_{xy} = \varepsilon_{yx} \\[2mm] \dfrac{\partial \eta}{\partial x} - \dfrac{\partial \xi}{\partial y} = 2\Omega_z \\[2mm] \dfrac{\partial \xi}{\partial z} - \dfrac{\partial \zeta}{\partial x} = 2\Omega_y \\[2mm] \dfrac{\partial \zeta}{\partial y} - \dfrac{\partial \eta}{\partial z} = 2\Omega_x \end{cases} \tag{2.76}$$

where

ε_{xx}, ε_{yy}, and ε_{zz} are the normal strains in the x, y, and z directions, respectively

ε_{xy}, ε_{yz}, and ε_{xz} are the shear strains in the xy, yz, and xz planes, respectively

Ω_x, Ω_y, and Ω_z are the rotation of the volume element with respect to the x-, y-, and z-axes, respectively

It is apparent that rotation of the volume element has no contribution to the distortion. Therefore, three normal strains and three shear strains are sufficient to describe the distortion in the solid.

In Figure 2.3, it is shown that there are nine stress components on the surface of a volume element:

$$\begin{vmatrix} T_{xx} & T_{xy} & T_{xz} \\ T_{yx} & T_{yy} & T_{yz} \\ T_{zx} & T_{zy} & T_{zz} \end{vmatrix} \tag{2.77}$$

Usually, stress is expressed as T_{ij} ($i, j = x, y, z$) where i indicates that the stress acts on a plane normal to the i-axis, j denotes the direction in which the stress acts; normal stress is when $i = j$, and shear stress is when $i \neq j$. It can be proven that the stress is symmetric, $T_{ij} = T_{ji}$. Therefore, six strain

tensor components are used to describe the distortion on a solid medium, while six stress components can show the corresponding stress. If the distortion is very small, there is a linear relationship between strain and stress. Each stress should be expressed as a function of six strain tensor components in the following format:

$$\begin{cases} T_{xx} = C_{11}\varepsilon_{xx} + C_{12}\varepsilon_{yy} + C_{13}\varepsilon_{zz} + C_{14}\varepsilon_{yz} + C_{15}\varepsilon_{zx} + C_{16}\varepsilon_{xy} \\ T_{yy} = C_{21}\varepsilon_{xx} + C_{22}\varepsilon_{yy} + C_{23}\varepsilon_{zz} + C_{24}\varepsilon_{yz} + C_{25}\varepsilon_{zx} + C_{26}\varepsilon_{xy} \\ T_{zz} = C_{31}\varepsilon_{xx} + C_{32}\varepsilon_{yy} + C_{33}\varepsilon_{zz} + C_{34}\varepsilon_{yz} + C_{35}\varepsilon_{zx} + C_{36}\varepsilon_{xy} \\ T_{yz} = C_{41}\varepsilon_{xx} + C_{42}\varepsilon_{yy} + C_{43}\varepsilon_{zz} + C_{44}\varepsilon_{yz} + C_{45}\varepsilon_{zx} + C_{46}\varepsilon_{xy} \\ T_{zx} = C_{51}\varepsilon_{xx} + C_{52}\varepsilon_{yy} + C_{53}\varepsilon_{zz} + C_{54}\varepsilon_{yz} + C_{55}\varepsilon_{zx} + C_{56}\varepsilon_{xy} \\ T_{xy} = C_{61}\varepsilon_{xx} + C_{62}\varepsilon_{yy} + C_{63}\varepsilon_{zz} + C_{64}\varepsilon_{yz} + C_{65}\varepsilon_{zx} + C_{66}\varepsilon_{xy} \end{cases} \tag{2.78}$$

where C_{ij} ($i, j = 1, 2, 3, 4, 5, 6$) are the elastic components, similar to the Hooke law in a spring. The elasticity characteristics of solid have in total 36 elastic components, making it much more complex than a fluid. However, all of these 36 elastic components are not completely independent. Because the elastic energy is a monotonic function of strain, it can be proven that the elastic components are also symmetric, $C_{ij} = C_{ji}$. For a symmetric crystal, the number of independent elastic components can be further reduced. For example, there are five, three, and two independent elastic components of a rhombohedral lattice crystal (e.g., quartz), a cubic crystal (e.g., sodium chloride), and an isotropic solid (e.g., metal and glass), respectively. For an isotropic solid, the stress–strain relations take the form

$$\begin{cases} T_{xx} = \lambda(\varepsilon_{xx} + \varepsilon_{yy} + \varepsilon_{zz}) + 2\mu\varepsilon_{xx} \\ T_{yy} = \lambda(\varepsilon_{xx} + \varepsilon_{yy} + \varepsilon_{zz}) + 2\mu\varepsilon_{yy} \\ T_{zz} = \lambda(\varepsilon_{xx} + \varepsilon_{yy} + \varepsilon_{zz}) + 2\mu\varepsilon_{zz} \\ \quad T_{yz} = \mu\varepsilon_{yz} \\ \quad T_{zx} = \mu\varepsilon_{zx} \\ \quad T_{xy} = \mu\varepsilon_{xy} \end{cases} \tag{2.79}$$

where λ and μ are the Lamè constants, and their relationships with the elastic components being

$$\lambda = C_{12} = C_{13} = C_{21} = C_{23} = C_{31} = C_{32}$$

$$\mu = C_{44} = C_{55} = C_{66} = \frac{1}{2}(C_{11} - C_{12}) \tag{2.80}$$

$$\lambda + 2\mu = C_{11} + C_{22} + C_{33}$$

with the other elastic components equal to zero. The Lamè constants are related to the elastic modulus (or Young's module) E and Poisson's ratio σ by the relationship

$$\lambda = \frac{E\sigma}{(1+\sigma)(1-2\sigma)}$$

$$\mu = \frac{E}{2(1+\sigma)} \tag{2.81}$$

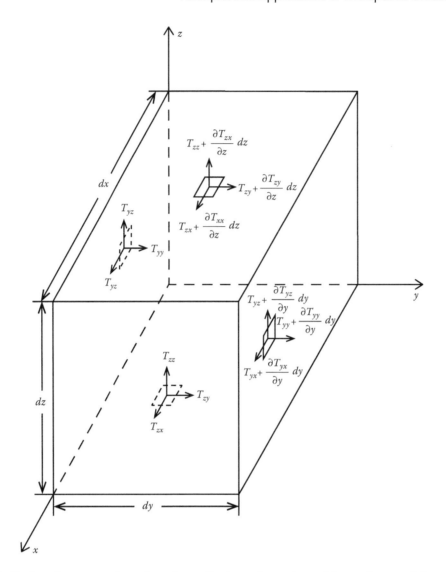

FIGURE 2.3   The distribution of stresses on the surface of a small volume element in a solid according to a deformation.

In order to derive the motion equation in the solid, the free body diagram on a volume element in the x direction is made. There are three forces applied to it:

$$F_x' = \left(T_{xx} + \frac{\partial T_{xx}}{\partial x} dx - T_{xx} \right) dy dz$$

$$F_x'' = \left(T_{yx} + \frac{\partial T_{yx}}{\partial y} dy - T_{yx} \right) dx dz \qquad (2.82)$$

$$F_x''' = \left(T_{zx} + \frac{\partial T_{zx}}{\partial z} dz - T_{zx} \right) dx dy$$

where F_x', F_x'', and F_x''' are the forces in the x direction applied on the surface normal to x-, y-, and z-axes, respectively. So, the net force in the x direction is

$$F_x = \left(\frac{\partial T_{xx}}{\partial x} + \frac{\partial T_{yx}}{\partial y} + \frac{\partial T_{zx}}{\partial z} \right) dx dy dz \qquad (2.83)$$

We can use Newton's second law to establish the motion equation of such a small cubic element in the x direction as well as those in the y and z directions as

$$\begin{cases} \rho \dfrac{\partial^2 \xi}{\partial t^2} = \dfrac{\partial T_{xx}}{\partial x} + \dfrac{\partial T_{yx}}{\partial y} + \dfrac{\partial T_{zx}}{\partial z} \\[2mm] \rho \dfrac{\partial^2 \eta}{\partial t^2} = \dfrac{\partial T_{xy}}{\partial x} + \dfrac{\partial T_{yy}}{\partial y} + \dfrac{\partial T_{zy}}{\partial z} \\[2mm] \rho \dfrac{\partial^2 \zeta}{\partial t^2} = \dfrac{\partial T_{xz}}{\partial x} + \dfrac{\partial T_{yz}}{\partial y} + \dfrac{\partial T_{zz}}{\partial z} \end{cases} \qquad (2.84)$$

where ρ is the density of the medium. Substituting Equations 2.78 and 2.75 into Equation 2.84 gives

$$\begin{cases} \rho \dfrac{\partial^2 \xi}{\partial t^2} = (\lambda + \mu) \dfrac{\partial \Delta}{\partial x} + \mu \nabla^2 \xi \\[2mm] \rho \dfrac{\partial^2 \eta}{\partial t^2} = (\lambda + \mu) \dfrac{\partial \Delta}{\partial y} + \mu \nabla^2 \eta \\[2mm] \rho \dfrac{\partial^2 \zeta}{\partial t^2} = (\lambda + \mu) \dfrac{\partial \Delta}{\partial z} + \mu \nabla^2 \zeta \end{cases} \qquad (2.85)$$

where $\Delta = \dfrac{\partial \xi}{\partial x} + \dfrac{\partial \eta}{\partial y} + \dfrac{\partial \zeta}{\partial z}$ and $\nabla^2 = \dfrac{\partial^2 \xi}{\partial x^2} + \dfrac{\partial^2 \eta}{\partial y^2} + \dfrac{\partial^2 \zeta}{\partial z^2}$. If the particle displacement is expressed as $\vec{S} = \xi \hat{i} + \eta \hat{j} + \zeta \hat{k}$, then

$$\rho \frac{\partial^2 \vec{S}}{\partial t^2} = (\lambda + 2\mu)\mathrm{grad}\Delta \vec{S} + \mu \nabla^2 \vec{S} \qquad (2.86)$$

Because $\Delta = \nabla \cdot \vec{S}$, the preceding equation can be rewritten as

$$\rho \frac{\partial^2 \vec{S}}{\partial t^2} = (\lambda + 2\mu)\nabla(\nabla \vec{S}) + \mu \nabla^2 \vec{S} \qquad (2.87)$$

Because $\nabla(\nabla \vec{S}) = \nabla^2 \vec{S} + \nabla \times (\nabla \times \vec{S})$, this equation can be rewritten as

$$\rho \frac{\partial^2 \vec{S}}{\partial t^2} = (\lambda + 2\mu)\nabla(\nabla \vec{S}) - \mu \nabla \times (\nabla \times \vec{S}) \qquad (2.88)$$

If the particle's velocity is expressed as $\vec{v} = v_x \hat{i} + v_y \hat{j} + v_z \hat{k}$, so $v_x = \partial \xi / \partial x$, $v_y = \partial \eta / \partial y$, and $v_z = \partial \zeta / \partial z$, Equation 2.87 can be rewritten as

$$\rho \frac{\partial^2 \vec{v}}{\partial t^2} = (\lambda + 2\mu)\nabla(\nabla \vec{v}) - \mu \nabla \times (\nabla \times \vec{v}) \qquad (2.89)$$

It is well known that a vector field can be expressed as the sum

$$\vec{v} = \nabla\varphi + \nabla \times \vec{\psi} \qquad (2.90)$$

where
  φ is the scalar potential
  $\vec{\psi}$ is the vector potential

The scalar potential φ is irrotational, so $\nabla \times \varphi = 0$, and φ is constant outside the acoustically perturbed region. The vector potential $\vec{\psi}$ is solenoidal, so $\nabla \cdot \vec{\psi} = 0$. For a fluid, $\vec{\psi} = 0$. The velocity potential can be written as

$$\begin{cases} v_x = \dfrac{\partial \varphi}{\partial x} + \dfrac{\partial \psi_z}{\partial y} - \dfrac{\partial \psi_y}{\partial z} \\[2mm] v_y = \dfrac{\partial \varphi}{\partial y} + \dfrac{\partial \psi_x}{\partial z} - \dfrac{\partial \psi_z}{\partial x} \\[2mm] v_z = \dfrac{\partial \varphi}{\partial z} + \dfrac{\partial \psi_y}{\partial x} - \dfrac{\partial \psi_x}{\partial y} \end{cases} \qquad (2.91)$$

We substitute this equation into Equation 2.90 to separate the scalar potential φ and the vector potential $\vec{\psi}$ as two independent equations:

$$\begin{cases} \rho \dfrac{\partial^2 \varphi}{\partial t^2} = (\lambda + 2\mu)\nabla^2 \varphi \\[3mm] \rho \dfrac{\partial^2 \vec{\psi}}{\partial t^2} = \mu \nabla^2 \vec{\psi} \end{cases} \qquad (2.92)$$

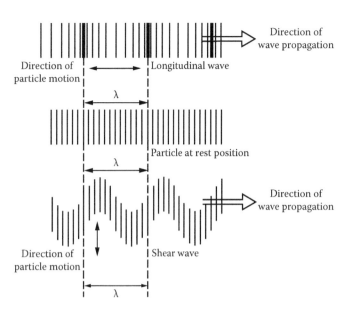

FIGURE 2.4 Vibration of particles with respect to longitudinal and shear waves.

So, the scalar potential φ and the vector potential $\vec{\psi}$ represent the longitudinal and shear waves in the solid, respectively, with the speeds of sound

$$c_L = \sqrt{\frac{\lambda + 2\mu}{\rho}}, \qquad c_T = \sqrt{\frac{\mu}{\rho}} \qquad (2.93)$$

The subscripts L and T stand for the longitudinal and shear waves, respectively.

Figure 2.4 shows that, for longitudinal waves, the motion of the particles is in the same direction as the propagation of the wave, leading to adjacent areas of compression and rarefaction in the material, while for shear waves the particle motion is perpendicular to the propagation direction. Longitudinal waves are the most important kind of ultrasound waves for biological purposes because shear waves damp out quickly in all tissues except bones.

ASSIGNMENT

Q1: There is a volumetric fluid source in the propagation medium, and the mass flow into the unit volume in unit time is $\rho_0 q(x, y, z, t)$. Derive the wave equation in terms of the medium velocity v in such a situation.

Q2: There is a volumetric force distribution in the propagation medium, and assume that the force applied on a unit volume of the medium is $F(x, y, z, t)$. Derive the wave equation in such a situation.

Q3: Assume the medium density to be inhomogeneous in the equilibrium state, that is, $\rho_0 = \rho_0(x, y, z)$. Verify that the wave equation in this case is

$$\nabla^2 p - \frac{1}{c_0^2}\frac{\partial^2 p}{\partial t^2} = \mathrm{grad}\, p \cdot \mathrm{grad}(\ln \rho_0)$$

Q4: If a wavefront plane varies according to the relationship $S = S_0(1 + \alpha_n x)^n$, where S_0 is the area at $x = 0$, and α_n is a constant, derive the wave equation in this case.

Q5: There is an infinite cylindrical sound source uniformly vibrating in the radial direction, and its radiation wavefront is cylindrical given by $S = 2\pi r$ for unit length, where r is the radius. Derive the wave equation of this acoustic field.

Q6: The surface of a sphere of mean radius $a = 5$ cm oscillates in radial harmonic motion with a frequency of 1000 Hz and a uniform velocity amplitude of 0.1 cm. Neglecting sound absorption in the air, determine the distance at which the sound pressure amplitude will be equal to the threshold of human hearing (20 µPa, rms).

REFERENCES

Auld BA. *Acoustic Fields and Waves in Solids*. Malabar, FL: Krieger Publishing, 1990.

Bilaniuk N, Wong GSK. Speed of sound in pure water as a function of temperature. *Journal of the Acoustical Society of America* 1993;93:1609–1612.

Blackstock DT. Transient solution for sound radiated into a viscous fluid. *Journal of the Acoustical Society of America* 1967;41:1312–1319.

Blackstock DT. *Fundamentals of Physical Acoustics*. New York: John Wiley & Sons, 2000.

Brillouin L. *Wave Propagation and Group Velocity*. New York: Academic Press, 1960.

Cobbold RSC, Sushilov NV, Weathermon AC. Transient propagation in media with classical or power-law loss. *Journal of the Acoustical Society of America* 2004;116:3294–3303.

Cruikshank DB. Growth of distortion in a finite-amplitude sound wave in air. *Journal of the Acoustical Society of America* 1966;40:731–733.

Del Grosso VA, Mader CW. Speed of sound in pure water. *Journal of the Acoustical Society of America* 1972;52:1442–1446.

Du GH, Zhu ZM, Gong XF. *Fundamental of Acoustics*. Shanghai, China: Shanghai Science & Technology Publishing, 1995.

Duck FA. *Physical Properties of Tissue*. London, U.K.: Academic Press, 1990.

Gurumurthy KV, Arthur RM. A dispersive model for the propagation of ultrasound in soft tissue. *Ultrasonic Imaging* 1982;4:355–377.

Hanin M. Propagation of an aperiodic wave in a compressible viscous medium. *Journal of Mathematical Physics* 1957;37:234–249.

Hassan W, Nagy PB. Simplified expressions for the displacements and stresses produced by the Rayleigh wave. *Journal of the Acoustical Society of America* 1998;104:3107–3110.

Jongen HAH, Thijssen JM, van den Aarssen M, Verhoef WA. A general model for the absorption of ultrasound by biological tissues and experimental verification. *Journal of the Acoustical Society of America* 1986;79:535–540.

Kino S. *Acoustic Waves: Devices, Imaging and Analog Signal Processing*. Englewood Cliffs, NJ: Prentice-Hall, 1986.

Leander JL. On the relation between the wavefront speed and the group velocity concept. *Journal of the Acoustical Society of America* 1996;100:3503–3507.

Leeman S. Ultrasound pulse propagation in dispersive media. *Physics in Medicine and Biology* 1980;25:481–488.

Ludwig R, Levin PL. Analytical and numerical treatment of pulsed wave propagation into a viscous fluid. *IEEE Transactions on Ultrasonics, Ferroelectrics, and Frequency Control* 1995;42:789–792.

Malecki I. *Physical Foundations of Technical Acoustics*. Oxford, U.K.: Pergamon Press, 1969.

Morse PM, Ingard KU. *Theoretical Acoustics*. New York: McGraw-Hill, 1968.

Nachman AI, Smith JF, Waag RC. An equation for acoustic propagation in inhomogeneous media with relaxation losses. *Journal of the Acoustical Society of America* 1990;88:1584–1595.

Pierce AD. *Handbook of Acoustics*. New York: John Wiley & Sons, 1998.

Rayleigh FRS. On the pressure of vibrations. *Philosophical Magazine* 1902;3:338–346.

Rayleigh FRS. On the momentum and pressure of gaseous vibrations, and on the connection with the virial theorem. *Philosophical Magazine* 1905;10:364–374.

Rayleigh JWS. *The Theory of Sound*. New York: Dover Publications, 1945.

Rose JL. *Ultrasonic Waves in Solid Media*. Cambridge, U.K.: Cambridge University Press, 1999.

Sushilov NV, Cobbold RSC. Frequency-domain wave equation and its time-domain solutions in attenuating media. *Journal of the Acoustical Society of America* 2004;115:1431–1436.

Szabo TL. *Diagnostic Ultrasound Imaging: Inside Out*. Burlington, MA: Elsevier, 2004.

Truesdell C. Mechanical foundation of elasticity and fluid dynamics. *Journal of Rational Mechanics and Analysis* 1952;1:125–171.

Truesdell C. Precise theory of the absorption and dispersion of forced plane infinitesimal waves according to the Navier-Stokes equations. *Journal of Rational Mechanics and Analysis* 1953;2:643–741.

Vincenti WG, Traugott SG. The coupling of radiative transfer and gas motion. *Annual Review of Fluid Mechanics* 1971;3:89–116.

Wang SH, Lee LP, Lee JS. A linear relation between the compressibility and density of blood. *Journal of the Acoustical Society of America* 2001;109:390–396.

Westervelt PJ. Parametric acoustic array. *Journal of the Acoustical Society of America* 1963;35:535–537.

Wood RW, Loomis AL. The physical and biological effects of high frequency sound waves of great intensity. *Philosophical Magazine* 1927;4:417–436.

Ziomek LJ. *Fundamentals of Acoustic Field Theory and Space-Time Signal Processing*. Boca Raton, FL: CRC Press, 1995.

3 Sound Reflection, Refraction, and Transmission

Using wave equation, the propagation of acoustic wave can be described. However, in the propagation pathway, the medium may not always be homogenous. In this chapter, we discuss what happens at the interface.

3.1 ACOUSTIC ENERGY AND INTENSITY

The presence of an acoustic wave causes the vibration of a particle around its equilibrium position and causes the compression or dilation in the medium (Ando 1969/70, Beissner 1974, Connor and Hynynen 2004, Ingard 2008). The former provides the kinetic energy of the particle, while the latter introduces potential energy. The sum of these two parts is the total acoustic energy due to the acoustic disturbance. Therefore, the acoustic wave actually causes the propagation of acoustic vibration energy.

If there is a sufficiently small volume element in the acoustic field with an initial volume of V_0, pressure of P_0, density of ρ_0, and velocity of v, the kinetic energy of this volume element due to the acoustic disturbance is

$$\Delta E_K = \frac{1}{2}(\rho_0 V_0)v^2 \tag{3.1}$$

In addition, because of the acoustic disturbance, the pressure of this volume element increases from P_0 to $P_0 + p$. So, the potential energy of such a volume element is

$$\Delta E_P = -\int_0^p p\,dV \tag{3.2}$$

where the negative sign represents the opposite directions of variation in pressure and volume of the element. Applying thermodynamic principle gives

$$dp = c_0^2 d\rho' \tag{3.3}$$

Because the mass of the volume element remains the same during the compression and dilation processes, the relationship between the variation in the volume and density of the element is

$$\frac{d\rho}{\rho} = -\frac{dV}{V} \quad \text{or} \quad \frac{d\rho'}{\rho} = -\frac{dV}{V} \tag{3.4}$$

For a small-amplitude wave, it can be simplified as

$$\frac{d\rho'}{\rho_0} = -\frac{dV}{V_0} \tag{3.5}$$

Substituting into Equation 3.3 gives

$$dp = -\frac{\rho_0 c_0^2}{V_0} dV \tag{3.6}$$

Substituting into Equation 3.2 gives

$$\Delta E_P = -\frac{V_0}{\rho_0 c_0^2}\int_0^p p\,dp = \frac{V_0}{2\rho_0 c_0^2} p^2 \tag{3.7}$$

So, the total acoustic energy of the volume element is

$$\Delta E = \Delta E_K + \Delta E_P = \frac{V_0}{2}\rho_0\left(v^2 + \frac{1}{\rho_0^2 c_0^2} p^2\right) \tag{3.8}$$

The acoustic energy density w is defined as the acoustic energy in unit volume:

$$w = \frac{\Delta E}{V_0} = \frac{1}{2}\rho_0\left(v^2 + \frac{1}{\rho_0^2 c_0^2} p^2\right) \tag{3.9}$$

Thus, the expression for the total acoustic energy for a plane wave is

$$\Delta E = \frac{V_0}{2}\rho_0\left[\frac{p_A^2}{\rho_0^2 c_0^2}\cos^2(\omega t - kx) + \frac{p_A^2}{\rho_0^2 c_0^2}\cos^2(\omega t - kx)\right]$$

$$= V_0 \frac{p_A^2}{\rho_0 c_0^2}\cos^2(\omega t - kx) \tag{3.10}$$

It can be shown that the variations in the kinetic and potential energies at any position in the planar acoustic field are in phase. So, the total acoustic energy increases to its maximum value of $V_0\left(p_A^2/\rho_0 c_0^2\right)$, which is twice the maximum value of the kinetic or potential energy. Because the acoustic field is a nonconservative system, energy is transferable.

Equation 3.10 is the instantaneous acoustic energy of the volume, and its time-averaged value is

$$\overline{\Delta E} = \frac{1}{T}\int_0^T \Delta E\,dt = \frac{1}{2}V_0 \frac{p_A^2}{\rho_0 c_0^2} \tag{3.11}$$

and the time-averaged acoustic energy density in a unit volume is

$$\bar{w} = \frac{\overline{\Delta E}}{V_0} = \frac{1}{2} \frac{p_A^2}{\rho_0 c_0^2} = \frac{p_e^2}{\rho_0 c_0^2} \quad (3.12)$$

where $p_e = p_A/\sqrt{2}$ is the effective acoustic pressure. In an ideal planar acoustic field, the amplitude of acoustic pressure is a distance-invariant constant. So, the average acoustic energy density is equal everywhere, which is another characteristic of a planar acoustic field in an ideal medium.

Using the expressions of both particle velocity and pressure, the power density (intensity) carried by the wave can be obtained. The instantaneous power flowing through a unit area perpendicular to the direction of wave propagation as one elemental volume of the fluid acts on a neighboring element is defined as the wave's power density:

$$\text{Power density} = \frac{\text{power}}{\text{area}} = \frac{\text{work}}{\text{area} \cdot \text{time}} = \frac{\text{force} \cdot \text{distance}}{\text{area} \cdot \text{time}}$$

$$= \text{pressure} \cdot \text{velocity}$$

$$I = pu = \frac{p^2}{Z} = Zu^2 \quad (3.13)$$

The average acoustic power is defined as the average acoustic energy flowing normally through an area in unit time, that is,

$$\overline{W} = \bar{w} c_0 S \quad (3.14)$$

where
  c_0 is the speed of sound
  S is the area

The unit of acoustic power is watts. Acoustic intensity or the average energy density is the average acoustic energy flowing through a unit area normal to the sound propagation direction:

$$I = \frac{\overline{W}}{S} = \bar{w} c_0 \quad (3.15)$$

According to the acoustic intensity, it can also be rewritten as the work done to the neighboring medium in the acoustic wave propagating direction at a unit area in unit time:

$$I = \frac{1}{T} \int_0^T \text{Re}(p) \text{Re}(v) dt \quad (3.16)$$

where Re(*) means the real part. The unit of acoustic intensity is W/m².

For a planar acoustic wave along the positive x direction, the acoustic intensity is

$$I = \frac{p_A^2}{2\rho_0 c_0} = \frac{p_e^2}{\rho_0 c_0} = \frac{1}{2} \rho_0 c_0 v_A^2 = \rho_0 c_0 v_e^2 = \frac{1}{2} p_A v_A = p_e v_e \quad (3.17)$$

where v_e is the effective particle velocity, $v_e = v_A/\sqrt{2}$. In contrast, for a planar acoustic wave along the positive x direction, the acoustic intensity is

$$I = -\bar{\varepsilon} c_0 = -\frac{p_A^2}{2\rho_0 c_0} = -\frac{1}{2} \rho_0 c_0 v_A^2 \quad (3.18)$$

It is seen that the sign of acoustic intensity is that of the acoustic wave propagation direction. Thus, the average power density (or intensity) for a sinusoidal wave is just one-half of its peak value, and a time average of I will yield the same result. It is also seen that doubling the magnitude of pressure will quadruple the acoustic power, which is analogous to the relationship between voltage and power in an electrical circuit.

3.2   SOUND PRESSURE AND INTENSITY LEVEL

There is great variation in the acoustic vibration energy. For example, the acoustic powers in human speech and a rocket are 10^{-5} and 10^9 W, respectively, differing by several orders. For easy comparison, the logarithmic expression of sound pressure and intensity is usually used in acoustics because it compresses the large range of typical sound pressures into a smaller, more practical scale, which incidentally also more closely parallels the human ear's ability to judge the relative loudness of sounds according to the ratio of their pressure.

Sound pressure level is represented by SPL and defined as

$$SPL = 20 \log_{10} \frac{p_e}{p_{\text{ref}}} (\text{dB}) \quad (3.19)$$

where $p_{\text{ref}} = 2 \times 10^{-5}$ Pa, which is the hearing threshold of humans at 1 kHz. In other words, the sound pressure level at such a hearing threshold is 0 dB.

Sound intensity level is represented by SIL and defined as

$$SIL = 10 \log_{10} \frac{I}{I_{\text{ref}}} (\text{dB}) \quad (3.20)$$

where $I_{\text{ref}} = 10^{-12}$ W/m², which is the corresponding acoustic intensity of pressure at 2×10^{-5} Pa in air:

$$SIL = 10 \log_{10} \frac{I}{I_{\text{ref}}} = 10 \log_{10} \left(\frac{p_e^2}{\rho_0 c_0} \cdot \frac{400}{p_{\text{ref}}^2} \right)$$

$$= SPL + 10 \log_{10} \frac{400}{\rho_0 c_0} \quad (3.21)$$

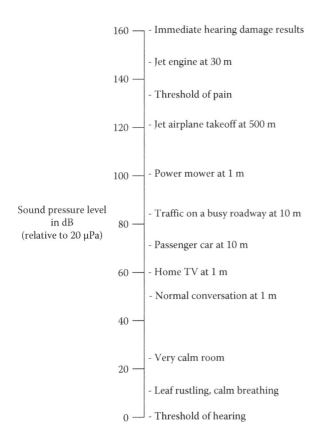

FIGURE 3.1 Sound pressure level in decibels (dB) in common environments.

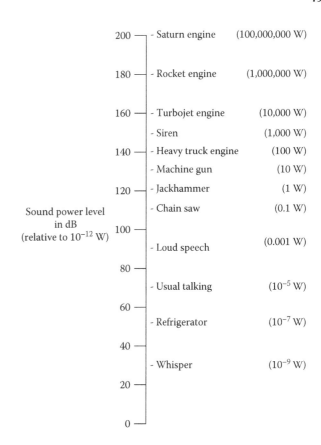

FIGURE 3.2 Sound intensity level in dB produced by objects.

It is clear that there is a very small modification term, $10\log_{10}(400/\rho_0 c_0)$, between sound pressure level and sound intensity level. So, if $\rho_0 c_0 = 400$ in the measurement, such as in air, $SIL = SPL$. The sound pressures and intensity levels encountered in everyday life are shown in Figures 3.1 and 3.2, respectively.

3.3 BOUNDARY CONDITIONS

Biomedical ultrasound application, either diagnosis or therapy, is based on the penetration of ultrasound waves into biological tissue or cells. Reflection of ultrasound waves at the various internal interfaces of the object is of importance. Whenever a wave passes from one region to a neighboring one with a different acoustic impedance, a certain amount of the incident power is reflected at the boundary and the remainder continues as a transmitted wave. The reflected wave serves as an indicator of the boundary position and shape, whereas the transmitted portion penetrates deeper into the interfaces. Such reflection and transmission are important characteristics of acoustic wave propagation.

The reflection, refraction, and transmission of sound waves occur at the interface of two media, so the boundary conditions determine such an acoustic phenomenon. Assume that there are two semi-infinite ideal fluids with acoustic impedance $\rho_1 c_1$ and $\rho_2 c_2$, respectively. Assume that at the interface,

there is an element with area A, weight ΔM, and negligible thickness. Its motion equation using Newton's second law is

$$[p(1) - p(2)]A = \Delta M \frac{d^2 x}{dt^2} = \Delta M \frac{dv}{dt} \qquad (3.22)$$

Because the thickness of this element is nearly zero and the acceleration cannot be infinite, there must be such a relationship in order to satisfy Equation 3.22:

$$p(1) - p(2) = 0 \qquad (3.23)$$

Such a relationship also exists in the absence of the sound wave, that is, when the static pressure at the interface of two media is continuous:

$$p_0(1) = p_0(2) \qquad (3.24)$$

When an acoustic wave propagates through this interface, $p(1) = p_0(1) + p_1$ and $p(2) = p_0(2) + p_2$, then

$$p_1 = p_2 \qquad (3.25)$$

It seems that the acoustic pressure at the interface of two media is continuous. In addition, it is also required that the net velocity components of the particles perpendicular to the left boundary side, v_1, must be equal to the perpendicular velocity component on the right side, v_2. Otherwise, the media would pull apart from each other at the boundary:

$$v_1 = v_2 \qquad (3.26)$$

Together, Equations 3.25 and 3.26 are the acoustic boundary conditions between two media.

3.4 NORMAL INCIDENCE

Consider the case shown in Figure 3.3 for a semi-infinite fluid interface at $x = 0$, and the acoustic impedances of the medium I and medium II are $Z_1 = \rho_1 c_1$ and $Z_2 = \rho_2 c_2$, respectively. A planar acoustic wave $p_i = p_{iA} e^{j(\omega t - kx)}$ is incident normally on the interface from medium I. Because of the impedance mismatch, some part of the sound wave will be reflected, while the other will be transmitted into medium II. The one-dimensional sound pressure p_1 in medium I can be expressed as

$$p_1 = A e^{j(\omega t - kx)} + B e^{j(\omega t + kx)} \qquad (3.27)$$

The first item represents the acoustic wave propagating in the x direction, while the second one is the wave propagating in the $-x$ direction. In this case, these two terms correspond to the incident wave and reflected wave $p_r = p_{rA} e^{j(\omega t + kx)}$, respectively. So, Equation 3.27 can be rewritten as

$$p_1 = p_i + p_r = p_{iA} e^{j(\omega t - kx)} + p_{rA} e^{j(\omega t + kx)} \qquad (3.28)$$

However, in the semi-infinite medium II, there will be no wave propagating in the $-x$ direction. So, the transmitted wave can be expressed as

$$p_2 = p_t = p_{tA} e^{j(\omega t - kx)} \qquad (3.29)$$

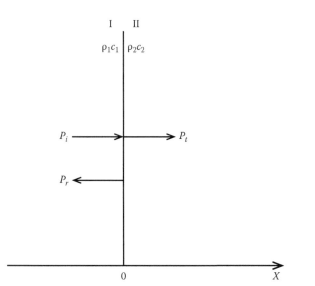

FIGURE 3.3 Normal incidence of an acoustic wave at the interface of two media.

The particle velocities in media I and II, v_1 and v_2, are

$$\begin{cases} v_1 = v_{iA} e^{j(\omega t - kx)} + v_{rA} e^{j(\omega t + kx)} \\ v_2 = v_{tA} e^{j(\omega t - kx)} \end{cases} \qquad (3.30)$$

where

$$v_{iA} = \frac{p_{iA}}{Z_1} = \frac{p_{iA}}{\rho_1 c_1}, \qquad v_{rA} = \frac{p_{rA}}{Z_1} = \frac{p_{rA}}{\rho_1 c_1}, \qquad v_{tA} = \frac{p_{tA}}{Z_2} = \frac{p_{tA}}{\rho_2 c_2}$$

Applying the acoustic boundary conditions at the interface, with the acoustic pressure and normal particle velocity being continuous, gives

$$p_1 \big|_{x=0} = p_2 \big|_{x=0} \quad \Rightarrow \quad p_{iA} + p_{rA} = p_{tA} \qquad (3.31)$$

$$v_1 \big|_{x=0} = v_2 \big|_{x=0} \quad \Rightarrow \quad v_{iA} + v_{rA} = v_{tA} \qquad (3.32)$$

From these equations, the reflection and transmission coefficients for sound pressure and velocity can be derived as

$$\begin{cases} R_p = \dfrac{p_{rA}}{p_{iA}} = \dfrac{Z_2 - Z_1}{Z_1 + Z_2} = \dfrac{Z_{12} - 1}{Z_{12} + 1} \\[2mm] R_v = \dfrac{v_{rA}}{v_{iA}} = \dfrac{Z_1 - Z_2}{Z_1 + Z_2} = \dfrac{1 - Z_{12}}{1 + Z_{12}} \\[2mm] T_p = \dfrac{p_{tA}}{p_{iA}} = \dfrac{2Z_2}{Z_1 + Z_2} = \dfrac{2Z_{12}}{1 + Z_{12}} \\[2mm] T_v = \dfrac{v_{tA}}{v_{iA}} = \dfrac{2Z_1}{Z_1 + Z_2} = \dfrac{2}{1 + Z_{12}} \end{cases} \qquad (3.33)$$

where $Z_{12} = \dfrac{Z_2}{Z_1}$ and $Z_{21} = \dfrac{Z_1}{Z_2}$. The relationship of acoustic energy through the interface can be expressed as

$$R_I = \frac{I_r}{I_i} = \frac{|p_{rA}|^2}{2\rho_1 c_1} \bigg/ \frac{|p_{iA}|^2}{2\rho_1 c_1} = \left(\frac{Z_2 - Z_1}{Z_2 + Z_1}\right)^2 = \left(\frac{Z_{12} - 1}{Z_{12} + 1}\right)^2 \qquad (3.34)$$

$$T_I = \frac{I_t}{I_i} = \frac{|p_{tA}|^2}{2\rho_2 c_2} \bigg/ \frac{|p_{iA}|^2}{2\rho_1 c_1} = \frac{4Z_1 Z_2}{(Z_1 + Z_2)^2} = \frac{4Z_{12}}{(1 + Z_{12})^2}$$

$$= (1 + R_I)^2 \frac{Z_2}{Z_1} \qquad (3.35)$$

where

R_I is the ratio of the acoustic intensity of the reflected wave to that of the incident wave

T_I is the ratio of the acoustic intensity of the transmitted wave to that of the incident wave

According to the principle of energy conservation, the total amount of power entering the boundary must be equal to the total amount of power leaving it. Thus,

$$I_i A_i = I_t A_t + I_r A_r \qquad (3.36)$$

where A_i, A_t, and A_r are the cross-sectional areas of the incident, transmitted, and reflected beams, respectively. Since $A_i = A_r$, the ratio of transmitted beam power to that of incident beam power is

$$\frac{\text{Transmitted beam power}}{\text{Incident beam power}} = \frac{I_t A_t}{I_i A_i} = 1 - \frac{I_r}{I_i} = 1 - R_I^2 \quad (3.37)$$

So, it is found that

$$\frac{I_r}{I_i} + \frac{I_t}{I_i} = R_1 + T_1 = 1 \qquad (3.38)$$

This equation derived from the conservation of total power using the ratios of intensity is valid only when the incident beam is normal to the interface. Although the conservation of power must always hold, intensity is not conserved when the cross-sectional area of the beam changes as the beam crosses the boundary. For oblique angles of incidence, the area of the beam will be altered as it passes across the interface, so $A_i \neq A_t$, which will be mentioned later.

It is clear that the reflection and transmission at the interface depend on the acoustic impedances of the media. Several usual situations will be discussed individually in the following:

1. $Z_1 = Z_2$ or $Z_{12} = 1$

Since media I and II are acoustically identical, it can be derived easily from Equation 3.33 that

$$R_p = R_v = 0, \quad T_p = T_v = 1 \qquad (3.39)$$

which illustrates that there is no reflection at the interface, and all acoustic energy is transmitted into medium II. In the other words, the interface seems to be nonexistent in the acoustic propagation pathway:

2. $Z_2 > Z_1$ or $Z_{12} > 1$, so

$$R_p > 0, \quad R_v < 0, \quad T_p > 0, \quad T_v > 0 \qquad (3.40)$$

This boundary is usually called the "hard" boundary since medium II is "harder" than medium I in acoustics. Similar to a ball bouncing back from a rigid wall, the particle velocity of the reflected wave has a phase change of 180° or π in comparison to the incident wave. However, the acoustic pressures of the reflected and incident waves have the same phase.

3. $Z_2 < Z_1$ or $Z_{12} < 1$, so

$$R_p < 0, \quad R_v > 0, \quad T_p > 0, \quad T_v > 0 \qquad (3.41)$$

This is a "soft" boundary because medium II is acoustically "softer" than medium I. So, there is a phase change of 180° or π between the reflected and incident acoustic pressures, but not the particle velocity.

4. $Z_2 \gg Z_1$ or $Z_{12} \approx \infty$; it can be easily derived that

$$R_p \approx 1, \quad R_v \approx -1, \quad T_p \approx 2, \quad T_v \approx 0 \qquad (3.42)$$

Because medium II is much "harder" than medium I, such as when a sound wave is incident from air to water, complete reflection occurs at this interface. Particle velocities of the reflected wave, v_r, and the incident wave, v_i, have the same magnitude but opposite phases. So, at the interface, the net particle velocity is 0. In contrast, the acoustic pressures of the reflected wave and incident wave have the same magnitude as well as phase. In medium II, there is no acoustic wave propagation. The pressure at the interface, $p_t = 2p_i$, is statically transferred to medium II and not in an oscillatory manner as an acoustic wave. In addition, there exists an acoustic standing wave in medium I as a result of the superposition of the incident wave and the reflected wave. The detailed characteristics of standing waves will be discussed in the next chapter. Therefore, short ultrasound pulses are usually used in the transcranial applications to avoid the unintentional generation of standing waves inside the skull (Connor and Hynynen 2004). However, the power transmission coefficient $\tau \approx 0$, so large and thick bones would block the propagation of ultrasound waves.

5. $Z_2 \ll Z_1$ or $Z_{12} \approx 0$; one can easily derive

$$R_p \approx -1, \quad R_v \approx 1, \quad T_p \approx 0, \quad T_v \approx 2 \qquad (3.43)$$

Full reflection will also occur at this very "soft" boundary, such as when a sound wave is incident from water to air. Therefore, ultrasound transmission into a gaseous medium (such as the lung or intestinal gas) is difficult. In order to achieve an excellent coupling effect, such gas should be carefully removed in both ultrasound imaging and therapy.

3.5 OBLIQUE INCIDENCE

The situation of an incident plane wave impinging upon a plane boundary between two media with different acoustic impedances is shown in Figure 3.4. In response to the vibrations caused by the incident wave at the interface, a reflected wave and a transmitted (refracted) wave are generated. If the perpendicular spacing between the phase fronts in medium I is defined as the wavelength λ_1, the corresponding

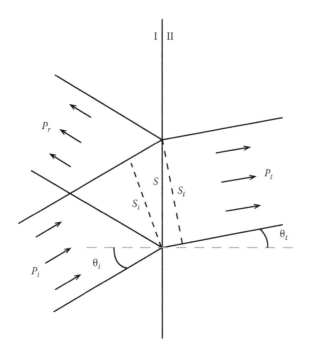

FIGURE 3.4 Three waves at the interface with angles such that the projections of their phase front distance d onto the boundary are equally spaced.

which is known as Snell's law, which applies as well to planar electromagnetic waves as to planar acoustic waves. It shows that whenever a wave strikes the boundary between two media with different phase velocities at a nonperpendicular angle, the transmitted wave will propagate at an angle that is different from the angle of incidence. This phenomenon is known as refraction of the transmitted wave, and the ratio on the right side of Snell's law is known as the index of refraction of medium II with respect to medium I. Note that for the special case when $\theta_i = 0$ (normal incidence) $\theta_t = 0$, and there is no refraction in this case regardless of the ratio of the phase velocities in the two regions.

Snell's law gives the precise angle of refraction for any interface. In practice, it is frequently required to quickly discern the direction of refraction rather than the exact angle or estimate rapidly the direction of bending of the wave propagation. As a simple memory aid for such a purpose, on entering a region of low velocity from a region of high velocity, the acoustic wave will bend such that the transmitted wave will propagate at an angle smaller than the angle of incidence, $\theta_i > \theta_t$. In the opposite case, when the acoustic wave passes from a region of lower velocity into a region of higher velocity, the transmitted wave will bend further away from the normal direction, $\theta_i < \theta_t$.

Assume that the angle between the propagation direction of the incident plane wave and the x-axis is θ_i and the interface between medium I and medium II is at $x = 0$. $\vec{n} = \cos\alpha \hat{i} + \cos\beta \hat{j}$, $\vec{r} = x\hat{i} + y\hat{j}$, $x = \vec{n} \cdot \vec{r}$, α and β are the angle between normal line of wave front and the x- and y-axes, respectively (Figure 3.5):

spacing between these phase fronts along the interface can be given by simple geometry as

$$d = \frac{\lambda_1}{\sin\theta_i} \tag{3.44}$$

$$p = p_A e^{j(\omega t - k\vec{n}\cdot\vec{r})} \tag{3.47}$$

Since the vibration at the interface caused by the incident wave generates the radiation of both a reflected and a transmitted waves, the projected distance between the phase fronts of these waves, d, must exactly match those of the incident wave along the entire boundary. It is easy to derive that, in order to match the spacing d, the reflected wave must be radiated back at a reflection angle equal to the angle of incidence because of the same wavelength of the incident and reflected waves:

$$\theta_r = \theta_i \tag{3.45}$$

which is the law of specular or "mirrorlike" reflection and is consistent with the similar phenomenon in optics, for light reflection from a mirror, because $\lambda = c/f$ and $c_1 = c_2$. However, the transmitted wave has a different wavelength in medium II than in medium I. In order for the transmitted wave to match the spacing of its phase fronts along the boundary to d, it must be tilted at an angle different from the angle of the incident wave. Geometry shows that the spacing d will be matched when

$$\frac{\sin\theta_i}{\sin\theta_t} = \frac{\lambda_1}{\lambda_2} = \frac{c_1}{c_2} \tag{3.46}$$

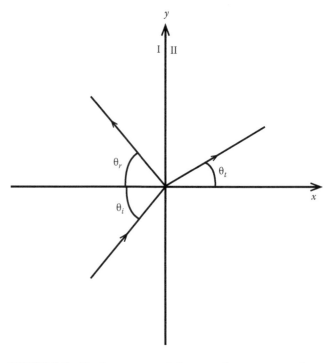

FIGURE 3.5 Total pressure and the normal components of particle velocity must be continuous across the boundary for an oblique incident wave.

If $k\vec{n} = \vec{k}$, the wave vector, $\vec{k} \cdot \vec{r} = k\vec{n} \cdot \vec{r} = k\cos\alpha x + k\cos\beta y$. Then the earlier equation can be rewritten as

$$p = p_A e^{j(\omega t - \vec{k} \cdot \vec{r})} = p_A e^{j(\omega t - kx\cos\alpha - ky\cos\beta)} \qquad (3.48)$$

The velocity along the x- and y-axes can be obtained as

$$\begin{cases} v_x = -\dfrac{1}{\rho_0} \displaystyle\int \dfrac{\partial p}{\partial x} dt = \dfrac{\cos\alpha}{\rho_0 c_0} p \\[3mm] v_y = -\dfrac{1}{\rho_0} \displaystyle\int \dfrac{\partial p}{\partial y} dt = \dfrac{\cos\beta}{\rho_0 c_0} p \end{cases} \qquad (3.49)$$

In Figure 3.5, $\alpha = \theta_i$ and $\beta = 90° - \theta_i$, so the sound pressure p and particle velocity in the x direction are

$$\begin{cases} p_i = p_{iA} e^{j(\omega t - k_1 x\cos\theta_i - k_1 y\sin\theta_i)} \\[3mm] v_{ix} = \dfrac{\cos\theta_i}{\rho_1 c_1} p_i \end{cases} \qquad (3.50)$$

where $k_1 = \omega/c_1$. The reflected wave is also in the xy plane with the angles $\alpha = 180° - \theta_r$ and $\beta = 90° - \theta_r$, so the sound pressure and particle velocity in the x direction are

$$\begin{cases} p_r = p_{rA} e^{j(\omega t + k_1 x\cos\theta_r - k_1 y\sin\theta_r)} \\[3mm] v_{rx} = -\dfrac{\cos\theta_r}{\rho_1 c_1} p_r \end{cases} \qquad (3.51)$$

Therefore, the acoustic field in medium I is the sum of the incident and reflected waves:

$$\begin{cases} p_1 = p_i + p_r = p_{iA} e^{j(\omega t - k_1 x\cos\theta_i - k_1 y\sin\theta_i)} + p_{rA} e^{j(\omega t + k_1 x\cos\theta_r - k_1 y\sin\theta_r)} \\[3mm] v_{1x} = v_{ix} + v_{rx} = \dfrac{\cos\theta_i}{\rho_1 c_1} p_{iA} e^{j(\omega t - k_1 x\cos\theta_i - k_1 y\sin\theta_i)} \\[3mm] \qquad - \dfrac{\cos\theta_r}{\rho_1 c_1} p_{rA} e^{j(\omega t + k_1 x\cos\theta_r - k_1 y\sin\theta_r)} \end{cases} \qquad (3.52)$$

However, there is only one refracted wave in medium II. Assume that the angle between the propagation direction of the refracted wave and the x-axis is θ_t; then $\alpha = \theta_t$ and $\beta = 90° - \theta_t$:

$$\begin{cases} p_t = p_{tA} e^{j(\omega t - k_2 x\cos\theta_t - k_2 y\sin\theta_t)} \\[3mm] v_{tx} = \dfrac{\cos\theta_t}{\rho_2 c_2} p_t \end{cases} \qquad (3.53)$$

where $k_2 = \omega/c_2$. At $x = 0$,

$$\begin{cases} p_{iA} e^{-jk_1 y\sin\theta_i} + p_{rA} e^{-jk_1 y\sin\theta_r} = p_{tA} e^{-jk_2 y\sin\theta_t} \\[3mm] \dfrac{\cos\theta_i}{\rho_1 c_1} p_{iA} e^{-jk_1 y\sin\theta_i} - \dfrac{\cos\theta_r}{\rho_1 c_1} p_{rA} e^{-jk_1 y\sin\theta_r} = \dfrac{\cos\theta_t}{\rho_2 c_2} p_{tA} e^{-jk_2 y\sin\theta_t} \end{cases}$$

$$(3.54)$$

This equation should be valid for any y value at $x = 0$. In order to satisfy this condition, the exponential factor of every term should be equal:

$$k_1 \sin\theta_i = k_1 \sin\theta_r = k_2 \sin\theta_t \qquad (3.55)$$

So,

$$\begin{cases} \theta_i = \theta_r \\[3mm] \dfrac{\sin\theta_i}{\sin\theta_t} = \dfrac{k_2}{k_1} = \dfrac{c_1}{c_2} = \dfrac{\lambda_1}{\lambda_2} \end{cases} \qquad (3.56)$$

which is an alternative derivation of Snell's law for reflection and refraction. It can be seen that at the medium interface, the reflection angle is equal to the incident angle and that the refraction angle depends on the ratio of sound speed in the two media. The higher the sound speed in medium II, the larger the refraction angle. Equation 3.54 can be simplified as

$$\begin{cases} p_{iA} + p_{rA} = p_{tA} \\[3mm] \dfrac{\cos\theta_i}{\rho_1 c_1} p_{iA} - \dfrac{\cos\theta_r}{\rho_1 c_1} p_{rA} = \dfrac{\cos\theta_t}{\rho_2 c_2} p_{tA} \end{cases} \qquad (3.57)$$

Then, the ratios of the reflected wave and transmitted wave to the incident wave, r_p and t_p, are

$$\begin{cases} r_p = \dfrac{p_{rA}}{p_{iA}} = \dfrac{\rho_2 c_2 \cos\theta_i - \rho_1 c_1 \cos\theta_t}{\rho_2 c_2 \cos\theta_i + \rho_1 c_1 \cos\theta_t} = \dfrac{\dfrac{\rho_2 c_2}{\cos\theta_t} - \dfrac{\rho_1 c_1}{\cos\theta_i}}{\dfrac{\rho_2 c_2}{\cos\theta_t} + \dfrac{\rho_1 c_1}{\cos\theta_i}} \\[6mm] t_p = \dfrac{p_{tA}}{p_{iA}} = \dfrac{2\rho_2 c_2 \cos\theta_i}{\rho_2 c_2 \cos\theta_i + \rho_1 c_1 \cos\theta_t} = \dfrac{2\dfrac{\rho_2 c_2}{\cos\theta_t}}{\dfrac{\rho_2 c_2}{\cos\theta_t} + \dfrac{\rho_1 c_1}{\cos\theta_i}} \end{cases} \qquad (3.58)$$

If $z_1 = \dfrac{p_i}{v_{ix}} = \dfrac{\rho_1 c_1}{\cos\theta_i}$ and $z_2 = \dfrac{p_2}{v_{ix}} = \dfrac{\rho_2 c_2}{\cos\theta_t}$ are the normal acoustic impedances, then Equation 3.58 can be rewritten as

$$\begin{cases} r_p = \dfrac{z_2 - z_1}{z_2 + z_1} \\[3mm] t_p = \dfrac{2z_2}{z_2 + z_1} \end{cases} \qquad (3.59)$$

So, they have similar expression as the normal incidence case, substituting the acoustic impedance Z by the normal acoustic impedance z. If the ratios of the density and wave number are $m = \dfrac{\rho_2}{\rho_1}$ and $n = \dfrac{k_2}{k_1} = \dfrac{c_1}{c_2}$, respectively, then Equation 3.59 can be rewritten as

$$\begin{cases} r_p = \dfrac{m\cos\theta_i - \sqrt{n^2 - \sin^2\theta_i}}{m\cos\theta_i + \sqrt{n^2 - \sin^2\theta_i}} \\[4mm] t_p = \dfrac{p_{tA}}{p_{iA}} = \dfrac{2m\cos\theta_i}{m\cos\theta_i + \sqrt{n^2 - \sin^2\theta_i}} \end{cases} \tag{3.60}$$

The ratios of the acoustic intensities are

$$\begin{cases} r_I = \dfrac{|p_{rA}|^2 / 2\rho_1 c_1}{|p_{iA}|^2 / 2\rho_1 c_1} = \dfrac{(\rho_2 c_2 \cos\theta_i - \rho_1 c_1 \cos\theta_t)^2}{(\rho_2 c_2 \cos\theta_i + \rho_1 c_1 \cos\theta_t)^2} \\[4mm] t_I = \dfrac{|p_{tA}|^2 / 2\rho_2 c_2}{|p_{iA}|^2 / 2\rho_1 c_1} = \dfrac{4\rho_1 c_1 \rho_2 c_2 \cos^2\theta_i}{(\rho_2 c_2 \cos\theta_i + \rho_1 c_1 \cos\theta_t)^2} \end{cases} \tag{3.61}$$

It is found that $r_I + t_I \neq 1$. When considering the sound beam of the incident and refracted waves, the ratio of the acoustic energy of the transmitted wave to that of the incident wave is

$$t_{\bar{W}} = \frac{I_t S_t}{I_i S_i} = \frac{t_I \cos\theta_t}{\cos\theta_i} = \frac{4\rho_1 c_1 \rho_2 c_2 \cos\theta_i \cos\theta_t}{(\rho_2 c_2 \cos\theta_i + \rho_1 c_1 \cos\theta_t)^2} \tag{3.62}$$

Since the reflected angle is equal to the incident angle, the sound beam of the reflected wave is same as that of the incident one. Therefore,

$$r_{\bar{W}} = r_I \tag{3.63}$$

It can be proven that $r_{\bar{W}} + t_{\bar{W}} = 1$, or the sum of the acoustic energies of reflected and refracted waves is equal to the that of the incident wave, which is expected from the energy conservation law.

We now discuss the various situations.

3.5.1 COMPLETE TRANSMISSION

When the incident angle θ_i satisfies $m\cos\theta_i - \sqrt{n^2 - \sin^2\theta_i} = 0$, the incident angle is

$$\sin\theta_{i0} = \sqrt{\frac{m^2 - n^2}{m^2 - 1}} \tag{3.64}$$

So, $r_p = 0$ and $t_p = 1$. That is, when the acoustic wave is incident at the interface at the angle θ_{i0}, there will be no reflection and the entire wave is transmitted into medium II. So, θ_{i0} is

called the complete transmission angle. In order to validate Equation 3.61, $0 \le \dfrac{m^2 - n^2}{m^2 - 1} \le 1$, or

$$\begin{cases} m > n > 1, & \text{for } m > 1 \\ m < n < 1, & \text{for } m < 1 \end{cases} \tag{3.65}$$

The former corresponds to $Z_2 > Z_1$, $c_1 > c_2$, while the latter to $Z_2 < Z_1$, $c_1 < c_2$.

3.5.2 COMPLETE REFLECTION

From Snell's law, it can be seen that, when $c_2 \le c_1$ ($n \ge 1$), always $\theta_t \le \theta_i$. However, when $c_2 > c_1$, $\theta_t > \theta_i$, or when the sound speed in medium II is larger than that in medium I, the refraction angle is always larger than the incident angle. When the incident angle is larger than a certain value (θ_{i0}), $\theta_t = 90°$ (the refracted wave propagates along the interface):

$$\theta_{i0} = \sin^{-1}\frac{c_1}{c_2} \tag{3.66}$$

where θ_{i0} is the complete reflection angle. When $\theta_i > \theta_{i0}$, $\sin\theta_t > 1$, meaning θ_t is not a real number, and there is no normal refracted wave in medium II. In this case, the reflection angle is equal to the incident angle, and the reflection coefficient is a complex number with the magnitude of 1. All acoustic energy of the incident wave is completely reflected into medium I but with a phase change.

3.5.3 GLANCING INCIDENCE

When $\theta_i = 90°$ (incident), $|r_p| \approx 1$ no matter what the acoustic impedances of medium I and medium II. If $c_2 > c_1$ ($n < 1$), the incident angle is definitely larger than the critical reflection angle θ_{i0}. However, when $c_2 < c_1$ ($n > 1$), complete reflection occurs only at $\theta_i = 90°$.

3.5.4 PERPENDICULAR TRANSMISSION

If the speed of sound in medium II is much smaller than that in medium I, $c_2 \ll c_1$, for an arbitrary incident angle $\theta_t \approx 0°$, or the refracted wave is always perpendicular to the interface.

Note that the boundary conditions used in this section have not attempted to match wave values tangential to the boundary, which is possible in the case of longitudinal waves and shear waves. Reflected and transmitted shear waves will therefore generally be set up in addition to reflected and transmitted longitudinal waves by an incident longitudinal wave when it strikes a boundary at a nonzero angle. This phenomenon is known as mode conversion. However, except for propagation in bones, shear waves decay very rapidly, and only the longitudinal waves remain at distances useful for biomedical applications.

3.6 MIDDLE-LAYER TRANSMISSION

In the practice, the acoustic wave propagates through a number of layers. For example, in abdominal ultrasound diagnosis, skin, fat, and muscle are the layers lying between the ultrasound transducer and the target. In Figure 3.6, it is assumed that there is a layer of medium with a thickness of D and acoustic impedance $Z_2 = \rho_2 c_2$ in an infinite medium with $Z_1 = \rho_1 c_1$. When a planar acoustic wave (p_i, v_i) is incident normally to the anterior surface of the middle layer, a part of the energy is reflected to medium I as reflected wave (p_{1r}, v_{1r}) and the other part is transmitted into the middle layer as (p_{2t}, v_{2t}). When the acoustic wave (p_{2t}, v_{2t}) meets the posterior surface because of the changes in acoustic impedance, some of it is reflected as (p_{2r}, v_{2r}) and the other is transmitted into medium I as (p_t, v_t). Using the coordinates in Figure 3.6, all wave components are described as

$$
\begin{cases}
p_i = p_{iA} e^{j(\omega t - k_1 x)} \\
v_i = v_{iA} e^{j(\omega t - k_1 x)} \\
p_{1r} = p_{1rA} e^{j(\omega t + k_1 x)} \\
v_{1r} = v_{1rA} e^{j(\omega t + k_1 x)} \\
p_{2t} = p_{2tA} e^{j(\omega t - k_1 x)} \\
v_{2t} = v_{2tA} e^{j(\omega t - k_1 x)} \\
p_{2r} = p_{2rA} e^{j(\omega t + k_2 x)} \\
v_{2r} = v_{2rA} e^{j(\omega t + k_2 x)}
\end{cases}
\tag{3.67}
$$

where $k_1 = \omega/c_1$ and $k_2 = \omega/c_2$. At $x = D$, the acoustic pressure and particle velocity are

$$
\begin{cases}
p_t = p_{tA} e^{j[\omega t - k_1(x-D)]} \\
v_t = v_{tA} e^{j[\omega t - k_1(x-D)]}
\end{cases}
\tag{3.68}
$$

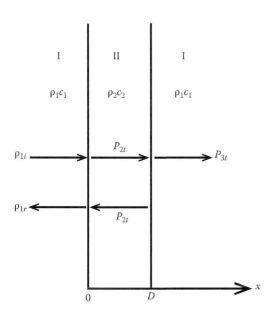

FIGURE 3.6 Acoustic wave transmission through a layer with thickness D.

The acoustic fields at $x < 0$ and $0 < x < D$ are superpositions of (p_i, v_i) with (p_{1r}, v_{1r}) and (p_{2t}, v_{2t}) with (p_{2r}, v_{2r}), respectively. At $x = 0$, the acoustic continuity conditions are

$$
\begin{cases}
p_{iA} + p_{1rA} = p_{2tA} + p_{2rA} \\
v_{iA} + v_{1rA} = v_{2tA} + v_{2rA}
\end{cases}
\tag{3.69}
$$

Similarly, the acoustic pressure and normal particle velocity are continuous at $x = D$:

$$
\begin{cases}
p_{2tA} e^{-jk_2 D} + p_{2rA} e^{jk_2 D} = p_{tA} \\
v_{2tA} e^{-jk_2 D} + v_{2rA} e^{jk_2 D} = v_{tA}
\end{cases}
\tag{3.70}
$$

Since all waves are planar,

$$
\begin{cases}
v_{iA} = \dfrac{p_{iA}}{Z_1}, \qquad v_{1rA} = -\dfrac{p_{1rA}}{Z_1} \\[2mm]
v_{2tA} = \dfrac{p_{2tA}}{Z_2}, \qquad v_{2rA} = -\dfrac{p_{2rA}}{Z_2} \\[2mm]
v_{tA} = \dfrac{p_{tA}}{Z_1}
\end{cases}
\tag{3.71}
$$

Substituting Equation 3.71 into Equations 3.69 and 3.70, the ratio of acoustic pressure of the transmitted wave at $x = D$ to that of the incident wave at $x = 0$ is

$$
t_p = \frac{p_{tA}}{p_{iA}} = \frac{2}{\sqrt{4\cos^2 k_2 D + (Z_{12} + Z_{21})^2 \sin^2 k_2 D}}
\tag{3.72}
$$

Therefore, the acoustic intensity transmission coefficient is

$$
t_I = \frac{I_t}{I_i} = \frac{|p_{tA}|^2 / 2\rho_1 c_1}{|p_{iA}|^2 / 2\rho_1 c_1} = \frac{4}{4\cos^2 k_2 D + (Z_{12} + Z_{21})^2 \sin^2 k_2 D}
\tag{3.73}
$$

The corresponding acoustic intensity reflection coefficient is

$$
r_I = \frac{|p_{1rA}|^2 / 2\rho_1 c_1}{|p_{iA}|^2 / 2\rho_1 c_1} = 1 - t_I
\tag{3.74}
$$

Equations 3.73 and 3.74 illustrate that the reflection and transmission of acoustic waves through a middle layer depend not only on the acoustic impedances of these two media (Z_1, Z_2) but also on the ratio of layer thickness to the wavelength (D/λ_2). Common situations are discussed individually in the following.

a. $k_2 D = \dfrac{2\pi D}{\lambda_2} \ll 1$, so $\cos k_2 D \approx 1$, $\sin k_2 D \approx 0$

From Equation 3.70, it is seen that $r_I \approx 1$, which suggests that, if the thickness of the middle layer is much smaller than the wavelength inside it, an acoustic wave would mostly transmit through it.

    b. $k_2D = n\pi$ ($n = 1, 2, 3, \ldots$), that is, $D = (\lambda_2/2)n$

It is seen that $r_I \approx 1$ from Equation 3.67, indicating no existence of the middle layer. In the ultrasound transducer, a matching layer with half-wavelength is usually applied in front of the piezoelectric material.

    c. $k_2D = (2n-1)\dfrac{\pi}{2}, Z_1 \ll Z_2$, that is,

$$D = (2n-1)\frac{\lambda_2}{4} \ (n = 1, 2, 3, \ldots)$$

It is seen that $r_I \approx 0$, which demonstrates a complete blockage on the acoustic wave.

For the oblique incidence case, the acoustic impedance in the normal direction and wave vector in the x direction $k_2' = k_2 \cos\theta_{2t}$ are used in Equations 3.72 and 3.73 to derive the acoustic intensity reflection and transmission coefficients as

$$\begin{cases} t_p = \dfrac{p_{tA}}{p_{iA}} = \dfrac{2}{\sqrt{4\cos^2 k_2'D + \left(\dfrac{z_2}{z_1} + \dfrac{z_1}{z_2}\right)^2 \sin^2 k_2'D}} \\[4ex] t_p = \dfrac{I_t}{I_i} = \dfrac{4}{4\cos^2 k_2'D + \left(\dfrac{z_2}{z_1} + \dfrac{z_1}{z_2}\right)^2 \sin^2 k_2'D} \end{cases} \tag{3.75}$$

ASSIGNMENT

Q1: The acoustic impedances of air and water at 20°C are 415×10^6 and 1.48×10^6 Rayleighs, respectively. Calculate the reflected pressure and acoustic transmission coefficient when a planar acoustic wave is incident normally from air to water. Recalculate when the planar acoustic wave is incident normally from water to air.

Q2: When an acoustic wave is incident from water to mud at an angle of 30°, what is the transmitted angle, the ratio of the reflected to the incident waves at the interface, and the transmission coefficient of acoustic energy? Acoustic impedances of water and mud are 1.48×10^6 and 3.2×10^6 Rayleighs, respectively.

Q3: Calculate the acoustic intensity transmission coefficient of waves with frequencies 200 and 2000 Hz incident on a mental plate in the air with thicknesses of 1 and 5 mm, respectively. Consider normal incidence. The acoustic impedance of the mental is 40×10^6 Rayleighs.

Q4: There is a wood plate in air with thickness 1 cm. Calculate the transmission coefficient (sound insulation in dB) for a 2000 Hz sound wave. If the plate is made of aluminum, what is the increase in transmission loss of the sound wave?

Q5: A room wall has a density of 2000 kg/m³ and a thickness of 20 cm. Calculate the sound transmission loss for sound wave of frequencies of 100 Hz, 500 Hz, 1000 Hz, 2000 Hz, 5000 Hz, and 10 kHz. If the wall thickness is doubled, what are the corresponding values? If a double wall is used with the distance between them 10 cm, what are the sound transmission losses?

Q6: In an anechoic room, the walls are usually treated with wedges of porous material, typically 0.5–1 m deep, to provide absorption over a wide range of frequencies (from 100 to 8000 Hz). In many measurements in such a room, it is desirable that the reflected amplitude is 20 dB lower than the incident amplitude. Calculate the absorption coefficient of the material in order to achieve this goal.

Q7: The absorption coefficient of a 1 m thick layer of fiberglass mounted on a rigid wall is 0.7 at 1000 Hz. An incident wave with a sound pressure level of 120 dB is incident on the absorber. Determine
    a. The magnitude of the pressure reflection coefficient
    b. The intensity of the incident and reflected waves in W/cm²
    c. The reduction in sound pressure level after one reflection

Q8: The normalized impedance of a porous layer of thickness L at sufficiently low frequencies (wavelength λ much larger than the layer thickness) is $\zeta = \Theta/3 + i/(H\gamma kL)$, where Θ is the normalized value of the total flow resistance of the layer, H is the porosity, $\gamma \approx 1.4$ is the specific heat ratio for air, and $k = \omega/c = 2\pi/\lambda$.
    a. What is the magnitude and phase angle of the complex pressure reflection coefficient in terms of the given parameters. In particular, calculate them for $R = 4$, $H = 0.95$, $L = 10$ cm, and $L/\lambda = 0.05$.
    b. Show that the absorption coefficient at normal incidence can be expressed as

$$\alpha \approx \frac{4\theta(kL')^2}{1 + (kL')^2(1+\theta)^2}$$

    where
      $\theta = \Theta/3$
      $L' = H\gamma L$

Q9: A pulsating sphere of mean radius a is surrounded by a concentric spherical enclosure of radius b and with totally reflecting walls. The radial velocity of the pulsating sphere is $u_r = |u|\cos(\omega t)$.
    a. Determine the pressure and velocity fields in the enclosure.
    b. What is the impedance at the source?

REFERENCES

Ando Y. On the sound radiation from semi-infinite circular pipe of certain wall thickness. *Acustica* 1969/70;22:219–225.

Beissner K. On the plane-wave approximation of acoustic intensity. *Journal of Acoustical Society of America* 1974;56:1043–1048.

Connor CW, Hynynen K. Patterns of thermal deposition in the skull during transcranial focused ultrasound surgery. *IEEE Transactions on Biomedical Engineering* 2004;51:1693–1706.

Ingard U. *Notes on Acoustics.* Hingham, MA: Infinity Science Press, 2008.

4 Acoustic Field and Wave Radiation

In Chapter 3, we have already seen the characteristics of acoustic wave propagation, including reflection and transmission. We discuss in this chapter how to synthesize the acoustic field when multiple acoustic sources are present.

4.1 ACOUSTIC INTERFERENCE

4.1.1 SUPERPOSITION PRINCIPLE

The propagation of a small-amplitude acoustic wave is governed by the wave Equations 2.24 and 2.27. In mathematics, the wave equations are linear ones, which requires that the small-amplitude acoustic waves satisfy the superposition principle. Assume that there are two acoustic waves with sound pressure of p_1 and p_2, respectively. They satisfy the following wave equations:

$$\begin{cases} \nabla^2 p_1 = \dfrac{1}{c_0^2} \dfrac{\partial^2 p_1}{\partial t^2} \\ \nabla^2 p_2 = \dfrac{1}{c_0^2} \dfrac{\partial^2 p_2}{\partial t^2} \end{cases} \quad (4.1)$$

The synthesized acoustic field p should also satisfy the wave equation:

$$\nabla^2 p = \frac{1}{c_0^2} \frac{\partial^2 p}{\partial t^2} \quad (4.2)$$

From Equation 4.1, we can get

$$\nabla^2 (p_1 + p_2) = \frac{1}{c_0^2} \frac{\partial^2 (p_1 + p_2)}{\partial t^2} \quad (4.3)$$

Comparing Equations 4.2 and 4.3, and considering that the acoustic boundary conditions are also linear, we can obtain

$$p = p_1 + p_2 \quad (4.4)$$

which means that the sound pressure of the synthesized acoustic field is the summation of two acoustic pressures. It is clear that the superposition principle of acoustic waves can be applied to the situation with multiple waves.

4.1.2 STANDING WAVE

Suppose that there are two plane waves with the same frequency but opposite propagation direction, given by

$$\begin{aligned} p_i &= p_{iA} e^{j(\omega t - kx)} \\ p_r &= p_{rA} e^{j(\omega t + kx)} \end{aligned} \quad (4.5)$$

Based on the superposition principle, the synthesized sound pressure is

$$p = p_i + p_r = 2 p_{rA} \cos kx\, e^{j\omega t} + (p_{iA} - p_{rA}) e^{j(\omega t - kx)} \quad (4.6)$$

It can be figured out that the two items in Equation 4.6 represent the standing wave and the plane wave propagating in the x direction with the amplitude of the difference between those of two plane waves, respectively. In the standing wave, every particle has the same phase in vibration, but its vibration amplitude varies with the location. Since $\cos(\omega t)$ alternates between $+1$ and -1 as a function of time, the envelope of the resultant wave has a spatial variation given by $\pm 2 p_{rA} \cos(kx)$. The resultant wave is fixed in space yet oscillates in time. However, no net power propagates. When $kx = n\pi$ or $x = n \dfrac{\lambda}{2}$ $(n = 1, 2, \ldots)$, called the anti-node, the sound pressure reaches its maximum value. In contrast, when $kx = (2n-1) \dfrac{\pi}{2}$ or $x = (2n-1) \dfrac{\lambda}{4}$ $(n = 1, 2, \ldots)$, the sound pressure is always 0. So, these locations are called nodes. The term "standing wave" is derived from the fact that the envelope is stationary in space; the positions of the envelope's maxima and minima do not propagate. However, when the relative amplitudes of the incident and reflected waves are unequal, the standing wave envelope has only partial minima that do not reach zero.

4.1.3 INTERFERENCE

Interference phenomenon will occur when two acoustic waves with the same frequency and a constant phase difference $\psi = \varphi_2 - \varphi_1$ combine, as in the following:

$$\begin{aligned} p_1 &= p_{1A} \cos(\omega t - \varphi_1) \\ p_2 &= p_{2A} \cos(\omega t - \varphi_2) \end{aligned} \quad (4.7)$$

We apply the superposition principle to calculate the synthesized sound pressure as

$$\begin{aligned} p &= p_1 + p_2 = p_{1A} \cos(\omega t - \varphi_1) + p_{2A} \cos(\omega t - \varphi_2) \\ &= p_A \cos(\omega t - \varphi) \end{aligned} \quad (4.8)$$

where

$$p_A^2 = p_{1A}^2 + p_{2A}^2 + 2p_{1A}p_{2A}\cos(\varphi_2 - \varphi_1)$$

$$\varphi = \tan^{-1}\frac{p_{1A}\sin\varphi_1 + p_{2A}\sin\varphi_2}{p_{1A}\cos\varphi_1 + p_{2A}\cos\varphi_2} \quad (4.9)$$

This equation demonstrates that the synthesized wave has the same frequency, but its amplitude is not the simple sum of the sound pressure of these waves but a function of the phase difference ψ. The average energy density of the synthesized sound field is expressed as

$$\bar{w} = \bar{w}_1 + \bar{w}_2 + \frac{p_{1A}p_{2A}}{\rho_0 c_0^2}\cos\psi \quad (4.10)$$

where $\bar{w}_1$ and $\bar{w}_2$ correlates to the average energy density of p_1 and p_2, respectively. If $\psi = 0, \pm 2\pi, \pm 4\pi,\ldots$, these two acoustic waves are in phase, and then

$$\begin{cases} p_A = p_{1A} + p_{2A} \\ \bar{w} = \bar{w}_1 + \bar{w}_2 + \dfrac{p_{1A}p_{2A}}{\rho_0 c_0^2} \end{cases} \quad (4.11)$$

In contrast, if $\psi = \pm\pi, \pm 3\pi,\ldots$, these two waves are out of phase, and then

$$\begin{cases} p_A = p_{1A} - p_{2A} \\ \bar{w} = \bar{w}_1 + \bar{w}_2 - \dfrac{p_{1A}p_{2A}}{\rho_0 c_0^2} \end{cases} \quad (4.12)$$

It can be seen that the average energy density in the acoustic field consisting of two waves with the same frequency and a fixed phase is not the simple summation of the average energy densities of these two waves but is related to the phase difference ψ when these two acoustic waves arrive a certain position. If $p_{1A} = p_{2A}$ and these two waves are in phase at a certain position, the synthesized sound pressure and average energy density are two and four times as those of each wave, respectively. In contrast, if they are out of phase, the synthesized sound pressure and average energy density are both zero. If $\psi = 0$ or $\pm 2\pi$, $\pm 4\pi,\ldots$, $\bar{w} = 4\bar{w}_1$, it seems that the two acoustic sources are close to each other.

However, it can be noticed that, if the frequencies of these two acoustic waves are not the same, but their phase difference is fixed, there will be no interference phenomenon. Assume that these two acoustic waves are

$$\begin{cases} p_1 = p_{1A}\cos(\omega_1 t - \varphi_1) \\ p_2 = p_{2A}\cos(\omega_2 t - \varphi_2) \end{cases} \quad (4.13)$$

The average acoustic energy density of the synthesized field is calculated as

$$\bar{w} = \bar{w}_1 + \bar{w}_2 + \frac{2p_{1A}p_{2A}}{\rho_0 c_0^2}\overline{\cos(\omega_1 t - \varphi_1)\cos(\omega_2 t - \varphi_2)} \quad (4.14)$$

It can be proven that, for a sufficiently long time, the value of the third item on the right of Equation 4.14 is zero. Then, the equation can be rewritten as

$$\bar{w} = \bar{w}_1 + \bar{w}_2 \quad (4.15)$$

Therefore, it can easily be figured out that these acoustic waves with different frequencies are not interfering.

4.1.4 ACOUSTIC WAVES WITH VARYING PHASES

In some situations, after a number of reflections, such as inside a room or in the skull, the acoustic field is the summation of all reflected waves from different directions with varying phases. Assume two such waves

$$\begin{cases} p_1 = p_{1A}\cos(\omega t - \varphi_1) \\ p_2 = p_{2A}\cos(\omega t - \varphi_2) \end{cases} \quad (4.16)$$

where ϕ_1 and ϕ_2 are the instantaneous phase of the acoustic waves arriving at a certain position. So, the combined sound pressure can be expressed as

$$p = p_1 + p_2 = p_{rA}\cos(\omega t - \varphi) \quad (4.17)$$

where

$$\begin{cases} p_A^2 = p_{1A}^2 + p_{2A}^2 + 2p_{1A}p_{2A}\cos(\varphi_2 - \varphi_1) \\ \varphi = \tan^{-1}\dfrac{p_{1A}\sin\varphi_1 + p_{2A}\sin\varphi_2}{p_{1A}\cos\varphi_1 + p_{2A}\cos\varphi_2} \end{cases} \quad (4.18)$$

Here ϕ_1, ϕ_2, and ϕ are all varying randomly within the range of 0–2π. The average acoustic energy density of the synthesized field is

$$\bar{w} = \frac{p_A^2}{2\rho_0 c_0^2} = \bar{w}_1 + \bar{w}_2 + \frac{p_{1A}p_{2A}}{\rho_0 c_0^2}\overline{\cos(\varphi_2 - \varphi_1)} \quad (4.19)$$

If the time is sufficiently long, it can be proven that

$$\bar{w} = \bar{w}_1 + \bar{w}_2 \quad (4.20)$$

which illustrates that these acoustic waves with the same frequency but randomly varying phases are also noninterfering ones. Therefore, for multiple waves, the average acoustic energy field of the synthesized field is the summation of the individual ones. The expression of sound pressure is

$$p_e^2 = p_{1e}^2 + p_{2e}^2 + \cdots + p_{ne}^2 \tag{4.21}$$

where
p_e is the effective sound pressure of the synthesized field
p_{je} ($j = 1, 2,\ldots, n$) is the effective sound pressure of each wave

4.2 ACOUSTIC WAVE RADIATION

4.2.1 PULSING SPHERICAL SOURCE

A pulsing spherical source is a spherical acoustic source undergoing uniform vibration; that is, every point on the surface of the spherical source vibrates in the radial direction at the same amplitude and phase. Although it is an ideal case, an arbitrary acoustic source can be considered as the summation of many small pulsing spherical sources using the superposition principle.

Assume that there are a sphere with radius r_0 and a tiny harmonic vibration of $\xi = dr$ at the surface. As a result, a spherical wave will radiate outward. The spherical coordinate and origin are set as in Figure 4.1. The spherical surface area at a distance r is $S = 4\pi r^2$. The wave equation can be expressed as

$$\frac{\partial^2 p}{\partial r^2} + \frac{\partial p}{\partial r} \cdot \frac{\partial \ln S}{\partial r} = \frac{1}{c_0^2}\frac{\partial^2 p}{\partial t^2} \tag{4.22}$$

Substituting $S = 4\pi r^2$ into the preceding equation, we get

$$\frac{\partial^2 p}{\partial r^2} + \frac{2}{r}\frac{\partial p}{\partial r} = \frac{1}{c_0^2}\frac{\partial^2 p}{\partial t^2} \tag{4.23}$$

If $Y = pr$, this equation can be rewritten as

$$\frac{\partial^2 Y}{\partial r^2} = \frac{1}{c_0^2}\frac{\partial^2 Y}{\partial t^2} \tag{4.24}$$

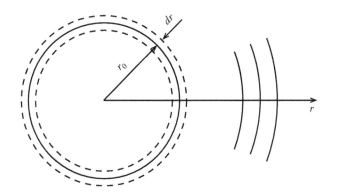

FIGURE 4.1 Diagram of a homogenous spherical wave.

And, its solution is

$$Y = Ae^{j(\omega t - kr)} + Be^{j(\omega t + kr)} \tag{4.25}$$

where A and B are undetermined constants. Then,

$$p = \frac{A}{r}e^{j(\omega t - kr)} + \frac{B}{r}e^{j(\omega t + kr)} \tag{4.26}$$

These two items represent the spherical wave radiating outward and inward, respectively. It is well known that there is no reflected wave at the source's surface. So, $B = 0$. Equation 4.26 can then be simplified as

$$p = \frac{A}{r}e^{j(\omega t - kr)} \tag{4.27}$$

where
A can be a complex number
Absolute value of A/r is the amplitude of sound pressure

The radial particle velocity is calculated as

$$v_r = -\frac{1}{j\omega\rho_0}\frac{\partial p}{\partial r} = \frac{A}{r\rho_0 c_0}\left(1 + \frac{1}{jkr}\right)e^{j(\omega t - kr)} \tag{4.28}$$

where the absolute value of $\dfrac{A}{r\rho_0 c_0}\left(1 + \dfrac{1}{jkr}\right)$ is the amplitude of the particle velocity.

The particle velocity at the surface of spherical source is $u = u_A e^{j(\omega t - kr_0)}$, where u_A is the velocity amplitude, and $-kr_0$ is the initial phase. At the surface of spherical source, the particle velocity is equal to the vibration velocity. The boundary condition is

$$(v_r)_{r=r_0} = u \tag{4.29}$$

Then,

$$A = \frac{\rho_0 c_0 k r_0^2}{1 + (kr_0)^2}u_A(kr_0 + j) = |A|e^{j\theta} \tag{4.30}$$

where

$$|A| = \frac{\rho_0 c_0 k r_0^2 u_A}{\sqrt{1 + (kr_0)^2}}, \qquad \theta = \tan^{-1}\left(\frac{1}{kr_0}\right) \tag{4.31}$$

The radiation pressure of the pulsing spherical source is

$$p = p_A e^{j(\omega t - kr + \theta)} \tag{4.32}$$

where $p_A = \dfrac{|A|}{r}$. Similarly, the particle velocity is

$$v_r = v_{rA}e^{j(\omega t - kr + \theta + \theta')} \qquad (4.33)$$

where v_A is the amplitude of radial velocity

$$v_{rA} = p_A \frac{\sqrt{1 + (kr)^2}}{\rho_0 c_0 kr}, \qquad \theta' = \tan^{-1}\left(\frac{-1}{kr}\right) \qquad (4.34)$$

Figure 4.2 shows the distribution of the sound pressure in the radial direction, which is inversely proportional relationship indicated in Equation 4.32. If r is sufficiently large or $\dfrac{dr}{r} \ll 1$, $\dfrac{dp_A}{p_A} \approx 0$, meaning that the change in the sound pressure is very small or the sound pressure is almost constant. In the other words, the plane of the wave front is very large in such a situation. So, it can be approximately considered as a plane wave.

In a spherical acoustic field, the acoustic intensity is calculated from Equations 4.32 and 4.33 as

$$\begin{aligned} I &= \frac{1}{T}\int_0^T \operatorname{Re} p \operatorname{Re} v_r\, dt \\ &= \frac{1}{T}\int_0^T p_A^2 \frac{\sqrt{1+(kr)^2}}{\rho_0 c_0 kr}\cos(\omega t - kr + \theta)\cos(\omega t - kr + \theta + \theta')\, dt \\ &= p_A^2 \frac{\sqrt{1+(kr)^2}}{\rho_0 c_0 kr}\frac{\cos\theta'}{2} \end{aligned} \qquad (4.35)$$

From Equation 4.34, it can be found that $\cos\theta' = \dfrac{kr}{\sqrt{1+(kr)^2}}$, so

$$I = \frac{p_A^2}{2\rho_0 c_0} = \frac{p_e^2}{\rho_0 c_0} \qquad (4.36)$$

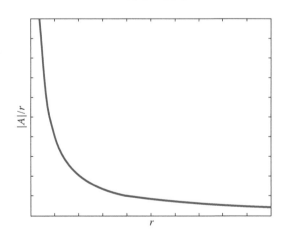

FIGURE 4.2 Acoustic pressure radiating from a spherical source along the propagation path.

It can be seen that Equation 4.36 has the same expression as that of a planar wave. The average acoustic power through such a spherical surface is

$$\overline{w} = I 4\pi r^2 = 4\pi r^2 \frac{p_e^2}{\rho_0 c_0} = \frac{2\pi}{\rho_0 c_0}|A|^2 \qquad (4.37)$$

The average acoustic power from a spherical source is not related to the distance to the source surface, which is in accordance with the law of energy conservation.

4.2.2 ACOUSTIC DIPOLE

An acoustic dipole consists of two tiny pulsing spherical sources (or point sources) with the same amplitude, phase difference of 180°, and a very small interval distance, as shown in Figure 4.3. The synthesized sound pressure is

$$p = \frac{A}{r_+}e^{j(\omega t - kr_+)} - \frac{A}{r_-}e^{j(\omega t - kr_-)} \qquad (4.38)$$

Since $r \gg l$, r_+ and r_- can be approximately rewritten using a Taylor expansion series as

$$r_+ \approx r + \frac{l}{2}\cos\theta, \qquad r_- \approx r - \frac{l}{2}\cos\theta \qquad (4.39)$$

Substituting Equation 4.39 into Equation 4.38 gives

$$\begin{aligned} p &\approx \frac{A}{r}e^{j(\omega t - kr)}\left(e^{j\frac{kl\cos\theta}{2}} - e^{-j\frac{kl\cos\theta}{2}}\right) \\ &= \frac{A}{r}e^{j(\omega t - kr)}\left(2j\sin\frac{kl\cos\theta}{2}\right) \end{aligned} \qquad (4.40)$$

If the interval distance is smaller than the wavelength, or $kl < 1$, Equation 4.40 can be simplified as

$$p \approx j\frac{kAl}{r}\cos\theta e^{j(\omega t - kr)} \qquad (4.41)$$

It is interesting to note that the form of the radiated acoustic field depends on the angle θ. In order to describe this characteristic of the direction-dependent radiation, the directivity is defined and calculated as

$$D(\theta) = \frac{(p_A)_\theta}{(p_A)_{\theta=0}} = |\cos\theta| \qquad (4.42)$$

The particle velocity in the radial direction is

$$v_r \approx j\frac{kAl}{\rho_0 c_0 r}\left(1 + \frac{1}{jkr}\right)\cos\theta e^{j(\omega t - kr)} \qquad (4.43)$$

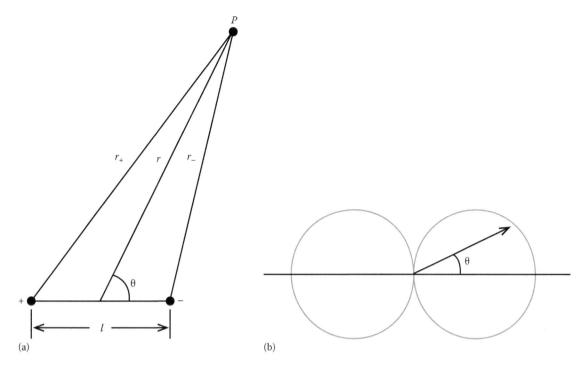

FIGURE 4.3 (a) Schematic of an acoustic dipole and (b) the corresponding directivity.

And, the acoustic intensity is

$$I = \frac{1}{T}\int_0^T \operatorname{Re} p \operatorname{Re} v_r dt = \frac{|A|^2 k^2 l^2}{2\rho_0 c_0 r^2}\cos^2\theta \qquad (4.44)$$

The average acoustic power is

$$\bar{w} = \iint_S I dS = \iint I r^2 \sin\theta d\theta d\phi = \frac{2\pi}{3\rho_0 c_0}|A|^2 k^2 l^2 \qquad (4.45)$$

Substituting Equation 4.30 into Equation 4.45 gives

$$\bar{w} = \frac{2}{3}\pi\rho_0 c_0 k^4 r_0^4 l^2 u_A^2 \qquad (4.46)$$

It can be concluded that the average acoustic power is independent of r, which is in accordance with the law of energy conservation.

4.2.3 Two In-Phase Sources

If there are two tiny pulsing spherical sources with the same frequency and phase with a distance l between them, the synthesized sound pressure is

$$p = \frac{A}{r_1}e^{j(\omega t - k r_1)} + \frac{A}{r_2}e^{j(\omega t - k r_2)} \qquad (4.47)$$

In the far field of $r \gg l$, the approximation is

$$r_1 \approx r - \Delta, \qquad r_2 \approx r + \Delta \qquad (4.48)$$

where $\Delta = \frac{l}{2}\sin\theta$, so

$$p = \frac{A}{r}e^{j(\omega t - kr)}[e^{-jk\Delta} + e^{jk\Delta}] = \frac{A}{r}e^{j(\omega t - kr)}\cdot 2\cos k\Delta$$

$$= \frac{A}{r}e^{j(\omega t - kr)}\cdot\frac{\sin 2k\Delta}{\sin k\Delta} \qquad (4.49)$$

The directivity is

$$D(\theta) = \frac{(p_A)_\theta}{(p_A)_{\theta=0}} = \left|\frac{\sin 2k\Delta}{2\sin k\Delta}\right| \qquad (4.50)$$

1. When $k\Delta = m\pi$ or $l\sin\theta = m\lambda$ $(m = 0, 1, 2, \ldots)$, $D(\theta) = 1$. It means that, at specific directions, the phase difference from two spherical sources is an integer times the wavelength or in phase so that the synthesized sound pressure is the maximum. The corresponding angle is

$$\theta = \sin^{-1}\frac{m\lambda}{l} \quad (m = 0, 1, 2, \ldots) \qquad (4.51)$$

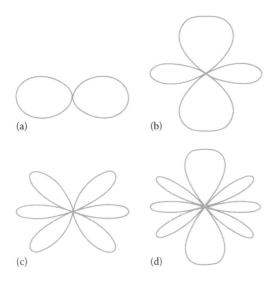

FIGURE 4.4 Directivity of two in-phase sources at different ratios of l/λ: (a) $l = (1/2)\lambda$, (b) $l = \lambda$, (c) $l = (3/2)\lambda$, and (d) $l = 2\lambda$.

The maximum value at $\theta = 0°$ is the main lobe, and the others are side lobes. It is known from Equation 4.51 that the number of side lobes within the range of $\left(0, \dfrac{\pi}{2}\right)$ is exactly the integer part of ratio $\dfrac{l}{\lambda}$. Because the acoustic energy at the main and side lobes are equal, such an energy distribution is not expected. So, the distance between these two sources should be smaller than wavelength for the non-appearance of any side lobes.

2. When $2k\Delta = m'\pi$ or $l\sin\theta = m'\dfrac{\lambda}{2}$ $(m' = 1, 3, 5, \ldots)$, $D(\theta) = 0$. It can be seen that, when the difference of acoustic travel is an odd number times the half-wavelength, the synthesized sound pressure is 0. The corresponding directions are

$$\theta = \sin^{-1}\frac{m'\lambda}{2l} \quad (m' = 1, 3, 5, \ldots) \tag{4.52}$$

The angular width of the main beam is determined as twice the angle for the first zero irradiation:

$$\overline{\theta} = 2\sin^{-1}\frac{\lambda}{2l} \tag{4.53}$$

It is noted that if $l < \dfrac{\lambda}{2}$, $\overline{\theta}$ has no solution. In other words, there is no zero radiation in any direction.

3. When $kl \ll 1$, $k\Delta = k\dfrac{l}{2}\sin\theta \ll 1$. So, $D(\theta) = 1$. That is, when these two tiny spherical sources are close to each other, acoustic radiation has no directivity. The synthesized sound pressure is $p \approx \dfrac{2A}{r}e^{j(\omega t - kr)}$.

Altogether, suppressing the side lobes and reducing the main beam width contradict each other. For example, if $l < \lambda$, there is no side lobes but the main beam width is quite large. In contrast, with the increase of l, the main beam becomes narrow but more side lobes appear. The directivities under different conditions are shown in Figure 4.4.

4.3 TRANSDUCER FIELD

4.3.1 PISTON ON A RIGID BAFFLE

The shape of the radiating beam from the transducer is important in ultrasound applications. The acoustic pressure wave that propagates from the face of an unfocused transducer generally maintains the approximate lateral dimensions of the transducer for a certain distance, but natural divergence begins to spread the transverse extent of the beam at larger distances so that the beam takes on a diverging nature. In the region near the transducer (the "near field"), the beam has many amplitude and phase irregularities due to interference between the contributing waves from all parts of the transducer's face, whereas in the region further from the transducer (the "far field"), the beam profile is much more uniform and well behaved. The radiation pattern of an ultrasound transducer is illustrated in the following, from which the transition distance between near-field and far-field regions and the beam spreading in the far field can be described.

4.3.1.1 General Acoustic Field

A piston source is a planar vibrator with the same velocity and phase in the normal direction of the surface. Assume that the radius is a and the velocity is $u = u_A{}^{ej\omega t}$. Select the center of the piston as the coordinate origin, as shown in Figure 4.5. The piston will be divided into a number of

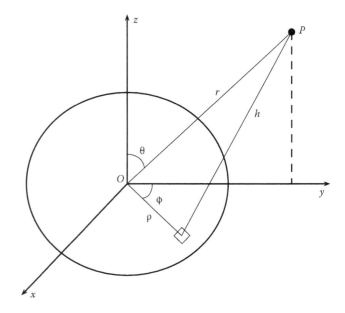

FIGURE 4.5 Schematic diagram of a piston on a baffle.

finite elements, and each one can be considered as a point source. Huygen's principle states that the radiation pattern from a general extended source can be constructed by considering the source as an appropriately weighted collection of point sources, each radiating outwardly propagating spherical waves. This decomposition of the complex problem into a summation of simpler parts (i.e., spherical waves radiating from point sources) is allowed because the wave equation is linear in the pressure variable. If the area of a finite element is dS and the source intensity is $dQ_0 = u_A dS$, the sound pressure at the position of P is

$$dp = j\frac{k\rho_0 c_0}{2\pi h}u_A dS e^{j(\omega t - kh)} \qquad (4.54)$$

where h is the distance between the finite element dS and the position P. The total radiation pressure of the piston at the observation point is the integral of the incremental pressures:

$$p = \iint dp = \iint_S j\frac{k\rho_0 c_0}{2\pi h}u_A e^{j(\omega t - kh)} dS \qquad (4.55)$$

It can be noted that for an arbitrary transducer shape and for an arbitrary observation point, this equation is quite difficult to evaluate and usually requires a computer solution. Here, several special shapes with certain symmetry (i.e., popular transducer types) are considered. For a piston, $dS = \rho d\rho d\phi$ and h is a function of ρ and ϕ:

$$h^2 = r^2 + \rho^2 - 2r\rho\cos(\vec{\rho}, \vec{r}) \;\Rightarrow\; h = r\sqrt{1 - \frac{2\rho}{r}\cos(\vec{\rho}, \vec{r}) + \frac{\rho^2}{r^2}} \qquad (4.56)$$

When $r \gg \rho$, this equation can be approximated as

$$h \approx r - \rho\cos(\vec{\rho}, \vec{r}) \qquad (4.57)$$

Substituting Equation 4.57 into Equation 4.55 gives

$$p = j\frac{\omega\rho_0 u_A}{2\pi r}e^{j(\omega t - kr)}\iint e^{jk\rho\cos(\vec{\rho}, \vec{r})}\rho d\rho d\phi \qquad (4.58)$$

Because $\vec{\rho} = |\rho|(\cos\phi\hat{i} + \sin\phi\hat{j})$ and $\vec{r} = |r|(\sin\theta\hat{i} + \cos\theta\hat{j})$, the angle can be calculated as $\cos(\vec{\rho}, \vec{r}) = \frac{\vec{\rho}, \vec{r}}{|\rho||r|}\sin\theta\cos\phi$. Then, Equation 4.58 can be rewritten as

$$p = j\frac{\omega\rho_0 u_A}{2\pi r}e^{j(\omega t - kr)}\int_0^a \rho d\rho \int_0^{2\pi} e^{jk\rho\sin\theta\cos\phi}d\phi \qquad (4.59)$$

Using the cylindrical Bessel function $J_0(x) = \frac{1}{2\pi}\int_0^{2\pi} e^{jx\cos\phi}d\phi$, $\int xJ_0(x)dx = xJ_1(x)$

$$p = j\omega\frac{\rho_0 u_A a^2}{2r}\left[\frac{2J_1(ka\sin\theta)}{ka\sin\theta}\right]e^{j(\omega t - kr)} \qquad (4.60)$$

The corresponding radial particle velocity is

$$v_r = -\frac{1}{j\omega\rho_0}\frac{\partial p}{\partial r} = \frac{1}{\rho_0 c_0}\left(1 + \frac{1}{jkr}\right)p \qquad (4.61)$$

The acoustic intensity is

$$I = \frac{1}{T}\int_0^T \mathrm{Re}\,p\,\mathrm{Re}\,v_r dt = \frac{1}{8}\rho_0 c_0 u_A^2 (ka)^2 \frac{a^2}{r^2}\left[\frac{2J_1(ka\sin\theta)}{ka\sin\theta}\right]^2 \qquad (4.62)$$

So, the power density decreases as $1/r^2$ in this region, as would be expected when the measurements are made sufficiently far away so that the source appears as a small radiator of diverging waves. The directivity of the piston is

$$D(\theta) = \frac{(p_A)_\theta}{(p_A)_{\theta=0}} = \left|\frac{2J_1(ka\sin\theta)}{ka\sin\theta}\right| \qquad (4.63)$$

So, the directivity of the piston depends on the ratio of the piston size to the wavelength. Figure 4.6 shows the directivity at $ka = 1$, $ka = 3$, $ka = 4$, and $ka = 10$. When $ka < 1$, $J_1(x) \approx (x/2)$ and $D \approx 1$, which means that the radiation is almost uniform

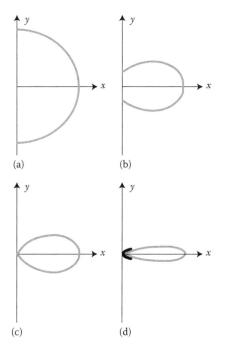

(a)

(b)

(c)

(d)

FIGURE 4.6 Directivity of a piston with different values of ka: (a) $ka = 1$, (b) $ka = 3$, (c) $ka = 4$, and (d) $ka = 10$.

if the piston is smaller than the wavelength. Equation 4.60 can be simplified as

$$p_L \approx j\omega \frac{\rho_0 u_A a^2}{2r} e^{j(\omega t - kr)} \quad (4.64)$$

which is approximated as a point source. The corresponding acoustic intensity at the low frequency is

$$I_L \approx \frac{1}{8} \rho_0 c_0 u_A^2 (ka)^2 \frac{a^2}{r^2} \quad (4.65)$$

Substituting Equation 4.64 into Equation 4.65 gives

$$I_L = \frac{p_{LA}^2}{2\rho_0 c_0} = \frac{p_{Le}^2}{2\rho_0 c_0} \quad (4.66)$$

where

p_{LA} is the amplitude of sound pressure radiated from piston at $ka < 1$

p_{Le} is the corresponding effective sound pressure

With the increase of ka (piston size or radiation frequency), the radiation directivity becomes sharper. When ka is equal to the first root of the first-order Bessel function, or in the direction

$$\theta_d = \sin^{-1} \frac{3.83}{ka} = \sin^{-1} 0.61 \frac{\lambda}{a} \quad (4.67)$$

$D = 0$ or the radiation is zero (see Figure 4.7). There are two side lobes and two more roots of $ka \sin \theta = 7.02$, 10.2 with corresponding $D = 0$ in the plot. The amplitude of the first side lobe is 0.14, so the acoustic intensity at such a location is about 0.02 times that in the main lobe ($\theta = 0°$).

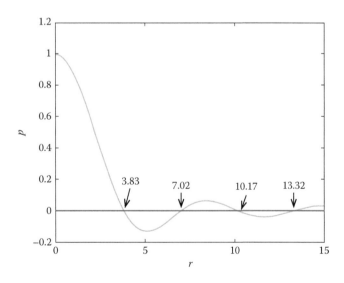

FIGURE 4.7 Lateral distribution of acoustic pressure generated by a piston.

4.3.1.2 Near Field: Fresnel Zone

However, if the observation point is close to the piston, the interference will be more complicated because the amplitude and phase of the acoustic waves from different elements on the piston vary. Assume that there is an annular element with inner diameter ρ and outer diameter $\rho + d\rho$ and that $d\rho$ is very small. The distance between every point on such an annular element to the piston's axis is $h = \sqrt{\rho^2 + z^2}$ (see Figure 4.8). The produced sound pressure is

$$dp = j \frac{k\rho_0 c_0}{2\pi h} u_A dS e^{j(\omega t - kh)} \quad (4.68)$$

where $dS = 2\pi\rho d\rho$ is the area of the annular element. After integrating with respect to ρ, the overall sound pressure at position of z is

$$p = jk\rho_0 c_0 u_A e^{j\omega t} \int_0^a \frac{e^{-kh}}{2\pi h} 2\pi\rho d\rho \quad (4.69)$$

Since $2\rho d\rho = 2h dh$

$$p = jk\rho_0 c_0 u_A e^{j\omega t} \int_z^R e^{-kh} dh$$

$$= -\rho_0 c_0 u_A e^{j\omega t} e^{-jk\frac{R+z}{2}} \left[e^{jk\frac{z-R}{2}} - e^{-jk\frac{z-R}{2}} \right]$$

$$= 2\rho_0 c_0 u_A \sin \frac{k}{2}(R-z) e^{j\left[\omega t - \frac{k}{2}(R+z) + \frac{\pi}{2}\right]} \quad (4.70)$$

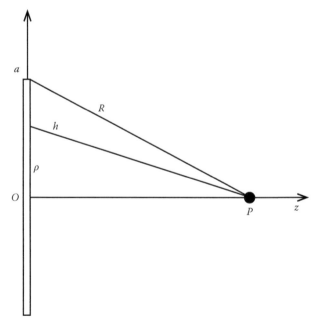

FIGURE 4.8 Schematic diagram of a piston in its near field.

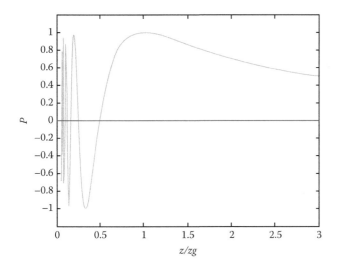

FIGURE 4.9 Acoustic pressure distribution on axis by a piston.

where $R = \sqrt{a^2 + z^2}$. The distribution of the sound pressure on the piston's axis is shown in Figure 4.9 and reveals the on-axis pressure variation in the near field of a circular transducer.

When z is very small or close to the source, at the position $\frac{k}{2}(R - z) = n\pi \quad (n = 1, 2, \ldots)$, the sound pressure is zero, while at the position of $\frac{k}{2}(R - z) = \left(n + \frac{1}{2}\right)\pi \quad (n = 0, 1, 2, \ldots)$, the sound pressure is maximum. It is clear that the variation of phase along the propagation direction provides the destructive and constructive interference patterns in the near field. Furthermore, with the increase in distance between the observation point and the piston on the axis, the spacing between the maximum and minimum positions will become larger.

When z is sufficiently large, or $z > 2a$, the sine function can be expanded as

$$\sin \frac{k}{2}(R - z) \approx \sin \frac{ka^2}{4z} = \sin \frac{\pi}{2}\frac{z_R}{z} \tag{4.71}$$

where

$$z_R = \frac{a^2}{\lambda} \tag{4.72}$$

At the position $z = z_R$, the sound pressure is maximum; when $z > z_R$, the sine function can be approximated as

$$\sin \frac{\pi}{2}\frac{z_R}{z} \cong \frac{\pi z_R}{2z} \tag{4.73}$$

So, the sound pressure is inversely proportional to the distance z, as in the case of the spherical wave. The position of the last maximum, z_R, is considered as the transition from the near field to the far field.

Substituting $z = 0$ in Equation 4.70 gives

$$p_N = 2\rho_0 c_0 u_A \sin \frac{ka}{2} e^{j\left(\omega t - \frac{ka}{2} + \frac{\pi}{2}\right)} \tag{4.74}$$

When $ka < 1$, it yields

$$p_N = 2\rho_0 c_0 u_A \frac{ka}{2} e^{j\left(\omega t - \frac{ka}{2} + \frac{\pi}{2}\right)} \tag{4.75}$$

So, the amplitude of acoustic pressure in the vicinity of the piston's center at low frequency is

$$p_{NA} = \rho_0 c_0 ka u_A \tag{4.76}$$

4.3.1.3 Far Field: Fraunhofer Zone

At a long distance away from the transducer, simplifications can be made to the general Equation 4.55 that allow the field to be calculated at any point (off-axis as well as on-axis) in the plane of observation (Figure 4.10). In regions $z > 2a$ and $z > z_g$, Equation 4.70 can be expressed using the Fraunhofer approximation as

$$p_{FA} \approx 2\rho_0 c_0 u_A \frac{ka^2}{4z} = \frac{\rho_0 c_0 u_A}{2z} ka^2 \tag{4.77}$$

Comparing Equations 4.76 and 4.77 gives

$$\frac{p_{FA}}{p_{NA}} = \frac{a}{2z} \tag{4.78}$$

It is clear that at a low frequency, such as when $ka < 1$, the ratio of the acoustic pressure in the far field on the axis to that close to the piston center is a constant. It is convenient to use this angle in Equation 4.67 as a measure of the divergence of the beam from a circular transducer, although some consider it too conservative. The smaller angular width of the half-width at the half-power points (–3 dB) rather than to the zeros is sometimes used; twice this angle is known as the full-width at half-maximum (FWHM) of intensity. Note the inverse relationship in Equation 4.67 between θ_d and the transducer radius a. When a comprises a large number of wavelengths

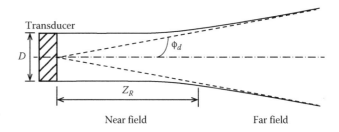

FIGURE 4.10 Shape of a beam as it propagates away from a piston.

(as measured in the tissue), the far-field beam is highly directed; conversely, when a is small, the beam spreads considerably as it propagates from the transducer. In fact, when a is approximately one-half of the wavelength (i.e., the diameter is one wavelength) or smaller, the half-angle of divergence is greater than 90° and the beam appears to be radiating hemispherically more or less isotropically from a point source.

4.3.2 RING TRANSDUCER

The equation derived for a circular transducer is also valid for the case of a ring transducer with negligible thickness T. So, the integral of Equation 4.59 is simply equal to the integration value radius a, and the corresponding acoustic field is given by

$$p_{ring} = \left[2\pi a T \cdot A_0 \cdot \frac{e^{j(\omega t - k \cdot R)}}{R} \right] \cdot J_0(ka \sin \theta) \qquad (4.79)$$

It can be seen that the directivity of a ring transducer is a Bessel function of order 0, as shown in Figure 4.11. It is found that the main lobe is narrow and the side lobes are enhanced for the ring transducer.

4.3.3 RECTANGULAR TRANSDUCER

The analysis of the far-field radiation from a rectangular transducer with width b in the x_0 direction and height h in the y_0 direction proceeds in a manner similar to that outlined earlier for a circular one as shown in Figure 4.12. Note that in the source plane, $\rho \cos \theta = x_0$. If we initially restrict the observation to be along the x_1 axis ($\varphi = \varphi_x$), Equation 4.59 may be integrated over the rectangular source to give

$$p(\varphi_x, r, t) = bh \frac{k \rho_0 c_0 u_A}{2\pi r} \cos \Psi(t) \left[\frac{\sin\left[(kb \sin \varphi_x)/2\right]}{(kb \sin \varphi_x)/2} \right] \qquad (4.80)$$

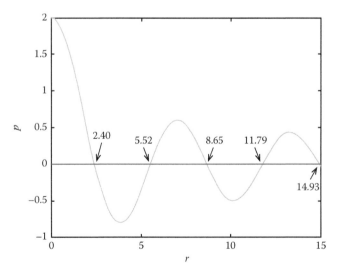

FIGURE 4.11 Lateral distribution of acoustic pressure generated by a ring.

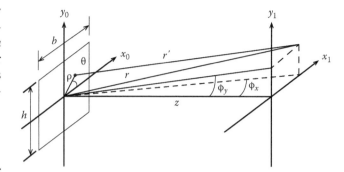

FIGURE 4.12 Schematic diagram of a rectangular transducer.

A similar expression holds for observations along the y_1 axis ($\varphi = \varphi_y$), and since the source is rectangular whose boundaries may be expressed by equations that are mathematically separable in x_0 and y_0, the complete expression for far-field intensity from a rectangular transducer is also separable in φ_x and φ_y:

$$I(\varphi_x, \varphi_y, r, t) = \left(\frac{bhu_A \sin(\omega t - kr)}{\lambda r} \right)^2$$

$$\rho_0 c_0 \left[\frac{\sin[(kb \sin \varphi_x)/2]}{(kb \sin \varphi_x)/2} \cdot \frac{\sin[(kh \sin \varphi_y)/2]}{(kh \sin \varphi_y)/2} \right]^2 \qquad (4.81)$$

The corresponding directionality is

$$D(\varphi_x, \varphi_y) = \left[\frac{\sin[(kb \sin \varphi_x)/2]}{(kb \sin \varphi_x)/2} \cdot \frac{\sin[(kh \sin \varphi_y)/2]}{(kh \sin \varphi_y)/2} \right] \qquad (4.82)$$

The far-field beam pattern from a rectangular transducer has the same qualitative features as those of a circular source, such as the main lobe and side lobes, except that the directionality has the form $\sin(x)/x$. Note that the half-angle to the first zero, marking the extent of the main lobe in the x_1 direction, is now given by

$$\sin\left(\frac{kb \sin \varphi_{xd}}{2} \right) = 0$$

or

$$kb \sin \varphi_{xd} = 2\pi \qquad (4.83)$$

or

$$\varphi_{xd} = \sin^{-1}\left(\frac{2\pi}{kb} \right) = \sin^{-1}\left(\frac{\lambda}{b} \right)$$

A similar equation describes the divergence as measured in the y_1 direction:

$$\varphi_{yd} = \sin^{-1}\left(\frac{\lambda}{h} \right) \qquad (4.84)$$

In contrast to the case of a circular transducer, the radiation pattern here is asymmetric. The inverse relationship between size and the divergence angle still applies. For a rectangular transducer that is taller than its width (i.e., $h > b$), the far-field radiation pattern of the transducer will be wider (i.e., $\varphi_{xd} > \varphi_{yd}$).

The beam pattern in the near field is very irregular, with many peaks and valleys close to the transducer surface. The near field in the lateral direction is mostly confined to the size of the transducer, but it is difficult to precisely define the edge of such an irregular field. When propagating toward the far field, the radiation pattern evolves into a well-defined single main lobe with some much smaller side lobes. The width of the main lobe increases linearly with distance, and the half-angle φ_d to the first zero on each side is inversely proportional to the transducer diameter. This progressive spreading pattern can be diagrammed qualitatively. In the near field, the beam is nearly collimated before reaching the transition distance z_R, where the beam spreading occurs. As the wave propagates toward the far field, the beam becomes widened further and eventually the divergence angle φ_d becomes constant. It is interesting to note that the beam from a small transducer is initially small but eventually becomes larger since it has a shorter z_R and a greater φ_d.

The irregularities within the near field and the presence of side lobes in the far field are sometimes an inconvenience when attempting to predict the beam pattern in the presence of reflectors. There is a technique for smoothing out these irregularities. It is based on the fact that, if the beam's amplitude profile as a function of radius was Gaussian-shaped at the transducer face (peaked at the center and decreasing to zero as $\exp(-\rho^2/a_1^2)$ toward the edges) rather than being of uniform amplitude across the transducer face, then the radiated beam's profile would be smoothly Gaussian-shaped everywhere in the near field as well as in the far field. Therefore, if by some means the transducer excitation profile approximates a Gaussian form with decreasing activity away from the center, the beam would be expected to be more uniform in its transverse behavior. Various ways of achieving a shaded profile at the radiating surface include placing a radially varying absorber in front of the transducer, designing the transducer face to have a star shape with some unexcited areas near the edges, or using a concentric ring transducer ("bull's eye") and exciting the outer rings with progressively less drive voltage than the center rings. All these techniques, known as apodization because they reduce the side lobes in the radiated beam pattern, will produce an overall smoother beam profile. The disadvantages, however, are that less total power is radiated, the transducer is more complex, and, as Equation 4.67 shows, the beam diverges at a greater angle since the effective transducer diameter is smaller.

4.3.4 Focused Transducer

The beam width from an unfocused transducer is generally too wide to give an adequate definition of the fine lateral features of objects being imaged or treated. Therefore, a lens or other focusing element such as a spherical reflector is usually employed to converge the radiating beam into a spot at

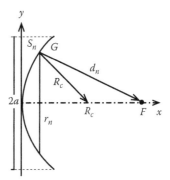

FIGURE 4.13 Schematic diagram of a focused transducer.

the focal plane of the lens. However, the size of the focused beam cannot be infinitely small since the natural divergence of a propagating wave as described in the previous section will attempt to spread even a converging beam, reducing the focusing effect of the lens. The further away the focal plane is from the lens, the larger will be the focused spot (Figure 4.13).

As opposed to optics, the acoustic lens material generally possesses an acoustic phase velocity that is greater than that of the material surrounding it (water or tissue). Thus, a converging (positive) lens will have a concave face. One can use Snell's law for ray tracing to show that a planoconcave lens having a surface with radius of curvature R_l will produce focusing at a focal length equal to

$$F = \frac{R_l}{1 - \dfrac{c_m}{c_l}} \tag{4.85}$$

where

c_l is the phase velocity of the lens material
c_m is the phase velocity of the medium into which the wave is focused

Lenses have the property of transforming angles into position. That is, all rays entering the lens at a common angle φ will get directed to a radius x_l in the focal plane. Under the small-angle approximation, geometry gives the transformation relationship as

$$\sin \varphi \approx \frac{x_l}{F} \tag{4.86}$$

Therefore, a lens of focal length F placed in front of the beam from a circular transducer whose radiation pattern is given by Equation 4.63 will transform the far-field angular distribution into a spatial distribution in the focal plane via the transformation of Equation 4.84. Making this substitution into Equation 4.63 yields the spatial distribution of the pressure at a focused spot from a circular transducer of radius a:

$$D(x_l) = \left[\frac{2J_1(kax_l/F)}{kax_l/F} \right] \tag{4.87}$$

and the focused pattern looks exactly like the far-field pattern, except that it is scaled down by an amount F/z. The focused spot has a dense central portion (corresponding to the main lobe) surrounded by minor rings (the side lobes). The diameter of the central portion is defined as previously: the distance between the first zero bounding the main lobe. From Equation 4.86, the radius of the first zero is found at

$$x_l = \frac{3.83F}{ka} = \frac{0.61F\lambda}{a} \qquad (4.88)$$

or, in terms of the transducer's size $D = 2a$, the diameter between first zeros is the focused spot diameter $d = 2x_1$:

$$d = 2.44\left(\frac{F}{D}\right)\lambda \qquad (4.89)$$

In the case of a rectangular transducer of width b, an analog can be used to find the width w of the focused spot in the direction parallel to b. Using Equations 4.80 and 4.84, the result for a rectangular transducer is

$$w = 2\left(\frac{F}{b}\right)\lambda \qquad (4.90)$$

The effects of divergence manifest in these relationships. The larger the diameter D of the transducer, the smaller the tendency of divergence to expand the beam, and the smaller the spot of the focus. Also, the further away the position of focus F, the greater is the effect of divergence, leading to a larger d. However, it may be impossible to get any narrowing at all in the beam diameter if the focal length is too large. d will be greater than D if

$$F > \frac{D^2}{2.44\lambda} \approx z_R \qquad (4.91)$$

In other words, no focusing occurs if the focal length of the lens is greater than the transition distance. Thus, it can be said that focusing is only possible at a distance within the near field and not in the far field of a transducer.

The spot size of the focused beam determines the transverse spatial resolution of a medical ultrasound imaging probe or the region of therapy. The focused beam is swept uniformly past a pair of point reflectors. The waveform of the envelope of the received echoes depends on the lateral spacing of the points. When far apart, the echoes from each point are distinct to form two separate and clear points in sonography. As they move closer, however, approaching the spacing d, the separate echoes start to blend together, and at some stage, the points become so close that their echoes cannot be separately resolved, which means that they appear as one reflecting object.

The spacing in the transverse plane at which the points are just separately resolvable is known as the lateral resolution (LR), which is approximately the diameter of the focused spot d:

$$LR \approx R = 2.44\left(\frac{F}{D}\right)\lambda \qquad (4.92)$$

and is the motivation for focusing the beam to reduce the focused spot size in ultrasound applications.

There is one disadvantage to tightly focusing of the beam. Although it improves the lateral resolution for reflecting objects located in the focal plane, points in planes either nearer or further away than the focal length are compromised because the beam is somewhat larger than d on either side of the focal plane. The problem gets worse with a decrease in the focused size. The axial distance over which the beam maintains its approximate focused size is termed as the depth of focus. It is noted that the beam shapes on both sides of the focal point are mirror images of each other about the focal plane. The beam behavior on one side of this plane, from the focus outward, has the same general characteristics of a beam propagating from an initially planar wavefront of a given diameter. Therefore, it stays approximately collimated within the transition distance z_R. Applying the transition distance of Equation 4.72 to the situation here, the depth of focus may be estimated to be twice this distance due to symmetry about the focal plane:

$$\text{Depth of focus} \approx \frac{d^2}{2\lambda} \approx 3\left(\frac{F}{D}\right)^2\lambda \qquad (4.93)$$

The tradeoff between focused size and depth of focus sometimes dictates a compromise in lens design. For example, in fixed-focus systems, the lens may be purposely given a nonspherical surface to cause the focused spot size to be larger than the theoretical limit, thereby increasing the depth of focus.

4.3.5 Line Transducer

Although circular transducers are used widely in ultrasound applications due to their easy fabrication, other geometrical configurations have certain advantages in some situations. Consider a line transducer with length L, width T, and negligible thickness. The coordinate system shown in Figure 4.14 is used to calculate the acoustic radiation field.

Using a similar approach as described in the previous sections, the field at point Q is given by

$$p_Q(R,\theta) = A_0 \int_{-L/2}^{L/2} \frac{e^{j(\omega t - kd)}}{d} \cdot T\,dx \qquad (4.94)$$

The distance between an infinitesimal surface element $T\,dx$ on the transmitter and point Q is given by

$$d^2 = (R\sin\theta - x)^2 + (R\cos\theta)^2 \qquad (4.95)$$

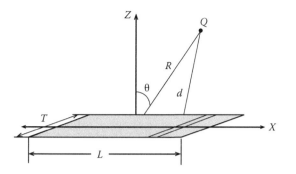

FIGURE 4.14 Schematic diagram of a line transducer.

Using the same approximation, it is assumed that

$$d \approx R - x \cdot \sin \theta \qquad (4.96)$$

The acoustic pressure at point Q is therefore given by

$$p_Q(R,\theta) = A_0 \cdot T \cdot \int_{-L/2}^{L/2} \frac{e^{j[\omega t - k(R - x\sin\theta)]}}{d} dx$$

$$= A_0 \cdot T \cdot \frac{e^{j(\omega t - kR)}}{R} \cdot \int_{-L/2}^{L/2} e^{jkx\sin\theta} dx \qquad (4.97)$$

Then, the solution is given by

$$p_Q(R,\theta) = \left[A_0 \cdot T \cdot L \cdot \frac{e^{j(\omega t - k \cdot R)}}{R} \right] \cdot \frac{\sin\left(k\frac{L}{2}\sin\theta \right)}{\left(k\frac{L}{2}\sin\theta \right)}$$

$$= \left[A_0 \cdot T \cdot L \cdot \frac{e^{j(\omega t - k \cdot R)}}{R} \right] \cdot \mathrm{sinc}\left(k\frac{L}{2}\sin\theta \right) \quad (4.98)$$

From this equation, it is clear that the directivity function in the plane containing the transducer is described by the sinc function.

4.4 ARRAY TRANSDUCER FIELD

4.4.1 LINEAR-ARRAY TRANSDUCER

4.4.1.1 General Acoustic Field

The transducer composed of a linear array of closely spaced elements, usually in rectangular shape as shown in Figure 4.15, is very common in medical ultrasound applications. Sometimes, it is manufactured by simply slicing a single, long piezoelectric bar. Although each element may be small in comparison to the wavelength, the overall width L can be appreciable (5–10 cm). If each element is excited in a sequential manner, the acoustic pattern is the same as that of a single element, that is, relatively broad in the horizontal or azimuth direction due to the limited dimensions of the elements. If the elements are excited simultaneously and coherently, the effective transducer width is L and the far-field divergence is much narrower as described by Equation 4.83. However, there is no change in the beam divergence in the elevation direction, no matter the way of element excitation, independently or coherently, because of the same effective array height h.

4.4.1.2 Grating Lobes

The radiation pattern of an array of elements is more complicated because of the appearance of grating lobes, which are the replicates of the main beam and side lobes with reduced amplitude positioned at around one or more discrete angles. The angles of the grating lobes, φ_{gn}, are found to be equal to those for which rays from two neighboring elements are in phase with each other by a multiple of 2π as constructive interference:

$$\sin\varphi_{gn} = \frac{l}{s} = \frac{n\lambda}{s} \quad \text{or} \quad \varphi_{gn} = \sin^{-1}\left(\frac{n\lambda}{s} \right), \quad n = \pm 1, \pm 2, \ldots$$

$$(4.99)$$

where l is the difference of path length between two neighboring elements. As the spacing s increases with respect to the wavelength λ, the grating lobes get closer, and their total number increases. The amplitude of the grating lobes is

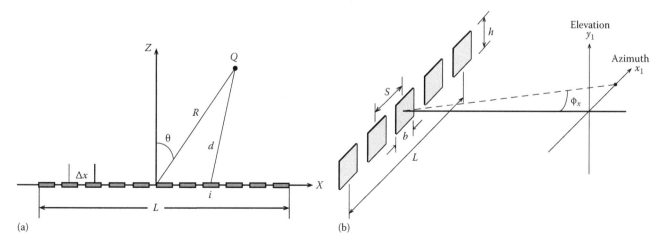

(a)

(b)

FIGURE 4.15 Schematic diagram of a linear-array transducer in (a) XZ plane and (b) 3D space.

determined by the directional factor H_r of the individual element and $\cos \varphi_x$:

$$\frac{\sin[(kb\sin\varphi_x)/2]}{(kb\sin\varphi_x)/2}\cos\varphi_x \qquad (4.100)$$

This envelope also has values of zeros like the main-beam directional factor, but at much larger angles since $b \ll L$. The positions of these zeros are

$$\varphi_{xe} = \sin^{-1}\left(\frac{m\lambda}{b}\right), \qquad m = \pm1, \pm2, \ldots \qquad (4.101)$$

In the special case of $s = 2b$, the angle of the second grating lobe $\varphi_{g2} = \sin^{-1}(2\lambda/b)$ falls at the first zero in the envelope $\varphi_{xe} = \sin^{-1}(\lambda/b)$ as well as the other even-order grating lobes. So, there are only the main beam and odd-order grating lobes. Altogether, the array parameters s, b, and L can be manipulated synergistically for the optimal acoustic field, depending on the desired application.

Reduction in the number and amplitude of the grating lobes is desirable in not only diagnosis to reduce potential sources of ambiguity in determining the direction of the echoes returned to the transducer but also therapy to focus ultrasound energy in a determined region and to avoid the unintended damage to the surrounding tissue. Although the magnitude of the grating lobes can be partially reduced in the transmitted pattern, such as by using very short transmitter pulses, it is not feasible in most of therapeutic ultrasound applications since its pulse duration is much longer than that of the diagnostic one. Another way to reduce grating lobe is by using nonuniform spacing between elements, defeating some of the constructive interference effects at off-axis angles.

4.4.1.3 Beam Steering

One of the main advantages of a linear-phased array is its ability to generate the desired acoustic field, which steers the acoustic beam electrically at arbitrarily chosen focal lengths. If all the elements transmit in phase simultaneously, the constructive interference will form a linear wavefront parallel to the transducer's surface. The phase of a spherical wave propagating a distance of $s = c \cdot t$ at a time interval of t is $\phi = k \cdot s$. If each element is small enough to be considered as a point source (element size relative to the wavelength), in order to form a straight wavefront titled at an angle β, all the emitted waves from each element reach this wavefront at the same phase, that is,

$$\omega \cdot t - k \cdot s_i + \phi_i = \omega \cdot t + \text{const} \qquad (4.102)$$

where
    s_i is the traveling distance of the spherical wave transmitted from the ith element
    ϕ_i is the initial phase of the ith element

For simplicity, the constant on the right-hand side of Equation 4.102 is set to zero:

$$-k \cdot x_i + \phi_i = 0 \quad \Rightarrow \quad \phi_i = k \cdot s_i \qquad (4.103)$$

Thus, the initial phase of the ith element is

$$\phi_i = k \cdot x_i \cdot \sin\beta \quad \Rightarrow \quad \phi_i = \left[(i-1)\cdot\Delta x - \frac{L}{2}\right]\cdot k \cdot \sin\beta \qquad (4.104)$$

which applies to a continuous wave transmission. However, in sonography, short pulses are implemented and the phase is replaced by the time delay for technical convenience. The time delay for the ith element (i.e., Δt_i) is calculated as

$$\phi_i = 2\pi f \cdot \Delta t_i = k \cdot s_i = \frac{2\pi}{\lambda} \cdot s_i$$

$$\Rightarrow \Delta t_i = \frac{2\pi}{2\pi \cdot f \cdot \lambda} = \frac{s_i}{c} \qquad (4.105)$$

$$\Rightarrow \Delta t_i = \left[(i-1)\cdot\Delta x - \frac{L}{2}\right]\cdot\frac{\sin\beta}{c}$$

To avoid the grating lobes, the element size of a linear-phased array is constrained to be relative to the wavelength. With such a constraint, the corresponding far-field pressure approximation for a beam steered at an angle of β is

$$P(R,\theta) = a \cdot \frac{e^{j(\omega t - kR)}}{R}\sum_{i=1}^{N} A_{0i} \cdot e^{j(k\cdot x_i \cdot \sin\theta + k\cdot x_i \cdot \sin\beta)} \qquad (4.106)$$

If the amplitudes of all elements are equal, this equation can be simplified as

$$P(R,\theta) = N \cdot A_0 \cdot a \cdot \frac{e^{j(\omega t - kR)}}{R}\cdot\left\{\frac{\sin\left[k\dfrac{N\cdot\Delta x}{2}\cdot(\sin\theta+\sin\beta)\right]}{N\sin\left[k\dfrac{\Delta x}{2}(\sin\theta+\sin\beta)\right]}\right\} \qquad (4.107)$$

whose maximum value is reached at the condition

$$k\frac{\Delta x}{2}(\sin\theta+\sin\beta) = m\pi, \quad m = 0, \pm1, \pm2, \ldots \qquad (4.108)$$

This implies the presence of undesired strong lobes in the acoustic field. If the first lobe for which $m = 1$ appears only at $\theta = \pi/2$, the constraint for the maximum steering angle $\beta_{\max}$ is given by

$$\frac{2\pi}{\lambda}\cdot\frac{\Delta x}{2}\left(\sin\frac{\pi}{2}+\sin\beta_{\max}\right) = \pi \quad \Rightarrow \quad \sin\beta_{\max} = \left(\frac{\lambda}{\Delta x}-1\right) \qquad (4.109)$$

or more explicitly

$$\beta_{\max} = \sin^{-1}\left(\frac{\lambda}{\Delta x} - 1\right) \tag{4.110}$$

It is noted that this condition must be kept in addition to $1 \le \lambda/\Delta x$. Otherwise, side lobes will occur anyway.

4.4.2 ANNULAR-PHASED ARRAYS

4.4.2.1 General Acoustic Field

Another popular phased-array transducer is the annular array type, which consists of multiple concentric rings arranged according to the requirement (e.g., equal area for each element in order to design the driving circuit conveniently and achieve the same power for all the activated elements). The gap between the rings is usually negligible. Annular arrays can be used to transmit a subset of the rings in order to achieve a sharp main lobe and receive with different subsets to avoid echoes from the side lobes. The focal point of a high-intensity focused ultrasound (HIFU) during a noninvasive and minimally invasive surgical procedure can be steered electrically on axis by using annular-phased arrays, which provides more flexibility than conventional transducers with the fixed focal distance.

A schematic depiction of an annular array is shown in Figure 4.16. The second ring has the same area as the first one:

$$\pi\left(r_2^2 - r_1^2\right) = \pi r_1^2 \quad \Rightarrow \quad r_2^2 = 2r_1^2 \quad \Rightarrow \quad r_2 = \sqrt{2}\cdot r_1 \tag{4.111}$$

Proceeding to the other rings, the general condition can be easily derived as

$$r_n = \sqrt{n}\cdot r_1 \tag{4.112}$$

And, if the outer radius and the number of rings are constrained as a and N, respectively, the radius of nth ring is given by

$$r_n = a\cdot\sqrt{\frac{n}{N}}, \quad n = 1,2,...,N \tag{4.113}$$

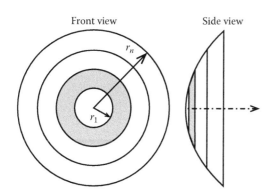

FIGURE 4.16 Schematic diagram of an annular-array transducer.

In general, the field pressure p_n transmitted from the nth ring is given by

$$p_n = A_n \cdot \frac{e^{j(\omega t - k\cdot R)}}{R} \int\limits_{r=r_n-1}^{r_n} 2\pi\cdot J_0(kr\sin\theta)\cdot r\cdot dr \tag{4.114}$$

where A_n is the corresponding amplitude for this particular ring. The overall field is the sum of the contributions of all the rings:

$$p(R,\theta) = \sum_{n=1}^{N} p_n(R,\theta) \tag{4.115}$$

4.4.2.2 Beam Steering

In the case of a concave array with curvature R_c, as illustrated in Figure 4.16, there is a point G located on the surface of the nth ring. Its distances to the x- and y-axes are r_n and S_n, respectively. The propagation distance from the point G to the focal point F, d_n, is given by

$$d_n = \sqrt{(F - S_n)^2 + r_n^2} \tag{4.116}$$

To focus an acoustic wave at point F, the phase of every ring should be the same. However, the distance from the inner and outer sides of the ring to the focal point varies. Hence, as a compromise, the average phase of waves reaching the focal point from the inner and outer sides of the ring will be used as the phase of the whole ring:

$$\phi = k\cdot F = k\cdot\left(\frac{d_n + d_{n-1}}{2}\right) + \phi_n \quad \text{or} \quad \phi_n = k\cdot\left[F - \left(\frac{d_n + d_{n-1}}{2}\right)\right] \tag{4.117}$$

It is noted that this generic expression does not assume equal area for every ring.

For an annular array with equal-area rings, the approximate formula is

$$\phi_n = k\cdot S_N\cdot\frac{2n-1}{2N}\cdot\frac{F - R_c}{F} \tag{4.118}$$

where S_N is the distance from the y-axis to the outermost ring. Meanwhile, the radius of an annular array with N equal-area rings is

$$R_c = \frac{2F_{\text{near}}\cdot F_{\text{far}}}{F_{\text{near}} + F_{\text{far}}} \tag{4.119}$$

where F_{near} and F_{far} are the required nearest and farthest focal distances steered along the array's axis of symmetry, respectively.

4.4.2.3 Bessel Beam

Another special application of an annular array is transmitting a "Bessel beam," a family of nondiffractive solutions to the wave equation. Using annular arrays, a single defined focal zone is obtained in each transmission. Moreover, the regions located outside the focal zone have poor spatial resolution. However, the main lobe is narrow over a very long range if activated with the Bessel beam. For example, a Bessel beam from a 10-ring annular array with an outside diameter of 25 mm has a beam width of only 1.27 mm over 200 mm on the axis. The acoustic pressure field for a Bessel beam of order 0 is described as

$$U(x,y,z,t) = J_0(\alpha\rho) \cdot e^{j(\beta Z - \omega t)}, \quad \rho = \sqrt{x^2 + y^2}, \quad \alpha^2 + \beta^2 = k^2 \tag{4.120}$$

where

J_0 is a Bessel function of order 0
k is the wave number
α is positive and real $\alpha \le k$
β is also real
$\alpha = 0$ means a regular planar wave

For an infinite aperture transducer, the -6 dB width of the Bessel beam, B_{6dB}, is given by

$$B_{6dB} = \frac{3.04}{\alpha} \tag{4.121}$$

The width between two zero pressures, B_{0-0}, is given by

$$B_{0-0} = \frac{4.81}{\alpha} \tag{4.122}$$

In comparison, for a finite aperture with radius R, the maximum range of the beam, Z_{max}, is given by

$$Z_{max} = R \cdot \sqrt{\left(\frac{k}{\alpha}\right)^2 - 1} \tag{4.123}$$

For an annular array containing N rings with equal area, the error in the phase at an infinite distance, ϕ_∞, from the array is given by

$$\phi_\infty = \frac{\pi}{N \cdot S}, \quad S = \frac{F \cdot \lambda}{R^2} \tag{4.124}$$

where
F is the focal length
S is the Fresnel number

In order to keep the phase error less than $\pi/2$, the radius of the innermost ring needs to match Bessel's function zero value. Hence, the parameter α is given by

$$\alpha = \frac{2.405\sqrt{N}}{R} \tag{4.125}$$

4.4.3 Sector-Vortex Array

The sector-vortex array consists of a concave transducer divided into N sectors of equal size, as shown in Figure 4.17. In some situations, there is a central hole, for example, for the incorporation of an ultrasound imaging probe. A cylindrical coordinate system (r, θ) on the transducer and (R, ψ) in the focal plane is assumed. The phase ϕ_i for each sector is

$$\phi_i = m[\theta_i + \beta(\theta_i)], \quad i = 1, 2, \dots, N \tag{4.126}$$

where
m is the vortex mode number
$$\theta_i = \frac{i2\pi}{N}$$
$\beta(\theta)$ is the phase modulation function

From symmetry considerations, such a phased transducer will always produce zero intensity along the central axis.

The complex amplitude of the driving signal for the ith element can be written as

$$A(\theta_i) = A_0 \exp j\left\{m\left[\theta_i + \beta(\theta_i)\right] - \omega_0 t\right\} \tag{4.127}$$

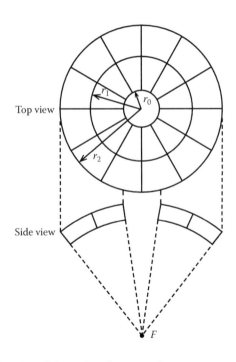

FIGURE 4.17 Schematic diagram of a sector-vortex array transducer.

where A_0 is the amplitude. The phase distribution over the sectors leads to an exciting field rotating at a phase velocity ω_p, where

$$\omega_p = \frac{\omega_0}{m}\left[1 + \frac{d}{d\theta}\beta(\theta)\right]^{-1} \qquad (4.128)$$

and ω_0 is the angular frequency of the driving signals.

An approximate analytic expression of the acoustic field in the focal plane for a simple case without modulation ($\beta = 0$), but sufficiently large N ($N \geq 4m$) can be derived as

$$B = \int_{r_0}^{r_1}\int_0^{2\pi} A(\theta)\exp j[k_f r\cos(\theta - \psi)]rd\theta dr$$

$$= A_0 \exp j\left[m\left(\psi + \frac{\pi}{2}\right) - \omega t\right],$$

$$\int_{r_0}^{r_1}\int_0^{2\pi} \exp j(m\theta - k_f r\sin\theta)rd\theta dr \qquad (4.129)$$

$$= 2\pi A_0 \exp j\left[m\left(\psi + \frac{\pi}{2}\right) - \omega t\right]\int_{r_0}^{r_1} J_m(k_f r)rdr$$

where $k_f = kR/F$. Equation 4.129 can be approximated as

$$B = A_0 \exp j\left[m\left(\psi + \frac{\pi}{2}\right) - \omega t\right]\pi\left(r_1^2 - r_0^2\right)J_m(k_f r_2) \qquad (4.130)$$

where $r_2 = \dfrac{2}{3}\dfrac{r_1^3 - r_0^3}{r_1^2 - r_0^2}$. It is seen that the field has a shape determined by the mth-order Bessel function. Thus, the vortex-shaped field is zero along the central axis for $m \neq 0$ and has a diameter proportional to the vortex mode m and phase velocity ω_p. When phase modulation function β is nonzero, there is a kind of spatial frequency modulation around the transducer. For certain β, the focal annuli are elongated along one axis, producing heating patterns in the shape of an ellipse.

Assume that the vortex array is divided into I tracks of radius r_i ($i = 1, ..., I - 1$), and each track has N sectors of equal size. The nth sector on the ith track is driven by the complex signal $A_i(\theta_n)$

$$A_i(\theta_n) = A_0 \exp j\left\{m\left[\theta_n + \beta(\theta_n)\right] + \gamma_i - \omega_0 t\right\},$$

$$n = 1, 2, ..., N, \quad i = 1, 2, ..., I \qquad (4.131)$$

where
$\theta_n = 2\pi n/N$ is the nth sector's angle
$m \in [-N/2, N/2]$ is the vortex mode number
$\beta(\theta)$ is the modulation function
ω is the angular frequency
A_0 and γ_i are the amplitude and phase, respectively

It shows that the driving signals allow the phase swivel m times per rotation along the track (the sign of m denoting the direction of the phase rotation).

An approximate analytic expression for the acoustic field $B_i(R, \psi)$ in the geometrical focal plane produced by the ith track can be derived as

$$B_i(R, \psi) = \int_{r_{i-1}}^{r_i}\sum_{n=1}^{N} A_i(\theta_n)\exp\left[k_R r\cos(\theta_n - \psi)\right]\cdot\left(\frac{2\pi}{N}\right)rdr \qquad (4.132)$$

where
$k_R = kR/F$
k is the wave number

When N is sufficiently larger than m, Equation 4.132 can be approximated as

$$B_i = \int_{r_{i-1}}^{r_i}\int_0^{2\pi} A_i(\theta)\exp\left[k_R r\cos(\theta - \psi)\right]rd\theta dr$$

$$= \begin{cases} A_0 \exp j[m(\psi + \pi/2) - \omega_0 t + \gamma_i]\cdot 2\pi\displaystyle\int_{r_{i-1}}^{r_i} J_m(k_R r)dr \quad m \geq 0 \\[4mm] A_0 \exp j[m(\psi + \pi/2) - \omega_0 t + \gamma_i]\cdot 2\pi\displaystyle\int_{r_{i-1}}^{r_i} J_{-m}(k_R r)dr \quad m \leq 0 \end{cases} \qquad (4.133)$$

The phase error of $A_i(\theta)$ is not larger than $\pi/8$ and can be ignored when $N \geq m$. Therefore,

$$2\pi\int_{r_{i-1}}^{r_i} J_{|m|}(k_R r)dr = \pi(r_i^2 - r_{i-1}^2)J_{|m|}\left(k_R\frac{2}{3}\cdot\frac{r_i^3 - r_{i-1}^3}{r_i^2 - r_{i-1}^2}\right) \qquad (4.134)$$

Then, the acoustic field $B(R, \psi)$ synthesized by the whole array is the summation of B_i:

$$B(R, \psi) = \sum_{i=1}^{I} B_i(R, \psi) \qquad (4.135)$$

Therefore, the acoustic field is approximated by the summation of a ring-shaped array with the $|m|$th-order Bessel function. The pressure distribution is cylindrically symmetric and controlled by both the mode number m and the phase difference between tracks, γ_i. If the widths of all the tracks are equal to w,

$$r_i = (i + \alpha)w, \quad i = 0, 1, ..., I \qquad (4.136)$$

where $\alpha = r_0/w$. Substitution of Equation 4.136 into Equation 4.135 gives the radial distribution B_R of the amplitude $B(R, \psi)$ as a function of $k_R w$:

$$B_R = 2\pi w^2 A_0 \sum_{i=1}^{I} b_i \qquad (4.137)$$

where

$$b_i = i_I J_{|m|}\left(k_R w\left[i_I + 1/12 i_I\right]\right) \exp j\gamma_i \qquad (4.138)$$

$$i_I = i + \alpha - 1/2 \qquad (4.139)$$

$$k_R w = \left(\frac{kw}{F}\right) R \qquad (4.140)$$

The power distribution patterns are determined by the interference between the annular elements, the radial distribution of the individual track b_i, and the overall power $\|B_R\|^2$. Figure 4.18 shows the pressure distribution of a two-track array ($I = 2$ and $\alpha = 1$) with mode $m = 5$. Since the primary and secondary lobes of an mth-order Bessel function are opposite in amplitude, the phase difference $\gamma_2 - \gamma_1$ is chosen to be π so that the primary lobe ($i = 1$) and the secondary lobe ($i = 2$) enhance each other, giving a high-intensity peak in the superimposed field. Figure 4.19 shows the pattern for a four-track array ($I = 4$ and $\alpha = 2$) driven with $m = 8$. The driving phases are chosen as $\gamma_1 = 0$, $\gamma_2 = \pi$, $\gamma_3 = 0$, and $\gamma_4 = \pi$. The ith lobe ($i = 1, 2, 3, 4$) match with each other at $k_R w \approx 4$, also producing a high-intensity peak.

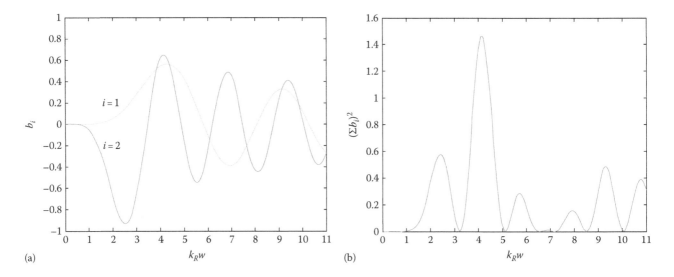

FIGURE 4.18   (a) Fields from each track in amplitude and (b) intensity profile of summation of Figure 4.19.

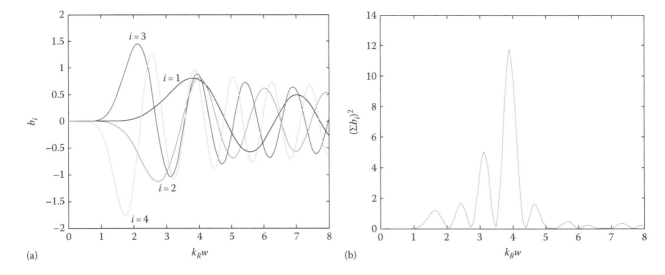

FIGURE 4.19   (a) Fields from each track in amplitude and (b) intensity profile of summation of Figure 4.20.

A multitrack array is more versatile in selecting annular foci dimensions than a single-track one, such as a larger annular foci from the same aperture with the same mode number. Moreover, in order to approximate a spherical shell for geometric focusing, multiple tracks are usually needed.

4.4.4 TWO-DIMENSIONAL PHASED ARRAY

To calculate the field from an array source, the Huygens principle—"Every source of waves for which the size of the source a is much less than the corresponding wavelength λ may be considered as a source for a spherical wave!"—is utilized as the basis for computing the field. Thus, the acoustic field stemming from any element with an area of ds in the array is given by

$$p(x,y,z) = \iint_{source} p_s ds = \iint_{source} A_0(\mu,\eta) \frac{e^{j[\omega t - kd(\mu,\eta,x,y,z) + \phi(\mu,\eta)]}}{d(\mu,\eta,x,y,z)} ds$$

$$(4.141)$$

where $d(x,y,z)$ is the distance between source (μ,η) on the transmitting surface and the point Q in the medium. $A_0(\mu,\eta)$ is the amplitude, and $\phi(\mu,\eta)$ is the corresponding phase. If the array consists of N elements, the total field is the summation of each element:

$$p(x,y,z) = \sum_{i=1}^{N} \iint_{source-i} p_s ds_i$$

$$= \sum_{i=1}^{N} \iint_{source-i} A_{0i}(\mu,\eta) \frac{e^{j[\omega t - kd(\mu,\eta,x,y,z) + \phi_i(\mu,\eta)]}}{d(\mu,\eta,x,y,z)} ds_i \quad (4.142)$$

where i is an index designating a certain element in the array. There is no limitation on the number of miniature elements (Figure 4.20).

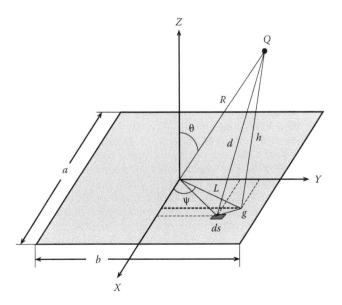

FIGURE 4.20   The schematic diagram of a 2D-array transducer.

ASSIGNMENT

Q1: There are two planar acoustic waves with different frequencies. Their acoustic pressures are $p_1 = p_{1A} \cos(\omega_1 t - k_1 x - \varphi_1)$ and $p_2 = p_{2A}\cos(\omega_2 t - k_2 x - \varphi_2)$, where the phases φ_1 and φ_2 are constants. Calculate the energy density of the acoustic field.

Q2: There are two point sources separated by a distance of l with the same frequency and with acoustic intensity Q_{01} and Q_{02}, respectively. Prove that the sound pressure in the far field can be expressed as

$$p = j\frac{k\rho_0 c_0}{4\pi r} e^{j(\omega t - kr)} \left[\begin{array}{l} (Q_{01}+Q_{02})\cos\left(\dfrac{\pi l\cos\theta}{\lambda}\right) \\[2mm] + j(Q_{01}-Q_{02})\sin\left(\dfrac{\pi l\cos\theta}{\lambda}\right) \end{array} \right]$$

Q3: If the vibration velocity of a piston surface is $u(t,\rho) = u_0\left(1-\dfrac{\rho^2}{a^2}\right)e^{j\omega t}$, determine the sound pressure in the far field.

Q4: If the vibration velocity of a piston surface is $u(t,\rho) = u_0\left(1-\dfrac{\rho^2}{a^2}\right)^n e^{j\omega t}$, prove that the radiation sound pressure in the far field is

$$p = j\frac{\omega\rho_0 u_0}{r} e^{j(\omega t - kr)} 2^n n! \frac{a^{1-n} J_{n+1}(ka\sin\theta)}{(k\sin\theta)^{n+1}}$$

REFERENCES

Archer-Hall JA, Gee D. A single integral computer method for axisymmetric transducers with various boundary conditions. *NDT International* 1980;13:95–101.

Arditi M, Foster FS, Hunt JW. Transient fields of concave annular arrays. *Ultrasonic Imaging* 1981;3:37–61.

Baker BB, Copson ET. *The Mathematical Theory of Huygens' Principle.* Oxford, U.K.: Clarendon Press, 1950.

Barton G. *Elements of Green's Functions and Propagation.* Oxford, U.K.: Oxford University Press, 1989.

Beaver WL. Sonic nearfields of a pulsed piston radiator. *Journal of the Acoustical Society of America* 1982;71:1406–1411.

Blackstock DT. Transient solution for sound radiated into a viscous fluid. *Journal of the Acoustical Society of America* 1967;41:1312–1319.

Bouwkamp CJ. A contribution to the theory of acoustic radiation. *Philips Research Reports* 1945/46;1:251–277.

Burckardt CB, Grandchamp PA, Hoffmann H. Focussing ultrasound over a large depth with an annular aperture: An alternative method. *IEEE Transactions on Sonics and Ultrasonics* 1975;SU-22:11–15.

Burckardt CB, Hoffmann H, Grandchamp PA. Ultrasound axicon: A device for focusing over a large depth. *Journal of the Acoustical Society of America* 1973;54:1628–1630.

Cain CA, Umemura S-I. Concentric-ring and sector-vortex phased-array applicators for ultrasound hyperthermia. *IEEE Transactions on Microwave Theory and Techniques* 1986;34:542–551.

Campbell JA, Soloway S. Generation of a nondiffracting beam with frequency-independent beamwidth. *Journal of the Acoustical Society of America* 1990;88:2467–2477.

Cathignol D, Faure P, Chavrier F. Acoustic field of plane or spherical transducer. *Acustica* 1997;83:410–418.

Cathignol D, Sapozhnikov OA, Theillere Y. Comparison of acoustic fields radiated from piezoceramic and piezocomposite focused transducers. *Journal of the Acoustical Society of America* 1999;105:2612–2617.

Coulouvrat F. Continuous field radiated by a geometrically focused transducer: Numerical investigation and comparison with an approximation model. *Journal of the Acoustical Society of America* 1993;94:1663–1675.

Crombie P, Bascom PAJ, Cobbold RSC. Calculating the pulsed response of linear arrays: Accuracy versus computational efficiency. *IEEE Transactions on Ultrasonics, Ferroelectrics, and Frequency Control* 1997;44:997–1009.

de Hoop AT, Zeroug S, Kostek S. Transient analysis of the transmitting properties of a focused acoustic transducer with an arbitrary rim. *Journal of the Acoustical Society of America* 1995;101:1269–1277.

Delannoy B, Lasota H, Bruneel C, Torquet R, Bridoux E. The infinite planar baffles problem in acoustic radiation and its experimental verification. *Journal of Applied Physics* 1979;50:5189–5195.

Dietz DR. Apodixed conical focusing for ultrasound imaging. *IEEE Transactions on Sonics and Ultrasonics* 1982;SU-29:128–138.

Doak PE. An introduction to sound radiation and its sources. In: Richards EJ, Mead DJ, eds., *Noise and Fatigue in Aeronautics.* New York: John Wiley & Sons, 1968, pp. 1–42.

Donnelly R, Power D, Templeman G, Whalen A. Graphical simulation of superluminal acoustic localized wave pulses. *IEEE Transactions on Ultrasonics, Ferroelectrics, and Frequency Control* 1994;41:7–12.

Du GH, Breazeale MA. The ultrasonic field of a Gaussian transducer. *Journal of the Acoustical Society of America* 1985;76:2083–2088.

Durnin J. Exact solutions for nondiffracting beams. I. The scalar theory. *Journal of the Optical Society of America A* 1987;4:651–654.

Durnin J, Miceli JJ, Eberly JH. Diffraction-free beams. *Physical Review Letters* 1987;58:1499–1501.

Filipczynski L, Etienne J. Theoretical study and experiments on spherical focusing transducers with Gaussian surface velocity distribution. *Acustica* 1973;28:121–128.

Freedman A. Transient fields of acoustic radiators. *Journal of the Acoustical Society of America* 1970;48:135–138.

Fung CCW, Cobbold RSC, Bascom PAJ. Radiation coupling of a transducer-target system using the angular spectrum method. *Journal of the Acoustical Society of America* 1992;92:2239–2247.

Gibson WGR, Cobbold RSC, Foster FS. Ultrasonic fields of a convex semi-spherical transducer. *Journal of the Acoustical Society of America* 1993;94:1923–1929.

Greenspan M. Piston radiator: Some extensions of the theory. *Journal of the Acoustical Society of America* 1979;65:608–621.

Guyomar D, Powers J. A Fourier approach to diffraction of pulsed ultrasonic waves in a lossless medium. *Journal of the Acoustical Society of America* 1987;92:354–359.

Harris GR. Review of transient field theory for a baffled planar piston. *Journal of the Acoustical Society of America* 1981a;70:10–20.

Harris GR. Transient field of a baffled planar piston having an arbitrary vibration amplitude distribution. *Journal of the Acoustical Society of America* 1981b;70:186–204.

Holm S. Bessel and conical beams and approximations with annular arrays. *IEEE Transactions on Ultrasonics, Ferroelectrics, and Frequency Control* 1998;45:712–718.

Hsu DK, Margeten FJ, Thompson DO. Bessel beam ultrasonic transducer: Fabrication method and experimental results. *Applied Physics Letters* 1989;55:2066–2068.

Hutchins DA, Hayward G. Radiated fields of ultrasonic transducer. In: Thurston RN, Pierce AD, eds., *Physical Acoustics: Ultrasonic Measurement Methods.* New York: Academic Press, 1990, pp. 1–80.

Jensen JA, Gandhi D, O'Brien WD. Ultrasound fields in an attenuating medium. *IEEE Ultrasonics Symposium Proceedings 1993*, Baltimore, MD, October 31, 1993 to November 3, 1993, pp. 943–946.

Kikuchi Y. Transducers for ultrasonic systems. In: Fry FF, ed., *Ultrasound: Its Applications in Medicine and Biology.* Amsterdam, the Netherlands: Elsevier, 1978.

Kozina OG, Makarov GI. Transient processes in the acoustic fields generated by a piston membrane of arbitrary shape. *Soviet Physics Acoustics* 1961;7:39–43.

Lasota H, Salamon R, Delannoy B. Acoustic diffraction analysis by the impulse response method: A line impulse response approach. *Journal of the Acoustical Society of America* 1984;76:280–290.

Leeman S, Healey AJ. Field propagation via the angular spectrum method. In: Lees S, Ferrari LA, eds., *Acoustical Imaging.* New York: Plenum Press, 1997, 363–368.

Li G, Bharath AA. Numerical study of limited diffraction, band limited, acoustic waves: A novel solution family of the homogeneous scalar wave equation. *Wave Motion* 1998;28:203–213.

Liu DL, Waag RC. Propagation and backpropagation for ultrasonic wavefront design. *IEEE Transactions on Ultrasonics, Ferroelectrics, and Frequency Control* 1997;44:1–13.

Lockwood JC, Willette JG. High speed method for computing the exact solution for the pressure variations in the nearfield of a baffled piston. *Journal of the Acoustical Society of America* 1973;3:735–741.

Lu JY, Greenleaf JF. Producing deep depth of field and depth-independent resolution in NDE with limited diffraction beams. *Ultrasonic Imaging* 1993;15:134–149.

Ludwig R, Levin PL. Analytical and numerical treatment of pulsed wave propagation into a viscous fluid. *IEEE Transactions on Ultrasonics, Ferroelectrics, and Frequency Control* 1995;42:789–792.

Madsen EL, Goodsitt MM, Zagzebski JA. Continuous waves generated by focused radiators. *Journal of the Acoustical Society of America* 1981;70:1508–1517.

Mair HD, Hutchins DA. Axial focusing by phased concentric annuli. In: Merklinger HM, ed., *Progress in Underwater Acoustics.* New York: Plenum Press, 1986, 619–626.

Markiewicz A, Chivers RC. Effects of baffle conditions on the nearfield of piston disk radiators. *Acustica* 1986;60:289–294.

O'Neil HT. Theory of focusing radiators. *Journal of the Acoustical Society of America* 1949;21:516–526.

Oberhettinger F. Note on the baffled piston problem. *Journal of Research of the National Bureau of Standards-B. Mathematics and Mathematical Physics* 1961a;65B:203–204.

Oberhettinger F. On transient solutions to the baffled piston problem. *Journal of Research of the National Bureau of Standards-B. Mathematics and Mathematical Physics* 1961b;65B:1–6.

Ohtsuki S. Ring function method for calculating nearfield of sound source. *Bulletin of the Tokyo Institute of Technology* 1974;123:23–27.

Patterson MS, Foster FS. Acoustic fields of conical radiators. *IEEE Transactions on Sonics and Ultrasonics* 1982;SU-29:83–92.

Penttinen A, Luukkala M. The impulse response and pressure nearfield of a curved ultrasonic radiator. *Journal of Physics D: Applied Physics* 1976;9:1547–1557.

Rayleigh FRS. On the passage of waves through apertures in plane screens and allied problems. *Philosophical Magazine* 1897;43:259–272.

Rayleigh JWS. *The Theory of Sound.* New York: Dover Publications, 1945.

San Emeterio JL, Ullate LG. Diffraction impulse response of rectangular transducers. *Journal of the Acoustical Society of America* 1992;92:651–662.

Schmerr LW, Sedov A, Lerch TP. A boundary diffraction wave model for a spherically focused ultrasonic transducer. *Journal of the Acoustical Society of America* 1997;101:1269–1277.

Skudrzyk E. *The Foundations of Acoustics: Basic Mathematics and Basic Acoustics.* New York: Springer-Verlag, 1971.

Sommerfeld A. *Optics: Lectures on Theoretical Physics.* New York: Academic Press, 1954.

Stepanishen PR. The time-dependent force and radiation impedance on a piston in a rigid infinite baffle. *Journal of the Acoustical Society of America* 1971a;49:841–849.

Stepanishen PR. Transient radiation from pistons in an infinite baffle. *Journal of the Acoustical Society of America* 1971b;49:1629–1638.

Stepanishen PR. Acoustic transients in the far field of a baffled circular piston using the impulse response approach. *Journal of Sound and Vibration* 1974;32:295–310.

Stepanishen PR, Forbes M, Letcher S. The relationship between the impulse response and angular spectrum methods to evaluate acoustic transient fields. *Journal of the Acoustical Society of America* 1991;90:2794–2798.

Stokes GG. Dynamical theory of diffraction. *Cambridge Philosophical Society Transactions* 1849;9:1–62.

Sushilov NV, Cobbold RSC. Frequency-domain wave equation and its time-domain solutions in attenuating media. *Journal of the Acoustical Society of America* 2004;115:1431–1436.

Tjotta JN, Tjotta S. Nearfield and farfield of pulsed acoustic radiators. *Journal of the Acoustical Society of America* 1982;71:824–834.

Tupholme GE. Generation of acoustic pulses by baffled plane pistons. *Mathematika* 1969;16:209–224.

Turnbull DH, Foster FS. Beam steering with pulsed two-dimensional transducer arrays. *IEEE Transactions on Ultrasonics, Ferroelectrics, and Frequency Control* 1991;38:320–333.

Umemura S, Cain C. The sector-vortex phased array: Acoustic field synthesis for hyperthermia. *IEEE Transactions on Ultrasonics, Ferroelectrics and Frequency Control* 1989;36:249–257.

Vecchio CJ, Schafer ME, Lewin PA. Prediction of ultrasonic field propagation through layered media using the extended angular spectrum method. *Ultrasound in Medicine and Biology* 1994;20:611–622.

Weight JP. Ultrasonic beam structures in fluid media. *Journal of the Acoustical Society of America* 1984;76:1184–1191.

Wells PNT. *Biomedical Ultrasonics.* New York: Academic Press, 1977.

Williams AO. Acoustic intensity distribution from a "piston" source II. The concave piston. *Journal of the Acoustical Society of America* 1946;17:219–227.

Williams EG. Numerical evaluation of the radiation from unbaffled finite planes using the FFT. *Journal of the Acoustical Society of America* 1983;74:343–347.

Williams EG, Maynard JD. Numerical evaluation of the Rayleigh integral for planar radiators sing the FFT. *Journal of the Acoustical Society of America* 1982;72:2020–2030.

Wolfe E, Marchand EW. Comparison of the Kirchhoff and the Rayleigh-Sommerfeld theories of diffraction at an aperture. *Journal of the Optical Society of America* 1964;54:587–594.

Wright FJ, Berry MV. Wave-front dislocations in the sound field of a pulsed circular piston radiator. *Journal of the Acoustical Society of America* 1984;75:733–748.

Wright WM, Medendorp NW. Acoustic radiation from a finite line source with N-wave excitation. *Journal of the Acoustical Society of America* 1968;43:966–971.

Wu P, Stepinski T. Extension of the angular spectrum approach to curved radiators. *Journal of the Acoustical Society of America* 1999;105:2618–2627.

Zemanek J. Beam behavior within the nearfield of a vibrating piston. *Journal of the Acoustical Society of America* 1970;49:181–191.

Ziolkowski RW. Exact solutions of the wave equation with complex source locations. *Journal of Mathematical Physics* 1985;26:861–863.

Ziomek LJ. *Fundamentals of Acoustic Field Theory and Space-Time Signal Processing.* Boca Raton, FL: CRC Press, 1995.

5 Acoustical Properties of Biological Tissue

In biomedical ultrasound application, acoustic properties of tissue and their influence on the acoustic field is important. In this chapter, acoustic attenuation, wave equation with attenuation, velocity relaxation and attenuation measurement are discussed.

5.1 CELL AND TISSUE

The cell is the fundamental and the smallest structural and functional unit of all living organisms (Figure 5.1), which can be classified as prokaryotic (unicellular and independent, such as bacteria and archaea) or eukaryotic (multicellular organisms, such as all plants and animals). The major difference between prokaryotes and eukaryotes is that eukaryotic cells contain membrane-bound compartments in which specific metabolic activities occur. All eukaryotic cells have a membrane 75–100 Å thick that envelops the cell, separates its interior from its environment, regulates what moves in and out (selectively permeable), and maintains the electric potential of the cell. Inside the membrane, a salty and gel-like cytoplasm takes up most of the cell volume, in which specialized organelles are located to perform the tasks necessary for the proper functioning of the cell. All cells possess deoxyribonucleic acid (DNA), the hereditary material of genes, and ribonucleic acid (RNA), containing the information necessary to build the various proteins such as enzymes, the cell's primary machinery. Oblong mitochondria are also seen in many cells, especially those of energy-related tissue, since the function of the mitochondria is to metabolize food substances by oxidation, yielding energy for the cell substance. In addition, other organelles useful in reproduction, transport, and secretion are sometimes found, along with inclusions or bits of waste material, food, enzymes, lipids, pigment, filaments, and so on. There are about 10^{13} cells in humans with size between 1 and 100 μm, which are visible only under a microscope.

Ultrasound-induced bioeffects in the cell depend on the parameters of the exposure beam. Little mechanical trauma or temperature elevation is found at low power output, such as in diagnosis. However, serious damage or rupture of the cell may occur at very high power levels, releasing the cell contents and causing cell death (apoptosis). If too many cells die, the tissue may not be able to repair itself sufficiently and consequently results in severe damage.

A large variety of cells from the same origin are assembled to constitute the different tissues to carry out a specific body function. The main constituents of the soft tissue are water (60%), protein (17%), and lipids (15%). An organ is a collection of multiple tissues grouping together in a structural unit to serve a common function. Animal tissues can be grouped into four fundamental types: connective, muscular, nervous, and epithelial, and their manifestation depends on the type of organism.

Tissue is continually adapting and self-regulating, growing and reproducing, becoming diseased, healing and repairing, altering metabolism, and interacting with other organs.

5.1.1 CONNECTIVE TISSUE

Connective tissues are fibrous tissues that are made up of cells separated by a nonliving material called the extracellular matrix. As the name implies, connective tissue provides supporting and connecting functions of various parts. Connective tissue, such as bone and blood, gives shape to organs and holds them in place by supporting and binding other tissues. It also serves as a background material through which cells wander, such as plasma cells for producing antibodies (important for fighting infection) and macrophages (for ridding the body of bits of foreign substances). Because of its ubiquitous presence, connective tissue is found in a vast variety of density, structure, and composition.

5.1.1.1 Loose Connective Tissue

It is a weblike structure composed of widely spaced fibroblast cells responsible for the generation of the loose and generally unorganized fibers in the tissue, which is often found between organs and filling other anatomical space. The fibers are not dense, but they still give the tissue its basic mechanical properties. Collagenous fibers have high tensile strength for toughness, while elastic fibers produce resilience in the tissue.

5.1.1.2 Dense Connective Tissue

There is a greater abundance of fibers, which are often organized to provide more strength or elasticity where needed. Tendons, organ capsules, and nerve sheaths are all dense connective tissue.

5.1.1.3 Bone

The extracellular substance of bone has become hardened through calcification, making the bone strong and rigid. Its principal role in the body is one of skeletal support, compartmentalization, and protection, as seen on the limbs and skull. Bone also provides a store of calcium to help in calcium regulation in the blood and other fluids and forms a compartment for bone marrow, which generates red blood cells. However, bone is a living, adaptable tissue with ample blood supply and cells for generation, protection, and repair, although its matrix is hardened calcium. There are numerous canals that infiltrate the bone, carrying metabolites to the osteocytes and containing small blood vessels in the large canals (Figure 5.2). Bone has a significantly higher acoustic phase velocity and impedance than soft tissue. The large mismatch in impedance at a soft tissue/bone interface results in a high reflection coefficient, making it difficult to penetrate bony areas with ultrasound. Meanwhile, shear waves will propagate in bone due to the rigidity of its matrix. Mode conversion to shear waves and scattering contribute to the high absorption coefficient for compressional waves in bone.

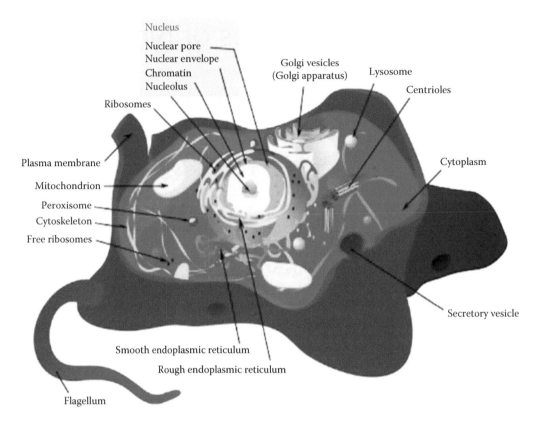

FIGURE 5.1 Diagrammatic structure of a representative animal cell.

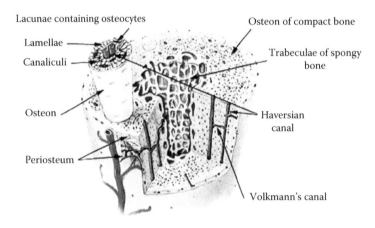

FIGURE 5.2 Diagrammatic structure of a compact bone and a spongy bone.

5.1.1.4 Blood

Blood is a fluid that carries nutrients to and waste products from all regions of the body (Figure 5.3). By means of its red blood cells (RBCs), or erythrocytes, blood transports oxygen from the lungs to all tissues to satisfy their metabolic requirements. In addition, blood can carry heat from or to a region as an important contributor to thermal regulation of the body. In vertebrates, it is composed of blood cells suspended in a liquid called blood plasma, which constitutes 54.3% of blood fluid, is mostly water (92% by volume), and contains dissipated proteins, glucose, mineral ions, hormones, carbon dioxide (plasma being the main medium for excretory product transportation), and blood cells themselves. The blood cells are mainly RBCs (or erythrocytes, 45% of whole blood), white blood cells (leukocytes, 0.7% of whole blood), and platelets (thrombocytes). The most abundant cells in vertebrate blood are RBCs. RBCs contain hemoglobin, an iron-containing protein that facilitates transportation of oxygen by reversibly binding to this respiratory gas and greatly increasing its solubility in blood. In contrast, carbon dioxide is almost entirely transported extracellularly dissolved in plasma as bicarbonate ion. Blood accounts for 8% of the human body weight, with an average density of ~1060 kg/m^3. Whole blood (plasma and cells) exhibits non-Newtonian fluid dynamics; its flow properties are adapted to flow effectively through tiny capillary blood vessels with less resistance than plasma by

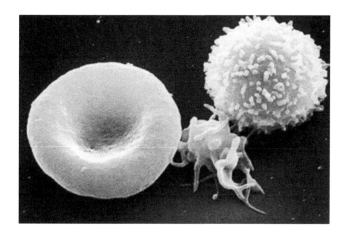

FIGURE 5.3 Scanning electron microscopy (SEM) images of a normal red blood cell, a platelet, and a white blood cell.

itself. Acoustically, the RBCs act as small point scatterers to an incident ultrasound wave, thus allowing the measurement of blood flow velocity by the Doppler shift technique. The acoustic characteristics of the internal material in a red blood cell are not greatly different from those of the surrounding plasma, and the cell membrane is too thin to appreciably affect the wave propagation. Also, since the size of an erythrocyte is much smaller than a wavelength, the scattering by each cell is classified as the Rayleigh scattering, which is generally weak and isotropic. However, the concentration of RBCs in healthy human blood is so great, $\sim 5 \times 10^6$ per mm^3, that the total scattered power from an irradiated volume may be detectable, even though each cell scatters weakly, making the Doppler blood flow measurements possible.

5.1.2 Muscular Tissue

Muscle cells form the active contractile tissue of the body, known as muscle tissue, which functions to produce force and cause motion, either locomotion or movement within internal organs (Figure 5.4). Muscle tissue is divided into three distinct categories: visceral or smooth muscle, skeletal muscle, and cardiac muscle. Smooth muscle is involuntarily controlled and is found in the inner linings of organs (e.g., digestive tract,

ducts of glands, arteries, and veins). Striated muscle occurs as voluntarily controlled skeletal muscle, whose fibers are organized into bundles attached to the bone for gross movement and force generation especially of the limbs. Cardiac muscle provides the force of the heart wall to contract and pump blood throughout an organism.

To be effective in force generation, the muscle bundle is generally tightly packed with fibers, and this affects its acoustical properties. The density, phase velocity, impedance, and attenuation values of muscle are all higher than those of water and other loose soft tissue. These values depend somewhat on whether the wave propagation direction is parallel or transverse to the muscle fiber's longitudinal axis. For example, attenuation in a direction parallel to the fibers is about twice that encountered when propagating perpendicular to them. However, the orientation effect is usually neglected when there is a lack of specific information regarding a muscle's orientation with regard to the beam's direction.

5.1.3 Nervous Tissue

Nervous tissue comprises the central nervous system (the brain and the spinal cord) and the peripheral nervous system (cranial nerves and spinal nerves, including the motor neurons). There are about 14 billion nerve cells (neurons) in humans, infusing almost every portion of the body and acting as transmission lines for collecting, transmitting, and distributing nerve impulses for local control or communication with the brain through the spinal cord. The dendrites project from the end of neurons where the cell body is located and function as multiple receivers or excitation sites in synaptic contact with the endings usually of several other prior neurons. A long projection (axon) carries the nerve's electrical impulses to its end, which may be further differentiated into fine endings, where it meets the dendrites of following neurons, or muscle, glandular, or epithelial cells (Figure 5.5).

If the neuron is involved in the transmission of impulses over long distances, its axon is generally myelinated, that is, coated with an insulating layer of myelin. Nerve bundles of myelinated axons appear white and form the white matter of the brain and the spinal cord. Collections of cell bodies

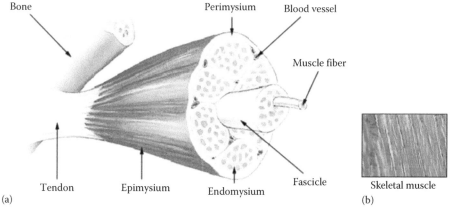

Bone · Perimysium · Blood vessel · Muscle fiber · Tendon · Epimysium · Endomysium · Fascicle

(a)

Skeletal muscle · Smooth muscle · Cardiac muscle

(b)

FIGURE 5.4 (a) Diagrammatic structure of muscle. (b) Photos of different muscle types at the appropriate magnification.

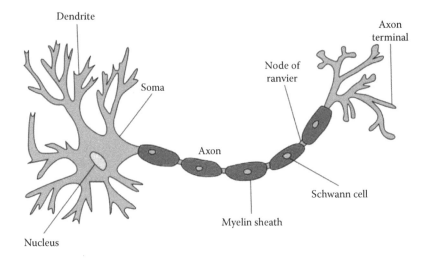

FIGURE 5.5 Diagrammatic structure of a nerve.

and bundles of unmyelinated axons comprise the gray matter of the brain and the spinal cord. However, only in the brain and the spinal cord, the neuron density is large enough to affect the propagation of acoustic waves.

5.1.4 EPITHELIAL TISSUE

Epithelial tissues are formed by cells in sheets or layers to cover the organ surfaces, such as the surface of the skin, the airways, the reproductive tract, and the inner lining of the digestive tract (Figure 5.6). The fundamental role of epithelial tissue is the protection, compartmentalization, and regulation of secretion and intake of substances by the organ it covers. Thus, the cells comprising an epithelial layer are linked via semipermeable, tight junctions. If only one layer of cells is present, the tissue is classified as *simple*. On the basis of the height of the cells in the layer, the tissue is classified as *squamous*, *cuboidal*, or *columnar*, in the order of increasing height. The simple squamous epithelium is found as a lining of blood vessels, kidney ducts, and other places where rapid diffusion through the tissue of vital substances such as oxygen or fluid molecules is important. The thicker and simpler cuboidal and columnar tissues are found in the digestive tract, in organ linings, and in glands where greater lining thickness may be useful but secretion or absorption is still required. When more protection or mechanical strength is required, there may be several layers of cells classified

as *stratified*. The stratified squamous epithelium is found on the inner and outer coverings of the body, including the skin, where mechanical wear and tear regularly removes cells from the outermost layer. Much of the mucous membrane lining of the respiratory tract is composed of epithelial tissue where the cells are tall and appear multilayered, but they actually form only a single layer. This tissue is called *pseudostratified columnar epithelium*. Hair-like projections known as *cilia* are found on the cell's outer surfaces to help transport foreign substance and to increase surface area for secretion or absorption. Transitional epithelial tissue is a combination of stratified squamous and columnar tissues, which allows the tissue to accommodate stretching while maintaining strength, such as the bladder lining. Being a relatively thin lining, epithelial tissue is normally not a major factor in determining the acoustical properties of most regions of the body. Connective tissue and muscle form a much larger percentage of the volume of most organs and viscera.

5.2 ACOUSTIC ATTENUATION

The pressure or intensity or power of an ultrasound wave decreases after propagating through tissue because of several reasons: divergence of the wavefront, elastic reflection at a planar interface, elastic scattering from irregularities or point scatterers, and sound absorption. The divergence of the wavefront leads to a dilution of the wave's energy into

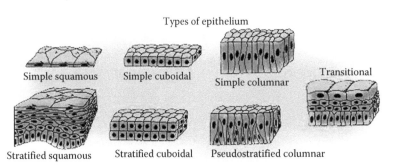

FIGURE 5.6 Diagramatic structure of different types of endothelial tissue.

an expanding cross-sectional area. Because of the acoustic impedance mismatch at intervening regions, wave reflection and transmission occur, as illustrated in Chapter 3. Even within an ideally homogeneous tissue, small particles, such as RBCs in the blood and multiple air-filled alveoli of lung tissue, cause broad-angle scattering of the incident wave and consequent reduction in forward intensity. The major cause of attenuation in most non-lung tissues is the absorption of the wave's energy, which transfers a portion of the originally organized acoustic energy to other energy forms, such as heat (through the effects of viscosity and heat conduction), chemical energy (i.e., sonochemistry), or light (i.e., sonoluminescence), while scattering represents the redirection of acoustic energy in the three-dimensional space (including reflection and refraction). It is noted that interference between the scattered wave and incident wave will occur. The exact cause of absorption by the molecules of biological tissue is still unknown and, in fact, is probably due to a complex variety of interactions.

Consider a plane wave incident on a specimen. Acoustic attenuation is commonly modeled under the assumption that the change in pressure is proportional to the distance from the source. Consequently,

$$\frac{dp(x)}{dx} = \alpha \cdot p(x) \qquad (5.1)$$

where
p is the acoustic pressure
α is the acoustic attenuation coefficient
x is the direction of wave propagation

This first-order differential equation has the solution

$$p(x) = p_0 e^{-\alpha \Delta x} \qquad (5.2)$$

where
p_0 is the initial pressure magnitude
Δx is the distance from the source

Since

$$I(x) = \frac{p^2(x)}{\rho_0 c_0} \qquad (5.3)$$

If the incident acoustic intensity at $x = 0$ is $I(0)$, the value in the specimen after propagating a distance of $x = d$ is

$$I(x)\big|_{x=d} = I(0)e^{-2\alpha x}\big|_{x=d} = I(0)e^{-2\alpha d} \qquad (5.4)$$

The units for α are m^{-1}, which is usually expressed as Np/cm in practice. It is also helpful to describe the change in acoustic intensity in a logarithmic scale:

$$10\log\left[\frac{I(0)}{I(x)}\right] = 20\alpha x \log(e) = 8.686\alpha x = \alpha_{dB}x \qquad (5.5)$$

Because of the relationship between the intensity and acoustic pressure, as shown in Equation 5.3,

$$\alpha_{dB} = \frac{10}{x}\log\left[\frac{I(0)}{I(x)}\right] = \frac{20}{x}\log\left[\frac{p(0)}{p(x)}\right] \qquad (5.6)$$

In most of the literature, taking care of the unit (Np/cm or dB/cm) of α for a correct

$$\alpha_{dB} = 8.686\alpha_{Np}, \qquad \alpha_{Np} = 0.1151\alpha_{dB} \qquad (5.7)$$

α is the effective attenuation coefficient, which is the sum of the scattering and absorption coefficients:

$$\alpha = \alpha_a + \alpha_s \qquad (5.8)$$

where α_s and α_a are the scattering and absorption coefficient, respectively.

For a plane wave with a single frequency f_c (or ω_c), its amplitude can be written as

$$p(x, t) = p_0 \exp(-\alpha x)\exp[-i\omega_c(t - x/c_0)] \qquad (5.9)$$

Its corresponding Fourier transform is

$$p(x, f) = p_0 \exp(-\alpha x - i\omega_c x/c_0)\delta(f - f_c) \qquad (5.10)$$

Usually, the loss is much smaller than the wavenumber, that is, $\alpha/k \ll 1$. Since $k = 2\pi/\lambda$, then $\alpha\lambda \ll 2\pi$, and there is little decay per wavelength. However, absorption will significantly add up over many wavelengths, and there may be an appreciable loss over the propagation distances encountered in the body (hundreds of wavelengths, i.e., 1.5 mm at 1 MHz, involved in a path of several centimeters).

In pure fluids, there is no scattering, so attenuation and absorption are identical. The effects of viscosity, which are represented by the shear and bulk viscosity coefficients μ and μ_B, govern the attenuation and relaxation losses. Therefore, the attenuation varies as the square of the frequency:

$$\alpha = \alpha_0 f^2 \qquad (5.11)$$

where
α_0 is a temperature-dependent factor in dB/cm/MHz
f is in megahertz (MHz)

Because deionized water is usually used as a reference medium in practice, the values of sound speed and attenuation are of importance for measurement. In experiments, it is found that the square-law frequency dependence is accurately true over a very wide range of temperatures up to at least 3 GHz (Figure 5.7) (Pinkerton 1949, Davidovich et al. 1972).

As a result of this frequency dependence on attenuation, acoustic pulses not only become smaller in amplitude but also change their shape during propagation. Absorption in the human body limits the detection of deep sites in sonography

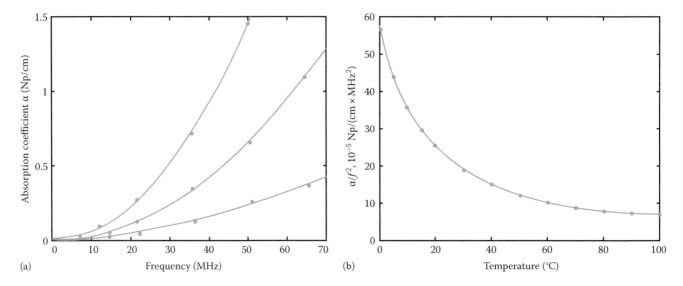

FIGURE 5.7 (a) Variation of absorption coefficient for water with frequency at temperatures 0°C, 20°C, and 60°C. (b) Temperature dependence of the attenuation factor for water. The attenuation factors up to 60°C were average measurements in frequency range 7.5–67.5 MHz; values of 60°C–100°C were from measurements at 52.5 MHz. (From Pinkerton, J.M.M., *Proc. Phys. Soc.*, B62, 129, 1949.)

as well as sufficient sonication in therapy. Acoustic loss in tissue is a function of frequency following a power law:

$$\alpha(f) = \alpha_0 + \alpha_1 f^\gamma \tag{5.12}$$

where
  α_0 is almost zero
  γ is the power law exponent

At low-MHz frequencies, scattering typically accounts for 10%–15% of the total attenuation. Meanwhile, the phase velocity usually changes with frequency during propagation, which is known as the *dispersion effect*:

$$c(f) = c_0 + \Delta c(f) \tag{5.13}$$

Although the dispersion effect is usually quite small in most tissues because of the high concentration of water (e.g., 60% in soft tissues), neglecting it may result in significant discrepancy, especially for broad-bandwidth pulses.

5.3 WAVE EQUATION WITH ATTENUATION

In Chapter 3, we assumed that the fluids for acoustic wave propagation are nonviscosous. The acoustic wave could be considered as a force source that causes a temporal momentum change of $\rho_0 u$ of the particles in the medium:

$$\frac{\partial p}{\partial x} + \rho_0 \frac{\partial u}{\partial t} = 0 \tag{5.14}$$

As a result, particle momentum causes the motion away from equilibrium and a subsequent restoring force due to the finite compressibility of the elastic materials, through which the pressure and particle velocity are coupled together to derive the wave equation. Energy, the sum of kinetic energy due to

particle velocity and potential energy due to the elastic force in the pressure wave, is transferred by the acoustic wave in an oscillating manner. No energy is lost if the force is purely elastic. In actual fluids or soft tissue, the viscous drag of the particles in the medium should be overcome in the propagation:

$$\frac{\partial}{\partial x}(p - p') + \rho_0 \frac{\partial u}{\partial t} = 0 \tag{5.15}$$

where p' is a viscosity-related pressure:

$$p' = \left(\frac{4\mu}{3} + \sigma\right)\frac{\partial u}{\partial x} \tag{5.16}$$

where
  μ is the dynamic coefficient of shear viscosity
  σ is the dynamic coefficient of bulk (compressional) viscosity

Substituting for p' yields

$$\frac{\partial}{\partial x}\left[p - \left(\frac{4\mu}{3} + \sigma\right)\frac{\partial u}{\partial x}\right] + \rho_0 \frac{\partial u}{\partial t} = 0 \tag{5.17}$$

The presence of p', although not large in magnitude, leads to an attenuation of the waves by power loss:

$$\frac{\partial p}{\partial t} + \frac{1}{K} \cdot \frac{\partial u}{\partial x} = 0 \tag{5.18}$$

where K is the compressibility of the medium. Substituting Equation 5.18 into Equation 5.17 gives

$$\frac{\partial}{\partial x}\left[p + \left(\frac{4\mu}{3} + \sigma\right)K\frac{\partial p}{\partial t}\right] + \rho_0 \frac{\partial u}{\partial t} = 0 \tag{5.19}$$

Carrying out the operation $\partial/\partial x$ on Equation 5.19 and $\partial/\partial t$ on Equation 5.14 and combining them allow the elimination of terms involving the variable u and derivation of a modified wave equation:

$$\frac{\partial^2 p}{\partial x^2} + \left(\frac{4\mu}{3} + \sigma\right) K \frac{\partial^3 p}{\partial x^2 \partial t} - \rho_0 K \frac{\partial^2 p}{\partial t^2} = 0 \quad (5.20)$$

A comparison of this equation with Equation 5.14 illustrates the introduction of an extra term (the second one with three levels of differentiation rather than two) to the wave equation. Therefore, all solutions will now have an exponential decay as a function of the travel distance:

$$p = p_0 \cdot e^{-\alpha t} \cdot e^{-i(\omega t - kx)} \quad (5.21)$$

Substituting Equation 5.21 into Equation 5.20, we get the expression for the attenuation coefficient:

$$\alpha = \frac{\left[(4\mu/3) + \sigma\right]\omega^2}{2\rho_0 c^3} \quad (5.22)$$

which is valid for small attenuations, $\alpha/k \ll 1$, as in normal tissue. As expected, the attenuation term is proportional to the viscosity coefficients μ and σ of the fluid and ω^2.

Furthermore, the phase velocity $c = \omega/k$ is increased slightly due to viscosity. However, the change is negligibly small for $\alpha/k \ll 1$, and

$$c \approx \frac{1}{\sqrt{\rho_0 K}} \quad (5.23)$$

Since the particle velocity and pressure waves are not exactly in phase, the acoustic impedance is changed slightly from a real number to a complex value. Under the assumption of small attenuation, this change can also be neglected:

$$Z \approx \sqrt{\frac{\rho_0}{K}} \quad (5.24)$$

In most practical simulations of a biomedical ultrasound device, the approximations of Equations 5.9 and 5.10 are valid within the accuracy required and simplify the calculations of propagation time, reflection coefficient, and angle of transmission.

There are two causes of acoustic absorption in fluids or soft tissue. One is the loss of power due to thermal diffusion. As a longitudinal wave propagates through a medium, the density fluctuations of the wave are accompanied by corresponding temperature variations. Within each cycle, heat will diffuse away from that portion of the wave with higher density toward the less dense region, which dissipates the acoustic wave and leads to a power loss proportional to the thermal conductivity of the medium. The other loss factor is the transfer of energy from the wave into the excitation of molecular vibration levels of the fluid, which is especially significant at certain frequencies. In any case, however, both of these absorption causes may be considered as small additions to the viscosity loss modeled in this section for tissues at medical ultrasound frequency.

5.4 VISCOSITY RELAXATION

Because biological tissues are not ideal viscous fluids, the absorption equations do not strictly apply. Attenuation measurements of many soft tissues show that the actual attenuation is more likely a linear function of frequency rather than proportional to the square of the frequency as predicted by Equation 5.8 at least in the range of medical ultrasound frequencies except for the bone and the blood (whose frequency-dependent power of attenuation is between 1 and 2) and for the lung tissue.

An important characteristic of relaxation is the postponed behavior in the medium with an associated time constant (relaxation time). If viscosity is modeled as a relaxation phenomenon, the effective viscosity μ_{eff} can then be written as

$$\mu_{eff} = \frac{\mu_0}{1 + (\omega/\omega_i)^2} \quad (5.25)$$

where
   μ_0 is the low-frequency viscosity
   ω_i is the -6 dB cut-off frequency at which the effective viscosity is reduced to one-half of its low-frequency value

The magnitude of μ_{eff} will drop off with the increase in frequency. Substituting Equation 5.25 into Equation 5.22 for μ and σ (i.e., due to molecular excitation, slight chemical changes, rearrangement of structural disorder, and temperature elevation), respectively, shows that the rate of increase of α with frequency will slow down significantly at higher frequencies. Thus, the frequency-dependent power for attenuation will be smaller than 2.

5.5 ATTENUATION MEASUREMENT

Measurement of the acoustic parameters for various biological tissues is complicated because of the significant variability in samples, such as the species of animal involved, the region of the body from where the sample is obtained, the homogeneity of the tissue, the temperature of the samples, the condition of sample (alive, perfused, or fixed), and the experimental technique. There are many approaches to characterize the ultrasonic properties of tissue (Bamber and Dunn 1997). Generally, the tissue preparation procedure should be designed to keep the tissue fresh and avoid trapping air bubbles for strong reflection at the interface and introduction of artifacts in the measurement. It is suggested to obtain the tissue immediately after the animal is sacrificed, immerse it in degassed phosphate-buffered saline (PBS), and store it in a cooler or refrigerator. The measurement should be carried out within a few hours after degassing the tissue for about 20–30 min in order to keep the conditions close to physiological ones.

There are two possible setups for characterizing the acoustic properties of tissue specimens: transmission (Bamber 1979) and reflection methods (Nassiri et al. 1979) (Figure 5.8). In the transmission setup (Figure 5.8a), a standard transducer for transmission is aimed at a broadband hydrophone acting as a point receiver placed coaxially (Bamber 1979). Several measurements of the signal with only degassed and deionized water between the transducer and hydrophone are used as the reference. The sample is then inserted in the path between the transducer and the hydrophone. Anterior and posterior surfaces of sample are better kept parallel and aligned perpendicular to the transducer axis in order to minimize the reflection from the water/tissue interface. If the transmitter and receiver contact with the sample firmly, as shown in Figure 5.8b, reflection at the interface will not be serious. In the reflection approach, a wideband transducer operates in the pulse-echo mode as both transmitter and receiver (Chivers and Hill 1975). A planar metal surface oriented normal to the beam axis and positioned at the last axial maximum is used as the target. The echo from this target can be isolated in the time domain from the excitation pulse, amplified to increase the signal-to-noise ratio, and then fed into a spectrum analyzer or a digital oscilloscope. Pressure waveforms in the reference medium (water) and through the tissue sample are used to calculate the propagation time and spectra in logarithmic scale by the processing software. Because of the double transmission through the specimen in the reflection mode, the actual propagation distance is twice of the tissue thickness.

5.5.1 Speed of Sound

The speed of sound (SOS) in the tissue is determined by comparing the time of flight through the sample, t_s, and through the water path alone, t_w. Since the pulse has a finite duration, a marker (e.g., the summit, valley, or zero-crossing) is designated for the measurements of the arrival times. For the sake of consistency, the same marker must be chosen for both reference and sample measurements. In addition, a cross-correlation technique can also be used to calculate the difference in traveling times, Δt, of the sample and reference measurement pulses, which is the time position of the maximum value of the cross-correlation of these two pulses:

$$\Delta t = t_s - t_w = \frac{d}{c_t} - \frac{d}{c_w} \qquad (5.26)$$

The uncertainty in this estimation is about half the sampling period. Then, SOS of the tissue sample, c_t, is determined as

$$c_t = \frac{1}{\dfrac{\Delta t}{d} + \dfrac{1}{c_w}} = \frac{c_w}{1 + \dfrac{c_w \Delta t}{d}} \qquad (5.27)$$

where d is the thickness of the sample measured by a digital caliper. SOS of water at a given temperature is determined by the fifth-order polynomial derived from

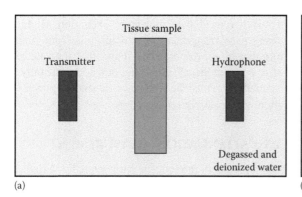

(a)

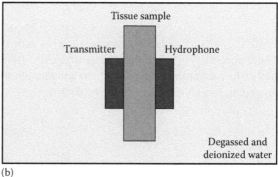

(b)

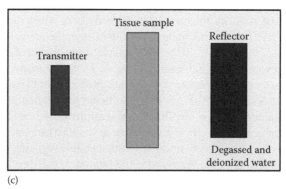

(c)

FIGURE 5.8 Diagram of the experimental setup of measuring acoustic properties of tissue specimen in (a) transmission mode, (b) contact mode, and (c) reflection mode.

extremely precise measurements of deionized water (Del Grosso and Mader 1972):

$$c_w = \sum_{i=0}^{5} k_i T^i \qquad (5.28)$$

where $k_0 = 1402.38754$, $k_1 = 5.03711129$, $k_2 = -5.80852166 \times 10^{-2}$, $k_3 = 3.34198834 \times 10^{-4}$, $k_4 = -1.47800417 \times 10^{-6}$, and $k_5 = 3.14643091 \times 10^{-9}$. The standard deviation is about 0.0026 m/s over the temperature range of $0.001°C \leq T \leq 95.126°C$.

Suppose that two markers, n and m, yield Δt values Δt_n and Δt_m, then the resulting disparity in SOS estimates is

$$\Delta SOS_{n,m} = \frac{c_w}{1 + \frac{c_w \Delta t_n}{d}} - \frac{c_w}{1 + \frac{c_w \Delta t_m}{d}} = \frac{c_w^2}{d} \cdot \frac{\Delta t_m - \Delta t_n}{\left(1 + \frac{c_w \Delta t_n}{d}\right)\left(1 + \frac{c_w \Delta t_m}{d}\right)} \qquad (5.29)$$

If the reference signal $x_w(t)$ is assumed to be a Gaussian modulated sinusoid,

$$x_w(t) = Ae^{-t^2/2\sigma_w^2} e^{i2\pi f_w t} \qquad (5.30)$$

where

A is the pulse magnitude
σ_w is a measure of the pulse duration
f_w is the center frequency
t is the time

The corresponding Fourier transform is

$$X_w(f) = A\sigma_w \sqrt{2\pi} e^{-(f-f_w)^2/2\sigma_f^2} \qquad (5.31)$$

where

f is the frequency
$\sigma_f = 1/(2\pi\sigma_w)$ presents the bandwidth

Both f_w and σ_f are estimated from Gaussian fits to the reference spectra. The linearly frequency-dependent attenuation of a medium can be described as

$$H(f) = e^{-\alpha fd} e^{-i2\pi f\Delta t} \qquad (5.32)$$

Then, the measured signal through medium, $x_s(t)$, is

$$x_s(t) = Be^{-(t-\Delta t)^2/2\sigma_s^2} e^{i2\pi f_s(t-\Delta t)} \qquad (5.33)$$

where

B is the pulse magnitude
σ_s is the pulse duration of the received signal

The downshifted center frequency f_s of the attenuated wave is given by

$$f_s = f_w - \sigma_f^2 \alpha d \qquad (5.34)$$

Suppose that locations of zero-crossings in the reference and measured signals are t_{n0} and t_{n1}, respectively:

$$\Delta t_n = t_{ns} - t_{nw} = \left(\frac{f_w}{f_s} - 1\right) t_{nw} + \Delta t = a t_{nw} + \Delta t \qquad (5.35)$$

Substituting Equation 5.34 gives

$$\Delta t_n = \frac{\sigma_f^2 \alpha d}{f_w - \sigma_f^2 \alpha d} \cdot t_{nw} + \Delta t \qquad (5.36)$$

Suppose that the envelope maximum is chosen as the second marker m so that the SOS estimate corresponds to group velocity $\Delta t_m = \Delta t$. The difference of transit-time differentials for these two markers is

$$\Delta t_m - \Delta t_n = -a t_{nw} = -a\frac{\tau}{f_w} \qquad (5.37)$$

where
$a = \sigma_f^2 \alpha d/(f_w - \sigma_f^2 \alpha d)$
τ is the interval between calibration marker n and m, $\tau = f_w t_{nw}$

Now, the denominator of Equation 5.29 can be simplified as

$$\left(1 + \frac{c_w s}{d}\right)\left[1 + \frac{c_w}{d}\left(\frac{a\tau}{f_0} + \Delta t\right)\right] = \frac{c_w}{c_t}\left[\frac{c_w}{c_t} + \frac{c_w a\tau}{df_w}\right] = \frac{c_w^2}{c_t^2} + \frac{c_w^2 a\tau}{c_t df_w} \qquad (5.38)$$

Combining Equations 5.29, 5.36, and 5.38 yields

$$SOS_n - c_t = -c_t^2 \cdot \frac{a\tau}{f_w d + a\tau c_t} \qquad (5.39)$$

If the second term in its denominator is negligible, Equation 5.39 simplifies to

$$SOS_n - c_t \approx -\frac{c_t^2 a\tau}{f_0 d} = -\frac{\tau c_t^2 \sigma_f^2 \alpha}{f_0^2} \cdot \frac{1}{1 - \left(\sigma_f^2 \alpha d/f_0\right)} \qquad (5.40)$$

which shows that the SOS estimated from group velocity has a strong dependence on the fractional bandwidth (σ_f/f_w). As a result, SOS values can be compensated using Equation 5.39 or 5.40.

5.5.2 Acoustic Attenuation

The captured waveform is subjected to fast Fourier transform (FFT) for obtaining the power spectrum. The discrepancy between the power spectra of the sample and the reference

TABLE 5.1

Typical Values of Acoustic Parameters for Mammalian Tissues

Tissue Type	Density ρ_0 (g/cm³)	Phase Velocity c (m/s)	Attenuation α (cm⁻¹)	Remarks
Human fresh and heparinized whole blood	1.055	1575, 1580	0.034 @ 2 MHz	Attenuation dependence approx. $f^{1.25}$, $0.017 \cdot f^{1.21}$
Human skull bone	1.738	2770 ± 185	1.5 @ 1 MHz	$f^{1.7}$
Human fresh brain	1.03	1460	0.06 @ 1 MHz	Pathology free, $0.09 \cdot f^{1.35}$
Human fresh breast		1510 ± 5	0.22 @ 1 MHz	Pathology free
Porcine fresh or refrigerated fat	0.937, 0.95	1479, 1450	0.07 ± 0.02 @ 1 MHz	@ 37°C, $0.07 \cdot f^1$
Bovine heart muscle	1.048 ± 0.0036, 1.045	1546 ± 4.7, 1570	0.185, 0.23 @ 1 MHz	
Bovine kidney	1.040	1572	0.09 @ 1 MHz	$0.12 \cdot f^1$
Porcine fresh liver	1.064, 1.06	1569.5 ± 4, 1590	0.149 @ 1 MHz	Pathology free, $0.1 \cdot f^{1.1}$
Canine fresh lung	0.4	658	4.3 @ 1 MHz	Inflated; $f^{0.6}$
Porcine muscle (along the fibers)	1.07, 1.065	1566, 1575	0.15 @ 1 MHz	Attenuation perpendicular to fibers
Porcine muscle (across the fibers)	1.065	1590	0.38 @ 1 MHz	
Skin	1.15	1730	1.06 @ 5 MHz	
Bone axial (trabecular)	1.9	4080 (longitudinal wave) 2800 (shear waves)	0.23–1.73 @ 0.2–1 MHz	
Eye (lens)	1.04	1650	0.9 @ 10 MHz	
Eye (vitreous humor)	1.01	1525	0.07 @ 6 MHz	
Teeth (dentine)	2.2	3600	9.21 @ 18 MHz	
Teeth (enamel)	2.9	5500	13.82 @ 18 MHz	
Water	1.0	1480–1500		$0.0002 f^2$

(water) is the *insertion loss*, and the corresponding attenuation at a specific frequency is given by

$$\alpha(f) = \frac{X_w(f)/X_s(f)}{d} + \alpha_w(f) \qquad (5.41)$$

where

$X_w(f)$ and $X_s(f)$ represent the logarithmic power spectra of the reference and sample signals at frequency f, respectively

d is the thickness of the sample

$\alpha_w(f)$ is the attenuation of water as interpolated (Duck 1990)

The emitting ultrasound is a frequency sweeping burst, with the upper and lower bounds determined by the bandwidth of the transmitting transducer (usually the bandwidth of the receiving hydrophone is higher than that of transmitting transducer). The tissue sample is moved laterally, and all data are used to calculate the average attenuation value and standard deviation in order to reduce the influence of tissue inhomogeneity. Then, a line is used to fit the experimental data at the sweeping frequencies by the least-squares method, and the frequency-dependent attenuation is determined from the slope. If the transducers do not contact with the tissue directly, reflections at the tissue/water interfaces due to impedance mismatch would occur, although the correction needed is quite small.

5.5.3 BACKSCATTER COEFFICIENT

The backscatter coefficient of a material to an acoustic wave is defined as the differential scattering cross section per

TABLE 5.2

Categories of Tissue Structure

Homogeneous	Same properties everywhere.
Inhomogenous	Predominantly same tissue type, but with small fluctuation in properties.
Heterogeneous	Large variations from point to point.
Isotropic	Properties do not vary with angle.
Anisotropic	Tissue properties vary with angle and have a preferential structural orientation.

unit solid angle (at 180°) per unit volume. The power spectrum of pulse-echo signals from a sample is compared with that of a perfect plane reflector (i.e., a polished metal plate as hard boundary, or the water/air interface as soft boundary) placed exactly at the position (Insana and Hall 1990). Since the attenuation of the sample is to be considered, its value should be measured before the experiment to measure the backscatter coefficient. Meanwhile, the diffraction of the sound beam should also be corrected for. There is usually a power law governing the backscatter coefficients and frequency. For example, a fourth-power dependence indicates a Rayleigh scattering from objects much smaller than the sound wavelength.

Acoustic properties of representative tissues are listed in Table 5.1. It is important to note that the attenuation factor, which is often given for a frequency $f = 1$ MHz and must be appropriately adjusted for other frequencies, is the combination of both absorption and point scattering effects. Tissue structure can be categorized as Table 5.2.

ASSIGNMENT

Q1: Assume that there is a fluid medium perpendicular to the z-axis with density ρ_0 and shear viscosity η'. A transverse vibration with velocity $U = U_0 e^{j\omega t}$ is applied to the fluid. Prove that the propagation of such a vibration is given by

$$\frac{\partial^2 v}{\partial z^2} = \left(\frac{\rho_0}{\eta'}\right)\frac{\partial v}{\partial t}$$

where v is the particle velocity normal to the z-axis. Find the solution of this equation and discuss its physics.

Q2: The attenuation coefficients of fresh and sea water (35% saline, 5°C) are 7×10^{-5} and 10^{-3} (dB/m), respectively. If an acoustic wave propagates through a distance of 2000 m in a planar way at a frequency of 5 kHz from the same source, what is the difference in dB in these two media? If the frequency is 20 kHz, what is the difference?

Q3: Prove that the 3D wave equation in a viscous fluid is

$$\frac{\partial^2 p}{\partial t^2} = c_0^2 \nabla^2 p + \eta \nabla^2 \frac{\partial p}{\partial t}$$

Derive the 1D wave equation in the radial direction r, and obtain its solution for a planar wave.

Q4: A planar ultrasound wave with acoustic intensity 720 mW/cm^2 and frequency 2 MHz propagates in a soft tissue, whose physical properties are $\rho_0 = 985$ kg/m^3, $c_0 = 1540$ m/s, and $\alpha = 0.6$ dB/cm/MHz. Calculate the amplitudes of particle displacement, particle velocity, acoustic pressure, and acoustic intensity at a location of close to the source, and 2.0 and 10 cm away from the source. Repeat the calculation for the ocular diagnosis frequency, for example, 10 MHz.

Q5: A continuous small-signal planar wave modulated by a rectangular wave with duty cycle 25% and fundamental frequency 5 MHz hits the interface of a gel phantom, whose attenuation coefficient is 0.8 dB/cm/MHz$^{1.5}$, $c_0 = 1560$ m/s, and $Z = 2.1$ MRay. Plot the power spectrum and acoustic intensity at the penetration depth in the gel, and determine the distance at which the amplitude of the acoustic pressure is reduced to 10% of the value at the launching site and the corresponding waveform.

REFERENCES

Anderson M, Trahey GE. The direct estimation of sound speed using pulse-echo ultrasound. *Journal of the Acoustical Society of America* 1986;104:3099–3106.

Auld BA. *Acoustic Fields and Waves in Solids*. Malabar, FL: Krieger Publishing, 1990.

Azhari H. *Basics of Biomedical Ultrasound for Engineers*. New York: Hoboken, New Jersey, John Wiley & Sons, 2010.

Bacon DR. Primary calibration of ultrasonic hydrophone using optical interferometry. *IEEE Transactions on Ultrasonics, Ferroelectrics, and Frequency Control* 1988;35:152–161.

Bamber JC. *Ultrasonic Characterisation of Structure and Pathology in Human Soft Tissue*. London, U.K.: London University, 1979.

Bamber JC. Attenuation and absorption In: Hill CR, ed., *Physical Principles of Medical Ultrasound*. Chichester, U.K.: Ellis Horwood, 1986. pp. 118–199.

Bamber JC. Acoustical characteristics of biological media. In: Crocker MJ, ed., *Encyclopedia of Acoustics*. New York: Wiley, 1997, pp. 1703–1726.

Bamber JC. Ultrasonic properties of tissues. In: Duck FA, Baker AC, Starritt HC, eds., *Ultrasound in Medicine*. Bristol, U.K.: Institute of Physics Publication, 1998, pp. 57–88.

Bamber JC. Attenuation and absorption. In: Bamber JC, ter Haar GR, eds., *Physical Principles of Medical Diagnosis*. Chichester, U.K.: Wiley, 2004a, pp. 118–199.

Bamber JC. Speed of sound. In: Bamber JC, ter Haar GR, eds., *Physical Principles of Medical Diagnosis*. Chichester, U.K.: Wiley, 2004b, pp. 167–190.

Bamber JC, Dunn F. Bioacoustics: Introduction and acoustical characteristics of biological media. In: Crocker MJ, ed., *Encyclopedia of Acoustics*. New York: John Wiley & Sons, 1997, pp. 1699–1726.

Bloch S. *Ultrasonic Tissue Characterization: Towards High-Intensity Focused Ultrasound Treatment Monitoring*. Seattle, WA: University of Washington, 1998, Seattle, WA, PhD dissertation.

Carstensen EL, Schwan HP. Acoustic properties of hemoglobin solutions. *Journal of the Acoustical Society of America* 1959;31:305–311.

Chaffai S, Padilla F, Berger G, Laugier P. *In vitro* measurement of the frequency-dependent attenuation in cancellous bone between 0.2 and 2 MHz. *Journal of the Acoustical Society of America* 2000;108:1281–1289.

Chen JF, Zagzebski JA, Madsen EL. Tests of backscatter coefficient measurement using broadband pulses. *IEEE Transactions on Ultrasonics, Ferroelectrics, and Frequency Control* 1993;40:603–607.

Chen X, Phillips D, Schwarz KQ, Mottley JG, Parker KJ. The measurement of backscatter coefficient from a broadband pulse-echo system: A new formulation. *IEEE Transactions on Ultrasonics, Ferroelectrics, and Frequency Control* 1997;44: 515–525.

Chivers RC, Hill CR. Ultrasonic attenuation in human tissue. *Ultrasound in Medicine and Biology* 1975;2:25–29.

D'Astous FT, Foster FS. Frequency dependence of ultrasound attenuation and backscatter in breast tissue. *Ultrasound in Medicine and Biology* 1986;12:795–808.

Davidovich LA, Makhkamov S, Pulatova L, Khabibullaev PK, Khaliulin MG. Acoustic properties of certain organic liquids at frequencies 0.3 to 3 GHz. *Soviet Physics: Acoustics* 1972;18:264–266.

Del Grosso VA, Mader CW. Speed of sound in pure water. *Journal of the Acoustical Society of America* 1972;52:1442–1446.

Duck FA. *Physical Properties of Tissue*. London, U.K.: Academic Press, 1990.

Duck FA. Propagation of sound through tissue. In: ter Haar G, Duck FA, eds., *The Safe Use of Ultrasound in Medical Diagnosis*. London, U.K.: British Institute of Radiology, 2000, pp. 4–15.

Dunn F. Attenuation and speed of ultrasound in lung: Dependence on frequency and inflation. *Journal of the Acoustical Society of America* 1986;80:1248–1250.

Fung YC. *Biomechanical Properties of Living Tissues*. New York: Springer Verlag, 1981.

Ginzberg VL. Concerning the general relationship between absorption and dispersion of sound waves. *Soviet Physics Acoustics* 1955;1 (1), 32–41.

Greenleaf JF, Sehgal CM. *Biologic System Evaluation with Ultrasound.* New York: Springer-Verlag, 1992.

Greenspan M, Tschigg CE. Speed of sound in water by a direct method. *Journal of Research of the National Bureau of Standards* 1957;59:249–254.

Gross SA, Johnston RL, Dunn F. Comprehensive compilation of empirical ultrasonic properties of mammalian tissues. *Journal of the Acoustical Society of America* 1978;64:423–457.

Gross SA, Johnston RL, Dunn F. Comprehensive compilation of empirical ultrasonic properties of mammalian tissues II. *Journal of the Acoustical Society of America* 1980;68:93–108.

Hall CS, Marsh JN, Hughes MS, Mobley J, Wallace KD, Miller JG, Brandenbeurger GH. Broadband measurements of the attenuation coefficient and backscatter coefficient for suspensions: A potential calibration tool. *Journal of the Acoustical Society of America* 1997;101:1162–1171.

Hall L. The origin of ultrasonic absorption in water. *Physical Review* 1948;73:773–781.

He P. Determination of ultrasonic parameters based on attenuation and dispersion measurements. *Ultrasonic Imaging* 1998a;20:275–287.

He P. Simulation of ultrasound pulse propagation in lossy media obeying a frequency power law. *IEEE Transactions on Ultrasonics, Ferroelectrics, and Frequency Control* 1998b;45:114–125.

Hertzfield KF, Litovitz TA. *Absorption and Dispersion of Ultrasonic Wave.* New York: Academic Press, 1959.

Insana MF, Hall TJ. Parametric ultrasound imaging from backscatter coefficient measurements: Image formation and interpretation. *Ultrasonic Imaging* 1990;12:245–267.

Kinsler LE, Frey AR, Coppens AB, Sanders JV. *Fundamentals of Acoustics.* New York: John Wiley & Sons, 2000.

Kuc R. Modeling acoustic attenuation of soft tissue with a minimum-phase filter. *Ultrasonic Imaging* 1984;6:24–36.

Langton CF, Ali AV, Riggs CM, Evans GP, Barfield WA. A contact method for the assessment of ultrasonic velocity and broadband attenuation in cortical and cancellous bone. *Clinical Physics and Physiological Measurement* 1990;11:243–249.

Langton CF, Palmer SB, Porter RW. Measurement of broadband ultrasonic attenuation in cancellous bone. *Engineering in Medicine* 1984;13:89–91.

Langton CM, Njeh CF. Acoustic and ultrasonic tissue characterization: Assessment of osteoporosis. *Proceedings of the Institution of Mechanical Engineers: Part H: Journal of Engineering in Medicine* 1999;213:261–269.

Madsen EL, Sathoff HJ, Zagzebski JA. Ultrasonic shear wave properties of soft tissues and tissue like materials. *Journal of the Acoustical Society of America* 1983;74:1346–1355.

Markham JJ, Beyer RT, Lindsay RB. Absorption of sound in fluids. *Reviews in Modern Physics* 1951;23:353–411.

Mast TD. Empirical relationships between acoustic parameters in human soft tissues. *Acoustics Research Letters* 2000;1:37–42.

Miller JG, Barzilai B, Milunski MR, Mohr GA. Myocardial tissue characterization: Clinical confirmation of laboratory results. *IEEE Ultrasonics Symposium Proceedings*, Montreal, Quebec, Canada, October 3 to 6, 1989, pp. 1029–1036.

Mottley JG, Miller JG. Anisotropy of the ultrasonic attenuation in soft tissues: Measurements in vitro. *Journal of the Acoustical Society of America* 1990;88:1203–1210.

Nassiri DK, Hill CR. The differential and total bulk acoustic scattering cross sections of some human and animal tissues. *Journal of the Acoustical Society of America* 1986;79:2034–2047.

Nassiri DK, Nicholas D, Hill CR. Attenuation of ultrasound in skeletal muscle. *Ultrasonics* 1979;17:230–232.

Norton GV, Novarini JG. Including dispersion and attenuation directly in the time domain for wave propagation in isotropic media. *Journal of the Acoustical Society of America* 2003;113:3024–3031.

O'Donnell M, Jaynes ET, Miller JG. Kramers–Kronig relationship between ultrasonic attenuation and phase velocity. *Journal of the Acoustical Society of America* 1981;69:696–701.

Pierce AD. *Acoustics.* Woodbury, NY: Acoustical Society of America, 1991.

Pinkerton JMM. A pulse method for the measurement of ultrasonic absorption in liquids: Results for water. *Nature* 1947;160:128–112.

Pinkerton JMM. The absorption of ultrasonic waves in liquids and its relation to molecular constitution. *Proceedings of the Physical Society* 1949;B62:129–141.

Reid JM. Standard substitution methods for measuring ultrasonic scattering in tissues. In: Shung KK, Thieme GA, eds., *Ultrasonic Scattering in Biological Tissues*. Boca Raton, FL: CRC Press, 1993, pp. 171–204.

Robinson DE, Ophir J, Wilson LS, Chen CF. Pulse-echo ultrasound speed measurements: Progress and prospects. *Ultrasound in Medicine and Biology* 1991;17:633–646.

Shung KK, Sigelmann RA, Reid JM. Scattering of ultrasound by blood. *IEEE Transactions on Biomedical Engineering* 1976;BME-23:460–467.

Shung KK, Thieme GA. *Ultrasonic Scattering in Biological Tissues.* Boca Raton, FL: CRC Press, 1993.

Szabo TL. Time domain wave equations for lossy media obeying a frequency power law. *Journal of the Acoustical Society of America* 1994;96:491–500.

Szabo TL. Causal theories and data for acoustic attenuation obeying a frequency power law. *Journal of the Acoustical Society of America* 1995;97:14–24.

Taylor KW, Wells PNT. Tissue characterization. *Ultrasound in Medicine and Biology* 1989;15:421–428.

Waag RC, Astheimer JP. Measurement system effects in ultrasonic scattering experiments. In: Shung KK, Thieme GA *Ultrasonic Scattering in Biological Tissues*. Boca Raton, FL: CRC Press, 1993, pp. 251–290.

Wang SH, Shung KK. An approach for measuring ultrasonic backscattering from biological tissues with focused transducers. *IEEE Transactions on Biomedical Engineering* 1997;44:549–554.

Waters KR, Hughes MS, Mobley J, Brandenbeurger GH, Miller JG. On the applicability of Kramers–Kronig relations for ultrasonic attenuation obeying a frequency power law. *Journal of the Acoustical Society of America* 2000;108:556–563.

Wear KA. Measurement of phase velocity and group velocity in human calcaneus. *Ultrasound in Medicine and Biology* 2000;26:641–646.

Wear KA. Fundamental mechanisms underlying broadband ultrasonic attenuation in calcaneus. In: Insana MF, Shung KK, eds., *Ultrasonic Imaging and Signal Processing*. Bellingham, WA: San Diego, CA, SPIE, 2001, pp. 427–429.

Wear KA. The dependence of time-domain speed-of-sound measurements on center frequency, bandwidth, and transit-time marker in human calcaneus *in vitro*. *Journal of the Acoustical Society of America* 2007;122:636–644.

Wells PNT. Absorption and dispersion of ultrasound in biological tissue. *Ultrasound in Medicine and Biology* 1975;1:369–376.

Wells PNT. *Biomedical Ultrasonics.* New York: Academic Press, 1977.

Wu JR. Effects of nonlinear interaction on measurements of frequency dependent attenuation coefficients. *Journal of the Acoustical Society of America* 1996;99:3380–3384.

Zeqiri B. An intercomparison of discrete-frequency and broad-band techniques for the determination of ultrasonic attenuation. In: Evans DH, Martin K, eds., *Physics in Medical Ultrasound.* London, U.K.: IPSM, 1988, pp. 27–35.

Zeqiri B. Reference liquid for ultrasonic attenuation. *Ultrasonics* 1989;27:314–315.

6 Nonlinear Acoustics

In linear acoustics, which is characterized by the sound pressures much smaller than the static pressure, the time-averaged value of the sound pressure or any other acoustic variable within a period is zero for most practical purposes. However, at sufficiently large sound pressures and corresponding fluid velocities, the time-averaged or mean values might become nonzero. Nonlinear acoustics contributes to a distortion of the waveform during propagation and can be described by various parameters, such as B/A (the ratio of the second to the first terms in a Taylor series expansion of the pressure as a function of density); the coefficient of non-linearity β, which takes self-convection into account; and the acoustic Mach number, which differentiates the linear regimes from the nonlinear regimes. During acoustic wave propagation, energy initially carried at the fundamental frequency gradually moves to the higher harmonics because of the accumulative effects of nonlinearity. As a result, the wave-front becomes steepened. Such a nonlinearity can be used to infer the properties of the medium or generate useful acoustic sources. Nonlinearity parameters may be determined by measuring the waveform distortion. Furthermore, laws have also been developed for the calculation of nonlinearity parameters of mixtures of media with known nonlinearities.

6.1 ONE-DIMENSIONAL NONLINEAR DYNAMIC EQUATION

Nonlinear acoustics begins with the simplest case, namely the propagation of finite-amplitude planar wave in an ideal fluid along the x direction. The corresponding dynamic equation can be described by

$$\frac{\partial v}{\partial t} + v\frac{\partial v}{\partial x} = -\frac{1}{\rho}\frac{\partial p}{\partial x} \quad (6.1)$$

And, the continuity equation is

$$\frac{\partial(\rho v)}{\partial x} = -\frac{\partial \rho}{\partial t} \quad (6.2)$$

For an ideal medium, the particle velocity v and sound speed c are both monotonic functions of density ρ:

$$v = v(\rho), \quad c = c(\rho) \quad (6.3)$$

Then, Equations 6.1 and 6.2 can be rewritten as

$$\frac{\partial v}{\partial t} + \left(v + \frac{1}{\rho}\frac{dp}{dv}\right)\frac{\partial v}{\partial x} = 0 \quad (6.4)$$

$$\frac{\partial \rho}{\partial t} + \frac{d(\rho v)}{d\rho}\frac{\partial \rho}{\partial x} = 0 \quad (6.5)$$

or

$$-\frac{\left(\dfrac{\partial v}{\partial t}\right)}{\left(\dfrac{\partial v}{\partial x}\right)} = -\left(\frac{\partial x}{\partial t}\right)_v = v + \frac{1}{\rho}\left(\frac{dp}{dv}\right) \quad (6.6)$$

$$-\frac{\left(\dfrac{\partial \rho}{\partial t}\right)}{\left(\dfrac{\partial \rho}{\partial x}\right)} = -\left(\frac{\partial x}{\partial t}\right)_\rho = v + \rho\left(\frac{dv}{d\rho}\right) \quad (6.7)$$

Since both ρ and v are monotonic variables,

$$\left(\frac{\partial x}{\partial t}\right)_\rho = \left(\frac{\partial x}{\partial t}\right)_v \quad (6.8)$$

Then, it can be derived from Equations 6.6 and 6.7 that

$$\rho^2\left(\frac{dv}{d\rho}\right) = \frac{dp}{dv} = c^2\left(\frac{d\rho}{dv}\right) \quad (6.9)$$

$$\left(\frac{dv}{d\rho}\right)^2 = \frac{c^2}{\rho^2} \quad (6.10)$$

where $c^2 = dp/d\rho$
so,

$$v = \pm\int \frac{c}{\rho}d\rho = \pm\int \frac{dp}{\rho c} \quad (6.11)$$

$$\left(\frac{\partial x}{\partial t}\right)_v = \left(\frac{\partial x}{\partial t}\right)_\rho = v \pm c \quad (6.12)$$

After integration, Equation 6.12 is rewritten as

$$x = (v \pm c)t + f(v) \quad \text{or} \quad v = F[x - (v \pm c)t] \quad (6.13)$$

where $f(*)$ and $F(*)$ are arbitrary functions. The $(\pm)$ sign means that, along the direction of x_+ and x_-, the speed of finite-amplitude sound wave is $v \pm c$, respectively.

6.2 NONLINEAR EQUATION OF STATE

The equation of state of an ideal gas in the adiabatic process is

$$P = P_0 \left(\frac{\rho}{\rho_0} \right)^\gamma \tag{6.14}$$

where
P is the pressure
P_0 is the pressure at the equilibrium state
ρ is the gas density
ρ_0 is the density at the equilibrium state
γ is the adiabatic index

Using Taylor's expansion up to the second-order item, we have

$$P = P_0 \left(\frac{\rho}{\rho_0} \right)^\gamma = P_0 + \gamma P_0 \left(\frac{\rho - \rho_0}{\rho_0} \right) + \frac{\gamma(\gamma-1)}{2} P_0 \left(\frac{\rho - \rho_0}{\rho_0} \right)^2 + \cdots \tag{6.15}$$

As shown in Chapter 2, for the small-amplitude acoustic wave only first-order item is retained, so

$$c^2 = \left(\frac{dP}{d\rho} \right)_{\rho=\rho_0} = \frac{\gamma P_0}{\rho_0} = c_0^2 \tag{6.16}$$

Therefore, the speed c_0 of small-amplitude sound in an ideal gas is a constant, which is determined by the physical properties of the medium. However, for most fluids, the equation of state is expanded around $\rho = \rho_0$ up to the second-order term:

$$P = P' + \left(\frac{\partial P}{\partial \rho} \right)_{\rho_0} (\rho - \rho_0) + \frac{1}{2} \left(\frac{\partial^2 P}{\partial \rho^2} \right) (\rho - \rho_0)^2 + \cdots \tag{6.17}$$

where P' is the pressure inside the fluid, and the coefficients of the first- and second-order items are constants relevant to temperature and pressure, respectively.

6.3 NONLINEARITY PARAMETERS

6.3.1 ACOUSTIC MACH NUMBER

The acoustic Mach number is defined as $M = u_{max}/c_0$, where u_{max} and c_0 are the maximum particle velocity and the small-signal acoustic phase speed in the propagation medium, respectively. For plane-wave propagation in a gas, $M = p_{max}/\gamma P_0$, where p_{max} is the maximum deviation of pressure from the equilibrium pressure P_0, and γ is the ratio of specific heats C_p/C_v for the gas. For linear acoustics, $M \ll 1$.

6.3.2 B/A

For small deviations from P_0, the pressure P may be expressed in terms of the density ρ and specific entropy s. In the adiabatic

(small induced temperature gradient with no appreciable heat flow occurring during a fraction of an acoustic period) or isentropic (thermodynamically reversible change with no change from its equilibrium entropy s_0) processes, the equation of state (both gas and fluid) can be expanded into the Taylor series:

$$P = P_0 + \left(\frac{\partial P}{\partial \rho} \right)_{s,\rho_0} (\rho - \rho_0) + \frac{1}{2!} \left(\frac{\partial^2 P}{\partial \rho^2} \right)_{s,\rho_0}$$

$$(\rho - \rho_0)^2 + \frac{1}{3!} \left(\frac{\partial^3 P}{\partial \rho^3} \right)_{s,\rho_0} (\rho - \rho_0)^3 + \cdots$$

$$= P_0 + \rho_0 \left(\frac{\partial P}{\partial \rho} \right)_{s,\rho_0} \left(\frac{\rho - \rho_0}{\rho_0} \right) + \frac{1}{2!} \rho_0^2 \left(\frac{\partial^2 P}{\partial \rho^2} \right)_{s,\rho_0}$$

$$\left(\frac{\rho - \rho_0}{\rho_0} \right)^2 + \frac{1}{3!} \rho_0^3 \left(\frac{\partial^3 P}{\partial \rho^3} \right)_{s,\rho_0} \left(\frac{\rho - \rho_0}{\rho_0} \right)^3 + \cdots$$

$$= P_0 + A \left(\frac{\rho - \rho_0}{\rho_0} \right) + \frac{1}{2} B \left(\frac{\rho - \rho_0}{\rho_0} \right)^2 + \frac{1}{6} C \left(\frac{\rho - \rho_0}{\rho_0} \right)^3 + \cdots \tag{6.18}$$

where $p = P - P_0$ is the excess or acoustic pressure, and $(\rho - \rho_0)/\rho_0$ is the condensation or fractional change in density of the propagation medium. Thus, for small deviations from equilibrium

$$A = \rho_0 \left(\frac{\partial P}{\partial \rho} \right)_{s,\rho_0} = \rho_0 c_0^2, \quad B = \rho_0^2 \left(\frac{\partial^2 P}{\partial \rho^2} \right)_{s,\rho_0}, \quad C = \rho_0^3 \left(\frac{\partial^3 P}{\partial \rho^3} \right)_{s,\rho_0} \tag{6.19}$$

and so on to higher orders. When the condensation is infinitesimal, the higher-order terms are negligible and the acoustic waves will propagate at the constant speed c_0. For finite-amplitude waves, the higher-order terms $B, C, \ldots$ become increasingly more important as the amplitude increases. Alternatively, for waves of a given amplitude, materials with larger values of $B, C, \ldots$ will cause the local sound speed to increase from c_0. The acoustic nonlinearity parameter B/A is the ratio of the quadratic coefficient in this expression to the linear coefficient and therefore provides a measure of the degree to which the local sound speed deviates from the small-signal (linear) case:

$$\frac{B}{A} = \frac{\rho_0}{c_0^2} \left(\frac{\partial^2 P}{\partial \rho^2} \right)_{s,\rho_0} = 2\rho_0 c_0 \left(\frac{\partial c}{\partial P} \right)_{s,\rho_0} \tag{6.20}$$

or

$$\frac{B}{A} = 2\rho_0 c_0 \left[\left(\frac{\partial c}{\partial p} \right)_T \right]_{\rho=\rho_0} + \frac{2c_0 T \kappa_e}{C_p} \left[\left(\frac{\partial c}{\partial T} \right)_p \right]_{\rho=\rho_0} \tag{6.21}$$

or

$$\frac{B}{A} = \left[\frac{\partial(1/k)}{\partial P}\right]_s - 1 \qquad (6.22)$$

where
  c is the local sound speed
  T is the temperature in kelvin

κ_e, C_p, and k are the volume coefficient of thermal expansion, the specific heat at constant pressure, and the adiabatic compressibility (the reciprocal of stiffness) for the propagation medium, respectively. Equation 6.20 shows that B/A is proportional to the change in sound speed for a change in pressure, provided that the pressure change is so rapid and smooth that isentropic conditions continue to hold. Equation 6.21 shows that B/A may be written as the sum of isothermal and isobaric components. However, the isothermal component of B/A typically dominates for most materials. Equation 6.23 suggests that materials whose stiffness changes greatly with changes in pressure, such as bubbly liquids, will have a large value of the acoustic nonlinearity parameter. Other physical characteristics described by B/A include the geometric packing of molecules and the form of the potential energy function that governs the forces between adjacent molecules in a material. The connection between molecular structure and B/A has been exploited by several investigators to determine the relative concentrations of "bound water" and "free water" in water/alcohol mixtures or the contribution of molecular subgroups to the overall nonlinearity of larger molecules in solution. It also suggests a connection between the level of macroscopic structure in soft tissue and its B/A value. Fluids in which B/A can be zero or negative are called "retrograde" fluids and have unusual thermodynamic properties. For an ideal gas, substituting Equation 6.14 to Equation 6.20 gives

$$\frac{B}{A} = \gamma - 1 \qquad (6.23)$$

The B/A values of common media are listed in Table 6.1. For most fluids, the B/A values are usually within the range of 5–10.
  The speed of sound is given by

$$c^2 = \gamma \frac{P_0}{\rho_0}\left(\frac{\rho}{\rho_0}\right)^{\gamma-1} = c_0^2\left(\frac{\rho}{\rho_0}\right)^{\gamma-1}$$

$$\therefore c = c_0\left(\frac{\rho}{\rho_0}\right)^{\frac{\gamma-1}{2}} \qquad (6.24)$$

TABLE 6.1

Summary of Nonlinearities of Common Media and Tissues

Substance	T (°C)	B/A
Distilled water	0	4.2
	20	5.0, 4.985 ± 0.063
	25	5.11 ± 0.20
	26	5.1
	30	5.31, 5.18 ± 0.033
	40	5.4
	60	5.7
	80	6.1
	100 (liquid)	6.1
Sea water (3.5% NaCl)	20	5.25
Isotonic saline	20	5.54 ± 0.032
	30	5.559 ± 0.018
Bovine serum albumin		
20 g/100 mL H_2O	25	6.23 ± 0.25
38.8 g/100 mL H_2O	30	6.68
Bovine liver	23	7.5–8.0
	30	7.23–8.9, 6.88
Bovine brain	30	7.6
Bovine heart	30	6.8–7.4
Bovine milk	26	5.1
Bovine whole blood	26	5.5
Chicken fat	30	11.27 ± 0.09
Porcine liver	26	6.9
Porcine heart	26	6.8
Porcine kidney	26	6.3
Porcine spleen	26	6.3
Porcine brain	26	6.7–7.0
Porcine muscle	26	6.5–6.6
	30	7.5–8.1
Porcine tongue	26	6.8
Porcine fat	26	9.5–10.9
	30	10.9–11.3
Porcine whole blood	26	5.8
Human liver	30	6.54
Human breast fat	22	9.206
	30	9.909
	37	9.633
Human multiple	22	5.603
	30	5.796
	37	6.178

Substituting Equation 6.24 for Equation 6.11 and then integrating give

$$v = \pm\int \frac{c}{\rho}\,d\rho = \pm c_0\int_{\rho_0}^{\rho}\left(\frac{\rho}{\rho_0}\right)^{(\gamma-1)/2}\,d\rho$$

$$= \pm\frac{2c_0}{\gamma-1}\left[\left(\frac{\rho}{\rho_0}\right)^{(\gamma-1)/2} - 1\right] = \pm\frac{2c_0}{\gamma-1}\left[\frac{c}{c_0} - 1\right] \qquad (6.25)$$

So,

$$c = c_0 \pm \frac{\gamma - 1}{2} v \qquad (6.26)$$

It illustrates that, when the particle velocity v is not negligible, the speed of sound at different locations of the waveform profile will be different.

6.3.3 COEFFICIENT OF NONLINEARITY (β)

The acoustic nonlinearity parameter describes the steepening of acoustic waves as they propagate through a material. A given point on a traveling acoustic waveform will propagate at the local sound speed B/A, given by Equation 6.22, where the so-called coefficient of nonlinearity $\beta \equiv 1 + B/2A$. Equation 6.22 shows that wave steepening has two distinct causes that are assumed to act independently: nonlinearities inherent in the material's properties, which are described by B/A, and those due to convection. The convective term arises from purely kinematic considerations and would therefore exist even if no material nonlinearities were present ($B/A = 0$). Equation 6.22 also shows that nonlinear effects are cumulative. The amount of distortion also depends on the time or distance the wave travels.

6.3.4 SPECIFIC INCREMENT OF B/A

Because the increase in B/A value with solute concentration χ of solutions of biological materials is often linear, the specific increment of B/A with concentration, that is, $\Delta(B/A)/\chi$, can be used to determine the relative contribution of the solute to the total nonlinearity of the solution. The specific increment is defined by differentiating Equation 6.20:

$$\frac{\Delta(B/A)}{\chi} = 2\rho_0 c_0 \left\{ \begin{array}{l} \frac{1}{\chi}\Delta\left(\frac{\partial c}{\partial P}\right)_{T_0} + ([c]+[\rho])\left(\frac{\partial c}{\partial P}\right)_{T_0} \\ + \frac{\kappa_e T}{\rho_0 C_p} \left\{ \begin{array}{l} \frac{1}{\chi}\Delta\left(\frac{\partial c}{\partial T}\right)_{P_0} \\ +([c]+[\kappa_e]-[C_p])\left(\frac{\partial c}{\partial T}\right)_{P_0} \end{array} \right\} \end{array} \right\} \qquad (6.27)$$

where $[c] = \Delta c/\chi c_0$, $[\rho] = \Delta\rho/\chi\rho_0$, $[\kappa_e] = \Delta\kappa_e/\chi\kappa_{e0}$, and $[C_p] = C_{p0}/\chi C_p$ with the subscript 0 denoting the solvent, and Δ meaning the difference between the solution and solvent for the corresponding parameter.

6.3.5 HIGH-ORDER NONLINEARITY PARAMETERS

The higher-order parameters C/A, D/A, …, become important at large acoustic pressure amplitudes or in extremely nonlinear media. From Equation 6.19, the third-order constant C/A may be written as

$$\frac{C}{A} = \frac{3}{2}\left(\frac{B}{A}\right)^2 + \frac{1}{\kappa}\frac{\partial[B/A]}{\partial P} \qquad (6.28)$$

For most materials, the first term in this equation exceeds the second term by several orders of magnitude.

6.3.6 MIXTURE LAWS FOR NONLINEARITY PARAMETERS

If the nonlinearity parameters of each of a n-component mixture of mutually immiscible materials are known, it is possible to derive expressions for the corresponding nonlinear parameters of the mixture as a whole:

$$k^2\beta = \sum_{i=1}^{n}\frac{\rho Y_i k_i^2 \beta_i}{\rho_i} = \sum_{i=1}^{n} X_i k_i^2 \beta_i \qquad (6.29)$$

where Y_i and X_i are the mass and volume fractions of component i, respectively. In Equation 6.29, k is the adiabatic compressibility and ρ the density of the mixture as a whole, given by $k = \sum_{i=1}^{n} k_i X_i$ and $\rho = \sum_{i=1}^{n} \rho_i X_i$, respectively, where the subscripted quantities refer to the component properties. Similar expressions may be developed for C/A, D/A, ….

6.4 NONLINEAR ACOUSTIC WAVE PROPAGATION

6.4.1 WAVEFORM DISTORTION

Substituting Equation 6.26 into Equation 6.13, we get

$$v = F\left[x - \left(\frac{\gamma+1}{2} v \pm c_0 \right) t \right] \qquad (6.30)$$

Assume that there is a sinusoidal wave $v = v_0 \sin \omega t$ propagating in the positive direction; then, the acoustic wave at x is

$$v = v_0 \sin\left(\omega t - \frac{\omega x}{c+v} \right) = v_0 \sin\left(\omega t - \frac{\omega x}{c_0 + \frac{\gamma+1}{2}v} \right)$$

$$= v_0 \sin\left(\omega t - \frac{\omega x}{c_0 + \beta v} \right) \qquad (6.31)$$

where $\beta = (\gamma + 1)/2$. It is clear that the speed of finite-amplitude sound is $c_0 + \beta v$. So, if $v > 0$, $c_0 + \beta v > c_0$; if $v < 0$, $c_0 + \beta v < c_0$; and if $v = 0$, the propagation speed is c_0. It is shown that the waveform distorts during its propagation, and such a distortion accumulates with the propagation distance.

6.4.2 DISCONTINUITY

Finally, the sinusoidal wave will become saw tooth shaped, which causes the discontinuity of the medium:

$$\frac{\partial v}{\partial x} = \infty \qquad \text{or} \qquad \frac{\partial x}{\partial v} = 0 \qquad (6.32)$$

Based on Equation 6.31, the critical distance of discontinuity is

$$x_k = \frac{\lambda c_0}{\pi(\gamma+1)v_0} = \frac{2\rho_0 c_0^3}{(\gamma+1)\omega p_0} \quad (6.33)$$

which shows that the discontinuity distance is inversely proportional to the initial sound pressure or particle velocity. M is the Mach number. So, Equation 6.33 can be rewritten as

$$x_k = \frac{\lambda}{\pi(\gamma+1)M} = \frac{1}{\beta M k} \quad (6.34)$$

Thus, x_k is inversely proportional to the Mach number as well as the wave number.

6.4.3 BESSEL–FUBINI SOLUTION

If the source is a pure sinusoidal wave, that is, $v = v_0 \sin \omega t$, at the position of x, the distortion of the acoustic pressure waveform is

$$v = v_0 \sin\left[\omega t - kx\left(1-\beta\frac{v}{c_0}\right)\right] = v_0 \sin\left(\omega t - kx + k\beta M x\frac{v}{v_0}\right)$$

$$= v_0 \sin\left(\omega t - kx + \frac{x}{x_k}\frac{v}{v_0}\right) \quad (6.35)$$

Thus, it can be expanded in a Fourier series as

$$\frac{v}{v_0} = \sum_{n=1}^{\infty} A_n \sin n(\omega t - kx) \quad (6.36)$$

where

$$A_n = \frac{1}{\pi}\int_0^{2\pi} \frac{v}{v_0} \sin n(\omega t - kx)d(\omega t - kx) = \frac{2J_n(n\sigma)}{n\sigma} \quad (6.37)$$

with $\sigma = \frac{x}{x_k}$, $J_n(n\sigma)$ the nth cylindrical Bessel function. So, Equation 6.35 can be expressed as

$$v = 2v_0 \sum_n \frac{J_n(n\sigma)}{n\sigma} \sin n(\omega t - kx) \quad (6.38)$$

which is the famous Bessel–Fubini solution and valid for $x < x_k$.
For the second harmonic

$$v_{2a} = v_0 \frac{J_2(2\sigma)}{\sigma} \quad (6.39)$$

Since $\frac{x}{x_k} < 1$, $J_2(2\sigma) \approx \frac{\sigma_2}{2}$

$$v_{2a} = \frac{v_0\sigma}{\sigma} = \left(\frac{\gamma+1}{4}\right)\left(\frac{v_0}{c_0}\right)^2 \omega x \quad (6.40)$$

The corresponding sound pressure of the second harmonic is

$$p_{2a} = \frac{\gamma+1}{4}\left(\frac{v_0}{c_0}\right)^2 \rho_0 c_0 \omega x = \frac{\omega x p_{1a}^2}{4\rho_0 c_0^3}(\gamma+1) = \frac{\omega x p_{1a}^2}{4\rho_0 c_0^3}\left(\frac{B}{A}+2\right)$$

$$(6.41)$$

where p_{1a} is the amplitude of the fundamental sound pressure. It is clear that the relationship between the critical distance and the second harmonic is

$$\frac{p_{2a}}{p_{1a}} = \frac{x}{2x_k} \quad (6.42)$$

When $x = x_k$ and $p_{2a} = p_{1a}/2$. When $x < x_k$, the amplitude of the second harmonic sound pressure is calculated from Equation 6.41 using the nonlinearity B/A. P_{2a} is proportional to ω, p_{1a}^2, x, and $(\gamma + 1)$. With the distortion of the waveform, the energy of the fundamental component will be decreased and transferred to the high-order harmonics. When the amplitude of the second harmonic reaches its maximum value, some of its energy will be transferred to the other harmonics $(n > 2)$.

6.5 INTERACTION OF FINITE-AMPLITUDE SOUND WAVES

In linear acoustics, sound waves will not have interaction with each other because of the linear superposition for each small-amplitude acoustic field. However, such a phenomenon will not be valid any more.

6.5.1 NONLINEAR WAVE EQUATION

For one-dimensional wave propagation along the x direction, the dynamic and continuity equations with nonlinear terms are

$$\frac{\partial v}{\partial t} + \frac{1}{2}\frac{\partial v^2}{\partial x} = -\frac{1}{\rho}\frac{\partial p}{\partial x} \quad (6.43)$$

$$\frac{\partial \rho}{\partial t} + \frac{\partial}{\partial x}(\rho v) = 0 \quad (6.44)$$

It is assumed that

$$\frac{\rho - \rho_0}{\rho_0} \approx \mu, \quad \frac{P - P_0}{p_0} \approx \mu, \quad \frac{v}{c_0} \approx \mu, \quad \mu < 1 \quad (6.45)$$

Ignoring the higher items than μ^2, the state equation (6.15) can be expressed as

$$p = P - P_0 = c_0^2(\rho - \rho_0) + \frac{\gamma - 1}{2\rho_0} c_0^2(\rho - \rho_0)^2 + \cdots \quad (6.46)$$

Substituting Equation 6.46 to Equation 6.43 gives

$$\frac{\partial v}{\partial t} = -\frac{1}{2} \frac{\partial v^2}{\partial x} - \frac{c_0^2}{\rho_0} \frac{\partial}{\partial x} \left[(\rho - \rho_0) + \frac{\gamma - 2}{2\rho_0}(\rho - \rho_0)^2 \right] \quad (6.47)$$

If Φ is defined as the velocity potential, $v = -(\partial\Phi/\partial x)$, and taking derivatives on both sides of Equation 6.47 gives

$$\frac{\partial^2 \Phi}{\partial t^2} = \frac{1}{2} \frac{\partial}{\partial t} \left(\frac{\partial \Phi}{\partial x} \right)^2 + \frac{c_0^2}{\rho_0} \left[\frac{\partial \rho}{\partial t} + \frac{\gamma - 2}{2\rho_0} \frac{\partial}{\partial t}(\rho - \rho_0)^2 \right] \quad (6.48)$$

Equation 6.44 can then be rewritten as

$$\frac{\partial \rho}{\partial t} = \frac{\partial \rho}{\partial x} \frac{\partial \Phi}{\partial x} + \rho \frac{\partial^2 \Phi}{\partial x^2} \quad (6.49)$$

Combining Equations 6.43, 6.47, and 6.48, the nonlinear wave equation is

$$\frac{\partial^2 \Phi}{\partial t^2} - c_0^2 \frac{\partial^2 \Phi}{\partial x^2} = \frac{\partial}{\partial t} \left[\left(\frac{\partial \Phi}{\partial x} \right)^2 + a \left(\frac{\partial \Phi}{\partial t} \right)^2 \right] \quad (6.50)$$

where $a = (\gamma - 1)/2c_0^2$.

6.5.2 Approximate Solution

An accurate solution of the wave equation is difficult, but the approximate one could be obtained using the method of small disturbance. Assume $\Phi = \Phi_1 + \Phi_2$, where Φ_1 and Φ_2 are first and second orders of approximations; then, the corresponding approximate equations are expressed as

$$\frac{\partial^2 \Phi_1}{\partial t^2} - c_0^2 \frac{\partial^2 \Phi_1}{\partial x^2} = 0 \quad (6.51)$$

$$\frac{\partial^2 \Phi_2}{\partial t^2} - c_0^2 \frac{\partial^2 \Phi_2}{\partial x^2} = \frac{\partial}{\partial t} \left[\left(\frac{\partial \Phi_1}{\partial x} \right)^2 + a \left(\frac{\partial \Phi_1}{\partial t} \right)^2 \right] \quad (6.52)$$

The solution of Equation 6.51 is

$$\Phi_1 = \Phi_a \left[1 - \cos \omega \left(t - \frac{x}{c_0} \right) \right] \quad (6.53)$$

So, Equation 6.50 can be rewritten as

$$\frac{\partial^2 \Phi_2}{\partial t^2} - c_0^2 \frac{\partial^2 \Phi_2}{\partial x^2} = Q \sin 2\omega \left(t - \frac{x}{c_0} \right) \quad (6.54)$$

where $Q = \frac{\gamma + 1}{2c_0^2} \Phi_a^2 \omega^3$. The solution of Equation 6.54 is

$$\Phi_2 = -\frac{Qx}{4\omega c_0} \cos 2\omega \left(t - \frac{x}{c_0} \right) = -\frac{\gamma + 1}{8c_0^3} \Phi_a^2 \omega^2 x \cos 2\omega \left(t - \frac{x}{c_0} \right) \quad (6.55)$$

which is the second harmonic of a distorted waveform. Therefore, the fundamental and second harmonic sound pressures are

$$p_1 = \rho_0 \frac{\partial \Phi_1}{\partial t} = p_{1a} \sin \omega \left(t - \frac{x}{c_0} \right) \quad (6.56)$$

$$p_2 = \rho_0 \frac{\partial \Phi_2}{\partial t} = \frac{(\gamma + 1)\omega x p_{1a}^2}{4\rho_0 c_0^3} \sin 2\omega \left(t - \frac{x}{c_0} \right) \quad (6.57)$$

where $p_{1a} = \rho_0 \Phi_a \omega$ and $p_{2a} = \frac{(\gamma + 1)\omega x}{4\rho_0 c_0^3} p_{1a}^2$.

6.5.3 Interaction of Sources at Different Frequencies

If there are two sources with angular frequencies ω_1 and ω_2 ($\omega_2 > \omega_1$) at $x = 0$,

$$\Phi = -\Phi_{1a} \cos \omega_1 t - \Phi_{2a} \cos \omega_2 t \quad (6.58)$$

The first-order approximate solution is

$$\Phi_1 = -\Phi_{1a} \cos \omega_1 \left(t - \frac{x}{c_0} \right) - \Phi_{2a} \cos \omega_2 \left(t - \frac{x}{c_0} \right) \quad (6.59)$$

So,

$$p_1 = p_{1a} \sin \omega_1 \left(t - \frac{x}{c_0} \right) + p_{2a} \sin \omega_2 \left(t - \frac{x}{c_0} \right) \quad (6.60)$$

where $p_{1a} = \rho_0 \omega_1 \Phi_{1a}$ and $p_{2a} = \rho_0 \omega_2 \Phi_{2a}$. The corresponding second-order approximate solution is

$$\frac{\partial^2 \Phi_2}{\partial t^2} - c_0^2 \frac{\partial^2 \Phi_2}{\partial x^2} = Q_1 \sin 2\omega_1 \left(t - \frac{x}{c_0} \right) + Q_2 \sin 2\omega_2 \left(t - \frac{x}{c_0} \right)$$

$$+ Q_3 \left\{ \begin{array}{l} (\omega_2 - \omega_1)\sin\left[(\omega_1 - \omega_2)\left(t - \frac{x}{c_0} \right) \right] \\ \\ +(\omega_2 + \omega_1)\sin\left[(\omega_1 + \omega_2)\left(t - \frac{x}{c_0} \right) \right] \end{array} \right\} \quad (6.61)$$

where

$$Q_1 = \frac{(\gamma+1)\omega_1^3\Phi_{1a}^2}{2c_0^2}, \qquad Q_2 = \frac{(\gamma+1)\omega_2^3\Phi_{2a}^2}{2c_0^2}, \qquad \text{and}$$

$$Q_3 = \frac{(\gamma+1)\omega_1\omega_2\Phi_{1a}\Phi_{2a}}{2c_0^2}$$

Thus, the linear inhomogenous differential equation is equivalent to the summation of the three independent equations:

$$\frac{\partial^2\Phi_2}{\partial t^2} - c_0^2\frac{\partial^2\Phi_2}{\partial x^2} = Q_1\sin 2\omega_1\left(t-\frac{x}{c_0}\right) \qquad (6.62)$$

$$\frac{\partial^2\Phi_2}{\partial t^2} - c_0^2\frac{\partial^2\Phi_2}{\partial x^2} = Q_2\sin 2\omega_2\left(t-\frac{x}{c_0}\right) \qquad (6.63)$$

$$\frac{\partial^2\Phi_2}{\partial t^2} - c_0^2\frac{\partial^2\Phi_2}{\partial x^2} = Q_3\left\{\begin{array}{l}(\omega_2-\omega_1)\sin\left[(\omega_1-\omega_2)\left(t-\dfrac{x}{c_0}\right)\right]+ \\[2mm] (\omega_2+\omega_1)\sin\left[(\omega_1+\omega_2)\left(t-\dfrac{x}{c_0}\right)\right]\end{array}\right\}$$

$$(6.64)$$

The corresponding solutions are

$$\Phi_{2a} = -\frac{(\gamma+1)\omega_1^2x\Phi_{1a}^2}{8c_0^3}\cos 2\omega_1\left(t-\frac{x}{c_0}\right) \qquad (6.65)$$

$$\Phi_{2b} = -\frac{(\gamma+1)\omega_2^2x\Phi_{2a}^2}{8c_0^3}\cos 2\omega_2\left(t-\frac{x}{c_0}\right) \qquad (6.66)$$

$$\Phi_{2c} = -\frac{(\gamma+1)\omega_1\omega_2x\Phi_{1a}\Phi_{2a}}{4c_0^3}\left\{\begin{array}{l}\cos\left[(\omega_1-\omega_2)\left(t-\dfrac{x}{c_0}\right)\right]- \\[2mm] \cos\left[(\omega_1+\omega_2)\left(t-\dfrac{x}{c_0}\right)\right]\end{array}\right\}$$

$$(6.67)$$

Therefore, the second-order approximate solution is

$$\Phi_2 = \Phi_{2a} + \Phi_{2b} + \Phi_{2c}$$

$$= -\frac{(\gamma+1)x}{8c_0^3}\left\{\begin{array}{l}-\omega_1^2\Phi_{1a}^2\cos 2\omega_1\left(t-\dfrac{x}{c_0}\right) \\[3mm] -\omega_2^2\Phi_{2a}^2\cos 2\omega_2\left(t-\dfrac{x}{c_0}\right) \\[3mm] 2\omega_1\omega_2\Phi_{1a}\Phi_{2a}\cos\left[(\omega_1-\omega_2)\left(t-\dfrac{x}{c_0}\right)\right] \\[3mm] -2\omega_1\omega_2\Phi_{1a}\Phi_{2a}\cos\left[(\omega_1+\omega_2)\left(t-\dfrac{x}{c_0}\right)\right]\end{array}\right\} \qquad (6.68)$$

Altogether, the sound pressure in the second-order approximation is

$$p_2 = \frac{(\gamma+1)\omega_1xp_{1a}^2}{4\rho_0c_0^3}\sin 2\omega_1\left(t-\frac{x}{c_0}\right)$$

$$+ \frac{(\gamma+1)\omega_2xp_{2a}^2}{4\rho_0c_0^3}\sin 2\omega_2\left(t-\frac{x}{c_0}\right)$$

$$+ \frac{(\gamma+1)xp_{1a}p_{2a}}{4\rho_0c_0^3}\left\{\begin{array}{l}(\omega_2-\omega_1)\sin\left[(\omega_2-\omega_1)\left(t-\dfrac{x}{c_0}\right)\right]+ \\[2mm] (\omega_2+\omega_1)\sin\left[(\omega_2+\omega_1)\left(t-\dfrac{x}{c_0}\right)\right]\end{array}\right\}$$

$$= p_{2\omega_1} + p_{2\omega_2} + p_{(\omega_1\pm\omega_2)} \qquad (6.69)$$

It shows that there are both second harmonics of the sources $(p_{2\omega_1}, p_{2\omega_2})$ and their sum $(p_{\omega_1+\omega_2})$ and difference $(p_{\omega_1-\omega_2})$, which further confirms that the linear superposition is invalid for finite-amplitude sound waves.

6.6 PROPAGATION OF FINITE-AMPLITUDE WAVE IN VISCOUS MEDIUM

6.6.1 BURGER'S EQUATION

It was shown in Chapter 5 that the attenuation in most soft tissue is proportional to the frequency with power 1–2. Although nonlinear propagation causes waveform distortion and energy transfer from the fundamental component to the higher harmonics, harmonics will have much higher attenuation. In a viscous medium, the dynamic equation is expressed as

$$\rho\left(\frac{\partial v}{\partial t} + \frac{1}{2}\frac{\partial v^2}{\partial x}\right) = -\frac{\partial p}{\partial x} + \left(\frac{4}{3}\eta' + \eta''\right)\frac{\partial^2 v}{\partial x^2} \qquad (6.70)$$

where η' and η'' are the shear and normal viscous coefficients, respectively. Assuming $\delta = (4/3)\eta' + \eta''$, $\rho = \rho_0 + \rho'$, the dynamic, continuity, and state equations can be rewritten as

$$(\rho_0+\rho')\frac{\partial v}{\partial t} + \rho_0 v\frac{\partial v}{\partial x} = -\frac{\partial p}{\partial x} + \delta\frac{\partial^2 v}{\partial x^2} \qquad (6.71)$$

$$\frac{\partial\rho'}{\partial t} + (\rho_0+\rho')\frac{\partial v}{\partial x} + v\frac{\partial\rho'}{\partial x} = 0 \qquad (6.72)$$

$$\frac{\partial p}{\partial x} = c_0^2\frac{\partial\rho'}{\partial x} + \frac{(\gamma-1)c_0^2}{\rho_0}\rho'\frac{\partial\rho'}{\partial x} \qquad (6.73)$$

Substituting Equation 6.73 to Equation 6.71 gives

$$\left(\rho_0 + \rho'\right)\frac{\partial v}{\partial t} + \rho_0 v \frac{\partial v}{\partial x} = -c_0^2 \frac{\partial \rho'}{\partial x} - \frac{(\gamma-1)c_0^2}{\rho_0}\rho'\frac{\partial \rho'}{\partial x} + \delta \frac{\partial^2 v}{\partial x^2} \quad (6.74)$$

If the coordinates are changed as

$$\tau = t - \frac{x}{c_0} \Rightarrow \frac{\partial}{\partial t} = \frac{\partial}{\partial \tau}, \quad x' = x \Rightarrow \frac{\partial}{\partial x} = \frac{\partial}{\partial x'} - \frac{1}{c_0}\frac{\partial}{\partial \tau} \quad (6.75)$$

Equations 6.74 and 6.72 can be rewritten as

$$\left(1 + \frac{\rho'}{\rho_0} - \frac{v}{c_0}\right)\frac{\partial v}{\partial \tau} = \frac{\delta}{\rho_0 c_0^2}\frac{\partial^2 v}{\partial \tau^2} + \frac{c_0}{\rho_0}\left[1 + (\gamma-1)\frac{\rho'}{\rho_0}\right]\frac{\partial \rho'}{\partial \tau} - \frac{c_0^2}{\rho_0}\frac{\partial \rho'}{\partial x} \quad (6.76)$$

$$\frac{1}{\rho_0}\left(1 - \frac{v}{c_0}\right)\frac{\partial \rho'}{\partial \tau} - \frac{1}{c_0}\left(1 + \frac{\rho'}{\rho_0}\right)\frac{\partial v}{\partial \tau} + \frac{\partial v}{\partial x} = 0 \quad (6.77)$$

If v/c_0 is replaced by ρ'/ρ_0, these two equations can be combined to remove ρ':

$$\frac{\partial v}{\partial x} - \frac{\beta}{c_0^3}v\frac{\partial v}{\partial \tau} = \frac{\delta}{2\rho_0 c_0^3}\frac{\partial^2 v}{\partial \tau^2} \quad (6.78)$$

which is Burger's equation. After the normalization, $W = v/v_0$, $\sigma = x/x_k$, $z = \omega\tau$, $\alpha = \omega^2 b/2\rho_0 c_0^3$, and $\Gamma = 1/\alpha x_k$, so Burger's equation can be normalized as

$$\frac{\partial W}{\partial \sigma} - W\frac{\partial W}{\partial z} = \frac{1}{\Gamma}\frac{\partial^2 W}{\partial z^2} \quad (6.79)$$

If expressed in p, the corresponding format is

$$\frac{\partial p}{\partial x} - \frac{\beta}{c_0^3}p\frac{\partial p}{\partial \tau} = \frac{\delta}{2\rho_0 c_0^3}\frac{\partial^2 p}{\partial \tau^2} \quad (6.80)$$

Burger's equation is the most widely used model for studying the combined effects of dissipation and nonlinearity on progressive planar waves.

6.6.2 SOLUTION OF BURGER'S EQUATION

Solution of Burger's equation is similar to that in an ideal medium, with the sum of first-order and second-order approximations, $v = v_1 + v_2$. The first-order approximation equation is

$$\frac{\partial v}{\partial x} - \frac{\delta}{2\rho_0 c_0^3}\frac{\partial^2 v}{\partial \tau^2} = 0 \quad (6.81)$$

Its solution is

$$v_1 = v_0 e^{-\alpha x}\sin\omega\tau \quad (6.82)$$

where $\alpha = \delta\omega^2/2\rho_0 c_0^3$ is the attenuation coefficient of a small-amplitude sound wave caused by viscosity and heat conduction loss. The second-order approximation equation is

$$\frac{\partial v_2}{\partial x} - \frac{\delta}{2\rho_0 c_0^3}\frac{\partial^2 v_2}{\partial \tau^2} = \frac{\beta}{c_0^2}v_1\frac{\partial v_1}{\partial \tau} = \frac{\beta\omega v_0^2}{2c_0^2}e^{-2\alpha x}\sin 2\omega\tau \quad (6.83)$$

Assuming that its solution has the form $v_2 = A(x)\sin 2\omega\tau$, Equation 6.83 can be rewritten as

$$\frac{\partial A}{\partial x} + 4\alpha A = \frac{\beta\omega v_0^2}{2c_0^2}e^{-2\alpha x} \quad (6.84)$$

This first order of an ordinary differential equation has the specific solution

$$v_2 = \frac{\beta\omega v_0^2}{4\alpha c_0^2}e^{-2\alpha x}\sin 2\omega\tau \quad (6.85)$$

The usual boundary condition would be that at $x = 0$ and $v_2 = 0$ (no second harmonic at the source). So, there would be another specific solution

$$v_2' = -\frac{\beta\omega v_0^2}{4\alpha c_0^2}e^{-4\alpha x}\sin 2\omega\tau \quad (6.86)$$

Altogether, the second-order approximate solution is

$$v_2 = \frac{\beta\omega v_0^2}{4\alpha c_0^2}(e^{-2\alpha x} - e^{-4\alpha x})\sin 2\omega\tau \quad (6.87)$$

The corresponding sound pressure of the second harmonic is

$$p_2(x,t) = \frac{(\gamma+1)p_a^2}{4b\omega}(e^{-2\alpha x} - e^{-4\alpha x})\sin 2\omega\left(t - \frac{x}{c_0}\right) \quad (6.88)$$

Here, the acoustic Reynold number is defined as $\text{Re} = p/\delta\omega$. The relationship between the sound pressure of the second harmonic and the propagation distance is not exponential as that of small-amplitude sound wave. The sound pressure of the second harmonic reaches its maximum when

$$\alpha x = \ln\sqrt{2} \quad \text{or} \quad x = \frac{\ln\sqrt{2}}{\alpha} \quad (6.89)$$

And, the maximum value is

$$(p_{2a})_{max} = \frac{(\gamma+1)p_{1a}^2}{16\delta\omega} = \frac{(\gamma+1)p_{1a}\text{Re}}{16} \quad (6.90)$$

So, when $x > (\ln \sqrt{2})/\alpha$, the increase of harmonics due to waveform distortion is smaller than the attenuation of the medium. So, the resultant harmonics decrease with the propagation distance. The ratio of the attenuation coefficients of finite-amplitude and small-amplitude waves is

$$\frac{\alpha'}{\alpha} \propto (e^{-\alpha x} - e^{-3\alpha x})^2 \qquad (6.91)$$

When $x = (\ln 3)/2\alpha$, α'/α is a maximum.

A useful method of obtaining a frequency-domain solution to Equation 6.78 is to assume a complex Fourier series solution of the form

$$v(x, \tau) = \frac{1}{2} \sum_{m=-\infty}^{\infty} v_m(x) e^{j2\pi m f_1 \tau} \qquad (6.92)$$

where $v_m(x)$ is the amplitude of the particle velocity of the mth harmonic. Then, Burger's equation can be rewritten as

$$\frac{\partial v_m(x)}{\partial x} = j \frac{\beta 2\pi f_1}{2c_0^2} \sum_{i=-\infty}^{\infty} (m-i) v_i v_{m-i} - \alpha_0 (m f_1)^2 v_m \quad (6.93)$$

where $\alpha \approx \frac{2\pi^2 f^2}{\rho_0 c_0^3} \left(\mu_B + \frac{4}{3}\mu \right) = \alpha_0 f^2$. To the first order

$$v_m(x + \Delta x) = v_m(x) + \frac{\partial v_m(x)}{\partial x} \Delta x \qquad (6.94)$$

which enables the iterative description of the propagation wave to be written as

$$v_m(x + \Delta x) = v_m(x) + \left[j \frac{\beta 2\pi f_1}{2c_0^2} \sum_{i=-\infty}^{\infty} (m-i) v_i v_{m-i} - \alpha_0 (m f_1)^2 v_m \right] \Delta x \qquad (6.95)$$

Then, the summation term can be rewritten in a form that is more convenient for computation, yielding

$$v_m(x + \Delta x) = v_m(x) + \begin{bmatrix} j \frac{\beta 2\pi f_1}{2c_0^2} \left(\sum_{i=1}^{m} i v_i v_{m-i} + \sum_{i=m+1}^{N} m v_i v_{m-1}^* \right) \\ -\alpha_0 (m f_1)^2 v_m \end{bmatrix} \Delta x \qquad (6.96)$$

where N is the number of harmonics to be retained in the computation process under the assumption of $v_{m=0} = 0$.

Note that, if there is no dissipation in the wave propagation, that is, $\delta = 0$ in Equation 6.80, the solution is

$$p = f \left[\tau + \left(\frac{\beta p}{\rho_0 c_0^3} \right) x \right] \qquad (6.97)$$

If there is no nonlinearity, $\beta = 0$ in Equation 6.80, and the solution has the form

$$p = p_0 \exp \left[j\omega\tau - \left(\frac{\delta\omega^2}{2c_0^3} \right) x \right] \qquad (6.98)$$

In both expressions, the coefficient of x is $O(\tilde{\varepsilon})$. Each solution thus exhibits the functional form

$$p = p(x_1, \tau), \quad x_1 = \tilde{\varepsilon} x, \quad \tau = t - x/c_0 \qquad (6.99)$$

The meaning of Equation 6.99 is that in a retarded time frame (i.e., for an observer in a reference frame that moves at speed c_0), nonlinearity and absorption separately produce only slow variations as functions of distance. Moreover, the relative order of the variations due to each effect is the same, that is, it is $O(\tilde{\varepsilon})$. We thus anticipate that the combined effects of nonlinearity and absorption will introduce variations of the same order. The coordinate x_1 is referred to as the "slow scale" corresponding to the retarded time frame τ.

6.6.3 Westervelt Equation

The general wave equation that accounts for nonlinearity up to the second order is given by the Westervelt equation (the lossy Westervelt equation for a thermoviscous fluid). The Westervelt equation is an appropriate approximation model of the full second-order wave equation with finite-amplitude and cumulative nonlinear effects, deriving from the equation of fluid motion by keeping up to quadratic-order terms:

$$\nabla^2 p - \frac{1}{c_0^2} \frac{\partial^2 p}{\partial t^2} + \frac{\delta}{c_0^4} \frac{\partial^3 p}{\partial t^3} = -\frac{\beta}{\rho_0 c_0^4} \frac{\partial^2 p^2}{\partial t^2} \qquad (6.100)$$

where

p is the sound pressure
c_0 is the small-signal sound speed
δ is the sound diffusivity
β is the nonlinearity coefficient
ρ_0 is the ambient density

The sound diffusivity is given by

$$\delta = \frac{1}{\rho_0} \left(\frac{4}{3}\mu + \mu_B \right) + \frac{k}{\rho_0} \left(\frac{1}{C_v} - \frac{1}{C_p} \right) \qquad (6.101)$$

where

μ is the shear viscosity
μ_B is the bulk viscosity
k is the thermal conductivity
C_v and C_p are the specific heat at constant volume and pressure, respectively

The right-hand side of the Westervelt equation may be regarded as a forcing function corresponding to a spatial distribution of virtual sources created by the sound wave itself. The integrated effect of these virtual sources on the sound wave accumulates with distance in the direction of propagation. Cumulative nonlinear effects produce waveform steepening and resonant harmonic interactions.

For the one-dimensional case, the Westervelt equation can be simplified as

$$\left(\frac{\partial^2}{\partial x^2} - \frac{1}{c_0^2}\frac{\partial^2}{\partial t^2}\right)p + \frac{\delta}{c_0^4}\frac{\partial^3 p}{\partial t^3} = -\frac{\beta}{\rho_0 c_0^4}\frac{\partial^2 p^2}{\partial t^2} \qquad (6.102)$$

6.6.4 GENERALIZED BURGER'S EQUATION

The generalized Burger's equation is an extension that takes into account the divergence (or convergence) of progressive spherical or cylindrical waves. The derivation begins by writing the Westervelt equation (6.91) as follows:

$$\left(\frac{\partial^2}{\partial r^2} + 2\frac{m}{r}\frac{\partial}{\partial r} - \frac{1}{c_0^2}\frac{\partial^2}{\partial t^2}\right)p + \frac{\delta}{c_0^4}\frac{\partial^3 p}{\partial t^3} = -\frac{\beta}{\rho_0 c_0^4}\frac{\partial^2 p^2}{\partial t^2} \qquad (6.103)$$

where the Laplacian operator is expressed in terms of the radial coordinate r for the wave field, with $m = 0, \frac{1}{2},$ and 1 for plane, cylindrical, and spherical waves, respectively. The field is assumed to depend only on r and t. Replacing r by x can recover the case of planar waves as Equation 6.102.

In linear theory, the pressure and particle velocity in time-harmonic spherical and cylindrical waves are also related by $p = \pm\rho_0 c_0 u$ for $kr \gg 1$ to the same order. Equation 6.103 is thus an appropriate model for the lowest-order nonlinear effects in spherical and cylindrical waves with weak attenuation ($\alpha \ll k$) provided $kr_0 \gg 1$, where r_0 is the source radius and k is the wavenumber corresponding to the lowest significant frequency component in the wave. Note that the restriction $kr_0 \gg 1$ on Equation 6.103 can be replaced by $r_0 > \lambda$, that is, the source radius is normally small in practice for diverging waves, because sources smaller than one wavelength are relatively inefficient and thus unlikely to radiate sound of finite amplitude. Conversely, Equation 6.103 cannot be used for distances closer than the order of one wavelength away from the focus in a converging wave field.

On the basis of the preceding discussion, it is reasonable to seek a solution having a functional form similar to Equation 6.80 for plane waves:

$$p = p(r_1, \tau), \qquad r_1 = \tilde{\varepsilon}r, \qquad \tau = t \mp \frac{r - r_0}{c_0} \qquad (6.104)$$

where $\tilde{\varepsilon}$ is simply an ordering parameter. As in the retarded time, the upper or lower sign is taken for a diverging (outgoing) or a converging (incoming) wave. The constant r_0/c_0 is

introduced for convenience, in order to obtain $\tau = t$ at the source $r = r_0$. The coordinate transformation from (r, t) to (r_1, τ) is similar to that for planar-wave case, and Equation 6.103 becomes, after $O(\tilde{\varepsilon}^3)$ terms are discarded and only $O(\tilde{\varepsilon})$ terms are retained,

$$\mp\tilde{\varepsilon}\frac{2}{c_0}\frac{\partial^2 p}{\partial r_1 \partial \tau} \mp\tilde{\varepsilon}\frac{2m}{c_0 r_1}\frac{\partial p}{\partial \tau} + \frac{\delta}{c_0^4}\frac{\partial^3 p}{\partial \tau^3} = -\frac{\beta}{\rho_0 c_0^4}\frac{\partial^2 p^2}{\partial \tau^2} \qquad (6.105)$$

Integration with respect to τ, multiplication by $\mp\frac{1}{2}c_0$, and transformation from the coordinates (r, t) to (r_1, τ) yield

$$\frac{\partial p}{\partial r} + \frac{m}{r}p \mp \frac{\delta}{2c_0^3}\frac{\partial^2 p}{\partial \tau^2} = \pm\frac{\beta p}{\rho_0 c_0^3}\frac{\partial p}{\partial \tau} \qquad (6.106)$$

This equation is the generalized Burger's equation, and it reduces to Burger's equation for $m = 0$ (in which case, one sets $r = x$ and $r_0 = 0$).

6.6.5 KHOKHLOV–ZABOLOTSKAYA–KUZNETSOV (KZK) EQUATION

To account for the combined effects of diffraction, absorption, and nonlinearity, Kuznetsov extended the work of Zabolotskaya and Khokhlov through the inclusion of the viscous loss term and obtained an approximate 3D equation for the velocity potential in an augmentation of Burger's equation in directional acoustic beams:

$$\frac{\partial^2 \varphi}{\partial t^2} - c_0^2 \nabla^2 \varphi = \frac{\partial}{\partial t}\left[\frac{1}{\rho_0}\left(\mu_B + \frac{4}{3}\mu\right)\nabla^2\varphi + (\nabla\varphi)^2 + \frac{\beta-1}{c_0^2}\left(\frac{\partial\varphi}{\partial t}\right)^2\right] \qquad (6.107)$$

where the right-hand side accounts for absorption and nonlinearity. In order to get a quasi-planar solution of a wave propagating in the x direction of a Cartesian coordinate system, this equation can be expressed in retarded time $\tau = t - z/c_0$:

$$\frac{\partial^2 \varphi}{\partial \tau \partial x} - \frac{c_0}{2}\nabla_\perp^2\varphi = \frac{\partial}{\partial \tau}\left[\frac{1}{2c_0^3\rho_0}\left(\mu_B + \frac{4}{3}\mu\right)\frac{\partial^2\varphi}{\partial \tau^2} + \frac{\beta}{2c_0^3}\left(\frac{\partial\varphi}{\partial \tau}\right)^2\right] \qquad (6.108)$$

where $\nabla_\perp^2 = \dfrac{\partial^2}{\partial x^2} + \dfrac{\partial^2}{\partial y^2}$ is a transverse Laplacian that operates in a plane normal to the propagation direction. Linear theory for directional beams reveals the existence of near-field and far-field regions, with the latter beginning roughly at the Rayleigh distance $\frac{1}{2}ka^2$, measured from the source along the beam axis. The near field is characterized by wavefronts that are approximately planar, and the far field is characterized by wavefronts that are spherical. Note that the relation $kz \gg 1$ is satisfied in the far field, where we have $z > \frac{1}{2}ka^2$ and therefore $kz > \frac{1}{2}(ka)^2 \gg 1$. With consideration of the viscosity term and frequency-dependent attenuation $\alpha = \alpha_0 f^2$, Equation 6.108 can have an alternative expression in terms of the pressure distribution using $p \approx \rho_0 \frac{\partial \varphi}{\partial t} = \rho_0 \frac{\partial \varphi}{\partial \tau}$:

$$\frac{\partial^2 p}{\partial \tau \partial x} - \frac{c_0}{2}\nabla_\perp^2 p = \frac{1}{2c_0^3 \rho_0}\frac{\partial}{\partial \tau}\left[\left(\mu_B + \frac{4}{3}\mu\right)\frac{\partial^2 p}{\partial \tau^2} + \beta\frac{\partial p^2}{\partial \tau}\right] \quad (6.109)$$

which is the well-known KZK equation. It is popular in calculating the radiated field when nonlinear effects must be accounted for. Moreover, it provides a reasonably good approximation to the field distribution for sources whose apertures are large compared to a wavelength ($ka \gg 1$), which ensures that the beam is fairly directional (quasi-plane wave) for observation points that are not too close to the source $z > a(ka)^{1/3}$ and lie within a reasonably narrow cone so that their off-axis locations are not too large. It should also be noted that someone should express the KZK equation in terms of the x component of the particle velocity under the assumption of $v_x \approx p/\rho_0 c_0$. For sources with cylindrical symmetry, cylindrical coordinates (r, z) can be used, in which case Equation 6.108 is still applicable, but the transverse Laplacian is given by $\nabla_\perp^2 = \frac{\partial^2}{\partial r^2} + \frac{1}{r}\frac{\partial}{\partial r}$.

In the absence of diffraction ($\nabla_\perp^2 p = 0$), the KZK equation reduces to Burger's equation.

6.7 MEASUREMENT OF NONLINEARITY B/A

Since the coefficients B and A are thermodynamic properties of the propagation medium, they can either be derived from the state equation of the material or be measured empirically if the state equation is unknown. The B/A value for a gas obeying the perfect gas law $P/P_0 = (\rho/\rho_0)^\gamma$ is $B/A = \gamma - 1$ and $C/A = (\gamma - 1)(\gamma - 2)$. For materials with no analytical equation of state, B/A must be measured. The finite-amplitude method is a technique in which the B/A value is inferred from the distorted pressure waveform in its propagation pathway via the growth of

the second harmonic. Another technique is the thermodynamic method, which relies on changes in the speed of sound that accompany changes in ambient pressure and temperature. The isentropic phase method makes use of Equation 6.20: the speed of sound is measured during a sufficiently rapid and smooth pressure change so that the system is considered thermodynamically reversible. This method has the advantage that a detailed knowledge of the thermodynamic properties of the propagation material and of the acoustic field of the transducer is unnecessary. The experimental precision in B/A measurement is about 10% for the finite-amplitude method and 5% for the thermodynamic method. More precise techniques have been necessitated by the use of the acoustic nonlinearity parameter in predictive models of tissue composition and nonlinear acoustic propagation in biological tissues, such as optical methods, parametric arrays, cavity resonance systems, and methods involving the measurement of volumetric effects. Values of B/A in typical biological tissue are listed in Table 6.1.

6.7.1 FINITE-AMPLITUDE METHOD

A schematic diagram of the experimental setup is shown in Figure 6.1. Both the transmitter and the receiver are immersed in deionized, degassed water. Either or both of them can be connected to a manipulator for 3D translational motion and/or rotation/titling. A small number of sinusoidal cycles generated from a function generator pass through a power amplifier before driving the transmitter. A broadband (at least covering several harmonics) hydrophone is used to pick up the acoustic burst, which is quantified by a digital oscilloscope and then transferred to a PC for further data analysis (i.e., spectrum). To ensure that the measured harmonic growth is only due to the nonlinearity in the medium, it is important to make certain that significant harmonics are not generated due to nonlinearities in the measurement system (i.e., source transducer and the driving electronics). The second harmonic level should be at least 40 dB below the fundamental for the highest electrical input used in the experiments. Since the determinations of B/A are made using much higher second harmonic levels, the nonlinearities of the equipment will not lead to errors in the B/A measurements.

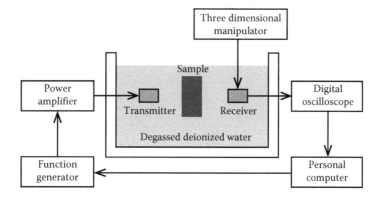

FIGURE 6.1 Schematic diagram of the experimental setup for measuring the nonlinearities of a material by the finite-amplitude method.

To have the experiment correspond to the theory discussed earlier, the transmitter and receiver must be aligned
along a common axis. The alignment procedure consists of
repeated reorientations of the transmitter and receiver until
an improvement in performance is no longer observed. For
example, the transmitter and receiver are aligned central to
each other for a small separation first. Before finite-amplitude
measurements, the attenuation coefficients at the fundamental and second harmonic are made as a function of distance
from the source. For each frequency, a best fit value for the
attenuation coefficient can be figured out by comparing these
results to the small-signal diffraction theory. The accuracy is
estimated to be ±3%. The speed of sound and density of the
liquid must also be determined in order to make the comparison between theory and experiment. The fundamental and
second harmonic spectrum levels are determined by a spectrum analysis method (e.g., fast Fourier transform [FFT]). For
each measurement distance, the pressure levels are recorded
in decibels relative to the fundamental level measured at a
reference distance, which is the minimum distance for which
standing waves between the transducers do not interfere with
the harmonic measurements. The experimentally determined
harmonics are compared with the theoretical prediction to
yield the B/A value. The fundamental ratios are normalized by
setting the measured ratio equal to the predicted ratio at the
reference distance. The measured second harmonic ratios are
then adjusted by the same correction factor used for the normalization. The peak source pressure p_0 is computed by applying the correction factor to the measured root-mean-squared
(rms) pressure at the reference distance. The predicted values
for the second harmonic ratios are then calculated with $\beta = 1$.
To determine B/A, the quotient of the measured and predicted
second harmonic ratios is averaged for all measurement distances. Since the predicted ratios are computed for $\beta = 1$, the
resulting average is just the measured β for the medium.
The B/A value is obtained using the relationship $B/A = 2(\beta-1)$.
The uncertainty of a B/A measurement is mainly due to that
of the receiver calibration.

6.7.2 Finite-Amplitude Insert-Substitution Method

The inset-substitution method is a comparative method that
measures the amplitude of the second harmonic p_{2o} of a reference liquid with known B/A (i.e., $B/A = 5.2$ for degassed, distilled water) and p_{2x} of a testing sample. For a finite-amplitude
planar wave in a lossless medium, the pressure amplitude of
its second harmonic at distance Z from the source is given by

$$p_2 = \frac{\omega p_1^2 Z}{4\rho_0 c_0^3}\left(\frac{B}{A}+2\right) \qquad (6.110)$$

Thus, the ratio of nonlinearities can be expressed as

$$\frac{(n+1)_x}{(n+1)_o} = \left(\frac{p_{2x}}{p_{2o}}\frac{L}{dD'D''} - \frac{L}{d} + 1\right)\frac{(\rho c^3)_x}{(\rho c^3)_o}\frac{1}{D''} \qquad (6.111)$$

where

$$D'' = \frac{(2\rho c)_x}{(\rho c)_o + (\rho c)_x}, \quad D' = \frac{(2\rho c)_o}{(\rho c)_o + (\rho c)_x}, \quad n = B/A + 1$$

and L is the length of the path between the transmitter and the
receiver. Therefore, $(B/A)_x$ can be calculated from p_{2x}/p_{2o}.
For a highly attenuating medium, the spatial rate of change
in the second harmonic amplitude is the sum of the change
due to the harmonic generation and sound absorption:

$$\frac{dp_2(Z)}{dZ} = Gp_1^2(Z) - \alpha_2 p_2(Z) \qquad (6.112)$$

where

$$G = [(n+1)\omega]/4\rho_0 c_0^3$$

$$p_1(Z) = p_1(0)e^{-\alpha_1 Z}$$

α_1 and α_2 are the attenuation coefficients for the fundamental and second harmonic components, respectively

The corresponding solution is

$$p_2(Z) = GZp_1^2(0)e^{-(\alpha_1 + \alpha_2/2)Z} \qquad (6.113)$$

Considering diffraction, the average pressure of the second
harmonic over the receiver is

$$\langle p_2(Z)\rangle = \int_0^z -GR_2(\mu)e^{ik\mu}\left[\left\langle p_1\left(Z - \frac{\mu}{2}\right)\right\rangle R_1(\mu)\right]^2 d\mu \qquad (6.114)$$

where

$$\langle p_1(Z)\rangle^2 = \langle p_1(0)\rangle^2 e^{i2k\mu}\left\{\frac{1 - 4e^{i\pi/4}\left[1 - \xi^2(Z)/2k^2a^2\right]}{/\sqrt{2\pi\xi(Z)}}\right\},$$

$$\xi(Z) = k\left(\sqrt{Z^2 + 4a^2} - Z\right)/2$$

and $R_1(\mu) = e^{-\alpha_1(Z-\mu)}$ is the correction for the attenuation of the
fundamental component from the source to the generating plane
of the second harmonic, and $R_2(\mu) = e^{-\alpha_2\mu}$ is for the attenuation
of the second harmonic from the generating plane to the receiver.
Therefore, a final expression of B/A taking into account both
attenuation and diffraction effects on the experiment is

$$\left(\frac{B}{A}\right)_x = \left\{\frac{\left[\frac{p_{2x}}{p_{2o}}\frac{L}{d}\frac{1}{R_1 R_2} - \left(\frac{L}{d} - 1\right)\frac{R_2}{R_1}\frac{F(k_o, d, L)}{F(k_o, 0, L)}\right]}{\frac{(\rho c^3)_x}{(\rho c^3)_o}\frac{F(k_o, 0, L)}{F(k_x, 0, d)}\sqrt{\frac{D''}{D'}}\left(\frac{B}{A} + 2\right)_o}\right\} - 2 \qquad (6.115)$$

where

$$R_1 = D'D''e^{-\alpha_1 d}$$

$$R_2 = \sqrt{D'D''}e^{-\alpha_2 d/2}$$

$F(k, o, Z)$ is the factor of diffraction correction

$$F(k,o,Z) = 1 - \frac{2}{r_2 - r_1}\int_{r_1}^{r_2}\frac{1-\xi^2(Z-\mu/2)/2k^2a^2}{\sqrt{\pi\xi(Z-\mu/2)}}\,d\mu \qquad (6.116)$$

The correction of diffraction of B/A values depends mainly on the difference between the sound speed in the sample and water. The main advantages of this method are that only a small amount of sample material is needed and absolute measurement of sound pressure is unnecessary. The sound speed and acoustic attenuation are measured by the pulse transmission method.

6.7.3 THERMODYNAMIC METHOD

The schematic diagram of the setup is shown in Figure 6.2 (the driving and receiving electronics are similar to those in Figure 6.1). The effect of standing wave should be minimized by reducing the output amplitude and pulse duration to the transmitter because it would cause the test medium to act as a fixed-pathlength interferometer and the amplitude versus frequency plot would show characteristic resonance peaks separated by a fundamental frequency, which is related to the speed of sound and the length of the medium. The signal from the function generator is fed into a network analyzer in which it is divided and amplified by an active power splitter into two output signals A and B with equal amplitude and phase. The signal B is transmitted through the interrogated medium by an ultrasound transducer. The signal picked up by the receiver is amplified and compared with the reference signal A by the network analyzer. The phase difference between the transmitted and the reference

signals is monitored by the digital oscilloscope. The steps involved are as follows:

1. The tested medium is allowed to equilibrate to the ambient temperature for at least 10 min until there is no further change in the phase.
2. The pressure is slowly increased by 1000 psi using a hydraulic pump, and the phase change during this process is determined. At this pressure, the system is allowed to equilibrate to the ambient temperature as indicated by no further change in the phase measurement.
3. The phase angle $\phi(p_0)$ is recorded, and the hydrostatic pressure is released to the atmospheric pressure. The phase decay accompanying the change in hydrostatic pressure is monitored as a function of time. By extrapolating, the phase angle $\phi(0)$ that corresponded to time zero can be determined. At this time point, the heat exchange between the sealed test vessel and the surroundings is negligible for all practical purposes, and B/A is estimated by the phase difference $\varphi(p_0) - \varphi(0)$ as in an adiabatic process:

$$\frac{B}{A} = -\frac{2\rho_0 c_0^3}{L\omega}\left(\frac{\partial\varphi}{\partial p}\right)_{s,0} \qquad (6.117)$$

where L is the length of the medium under investigation.

4. Because of the rapid change in hydrostatic pressure, the temperature of the medium contained in the vessel changes slightly. As the system equilibrates with the ambient temperature in 15–20 min, the phase angle reaches an equilibrium value of $\phi(e)$. The phase difference $\varphi(p_0) - \varphi(e)$ corresponds to an isothermal process to estimate $(B/A)'$:

$$\left(\frac{B}{A}\right)' = -\frac{2\rho_0 c_0^3}{L\omega}\left(\frac{\partial\varphi}{\partial p}\right)_{T,0} \qquad (6.118)$$

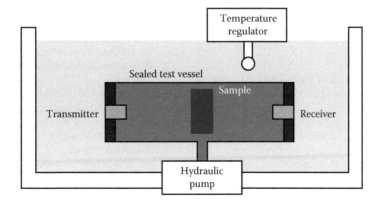

FIGURE 6.2 Schematic diagram of the experimental setup of measuring the nonlinearity of a material by the thermodynamic method.

5. The temperature of the bath is then raised to a known higher value. The accompanying change in phase $\Delta\theta$ is measured to determine the sound speed at the new temperature:

$$c_0(T) = \frac{c_0(R)}{1 + [\Delta\theta c_0(R)/L\omega]} \quad (6.119)$$

Steps 1–4 can be repeated to estimate B/A and $(B/A)'$.

6.7.4 Improved Thermodynamic Method

This method directly measures the speed of sound in an isentropic process:

$$\frac{B}{A} = \rho_0 c_0 \left(\frac{\partial c}{\partial p} \right)_s \quad (6.120)$$

When the distance from the transmitter to the receiver is fixed, the change in sound speed due to the variation in the static pressure is equivalent to the change in transmission time Δt. The corresponding phase shift is $\Delta\varphi = \omega\Delta t$, so

$$\frac{B}{A} = -\frac{2\rho_0 c_0^3}{\omega d} \left(\frac{\Delta\varphi}{\Delta p} \right)_s \quad (6.121)$$

The main advantage of this method is that it does not require calibration of the sound field and knowledge of thermodynamic parameters (i.e., the thermal expansion coefficient and specific heat at constant pressure), which are sometimes difficult with great precision. The accuracy of this method is estimated to be 4% for liquids and 7% for soft tissues.

ASSIGNMENT

Q1: Using transform $x' = x - c_0 t$ and $\tau = t$, prove that the corresponding Burger's equation is

$$\frac{\partial v}{\partial \tau} + \beta v \frac{\partial v}{\partial x} = b \frac{\partial^2 v}{\partial x^2}$$

Q2: Prove that superposition principle is invalid for nonlinear acoustic propagation.

REFERENCES

Bacon DR. Finite amplitude distortion of the pulsed fields used in diagnostic ultrasound. *Ultrasound in Medicine and Biology* 1984;10:189–195.

Bakhvalov NS, Zhileikin YM, Zabolotskaya EA. *Nonlinear Theory of Sound Beams.* New York: American Institute of Physics, 1987.

Berg AM, Naze Tjotta J. Numerical simulation of sound pressure field from finite amplitude, plane or focusing, rectangular apertures. In: Hobaek H, ed., *Advances in Nonlinear Acoustics.* London, U.K.: Elsevier, 1993, pp. 309–314.

Berntsen J. Numerical calculations of finite amplitude sound beam. In: Hamilton MF, Blackstock DT, eds., *Frontiers of Nonlinear Acoustics.* London, U.K.: Elsevier, 1990, pp. 191–196.

Beyer RT, Letcher SV. *Nonlinear Acoustics. Physical Ultrasonics.* New York: Academic Press, 1969.

Bjorno L. Forty years of nonlinear ultrasound. *Ultrasonics* 2002; 40(1–8):11–17.

Blackstock DT. Propagation of plane sound waves of finite amplitude in nondissipative fluids. *Journal of the Acoustical Society of America* 1962;34:9–30.

Blackstock DT. Connection between the Fay and Fubini solutions for plane sound waves of finite amplitude. *Journal of the Acoustical Society of America* 1966;39:1019–1026.

Blackstock DT. Generalized Burgers equation for plane waves. *Journal of the Acoustical Society of America* 1985;77:2050–2053.

Blackstock DT. Chapter 1, History of nonlinear acoustics: 1750's–1930's. In: Hamilton MF, Blackstock DT, eds., *Nonlinear Acoustics.* San Diego, CA: Academic Press, 1998, pp. 1–23.

Carstensen EL, Law WK, McKay ND, Muir TG. Demonstration of nonlinear acoustical effects at biomedical frequencies and intensities. *Ultrasound in Medicine and Biology* 1980;6:359–368.

Christopher T, Parker KJ. New approaches to nonlinear diffractive field propagation. *Journal of the Acoustical Society of America* 1991;90:488–499.

Cleveland RO, Hamilton MF, Blackstock DT. Time-domain modeling of finite-amplitude sound in relaxing fluids. *Journal of the Acoustical Society of America* 1996;99:3312–3318.

Cobb WN. finite amplitude method for the determination of the acoustic nonlinearity parameter B/A. *Journal of the Acoustical Society of America* 1983;73:1525–1531.

Dalecki D, Carstensen EL, Parker KJ. Absorption of finite amplitude focused ultrasound. *Journal of the Acoustical Society of America* 1991;89:2435–2447.

Duck FA. *Physical Properties of Tissue: A Comprehensive Reference Book.* London, U.K.: Academic Press, 2002.

Earnshaw S. On the mathematical theory of sound. *Philosophical Transactions of the Royal Society of London* 1860;150:133–148.

Everbach EC. Parameters of nonlinearity of acoustic media. In: Crocker MJ, ed., *Encyclopedia of Acoustics.* New York: John Wiley & Sons, 1997, pp. 219–226.

Froysa KE. Linear and weakly nonlinear propagation of a pulsed sound beam. PhD thesis, Department of Applied Mathematics, University of Bergen, Bergen, Norway, 1991.

Ginsberg JH, Hamilton MF. Computational methods. In: Hamilton MF, Blackstock DT, eds., *Nonlinear Acoustics.* San Diego, CA: Academic Press, 1998, pp. 309–342.

Gol'dberg ZA. On the propagation of plane waves of finite amplitude. *Soviet Physics. Acoustics* 1957;3:340–347.

Gong X-F, Zhu Z-M, Shi T, Huang J-H. Determination of the acoustic nonlinearity parameter in biological media using FAIS and ITD methods. *Journal of the Acoustical Society of America* 1989;86:1–5.

Gould RK, Smith CW, Williams Jr AO, Ryan RP. Measured structure of harmonics self-generated in an acoustic beam. *Journal of the Acoustical Society of America* 1997;40:2539–2548.

Hamilton MF. Sound beams. In: Hamilton MF, Blackstock DT, eds., *Nonlinear Acoustics.* San Diego, CA: Academic Press, 1998, pp. 233–261.

Haran ME, Cook BD. Distortion of finite amplitude amplitude ultrasound in lossy media. *Journal of the Acoustical Society of America* 1983;73:774–779.

Hart TS, Hamilton MF. Nonlinear effects in focused sound beams. *Journal of the Acoustical Society of America* 1988;84:1488–1496.

Humphrey VF. Nonlinear propagation in ultrasonic fields: Measurements, modelling and harmonic imaging. *Ultrasonics* 2000;38:267–272.

Khokhlov RV, Soluyan SI. Propagation of acoustic waves of moderate amplitude through dissipative and relaxing media. *Acoustica* 1964;14:241–247.

Khokhlov VA, Souchon R, Tavakkoli J, Sapozhnikov OA, Cathignol D. Numerical modeling of finite-amplitude sound beams: Shock formation in the near field of a cw plane piston source. *Journal of the Acoustical Society of America* 2001;110:95–108.

Kuznetsov VP. Equations of nonlinear acoustics. *Soviet Physics. Acoustics* 1971;16:467–470.

Labat V, Remenieras JP, Bou Matar O, Ouahabi A, Patat F. Harmonic propagation of pulsed finite amplitude sound beams: Experimental determination of the nonlinearity parameter B/A. *Ultrasonics* 2000;38:292–296.

Lamb H. *The Dynamical Theory of Sound.* London, U.K.: Edward Arnold & Co., 1925.

Law WK, Frizzell LA, Dunn F. Ultrasonic determination of the nonlinearity parameter B/A for biological media. *Journal of the Acoustical Society of America* 1981;108:906–917.

Law WK, Frizzell LA, Dunn F. Determination of the nonlinearity parameter B/A of biological media. *Ultrasound in Medicine and Biology* 1985;11:307–318.

Lee CP, Wang TG. Acoustic radiation pressure. *Journal of the Acoustical Society of America* 1993;94:1099–1109.

Lee YS, Hamilton MF. Time-domain modeling of pulsed finite-amplitude sound beams. *Journal of the Acoustical Society of America* 1995;97:906–917.

Li S, McDicken WN, Hoskins PR. Nonlinear propagation in Doppler ultrasound. *Ultrasound in Medicine and Biology* 1993;19:359–364.

Lighthill MJ. Viscosity effects in sound waves of finite amplitude. In: Batchelor GK, Davies RM, eds., *Surveys in Mechanics.* Cambridge, U.K.: University Press, 1956, pp. 250–351.

Muir TG, Carstensen EL. Prediction of nonlinear acoustic effects at biomedical frequencies and intensities. *Ultrasound in Medicine and Biology* 1980;6:345–357.

Nachef S, Cathignol D, Tjotta JN, Berg AM, Tjotta S. Investigation of a high-intensity sound beam from a plane transducer: Experimental and theoretical results. *Journal of the Acoustical Society of America* 1995;98:2303–2323.

Naugol'nykh KA, Soluyan SI, Khokhlov RV. Spherical waves of finite amplitude in a viscous thermally conducting medium. *Soviet Physics. Acoustics* 1963;9:42–46.

Naugolnykh K, Ostrovsky L. *Nonlinear Wave Processes in Acoustics.* Cambridge, U.K.: Cambridge University Press, 1998.

Naze Tjotta J, Tjotta S. Nonlinear equations of acoustics, with applications to parametric arrays. *Journal of the Acoustical Society of America* 1981;69:1644–1652.

Parker KJ. Observation of nonlinear acoustic effects in a B-scan imaging instrument. *IEEE Transactions on Sonics and Ultrasonics* 1985;32:4–8.

Pierce AD. *Acoustics: An Introduction to Its Physical Principles and Applications.* New York: Acoustical Society of American Publications, 1989.

Remenieras JP, Bou Matar O, Labat V, Patat F. Time-domain modeling of nonlinear distortion of pulsed finite amplitude sound beams. *Ultrasonics* 2000;38:305–311.

Rudenko OV, Soluyan SI. *Theoretical Foundations of Nonlinear Acoustics.* New York: Consultants Bureau, 1977.

Sehgal CM, Bahn RC, Greenleaf JF. Measurement of the acoustic nonlinearity parameter B/A in human tissues by a thermodynamic method. *Journal of the Acoustical Society of America* 1984;76:1023–1029.

Starritt HC, Duck FA, Hawkins AI, Humphrey VF. The development of harmonic distortion in pulsed finite-amplitude ultrasound passing through liver. *Physics in Medicine and Biology* 1986;31:1401–1409.

Starritt HC, Perkins MA, Duck FA, Humphrey VF. Evidence for ultrasonic finite-amplitude distortion in muscle using medical equipment. *Journal of the Acoustical Society of America* 1985;77:302–306.

Tavakkoli J, Cathignol D, Souchon R, Sapozhnikov OA. Modeling of pulsed finite-amplitude focused sound beams in time domain. *Journal of the Acoustical Society of America* 1998;104:2061–2072.

TenCate JA. An experimental investigation of the nonlinear pressure field produced by a plane circular piston. *Journal of the Acoustical Society of America* 1993;94:1084–1089.

Tjotta S. On some nonlinear effects in ultrasonic fields. *Ultrasonics* 2000;38:278–283.

Too GPJ, Ginsberg JH. Nonlinear progressive wave equation model for transient and steady-state sound beams. *Journal of the Acoustical Society of America* 1992;92:59–68.

Zabolotskaya EA, Khokhlov RV. Quasi-plane waves in the nonlinear acoustics of confined beams. *Soviet Physics. Acoustics* 1969;15:35–40.

Zemp RJ, Tavakkoli J, Cobbold RSC. Modeling of nonlinear ultrasound propagation in tissue from array transducers. *Journal of the Acoustical Society of America* 2003;113:139–152.

Zhang J, Dunn F. A small volume thermodynamic system for B/A measurement. *Journal of the Acoustical Society of America* 1991;89:73–79.

Zhang J, Kuhlenschmidt MS, Dunn F. Influences of structural factors of biological media on the acoustic nonlinearity parameter B/A. *Journal of the Acoustical Society of America* 1991;89:80–91.

7 Cavitation

Cavitation refers to the formation of bubbles and the subsequent bubble dynamics in liquids. Bubbles may be of a gas or vapors of liquids (i.e., water, organic solvents, biological tissue, liquid helium, and molten metals) under a wide range of conditions, such as under hydrodynamic, thermal, or acoustic excitation. Bubble cavitation can be categorized as transient (inertial) or stable. Transient cavitation involves significant changes in the bubble size within over only a few acoustic cycles and results in the violent collapse of the bubble wall. In comparison, stable cavitation usually is small-amplitude oscillations about an equilibrium radius over a time scale of thousands of acoustic cycles in most instances. However, these two phenomena do not have a strict borderline and can change from one form to the other. For example, accumulative stable cavitation of long duration can gradually become transient cavitation, and the smaller bubbles produced by the collapse of a transient cavity would undergo stable cavitation afterward.

The scientific research on cavitation accompanied the use of high-powered, high-speed steam turbines in the mid-1800s. Sir John Thornycroft found a decrease in efficiency of the propeller in torpedo destroyers in rapid motion as a result of loss of contact with the water and the significant erosion on the propeller (Figure 7.1). So, the British Admiralty appointed a special subcommittee, in which Lord Raleigh was involved, to investigate this phenomenon termed "cavitation" in 1917. At almost the same time (1915–1920), acoustic cavitation was observed by Langevin and his coworkers during their research on ultrasonics. The fundamental problems of cavitation were by now well formulated.

The influence of acoustic cavitation is related to the disruption of a liquid by the presence of bubbles. Consequently, the bubble's interior and the medium immediately surrounding the bubble undergo continuous change during cavitation, which include the shape and size of the bubble–liquid interface, molecular diffusion through the bubble wall, the gas concentration in the surrounding liquid, microscopic acoustic streaming near the bubble and the subsequent shear stresses, the interior pressure and temperature of bubble, the radiated acoustic energy, and hinderance to bubble oscillations by thermal and viscous damping. Each of these processes manifests itself differently, yet they work synergistically for the induced chemical, physical, and biological effects of ultrasound. Therefore, in order to understand the underlying mechanism, a general analysis of acoustic cavitation is required.

7.1 NUCLEATION

A liquid can support a considerable amount of tensile stress because of the intermolecular connection maintaining its liquid state. The tensile strength of water ranges from 1300 to 4000 bars. The fracture of the liquid and the subsequent collapse of the cavity in an acoustic field are one aspect of acoustic cavitation. However, numerous measurements of the acoustic cavitation threshold (tensile strength) of water show that it is several orders of magnitude smaller than the theoretical prediction, which is due to inhomogeneities (i.e., small pockets of gas) in the bulk of the fluid serving as preferential sites for liquid fracture. This phenomenon is similar to the initial crack formed at the weakest link (i.e., in the form of dislocations and grain boundaries) in a solid. Meanwhile, the surface tension of the liquid will tend to force the gas out of the bubble (unless the liquid is supersaturated with gas), and the bubble will dissolve. Consequently, nucleation depends on a mechanism for the stabilization of these gas pockets from dissolution.

In the bulk of a pure liquid at a given pressure, temperature and density fluctuations cause the rapid formation and destruction of microscopic vapor cavities. The pressure and temperature requirements for the onset of cavitation are given by

$$\frac{16\pi[\sigma(T)]^3}{3\kappa T[P(T)-P_L]^2} = \ln C(T) \tag{7.1}$$

where

T is the absolute temperature
κ is the Boltzmann constant
$\sigma(T)$ is the surface tension at the interface of liquid and vapor
$[P(T)-P_L]$ is the pressure differential across the cavity interface
$P(T)$ is the pressure of the vapor at the interface
P_L is the pressure in the liquid
$C(T)$ presents the complex kinetics of the nucleation process

In the homogeneous nucleation in a liquid, it is generally assumed that the liquid does not contain any foreign substances that may act as a reference nucleation site.

Since the cavitation thresholds of liquid measured in practice are significantly less than the homogenous nucleation threshold, it is reasonable to assume that most liquids contain foreign substances that stabilize pockets of gas and thus weaken the liquid to stress failure. The obvious and simplest model of a nucleus is that of a free bubble. However, because of the "Laplace pressure" (i.e., an internal pressure caused by surface tension and equal to $2\sigma/R$, where R is the bubble radius), these bubbles gradually dissolve as the gas is forced out of the interior of the bubble into the liquid. The differential

(a)

(b)

FIGURE 7.1 (a) Bubble produced when the propeller is rotating and (b) erosion on its surface by cavitation.

equation describing the change in equilibrium radius R_0 with time for an undisturbed liquid is given by

$$\frac{dR_0}{dt} = \frac{Dd}{R_0}\left[\frac{C_i/C_0 - 1}{1 + 4\sigma/3R_0P_\infty}\right]\left(1 + \frac{R_0}{\sqrt{\pi Dt}}\right) \qquad (7.2)$$

where
   D is the diffusion constant
   C_0 is the equilibrium or saturated concentration of the gas in the liquid (moles per unit volume)
   C_i is the concentration of dissolved gas in the liquid far away from the bubble
   P_∞ is the ambient pressure in the liquid far away from the bubble
   $d = R_gTC_0/P_\infty$, where R_g is the universal gas constant

Thus, a liquid containing free gas bubbles and is not supersaturated will rid itself of these bubbles as the small ones dissolve and the large ones rise to the surface. If the liquid is supersaturated (e.g., a carbonated beverage), bubbles can grow from small stabilized gas pockets on the walls of the vessel to a visible size in a short time.

7.2 CAVITATION THRESHOLD

If the amplitude of an acoustic field is sufficiently low, a nucleus undergoes no appreciable change in size and becomes stabilized. When the amplitude reaches a sufficiently high value, the nucleus becomes unstable and grows within a few acoustic cycles into a mostly vapor-filled bubble, which is detectable in the experiment. In theory, the cavitation threshold is the value of the acoustic pressure amplitude necessary for the nucleus to become unstable. In practice, it is the acoustic pressure amplitude that produces detectable cavitation events. Ideally, the two thresholds are identical, and the discrepancy is only due to the sensitivity of cavitation measurement approach.

Liquids saturated with gas have a transient cavitation threshold of about 1–2 bars. The bubbles contain an appreciable amount of gas that cushions the collapse, and its cavitation is associated with the presence of numerous gas bubbles that gradually appear in the bulk of the liquid. Usually, "streamers" or relatively long lines of bubbles are present. Continual hissing or frying noise, instead of a clearly distinguishable audible snap or pop, is heard. Cavitation threshold increases to about 7 bars linearly as the gas concentration is reduced to about 25% of saturation. In that situation, the cavitation becomes crisper and transient.

The cavitation threshold is proportional to the hydrostatic pressure applied to the liquid, and it increases with the decreasing number and size of solid contaminants but varies inversely with surface tension through an empirical relationship:

$$\cos\alpha_E = \frac{C}{\sigma} - 1 \qquad (7.3)$$

where
   α_E is the equilibrium contact angle
   $C \approx 50$ dyn/cm is a constant for surface tension >10 dyn/cm

It is different from the Laplace pressure that varies directly with surface tension. The cavitation threshold decreases with temperature almost in a linear manner except for liquids near the boiling point, where the threshold drops to zero.

7.3 RECTIFIED DIFFUSION

Rectified diffusion is the process by which dissolved gas in the liquid converts into free gas in the form of bubbles by the action of an acoustic wave. The rectification of mass transfer is important whenever the applied acoustic pressure exceeds the threshold for rectified diffusion, which is on the order of 1 bar or less. For liquids that are not supersaturated with gas, the Laplace pressure exerted on the gas contained within a bubble by the force of surface tension determines the cavitation threshold by

$$P_i = \frac{2\sigma}{R_0} \qquad (7.4)$$

where

P_i is the internal pressure
R_0 is the equilibrium radius of the bubble

For relatively large gas bubbles (i.e., 1 cm in diameter), the value is small (<0.001 bar). However, for bubbles on the order of 1 μm, this pressure can be quite large (>1 bar). Therefore, small, free gas bubbles in a liquid with a surface tension must be stabilized; otherwise, they will dissolve rapidly. In the acoustic field, a bubble can oscillate about an equilibrium radius and grow in size via the following mechanisms:

1. *Surface area effect*: In the contraction phase, the gas concentration inside the bubble increases, and gas diffuses outward through the bubble wall. On the contrary, when the gas concentration decreases, the gas diffuses inward during bubble expansion. Because the diffusion rate is proportional to the surface area, the inflow of gas is more than the outflow. Therefore, there is a net increase in the amount of gas in the bubble over a complete oscillation cycle.
2. *Shell effect*: The diffusion rate of gas in a liquid is proportional to the gradient of its concentration. When the bubble contracts, the complementary liquid shell expands, and the gas concentration near the bubble wall reduces. Thus, the diffusion rate of gas away from the bubble is greater than that at equilibrium. Conversely, the gas concentration near the bubble increases during bubble expansion, and the diffusion rate of gas toward the bubble is greater than average. This kind of convection can enhance the rectified diffusion.

In summary, both the surface area and shell effects are necessary in describing the dynamics of the bubble cavitation.

The well-known Rayleigh–Plesset equation for the motion of the gas bubble is given by

$$R\ddot{R} + \frac{3}{2}\dot{R}^2 + \frac{1}{\rho}\left\{ P_0\left[1 - \left(\frac{R_0}{R}\right)^{3\eta}\right] - P_A\cos\omega t + \rho R_0\omega_0 b\dot{R}\right\} = 0 \qquad (7.5)$$

where

R is the instantaneous bubble radius
R_0 is the equilibrium bubble radius
ρ is the density of the liquid
η is the polytropic exponent of the gas inside the bubble
P_A is the acoustic pressure
ω is the angular frequency
ω_0 is the small-amplitude resonance frequency
b is a damping factor only close to the bubble resonance

Furthermore, $P_0 = P_\infty + 2\sigma/R_0$, where P_∞ is the ambient pressure.

The small-amplitude resonance frequency of the bubble is

$$\omega_0^2 = \frac{1}{\rho R_0^2}\left(3\eta P_0 - \frac{2\sigma}{R_0}\right) \qquad (7.6)$$

The gas diffusion equation is governed by Fick's law of mass transfer

$$\frac{dC}{dt} = \frac{\partial C}{\partial t} + \vec{v}\cdot\nabla C = D\nabla^2 C \qquad (7.7)$$

where

C is the gas concentration in the liquid
$\vec{v}$ is the liquid velocity
D is the diffusion constant

The gradient of the number n of gas moles in a bubble is given by

$$\frac{dn}{dt} = 4\pi D R_0 C_0 \cdot \left[\langle R/R_0\rangle + R_0\sqrt{\frac{\langle(R/R_0)^4\rangle}{\pi D t}}\right]H \qquad (7.8)$$

where t is the time, and H is defined by

$$H = C_i/C_0 - \langle(R/R_0)^4(P_g/P_\infty)\rangle / \langle(R/R_0)^4\rangle \qquad (7.9)$$

where P_g is the instantaneous gas pressure inside the bubble given by

$$P_g = P_0\left(\frac{R_0}{R}\right)^{3\eta} \qquad (7.10)$$

This assumption of polytropic exponent is commonly made and has been shown to be reasonably accurate for small-amplitude pulsation. However, for moderate- to large-amplitude oscillations, it should be modified.

The value of R/R_0 is obtained by an expansion solution of the Rayleigh–Plesset equation as

$$\frac{R}{R_0} = 1 + \alpha\left(\frac{P_A}{P_\infty}\right)\cos(\omega t + \delta) + \alpha^2\kappa\left(\frac{P_A}{P_\infty}\right)^2 + \cdots \qquad (7.11)$$

where

$$\alpha^{-1} = \left(\frac{\rho R_0^2}{P_\infty}\right)\sqrt{\left(\omega^2 - \omega_0^2\right)^2 + (\omega\omega_0 b)^2} \qquad (7.12)$$

$$\kappa = \frac{(3\eta + 1 - \beta^2)/4 + (\sigma/4R_0 P_\infty)(6\eta + 2 - 4/3\eta)}{1 + (2\sigma/R_0 P_\infty)(1 - 1/3\eta)} \qquad (7.13)$$

$$\delta = \tan^{-1}\left[\frac{\omega\omega_0 b}{\omega^2 - \omega_0^2}\right] \qquad (7.14)$$

$$\beta^2 = \frac{\rho\omega^2 R_0^2}{3\eta P_\infty} \tag{7.15}$$

Since the damping of the bubble pulsations is only important near resonance (or at harmonics and subharmonics of the resonant frequency), it is estimated by using an effective viscosity in a nonrigorous way, which consists of three common damping mechanisms: thermal, viscous, and radiation:

$$b = b_t + b_v + b_r \tag{7.16}$$

where

$$b_t = 3(\gamma-1)\left[\frac{X(\sinh X + \sin X) - 2(\cosh X - \cos X)}{X^2(\cosh X - \cos X) + 3(\gamma-1)X(\sinh X - \sin X)}\right] \tag{7.17}$$

$$X = R_0\sqrt{\frac{2\omega}{D_1}} \tag{7.18}$$

$$b_v = \frac{4\omega\mu}{3\eta P_\infty} \tag{7.19}$$

$$b_r = \frac{\rho R_0^3 \omega^3}{3\eta P_\infty c} \tag{7.20}$$

$$\eta = \gamma\left(1 + b_t^2\right)^{-1}\left[1 + \frac{3(\gamma-1)}{X}\left(\frac{\sinh X - \sin X}{\cosh X - \cos X}\right)\right]^{-1} \tag{7.21}$$

where

    γ is the ratio of specific heats
    $D_1 = \kappa_1/\rho_1 C_{p1}$, where κ_1 is the thermal conductivity of the gas inside the bubble, ρ_1 is the gas density, and C_{p1} is the specific heat of gas at constant pressure
    μ is the viscosity of liquid
    c is the speed of sound in the liquid

The growth rate equation has an expression as

$$\langle R/R_0 \rangle = 1 + \kappa\alpha^2\left(\frac{P_A}{P_\infty}\right)^2 \tag{7.22}$$

$$\langle (R/R_0)^4 \rangle = 1 + (3+4\kappa)\alpha^2\left(\frac{P_A}{P_\infty}\right)^2 \tag{7.23}$$

$$\langle (R/R_0)^4(P_g/P_\infty) \rangle = \left[\begin{array}{c} 1 + \frac{3(\eta-1)(3\eta-4)}{4}\alpha^2\left(\frac{P_A}{P_\infty}\right)^2 \\ + (4-3\eta)\kappa\alpha^2\left(\frac{P_A}{P_\infty}\right)^2 \end{array}\right]\left(1 + \frac{2\sigma}{R_0 P_\infty}\right) \tag{7.24}$$

For an ideal gas, the relationship of the equilibrium radius and the number of moles of gas is

$$P_0 = \frac{3\eta R_g T}{4\pi R_0^3} \tag{7.25}$$

This equation is a good approximation in linear acoustics so that the density fluctuations do not become excessive. In addition, for low host-liquid temperatures, the ratio of vapor to gas within the bubble remains low, which is often met in most experimental situations.

The rate of change in the equilibrium bubble radius is expressed as

$$\frac{dR_0}{dt} = \frac{Dd}{R_0}\left[\langle R/R_0 \rangle + R_0\sqrt{\frac{\langle (R/R_0)^4 \rangle}{\pi Dt}}\right]$$

$$\times\left(1 + \frac{4\sigma}{3R_0 P_\infty}\right)^{-1}\left(\frac{C_i}{C_0} - \frac{\langle (R/R_0)^4(P_g/P_\infty) \rangle}{\langle (R/R_0)^4 \rangle}\right) \tag{7.26}$$

The threshold of acoustic pressure for bubble growth is obtained by setting $dR_0/dt = 0$ and given by

$$P_A^2 = \frac{(\rho R_0^2\omega_0^2)\left[(1-\omega^2/\omega_0^2)^2 + b^2(\omega^2/\omega_0^2)\right](1 + 2\sigma/R_0 P_g - C_i/C_0)}{(3+4\kappa)C_i/C_0 - [3(\eta-1)(3\eta-4)/4 + (4-3\eta)\kappa](1 + 2\sigma/R_0 P_\infty)} \tag{7.27}$$

The expressions here are the governing equations for rectified diffusion to calculate the threshold for inception of growth and the rate of this growth and also valid for diffusion in the absence of an applied sound field (i.e., $P_A = 0$) below the threshold.

7.4 BUBBLE DYNAMICS

In order to examine the effect of an acoustic field on a liquid, it is desirable first to understand the response of a single bubble to an oscillating pressure field (Figure 7.2).

7.4.1 STABLE CAVITATION

A stable cavity is one that oscillates nonlinearly about its equilibrium radius. To account adequately for the motion of the bubble interface, it is necessary to solve the equations of conservation of mass, momentum, and energy for both the gas and the liquid phases and match the boundary conditions.

Conservation of mass

$$\frac{1}{\rho}\left(\frac{\partial p}{\partial t} + u\frac{\partial p}{\partial r}\right) = -\left(\frac{\partial u}{\partial r} + \frac{2u}{r}\right) \tag{7.28}$$

Conservation of momentum

$$\frac{\partial u}{\partial t} + u\frac{\partial u}{\partial r} = -\frac{1}{\rho}\frac{\partial \rho}{\partial r} \tag{7.29}$$

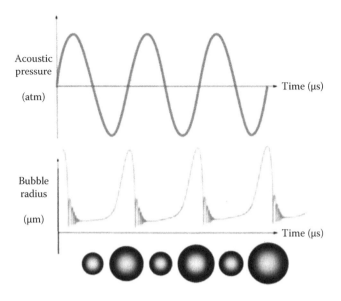

FIGURE 7.2 Stable bubble cavitation in response to acoustic pressure.

Conservation of energy

$$\rho c_v \left(\frac{\partial T}{\partial t} + u \frac{\partial T}{\partial r} \right) = \kappa \nabla^2 T - p \nabla \cdot u \qquad (7.30)$$

where

    u is the radial velocity
    c_v is the specific heat at a constant volume
    κ is the thermal conductivity of the fluid

The most widely used equation, the Rayleigh–Plesset equation, is written in a different form

$$\rho(R\ddot{R} + 3\dot{R}^2/2) == P_i - P_e = P_i(R,t) - P_A(t) - 2\sigma/R - 4\mu\dot{R}/R \qquad (7.31)$$

where

    R is the bubble radius
    ρ is the liquid density
    P_i and P_e are the internal pressure and the external (in the liquid including the surface to the bubble) pressure, respectively

$$P_i = P_\infty \left(\frac{R_0}{R} \right)^{3\kappa} \qquad (7.32)$$

In the polytropic approximation, the viscous damping is not adequately accounted for except for very small bubbles or at very high frequencies. Since the principal damping mechanism is thermal in origin, introduction of the internal pressure via the polytropic exponent will result in the absence of the phase information that accounts for it. However, retaining the internal pressure as a spatial and temporal variable introduces a nearly intractable complication. The damping problem was circumvented by introducing an effective viscosity, as shown previously, which could account for the contributions of

thermal and radiation damping. Thus, a linearized solution to the Rayleigh–Plesset equation in a first-order approximation to the behavior of an oscillating gas bubble in the liquid can be obtained based on the assumption that the internal pressure is given by the polytropic approximation, and the bubble radius can be written as

$$R = R_0(1 + \chi) \qquad (7.33)$$

This approach leads to an analytical expression that closely approximates the threshold, the growth rate of rectified diffusion, and indirect measurement of the variation in the polytropic exponent with the bubble radius.

A model taking the damping by sound radiation into account is the Gilmore model:

$$\left(1 - \frac{\dot{R}}{C}\right)R\ddot{R} + \frac{3}{2}\left(1 - \frac{1}{3}\frac{\dot{R}}{C}\right)\dot{R}^2 = \left(1 + \frac{\dot{R}}{C}\right)H + \left(1 - \frac{\dot{R}}{C}\right)\frac{R}{C}\dot{H} \quad (7.34)$$

where

    C is the sound velocity in the liquid at the bubble interface
    H is the enthalpy evaluated at the bubble wall, given by

$$H = \int_{P_\infty}^{P(R)} \frac{dp}{\rho} \qquad (7.35)$$

7.4.2 Transient Cavitation

A small gas bubble subject to acoustic pressure amplitude of more than a few bars experiences such violent radial pulsations that the wall velocity of the collapsing bubble approaches the speed of sound (Figure 7.3). A transient cavity grows rapidly to some maximum size and then collapses violently without further oscillation, except for a few rebounds.

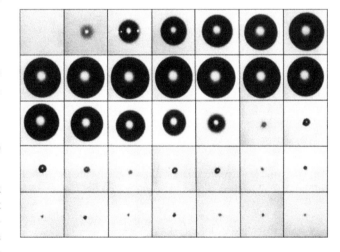

FIGURE 7.3 Free oscillation of a single laser-produced bubble in water. The framing rate is 75,000 frames/s, and the maximum radius is 2 mm. (Reprinted from *Ultrason. Sonochem.*, 14, Lauterborn, W., Kurz, T., Geisler, R., Schanz, D., and Lindau, O., Acoustic cavitation, bubble dynamics and sonoluminescence, 484–491, Copyright 2007, with permission from Elsevier.)

7.4.3 Bubble Dynamics inside the Tissue

If a microbubble oscillates inside the tissue, the boundary and initial conditions will become

$$r = R: \quad p = p_g - \frac{2\sigma}{R} + \tau_{rr}$$

$$r = \infty: \quad p = p_\infty \tag{7.36}$$

$$t = 0: \quad R = R_0, \dot{R} = 0$$

In the near field, bubble compression and expansion dominate, and the surrounding medium is considered incompressible. The pressure distribution is

$$u = -\frac{\dot{R}R^2}{r^2} \tag{7.37}$$

$$p_{in} = p_a - \rho_0 \left(R\ddot{R} + \frac{3}{2}\dot{R}^2 \right) + \frac{\rho_0}{r}(R^2\dot{R})' - \frac{\rho_0}{2}\frac{R^4\dot{R}^2}{r^4}$$

$$+ \tau_{rr}\big|_R^r + 3\int_R^r \frac{\tau_{rr}}{r} dr \tag{7.38}$$

The linear Voigt model can be used to describe the viscoelastic properties of an incompressible material (i.e., soft tissue) to determine the stress

$$3\int_R^\infty \frac{\tau_{rr}}{r} dr = -\left[\frac{4G}{3R^3}\left(R^3 - R_0^3\right) + \frac{4\mu\dot{R}}{R} \right] \tag{7.39}$$

Then, the equation of bubble dynamics can be rewritten as

$$\left(1 - \frac{\dot{R}}{c}\right)R\ddot{R} + \frac{3}{2}\left(1 - \frac{\dot{R}}{3c}\right)\dot{R}^2 = \left(1 + \frac{\dot{R}}{c}\right)\frac{p_a - p_1}{\rho} + \frac{R}{\rho c}\frac{d}{dt}[p_a - p_1] \tag{7.40}$$

where

$$p_a = p_g - \frac{2\sigma}{R} + \tau_{rr}(R,t) \tag{7.41}$$

$$\frac{d}{dt}(p_a - p_1) = \frac{dp_g}{dt} + \frac{2\sigma\dot{R}}{R^2} + P_A\frac{dg(t)}{dt} - 4G\frac{R_0^3\dot{R}}{R} - 4\mu\left(-\frac{\dot{R}^2}{R^2} + \frac{\ddot{R}}{R}\right) \tag{7.42}$$

Applying the polytropic relationship of gas inside the bubble, Equation 7.42 becomes

$$\frac{d}{dt}(p_a - p_1) = \left(\frac{2\sigma}{R} - 3\kappa p_g\right)\frac{\dot{R}}{R} + P_A\frac{dg(t)}{dt} - 4G\frac{R_0^3\dot{R}}{R}$$

$$-4\mu\left(-\frac{\dot{R}^2}{R^2} + \frac{\ddot{R}}{R}\right) \tag{7.43}$$

It is found that the effect of rigidity is much greater than that of surface tension, meaning that a larger bubble will exhibit

a much greater stiffness than a free bubble of equivalent size. The tissue elasticity causes the resonance peaks to shift to bubble radii 2–4 times larger than that in the free field and greatly reduces the amplitude and nonlinearity of the oscillation. However, the effect of elasticity will be less when the driving pressure is strong. The cross sections for bubbles surrounded by tissue are less than for resonant bubbles in water, which results in the difficulty in the acoustic detection of bubbles in tissue.

The inertial cavitation threshold increases as the elasticity increases. There is a V-shaped relationship between the bubble size and inertial threshold. Strong subharmonics occur above the corresponding inertial cavitation threshold when the fundamental frequency component has the maximum emission level, and the gap between them increases with the tissue elasticity. The strong subharmonic signal region becomes smaller when the viscosity or elasticity increases.

7.4.4 Acoustic Microstreaming

If the amplitude of the ultrasonic wave is large, a steady direct current (DC) flow, which is termed as acoustic streaming, will be produced in an acoustic field, which is a second-order phenomenon. If it is small-scale and boundary-associated with a trapped oscillating bubble, it becomes microstreaming. Close to an oscillating microbubble, the DC velocity of microstreaming drops rapidly across the viscous boundary layer with its thickness on the order of $\sqrt{2v/\omega}$, where v is the kinematic shear viscosity coefficient. Because of the viscosity of the medium, particles adhere to the boundary layer. Consequently, the oscillatory (AC) particle velocity drops from the value of the mainstream, v_0, to zero. The thickness of the AC layer is comparable to that of the DC layer. There is significant shear stress due to both the high DC and AC velocity gradients imposed upon the cells in this region. DC stress, rather than AC stress, is the primary mechanism for cell disruption. If both the displacement amplitude ξ_0 and the viscous boundary layer thickness δ are smaller than the bubble radius or the tip of an object, R_0, the maximum DC shear stress is given by

$$S_{max} = \frac{2\pi^{3/2}\xi_0^2(\rho f^3\eta)^{1/2}}{R_0} \tag{7.44}$$

$$\delta = \sqrt{\frac{\eta}{\pi f \rho}} \tag{7.45}$$

where η and ρ are viscosity and density of the medium, respectively.

The streaming fields have symmetric and orderly patterns about a vertical axis through the center of the bubble, which depend on the driving frequency, the liquid viscosity, and the amplitude of the bubble oscillation. Four different regimes of bubble-associated streaming patterns may be distinguished, as shown in Figure 7.4.

Regime I is observed near a surface-contaminated bubble in a liquid of low viscosity and associated with the presence of

FIGURE 7.4 Four different microstreaming patterns near an oscillating hemispherical bubble adhering to a solid boundary. (Reprinted with permission from Elder, S.A., Cavitation microstreaming, *J. Acoust. Soc. Am.*, 31, 54–64, Copyright 1959, American Chemical Society.)

an acoustic boundary layer on the bubble surface. The streaming at low amplitude consists of large and small vortex rings surrounding the bubble. The upper ring rotates in such a way as to move the liquid outward along the vertical axis, while the lower one rotating in the opposite sense is confined into a narrow region near the base of the bubble. Such a pattern is useful in verifying whether a small bubble possesses a surface skin that causes the bubble to collect particles from the liquid. However, the bubble surface remains stationary, indicating the existence of a thin boundary layer region.

Regime II receives the greatest emphasis in the investigation because of its frequent appearance over a wide range of amplitudes and viscosities. A surface mode appears, and the two vortex rings give way to a vigorous outward circulation near the top of the bubble (Figure 7.5). For moderate-viscosity liquids, in general, surface modes of vibration are not as readily excited as in the low-viscosity case. At higher amplitude, the bubble will break loose and disappear. The chaotic mode is not observed at all for kinematic viscosities greater than 0.07 cm²/s.

Regime III is observed most readily in liquids of low viscosity, although it is possible at higher viscosities with sufficiently large driving amplitudes. The bubble is seized with the surface mode vibration, and simultaneously streaming motion is visible near the top of the bubble. Its occurrence coincides with the onset of the first surface mode and the dissolution of Regime II at higher amplitude. The greatest acceleration of the streaming particles occurs very close to the two uppermost antinodes of the surface mode. The surface velocity is

proportional to the incident field amplitude for a given bubble under the assumption of constant damping of the bubble.

Regime IV occurs only in the least viscous solutions and seems to be a return to Regime II streaming as the amplitude becomes too large to permit the existence of a single stable surface mode. When the surface velocity reaches ~0.6 m/s, the stable surface mode dissolves into a chaotic surface agitation, the streaming pattern taking the form of a large vortex ring surrounding the bubble. Further increasing amplitude will only enhance the effect of rectified diffusion. Regimes II and IV can be accounted for by a theory of surface mode–radial mode coupling and are probably the most important for acoustic effects for large driving amplitudes at high and low viscosities, respectively.

Although microstreaming (a second-order nonlinear effect) is not as strong as the collapse force of cavitation (the first-order inertial cavitation flow), it can still lyse an artificial vesicle, which is relevant for therapeutic purposes. Using particle image velocimetry (PIV) and streak imaging technology, the pattern of microstreaming and the velocity vector can be visualized and calculated, as shown in Figure 7.6. The amplitude of bubble oscillation is 3–4 orders greater than the fluid displacement in linear acoustics. Thus, the horizontal translation of the bubble may be due to the secondary Bjerknes forces between the bubble and its virtual images on the surrounding vertical walls. Quadrupole microstreaming patterns (i.e., four vortices) usually occur in a variety of viscous fluids, with an object oscillating in a line. If the bubble motion takes a circular path, the microstreaming pattern is a single vortex

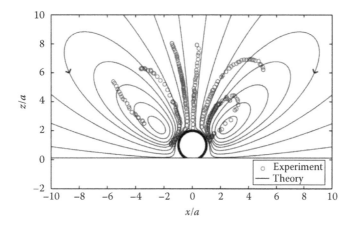

FIGURE 7.5 Comparison of the experimental measurement and theoretical modeling of the streaming flow in Regime II. (Reprinted by permission from Macmillan Publishers Ltd., *Nature* (Marmottant, P. and Hilgenfeldt, S., Controlled vesicle deformation and lysis by single oscillating bubbles, 423, 153–156, Copyright 2003.)

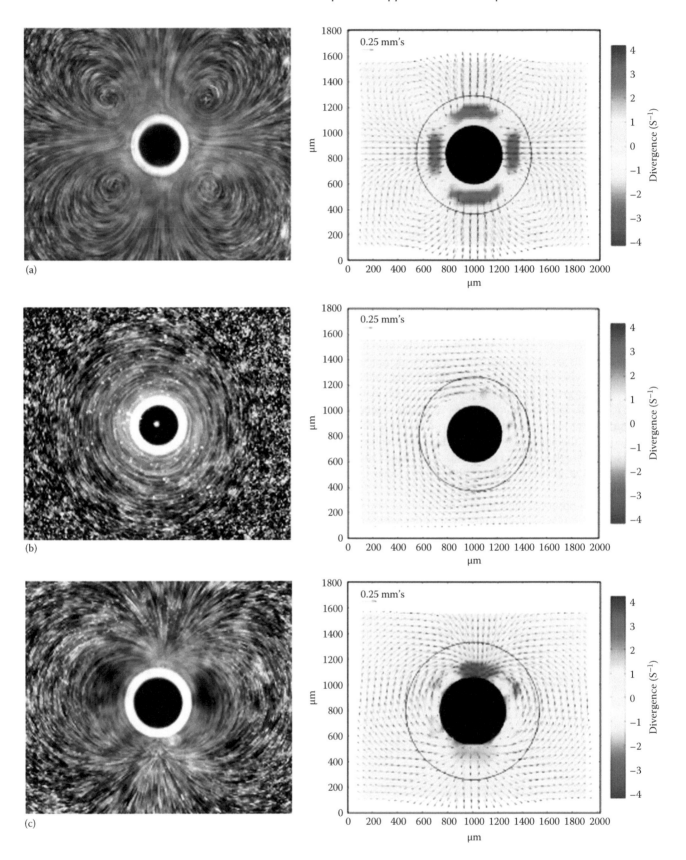

FIGURE 7.6   (a) Quadrupole, (b) circular vortex, and (c) dipole microstreaming patterns created by linear translation, circular translation, and radial oscillation of a ~230 μm radius bubble and forced at 2.42, 1.19, and 8.66 kHz, respectively. Left column is the streak image, and the right column is particle image velocimetry (PIV) velocity vector field and divergence around bubble circumference. (Reprinted from *Ultrasonics*, 50, Collis, J., Manasseh, R., Liovic, P., Tho, P., Ooi, A., Petkovic-Duran, K., and Zhu, Y., Cavitation microstreaming and stress fields created by microbubbles, 273–279, Copyright 2010, with permission from Elsevier.)

centered on the bubble. When bubbles are forced to undergo their natural (volumetric) mode of oscillation, similar to the situation of contrast agent microbubbles being driven by clinical ultrasound, the corresponding microstreaming is in the pattern of a "dipole."

7.4.5 ACOUSTIC RADIATION FORCE

When a nonfocused ultrasound beam passes through a fluid with the sound speed c and impinges on an ideal nonreflecting absorbing object, the object responds as if a steady force (radiation force) were acting on it, given by

$$F_{rad} = \frac{W}{c} \qquad (7.46)$$

where W is the total power absorbed by the object. If the object is a perfect reflector on which the acoustic wave is normally incident, F_{rad} should be doubled. If the incident ultrasound beam is a focused one, Equation 7.46 is replaced by

$$F_{rad} = \frac{W[1 + \cos(\alpha)]}{2c} \qquad (7.47)$$

where α is half the convergence angle.

The acoustic force on a spherical particle in a volume V is expressed by the Gor'kov equation

$$F_x = \frac{V d\Phi}{dx} \qquad (7.48)$$

$$\Phi = D\langle E_k \rangle - (1-\beta)\langle E_p \rangle$$
$$D = \frac{3(\rho_s - \rho_o)}{2\rho_s + \rho_o} \qquad (7.49)$$

where

$\langle E_k \rangle$ and $\langle E_p \rangle$ are the time-averaged densities of the kinetic energy and potential energy, respectively

β is the ratio of the compressibility of the particle to that of the surrounding medium

ρ_s is the particle density

ρ_o is the medium density

In a standing-wave field,

$$F_x = -VEk\Phi \sin(2kx) \qquad (7.50)$$

$$\Phi = \frac{3(\rho_s - \rho_o)}{2\rho_s + \rho_o} + \frac{\beta_o - \beta_s}{\beta_o} \qquad (7.51)$$

where

E is the acoustic energy density

k is the wave number

The lateral primary radiation force can be expressed as

$$F_L = V\nabla E_{ac}\left[\frac{3(\rho_s - \rho_o)}{2\rho_s + \rho_o}\cos^2(kx) - \frac{\beta_o - \beta_s}{\beta_o}\sin^2(kx)\right] \qquad (7.52)$$

The secondary acoustic force F_s will be significant if interparticle distances are short, because of the interaction of acoustic waves scattered from the particles. For two spherical particles of identical radius, F_s is expressed as

$$F_s = 4\pi a^6\left[\frac{(\rho_s - \rho_o)^2(3\cos^2\theta - 1)}{6\rho_m d^4}v^2(x) - \frac{\omega^2\rho_o(\beta_o - \beta_s)^2}{9d^2}p^2(x)\right] \qquad (7.53)$$

where

v is the velocity

p is the acoustic pressure

ω is the angular frequency

a is the radius of the particle

d is the interparticle distance

θ is the angle between the axis of the incident wave and the central line connecting these two particles

These two terms are dependent on the density and compressibility difference, respectively, and both of them increase with shorter interparticle distances. The compressibility-based term is an attractive force that is most relevant to intermediate interparticle distances. In comparison, the density-associated term is angle dependent, and it contributes to the attractive force perpendicular to the direction of the standing wave and to the repulsive one for particles aligned to the standing wave. It dominates for very small particles or at very short center–center distances.

For most microparticles in water ($D > 1$, $\beta < 1$), F_x is the time-averaged value of the product of the particle's volume and the local pressure and the local pressure gradient in the direction of increasing $\langle E_k \rangle$:

$$F_x = \left\langle -V\frac{\partial p}{\partial x}\right\rangle \qquad (7.54)$$

And, the general expression of the primary Bjerknes force is given by

$$F_p = -\langle V(t) \cdot \nabla p(r, t)\rangle \qquad (7.55)$$

Thus, small particles will be collected at pressure maxima, while large ones will go to pressure minima, which has been used to concentrate erythrocytes, DNA, and hybrodoma cells under ultrasound exposure (Figure 7.7). If the gas bubble is less than resonance size, it will be attracted toward the antinode; if greater, it will be forced away from the antinode toward a node. The magnitude of the secondary or mutual Bjerknes force between two bubbles of mean radii $\bar{R}_1$ and $\bar{R}_2$, which are pulsating in the vicinity of each other and separated by some distance r, is given by

$$F_s = -\frac{2\pi\rho\omega^2\left(\delta_1\bar{R}_1^2\right)\left(\delta_2\bar{R}_2^2\right)\cos\theta}{r^2} \qquad (7.56)$$

If the pulsations are in phase, the force is attractive, and if they are out of phase, it is repulsive. Small bubble coalescence is greatly aided by the secondary Bjerknes force (Figure 7.8).

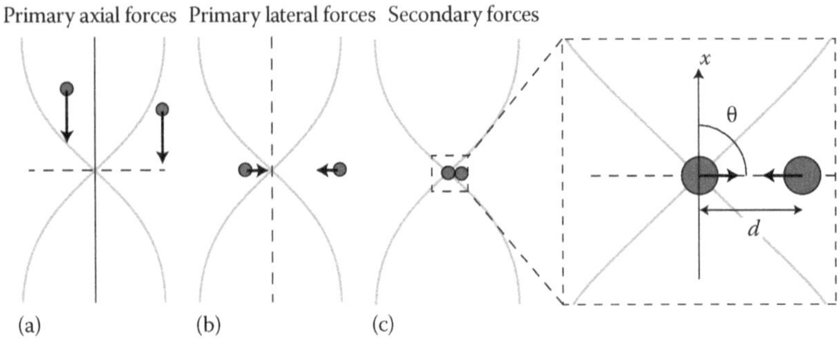

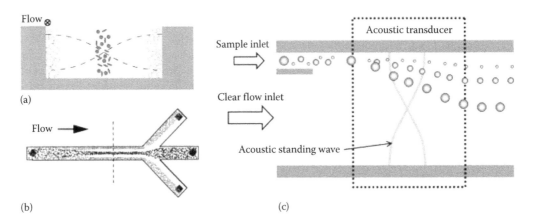

FIGURE 7.7 Schematic diagram of the mechanisms of particle capture and aggregation in an acoustic field. (a) Particle levitation to the nodal plane of a standing wave by primary axial radiation forces. (b) Retainment of the particles by lateral radiation forces due to the 3D gradient of the acoustic field. (c) Aggregation by the secondary acoustic forces if the interparticle distance is sufficiently small. (Hammarström, B., Laurell, T., and Nilsson, J., Seed particle-enabled acoustic trapping of bacteria and nanoparticles in continuous flow systems, *Lab. Chip*, 12, 4296–4304, 2012. Reproduced by permission of The Royal Society of Chemistry.)

FIGURE 7.8 Implementation of acoustic forces to separate particles. (a) Two types of particles are trapped in nodal and antinodal planes of a standing wave, respectively. (b) Top view of a continuous separation. (c) Free-flow acoustophoresis. (Reprinted from *Mech. Res. Commun.*, 36, Tsutsui, H. and Ho, C.-M., Cell separation by non-inertial force fields in microfluidic system, 92–103, Copyright 2009, with permission from Elsevier.)

7.4.6 Dynamics of Nonspherical Bubbles

If the spherical symmetry of bubble cannot be maintained, the corresponding bubble dynamics becomes exceedingly complex in view of both theoretical and experimental analyses because of deviations from the spherical model and the multitude of factors that promote such deviations (Figure 7.9). The specification of a nonspherical bubble shape can be expressed as an expansion in terms of spherical harmonics $Y_n^m(\theta, \phi)$:

$$S(r, \theta, \phi, t) = r - R(t) - \sum_{n,m} a_{nm}(t) Y_n^m(\theta, \phi) \quad (7.57)$$

where
    $S(r, \theta, \phi, t) = 0$ means the bubble surface
    $R(t)$ is the instantaneous radius
    $a_{nm}(t)$ is the spherical harmonic component of order n and degree m

It is to be noted that expansion of the spherical harmonics may not be the only choice and the most convenient for all practical problems.

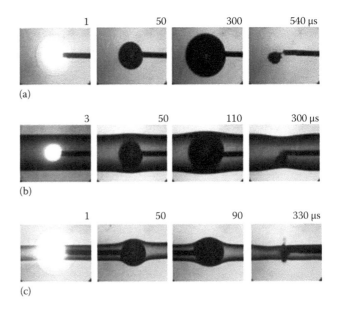

FIGURE 7.9 Bubble dynamics (a) in water, (b) inside a 3.5 mm silicon tube, and (c) inside a 1.5 mm silicon tube. (Courtesy of S.L. Zhu.)

In the small-amplitude approximation of $|a_{nm}|/R \ll 1$, the motion equation of an incompressible, inviscid, unbounded fluid with a free surface for a_{nm} that does not contain the index m is found to be

$$\ddot{a}_n + 3\left(\frac{\dot{R}}{R}\right)\dot{a}_n + (n-1)\left[\frac{(n+1)(n+2)\sigma}{\rho R^3} - \frac{\ddot{R}}{R}\right]a_n = 0 \quad (7.58)$$

$n = 1$ corresponds to a translation of the bubble center, whereupon this equation is simplified as $R^3 \dot{a}_1 = $ const. It expresses the constancy of the total liquid momentum, which is proportional to the product of bubble volume and the translational velocity. If all the others vanish, it happens to be an exact result irrespective of the small-amplitude approximation. However, for most cases, $n \geq 2$. The introduction of viscous effects complicates the matter substantially, and an integral–differential structure is derived except in the case where the viscous diffusion length is comparable to the bubble radius:

$$\ddot{a}_n + [3\dot{R}/R + 2(n+2)(2n+1)\mu/R^2]\dot{a}_n$$
$$+ (n-1)[(n+1)(n+2)\sigma/\rho R^3 + 2(n+2)\mu\dot{R}/R^3 - \ddot{R}/R]a_n = 0 \quad (7.59)$$

The inherent dynamic instability of contracting bubbles, the proximity of solid boundaries or of free surfaces, buoyancy effects in bubble collapse or coalescence, and cavitation damage are dominated by these effects, which also increase the heat- and mass-transfer rates.

7.4.7 STABILITY OF SPHERICAL GROWTH AND COLLAPSE

Even with the neglect of the surface tension term, which has a significant stabilizing effect only for very small bubbles, the qualitative behavior is not understood quite well. Indeed, the stability characteristics depend not only on the acceleration of the interface, as in the plane case of the Rayleigh–Taylor instability, but also on its velocity. It is clearly a consequence of the geometry, since the divergence of the streamlines during bubble growth has a stabilizing effect, while the reverse occurs during the collapse. In the case of very large acceleration but small velocity (i.e., at the early stage of bubble explosion), instability of the spherical shape predicted theoretically is indeed known to exist. A similar situation occurs in the early growth stage of bubble cavitation or boiling bubble in the liquid or soft tissue, in which the duration of large acceleration, but small velocity, is so short that there is no sufficient time to develop the instability.

The quantitative investigation of the stability for a bubble expanding or collapsing under a fixed pressure difference $P = p_i - p_\infty$ is available from the closed-form solutions of the equation above. For the bubble growth, $\dot{R}^2 \cong 2P/3\rho$, from which it is seen that $a_n \to$ const as $R \to \infty$. Although an unstable behavior is expected on the basis of the plane Rayleigh–Taylor case, no significant deviations from the spherical shape occur in practice because the large acceleration takes place only over a small portion and its destabilizing influence is

effectively counterbalanced by the stretching of the bubble surface due to the divergence of the streamlines. In addition, high growth rates for instabilities of larger-order n do not happen because of viscous effects. In the collapse of a cavitation bubble, $b_n = (R_i/R)^{3/2}a_n$, so Equation 7.38 takes the form

$$\dot{b}_n - \left[(3/4)(\dot{R}/R)^2 + (n+\tfrac{1}{2})\ddot{R}/R - (n-1)(n+1)(n+2)\sigma/\rho R^3\right]b_n = 0 \quad (7.60)$$

Since during the collapse $\dot{R}^2 \cong -(2/3)(P/\rho)(R_i/R)^3$, $b_n = -nc^2 R^{-5}$, where c is a constant.

7.4.8 COLLAPSE CLOSE TO A RIGID WALL

In principle, deviations from a spherical bubble in an unbounded liquid can occur only through the amplification of preexisting small perturbations. However, such initial perturbation is unnecessary for bubble deformation near a boundary, because the asymmetric flow induced by the boundary is large enough to generate highly distorted bubbles. With the neglect of viscosity, the virtual mirror approach, in which an imaginary bubble identical to the real one is located symmetrically with respect to the boundary, would provide an insight into these effects. It is found that the portion of the bubble farther from the wall acquires a greater velocity than the one near the wall, because the flow induced by the imaginary bubble adds to the collapse velocity. This asymmetry leads to a high-velocity jet toward the wall, while the sink-like flow also attracts the bubble toward the wall (Figure 7.10).

It is well known that when cavitation bubbles are permitted to collapse near a boundary, either soft or hard, the collapse is asymmetric and instabilities develop that grow without bound. Note that for an asymmetrical collapse, portion of the host liquid is delivered to the center of the bubble. Because the liquid is an immense heat reservoir, a typical temperature profile within the bubble indicates that, although the temperature at the center of the bubble may be several thousands of degrees, the temperature near the bubble wall must be near that of the liquid. Meanwhile, a liquid jet develops and penetrates into the interior of the gas bubble where the temperature is elevated. In this case, small droplets of liquid can be deposited within the interior of the bubble, which would then be heated much more effectively during a subsequent collapse than if the liquid were near the bubble interface. Thus, when an asymmetrical bubble collapse occurs, the liquid can be elevated to high temperatures and spectral characteristics of the liquid rather than the gas can be observed.

7.4.9 SONOLUMINESCENCE

Sonoluminescence (SL) can occur in a single, stable, pulsating gas bubble and be seen by the naked eye as a faint glow that is distributed throughout the bulk liquid (Figure 7.11). SL was first observed in 1934 by two German scientists at the University of Cologne during experiments with SONAR. But the observation was nearly overlooked and buried until 1989, when two American scientists were able to re-create the

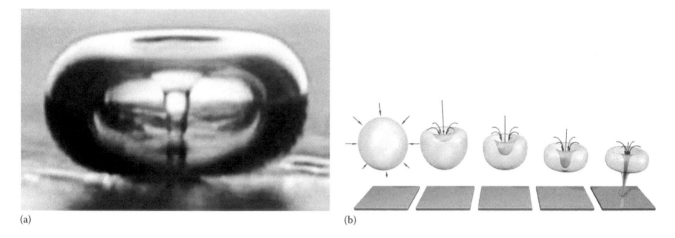

(a)   (b)

FIGURE 7.10   Micro-jet creation through collapsing cavitation bubbles near a solid boundary (a) by high-speed photography and (b) schematic illustration.

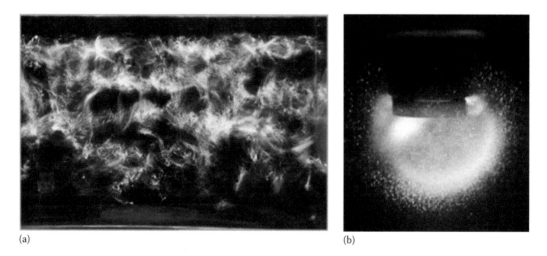

(a)   (b)

FIGURE 7.11   Sonoluminescence images produced by the ultrasonic horn taken at (a) the Physikalisches Institut III of Goettingen University and (b) at University of Illinois at Urbana-Champaign.

phenomenon through what later would become known as the single-bubble experiment. The SL spectrum of single bubble is devoid of the major peaks and can be fitted only by that of blackbody with temperatures as high as 40,000 K (surface of sun is ~7000 K) because of the symmetry of the collapse. A shock wave developed within the gas is also responsible for the enormous temperatures achieved. In contrast, there are several differences in the SL of bubble clouds. Because bubble clouds consist of individual oscillating bubbles, cavitation jet velocities and shock wave intensities in excess of those of individual bubbles can be produced as a result of cooperative interaction between the bubbles. First, the cavitation bubbles tend to collapse asymmetrically, thus introducing liquid into the interior of the bubble, which is heated by adiabatic compression. Second, the corresponding SL spectrum is dominated by the characteristics of the liquid rather than the gas. Although measurements of the temporal duration of the SL flashes by a single bubble are difficult because of its transient nature (<50 ps), the lifetime of bubble clouds can be several nanoseconds. Third, the duration of bubble-cloud SL is much longer, and the effects of heat conduction should be considered. Finally, because the symmetry of the collapse is destroyed, the final temperatures achieved in this case are relatively low.

Whenever bubble cavitation occurs in the vicinity of a surface, either hard or soft, the asymmetric hydrodynamic flow field has a preferential geometry to develop a jet. Consequently, SL is also produced from the vortex ring as a topological residue of the asymmetric collapse, which can result in free-radical production close to the solid surface, a rich region for sonochemical reactions.

7.4.10   Dynamics of Shelled Bubble

Microbubbles have been used in medical ultrasound diagnostics as a contrast agent because of their significant low acoustic impedance in comparison to the biological tissue for almost perfect reflection. In practice, a free air bubble lacks stability and has a limited lifetime in the circulation system. In order to overcome this shortcoming (high gas diffusion), a solid shell made from a lipid (i.e., serum albumin) or a protein or a polymer with low gas permeability is used to encapsulate the inertial and insoluble gas (i.e., perfluoropropane), as shown in Figure 7.12.

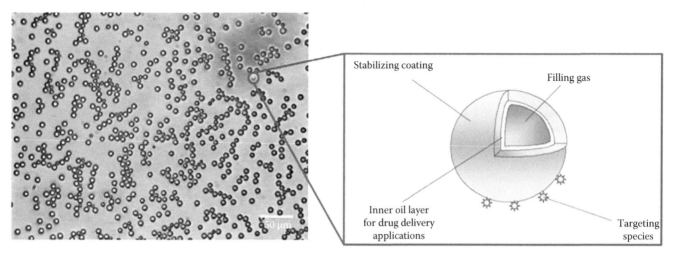

FIGURE 7.12 Photo and structure of an ultrasound contrast agent.

These shelled microbubbles are also used in ultrasound therapy to enhance the cavitation effects. The inclusion of shell adds complexity in the modeling of bubble dynamics.

7.4.11 Newtonian Interfacial Rheological Model

The extra stress on the surface and the abrupt change in the bulk viscous stress across the interface can be described as

$$\tau_s = \sigma I_s + (\kappa^s - \mu^s)(I_s : D_s)I_s + 2\mu^s D_s$$
$$[\tau \cdot n]_{\text{surface}} = \nabla_s \cdot \tau_s \qquad (7.61)$$

where

κ^s and μ^s are the interfacial dilatational and shear viscosities, respectively

I_s and D_s are the surface identity and surface strain rate tensors, respectively

Adsorption of a small amount of surface-active materials leads to a reduction in surface tension.

7.4.12 Viscoelastic Interfacial Rheological Model

Dilatational surface elasticity or Gibb's elasticity can also be treated as an effect due to surface tension gradients. Hence, the dilatational elasticity E^s is introduced as

$$E^s = \left(\frac{\partial \sigma}{\partial \alpha}\right)_{\beta=0}, \qquad \sigma = \sigma_0 + E^s\beta \qquad (7.62)$$

where $\beta = \delta A/A = \left[(R/R_E)^2 - 1\right]$ is the fractional change in the area from the equilibrium condition. Note that in Newtonian model, the dynamics does not have such a reference unstrained state. Because dilatational elasticity can be considered as a surface tension gradient effect, such elastic behavior does not necessarily violate the Newtonian model of interfacial rheological behavior. So, the dynamic boundary condition becomes

$$p_{r=R} = p_g - 4\mu\frac{\dot{R}}{R} - \frac{4\kappa^s\dot{R}}{R^2} - \frac{2\sigma_0}{R} - \frac{2E^s}{R}\left[\left(\frac{R}{R_E}\right)^2 - 1\right] \qquad (7.63)$$

At the initial zero-motion state, the interior pressure is

$$p_{g0} = p_0 + \frac{2\sigma_0}{R_0} + \frac{2E^s}{R_0}\left[\left(\frac{R_0}{R_E}\right)^2 - 1\right] \qquad (7.64)$$

In order to ensure the stability of microbubbles, the pressure inside and outside the bubble should be equal, that is, $p_{g0} = p_0$. So, the equilibrium radius is

$$R_E = R_0\left(1 - \frac{\sigma_0}{E^s}\right)^{-1/2} \qquad (7.65)$$

The interior bubble pressure is assumed to vary with bubble volume polytropically, as

$$p_g R^{3k} = p_{g0}R_0^{3k} \qquad (7.66)$$

For the viscoelastic interfacial rheology,

$$\rho\left(R\ddot{R} + \frac{3}{2}\dot{R}^2\right) = p_{g0}\left(\frac{R_0}{R}\right)^{3k} - 4\mu\frac{\dot{R}}{R} - \frac{4\kappa^s\dot{R}}{R^2}$$
$$- \frac{2\sigma_0}{R} - \frac{2E^s}{R}\left[\left(\frac{R}{R_E}\right)^2 - 1\right] - p_0 + p_A(t) \qquad (7.67)$$

For Newtonian rheology,

$$\rho\left(R\ddot{R} + \frac{3}{2}\dot{R}^2\right) = p_{g0}\left(\frac{R_0}{R}\right)^{3k} - 4\mu\frac{\dot{R}}{R} - \frac{4\kappa^s\dot{R}}{R^2} - \frac{2\sigma}{R} - p_0 + p_A(t) \qquad (7.68)$$

With the assumption of liquid incompressibility, the acoustic pressure scattered by a bubble is

$$p_s(r,t) = \rho\frac{R}{r}(2\dot{R}^2 + R\ddot{R}) \qquad (7.69)$$

The corresponding scattering cross section is

$$\sigma_s(t) = \frac{\langle r^2 p_s(t)^2 \rangle}{p_A^2} \tag{7.70}$$

The damping term has an additional interface term compared to the free bubble case:

$$\delta_{total} = \delta_{liquid} + \delta_{interface} + \delta_{radiation}$$

$$\delta_{liquid} = \frac{4\mu}{\rho\omega_0 R_0^2}, \quad \delta_{interface} = \frac{4\kappa^s}{\rho\omega_0 R_0^3}, \quad \delta_{radiation} = \frac{\omega^2 R_0}{\omega_0 c} \tag{7.71}$$

The extinction cross section for the linearized dynamics is

$$\sigma_e^{(l)} = 4\pi R_0^2 \frac{c\delta_{total}}{\omega_0 R_0} \frac{\Omega}{\left[(1-\Omega^2)^2 + \Omega^2 \delta_{total}^2\right]} \tag{7.72}$$

where $\Omega = \omega/\omega_0$. The absorption and scattering by microbubble in the frequency domain result in attenuation as

$$\alpha(\omega) = 10\log_{10} e \int_{a_{min}}^{a_{max}} \sigma_e(a;\omega)n(a)da \tag{7.73}$$

where

    $n(a)da$ is the number of bubbles per unit volume with radius in $(a, a + da)$

    $a_{max(min)}$ is the maximum (minimum) bubble radius

The Newtonian model performs best for collapse and predicts subharmonic response very well for most frequencies even with a high surface tension term. At high excitations with large excursion of the bubble surface, surface dilatational elasticity of the encapsulation will become quite low and is not included in the model. In contrast, the viscoelastic model consistently underpredicts the subharmonic response. The high value of surface dilatational elasticity, linearly interpolated from the low amplitude attenuation data, makes a stiff encapsulation for a lower response at high power activation.

Oscillation of contrast agent microbubbles by ultrasound produces large variations in the surface area. When the surface area is reduced to that covered by the monolayer lipid, the effective surface tension decreases sharply. Further compression leads to the occurrence of buckling and the vanishing of the surface tension. In contrast, surface tension increases when a slow expansion separates these molecules. Three parameters can be used to describe the surface tension: the buckling area of the bubble, $A_{buckling}$; an elastic modulus, χ; and a critical break-up tension for the elastic regime, $\sigma_{break-up}$. The buckling area of the bubble depends on the number of lipid molecules n at the interface and the molecular area at buckling, $a_{buckling}$ (e.g., 0.4 nm^2 for phospholipid molecules

normal to the interface), $A_{buckling} = na_{buckling}$. The tension of the monolayer model is

$$\sigma(R) = \begin{cases} 0 & R \le R_{buckling} \\ \chi\left(\frac{R^2}{R_{buckling}^2} - 1\right) & R_{buckling} \le R \le R_{break-up} \\ \sigma_{water} & R \ge R_{break-up} \end{cases} \tag{7.74}$$

The compression modulus K_V of the bubble is given by

$$K_V = \begin{cases} \kappa P & \text{buckled} \\ \kappa P + \frac{4}{3}\frac{\chi}{R} & \text{elastic} \\ \kappa P + \frac{3\kappa-1}{3}\frac{2\sigma_{water}}{R} & \text{free/broken} \end{cases} \tag{7.75}$$

where V is the bubble volume in the limit of $\chi \gg \sigma_{water}$ for phospholipids. The compression modulus is much higher when the bubble is in the elastic state. So, the eigenfrequency is

$$\omega_0^2 = \frac{3}{\rho_l R_0^2} K_V \tag{7.76}$$

Thus, elastic bubbles have a much higher resonance frequency than free or buckled ones. At the elastic state, $\sigma = \chi(A/A_{buckling}-1)$. The shell is in the elastic state only within a narrow area: $A_{break-up} = A_{buckling}\sqrt{1 + \sigma_{break-up}/\chi}$ before rupture of the shell and $A_{ruptured} = A_{buckling}\sqrt{1 + \sigma_{water}/\chi}$ after rupture.

Within this regime, the surface tension is a linear function of the area, or of the square of the radius, with small variations around the equilibrium radius R_0:

$$\sigma(R) = \sigma(R_0) + \chi\left(\frac{R^2}{R_0^2} - 1\right) \approx \sigma(R_0) + 2\chi\left(\frac{R}{R_0} - 1\right), \quad |R-R_0| \ll R_0 \tag{7.77}$$

During the oscillation, the balance of normal stresses at the interface is

$$p_g(t) - p_l(t) = \frac{2\sigma(R)}{R} + 4\mu\frac{\dot{R}}{R} + 4\kappa_s\frac{\dot{R}}{R^2} \tag{7.78}$$

where

    p_g is the gas pressure in the bubble

    p_l is the liquid pressure

    μ is the liquid viscosity

    $\kappa_s = 3\epsilon\mu_{lipid}$ is the surface dilatational viscosity from the monolayer, with μ_{lipid} the bulk lipid viscosity

Combining the Raleigh–Plesset equation and the polytropic gas law with boundary conditions, the bubble dynamics becomes

$$p_l\left(R\ddot{R} + \frac{3}{2}\dot{R}^2\right) = \left[p_0 + \frac{2\sigma(R_0)}{R_0}\right]\left(\frac{R}{R_0}\right)^{-3\kappa}\left(1 - \frac{3\kappa}{c}\dot{R}\right)$$

$$- p_0 - \frac{2\sigma(R)}{R} - 4\mu\frac{\dot{R}}{R} - 4\kappa_s\frac{\dot{R}}{R^2} - p_{ac}(t) \tag{7.79}$$

If the shell is assumed to be an elastic solid with damping, its stress $T_{s,rr}$ consists of two parts:

$$T_{s,rr} = (\lambda_s + 2G_s)\frac{\partial \varepsilon_{rr}}{\partial r} + 2\lambda_s \frac{\varepsilon_{rr}}{r} \qquad (7.80)$$

where λ_s and G_s are the Lamé constants, and ε_{rr} is the radial strain given by

$$\varepsilon_{rr} = \frac{R_1^2}{r^2}(R_1 - R_{e1}) \qquad (7.81)$$

where R_{e1} is the unstrained equilibrium position of interface 1. Then,

$$T_{s,rr} = -4\frac{R_1^2}{r^3}\left[G_s(R_1 - R_{e1}) + \mu_s U_1 \right] \qquad (7.82)$$

Therefore, the Rayleigh–Plesset-like equation for the dynamics of encapsulated bubbles of damped elastic solids is

$$R_1 U_1 \left[1 + \left(\frac{\rho_l - \rho_s}{\rho_s} \right)\frac{R_1}{R_2} \right] + U_1^2 \left[\frac{3}{2} + \left(\frac{\rho_l - \rho_s}{\rho_s} \right)\left(\frac{4R_2^3 - R_1^3}{2R_2^3} \right)\frac{R_1}{R_2} \right]$$

$$= \frac{1}{\rho_s}\left[\begin{array}{c} p_{g,eq}\left(\dfrac{R_{01}}{R_1} \right)^{3\kappa} - p_\infty(t) - \dfrac{2\sigma_1}{R_1} - \dfrac{2\sigma_2}{R_2} \\[2ex] -r\dfrac{U_1}{R_1}\left(\dfrac{V_s \mu_s + R_1^3 \mu_l}{R_2^3} \right) - 4\dfrac{V_s G_s}{R_2^3}\left(1 - \dfrac{R_{e1}}{R_1} \right) \end{array} \right] \qquad (7.83)$$

where $V_s = R_{02}^3 - R_{01}^3$. It is noted that the acceleration of interface 1 is increased by a factor proportional to the difference between the densities of shell and liquid. If $\rho_s > \rho_l$, the acceleration is reduced, and if $\rho_s < \rho_l$, it is increased.

The viscosity of the shell produces a remarkable decrease in both the linear and nonlinear pulsation amplitudes of encapsulated bubbles. In comparison, the effects of the density and surface tension of the shell are relatively minor. As the size of the bubble decreases, the effects of the shell become more significant. The effective absence of surface tension means that the pressure found within an encapsulated bubble is less than that in a free bubble. Only when the insonating acoustic frequency is much higher than the resonance frequency of a bubble do free and encapsulated bubbles behave in a similar manner. The increased stiffness by the shells reduces the change in the speed of encapsulated bubbles, even in the resonance range. At well below the resonant frequency, the speed of sound decreases with the shell rigidity and the shell thickness, but it increases with the volume fraction. Free bubbles produce the largest values of attenuation. The shell rigidity predominently determines the maximum attenuation in the frequency domain. The shape of the attenuation response is more dependent on shell viscosity than rigidity.

7.5 CAVITATION EFFECTS

The violent motion of cavities in an acoustical field produces a variety of unique phenomena, such as the peculiar noise emission, the destructive action on all kinds of material from soft tissue to hard steel, enhancement of chemical reaction, and therapeutic effects in the medicine.

7.5.1 MATERIAL DESTRUCTION

Destructive action, such as due to the high pressures and temperatures in the bubble, generation of the shock waves and liquid jet in the final stage of bubble collapse, and bubble-formed-pits on a solid surface, is unwanted and may lead to failure of the parts due to progressive removal of material by cavitation, such as the erosion on propellers in Figure 7.1. The damaging process is very involved and of a statistical nature even under highly controlled conditions because of the inhomogeneity of the cavitation nuclei. It seems that only a bubble collapsing in contact with the boundary leads to pit formation. Usually, aluminum, bronze, and titanium are proven to be quite cavitation resistant.

However, in a well-controlled form, cavitation has been utilized for material cleaning, such as removing grinding material from lenses. Bubble cavitation can generate vigorous shear forces by the oscillating bubbles, which can be used to remove surface contamination. Bubble-associated turbulence can cause destruction of paramecia (i.e., through simple mechanical buffeting) and is one of the important agents in ultrasonic cleaning. It is found that cavitation microstreaming is most pronounced for bubbles undergoing volume resonance, situated on solid boundaries, and even at low sound amplitudes.

7.5.2 SOUND EMISSION

Measurements of the noise emission have been done with subsequent application of the new methods from nonlinear dynamics to the sound output from the liquid. These are phase-space analysis, dimensional analysis, and Lyapunov analysis. A period-doubling route to chaos has been found. After a cascade of period doublings, a chaotic noise attractor appears. It is possible to determine the dimension of the noise attractors. Surprisingly small fractal dimensions between 2 and 3 are found. The appearance of fractal attractors as well as period-doubling sequences suggests that the system of oscillating bubbles in an acoustic field is a chaotic system. The definition of a chaotic system is that at least one of the Lyapunov exponents of the system making up the Lyapunov spectrum should be positive. A Lyapunov exponent is a measure of how fast two neighboring trajectories in phase-space separate. Thus, acoustic cavitation noise has been proven to be a chaotic system with only a small number of nonlinear degrees of freedom. According to the fractal dimension, only three variables should be sufficient to describe the dynamics of the system. This finding suggests that a high degree of cooperation must occur among the bubbles. The highly structured bubble ensemble confirms this viewpoint.

7.5.3 CHEMICAL REACTION

At the collapse stage, temperature inside a cavity can reach 10,000°C and pressures as high as 10,000 bars. Under such an environment, certain chemical effects can occur that may be of major importance, which are termed sonochemistry. It is found that cavitation can not only affect the rate of chemical reactions but also induce reactions that would not occur in the absence of cavitation. Sonochemistry would occur only within the vapor phase because both the thermal capacity and conductivity of a liquid are larger than those of most gases, and the total energy existing within a collapsed cavitation bubble is relatively high but low in absolute terms (measured in units of MeV, rather than joules). Accordingly, liquid temperature rises only in the tens or hundreds of kelvins are expected. In contrast, for microjets and droplets that are injected into the interior of a collapsed bubble, the temperature rise can be fairly large. Suppose a transient-cavitation bubble collapse can generate 300 MeV of total thermal energy. If 30 water droplets with the radius of 1.0 μm are contained within the collapsed bubble, and only 10% of this energy is used to raise the temperature of these droplets to superheated vapor, a temperature increase of 10,000 K is estimated by using Joules' law.

7.5.4 PHYSICAL EFFECTS

The Bjerknes forces exerted on gas bubbles in a liquid can remove bubbles from glass melts or effect liquid degassing. The cell disruption or formation of small vesicles by cavitation has become popular.

7.5.5 BIOEFFECTS

As cavitation is connected with high pressures and temperatures, with the possibility of damage, it normally has to be avoided for reasons of safety. Therefore, quite a number of experiments have been conducted to learn about the damage potential to biological material (tissue, blood cells, etc.). It has been found that, in the presence of gas bubbles, the damage potential of ultrasonic waves becomes higher. Even small (10 mW/cm²) peak intensities of pulsed ultrasound with pulse lengths of a few microseconds only are sufficient to induce damage. Unless the duty cycle is also low, however, damage may nevertheless occur, and even in one acoustic cycle, preexisting cavities may be set into violent motion if they meet the appropriate conditions.

7.6 CAVITATION MEASUREMENT

7.6.1 HIGH-SPEED IMAGING

Bubble dynamics can be characterized by using a high-speed shadowgraph imaging system. A pulsed Nd:YAG laser (pulse duration of only several nanoseconds) is expanded by a concave lens and collimated using a Schlieren mirror to form a parallel light beam through the test chamber. The image is projected through a combination of lenses and mirrors onto a charge-coupled device (CCD) camera, and an appropriate magnification is selected depending on the maximum bubble size. By adjusting the delay time of the trigger signals, a series of high-speed shadowgraph images can be recorded at various stages of the bubble oscillation under the assumption that bubble cavitation is consistent and repeatable. Then, a representative sequence of the whole event can be composed (Figure 7.13). Although the experimental setup is easy, the reconstruction of the whole process is time consuming and some important phenomena may be missed if the bubble lifespan is long and the time interval of imaging is short. Because of the stochastic nature of cavitation, several repetitions may be required for the whole process.

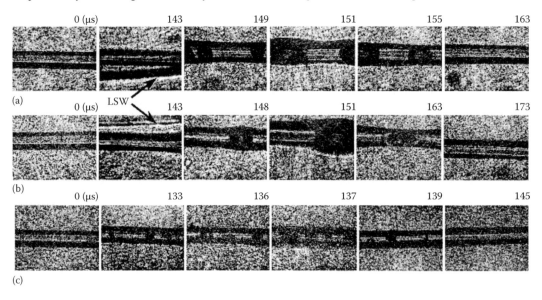

FIGURE 7.13 Representative sequences of bubble dynamics in the 200 mm hollow fiber produced by (a) standard lithotripter shock wave (LSW) at 16 kV, (b) LSW at 24 kV, and (c) the inverted LSW (ILSW) at 24 kV. The number above each image frame is the time delay in milliseconds after the spike discharge. (Reprinted from *Ultrasound Med. Biol.*, 27, Zhong, P., Zhou, Y., and Zhu, S., Dynamics of bubble oscillation in constrained media and mechanisms of vessel rupture in SWL, 119–134, Copyright 2001, with permission from Elsevier.)

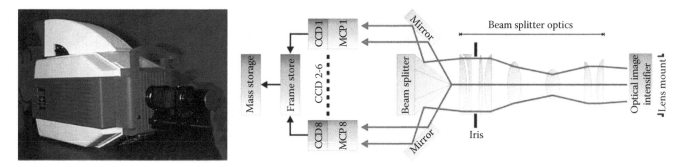

FIGURE 7.14 Photo and schematic structure of the Imacon 468 high-speed camera.

Multiple images captured in a single cavitation process are preferred. A commercially available high-speed digital camera is Imacon 468 (DRS Hadland, Inc., Cupertino, CA), which has eight individual CCD cameras (Figure 7.14). The image is collected by a common optical system and is then split into eight identical copies, and these are relayed to the individual cameras, which are then exposed in series. The camera has a capacity of 100 million frames/s and can generate up to 16 images (two images from each CCD camera) with variable shutter duration (10 ns to 1 ms) (Brujan et al. 2001, Dayton et al. 2001, Pishchalnikov et al. 2003, Sankin et al. 2005, Chen et al. 2011). The other is the Brandaris 128 developed at Thoraxcentre, Erasmus, MC, the Netherlands (Figure 7.15). It has 128 digital frames with 500×292 pixels in each frame with a speed of up to 25 million frames/s (40 ns interframe time) (Chin et al. 2003, Van der Meer et al. 2004, Bouakaz et al. 2005, Marmottant et al. 2006, De Jong et al. 2007, Garbin et al. 2007). A customized rotating mirror camera frame is used to project the image to the appropriate CCD. Full sequences can be repeated every 20 ms, and six full sequences can be stored in the memory buffer. However, these two cameras are quite expensive and are available only in a few laboratories in the world.

FIGURE 7.15 Photo of the Brandaris 128 high-speed camera.

Since the change in the bubble wall is usually symmetric, full imaging of a single bubble may not provide more information but only limited frame rate. Scattered laser light from the bubble is focused through one of the quartz windows onto the entrance slit of the streak camera (Figure 7.16). When the vertical line is positioned over the center of the bubble, this "streak" image represents the diameter of the bubble as a function of time, from which a radius–time curve of the bubble can be determined. The temporal and spatial resolutions of the streak image can be ~400 ps and 0.12 μm, respectively, which are limited mainly by the aberrations of the optical system, jitter of the trigger signal, and spatial instabilities of the bubble itself. The maximum error is estimated to be ~2 pixels due to the threshold in determining the bubble's boundary (Morgan et al. 2000, Pecha and Gompf 2000).

7.6.2 Acoustic Emission

Acoustic emission (AE) associated with shock-wave-induced bubble oscillation can be detected using a passive cavitation detection (PCD) system, such as a 1 MHz focused transducer that is aligned orthogonal to the lithotripter axis and confocally with the transducer (Figure 7.17). AE produced by the rapid oscillation of cavitation bubbles in a lithotripter field has a characteristic double-burst structure in each trace, with the first burst corresponding to the initial compression and ensuing expansion of cavitation nuclei by the incident lithotripter shock wave (LSW) and the second burst corresponding to the subsequent inertial collapse of the bubbles. The collapse time of the bubble cluster (*tc*) is conveniently defined as the time delay between the peak pressure of the first and second AE bursts (Church 1989, Coleman et al. 1992, 1996, Zhong et al. 1997, 1999). A dual PCD uses two perpendicular confocal transducers so that the effective focal dimension is proportional to the width, not the length, of each transducer. A coincidence detection algorithm, employing cross-correlation of two signals, can identify cavitation events originating from the effective focal region only (Cleveland et al. 2000, Bailey et al. 2005).

The sound emitted by the cavitation bubble field can be picked up by a broadband and focused transducer after appropriate amplification and filtering (i.e., a 20 MHz fourth-order passive bandpass filter). The root-mean-square (RMS) voltage of the signal picked up by the focused transducer is

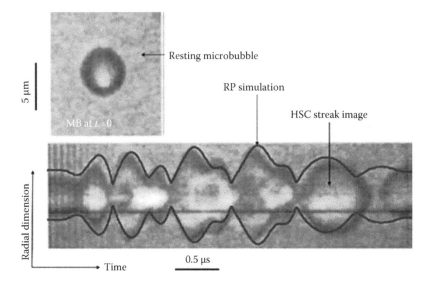

FIGURE 7.16 Radial oscillation of a 2.5 µm microbubble under the sonication of a seven-cycle burst with a peak negative pressure of 400 kPa at 2.25 MHz observed from the high-speed camera (HSC) streak image and predicted by the Rayleigh–Plesset equation. (Reprinted from *IEEE Trans. Ultrason., Ferroelectr. Freq. Control*, 47, Morgan, K.E., Allen, J.S., Dayton, P.A., Chomas, J.E., Klibaov, A., and Ferrara, K.W., Experimental and theoretical evaluation of microbubble behavior: Effect of transmitted phase and bubble size, 1494–1509, Copyright 2000, with permission from Elsevier.)

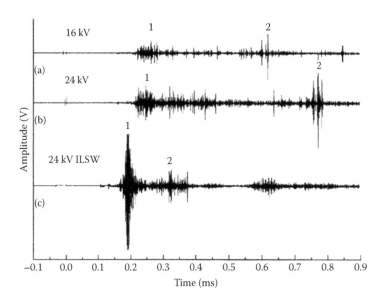

FIGURE 7.17 Representative acoustic emission signals produced by (a) standard lithotripter shock wave (LSW) at 16 kV, (b) LSW at 24 kV, and (c) the inverted LSW (ILSW) at 24 kV in free field. (Reprinted from *Ultrasound Med. Biol.*, 27, Zhong, P., Zhou, Y., and Zhu, S., Dynamics of bubble oscillation in constrained media and mechanisms of vessel rupture in SWL, 119–134, Copyright 2001, with permission from Elsevier.)

a measure of the cavitation events (Everbach et al. 1997). The averaged maximal peak-to-peak amplitudes from inertial cavitation (IC) present as "spikes" above the baseline in the recorded trace. Because the IC dose (ICD) is a relative value that depends highly on the experimental setup and especially on the sensitivity of the receiving hydrophone, results can be compared quantitatively only for experiments using exactly the same equipmental setup (Poliachik et al. 1999, Chen et al. 2003). Furthermore, each recorded waveform can be converted to the frequency domain using fast Fourier transform (FFT), which shows bursts of the various spectral lines

and broadband noise. Then, the spectra can be plotted versus time (Lauterborn and Cramer 1981, Cramer and Lauterborn 1982). An example is shown in Figure 7.18. A specific narrow-frequency window, whose central frequency is the mean value of the third and fourth harmonics, is chosen to evaluate the amount of IC-induced broadband noise by calculating its RMS value, which is then registered to the time trace. The overall ICD is calculated as the integrated the area under such curves over the effective exposure period and then normalized by a background value to minimize the effect of the experimental setup (Tu et al. 2006).

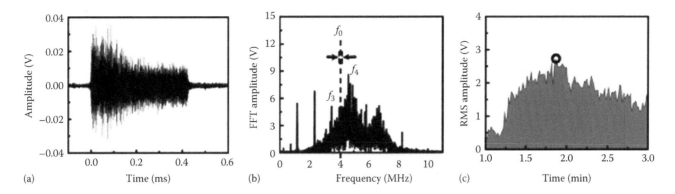

(a) Time (ms)

(b) Frequency (MHz)

(c) Time (min)

FIGURE 7.18 (a) Sample and (b) FFT spectrum of received amplitude signal, and (c) the plot as a time trace for ICD calculation from the waveforms using the PCD system. (Reprinted from *Ultrasound Med. Biol.*, 32, Tu, J., Hwang, J.H., Matula, T.J., Brayman, A.A., and Crum, L.A., Intravascular inertial cavitation activity detection and quantification in vivo with Optison, 1601–1609, Copyright 2006, with permission from Elsevier.)

Small-time Fourier transforms (STFT) can also be then performed on the collected PCD data for the corresponding spectrogram $F(t, f)$. A specific frequency window, for example, the mean value of the second and third harmonic frequencies, is chosen to evaluate the amount of inertial cavitation-induced broadband noise during the high-intensity focused ultrasound (HIFU) exposure by calculating the time-averaged and frequency-averaged amplitude

$$\bar{F} = \frac{\int_0^T \int_{f_1}^{f_2} F(t, f) \, df \, dt}{T \cdot (f_2 - f_1)} \tag{7.84}$$

where

T is the pulse duration time
f_2 and f_1 are the upper and lower bounds of the integral, respectively

Meanwhile, the amplitude of the higher-order harmonic frequency is used to represent the stable cavitation, as shown in Figure 7.19.

7.6.3 Sonoluminescence

The excellent match between the multiple-bubble sonoluminescence (MBSL) and the synthetic spectra demonstrates that SL is a thermal chemiluminescence process (Figure 7.20) (Jeffries et al. 1992). Thus, the cavitation temperature in sonochemical reactions can be determined by the spectroscopic approach about the emissivities of many metal-atom excited states as those used to monitor the surface temperature of stars (Suslick et al. 1986). Although single bubble sonoluminescence (SBSL) can provide much insight into the physics of cavitation, single bubbles simply do not contain sufficient

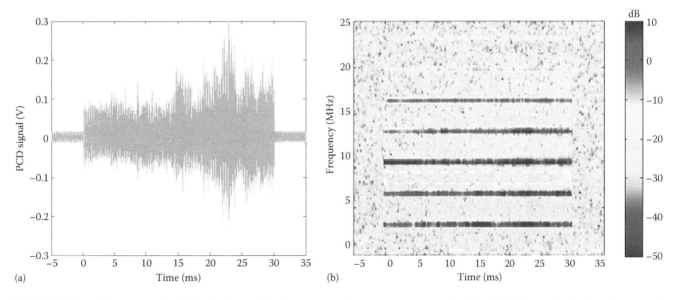

(a) Time (ms)

(b) Time (ms)

FIGURE 7.19 (a) Representative PCD signals and (b) the corresponding spectrogram produced by HIFU burst. Production of more and stronger harmonics is due to the stable cavitation, and the white noise in the spectrogram is associated with inertial cavitation. (Reprinted with permission from Zhou, Y. and Gao, X.W., Variations of bubble cavitation and temperature elevation during lesion formation by high-intensity focused ultrasound, *J. Acoust. Soc. Am.*, 134, 1683–1694, Copyright 2013, American Institute of Physics.)

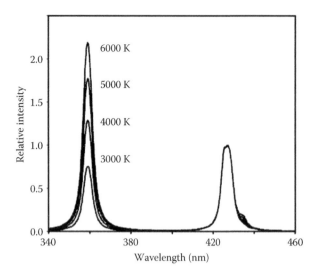

FIGURE 7.20 Calculated spectra of multi-bubble sonoluminescence (MBSL) emission from excited states of Cr atoms as a function of temperature. (Reprinted by permission from Macmillan Publishers Ltd., *Nature* (McNamara, W.B., Didenko, Y.T., and Suslick, K.S., Sonoluminescence temperatures during multi-bubble cavitation, 401, 772–775, Copyright 1999.)

material to drive chemical reactions in practice. The spectra of MBSL and SBSL are dramatically different (Matula et al. 1995). MBSL can be observed in essentially all liquids and is generally dominated by atomic and molecular emission lines, whereas SBSL has been observed primarily in aqueous liquids and its interpretation is much less clear (featureless emission that increases with decreasing wavelength). The difference between SBSL and MBSL spectra is likely related to the severity of collapse (Suslick et al. 1999).

The SL intensity can be measured with a photomultiplier tube (PMT), which can localize the arrival of a photon to a time much shorter than the response time to a pulse of light (Figure 7.21). The transit time spreads can be as small as

25 ps, whereas the rise times are about 150 ps. In order to ensure reproducibility and consistency in the experiments and to avoid the initial and variable effect of dissolved air in water, pretreatment to the solution is required (Gompf et al. 1997, Hiller et al. 1998, Kanthale et al. 2008).

7.6.4 Chemical Assessment

The total amount of peroxides produced during sonication is measured by using the Fricke dosimeter (Spinks and Woods 1990, von Sonntag et al. 1999). Fe^{2+} is known to be oxidized by H_2O_2, peroxy radicals ($RO_2^•$), hydroperoxides (RO_2H), and alkoxyradicals ($RO^•$). Organic peroxides ($ROOR'$) react with Fe^{2+} in a similar way as H_2O_2. The oxidation of Fe^{2+} is followed by the determination of $[Fe^{3+}]$ by UV spectrophotometry. Because the oxidation of Fe^{2+} by organic peroxides is slow, the solutions are kept for 24 h in the dark at room temperature for analysis after being mixed with the Fricke solution (2×10^{-2} M $FeSO_4$ and 0.9 M H_2SO_4) (Spinks and Woods 1990, Suslick et al. 1999, Segebarth et al. 2002).

$$
\begin{aligned}
H_2O_2 + Fe^{2+} + H^+ &\rightarrow Fe^{3+} + {}^•OH + H_2O \\
{}^•OH + Fe^{2+} + H^+ &\rightarrow Fe^{3+} + H_2O \\
RO_2^• + Fe^{2+} + H^+ &\rightarrow Fe^{3+} + RO_2H \\
RO_2H + Fe^{2+} + H^+ &\rightarrow Fe^{3+} + RO^• + H_2O \\
RO^• + Fe^{2+} + H^+ &\rightarrow Fe^{3+} + ROH \\
ROOR' + Fe^{2+} + H^+ &\rightarrow Fe^{3+} + RO^• + ROH
\end{aligned}
\tag{7.85}
$$

After each irradiation, the yield of H_2O_2 is measured using a UV–Vis absorption spectrophotometer. The iodide reagent is prepared by mixing equal volumes (1 mL) of solution A (0.4 M KI, 0.05 M NaOH, 0.00016 M $(NH_4)_6Mo_7O_{24} \cdot 4H_2O$) and solution B (0.1 M $KHC_8H_4O_4$). One milliliter of the sonicated

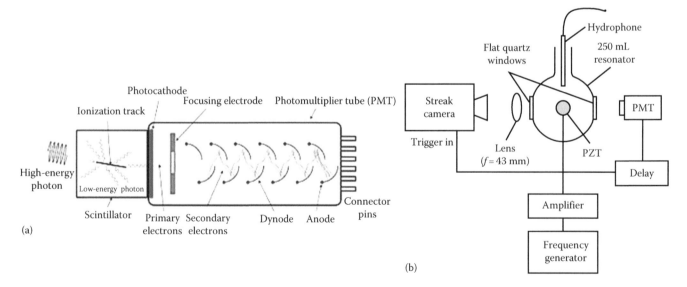

FIGURE 7.21 (a) Structure of a photomultiplier tube and (b) its application in the measurement of sonoluminescence during the bubble cavitation.

sample is added to the iodide reagent, and after mixing, the absorbance of I_3^- is measured spectrophotometrically using $\varepsilon = 26{,}400 \ M^{-1} \ cm^{-1}$ at 353 nm (Kanthale et al. 2008).

REFERENCES

Apfel RE. The role of impurities in cavitation-threshold determination. *Journal of the Acoustical Society of America* 1970;48:1179–1186.

Apfel RE. Vapor nucleation at a liquid–liquid interface. *Journal of Chemical Physics* 1971;54:62–63.

Apfel RE. Acoustic cavitation. In: Edmonds P, ed., *Methods of Experimental Physics Series*. New York: Academic Press, 1981a, p. 355.

Apfel RE. Acoustic cavitation prediction. *Journal of the Acoustical Society of America* 1981b;69:1624–1633.

Apfel RE. Acoustic cavitation: A possible consequence of biomedical uses of ultrasound. *British Journal of Cancer* 1982;45:140–146.

Ashokkumar M. The characterization of acoustic cavitation bubbles—An overview. *Ultrasonics Sonochemistry* 2011;18:864–872.

Ashokkumar M, Lee J, LIida Y, Yasui K, Kozuka T, Tuziuti T, Towata A. Spatial distribution of acoustic cavitation bubbles at different ultrasound frequencies. *ChemPhysChem* 2010;11:1680–1684.

Atchley AA, Crum LA. Acoustic cavitation and bubble dynamics. In: Suslick KS, ed., *Ultrasound: Its Chemical, Physical, and Biological Effects*. New York: VCH Publisher, 1988, pp. 1–64.

Atchley AA, Prosperetti A. The crevice mode of bubble nucleation. *Journal of the Acoustical Society of America* 1989;86:1065–1084.

Bailey MR, Pishchalnikov YA, Sapozhnikov OA, Cleveland RO, McAteer JA, Miller NA, Pishchalnikova IV, Connors BA, Crum LA, Evan AP. Cavitation detection during shockwave lithotripsy. *Ultrasound in Medicine and Biology* 2005;31:1245–1256.

Boteler JM, Sutherland GT. Tensile failure of water due to shock wave interactions. *Journal of Applied Physics* 2004;96:6919–6924.

Bouakaz A, Versluis M, de Jong N. High-speed optical observations of contrast agent destruction. *Ultrasound in Medicine and Biology* 2005;31:391–399.

Brennen CE. *Cavitation and Bubble Dynamics*. New York: Oxford University Press, 1995.

Brennen CE. Cavitation in biological and bioengineering contexts. *5th International Symposium on Cavitation*, Osaka, Japan, 2003, pp. 1–9.

Brenner MP, Hilgenfeldt S, Lohse D. Single-bubble sonoluminescence. *Reviews of Modern Physics* 2002;74:425–484.

Brett HWW, Jellinek HHG. Degradation of long chain molecules by ultrasonic waves. Part V. Cavitation and the effect of dissolved gases. *Journal of Polymer Science Part A* 1954;13:441–459.

Briggs LJ. Limiting negative pressure of water. *Journal of Applied Physics* 1950;21:721–722.

Brotchie A, Statham T, Zhou M, Dharmarathne L, Grieser F, Ashokkumar M. Acoustic bubble sizes, coalescence, and sonochemical activity in aqueous electrolyte solutions saturated with different gases. *Langmuir* 2010;26:12690–12695.

Brujan E-A, Nahen K, Schmidt P, Vogel A. Dynamics of laser-induced cavitation bubbles near an elastic boundary. *Journal of Fluid Mechanics* 2001;433:251–281.

Caupin F, Herbert E. Cavitation in water: A review. *Comptes Rendus Physique* 2006;7:1000–1017.

Ceccio SL, Brennen CE. Observations of the dynamics and acoustics of travelling bubble cavitation. *Journal of Fluid Mechanics* 1991;233:633–660.

Chen H, Kreider W, Brayman AA, Bailey MR, Matula TJ. Blood vessel deformations on microsecond time scales by ultrasonic cavitation. *Physical Review Letters* 2011;106:034301.

Chen W-S, Matula TJ, Brayman AA, Crum LA. A comparison of the fragmentation thresholds and inertial cavitation doses of different ultrasound contrast agents. *Journal of the Acoustical Society of America* 2003;113:643–651.

Chin CT, Lancée C, Borsboom J, Mastik F, Frijlink ME, de Jong N, Versluis M, Lohse D. Brandaris 128: A digital 25 million frames per second camera with 128 highly sensitive frames. *Review of Scientific Instruments* 2003;74:5026–5034.

Church CC. A theoretical study of cavitation generated by an extracorporeal shock wave lithotripter. *Journal of the Acoustical Society of America* 1989;86:215–227.

Church CC. The effects of an elastic solid surface layer on the radial pulsations of gas bubbles. *Journal of the Acoustical Society of America* 1995;97:1510–1521.

Cleveland RO, Sapozhnikov OA, Bailey MR, Crum LA. A dual passive cavitation detector for localized detection of lithotripsy-induced cavitation in vitro. *Journal of the Acoustical Society of America* 2000;107:1745–1758.

Coakely WT, Nyborg WL. Cavitation: Dynamics of gas bubbles; applications. In: Fry FJ, ed., *Ultrasound: Its Applications in Medicine and Biology*. New York: Elsevier, 1978, pp. 77–159.

Coleman A, Choi M, Saunders J. Detection of acoustic emission from cavitation in tissue during clinical extracorporeal lithotripsy. *Ultrasound in Medicine and Biology* 1996;22:1079–1087.

Coleman A, Choi M, Saunders J, Leighton T. Acoustic emission and sonoluminescence due to cavitation at the beam focus of an electrohydraulic shock wave lithotripter. *Ultrasound in Medicine and Biology* 1992;18:267–281.

Collis J, Manasseh R, Liovic P, Tho P, Ooi A, Petkovic-Duran K, Zhu Y. Cavitation microstreaming and stress fields created by microbubbles. *Ultrasonics* 2010;50:273–279.

Contamine RF, Wilhelm A, Berlan J, Delmas H. Power measurement in sonochemistry. *Ultrasonics Sonochemistry* 1995;2:S43–S47.

Cramer E, Lauterborn W. Acoustic cavitation noise spectra. *Applied Scientific Research* 1982;38:209–214.

Cravotto G, Cintas P. Power ultrasound in organic synthesis: Moving cavitational chemistry from academia to innovative and large-scale applications. *Chemical Society Reviews* 2006;35:180–196.

Crum LA. Tensile strength of water. *Nature* 1979;278:148–149.

Crum LA. Acoustic cavitation thresholds in water. In: Lauterborn W, ed., *Cavitation and Inhomogeneities in Underwater Acoustics*. New York: Springer-Verlag, 1980, pp. 84–89.

Crum LA. Acoustic cavitation. *IEEE Ultrasonics Symposium*, San Diego, CA, October 27 to 29, 1982a, pp. 1–11.

Crum LA. Nucleation and stabilization of microbubbles in liquids. *Applied Scientific Research* 1982b;38:101–115.

Crum LA. Sonoluminescence, sonochemistry and sonophysics. *Journal of the Acoustical Society of America* 1994;95:559–563.

Crum LA. Comments on the evolving field of sonochemistry by a cavitation physicist. *Ultrasonics Sonochemistry* 1995;2:S147–S152.

Crum LA, Hansen GM. Generalized equations for rectified diffusion. *Journal of the Acoustical Society of America* 1982;72:1586–1592.

Crum LA, Prosperetti A. Nonlinear oscillations of gas bubbles in liquids: An interpretation of some experimental results. *Journal of the Acoustical Society of America* 1983;73:121–127.

Crum LA, Roy RA. Sonoluminescence. *Physics Today* 1994;47:22–30.

Cupin F. Liquid–vapor interface, cavitation, and the phase diagram of water. *Physical Review E* 2005;71:051605.

Dayton PA, Chomas JE, Lum AF, Allen JS, Lindner JR, Simon SI, Ferrara KW. Optical and acoustical dynamics of microbubble contrast agents inside neutrophils. *Biophysical Journal* 2001;80:1547–1556.

De Jong N, Emmer M, Chin CT, Bouakaz A, Mastik F, Lohse D, Versluis M. "Compression-only" behavior of phospholipid-coated contrast bubbles. *Ultrasound in Medicine and Biology* 2007;33:653–656.

Didenko YT, Suslick KS. The energy efficiency of formation of photons, radicals and ions during single-bubble cavitation. *Nature* 2002;418:394–397.

Dixon HH. Note on the tensile strength of water. *Scientific Proceedings of the Royal Dublin Society* 1909;12:60–65.

Duck FA. Radiation pressure and acoustic streaming. In: Duck FA, Baker AC, Starritt HC, eds., *Ultrasound in Medicine*. Bristol, U.K.: Institute of Physics Publication, 1998, pp. 39–56.

Elder SA. Cavitation microstreaming. *Journal of the Acoustical Society of America* 1959;31:54–64.

Eller A, Flynn HG. Rectified diffusion during nonlinear pulsations of cavitation bubbles. *Journal of the Acoustical Society of America* 1965;37:493–503.

Everbach EC, Makin IRS, Azadniv M, Meltzer RS. Correlation of ultrasound-induced hemolysis with cavitation detector output in vitro. *Ultrasound in Medicine and Biology* 1997;23:619–624.

Finch RD. Sonoluminescence. *Ultrasonics* 1963;1:87–98.

Finch RD, Kagiwada R, Barmatz M, Rudnick I. Cavitation in liquid helium. *Physical Review* 1964;A134:1425–1438.

Fisher JC. The fracture of liquids. *Journal of Applied Physics* 1948;19:1062–1067.

Flannigan DJ, Suslick KS. Plasma formation and temperature measurement during single-bubble cavitation. *Nature* 2005;434:52–55.

Flynn HG. Physics of acoustic cavitation in liquids. In: Mason WP, ed., *Physical Acoustics*. New York: John Wiley & Sons, 1964, pp. 57–152.

Flynn HG. Generation of transient cavities in liquids by microsecond pulses of ultrasound. *Journal of the Acoustical Society of America* 1982;72:1926–1932.

Fox FE, Herzfeld KF. Gas bubbles with organic skin as cavitation nuclei. *Journal of the Acoustical Society of America* 1954;26:984–989.

Franc JP, Michel JM. *Fundamentals of Cavitation*. Dordrecht, Netherlands: Kluwer Academic Publishers, 2004.

Frinking PJA, de Jong N. Acoustic modeling of shell-encapsulated gas bubbles. *Ultrasound in Medicine and Biology* 1998;24:523–533.

Gaitan DF, Crum LA, Church CC, Roy RA. Sonoluminescence and bubble dynamics for a single, stable, cavitation bubble. *Journal of the Acoustical Society of America* 1992;91:3166–3183.

Galloway WJ. An experimental study of acoustically induced cavitation in liquids. *Journal of the Acoustical Society of America* 1954;36:2287–2292.

Garbin V, Cojoc D, Ferrari E, Di Fabrizio E, Overvelde M, Van Der Meer S, De Jong N, Lohse D, Versluis M. Changes in microbubble dynamics near a boundary revealed by combined optical micromanipulation and high-speed imaging. *Applied Physics Letters* 2007;90:114103.

Gompf B, Günther R, Nick G, Pecha R, Eisenmenger W. Resolving sonoluminescence pulse width with time-correlated single photon counting. *Physical Review Letters* 1997;79:1405.

Gong C, Hart DP. Ultrasound induced cavitation and sonochemical yields. *Journal of the Acoustical Society of America* 1998;104:2675–2682.

Hammarström B, Laurell T, Nilsson J. Seed particle-enabled acoustic trapping of bacteria and nanoparticles in continuous flow systems. *Lab on a Chip* 2012;12:4296–4304.

Hammitt FG. *Cavitation and Multiphase Flow Phenomena*. New York: McGraw-Hill, 1980.

Harvey EN, Barnes DK, McElroy WD, Whiteley AH, Pease DC, Cooper KW. Bubble formation in animals I: Physical factors. *Journal of Cellular and Comparative Physiology* 1944a;24:1–22.

Harvey EN, Cooper KW, Whiteley AH, Pease DC, McElroy WD. Bubble formation within single cells. *Biological Bulletin* 1946;91:236–237.

Harvey EN, McElroy WD, Whiteley AH. On cavity formation in water. *Journal of Applied Physics* 1947;18:162–172.

Harvey EN, Whiteley AH, WDMcElroy, Pease DC, Barnes DK. Bubble formation in animals II: Gas nuclei and their distribution in blood and tissues. *Journal of Cellular and Comparative Physiology* 1944b;24:23–24.

Hayward ATJ. The role of stabilized gas nuclei in hydrodynamic cavitation inception. *Journal of Physics D* 1970;3:574–579.

Hiller R, Putterman SJ, Barber BP. Spectrum of synchronous picosecond sonoluminescence. *Journal of the Acoustical Society of America* 1992;92:254–2454.

Hiller RA, Putterman SJ, Weninger KR. Time-resolved spectra of sonoluminescence. *Physical Review Letters* 1998;80:1090.

Jeffries JB, Copeland RA, Suslick KS, Flint EB. Thermal equilibration during cavitation. *Science* 1992;256:248–248.

Kanthale P, Ashokkumar M, Grieser F. Sonoluminescence, sonochemistry (H_2O_2 yield) and bubble dynamics: Frequency and power effects. *Ultrasonics Sonochemistry* 2008;15:143–150.

Kell GS. Early observations of negative pressures in liquids. *American Journal of Physics* 1983;51:1038–1041.

Knapp RT, Daily JW, Hammitt FG. *Cavitation*. New York: McGraw-Hill, 1970.

Kondo T, Gamson J, Mitchell JB, Riesz P. Free radical formation and cell lysis induced by ultrasound in the presence of different rare gases. *International Journal of Radiation Biology* 1988;54:955–962.

Kondo T, Kano E. Effect of free radicals induced by ultrasonic cavitation on cell killing. *International Journal of Radiation Biology* 1988;54:475–486.

Krefting D, Mettin R, Lauterborn W. High-speed observation of acoustic cavitation erosion in multibubble systems. *Ultrasonics Sonochemistry* 2004;11:119–123.

Lauterborn W, Cramer E. On the dynamics of acoustic cavitation noise spectra. *Acta Acustica United with Acustica* 1981;49:280–287.

Lauterborn W, Kurz T, Geisler R, Schanz D, Lindau O. Acoustic cavitation, bubble dynamics and sonoluminescence. *Ultrasonics Sonochemistry* 2007;14:484–491.

Leighton TG. *The Acoustic Bubble*. London, U.K.: Academic Press, 1994.

Leighton TG, Walton AJ, Pickworth MJW. Primary Bjerknes forces. *European Journal of Physics* 1990;11:47–50.

Lewin PA, Jensen LB. Acoustic pressure amplitude thresholds for rectified diffusion in gaseous microbubbles in biological tissue. *Journal of the Acoustical Society of America* 1981;69:846–852.

Liebermann L. Air bubbles in water. *Journal of Applied Physics* 1957;28:205–211.

Lifka J, Ondruschka B, Hofmann J. The use of ultrasound for the degradation of pollutants in water: Aquasonolysis—A review. *Engineering in Life Sciences* 2003;3:253–262.

Margulis M. Sonoluminescence and sonochemical reactions in cavitation fields. A review. *Ultrasonics* 1985;23:157–169.

Marmottant P, Hilgenfeldt S. Controlled vesicle deformation and lysis by single oscillating bubbles. *Nature* 2003;423: 153–156.

Marmottant P, van der Meer S, Emmer M, Versluis M, de Jong N, Hilgenfeldt S, Lohse D. A model for large amplitude oscillations of coated bubbles accounting for buckling and rupture. *Journal of the Acoustical Society of America* 2005;118:3499–3505.

Marmottant P, Versluis M, de Jong N, Hilgenfeldt S, Lohse D. High-speed imaging of an ultrasound-driven bubble in contact with a wall: "Narcissus" effect and resolved acoustic streaming. *Experiments in Fluids* 2006;41:147–153.

Mason TJ, Lorimer JP. *Applied Sonochemistry*. Weinheim, Germany: Wiley-VCH, 2002.

Matula TJ. Inertial cavitation and single-bubble sonoluminescence. *Philosophical Transactions of the Royal Society A: Mathematical, Physical and Engineering Sciences* 1999;357:225–249.

Matula TJ, Crum LA. Evidence for gas exchange in single-bubble sonoluminescence. *Physical Review Letters* 1998;80: 865–868.

Matula TJ, Roy RA, Mourad PD, McNamara III WB, Suslick KS. Comparison of multibubble and single-bubble sonoluminescence spectra. *Physical Review Letters* 1995;75:2602.

McNamara WB, Didenko YT, Suslick KS. Sonoluminescence temperatures during multi-bubble cavitation. *Nature* 1999;401:772–775.

Mead EL, Sutherland RG, Verrall RE. The effect of ultrasound on water in the presence of dissolved gases. *Canadian Journal of Chemistry* 1976;54:1114–1120.

Miller DL. Overview of experimental studies of biological effects of medical ultrasound caused by gas body activation and inertial cavitation. *Progress in Biophysics and Molecular Biology* 2007;93:314–330.

Miller DL, Nyborg WL, Whitcomb CC. Platelet aggregation induced by ultrasound under specialized conditions *in vitro*. *Science* 1979;205:505–507.

Miller MW, Miller DL, Brayman AA. A review of *in vitro* bioeffects of inertial ultrasonic cavitation from a mechanistic perspective. *Ultrasound in Medicine and Biology* 1996;22: 1131–1154.

Morgan KE, Allen JS, Dayton PA, Chomas JE, Klibaov A, Ferrara KW. Experimental and theoretical evaluation of microbubble behavior: Effect of transmitted phase and bubble size. *IEEE Transactions on Ultrasonics, Ferroelectrics and Frequency Control* 2000;47:1494–1509.

Moss WC, Levatin JL, Szeri AJ. A new damping mechanism in strongly collapsing bubbles. *Proceedings of the Royal Society A: Mathematical, Physical and Engineering Sciences* 2000;456:2983–2994.

Neppiras EA. Acoustic cavitation. *Physics Reports* 1980;61:160.

Neppiras EA, Noltingk BE. Cavitation produced by ultrasonics: Theoretical conditions for the onset of cavitation. *Proceedings of the Physical Society* 1951;B64:1032.

Noltingk BE, Neppiras EA. Cavitation produced by ultrasonics. *Proceedings of the Physical Society Section B* 1950;63:674–685.

Nyborg WL. Acoustic streaming. In: Hamilton MF, Blackstock DT, eds., *Nonlinear Acoustics*. New York: Academic Press, 1998, pp. 207–231.

Ohl CD, Kurz T, Geisler R, Lindau O, Lauterborn W. Bubble dynamics, shock waves and sonoluminescence. *Philosophical Transactions of the Royal Society A: Mathematical, Physical and Engineering Sciences* 1999;357:269–294.

Pecha R, Gompf B. Microimplosions: Cavitation collapse and shock wave emission on a nanosecond time scale. *Physical Review Letters* 2000;84:1328.

Petersson F, Nilsson A, Holm C, Jonsson H, Laurell T. Separation of lipids from blood utilizing ultrasonic standing waves in microfluidic channels. *Analyst* 2004;129:938–943.

Pishchalnikov YA, Sapozhnikov OA, Bailey MR, Williams Jr JC, Cleveland RO, Colonius T, Crum LA, Evan AP, McAteer JA. Cavitation bubble cluster activity in the breakage of kidney stones by lithotripter shockwaves. *Journal of Endourology* 2003;17:435–446.

Plesset MS, Prosperetti A. Bubble dynamics and cavitation. *Annual Review of Fluid Mechanics* 1977;99:145–185.

Poliachik SL, Chandler WL, Mourad PD, Bailey MR, Bloch S, Cleveland RO, Kaczkowski P, Keilman G, Porter T, Crum LA. Effect of high-intensity focused ultrasound on whole blood with and without microbubble contrast agent. *Ultrasound in Medicine and Biology* 1999;25:991–998.

Putterman SJ, Weninger KR. Sonoluminescence: How bubbles turn sound into light. *Annual Review of Fluid Mechanics* 2000;32:445–476.

Rooze J, Rebrov EV, Schouten JC, Keurentjes JTF. Dissolved gas and ultrasonic cavitation—A review. *Ultrasonics Sonochemistry* 2013;20:1–11.

Sankin G, Simmons W, Zhu S, Zhong P. Shock wave interaction with laser-generated single bubbles. *Physical Review Letters* 2005;95:034501.

Sarkar K, Shi WT, Chatterjee D, Forsberg F. Characterization of ultrasound contrast microbubbles using in vitro experiments and viscous and viscoelastic interface models for encapsulation. *Journal of the Acoustical Society of America* 2005;118:539–550.

Segebarth N, Eulaerts O, Reisse J, Crum LA, Matula TJ. Correlation between acoustic cavitation noise, bubble population, and sonochemistry. *Journal of Physical Chemistry* B 2002;106:9181–9190.

Shchukin DG, Möhwald H. Sonochemical nanosynthesis at the engineered interface of a cavitation microbubble. *Physical Chemistry Chemical Physics* 2006;8:3496–3506.

Spinks JWT, Woods RJ. An introduction to radiation chemistry. 1990, New York, John Wiley & Sons.

Suslick KS, Didenko Y, Fang MM, Hyeon T, Kolbeck KJ, McNamara WB, Mdleleni MM, Wong M. Acoustic cavitation and its chemical consequences. *Philosophical Transactions of the Royal Society of London. Series A: Mathematical, Physical and Engineering Sciences* 1999;357:335–353.

Suslick KS, Flannigan DJ. Inside a collapsing bubble: Sonoluminescence and the conditions during cavitation. *Annual Review of Physical Chemistry* 2008;59:659–683.

Suslick KS, Hammerton DA, Cline RE. Sonochemical hot spot. *Journal of the American Chemical Society* 1986;108:5641–5642.

Suslick KS, Mdleleni MM, Ries JT. Chemistry induced by hydrodynamic cavitation. *Journal of the American Chemical Society* 1997;119:9303–9304.

Temperley HNV. The behaviour of water under hydrostatic tension: II. *Proceedings of the Physical Society* 1946;58: 436–443.

Temperley HNV. The behaviour of water under hydrostatic tension: III. *Proceedings of the Physical Society* 1947;59:199–208.

Temperley HNV, Chambers LG. The behaviour of water under hydrostatic tension: I. *Proceedings of the Physical Society* 1946;58:420–436.

ter Haar GR, Daniels S, Eastaugh KC, Hill CR. Ultrasonically induced cavitation *in vivo*. *British Journal of Cancer* 1982; Suppl. V:151–155.

Trevena DH. *Cavitation and Tension in Liquids*. Bristol, U.K.: Adam Hilger, 1987.

Tsutsui H, Ho C-M. Cell separation by non-inertial force fields in microfluidic system. *Mechanics Research Communications* 2009;36:92–103.

Tu J, Hwang JH, Matula TJ, Brayman AA, Crum LA. Intravascular inertial cavitation activity detection and quantification in vivo with Optison. *Ultrasound in Medicine and Biology* 2006;32: 1601–1609.

Unger EC, Porter T, Culp W, Labell R, Matsunaga T, Zutshi R. Therapeutic application of lipid-coated microbubbles. *Advanced Drug Delivery Reviews* 2004;56:1291–1314.

Van der Meer S, Versluis M, Lohse D, Chin C, Bouakaz A, De Jong N. The resonance frequency of SonoVue™ as observed by high-speed optical imaging. *2004 IEEE Ultrasonics Symposium*, 2004, pp. 343–345, Montreal, Canada.

von Sonntag C, Mark G, Tauber A, Schuchmann H-P. OH radical formation and dosimetry in the sonolysis of aqueous solutions. *Advances in Sonochemistry* 1999;5:109–145.

Wang TG, Lee CP. Radiation pressure and acoustic levitation. In: Hamilton MF, Blackstock DT, eds., *Nonlinear Acoustics*. New York: Academic Press, 1998, pp. 177–205.

William AR. *Ultrasound: Biological Effects and Potential Hazards*. New York: Academic Press, 1983, London, UK.

Winnick J, Cho SJ. PVT behavior of water at negative pressures. *Journal of Chemical Physics* 1971;55:2092–2097.

Yanagida H. The effect of dissolved gas concentration in the initial growth stage of multi cavitation bubbles: Differences between vacuum degassing and ultrasound degassing. *Ultrasonics Sonochemistry* 2008;15:492–496.

Yang X, Church CC. A model for the dynamics of gas bubbles in soft tissue. *Journal of the Acoustical Society of America* 2005;118:3595–3606.

Young FR. *Cavitation*. London, U.K.: McGraw-Hill, 1989.

Yount DE, Gillary EQ, Hoffman DC. A microscopic investigation of bubble formation nuclei. *Journal of the Acoustical Society of America* 1984;76:1511–1521.

Zheng H, Dayton PA, Caskey C, Zhao S, Qin S, Ferrara KW. Ultrasound-driven microbubble oscillation and translation within small phantom vessels. *Ultrasound in Medicine and Biology* 2007;33:1978–1987.

Zhong P, Cioanta I, Cocks FH, Preminger GM. Inertial cavitation and associated acoustic emission produced during electro-hydraulic shock wave lithotripsy. *Journal of the Acoustical Society of America* 1997;101:2940–2950.

Zhong P, Lin H, Xi X, Zhu S, Bhogte ES. Shock wave–inertial microbubble interaction: Methodology, physical characterization, and bioeffect study. *Journal of the Acoustical Society of America* 1999;105:1997–2009.

Zhong P, Zhou Y, Zhu S. Dynamics of bubble oscillation in constrained media and mechanisms of vessel rupture in SWL. *Ultrasound in Medicine and Biology* 2001;27:119–134.

Zhou Y, Gao XW. Variations of bubble cavitation and temperature elevation during lesion formation by high-intensity focused ultrasound. *Journal of the Acoustical Society of America* 2013;134:1683–1694.

8 Transducer

A key component of a therapeutic ultrasound system is the transducer, which generates the acoustic waves and sometimes also picks up the echoes as the diagnostic probes. Piezoelectricity is the electric charge that accumulates in certain solid materials in response to an applied mechanical stress, and it was discovered in 1880 by French physicists Jacques Curie and Pierre Curie (Curie and Curie 1880). The piezoelectric effect is a reversible process: an applied electrical field generates a mechanical strain. Biological tissues, such as bone, DNA, and various proteins, also have piezoelectricity (Fukada and Yasuda 1957, Storri et al. 1998, Fukada and Ando 2003, Tombelli et al. 2006, Behari 2009, Lemanov et al. 2011). Above the Curie point, a phase transition occurs and the crystal structure can change to a crystallographic class in which piezoelectricity ceases to exist.

During World War I, the quartz (SiO_2) transducer was first used in the sonar system to detect submarines. Among the 20 crystal classes with piezoelectricity, 10 are pyroelectric, which show a spontaneous polarization that changes with temperature. If the dipole moment of a pyroelectric medium can be redirected by an externally applied electric field, the material is called ferroelectric. Around the end of World War II, piezoelectric ceramics were developed by making a suitably balanced mixture that included an appropriate binder, followed by pressing and then firing at a high temperature. Domains of the same polarization direction are formed within the polycrystallites, and these are bound by other domains with different polarization directions. Overall, because the domains are randomly oriented, the macroscopic behavior of the polycrystalline ceramic is approximately isotropic. By applying a high DC electric field at a temperature close to the Curie point and with the field present when the temperature is lowered, many of the domains can be made to align with the applied field and some grow in volume, causing the piezoelectric properties to be greatly enhanced, a process called poling. Piezoceramics have the important advantage of easy fabrication into a variety of shapes with controlled directions of polarization. Relaxor-based ferroelectric materials can be grown in single-crystalline form to a sufficiently large size for fabricating arrays and have a higher piezoelectric coupling factor than lead zirconate titanate (PZT) ceramics. The performance of ultrasound transducers can also be improved through the use of a composite arrangement of piezoelectric and non-piezoelectric (i.e., polymer and epoxy) materials. A piezocomposite combines the superior piezoelectric properties of the ceramic with the much lower acoustic impedance of the polymer, resulting in an effective acoustic propagation. Relaxor-based single crystals can also be used in building piezocomposites. Polyvinylidene fluoride (PVDF), a semicrystalline polymer with long molecular chains and structure, exhibits ferroelectricity and,

therefore, piezoelectric properties. It is used in the fabrication of broadband hydrophones, high-frequency imaging (i.e., 50–200 MHz), and the acoustic microscope.

In this chapter, the principle of piezoelectricity and the configuration and characteristics of popular ultrasound transducers are presented.

8.1 PIEZOELECTRICITY

Suppose a force $\vec{F}$ is acting on one of the faces of an elementary cube in Cartesian coordinates. The generated stress $[T]$ can be expressed as

$$\left[T_{ij} \right] = \begin{bmatrix} T_{xx} & T_{xy} & T_{xz} \\ T_{yx} & T_{yy} & T_{yz} \\ T_{zx} & T_{zy} & T_{zz} \end{bmatrix} \qquad (8.1)$$

where the components T_{xx}, T_{yy}, and T_{zz} are the normal stresses and the others are the shear stresses. Since there is no rotation in a homogeneous medium in equilibrium, which requires $T_{xy} = T_{yx}$, $T_{xz} = T_{zx}$, and $T_{zy} = T_{yz}$, only six of the nine stress components are independent, which can be rewritten as

$$\text{normal:} \quad T_1 = T_{xx}, \quad T_2 = T_{yy}, \quad T_3 = T_{zz}$$
$$\text{shear:} \quad T_4 = T_{yz} = T_{zy}, \quad T_5 = T_{xz} = T_{zx}, \quad T_6 = T_{xy} = T_{yx} \qquad (8.2)$$

So, the stress has the matrix expression

$$\{T\} = \begin{bmatrix} T_1 & T_2 & T_3 & T_4 & T_5 & T_6 \end{bmatrix}' \qquad (8.3)$$

Similarly, the strain can be defined as

$$\{S\} = \begin{bmatrix} S_1 \\ S_2 \\ S_3 \\ S_4 \\ S_5 \\ S_6 \end{bmatrix} = \begin{bmatrix} S_{xx} & S_{xy} & S_{xz} \\ S_{yx} & S_{yy} & S_{yz} \\ S_{zx} & S_{zy} & S_{zz} \end{bmatrix} \qquad (8.4)$$

where S_1, S_2, and S_3 are the tensile strains along the x, y, and z directions, respectively, and the others are pure shear strains. In the polarized state, the produced electric displacement is

$$\{D\} = \varepsilon_0 \{E\} + \{P\} \qquad (8.5)$$

where

$\{D\}$, $\{E\}$, and $\{P\}$ are three-element vectors of the electric field, electric displacement, and polarization, respectively

ε_0 is the permittivity of free space

In a piezoelectric dielectric medium, there is no free charge and all charges are bound dipolar ones. Thus,

$$\nabla \cdot \{D\} = 0 \tag{8.6}$$

When a piezoelectric material is subjected to an elastic deformation, the stress will be linearly related to the strain and the electric field:

$$\{T\} = [c^E]\{S\} - [e]'\{E\}$$

$$= \begin{bmatrix} c_{11}^E & c_{12}^E & c_{13}^E & c_{14}^E & c_{15}^E & c_{16}^E \\ c_{21}^E & c_{22}^E & c_{23}^E & c_{24}^E & c_{25}^E & c_{26}^E \\ c_{31}^E & c_{32}^E & c_{33}^E & c_{34}^E & c_{35}^E & c_{36}^E \\ c_{41}^E & c_{42}^E & c_{43}^E & c_{44}^E & c_{45}^E & c_{46}^E \\ c_{51}^E & c_{52}^E & c_{53}^E & c_{54}^E & c_{55}^E & c_{56}^E \\ c_{61}^E & c_{62}^E & c_{63}^E & c_{64}^E & c_{65}^E & c_{66}^E \end{bmatrix} \begin{bmatrix} S_1 \\ S_2 \\ S_3 \\ S_4 \\ S_5 \\ S_6 \end{bmatrix} - \begin{bmatrix} e_{11} & e_{21} & e_{31} \\ e_{12} & e_{22} & e_{31} \\ e_{13} & e_{23} & e_{33} \\ e_{14} & e_{24} & e_{34} \\ e_{15} & e_{25} & e_{35} \\ e_{16} & e_{26} & e_{36} \end{bmatrix} \begin{bmatrix} E_1 \\ E_2 \\ E_3 \end{bmatrix} \tag{8.7}$$

where
   [c] is the stiffness matrix
   [e] is the piezoelectric stress matrix

In a similar way, the electric displacement can be expressed in terms of the strain and electric field as

$$\{D\} = [e]\{S\} + [\varepsilon^S]\{E\} \tag{8.8}$$

where $[\varepsilon^S]$ is a constant strain (clamped) 3×3 permittivity matrix, which is different from those measured under constant stress conditions $[\varepsilon^T]$. The whole sets of relationships of $\{T\}$, $\{S\}$, $\{D\}$, and $\{E\}$ are, then

$$\{T\} = [c^E]\{S\} - [e]'\{E\}, \quad \{D\} = [e]\{S\} + [\varepsilon^S]\{E\} \tag{8.9}$$

$$\{S\} = [s^E]\{T\} + [d]'\{E\}, \quad \{D\} = [d]\{T\} + [\varepsilon^T]\{E\} \tag{8.10}$$

$$\{S\} = [s^D]\{T\} + [g]'\{E\}, \quad \{E\} = -[g]\{T\} + [\beta^T]\{D\} \tag{8.11}$$

$$\{T\} = [c^D]\{S\} - [h]'\{D\}, \quad \{E\} = -[h]\{S\} + [\beta^S]\{D\} \tag{8.12}$$

where [d], [e], [g], and [h] are 3×6, $[\varepsilon]$ and $[\beta]$ are 3×3, and [c] and [s] are 6×6 matrices. They are related to each other by

$$[s^D] = [c^D]^{-1}, \quad [s^E] = [c^E]^{-1}, \quad [\varepsilon^T] = [\beta^T]^{-1}, \quad [\varepsilon^S] = [\beta^S]^{-1}$$
$$[d] = [\varepsilon^T][g] = [e][s^E], \quad [e] = [d][c^E] = [\varepsilon^S][h]$$
$$[g] = [h][s^D] = [\beta^T][d], \quad [h] = [\beta^S][e] = [g][c^D]$$
$$[s^E] - [s^D] = [g]'[d] = [d]'[g], \quad [c^D] - [c^E] = [h]'[e] = [e]'[h]$$
$$[\varepsilon^T] - [\varepsilon^S] = [e][d]' = [d][e]', \quad [\beta^S] - [\beta^T] = [g][h]' = [h][g]' \tag{8.13}$$

The piezoelectric strain coefficient d_{ii} is related to the stress coefficient e_{ii} by

$$e_{ii} = d_{ii}c_{ii} \tag{8.14}$$

where c_{ii} is the material's elastic stiffness constant under conditions of constant electric field. The piezoelectric coefficient g_{ii} is related to d_{ii} by

$$d_{ii} = g_{ii}\varepsilon_r\varepsilon_0 \tag{8.15}$$

where
   ε_r is the relative dielectric constant of the transducer material (under unrestrained or free conditions)
   ε_0 is the permittivity of free space ($\varepsilon_0 = 8.85 \times 10^{-12}$ F/m)

If the applied electric field is also in the same direction, the piezoelectric modulus for a longitudinal wave, d_{33}, is given by

$$d_{33} = \frac{\Delta x_t}{V_t} \tag{8.16}$$

where
   V_t is the applied voltage
   Δx_t is the change in the thickness direction

It is evident that, if there are no stress components and all [d] coefficients are zero except d_{33}, then d_{33} expresses the deformation due to a given applied electric field, corresponding to the inverse piezoelectric effect. For an ultrasound transmitter, the value of the piezoelectric modulus is expected to be high. With an external change in thickness Δx_r, the resultant voltage V_r across the piezoelectric element is

$$V_r = h_{33}\Delta x_r \tag{8.17}$$

where h_{33} is the piezoelectric deformation constant. The [g] coefficient expresses the change in electric field with pressure. If a pressure P is applied to the piezoelectric material, the voltage across it is

$$V_r = g_{33} \cdot l \cdot P \tag{8.18}$$

where
   l is the thickness
   g_{33} is the piezoelectric pressure constant

When the transducer works as a receiver, the value of g_{33} should be high for obtaining sufficient sensitivity. Therefore, although d_{33} may be small, piezoceramics may still have a high g_{33}.

If the permittivity is written in a complex form as $\varepsilon = \varepsilon' - j\varepsilon''$, the imaginary term corresponds to real power loss. The tangent of the material, defined as $\tan \delta = \varepsilon''/\varepsilon'$, gives the angle by which the current through an ideal capacitor. $Q_e = \varepsilon'/\varepsilon''$ is the maximum energy stored over a cycle divided by the energy dissipated per cycle. Meanwhile, $Q_m = c'/c''$ accounts for viscous losses.

The piezoelectric coupling factor represents the energy conversion efficiency from mechanical to electrical, or vice versa, and is used to assess and compare the performance of different piezoelectric materials. It is defined as the square root of electrical (mechanical) work done under ideal conditions in comparison to the total energy stored from a mechanical (electrical) source in the piezoelectric and inverse piezoelectric effects, respectively:

$$k_{ij}^2 = h_{ji} d_{ji} \qquad (8.19)$$

To derive the coupling factor, a compressive force ($T_3 \neq 0$, $T_1 = T_2 = 0$) is applied to a piezoelectric material with electrodes on its top and bottom. If the electrodes are shortened, then the resulting strain is $S_{zz} = s_{zz}^E T_3$ and the change takes the path from (a) to (b) as shown in Figure 8.1. If the electrodes are opened after the cessation of the compressive force, then the strain will decrease from (b) to (c) at a constant $[D]$. If the electrodes are then connected to an electrical load, then the strain reduces to 0 along the path from (c) to (d). As a result, energy will be delivered to the load. During the compression process, the energy stored is $W_1 + W_2$. With the electrical load, the work done is W_1. So, the piezoelectric coupling factor is

$$k_{33}^l = \sqrt{\frac{W_1}{W_1 + W_2}} = \sqrt{\frac{s_{33}^E - s_{33}^D}{s_{33}^E}} \qquad (8.20)$$

Since $s_{33}^E - s_{33}^D = d_{33}^2 / \varepsilon_{33}^T$, this equation can be rewritten as

$$k_{33}^l = \frac{d_{33}}{\sqrt{s_{33}^E \varepsilon_{33}^T}} \qquad (8.21)$$

If the piezoelectric material is clamped in the lateral direction ($S_3 \neq 0$, $S_1 = S_2 = 0$), then the coupling factor can also be derived using a similar method as

$$k_{33}^t = \frac{e_{33}}{\sqrt{c_{33}^D \varepsilon_{33}^S}} \qquad (8.22)$$

Then, the overall electromechanical coupling coefficient is given by

$$1 - k_{33}^2 \approx \left[1 - \left(k_{33}^l \right)^2 \right] \left[1 - \left(k_{33}^t \right)^2 \right] \qquad (8.23)$$

It is clear that if $k_{33}^l = k_{33}^t$, then the efficiency of the piezoelectric element will be doubled.

The matrices used for a piezoelectric material can be greatly simplified by taking into consideration the symmetries in the crystal structure. The poling process causes the polycrystallites to become anisotropic, with an axis of symmetry that coincides with the poling direction. Because of the high degree of symmetry, many coefficients become zero and the number of independent ones is significantly reduced. The elastic stiffness matrix then becomes

$$[c^E] = \begin{bmatrix} c_{11}^E & c_{12}^E & c_{13}^E & 0 & 0 & 0 \\ c_{12}^E & c_{11}^E & c_{13}^E & 0 & 0 & 0 \\ c_{13}^E & c_{13}^E & c_{11}^E & 0 & 0 & 0 \\ 0 & 0 & 0 & c_{44}^E & 0 & 0 \\ 0 & 0 & 0 & 0 & c_{44}^E & 0 \\ 0 & 0 & 0 & 0 & 0 & c_{66}^E \end{bmatrix} \qquad (8.24)$$

where $c_{66}^E = \frac{1}{2} \left(c_{11}^E - c_{12}^E \right)$.

$$[e] = \begin{bmatrix} 0 & 0 & 0 & 0 & e_{15} & 0 \\ 0 & 0 & 0 & e_{15} & 0 & 0 \\ e_{31} & e_{31} & e_{33} & 0 & 0 & 0 \end{bmatrix} \qquad (8.25)$$

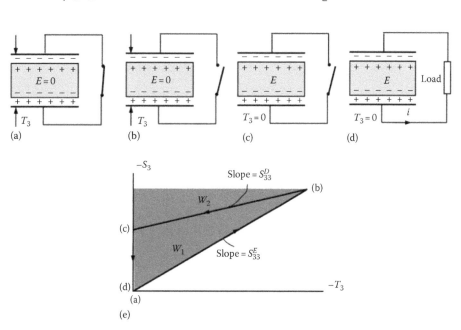

FIGURE 8.1 Derivation of the piezoelectric coupling factor for an ultrasound transducer by considering a mechanical to electrical work conversion cycle $T_3 \neq 0$, $T_1 = T_2 = 0$. (a) A compressive stress is applied to the electrodes shorted. (b) The electrodes are open while compressed. (c) The compressive stress is removed while the electric displacement D_3 remains unchanged but not the electric field. (d) The electrodes are connected to an electrical load. (e) Work cycle is used for calculating the coupling factor k_{33}^l by using the mechanical source, $W_1 + W_2$, where W_1 is the electrical work done.

$$[\varepsilon^S] = \begin{bmatrix} \varepsilon_{11}^S & 0 & 0 \\ 0 & \varepsilon_{11}^S & 0 \\ 0 & 0 & \varepsilon_{33}^S \end{bmatrix} \qquad (8.26)$$

Therefore, there are in total only 10 independent coefficients to describe the properties of piezoceramics. A single matrix can be used to express each pair of constitutive equations, such as

$$\begin{bmatrix} S_1 \\ S_2 \\ S_3 \\ S_4 \\ S_5 \\ S_6 \\ D_1 \\ D_2 \\ D_3 \end{bmatrix} = \begin{bmatrix} s_{11}^E & s_{12}^E & s_{13}^E & 0 & 0 & 0 & 0 & 0 & d_{31} \\ s_{12}^E & s_{11}^E & s_{13}^E & 0 & 0 & 0 & 0 & 0 & d_{31} \\ s_{13}^E & s_{13}^E & s_{33}^E & 0 & 0 & 0 & 0 & 0 & d_{33} \\ 0 & 0 & 0 & s_{44}^E & 0 & 0 & 0 & d_{15} & 0 \\ 0 & 0 & 0 & 0 & s_{44}^E & 0 & d_{15} & 0 & 0 \\ 0 & 0 & 0 & 0 & 0 & s_{66}^E & 0 & 0 & 0 \\ 0 & 0 & 0 & 0 & d_{15} & 0 & \varepsilon_{11}^T & 0 & 0 \\ 0 & 0 & 0 & d_{15} & 0 & 0 & 0 & \varepsilon_{11}^T & 0 \\ d_{31} & d_{31} & d_{33} & 0 & 0 & 0 & 0 & 0 & \varepsilon_{33}^T \end{bmatrix} \begin{bmatrix} T_1 \\ T_2 \\ T_3 \\ T_4 \\ T_5 \\ T_6 \\ E_1 \\ E_2 \\ E_3 \end{bmatrix}$$

$$(8.27)$$

where $s_{66}^E = \frac{1}{2}\left(s_{11}^E - s_{12}^E\right)$.

In contrast, polymer-based piezoelectric films, such as PVDF with long-chain polar molecules embedded in an amorphous phase matrix, are semicrystalline. They are asymmetric about the polarization axis. Consequently, there are a higher number of independent mechanical, electrical, and piezoelectric coefficients:

$$\begin{bmatrix} S_1 \\ S_2 \\ S_3 \\ S_4 \\ S_5 \\ S_6 \\ D_1 \\ D_2 \\ D_3 \end{bmatrix} = \begin{bmatrix} s_{11}^E & s_{12}^E & s_{13}^E & 0 & 0 & 0 & 0 & 0 & d_{31} \\ s_{12}^E & s_{22}^E & s_{23}^E & 0 & 0 & 0 & 0 & 0 & d_{32} \\ s_{13}^E & s_{23}^E & s_{33}^E & 0 & 0 & 0 & 0 & 0 & d_{33} \\ 0 & 0 & 0 & s_{44}^E & 0 & 0 & 0 & d_{24} & 0 \\ 0 & 0 & 0 & 0 & s_{55}^E & 0 & d_{15} & 0 & 0 \\ 0 & 0 & 0 & 0 & 0 & s_{66}^E & 0 & 0 & 0 \\ 0 & 0 & 0 & 0 & d_{15} & 0 & \varepsilon_{11}^T & 0 & 0 \\ 0 & 0 & 0 & d_{24} & 0 & 0 & 0 & \varepsilon_{22}^T & 0 \\ d_{31} & d_{32} & d_{33} & 0 & 0 & 0 & 0 & 0 & \varepsilon_{33}^T \end{bmatrix} \begin{bmatrix} T_1 \\ T_2 \\ T_3 \\ T_4 \\ T_5 \\ T_6 \\ E_1 \\ E_2 \\ E_3 \end{bmatrix}$$

$$(8.28)$$

The properties of the most popular piezoelectric materials are listed in Table 8.1. It can be seen that barium titanate or PZT is used widely as radiators because of their high value of e_{ii}. But the other parameters, such as the coupling of the transducer to the electrical circuit, the internal losses, phase velocity, dielectric constant, and the Curie temperature that determines the working range of temperature, and flexibility and ease of fabrication, should also be considered in the transducer design. Quartz and piezoceramics have high acoustic impedances due to their dense and

---

TABLE 8.1
Properties of Piezoelectric Materials

Material	Quartz (X-Cut)	Barium Titanate (BaTiO₃)	Lead Zirconium Titanate (PZT)	Polyvinylidene Fluoride (PVDF)	Lead Metaniobate (PbNb₂O₆)	Lithium Sulfateobate	Lithium Niobate (LiNbO₃)
Density ρ_0 (kg/m³)	2650–2700	5300–5700	7500–7800	1300–1800	6200	2060	4640
Elastic stiffness C_{ii} (N/m²)	86×10^9	110×10^9	83×10^9	3×10^9			
Sound speed c (m/s)	5740–5800	5200–5300	4000–4200	1400–2600	3300	5460	7320
Acoustic impedance Z_c (ray)	15×10^6	30×10^6	30×10^6	2.5×10^6			
Relative dielectric constant: ε_r	4.5	960–1700	480–1200	11.5–12	300	10.3	30
Piezoelectric stress coeff. e_{ii} (N/V·m)	0.17	8.6	9.2	0.069			
d_{33} (m/V)	2.3×10^{-12}	125×10^{-12}	125×10^{-12}	25×10^{-12}	85×10^{-12}	15×10^{-12}	6×10^{-12}
Transmission	Poor	Excellent	Excellent	Fair	Good	Fair	Poor
g_{33} (V/m)	57×10^{-13}	14×10^{-13}	30×10^{-13}	230×10^{-13}	32×10^{-13}	156×10^{-13}	23×10^{-13}
Receiving	Good	Fair	Fair	Excellent	Fair	Excellent	Fair
h_{33} (V/m)	4.9×10^9	1.5×10^9	—	—	1.9×10^9	8.2×10^9	6.7×10^9
K_t	0.1	0.33	0.35	0.1–0.14	0.38	0.38	0.47
K_r	0.1	−0.25	−0.46	—	−0.07	—	—
K_{33}^2		0.20	0.45		0.16		
Q	>10,000	350	400	<15	15	>1,000	>1,000
T_C (K)	847	393	618	438–453	>673	403	1483
Temperature tolerance	Excellent	Poor	Poor	Good	Good	Poor	Excellent

relatively incompressible nature. The polymer material is much softer and less dense. The acoustic impedance of PVDF is close to that of tissue, and therefore more power can be coupled into the tissue. PVDF material has a broad bandwidth, but it suffers from large internal loss and low temperature tolerance, which limit the amount of emission power due to low piezoelectric transmission coefficients and heat generation by the internal loss. So, it is used more in the receiver mode.

8.2 EQUIVALENT CIRCUIT

For a usual ultrasound transducer, its two opposite faces are plated with conductive metal films. A voltage V is applied to the electrodes to produce an electric field E_3 across the thickness l of the transducer whose magnitude is given by

$$E_3 = \frac{V}{l} \tag{8.29}$$

The surface charge density at the transducer surface is

$$\sigma_3 = \varepsilon_r \varepsilon_0 E_3 - d_{33} p_3 \tag{8.30}$$

where p_3 is the generated acoustic pressure and determined by

$$p_3 = e_{33} E_3 - c_{33} \left(\frac{\partial \xi}{\partial z} \right) \tag{8.31}$$

The transducer can be considered as a parallel-plate capacitor, and its capacitance is given by

$$C_0 = \frac{q}{V} = \frac{\sigma_i A}{E_i l} = \frac{\varepsilon_e A}{l} \tag{8.32}$$

where
  q is the total charge on each plate
  A is the area of plate or the transducer area
  ε_e is the effective dielectric constant of material between the plates

If the transducer is clamped, $\partial \xi / \partial z = 0$, and

$$\sigma_3 = \varepsilon_r \varepsilon_0 E_3 - d_{33} e_{33} E_3 = \varepsilon_r \varepsilon_0 \left(1 - \frac{d_{33} e_{33}}{\varepsilon_0 \varepsilon_r} \right) E_3$$

$$= \varepsilon_r \varepsilon_0 (1 - g_{33} e_{33}) E_3 = \varepsilon_r \varepsilon_0 (1 - \kappa^2) E_3 \tag{8.33}$$

So, the effective dielectric constant under clamped conditions is given by

$$\varepsilon_e = \varepsilon_r \varepsilon_0 (1 - \kappa^2) \tag{8.34}$$

And, the corresponding capacitance of the transducer is

$$C_0 = \varepsilon_r \varepsilon_0 (1 - \kappa^2) \frac{A}{l} \tag{8.35}$$

where $\kappa = \sqrt{g_{33} e_{33}}$ is the coefficient of electromechanical coupling of the materials.

The velocity of the front face of the transducer at resonance is

$$u_f = \frac{2 e_{33} E_3}{Z_1 + Z_2} \tag{8.36}$$

where Z_1 and Z_2 are the acoustic impedance of the media on either side of the transducer (back and front sides, respectively). If the ultrasound transducer has air in the back region, as in most cases of therapy, $Z_1 \approx 0$ and then Equation 8.36 can be simplified as

$$u_f = \frac{2 e_{33} E_3}{Z_2} \tag{8.37}$$

The acoustic intensity transmitted is then easily found from velocity continuity and the relationship $I = Z u^2$:

$$I = \frac{4 e_{33}^2 E_3^2}{Z_2} \tag{8.38}$$

For the most popular configuration, where the transducer is excited by a sinusoidal voltage source, the average radiated intensity is given by

$$I_{ave} = \frac{2 e_{33}^2 V_0^2}{l^2 Z_2} \tag{8.39}$$

where V_0 is the peak sinusoidal exciting voltage and a factor ½ was used to give the time average of the intensity.

Thus, at the resonant frequency, the transducer is equivalent to the parallel connection of a capacitor C_0 and a resistor R_m, which represents the transformation of electrical power into radiated acoustic power (Figure 8.2):

$$R_m = \frac{V_0^2}{2 I_{ave}} = \frac{l^2 Z_2}{4 e_{33}^2 A} \tag{8.40}$$

The capacitance C_0 is moderately high due to the large value of ε_r for many piezoelectric materials, and the resistance R_m is inversely proportional to the radiated acoustic power. So, high radiation means a low value of R_m. To efficiently match the resonance frequency of the transducer to the driving generator, a parallel inductor L_0 is added between the transducer and the generator, and its value is chosen so that the electrical resonant frequency $\omega = \sqrt{1/L_0 C_0}$ is matched to that of acoustic resonance. However, this model works exactly only at the resonance frequency. In a general description, an inductor L and a capacitor C in series with R_m characterize the equivalent circuit. Their impedances cancel each other at resonance, and the equivalent circuit is capacitive below resonance but inductive above it. Two more resistors, a parallel resistor R_k and a series one R_a, can be included to account for the leakage current and internal absorption in the material, respectively.

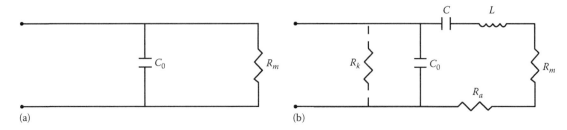

FIGURE 8.2 Equivalent electrical circuit for (a) a lossless transducer exactly at resonance and (b) in the neighborhood of resonance. C_0 is the parallel-plate capacitance of the device, R_m is a resistance representing the acoustic radiation power, L is the inductance, C is the capacitance, and R_k and R_a are possible nonradiative losses in the transducer. At resonance, the impedances of L and C cancel each other.

The behavior of transducer can also be represented as a three-port device: one electrical port for the current and voltage between the two parallel plates, and two acoustic ports for the force and particle velocity on both surfaces. In a first-order approximation, the strain within the elementary volume of the transducer is

$$S \approx \frac{\partial \xi}{\partial z} \tag{8.41}$$

Applying Newton's second law to that elementary element, one obtains

$$A \frac{\partial T}{\partial z} dz = \rho_0 A dz \frac{\partial^2 \xi}{\partial t^2} \tag{8.42}$$

which can be simplified as

$$\frac{\partial T}{\partial z} = \rho_0 \frac{\partial^2 \xi}{\partial t^2} \tag{8.43}$$

Using the constitutive relation in the spatial derivative format,

$$\frac{\partial T}{\partial z} = c^D \frac{\partial S}{\partial z} - h \frac{\partial D}{\partial z} \tag{8.44}$$

Since $\partial D/\partial z = 0$ in the absence of free charge in the piezoelectric material, the differential equation for the displacement is

$$c^D \frac{\partial^2 \xi}{\partial z^2} = \rho_0 \frac{\partial^2 \xi}{\partial t^2} \tag{8.45}$$

The forces on the two surfaces relate to the stresses by $F_1 = -AT(0)$ and $F_2 = -AT(l)$. If the wavenumber is denoted by $\beta = \omega/c_0^D = \omega/\sqrt{c^D/\rho_0}$, then the harmonic solution to the displacement differential equation is given by

$$\xi(z, t) = B_1 e^{j(\omega t - \beta z)} + B_2 e^{j(\omega t + \beta z)} = e^{j\omega t}[B_3 \sin(\beta z) + B_4 \cos(\beta z)] \tag{8.46}$$

Then, the velocities on both sides are

$$v_1 = j\omega B_4, \quad v_2 = -v(l) = -j\omega[B_3 \sin(\beta l) + B_4 \cos(\beta l)] \tag{8.47}$$

So, the forces are

$$F_1 = -A[c^D S - hD]_{z=0} = A\left[hD - \beta c^D B_3\right]e^{j\omega t}$$

$$F_2 = -A[c^D S - hD]_{z=l}$$

$$= A\left[hD - \beta c^D B_3 \cos(\beta l) + \beta c^D B_4 \cos(\beta l)\right]e^{j\omega t} \tag{8.48}$$

If the electric displacement phasor can be expressed in terms of the current phasor by $D = I/j\omega A$, and the acoustic radiation impedance is written as $Z_a = AZ_0 = A\rho_0 \sqrt{c^D/\rho_0} = A\sqrt{c^D \rho_0}$, the force equations can be further simplified as

$$F_1 = -j\left[Z_a \cot(\beta l)v_1 + Z_a \operatorname{cosec}(\beta l)v_2 + \frac{hI}{\omega}\right]$$

$$F_2 = -j\left[Z_a \operatorname{cosec}(\beta l)v_1 + Z_a \cot(\beta l)v_2 + \frac{hI}{\omega}\right] \tag{8.49}$$

The applied voltage is

$$V = \int_0^l E dz = \int_0^l \left(-hS + \frac{D}{\varepsilon^S}\right) dz$$

$$= \frac{Dl}{\varepsilon^S} - h \int_0^l \frac{\partial \xi}{\partial z} dz = \frac{Dl}{\varepsilon^S} - h\left[B_3 \sin(\beta z) + B_4 \cos(\beta z)\right]_0^l$$

$$= -\frac{j}{\omega}\left[\frac{Il}{A\varepsilon^S} + h(v_1 + v_2)\right] = -\frac{j}{\omega}\left(hv_1 + hv_2 + \frac{I}{C_0}\right) \tag{8.50}$$

Therefore, the three-port model can be written in the compact matrix form as

$$\begin{bmatrix} F_1 \\ F_2 \\ V \end{bmatrix} = -j \begin{bmatrix} Z_a \cot(\beta l) & Z_a \operatorname{cosec}(\beta l) & \dfrac{h}{\omega} \\ Z_a \operatorname{cosec}(\beta l) & Z_a \cot(\beta l) & \dfrac{h}{\omega} \\ \dfrac{h}{\omega} & \dfrac{h}{\omega} & \dfrac{1}{\omega C_0} \end{bmatrix} \begin{bmatrix} v_1 \\ v_2 \\ I \end{bmatrix} \tag{8.51}$$

If the transducer is working as a transmitter excited by a voltage V_s, the acoustic impedances of the transmitter and backing material can be written as $Z_B = -F_1/v_1$ and $Z_T = -F_2/v_2$, respectively, and the source impedance as $Z_S = -(V - V_S)/I$. If the propagation loss is considered, the waveform will be represented in a complex form as $\gamma = \alpha + j\beta$, where α is the attenuation constant. So, a more general form of Equation 8.51 is

$$\begin{bmatrix} F_1 \\ F_2 \\ V \end{bmatrix} = -j \begin{bmatrix} Z_a \cot(\gamma l) & Z_a \mathrm{cosec}(\gamma l) & \dfrac{h}{\omega} \\ Z_a \mathrm{cosec}(\gamma l) & Z_a \cot(\gamma l) & \dfrac{h}{\omega} \\ \dfrac{h}{\omega} & \dfrac{h}{\omega} & jR_S + \dfrac{1}{\omega C_0} \end{bmatrix} \begin{bmatrix} v_1 \\ v_2 \\ I \end{bmatrix}$$

(8.52)

The Mason model describes the equivalent circuit using the matrix equation as shown in Figure 8.3, where $Z_1 = Z_2 = jZ_a\tan(\beta l/2)$, $Z_3 = -jZ_a\mathrm{cosec}(\beta l)$, and $\varphi = 1/(C_0 h) = 1/(C_0 g c D)$ (Figure 8.4).

The Redwood model is more amenable to transmission analysis and physical interpretation by replacing Z_1, Z_2, and Z_3 by a transmission line model, which is half the resonant wavelength, $\lambda_0/2$. The transformer output drives this transmission line on the common (shield) connection, and consequently the transformer output appears as two sources, one at each end of the transmission line. The transient acoustic response of the transducer when a step of voltage is applied and the electrical response when a step function of force is applied can be determined (Figure 8.5).

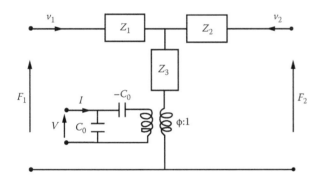

FIGURE 8.3 Mason equivalent circuit model of a piezoelectric material working in the thickness mode.

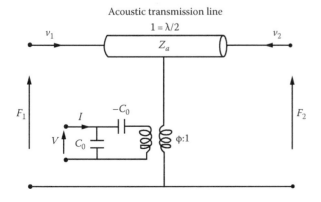

FIGURE 8.4 Redwood model obtained from the Mason model by using an equivalent acoustic transmission line.

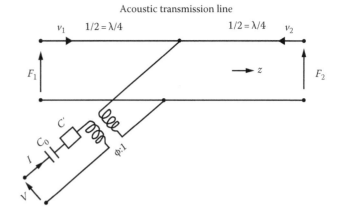

FIGURE 8.5 Equivalent circuit of the KLM model of a disk transducer operating in the thickness.

The KLM model is also based on the compact matrix form Equation 8.52 and governs the terminal properties of all three models by straightforward analysis for longitudinal thickness mode in a piezoelectric disk whose lateral dimensions are larger than its thickness. It contains two quarter-wave acoustic transmission lines with their common terminals connected to the transformer. The KLM model replaces the distributed coupling by a single coupling point, and the resultant discrepancy between this approximation and distributed coupling is included through the frequency-dependent transformer turns ratio and a series reactance. There are two capacitors in series at the electrical port: the clamped capacitor, and one whose capacitance is negative for $\omega < \omega_0$ and approaches $-\infty$ at the resonant frequency. In the neighborhood, $|C'| \gg C_0$, so that C' can be regarded as a short circuit:

$$C' = \frac{-C_0}{(k^t)^2} \frac{\pi\omega/\omega_0}{\sin(\pi\omega/\omega_0)}$$

$$\varphi = k^t \sqrt{\frac{\pi}{\omega_0 C_0 Z_a}} \frac{\sin(\pi\omega/2\omega_0)}{\pi\omega/2\omega_0}$$

(8.53)

In all three models, the transformer ratio is a dimensional parameter, which is due to the interface between the electrical and mechanical parts of the circuit. For the Mason and Redwood models, the transformer turns ratios are independent of the frequency, but not for the KLM model. The piezoelectric material behaves as a distributed acoustic delay line that is excited throughout its thickness by the time-varying electric field.

8.3 TRANSDUCER DESIGN

The lowest frequency that satisfies the resonance condition is called the fundamental frequency of the crystal (Figure 8.6). Usually, a standing wave of a single half-wavelength is established in the piezoelectric material with pressure nodes on both surfaces. For a transducer of thickness l, the fundamental frequency f_0 will have a wavelength λ_0 inside the transducer such that

$$\frac{\lambda_0}{2} = l$$

(8.54)

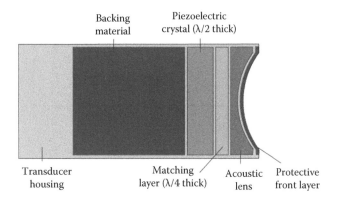

FIGURE 8.6 Schematic diagram of an ultrasound transducer.

Since $\lambda_0 = c_l/f_0$, where c_l is the longitudinal wave velocity in the piezoelectric material,

$$f_0 = \frac{c_l}{2l} \tag{8.55}$$

However, at a very high frequency, the wavelength is quite small, making some crystals fragile. So, transducer may be employed in the higher harmonic modes. For example, if working in the third harmonic mode, the thickness of the crystal can be increased by three times.

The narrowness or broadness of the resonance frequency of a transducer, as measured by the frequency width Δf to the half-power points or decreased by −6 dB, is defined by the quality factor Q:

$$\frac{f_0}{\Delta f} = Q \tag{8.56}$$

Thus, a high Q means a very narrowly peaked resonance, which is good for energy radiation, whereas a low-Q transducer has a broadband response, which is preferred as a receiver. The value of Q is determined by the internal loss in the transducer, where the largest contributor is the transmission of acoustic power through the faces into the neighboring regions. If there is air on both sides of the transducer, the significant impedance mismatch leads to difficulty in the escape of acoustic power and an extremely high Q, such as for the accurate frequency control in quartz watches. In biomedical ultrasound applications, tissue replaces air at the front surface to transmit the acoustic wave. Subsequently, the Q value decreases dramatically due to the reduction in the impedance mismatch.

In order to maximize the acoustic energy coupling to soft tissue or water, a quarter-wavelength layer at the resonant frequency and acoustic impedance of $Z_M = \sqrt{Z_a Z_T}$ will be attached to the front surface of piezoceramics, as was shown in Chapter 3. The radiation resistance becomes

$$R_{a0} = \frac{2(k^t)^2}{\pi^2 f_0 C_0} \frac{Z_a}{Z_B + Z_T} \approx \frac{2(k^t)^2}{\pi^2 f_0 C_0} \tag{8.57}$$

because $Z_B \approx 0$ in the air-backed case, and $Z_a = Z_T$ after applying the matching layer.

In the absence of any load on the transducer, the total electrical input impedance is determined using the KLM model as

$$Z_{in} = \frac{V}{I} = \frac{1}{j\omega C_0}\left[1 - \frac{(k^t)^2}{(\pi\omega/2\omega_0)}\tan\left(\frac{\pi\omega}{2\omega_0}\right)\right] = \frac{1}{j\omega C_0} + Z_m \tag{8.58}$$

where Z_m is the motional acoustic impedance and can be identified as the additional input impedance caused by the acoustic motion. It is obvious that at the odd harmonics, $\omega = (2n + 1)\omega_0$, anti-resonances (parallel resonances) occur, and $|Z_{in}| \to \infty$. In contrast, resonances occur at $|Z_{in}| = 0$ corresponding to

$$\frac{\pi\omega}{2\omega_0} = (k^t)^2 \tan\left(\frac{\pi\omega}{2\omega_0}\right) \tag{8.59}$$

So, the shift in the series resonance frequency can be attributed to the electromechanical coupling, and the series and parallel resonance frequencies (ω_s and ω_p, respectively) are related by

$$k^t = \sqrt{\frac{\pi\omega_s}{2\omega_p}\tan\frac{\pi(\omega_p - \omega_s)}{2\omega_p}} \tag{8.60}$$

Thus, by measuring the two resonance frequencies for an unloaded transducer, the coupling factor can be determined.

In comparison, the motional acoustic impedance of the transducer for any load or backing impedance is given by

$$Z_m = \frac{(k^t)^2\left\{-2Z_a^2\left[1-\cos\left(\frac{\pi\omega}{\omega_0}\right)\right] + jZ_a(Z_B + Z_T)\sin\left(\frac{\pi\omega}{\omega_0}\right)\right\}}{j\omega C_0\left(\frac{\pi\omega}{\omega_0}\right)\left[(Z_a^2 + Z_B Z_T)\sin\left(\frac{\pi\omega}{\omega_0}\right) - jZ_a(Z_B + Z_T)\cos\left(\frac{\pi\omega}{\omega_0}\right)\right]} \tag{8.61}$$

As expected, the condition is changed to the unloaded one when $Z_B = Z_T = 0$.

In order to increase the delivery of energy from an electrical source to an acoustic load, both the electrical matching between the source and transducer and the acoustic matching conditions at the front and back ports should be carefully considered.

By using an electrical matching network between the source and the transducer, the power transfer efficiency can be optimized based on the well-known fact that the maximum power transfer is obtained when the load impedance is equal to the complex conjugate of the source impedance, that is,

$$Z_S = Z_L^* \tag{8.62}$$

For a source with real output impedance, the network should cancel out the imaginary part of the load impedance. If the transducer is represented as the series combination of an acoustic radiation resistance $R_a(\omega)$ and a reactance $X(\omega)$, the input impedance of the network of the transducer should be equal to the source resistance R_S over the entire frequency range. So, the maximum electrical power delivered by the source is

$$P_e = \frac{|V_S|^2}{8R_S} \qquad (8.63)$$

The power transfer efficiency of both acoustic ports can be defined by

$$\eta = \frac{P_a}{P_e} = \frac{P_a}{|V_S|^2/(8R_S)} \qquad (8.64)$$

where P_a is the total acoustic output power at both ports. As a result, the transfer efficiency for the front port is reduced by a factor of $Z_T/(Z_T + Z_B)$. If the loads on both ports are equal, the transfer efficiency will be decreased by −3 dB (Figure 8.7).

The input impedance of transducer can be modeled by an acoustic radiation resistance R_{a0} in series with a reactance (usually a capacitor). The general optimization of the electrical matching network leads to the maximum power transfer over the required bandwidth. If $L = 1/(\omega_0 C_0)$, the input impedance seen by the source will be purely resistive and the power transfer efficiency is

$$\eta = \frac{4R_{a0}R_S}{(R_{a0} + R_S)^2} \qquad (8.65)$$

which is ideally unity when $R_{a0} = R_S$. In the general off-resonance case, the input impedance of the transducer becomes $Z_{\text{in}}(\omega) = R_a(\omega) + jX(\omega)$ and the inductance is $L = 1/(\omega_0^2 C_0)$. So, the efficiency is

$$\eta = \frac{4R_{a0}R_S}{(R_{a0} + R_S)^2 + \left[\omega/\left(\omega_0^2 C_0\right) + X(\omega)\right]^2} \qquad (8.66)$$

Using a matching layer thickness of exactly $\lambda/4$, the tolerated angular frequency range $\Delta\omega$ is

$$\frac{\Delta\omega}{\omega_0} = 2 - \frac{4}{\pi}\cos^{-1}\left|\frac{2R_m\sqrt{Z_a Z_T}}{(Z_T - Z_a)\sqrt{1 - R_m^2}}\right| \qquad (8.67)$$

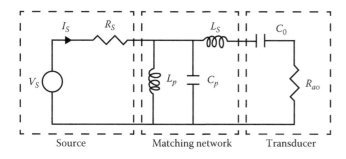

Source   Matching network   Transducer

FIGURE 8.7 General electrical diagram of the driving source, matching network, and ultrasound transducer.

where R_m is the maximum reflection coefficient. Consequently, the impedance seen by port 2 is

$$Z_i(\omega) = \sqrt{Z_a Z_T}\,\frac{Z_T + j\sqrt{Z_a Z_T}\,\tan(a_m\omega/\omega_0)}{\sqrt{Z_a Z_T} + jZ_T\tan(a_m\omega/\omega_0)} \qquad (8.68)$$

where $a_m = \alpha\pi/2$ and $\alpha = 1$ for $\lambda/4$ at ω_0.

Meanwhile, the acoustic impedance of the backing material has a major effect on the power transfer efficiency and bandwidth. If the transducer is air backed, $Z_B \approx 0$, and almost no power is lost through the backing. At the resonance frequency, the acoustic impedance R_{a0} is in series with C_0, so the acoustic power delivered to either port is

$$P_a(\omega_0) = \frac{|V_S|^2 R_{a0}/2}{(R_{a0} + R_S)^2 + 1/(\omega_0 C_0)^2} \qquad (8.69)$$

If $Z_B = Z_T$, the power transfer efficiency becomes

$$\eta = \frac{2R_S R_{a0}}{(R_{a0} + R_S)^2 + 1/(\omega_0 C_0)^2} \qquad (8.70)$$

So, the maximum power transfer efficiency is achieved at the condition $R_S = R_{a0}\sqrt{1 + 1/(\omega C_0 R_{a0})^2}$, and the optimal transfer efficiency is

$$\eta_F^{\text{opt}} = \frac{1}{1 + \sqrt{1 + \pi^2/(k^t\sqrt{2})^4}} \qquad (8.71)$$

8.4 CAPACITIVE TRANSDUCER

The idea of using the electrostatic force between different charges on a conducting membrane and a closely spaced metal plate as a means of both emitting and receiving sound wave can be traced back to the 1880s. With the development of silicon-based microfabrication technology in the integrated circuit (IC) industry, this method has taken a new life. For a simple structure of a parallel plate in which the moveable upper plate is supported by a spring and the applied voltage is V, the attractive electrostatic force between the plates can be found from the change in stored energy for an incremental change in plate separation:

$$F_C = -\frac{\partial W}{\partial z} = -\frac{\partial}{\partial z}\left(\frac{1}{2}CV^2\right) = -\frac{1}{2}V^2\frac{\partial}{\partial z}\left(\frac{\varepsilon A}{l-z}\right) = \frac{\varepsilon AV^2}{(l-z)^2} \qquad (8.72)$$

where

l is the thickness of the medium
ε is its permittivity
A is the plate area
C is the capacitance

In the absence of any DC bias but only a sinusoidal source of $V_{ac} \sin(\omega t)$, the attractive force is

$$F_C = \frac{\varepsilon A [V_{ac} \sin(\omega t)]^2}{(l-z)^2} = \frac{\varepsilon A V_{ac}^2 [1-\cos(2\omega t)]}{2(l-z)^2} \quad (8.73)$$

which corresponds to a DC component and a second harmonic component. In the case of $V = V_p + V_{ac} \sin(\omega t)$ and $V_{ac} \ll V_p$, the force is given by

$$F_C = \frac{\varepsilon A [V_p + V_{ac} \sin(\omega t)]^2}{(l-z)^2} \approx \frac{\varepsilon A \left[V_p^2 + 2V_p V_{ac} \sin(\omega t) \right]}{(l-z)^2} \quad (8.74)$$

Consequently, by applying a polarizing voltage or by incorporating a medium with a permanent polarization, the fundamental component can be retained.

The restoring force comes from the supported spring, so the equilibrium position is given by

$$F_C = F_R = kz = \frac{\varepsilon A V^2}{(l-z)^2} \quad (8.75)$$

If the electrostatic force can no longer be balanced by the spring's restoring force at a certain distance, an increment in the applied voltage will cause a catastrophic collapse of the moveable plate to the underlying one, of which the approximate critical voltage V_c is determined by using the condition $\partial z / \partial V = \infty$:

$$V_c = \sqrt{\frac{8kl^3}{27\varepsilon A}} \quad (8.76)$$

where the critical distance is $z_c = l/3$. As the critical voltage is approached, the sensitivity to an incident pressure on the membrane displacement increases, and the electromechanical coupling factor could be as high as 0.7.

8.5 TRANSDUCER ARRAY

8.5.1 Pseudo-Inverse Approach

The pseudo-inverse approach developed for multifocus synthesis has been widely utilized in phased-array applications. The theory of this approach is introduced briefly here. For an array consisting of N elements with an arbitrary geometry, the pressure at a specific point could be expressed as

$$p(r) = \frac{j\rho c k}{2\pi} \sum_{n=1}^{N} u_n \int_S \frac{e^{-jk(r-r')}}{|r-r'|} dS \quad (8.77)$$

where
  u_n and S are the velocity and surface area of the nth element, respectively
  r is the distance between the specific point and the element
  r' is the distance between the specific point and the center of the phased-array transducer

If the number of specific points is M, Equation 8.77 can be written as

$$p(r_m) = \frac{j\rho c k}{2\pi} \sum_{n=1}^{N} u_n \int_S \frac{e^{-jk(r_m-r')}}{|r_m-r'|} dS, \quad m = 1,2,...,M \quad (8.78)$$

This equation can be expressed in matrix form as

$$[H] \cdot \{u\} = \{p\} \quad (8.79)$$

where
  $\{u\} = [u_1, u_2, ..., u_n]'$ is the complex excitation vector
  The vector $\{p\} = [p(r_1), p(r_2), ..., p(r_M)]'$ denotes the complex pressure at the specific points in the field
  $[H]$ is the forward propagation operator with the elements given by

$$H(m,n) = \frac{j\rho c k}{2\pi} \sum_{n=1}^{N} u_n \int_S \frac{e^{-jk(r_m-r_n)}}{|r_m-r_n|} dS \quad (8.80)$$

When $[H]$ is full rank, Equation 8.79 has the minimum-norm solution as

$$\{u\} = [H^*]'([H] \cdot [H^*]')^{-1} \cdot \vec{p} \quad (8.81)$$

where
  $[H^*]'$ is the conjugate transpose of $[H]$
  $\hat{u}$ is the final driving signal

The array's excitation efficiency η_A is defined as

$$\eta_A = \frac{<u_e, u_e>}{NU_{max}^2} \times 100\% = \frac{\sum_{n=1}^{N} |u_{en}|^2}{NU_{max}^2} \times 100\% \quad (8.82)$$

where
  u_e is the actual excitation vector of the array
  $< *,* >$ defines the inner product of two complex column vectors
  U_{max} is the maximum amplitude of particle velocity attainable at the surface of the transducer elements used in fabricating the array

The elements of the complex excitation vector vary in both amplitude and phase. A wide dynamic range of the amplitudes of the driving signals means that the array operates with low excitation efficiency. In order to increase the efficiency, a weighting function is implemented to the pseudo-inverse phasing scheme. Subsequently, the weighted minimum-norm solution is given by

$$\{u\} = [H^*]'([H] \cdot [W] \cdot [H^*]')^{-1} \cdot \{p\} \quad (8.83)$$

where $[W]$ is an $N \times N$ real positive definite weighting matrix.

A judicious choice of [W] can significantly improve the array efficiency when compared with the unweighted solution. An iterative weighting algorithm based on Equation 8.83 was utilized to increase the array excitation efficiency to the desired level so that the specified power deposition at the control points could be achieved. The algorithm is summarized as follows:

Step 0: [W] = [I], where [I] is the identity matrix.
Step 1: Compute $\hat{u}_w$ using Equation.
Compute η_A using Equation 8.82.
If η_A is sufficient, go to Step 3. Otherwise, $[H^*]' = [W] \cdot [H^*]'$
Step 2: Construct the new weighting matrix [W]

$$W(m, n) = \begin{cases} \dfrac{1}{|\hat{u}_{wn}|} & m = n \\ 0 & m \neq 0 \end{cases} \qquad (8.84)$$

Step 3: Go to step 1.

In step 2, $\{u_{wn}; n = 1, 2, \ldots, N\}$ are the elements of the vector $\{u_w\}$. The choice of [W] in Equation 8.84 intuitively forces the vector $\hat{u}_w$ toward uniform amplitude distribution after each iteration. This weighting algorithm is straightforward and has demonstrated to be very useful in increasing the array excitation efficiency to nearly 100% efficacy in most cases.

REFERENCES

Auld BA. *Acoustic Fields and Waves in Solids*. Malabar, FL: Krieger Publishing, 1990.

Behari J. Piezoelectricity in bone. In: *Biophysical Bone Behavior: Principles and Applications*. Chichester, U.K.: John Wiley & Sons, 2009, 53–100 (Behari's own book).

Cady WG. *Piezoelectricity*. Mineola, NY: Dover, 1964.

Curie P, Curie J. Développement par pression de l'électricit polaire dans les cristaux hémièdres à faces enclinées. *Comptes Rendus* 1880;91:383.

de Jong N, Souquet J, Bom N. Vibration modes, matching layers, and grating lobes. *Ultrasonics* 1985;24:176–182.

Foster FS. Transducer materials and probe construction. *Ultrasound in Medicine and Biology* 2000;26:S2–S5.

Fukada E, Ando Y. Piezoelectricity in oriented DNA films. *Journal of Polymer Science Part A-2: Polymer Physics* 2003;10:565–567.

Fukada E, Yasuda I. On the piezoelectric effect of bone. *Journal of the Physical Society of Japan* 1957;12:1158–1162.

Hunt JW, Arditi M, Foster FS. Ultrasound transducers for pulse-echo medical imaging. *IEEE Transactions on Biomedical Engineering* 1983;BME-30:452–481.

Hutson AR, White DL. Elastic wave propagation in piezoelectric semiconductors. *Journal of Applied Physics* 1962;33:40–47.

Jaffe B, Cook WR, Jaffe H. *Piezoelectric Ceramics*. New York: Academic Press, 1971.

Kawai H. The piezoelectricity of polyvinylidene fluoride. *Japanese Journal of Applied Physics* 1969;8:975–976.

Kino GS. *Acoustic Waves: Devices, Imaging, and Analog Signal Processing*. Englewood Cliffs, NJ: Prentice-Hall, 1987.

Kojima T. Matrix array transducer and flexible matrix array transducer. *IEEE Ultrasonics Symposium Proceedings*, 1986, pp. 649–654, Nov. 17–19, Williamsburg, VA.

Kyame JJ. Conductivity and viscosity effects on the wave propagation in piezoelectric material. *Journal of the Acoustical Society of America* 1954;26:990–993.

Lemanov VV, Popov SN, Pankova GA. Piezoelectricity in protein amino acids. *Physics of the Solid State* 2011;53:1191–1193.

Lerch R. Simulation of piezoelectric devices by two- and three-dimensional finite elements. *IEEE Transactions on Ultrasonics, Ferroelectrics, and Frequency Control* 1990;41:225–230.

Mason WP. An electrochemical representation of a piezoelectric crystal used as a transducer. *Proceedings of the IRE* 1935;23:1252–1263.

Mason WP. *Piezoelectric Crystals and Their Application to Ultrasonics*. New York: Van Nostrand, 1950.

Mason WP. *Physical Acoustics*. New York: Academic Press, 1964.

McKeighen R. Finite element simulation and modeling of 2D arrays for 3D ultrasonic imaging. *IEEE Transactions on Ultrasonics, Ferroelectrics, and Frequency Control* 2001;48:1395–1405.

Mills DM, Smith SW. Finite element comparison of single crystal vs. multi-layer composite arrays for medical ultrasound. *IEEE Transactions on Ultrasonics, Ferroelectrics, and Frequency Control* 2002;49:1015–1020.

Oralkan O, Ergun AS, Johnson JA, Karaman M, Demirci U, Kaviani K, Lee TH, Khuri-Yakub BT. Capacitive micromachined ultrasonic transducers: Next-generation arrays for acoustic imaging? *IEEE Transactions on Ultrasonics, Ferroelectrics, and Frequency Control* 2002;49:1596–1610.

Reid JM, Wild JJ. Current developments in ultrasonic equipment for medical diagnosis. *Proceedings of the National Electronics Conference* 1958;12:1002–1015.

Selfridge AR, Baer R, Khuri-Yakub BT, Kino GS. Computer-optimized design of quarter-wave acoustic matching and electrical networks for acoustic transducers. *IEEE Ultrasonics Symposium Proceedings*, 1981, pp. 644–648, Stanford University, CA, Oct. 14–16.

Selfridge AR, Gehlbach S. KLM transducer model implementation using transfer matrices. *IEEE Ultrasonics Symposium Proceedings*, 1985, pp. 875–877, San Francisco, CA, Oct. 16–18.

Selfridge AR, Kino GS, Khuri-Yakub BT. Fundamental concepts in acoustic transducer array design. *IEEE Ultrasonics Symposium Proceedings*, 1980, pp. 989–993, Boston, MA, Nov. 5–7.

Storri S, Santoni T, Mascini M. A piezoelectric biosensor for DNA hybridisation detection. *Analytical Letters* 1998;31:1795–1808.

Szabo TL. Miniature phased-array transducer modeling and design. *IEEE Ultrasonics Symposium Proceedings*, 1982, pp. 810–814, San Diego, CA, Oct. 27–29.

Szabo TL. Transducer arrays for medical ultrasound imaging. In: Duck FA, Baker AC, Starritt HC, eds., *Ultrasound in Medicine, Medical Science*. Bristol, U.K.: Institute of Physics Publishing, 1998, 91–111.

Tombelli S, Minunni M, Santucci A, Spiriti MM, Mascini M. A DNA-based piezoelectric biosensor: Strategies for coupling nucleic acids to piezoelectric devices. *Talanta* 2006;68:806–812.

von Kervel SJH, Thijssen JM. A calculation scheme for the optimum design of ultrasonic transducers. *Ultrasonics* 1983;21: 134–140.

von Ramm OT. 2D arrays. *Ultrasound in Medicine and Biology* 2000;S1:S10–S12.

Wildes DG, Chiao RY, Daft CMW, Rigby KW, Smith LS, Thomenius KE. Elevation performance of 1.25D and 1.5D transducer array. *IEEE Transactions on Ultrasonics, Ferroelectrics, and Frequency Control* 1997;44:1027–1036.

9 Acoustic Field Calibration and Measurement

9.1 ACOUSTIC PRESSURE MEASUREMENT

Miniature hydrophones are the most frequently used devices for measuring acoustic pressure distribution of medical ultrasound equipment, either diagnostic or therapeutic type. There are two most prominent types of the piezoelectric polymer polyvinylidene fluoride (PVDF) miniature ultrasonic hydrophones configured in either a needle or spot-poled membrane type. In both types, the sensitive element thickness varies from 9 to 50 μm, and effective diameter is in the range of 0.2–1 mm. Reflection from the membrane surface can establish a standing wave for long tone-burst excitation. The most straightforward way is to rotate either the acoustic source or the hydrophone in order to avoid parallel reflecting surfaces. However, the measured signal will decrease and need correction because of its directional response. Another solution is to match the acoustic impedance of the film to that of water and to make it acoustically transparent. The characteristics of hydrophone that are important in the measurement are sensitivity, electrical impedance, frequency response, dynamic range, directional response, and effective size (Figure 9.1).

Frequency response describes the magnitude and phase of the hydrophone's sensitivity as a function of frequency. To reproduce waveforms faithfully, the hydrophone (and accompanying electronics) should have a uniform frequency response in the measurement range, and the −6 dB bandwidth should extend to at least three octaves above the center frequency. A PVDF hydrophone can have useful bandwidths extending beyond 50 MHz. However, the frequency response depends on the PVDF film thickness at higher frequencies and on the construction approach. Because of the nonlinear acoustic phenomenon in the therapeutic ultrasound field at high power, the requirement of broad bandwidth is more significant than that for diagnostic ultrasound measurement (Figure 9.2).

The effective dimensions of the hydrophone, which is different from its geometrical or physical dimensions, should be small compared to the acoustic wavelength and ultrasound beam dimensions for accurate waveform measurement. The measurement result of the hydrophone is the spatial integral of pressure over its active element. So, the more incident beam the hydrophone encompasses, the less the resultant spatial average effect. A hydrophone's effective diameter is usually determined from its directional response $D(f, \theta)$ at some frequency f and angle θ relative to normal incidence, which

is similar to the theoretical simulation of a uniform circular receiver for certain types of needle-type hydrophones, $2J_1(x)/x$

$$x = \frac{\pi d_h f \sin \theta}{c} \tag{9.1}$$

where
J_1 is the first-order Bessel function of the first kind
d_h is the effective hydrophone diameter
c is the speed of sound in the medium

However, in several unique directional response characteristics of the spot-poled membrane hydrophone, significant deviation from the earlier model is found. One reason is the presence of side lobes due to Lamb waves propagating across the PVDF film surface when sound is incident at the critical angle (~50°). Although these side lobes are rather large in comparison to the main lobe, especially at low frequencies (1–2 MHz), their presence has little influence to many acoustic characterization because of the large angles involved. The other problem is the apodized behavior and asymmetry of the sensitive region, which is due to the directional dependence of the piezoelectric sensitivity, and fringe poling fields. The fringe fields arise during electrical poling of the PVDF film, because the poling field is not confined to the volume of polymer film directly between the electrodes, and forms at the electrode boundaries. For electrodes in diameter of no less than 1 mm, the effect of the fringing field on the hydrophone's directional response is apparently small. Usually below 1 mm, the relatively gradual decrease in piezoelectric activity in the radial direction and a distortion of the fringe field at the junction of the electrical leads and the electrodes contribute to apodization and asymmetry in the directional response, respectively. The effective diameter decreases with increase in frequency. The smaller the geometrical electrode diameter, the greater will be the apodization and asymmetry effects. A needle or membrane hydrophone has a limited spatial resolution, which is determined by the size of its active element. To avoid the error due to spatial averaging, the effective radius of a needle hydrophone should be smaller than one-quarter of the acoustic wavelength or one-third of the −6 dB beamwidth of the focused acoustic field.

If the spatial averaging effect of the hydrophone is ignored, the exact nature of the receiving response is immaterial. If the directional response deviates significantly from the uniform receiver model, attention should be paid in interpreting the

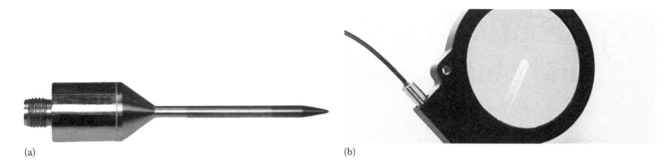

(a)   (b)

FIGURE 9.1 Needle (a) and membrane (b) PVDF hydrophone.

effective diameters. The appropriate compensation factors for effects of spatial averaging by the hydrophone are not universally available. In practice, the effective hydrophone size (or the largest dimension) d_h is recommended to be

$$d_h < \frac{z\lambda}{2d_s} \qquad (9.2)$$

where
  z is the distance between the source and the hydrophone
  λ is the acoustic wavelength
  d_s is the effective diameter of the transducer

A preamplifier is sometimes used together with the hydrophone in acoustic field measurements. For load capacitances ≥15 pF, the hydrophone impedance is <10 kΩ at 1 MHz and <5 kΩ for frequencies larger than 2 MHz. So, the preamplifier should have an input impedance of a few hundred kiloohms. The bandwidth in the frequency response of the preamplifier should not be less than that of the hydrophone. It should also have low noise and linear response to the hydrophone input. The maximum input and output voltage requirements for the preamplifier are determined by the hydrophone sensitivity, maximum expected pressure amplitude, and preamplifier voltage gain. The sensitivity M_I at the preamplifier input for a pressure wave of amplitude P is $M_I = V_I/P$, and the sensitivity M_O at the preamplifier output is $M_O = V_O/P$. If G is the gain of preamplifier, then $M_O = GM_I$, and the following conditions can be required for the maximum input and output voltages, $V_{I\max}$ and $V_{O\max}$:

$$V_{I\max} \ge M_I P_{\max} = \frac{M_O P_{\max}}{G} \qquad (9.3)$$

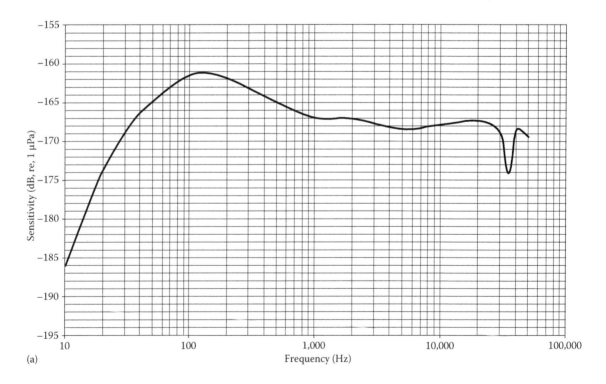

(a)

FIGURE 9.2 Frequency response (a) of a hydrophone.   (*Continued*)

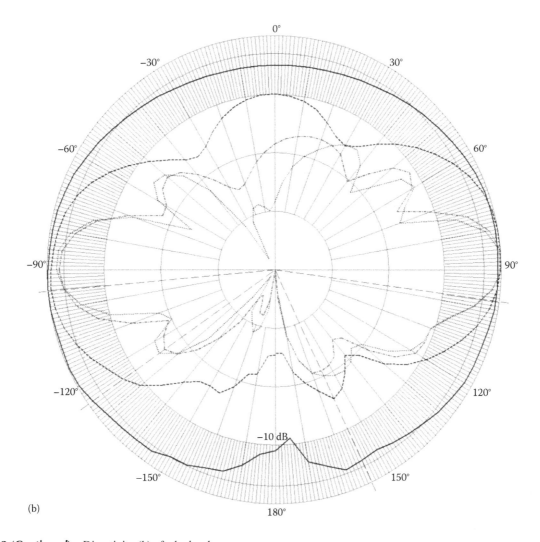

(b)

FIGURE 9.2 (*Continued*) Directivity (b) of a hydrophone.

$$V_{Omax} \geq M_O P_{max} = GM_I P_{max} \qquad (9.4)$$

where P_{max} is the maximum expected pressure amplitude. The maximum current across the load impedance Z_L is

$$I_{max} \geq \frac{V_{Omax}}{|Z_L|} = \frac{M_O P_{max}}{|Z_L|} \qquad (9.5)$$

Usually, Z_L is a 50 Ω resistor in parallel with a capacitor (100 pF from 1 m of 50 Ω coaxial cable). The product of sensitivity and the total capacitance of the measurement system, C_T, including the hydrophone, preamplifier, and oscilloscope, is almost constant:

$$M_I(\text{long}) \cdot C_T(\text{long}) = M_I(\text{short}) \cdot C_T(\text{short}) \qquad (9.6)$$

where "long" and "short" refer to the cable length. Usually, to suppress the effects of cable resonance, the cable length is recommended to be not more than 15 cm between the hydrophone housing and preamplifier. So, its capacitance should satisfy the following condition:

$$C_T(\text{short}) \geq \frac{G \cdot M_I(\text{long}) \cdot C_T(\text{long}) \cdot P_{max}}{V_{Omax}} \qquad (9.7)$$

Mapping of the acoustic field is carried out in a testing tank, whose design should consider the following items: the geometry of the source and hydrophone, the positional precision and mechanical stability of the scanning mechanism, and the alignment of the scanning plane with the plane of the acoustic source. The acoustic measurement tank should be filled with distilled, degassed water, and its walls should be acoustically treated (i.e., equipped with acoustic absorbing material) to suppress reflections that might interfere with the measurements. High conductivity of the tap water acts as a shunt across the exposed leads on the membrane, and consequently the sensitivity will decrease markedly. International Electrotechnical Commission (IEC) has recommended the water conductivity to be less than 10 µS/cm. Degassed water with oxygen concentration of 2–3 mg/L should be used in ultrasound measurement and application to minimize the effect of acoustically induced bubble cavitation, which can be achieved by using vacuum degassing, boiling, and addition of sodium sulfite. When not in use, the simplest way to maintain the degassing status of water is to seal the water surface by the ordinary kitchen plastic wrap. Even in degassed water, tiny bubbles may develop on the face of the transducer and the hydrophone and lead to erroneous measurement results. Thus, any possible air bubbles on the hydrophone face must be removed periodically throughout the acoustic field characterization using a water jet from a syringe because it is not advisable to touch the sensing element of the hydrophone. In comparison, the water temperature of 15°C–25°C has no significant effect on the measurement results.

To produce clean and unambiguous oscilloscope traces, it is necessary to have a stable trigger signal synchronized to the electrical excitation of the ultrasound transducer. It is suggested to check both the trigger signal in synchronization with the source excitation and the delayed pressure profile in an oscilloscope. There are several methods of obtaining the correct trigger signal: an external synchronization signal provided by ultrasound system, RF signal picked up by either an unshielded wire or a coil near the source when emitting acoustic wave, RF signal picked up by unshielded spot-pole membrane hydrophone (needle hydrophone is usually well shielded), acoustic signal measured by the hydrophone aligned close to the transducer, or light picked up by a fast photodetector as spike discharge (e.g., underwater discharge by a shock wave lithotripter). When using the source excitation signal, it should be stable, or jitter will be present in the measurement. When in continuous-wave (CW) field, the picked up RF by an unshielded hydrophone may interfere or overlap with the acoustic signal. However, for most types of therapeutic ultrasound devices, pulsed mode is used in the measurement in order to avoid bubble cavitation damage to the hydrophone.

Before measurement, the transducer and hydrophone should be cleaned and immersed in purified water for at least 30 min with ultrasound system warmed up for at least 15 min in order to ensure that there is no gas bubble attached to the surface of the transducer and hydrophone during the measurement. During warmup, the ultrasound system can be set at a small power output. Before the acoustic field mapping, the most important issue of measurement is that the tilt and rotation axes of the hydrophone (or that of the transducer) should be adjusted to pass through the active element of the hydrophone. Because the hydrophone is usually much lighter than the ultrasound transducer, the hydrophone is suggested to be connected with a 3D translational stage for scanning while the transducer is stationary. There are at least five degrees of freedom, x, y, and z, tilt, and rotation, which are totally independent for easy alignment. The transducer is initially aligned by eye so that the acoustic beam axis is parallel to needle hydrophone or perpendicular to the membrane hydrophone. With the hydrophone

and the transducer at their largest separation, the signal is maximized by manually altering the tilt and rotation of the transducer while monitoring the signal on an oscilloscope. Then, the separation is decreased toward the focal distance, and the position of the hydrophone is optimized again for the maximum signal. The shift of the location of the maximum signal is used for tilt and rotation adjustment. Such a process is repeated until the beam axis of the transducer is parallel to the z-axis of the coordinate positioning system or until the discrepancy of the maximum signal at these two distances is smaller than one-tenth of the wavelength according to GB/T 16846-2008. If the transducer is still in the development stage without being fully waterproof, the transducer's acoustic axis can face downward with only its face immersed in water, which facilitates positioning. The directivity of the hydrophone becomes narrower at higher frequency, so the misalignment may introduce significant measurement errors. However, sometimes an experienced operator can align the hydrophone to the focal point in a shorter time than using a computerized search (Figure 9.3).

After the alignment, the acoustic field will be mapped usually automatically. The measured signal profile at each position, may be after averaging, will be saved in a personal computer (PC) before moving to the next position. A program in the PC will process the data for acoustic intensity and pressure distribution in three dimensions. The spatial sampling interval is typically one-half the acoustic wavelength of the transmitting transducer and may be smaller than the active diameter of the hydrophone itself, whose averaging effect leads to an underestimation of the measurement but widening of the −6 dB beam diameter. If the beam shape is assumed to be a Bessel function, a correction factor is given as $(3 - \beta)/2$, where β is the ratio of the signal measured at one hydrophone radius from the axis to that on the axis. Although it is not intended to be a rigorous method, it can be readily and quickly applied and gives reliable results. The accuracy of the correction is estimated to be ±10% for $\beta > 0.8$, which corresponds to using a hydrophone with a radius less than 0.6 times the −6 dB beam radius.

The important characteristics of the acoustic field include, but is not limited to, the temporal peak positive and negative pressure amplitudes p^+ and p^-; the maximum intensity I_m (i.e., the temporal average intensity over the largest half-cycle in the pressure waveform); the spatial peak-pulse average intensity I_{SPPA}; and spatial peak-temporal average intensity I_{SPTA}; the −6 dB beam size; and the pulse duration. If the measured signal in voltage is $v(t)$, the corresponding acoustic pressure can be obtained using the hydrophone's sensitivity M:

$$p(t) = \frac{v(t)}{M} \tag{9.8}$$

In linear acoustics, the value of M is selected at the driving frequency of ultrasound in the frequency response plot. However, waveform distortion at high-power output due to the nonlinear phenomenon leads to the energy spreading

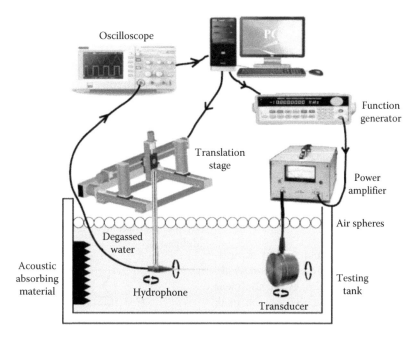

FIGURE 9.3 Schematic diagram of acoustic field mapping by a hydrophone under the control of a computer system.

from the fundamental component to the higher harmonics. In order to obtain accurate measurement, deconvolution could be performed:

$$p(t) = FFT^{-1}\left\{\frac{FFT[v(t)]}{M(f)}\right\} \tag{9.9}$$

where

FFT is the fast Fourier transform
FFT^{-1} is its inverse transform
$M(f)$ is the frequency response of the hydrophone

The frequency range should be selected appropriately to minimize the outcome noise by introducing small values in the denominator.

At the focal point, where the acoustic pressure reaches its maximum value, the spatial peak acoustic intensity I_{SP} is determined as

$$I_{SP}(t) = \frac{p^2(t)}{2\rho_0 c_0} \tag{9.10}$$

Then, the spatial-peak pulse-average acoustic intensity I_{SPPA} is

$$I_{SPPA} = \frac{1}{T_0}\int_0^{T_0} I_{SP}(t)dt \tag{9.11}$$

where T_0 is the pulse duration, and the starting point of the pulse is aligned with the 0. The maximum intensity I_m is

$$I_m = \frac{1}{T^+}\int_{t_0}^{t_0+T^+} I_{SP}(t)dt \tag{9.12}$$

where T^+ is the duration of the half-cycle. Because of the nonlinear acoustic effects, the compressional component is always larger than the tensile one. So, the positive half-cycle is selected for the maxim p^+. The spatial-peak temporal-average acoustic intensity I_{SPTA} is determined as

$$I_{SPTA} = \frac{1}{T}\int_0^T I_{SP}(t)dt = I_{SPPA} \cdot PD \cdot PRF \tag{9.13}$$

where

PD is the pulse duration
PRF is the pulse repetition frequency
T is the reciprocal of PRF

If the beam sweeps across the hydrophone (i.e., electrical steering of the HIFU focus), Equation 9.13 has the alternative expression

$$I_{SPTA} = \sum_{i=1}^{N}\int_0^T I_i(t)dt \tag{9.14}$$

where N is the number of beams produced per pulse. The spatial-average temporal-average acoustic intensity I_{SATA} is determined as

$$I_{SAPA} = \frac{\int_S I_{SPPA}dS}{\int_S dS} \tag{9.15}$$

where the mapping area S is selected to be large enough to accumulate all significant side lobes and grating lobes.

Scanning the hydrophone at the focal plane of HIFU transducer in two-dimensional space with the step size

$\Delta x = \Delta y \leq 0.1\lambda$ can be done to calculate spatial-averaged acoustic intensity I_{sal}:

$$I_{sal} = \frac{\sum_{i=1}^{N} p_{rms6i}^2(x_i, y_i, F_{pres})}{\rho c N} \qquad (9.16)$$

where

p_{rms6i} is the effective acoustic pressure no less than half of the maximum effective acoustic pressure $p_{rms,max}$

N is the number of the measurement points at the focal plane eligible for calculation

The maximum size of the −6 dB region at the focal plane is defined as Δd_r when measuring I_{sal}.

Lateral beam plots in the x or y direction for any acoustical parameter can be undertaken to determine the beamwidth. Because of the pressure waveform distortion by the nonlinear acoustic propagation, the beam profile based on the peak negative pressure is more reliable than on positive pressure.

We can use a similar approach to measuring Δd_r to find the second maximum $p_{rms,sm}$ at the acoustic pressure focal plane $p_{rms}(x_i, y_i, F_{pres})$, and the maximum lateral side-lobe level is determined as

$$L_{sm} = 20\log\frac{p_{rms,sm}}{p_{rms,max}} \qquad (9.17)$$

If the hydrophone is scanned along the acoustical axis (z direction), the distance of two points along the z direction whose acoustic pressure is equal to half of the maximum effective acoustic pressure is defined as Δd_z. Using a similar approach as in measuring Δd_z to find the second maximum $p_{rms,asm}$ along the acoustical axis, the axial side-lobe level is determined as

$$L_{asm} = 20\log\frac{p_{rms,asm}}{p_{rms,max}} \qquad (9.18)$$

The time delay between the transmitted pulse and the measured signal by the hydrophone at the acoustic focal point, Δt, is used to determine the acoustic focal length:

$$F_{pres} = c\Delta t \qquad (9.19)$$

The effective radius of a concave ultrasound transducer is determined by

$$a = \frac{F_{pres}}{k}\left(\frac{1.62}{W_{pb3}} + \frac{2.22}{W_{pb6}}\right) \qquad (9.20)$$

where W_{pb3} and W_{pb6} are the −3 and −6 dB sound beam widths of the main lobe measured at the focal plane $p_{rms}(x_i, y_i, F_{pres})$, respectively. So, the effective transmission of the transducer is

$$A = 2\pi F_{pres}^2(1 - \cos\varphi) \qquad (9.21)$$

where $\varphi = \arcsin(a/F_{pres})$ is half of the focusing angle. The acoustic pressure focusing gain is

$$G_{pfocal} = p_{rms,max}\sqrt{\frac{A}{\rho c P}} \qquad (9.22)$$

where P is the acoustic power. For a concave array consisting of N transducers with the same transmission condition, the total effective radiation area A is calculated as $A = \sum_{i=1}^{N} A_i$, where A_i is the effective radiation area of the ith transducer.

The amplitudes of acoustic pressure components are calculated by performing FFT on the waveform profile measured by the hydrophone. The harmonic distortion coefficient D is defined as

$$D = \sqrt{\frac{\sum_{i=2}^{n} p_i^2}{\sum_{i=1}^{n} p_i^2}} = \sqrt{\frac{\sum_{i=2}^{n} H_i^2}{1 + \sum_{i=2}^{n} H_i^2}} \times 100\% \qquad (9.23)$$

where

p_i is the acoustic pressure of the ith harmonic

$H_i = p_i/p_1$ is the ratio of the amplitude of the ith harmonic component to that of fundamental one

n is the largest number of harmonics in consideration that is determined by the signal-to-noise ratio ($H_i = (p_i/p_1)/N$) of p_i

The spatial-peak intensities as measured in water should be corrected for tissue attenuation (i.e., $\alpha = 0.6$ dB/MHz/cm). The compensation factor for the *in situ* spatial-peak intensities is $e^{-0.069fd}$, where f is the driving frequency [MHz], and d is the focal length [cm]. Random uncertainty is determined by repeating the measurements, preferably by removing the transducer and hydrophone from the water bath, replacing them, and re-optimizing their positions. Ideally, at least four such repeats should be made (Preston et al. 1988).

Acoustic waves in a liquid induce change in mass density, which modulates the density and optical refractive index of the medium. The fiber-optic probe hydrophone (FOPH) has been developed to measure the acoustic field by exploiting the piezo-optic effect, especially those with high intensity (Figure 9.4). The change in the refractive index can be measured by the light reflected at the tip of a glass fiber submerged in a testing medium. Laser light propagates to the optical fiber tip via a 2 × 2 or 1 × 2 optical coupler and is then partially transmitted through and partially reflected at the interface of fiber and water. The reflected signal is subsequently directed to a fast semiconductor photodetector via an optical coupler and converted to an electric signal after being amplified and filtered. Using the stationary light reflection signal and intrinsic parameter-based calibration of the FOPH system, the relation between pressure and photodiode signal can be determined. In comparison to conventional PVDF membrane and needle hydrophone, FOPH has the advantages of high-pressure measurements,

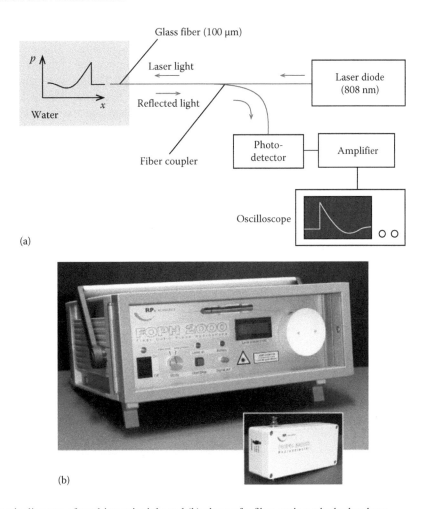

(a)

(b)

FIGURE 9.4 (a) Schematic diagram of working principle and (b) photo of a fiber-optic probe hydrophone.

self-calibration, immunity to electromagnetic interference, robustness to cavitation damage, and lifelong use of the 20 m measuring glass fiber. If the glass fiber is broken, a new fiber endface is cut perpendicular to the fiber axis immediately using professional cleaving tools. Transimpedance amplification shows a bandwidth from DC to 20 MHz. Finally, noise due to coherent fluctuation of the laser source could be compensated by subtracting the background noise signal, and an appropriate photodetector to minimize the stray light.

The reflection coefficient as a function of pressure is derived from the Fresnel formulas, the relationship between the refraction index and density, and the state equation of the media (Staudenraus and Eisenmenger 1993). Commercial step-index silica fibers have a core diameter of 100–200 μm and a light propagation angle of <15° to the fiber axis. The reflection coefficient R at a flat, perpendicular interface is well approximated by

$$R = \left(\frac{n_c - n_w}{n_c + n_w} \right)^2 \qquad (9.24)$$

where n_c and n_w are the refractive indexes of the fiber core and water, respectively.

The change of state may be approximated by Poisson's law. The piezo-optic relationship between the acoustic pressure and the change in refractive index in water up to 1 GPa can be determined via a modified isentropic version of the Tait equation:

$$\frac{P_0 + p(t) + Q}{\rho(t)^\gamma} = K_1 \qquad (9.25)$$

where
K_1 is a constant
P_0 is the ambient pressure
$p(t)$ is the acoustic pressure
$Q = 295.5$ MPa
$\gamma = 7.44$ under normal experimental conditions (i.e., $T = 20°C$, $P_0 = 100$ kPa, and $\rho_0 = 1000$ kg/m^3)

The Gladstone–Dale model relates the density and refractive index of water and is valid for compressional pressures up to approximately 500 MPa (Yadav et al. 1973, Davision and Graham 1979):

$$\frac{n_w(t) - 1}{\rho(t)} = K_2 \qquad (9.26)$$

where K_2 is another constant. The error is within 5% as confirmed from static compression. For water and a light beam

($\lambda = 800$ nm, $n_w = 1.329$), $\Delta n_w/\Delta p \approx 1.4 \times 10^{-4}$ MPa^{-1} (Yadav et al. 1973), and the pressure-dependent change of refractive index of a silica fiber is $\Delta n_c/\Delta p \approx 5 \times 10^{-6}$ MPa^{-1} (Barker and Hollenbach 1970). Hence, for simplicity, n_c is considered constant with regard to the fiber compressibility by a final pressure data correction of +3.6%.

The sensitivity H of the hydrophone is defined by the change in light reflectivity ΔR divided by the acoustic pressure p and the static reflectivity R_0:

$$H = \frac{\Delta R}{(R_0 + S)p} \qquad (9.27)$$

where S is the additional stray light due to nonideal experimental conditions. It is calculated that for acoustic pressure in the range of 5–30 MPa, the nonlinear aberration from an average sensitivity of -1.95×10^{-3} MPa^{-1} is below 5% assuming an incompressible fiber and zero stray light. At the temperature range of 20°C–40°C, the decrease in sensitivity of the hydrophone is about 3×10^{-3}/°C. In addition, the signal-to-noise ratio of the FOPH can be significantly improved by reducing the photon noise, in principle. So, the minimum detectable pressure p_{min} is given by

$$p_{min} = \left(\frac{2(R_0 + S)h\nu\Delta f}{qW_0} \right)^{1/2} \cdot \left(\frac{dR}{dp} \right) \qquad (9.28)$$

The low noise of the hydrophone at the desired large bandwidth Δf requires a light source with high photon flux $W_0/h\nu$ ($W_0 = 0.1–1$ W), a photodetector with high quantum yield q, and a small S compared to the static reflection R_0. The minimum detectable pressure is experimentally determined by the quotient of noise level and sensitivity, resulting in $P_{min} \approx 1$ MPa (Table 9.1).

The light intensity that is reflected at the glass/water interface depends on the acoustic pressure. The principle of this sensor is based on the pressure-dependent reflectivity of a laser beam at an interface between a solid silica glass block and water (Figure 9.5). The optical reflectivity at this interface is determined by the indices of refraction of glass and water. As these refractive indices are dependent on the density of the medium, the reflectivity itself is also pressure dependent.

In addition, the optical head can consist of a solid silica glass block (90 mm × 60 mm × 30 mm) on a displaeable mount, as shown in Figure 9.5. The other optical components are similar to that of FOPH. The measurement location is defined by the position of the light spot. This is approximately at the center of the optical head on the front side of the glass block. The relationship between acoustic pressure and light reflectivity is described by Fresnel's laws and the Tait equation. Unwanted reflections of the sound wave from the boundaries or the back side of the glass block do not influence the measurements since they are sufficiently delayed by the large dimensions of the glass block. The usable region on the front side of the glass block is the region that has a minimum distance of 20 mm from the boundaries of the glass block. In case of a local damage of the glass surface or volume, the glass block can be displaced by a spindle without significantly influencing the optical alignment. A new measurement can be done without new calibration. If all possible measurement positions on the glass surface are used and damaged, the glass block will have to be replaced.

Because of the simultaneous measurement of the baseline voltage level U_{DC} and referring the measured values U_{AC} after self-calibration, the pressure results are independent of a small misalignment of the optics or the quality of the fiber connectors. The differential change of the output voltage ΔU_{AC} by a change of the sound pressure $\Delta p(t)$ at a pressure p is given by

$$\left. \frac{\Delta U_{AC}}{\Delta p} \right|_p = \frac{Z_{T,AC}}{Z_{T,DC}} \cdot C_R \cdot \left. \frac{dR_{GW}(p)}{dp} \right|_{C_R \cdot p} \cdot \frac{1}{R_{GW}(p=0)} \cdot U_{DC} \qquad (9.29)$$

where

p is the sound pressure of the forward traveling sound wave toward the optical head
$C_R = 1.79$ is the factor due to the reflection of the acoustic wave at the glass/water interface
U_{AC} is the signal output
U_{DC} is the base level of direct current (DC) output voltage
$Z_{T,DC}$ is the transimpedance in the DC path of the amplifier
$Z_{T,AC}$ is the transimpedance in the AC path of the amplifier
$R_{GW}(p=0)$ is the optical base reflectivity of the water/air interface
$dR_{GW}(p)/dp$ is the dependence of the optical reflectivity at the water/air interface on the sound pressure p

TABLE 9.1

Comparison of the Hydrophone Specifications

	FOPH	Membrane	Needle	HGL
Spatial resolution (µm)	10–100	100–500	40–1500	85–1000
Rise time (ns)	3			
Bandwidth (MHz)	0–150	0.5–45	0.25–40	0.25–40
Sensitivity (mV/MPa)	2–150	12–350	6–1200	8–510
Acoustical pressure range (MPa)	−60 to 400			
Acceptance angle (°)			15–110	20–150
Water temperature range (°C)	5–95	<40	<50	<50
Accuracy	±5%			

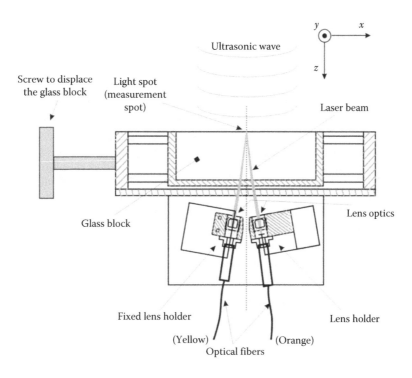

FIGURE 9.5 Schematic diagram of light spot block in the acoustic pressure measurement.

Optical reflectivity is determined from Fresnel's law:

$$n_w(p) = 1.3286 + \frac{1.4 \cdot 10^{-4}}{\text{MPa}} \cdot p \tag{9.30}$$

$$n_c(p) = 1.4538 + \frac{5 \cdot 10^{-6}}{\text{MPa}} \cdot p \tag{9.31}$$

Sensitivity of the optical block is determined as

$$S_{\text{norm}} = \left. \frac{\Delta U_{\text{AC}}}{\Delta p} / U_{\text{DC}} \right|_p \tag{9.32}$$

9.2 RADIATION FORCE BALANCE

The absorbing target should have an acoustic pressure reflection coefficient of $\leq 5\%$ and an acoustic pressure transmission coefficient of $<10\%$. Its minimum size should be at least 1.5 times as large as the -26 dB sound beam width in the plane. The support or suspension of the target should have sufficient stability for negligible measurement error due to horizontal displacement. The face of the absorbing target should be perpendicular to the axis of the sound beam. The resolution of force detection should be at least 10^{-3} N. The sensitivity of hydrophone should be not less than 10 nV/Pa. Its variation should be within ± 6 dB in the twice the working frequency range. At the working frequency, the -6 dB main beam along the acoustic axis should be $\geq 70°$.

The effective radius of the hydrophone, a, is determined as

$$a_{-3\text{dB}} = \frac{1.62}{k \sin \theta_{-3\text{dB}}} \tag{9.33}$$

$$a_{-6\text{dB}} = \frac{2.22}{k \sin \theta_{-6\text{dB}}} \tag{9.34}$$

$$a = \frac{1}{2}(a_{-3\text{dB}} + a_{-6\text{dB}}) \tag{9.35}$$

where

k is the wavenumber in water
$\theta_{-3\text{ dB}}$ and $\theta_{-6\text{ dB}}$ are half of the -3 and -6 dB beam width of the main lobe, respectively

Theoretically, the effective radius of the sensing element of the hydrophone should be no greater than a quarter wavelength of the ultrasound burst:

$$a_{\text{max}} = \frac{\lambda}{8a_1} \sqrt{l^2 + a_1^2} \tag{9.36}$$

where

λ is the wavelength at the working frequency
a_1 is the effective radius of the ultrasound transducer
l is the distance between the hydrophone and the transducer surface

The hydrophone should keep the linear electric voltage output under the instantaneous acoustic pressure of 10 MPa with a nonlinear distortion of $<10\%$.

When the hydrophone is scanning around the focal point of the ultrasound transducer, the scanning step size at the focal plane and along the axial direction should be less than 0.1λ and 0.2λ, respectively. The resolution of the step size should be better than 10 μm. For the ultrasound system whose focusing beam propagates upward, the transducer should be fixed at the bottom of the testing tank with its sound beam vertically upward. The absorbing target is suspended with a digital balance to measure the normal radiation force acting on it. The line connecting the absorbing target and the digital balance should be strong but sufficiently thin. In contrast, for the ultrasound system whose focusing beam propagates upward, force transferring mechanical parts or an underwater load cell should be used, as shown in Figure 9.6, to measure the radiation force acting on the absorbing target vertically downward. No matter which configuration is used, the testing tank should be sufficiently large with sound-absorbing tiles on the inner wall to avoid the generation of reflection at the boundary. The transducer or the transducer array should be totally immersed in water, and the produced sound beam should aim at the geometrical center of the absorbing target.

Similar to the requirement of the testing tank in measuring acoustic pressure, an acoustic absorber is placed on the inner surface, bottom, and the water/air interface to avoid the influence of potential reflections during the power measurement. Water in the testing tank should be purified with oxygen concentration ≤4 mg/L and temperature within the range of 23°C ± 3°C. After warming up for 15 min, the frequency and electrical power stability of the transducer should be $10^{-4}/4$ h and 10%/4 h, respectively. In order to guarantee the safety and lifespan of the hydrophone, HIFU should be working in the burst mode, with pulse duration no longer than 100 μs and pulse repetition frequency less than 1 kHz. The digital oscilloscope should have a bandwidth of at least 10 times that of the working frequency of the HIFU transducer. To avoid the influence of nonlinear and acoustic streaming on the measurement

of radiation force, the sound-absorbing target should be close to the sound source and normal to the acoustic axis, with the geometrical center aligned with the acoustic axis. The distance between the sound-absorbing target and the transducer should be no larger than 0.7 times of the focal length. Before the measurement, the absorbing target should be immersed for at least 30 min and the ultrasound system should warm up for at least 15 min. To reduce the effect of thermal shifting, the difference in the stable reading from a digital balance between HIFU on and off in a short time (i.e., 2–3 s) should be equal to the ratio of the normal radiation force acting on the absorbing target, F, to the gravitational acceleration g. It is to be noted that when the lever is used in connection with the digital balance, the force acting should be calibrated for the actual radiation force measured. During measurement, any small bubbles attached to the absorbing target and the surface of the HIFU transducer should be removed carefully. The following cased are considered:

1. Concave transducer

$$P = \frac{2Fc}{1 + \cos\beta} e^{2\alpha d} \tag{9.37}$$

where
P is the acoustic power
F is the normal radiation force acting on the absorbing target
c is the speed of sound in water
$β$ is half the focusing angle of the HIFU transducer
$α$ is the attenuation coefficient of water
d is the distance between the absorbing target and HIFU transducer

2. Annular concave transducer

$$P = \frac{2Fc}{\cos\beta_1 + \cos\beta_2} e^{2\alpha d} \tag{9.38}$$

where
$β_1$ is half the focusing angle of the outer diameter of the HIFU transducer
$β_2$ is half the focusing angle of the inner diameter of the HIFU transducer
d is the distance between the absorbing target and the effective center of the HIFU transducer

3. Array consisting of piston transducers
   If an array consists of N identical piston transducers distributed on a common spherical surface with all sound beam axes aligned with the spherical center and each piston has the same acoustic transmission power, the total acoustic power is calculated as

$$P = \frac{NFc(\text{corr})}{\sum_{i=1}^{n} \cos\theta_i} e^{2\alpha d} \tag{9.39}$$

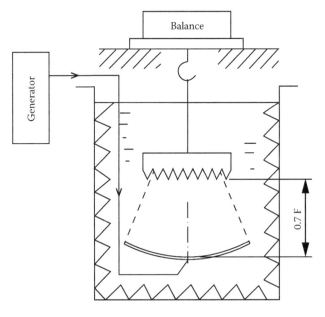

FIGURE 9.6 Schematic diagram of acoustic radiation force balance.

where

F is the total normal radiation force acting on the absorbing target from all piston transducers

θ_i is the angle between the ith piston element and the main axis of the array or the incident angle of such element to the absorbing target

corr is the correction factor for the divergence of each piston element, given by

$$\text{corr} = \frac{P}{cF} = \frac{1 - J_1(2ka)/ka}{1 - J_0^2(ka) - J_1^2(ka)} \quad (9.40)$$

where

k is the wavenumber

a is the radius of the acting element of the piston transducer

$J_0(*)$ is the zero-order Bessel function

$J_1(*)$ is the first-order Bessel function

4. Array consisting of self-focusing elements

If an array consists of N identical self-focusing transducers distributed on a common spherical surface with the focal point of each self-focusing transducer aligned with the center of the spherical surface and the acoustic transmission power of each self-focusing transducer is the same, the total acoustic power is calculated as

$$P = \frac{2NFc}{1 + \cos\beta} e^{2\alpha d} / \sum_{i=1}^{n} \cos\theta_i \quad (9.41)$$

where

β is half the focusing angle of each self-focusing transducer

θ_i is the angle between the sound beam of the ith self-focusing element and the main acoustic axis of the array or the incident angle to the absorbing target

At low power levels, the precision of the radiation force is limited by fluctuations and drifts by the other forces acting on the target (the Archimedes upthrust, forces due to convection currents, pressure gradients in the water due to building vibrations, and surface tension acting on the target suspension). Degassing of the water may occur at high *in situ* acoustic intensities, especially in the standing wave field in front of the target.

9.3 CALORIMETER

Using some form of calorimetry is an alternative approach to acoustic power measurement. Acoustic energy is converted into heat, which is then measured by the calorimeter. The operation principle of calorimetry is simple. The transducer faces the calorimeter at a certain distance to the tube. A fraction of the acoustic power radiating from the transducer is absorbed by the oil (castor oil), and the remaining part is reflected. The absorbed fraction is calculated from the acoustic absorption coefficient α of castor oil at various temperatures T and for various frequencies f. At 1 MHz and in the temperature range

of 15°C–30°C, $\alpha \propto f^{5/8}$. Constant-flow calorimetry is used measuring acoustic powers as low as 1 mW at frequencies of a few megahertz (Torr and Watmough 1977).

The calorimeter is submerged in a water-filled tank, and castor oil is pumped through the calorimeter at a constant rate with a peristaltic pump. The equilibrium temperature difference between the inflowing and outflowing oil of the calorimeter is monitored by calibrated thermometers, which form the arms of a DC Wheatstone bridge. The out-of-balance voltage of the bridge is amplified and recorded. The absorbed acoustic power qW is equal to the electrical power dissipated by the heater to produce the same temperature difference, and a knowledge of the absorbed fraction q allows the calculation of the total acoustic power W of the transducer. The minimum absorbed power that can be detected by the calorimeter is about 0.2 mW, below which the accuracy of temperature measurement is affected by random noise. The simple constant-flow calorimeter can measure the acoustic power in the range from 1 mW to 10 W.

The response time of the calorimeter to reach equilibrium from a power change is determined by the internal volume of the calorimeter and the flow rate of oil. The flow through the calorimeter is found to be nonturbulent, and the response time is well defined. For example, the response time of a calorimeter with 9 cm³ of oil and a flow rate of around 0.6 cm³/s is 15 s. The sensitivity of the calorimeter is inversely proportional to the oil flow rate. The main shortcoming of the calorimeter is the alignment. Accuracy of this approach also depends on the reliability of the following assumptions: negligible reflection at the interfaces of castor oil and water, negligible scattering from the heater wire, planar ultrasound wave for the calorimeter, equal heating by electrical and acoustic power, negligible heat flow from the transducer, and the equivalence of the power absorbed by the oil from pulsed and continuous beam.

The temporal temperature change of the calorimetric method can be analyzed in a comparative form (Margulis and Margulis 2003) in order to increase accuracy and sensitivity of static calorimetry by continuous registration of the temperature curve during the sonication, $T_u(t)$, and during heating, $T_h(t)$. If $T_u(t)$ and $T_h(t)$ are very close to each other, the heater power may be treated the same as the acoustic power absorbed by the liquid. The heat exchange is proportional to small temperature change ΔT. So, the differential equations are

$$\frac{d(\Delta T)}{dt} = \begin{cases} \dfrac{W}{Cm} - a\Delta T, & (t \le t_{us}) \\ -a\Delta T, & (t > t_{us}) \end{cases} \quad (9.42)$$

where

W is the heater power or absorbed acoustic power

C is the heat capacity of the liquid

m is a mass of the liquid

a characterizes heat exchange between liquid under ultrasound and outward objects

t_{us} is the sonication duration

Then, the explicit solution for the ordinary differential equation (ODE) given before is

$$\Delta T(t) = \begin{cases} \dfrac{W}{Cm}\left(\dfrac{1-e^{-at}}{at}\right), & (t \le t_{us}) \\[3mm] \dfrac{Wt_{us}}{Cm}\left(\dfrac{1-e^{-at_{us}}}{at_{us}}\right)e^{-at}, & (t > t_{us}) \end{cases} \quad (9.43)$$

The temperature rise and fall can be simply fitted by a line and an exponential decay curve, respectively:

$$\Delta T(t) = \begin{cases} At + B, & (t \le t_{us}) \\ \Delta T_{max}e^{-at}, & (t > t_{us}) \end{cases} \quad (9.44)$$

where ΔT_{max} corresponds to their intersection, as shown in Figure 9.7. During sonication, $\Delta T_{us\,max}$ is proportional to the absorbed acoustic power. First, set the power of the heater as

$$W_{h1} = \frac{C_v m \Delta T_{us\,max}}{t_{us}} \quad (9.45)$$

where C_v is the heat capacity of liquid. The corresponding $\Delta T_{h1\,max}$ is determined by the measurement curve. Then, set the heater working time equal to t_{us} and power to

$$W_{h2} = \frac{W_{h1}\Delta T_{us\,max}}{\Delta T_{h1\,max}} \quad (9.46)$$

$\Delta T_{h2\,max}$ can be determined from the experimental curve $\Delta T_{h2}(t)$. Finally, the absorbed acoustic power is given by

$$W_{us} = \frac{W_{h2}\Delta T_{us\,max}}{\Delta T_{h2\,max}} \quad (9.47)$$

Measurements can be performed at equilibrium liquid temperature in the range of 10°C–50°C at the absorbed acoustic power of 3–150 W. Accuracy is ~3% for a magnetostrictive or piezoelectric transducer. Resolution of temperature and sonication are 0.1°C and 0.01 s, respectively.

9.4 HOLOGRAPHY

Acoustic holography is a method that is used to estimate the sound field near a source by measuring the acoustic parameters away from the source via an array of pressure and/or

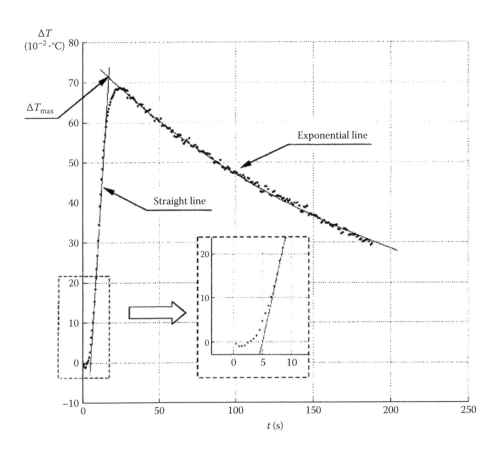

FIGURE 9.7 Experimental results of temperature rise and decay during calorimetric measurement and its approximation curves. (Reprinted from *Ultrason. Sonochem.*, 10, Margulis, M.A. and Margulis, I.M., Calorimetric method for measurement of acoustic power absorbed in a volume of a liquid, 343–345, Copyright 2003, with permission from Elsevier.)

particle velocity transducers. A hydrophone is first scanned in a water tank over a plane using a three-dimensional positioning system to measure the complex pressure field as a function of position. The field is then projected on to a series of new planes. The method is shown to be accurate for use with phase-controlled field patterns, providing a rapid and accurate method for obtaining field information over a large spatial volume, and can significantly simplify the characterization procedure required for phased-array application used for therapy. Most significantly, the wavefront propagated back to a phased array can be used to predict the field produced by different phase and amplitude settings of the array elements.

The acoustical pattern becomes more important with increase of transducer complexity. The angular spectrum or plane-wave decomposition technique was initially developed in the study of optical diffraction, but it can also be applied to acoustic wave propagation and was shown to be equivalent to the diffraction integral approach. Acoustic propagation between parallel planes is modeled using the two-dimensional (2D) Fourier transform of the acoustic field, with each element in the spatial frequency domain multiplied by the phase factor. The advantages of the method include high computational efficiency using the FFT and high spatial resolution, even in the near field of acoustic sources. The angular spectrum method has already been demonstrated to reconstruct the distribution of surface velocity at an acoustic radiator. Furthermore, it models both forward propagation from the source plane to a hydrophone and back propagation from the hydrophone plane back to the source. The acoustic propagation effects in the tissue, such as absorption, refraction, dispersion, and phase distortion, can also be modeled using the angular spectrum method. In practice, the restriction to planar source distributions does not significantly reduce its implementation since many ultrasound radiators are not planar.

The acoustic field is assumed to satisfy the linearized acoustic wave equation. A time-harmonic pressure is generally expressed as $P(r, t) = P(r)e^{i\omega t}$. A Helmholtz equation in Cartesian coordinates is obtained through the substitution of the 2D Fourier integral

$$P(x, y, z) = \frac{1}{2\pi} \int\int p(k_x, k_y, z)e^{i\omega t}e^{ik_x x}e^{ik_y y}dk_x dk_y \quad (9.48)$$

where the spatial frequency (k_x, k_y) represents a plane wave traveling with direction $(n_x = k_x/k, n_y = k_y/k)$. Then, the wave equation becomes

$$(\nabla^2 + k^2)P(r) = 0 \Rightarrow$$
$$\frac{\partial p(k_x, k_y, z)}{\partial z} + \left(k^2 - k_x^2 - k_y^2\right)p(k_x, k_y, z) = 0 \quad (9.49)$$

To propagate a planar field at a distance z_0 in front of a source to a new plane z, the advanced solution is

$$p(k_x, k_y, z) = p(k_x, k_y, z_0)e^{i(z-z_0)\sqrt{k^2 - k_x^2 - k_y^2}} \quad (9.50)$$

The propagation is then modeled by multiplying (k_x, k_y) with an appropriate and simple phase propagation factor as the transfer function:

$$G(k_x, k_y, z_0, z) = e^{i(z-z_0)\sqrt{k^2 - k_x^2 - k_y^2}} \quad (9.51)$$

In the wave vector space, the field recorded in a plane z_0 is thus related to the field at any other plane z by a simple transfer function. In order to obtain the pressure field at z, the inverse Fourier transform is performed to Equation 9.51.

Backward projection to the source could serve as a source function for a more realistic acoustic modeling. Although the hydrophone is not ideally omnidirectional, its directivity is nearly constant over the measurement planes. The hydrophone is aligned to the center of the acoustic source plane and then raster scanned. Because the measured acoustic pressure amplitude varies significantly within the acoustic field, the range of the oscilloscope is adjusted automatically to set the sensitivity as high as possible without causing an overload. The temporal shift of the pulse at different positions is also compensated. The signal is averaged to increase the signal-to-noise ratio. Subsequently, the hydrophone is moved to the next position, and the signal acquisition process is repeated until the entire plane is covered. The velocity field radiated by a transducer can also be measured using a laser interferometer. The reconstructed pressure field distributions are similar to the theoretical predictions, as shown in Figure 9.8.

The primary limitation to reconstruction of small features is the bandwidth of the spatial frequency. A small hydrophone and a high sampling rate are preferred. Planar projection allows acoustic field characterization in the focal region of a high-intensity beam and at a position close to the source where the hydrophone directionality can affect measurements. Formation of standing waves between the transducer and the hydrophone holder may produce oscillations in the measured data. Field backprojection toward the transducer can check the performance of transducer and provide source condition for theoretical modeling, such as field designs for beam steering as well as multiple foci patterns of a phased array. Interference in front of the transducer may affect the direct measurements, which can be minimized by projecting backward from the focus where a much smaller area is scanned. The spatial undersampling of the finite aperture and the spatial cutoff due to neglecting the evanescent component of the wave field can introduce errors in the reconstruction using the discrete Fourier transform, such as a "bump" in the center of the source in both magnitude and phase. If the time delay from the measurement plane to the reconstruction plane is greater than the time window, signal wrapping will occur. Therefore, the projection distance is limited. In order to avoid such a shortcoming, time padding the waveform is necessary and the amount of time padding required depends on the distance over which the field is propagated.

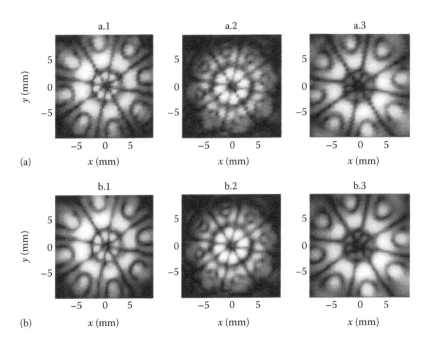

FIGURE 9.8   Comparison of (a) radial measurement of the sector vortex array at three different distances and (b) their projections. (Reprinted with permission from Clement, G.T. and Hynynen, K., Field characterization of therapeutic ultrasound phased arrays through forward and backward planar projection, *J. Acoust. Soc. Am.*, 108, 441–446, Copyright 2000, American Institute of Physics.)

9.5   INTERFEROMETRIC MEASUREMENT

The heterodyne interferometer used for measuring mechanical displacement consists of a reference beam at a wavelength λ and a probe beam that has a frequency shifted by a Bragg cell (f_B = 70 MHz) and focused on the inspected surface in a diameter of 20–50 μm (Gendreu et al. 1995). The beating between the reference and probe beams is detected by a photodetector so that the vibrating surface modulates the phase of light at frequency f_B. The modulation index $\Delta\Phi$ is proportional to the normal displacement, $\Delta\Phi = (4\pi/\lambda)u$. Phase demodulation is carried out in a broadband (20 kHz–30 MHz) electronics, whose bandwidth could be limited to reduce the acquisition noise. This approach can measure surface displacements as small as 0.3 Å.

The piezocomposite transducer is first immersed in water, with its reflective gold-coated surface in front of the optical probe. The transducer is scanned in two dimensions to measure the transient normal displacement $u(t)$ of the transducer's front surface. Then, the transmitted field can also be measured by an optical probe, such as a thin, reflective membrane of thickness ~15 μm, immersed in water. The motion of the membrane induced by the acoustic pressure is measured by the interferometer. At a frequency lower than 10 MHz, a 15 μm thick Mylar membrane follows the particle motion in the fluid. Furthermore, the electromagnetic coupling between the electrodes and the probe determines the location of field measurements. The high stability of the optical probe allows measurements to be performed over a long time (10 h or more if necessary) without requiring any alignment of the optical components. Moreover, the heterodyne process makes the measurements insensitive to the low-frequency thermal and mechanical disturbances occurring in the water tank.

The optic fiber measurement is an interferometric technique (Menssen et al. 1991). A single-mode fiber forms the measuring arm of a Michelson interferometer, and a 200 nm titanium layer is coated on the cut tip to improve the optical reflection at the interface of the fiber and the fluid. Although titanium has a lower optical reflection coefficient than aluminum or gold, it can bear shock-wave exposure. The fiber tip is placed in the focal region of the transducer toward the acoustic source. The variation in the optical path is measured by the interferometer, which consists of a He–Ne laser as an illuminating light source and a polarizing beam splitter for dividing and combining the optical field. To detect the output field, two balanced photodetectors with a −3 dB bandwidth at >100 MHz and a difference transimpedance amplifier that suppresses the noise produce two output signals with a phase difference of 90°. The displacement of the fiber tip ζ_F is obtained by an arctan operation applied to the quotient of these two signals (Menssen et al. 1991), and the acoustic pressure is obtained by time differentiation.

9.6   FIELD MAPPING USING VIBROMETER

When a vibrometer is used to detect velocities on a pellicle, its laser beam interacts with acoustic waves passing through the pellicle. As the vibrometer interprets local perturbations as velocities, the piezo-optic coefficient of the water must be derived to determine the velocity on the pellicle. Refractive index n_0 = 1.33 and piezo-optic coefficient $n_1 = (\partial n/\partial p)_s = 0.32$ are adopted under the conditions of normal atmospheric pressure, water temperature $T = 20°$C, and laser wavelength $\lambda = 632$ nm. A pellicle is placed at the plane x_1, and the acoustic wavefront travels to the position

x_2 at time t. Under the condition where the amplitude of the acoustic wave is much smaller than the acoustic wavelength, the optical path length $q(t)$ detected by the vibrometer at time t can be denoted by

$$q(t) = -2n_0 a(t) + \frac{2n_1}{\rho c^2} \int_{x_1}^{x_2} p(x,t) dx \qquad (9.52)$$

where

$a(t)$ is the amplitude of the acoustic wave at time t
$p(x, t)$ is the acoustic pressure at point x and time t
ρ is the density of water
c is the speed of sound in water

For linear plane waves, the velocity v detected by the vibrometer is given by

$$v(t) = -2(n_0 - n_1) u(x_1, t) \qquad (9.53)$$

where $u(x_1, t)$ is the velocity on the pellicle at time t. $n_p^* = n_0 - n_1$ is defined as the effective refractive index of the plane waves. When considering the spherical radiation of an acoustic wave, the equation can be rewritten as

$$q(t) = -2n_0 a(t) + \frac{2n_1}{\rho c^2} A_0 e^{i\omega t} \int_{x_1}^{x_2} \frac{e^{-ikx}}{x} dx \qquad (9.54)$$

where

A_0 is the amplitude of acoustic pressure
ω is the angular frequency
k is the wavenumber

So, the ratio of the measured velocity v to the pellicle velocity u is given by

$$n_s^* = n_0 - n_1 \frac{(kx_1)^2}{kx_1 - i} e^{ikx_1} \cdot \left(\frac{e^{-ikx_2}}{kx_2} + i \int_{x_1}^{x_2} \frac{e^{-ikx}}{x} dx \right) \qquad (9.55)$$

n_s^* can be regarded as the effective refractive index of the spherical wave and can be simplified as $n_s^* = n_0 - n_1$ when the pellicle is placed at three wavelengths away from the transducer ($x_1 > 3\lambda$). Furthermore, if the laser beam of the vibrometer scans an angle α from the x-axis and the acoustic incident angle is θ on the pellicle, the effective refractive index becomes

$$n_s^* = n_0 \frac{\cos\theta}{\cos\alpha} - n_1 \frac{(kr_0)^2}{kr_0 - i} e^{ikr_0}$$

$$\cdot \left(\frac{e^{-ikx_2}}{kx_2\sqrt{1-(d/x_2)^2}} + i \int_{x_1/\cos\theta}^{x_2} \frac{e^{-ikr-ikx}}{r\sqrt{1-(d/r)^2}} dr \right) \qquad (9.56)$$

where

$d = x_1 \cos\alpha (\tan\theta + \tan\alpha)$
r_0 is the distance from the projector to point p

To reduce the influence of piezo-optic effect on the output signals of a vibrometer, scanning measurements should be carried out at small angles.

For linear plane waves, the acoustic pressure p can be derived from the detected velocity v by

$$p(t) = -\frac{\rho c v(t)}{2 n_p^*} \qquad (9.57)$$

For small-amplitude spherical waves, the acoustic pressure p is given by

$$p(x, t) = -\frac{\rho c u(x,t)}{1 + i\lambda / 2\pi x} \qquad (9.58)$$

Hence, the acoustic pressure p can be expressed in terms of the detected velocity v:

$$p(x, t) = -\frac{\rho c v(t)}{2 n_s^* (1 + i\lambda / 2\pi x)} \qquad (9.59)$$

Assuming that a concave spherical transducer has a uniform velocity v_0, the velocity at the field point (x, r) can be given by

$$u(x, r) = \frac{v_0}{x} e^{-\frac{ikr^2}{2x}} \int_{r'=0}^{a} \exp\left[-\frac{ikr'^2}{2}\left(\frac{1}{x} - \frac{1}{R}\right) \right] J_0\left(\frac{krr'}{x}\right) r' dr' \qquad (9.60)$$

where

R is the curvature radius of the spherical shell
J_0 is the Bessel function of zeroth order

The schematic diagram of an optical method for detection of focused fields is shown in Figure 9.9. In the system, tone-burst signals are generated by a signal generator, amplified by a power amplifier, and then input to a focusing transducer that is fixed to the wall of a water tank to radiate its acoustic energy directly into the water. A bilaminar shielded membrane hydrophone coated on either side with gold is used to reflect the laser beams. A laser scanning vibrometer is placed on the wall opposite to the transducer, and its laser beam is coincident with the beam axis of the transducer. The vibrometer is placed 60 cm away from the pellicle, and the laser beam is 150 μm in diameter, so the largest deflection angle is less than 2.0° (Figure 9.10).

9.7 SCHLIEREN IMAGING

The fundamental principle and theory of schlieren optical diffraction were first described by Raman and Nath (1935). Ultrasound waves produce a periodic variation in the density

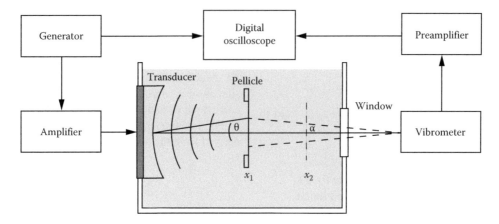

FIGURE 9.9 Schematic diagram of the focused field measurement using a scanning laser vibrometer.

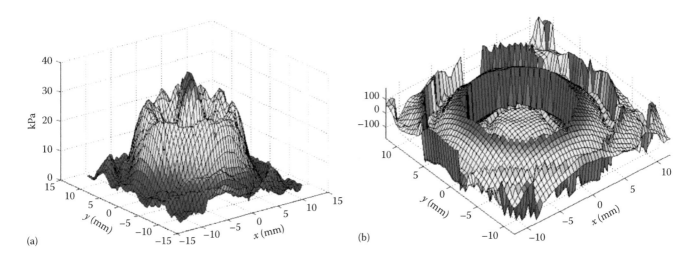

FIGURE 9.10 Distribution of measured (a) amplitude and (b) phase on the plane 35 mm away from the focusing transducer using a scanning vibrometer. (Reprinted with permission from Wang, Y., Tyrer, J., Ping, Z., and Wang, S., Measurement of focused ultrasonic fields using a scanning laser vibrometer, *J. Acoust. Soc. Am.*, 121, 2621–2627, Copyright 2007, American Institute of Physics.)

and refraction index of the medium and behave as a phase grating to a normally incident planar monochromatic light. An appropriate optical setup allows the construction of the phase grating graphically.

The basic configuration of a schlieren system is shown in Figure 9.11 (Schneider and Shung 1996). A monochromatic laser beam is normally incident upon the acoustic wave from an ultrasound transducer. The Fraunhofer diffraction pattern due to the phase shift of the laser beam is proportional to the pressure integral along the light beam. The nth-order resultant normalized intensities of the diffracted light are given by

$$I_n = J_n^2(\upsilon) \tag{9.61}$$

where
  J_n is the nth-order Bessel function
  υ is the Raman–Nath parameter representing the maximum optical phase retardation on the yz plane, given by

$$\upsilon(y,z) = A\sin(\omega_a t + k_a y)f_p(z) \tag{9.62}$$

where
  A is a constant
  ω_a is the angular frequency
  k_a is the acoustic wavenumber
  f_p is the line integral of the acoustic peak pressure p_i

$$f_p(z) = \int p_i(x,z)dx \tag{9.63}$$

The level in a gray image is related to the acoustic intensity integral along the optical path in a side view. Using tomographic reconstruction, a 2D gray map may be obtained, which shows the distribution of the acoustic intensity in the plane $y = y_0$.

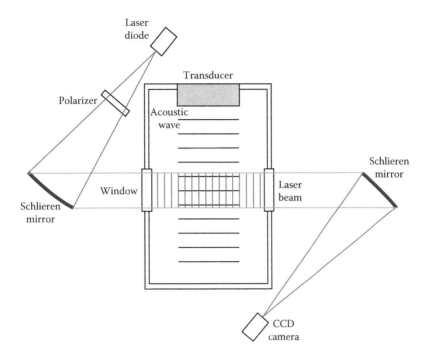

FIGURE 9.11 Schematic diagram of schlieren imaging for acoustic field characterization.

To implement the schlieren method in the characterization of ultrasound transducers, a water tank, a translational system driven by a step motor, a pulser, a function generator, a digital delay generator, a charge-coupled device (CCD) camera, a water circulation system, sophisticated optical lenses and mirrors for schlieren imaging, and a control PC are required. By setting the delay time between the laser firing triggered by a TTL trigger pulse and the ultrasonic pulse along its propagation path, the intensity of the resulting image can be captured at the required temporal points. The image at the focal plane of the optics is captured by the CCD camera, stored by a frame grabber, and then transferred to the PC for further processing. The high resolution of the schlieren system makes it possible to detect small particles in water. In order to minimize such background artifacts, a water circulation system with a filter and a pump is needed. In addition, any temperature gradient will change the refraction index of the medium, and the heat emitted by the transducer will be reduced by the circulating water and will have no influence on the resulting image. These lenses and mirrors are used to collimate the light beam, measure the changes in the refractive index, and project the schlieren pattern onto the CCD camera. The limited viewing field due to the size of the optical lenses used requires the alignment of the transducer for the specific region of interest. In addition, tomographic reconstruction could be performed by rotating the transducer over 180° by a stepper motor with a translator.

The protocols of a typical schieren image experiment may be concisely described as follows. First, the transducer is excited. Then, the laser is triggered at a predetermined delay time, controlled by the digital delay generator. The light beam is normally incident upon the acoustic beam. After the interaction, the schlieren image is focused on the CCD camera.

It is important to excite the transducer and fire the laser at high speed (thousands of pulses per second) so that the pulse seems stationary in the captured image. Tomographic reconstruction is required to obtain the acoustic field along the optical path. The synchronous TTL output from a digital delay generator is used to excite a transducer and trigger the laser, and appropriate and highly accurate delay is set for the firing of the laser because the speed of sound in water is much smaller. The variation of delay determines the propagation distance of the pulse from the transducer. The frame grabber is programmed to combine several images needed for postprocessing and reconstruction. The 3D reconstruction from the acquired images is accomplished by a well-established backprojection tomographic reconstruction algorithm. Afterward, a 2D rendering of the slice is performed. The gray level in the schlieren image is proportional to the time-averaged acoustic intensity, and the absolute values may be obtained by calibrating the gray levels with the hydrophone measurement data (Figure 9.12).

9.8 LIGHT DIFFRACTION TOMOGRAPHY

Light diffraction tomography enables the measurement of nonperturbing pressure and phase with high spatial resolution. Under certain circumstances, light diffraction tomography as an absolute measurement method can provide an accuracy of 5%–10% and satisfactory signal-to-noise ratio, reproducibility, spatial resolution, linearity, and frequency response (Figure 9.13).

The acoustic pressure from an ultrasonic transducer in the CW or tone-burst mode affects the refraction index. When a light beam intersects the ultrasound wave, it is diffracted and split into several beams propagating in different directions

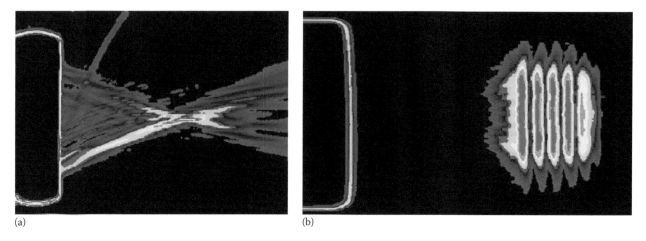

(a)   (b)

FIGURE 9.12 Representative schlieren images of the acoustic field excited by a transducer in the mode of (a) continuous wave and (b) pulsed wave.

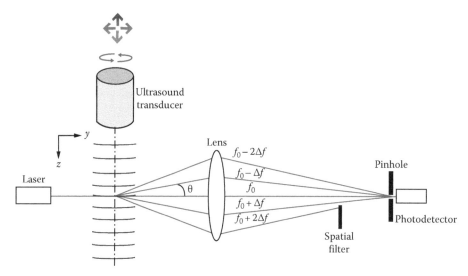

FIGURE 9.13 Schematic diagram and principle of light diffraction tomography. f_0 is the frequency of the laser, and Δf is the frequency of ultrasound burst.

(diffraction orders). The relative light intensity of the mth diffraction is

$$\frac{J_m^2(\hat{v})}{J_n^2(\hat{v})} \tag{9.64}$$

where
J_m is the Bessel function of the mth order
$\hat{v} = k_{e0}p_{op}P_{ave}L$ is the optical phase retardation (the Raman–Nath parameter), k_{e0} is the wavenumber of light, $p_{op} = \partial n/\partial p$ is the piezo-optic constant (2.0×10^{-9} m²/N at 22°C), P_{ave} is the average pressure amplitude, and L is the sound field depth:

$$\sin\theta = \pm\frac{n\lambda}{\Lambda} \tag{9.65}$$

where
θ is the angle between the diffracted beam of the nth order compared to the incident beam
λ is the light wavelength
Λ is the ultrasound wavelength

The frequency of diffracted light is shifted up or down in multiples of the ultrasound frequency.

If the ultrasound field is propagating in the z direction

$$p(x,y,z,t) = p_{max}(x,y,z)\sin\left[\omega t + kz + \phi(x,y,z)\right] \tag{9.66}$$

If the direction of the laser light is y, under Raman–Nath conditions

$$v(x,z,t) = k_{e0}p_{op}\int_{-\infty}^{\infty} p_{max}(x,y,z)\sin[\omega t + kz + \phi(x,y,z)]dy$$

$$= \hat{v}_c(x,z)\sin(\omega t + kz) + \hat{v}_s(x,z)\cos(\omega t + kz)$$

$$= \hat{v}(x,z)\sin[\omega t + kz + \overline{\phi}(x,z)] \tag{9.67}$$

where $\bar{\phi}(x, z)$ is the mean acoustic phase, and

$$\hat{v}_c(x, z) = \hat{v}(x, z)\cos[\bar{\phi}(x, z)]$$

$$= k_{e0} p_{op} \int_{-\infty}^{\infty} p_{max}(x, y, z)\cos[\phi(x, y, z)]dy \quad (9.68)$$

$$\hat{v}_s(x, z) = \hat{v}(x, z)\sin[\bar{\phi}(x, z)]$$

$$= k_{e0} p_{op} \int_{-\infty}^{\infty} p_{max}(x, y, z)\sin[\phi(x, y, z)]dy \quad (9.69)$$

If only the zeroth and the first positive or negative orders are kept, such as focusing the diffracted light and blocking the diffraction orders with a spatial filter, the light intensity picked up by a photodetector is

$$I(x, z, t) = I_0 \left\{ J_0^2[\hat{v}(x, z)] + J_1^2[\hat{v}(x, z)] \right\}$$

$$+ 2I_0 J_0^2[\hat{v}(x, z)] J_1^2[\hat{v}(x, z)]\cos[\omega t + kz + \bar{\phi}(x, z)] \quad (9.70)$$

where I_0 is the light intensity. If I_0 and the amplitude and the phase of the ultrasonic waveform are known, the AC part of the light intensity, $\hat{v}$, can be evaluated, from which $\hat{v}_c$ and $\hat{v}_s$ can be calculated to reconstruct the tomography of ultrasound beam cross section. The laser beam should be at least several acoustic wavelengths in width. The separation between the diffraction orders, r, is given by

$$r \approx F\lambda/\Lambda \quad (9.71)$$

where F is the focal length of the lens. The pinhole placed in front of the photodetector has to be small compared with the acoustic wavelength to avoid underestimation on the amplitude of the measured signal. To enable tomographic reconstruction of the pressure $p_{max}(x, y, z)$ and $\phi(x, y, z)$ in the measured plane using established algorithm, such as algebraic reconstruction technique, the transducer will be rotated at equal angular steps over 180° for several projections. The spatial resolution in light diffraction tomography is determined by the detection width, the sample distance in a projection, and the number of projections per 180°. However, there is a compromise between time and resolution, which determines the number of projections and samples. The phase is calculated by cross-correlating the present captured waveform and a reference that is collected at the center of the first projection in each ultrasonic exposure since the phase is very sensitive to temperature variations (Figure 9.14).

Both the thickness and amplitude the acoustic field should be sufficiently small to ignore natural diffraction and ray-bending effects, respectively. The total uncertainty is determined by combining the estimation of error, which is

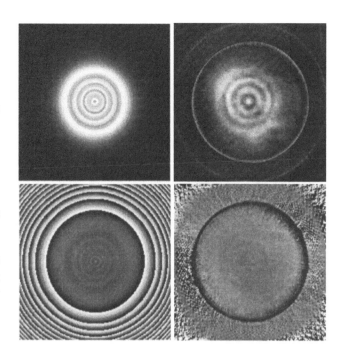

FIGURE 9.14 Comparison of the theoretical simulation (left column) and measurement with light diffraction tomography (right column) of the pressure map (upper row) and phase map (lower row) 10 mm from the surface of a 2-MHz air-coupled transducer of diameter of 8.6 mm. (Reprinted from *Ultrasonics*, 37, Almqvist, M., Holm, A., Persson, H.W., and Lindström, K., Characterization of air-coupled ultrasound transducers in the frequency range 40 kHz–2 MHz using light diffraction tomography, 565–575, Copyright 2000, with permission from Elsevier.)

assumed to be random and independent, in quadrature, reproducibility due to the variation in laser intensity, photodetector gain stability, temperature variation and air motion (±6%); theoretical considerations (±1% for $\hat{v} < 1$ and $Q < 0.5$); tomography algorithm (±5%); and piezooptic constant (±10%). So, the total uncertainty is about ±13% (Almqvist et al. 2000). Theoretically, this technique has an almost flat frequency response for very low pressure within the Raman–Nath region for both continuous and pulsed ultrasound for $\hat{v} < 0.4$. Its bandwidth is much larger than that of hydrophone. Light diffraction tomography offers much higher spatial resolution (i.e., 160 μm) and signal-to-noise ratio compared with the pulse-echo measurement.

9.9 LIGHT SCANNING OF ULTRASOUND FIELD

In a mechanical scanning system, acquisition speed is very low. Although the use of an array of detectors may overcome this problem, the large number of channels required to spatially sample with adequate resolution over a large region results in prohibitive expense. An alternative solution is encoding the spatial distribution of the acoustic pressure to an optical field and then addressing the output continuously by a sensitive optical sensor. Subsequently, effective element sizes and spatial sampling intervals are significantly smaller than the active element of the hydrophone and can be reduced to a few micrometers, which is the optical diffraction limit, in principle. Detecting the variations in the optical thickness of a thin polymer film, generated by an incident ultrasound wave, which acts as a Fabry–Perot interferometer (FPI), and consequently measuring the small optical phase shift (<150 mrad) are one of the solutions. The relationship between the reflected intensity modulation and optical phase is determined by the FPI transfer function (ITF). By mechanically scanning the photodetector over the reflected beam of the FPI, the distribution of the incident acoustic field as well as the pressure profile can be measured. This approach is an alternative for acoustic field characterization, ultrasonic nondestructive testing, and biomedical acoustic imaging and is immune to electromagnetic interference because it is passive electrically (Figure 9.15).

The FP sensor is inexpensive in batch fabrication with high repeatability, and is extremely rugged and provides many years of reliable operation despite long periods of immersion in water. The vertical-cavity surface-emitting laser (VCSEL) is a cheap alternative to external cavity lasers and other continuously tunable lasers used for interrogation. A normally incident focused light beam of 50 μm diameter, which could be reduced further for high-frequency applications by either increasing the diameter of the collimated beam or decreasing the focal length of the lens, is scanned over the region of interest in a line pathway. The output signal of the photodetector goes through a high-pass filter (>300 kHz) to remove the DC optical signal reflected from the sensor and the low-frequency fluctuations from the VCSEL (Zhang and Beard 2006). There are certain variations in the sensitivity of the FP sensor because of changes in the optical thickness of the polymer film and subsequent changes in the slope of ITF. Tuning the laser wavelength over the free spectral range (FSR) of the FPI (5.5 nm at 850 nm) and to the peak value of the ITF phase derivative can recover the ITF and obtain the maximum sensitivity. The point-to-point acquisition is, therefore, limited only by the response time of galvanometer, the wavelength tuning speed of the VCSEL, and the associated control electronics. Using the acoustic phase sensitivity of 0.06 rad/MPa, an upper limit of linear detection is about 1.2 MPa. Although this approach is used primarily for high frequencies (i.e., MHz), the measurement of low-frequency CW or quasi-CW is also applicable. Its bandwidth can be increased by reducing the polymer film thickness, which is limited by the tenability of the VCSEL, and decreasing the focused beam size and scanning step size accordingly to meet the requirement of the spatial Nyquist sampling theorem (Figure 9.16).

The detection sensitivity or noise-equivalent pressure (NEP) is defined as the acoustic pressure that has a signal-to-noise ratio of 1 at the low frequency, $\lambda_a \gg l$, where λ_a is the acoustic wavelength, and l is the thickness of FPI:

$$NEP = \frac{N}{S} \qquad (9.72)$$

where

N is the minimum detectable optical power modulation reflected from the sensor within a specific bandwidth

S is the sensor sensitivity defined as the reflected optical power per acoustic pressure (μW/MPa) at the FPI optimum phase bias φ_0 depending on the reflection coefficient of the FPI mirror, the thickness and elastic and photoelastic properties of the polymer film, and the acoustic impedance of the backing stub.

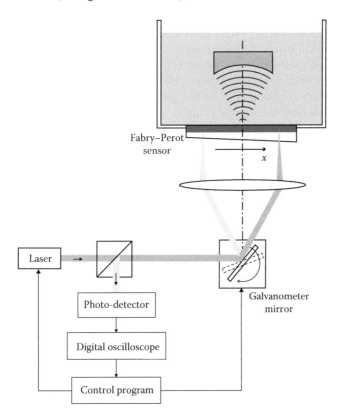

FIGURE 9.15 Schematic of the line-scanning ultrasound field mapping system.

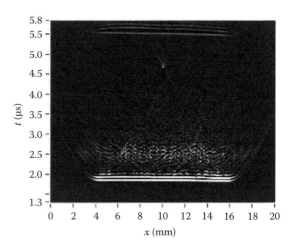

FIGURE 9.16 Line scan of a 15 MHz planar transducer in a diameter of 22 mm aligned 2.7 mm away from the FP sensor. (Zhang, E. and Beard, P., Broadband ultrasound field mapping system using a wavelength tuned, optically scanned focused laser beam to address a Fabry Perot polymer film sensor, *IEEE Trans. Ultrason. Ferroelectr. Freq. Control*, 53, 1330–1338 © 2006, IEEE.)

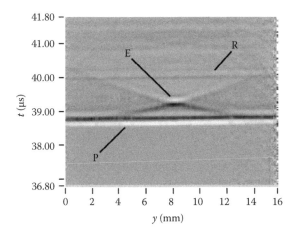

FIGURE 9.17 Vertical line scans of a pulsed 3.5 MHz planar transducer output 5.8 cm away from a 40 μm Parylene sensor. P, initial plane wave; E, edge wave component; R, reflection from back surface of a 4 mm glass backing stub. (Beard, P.C., Two-dimensional ultrasound receive array using an angle-tuned Fabry–Perot polymer film sensor for transducer field characterization and transmission ultrasound imaging, *IEEE Trans. Ultrason. Ferroelectr. Freq. Control*, 52, 1002–1012 © 2005, IEEE.)

For the 75 μm PET sensor, the peak *NEP* was 6.5 kPa over 25 MHz measurement bandwidth of a 200 μm PVDF hydrophone. The uniformity of response is determined by the acoustic impedance mismatches at the boundaries of the film due to the backing and the surrounding water on both ends.

An example of the acoustic field from a flat transducer is shown in Figure 9.17. This planar wavefront is followed by inverted edge waves originating from the circumference of the transducer, demonstrating the characteristic "X" shape. The edge waves around the transducer circumference arrive at the same time so that the center of X shape has the maximum pressure. As the detector moves off axis, asymmetry will occur. The difference between the time of flight of the edge wave and that of the initial planar component increases with decreasing distance between transducer and detector. However, it is to be noted that only the amplitude of the detected signal is required (Figure 9.18).

The availability of inexpensive and relatively long-exposure CCD cameras in conjunction with a pulsed interrogating laser, which determines the temporal resolution of measurement, is the alternative to piezoelectric arrays with their fabrication difficulties and high cost. By changing the geometry of the illuminating beam, any arbitrary transducer can be characterized. A large FPI reflected beam can be optically scanned over a small detector array using galvanometer mirrors to synthesize a larger array aperture quickly. In addition, the incident beam could also be focused to a diffraction-limited spot and scanned over the sensor while receiving the reflected beam with a single detector. By optimizing the ITF and increasing the interrogating laser power, detection sensitivities with a 50 μm optical element can be 0.1 kPa, which is advantageous over piezoelectric characterization in the acoustic performance.

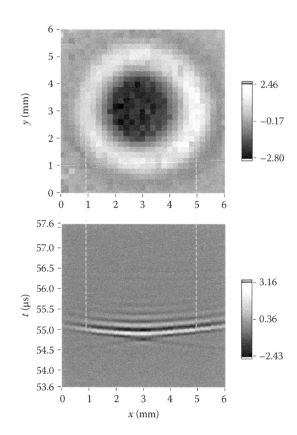

FIGURE 9.18 Two-dimensional scans of a pulse from a 5 MHz focused transducer 8.2 cm away from a 75 μm PET sensor. The upper and lower rows show the lateral pressure distribution and the corresponding pressure time records $p(x, t)$ through the center of each of the xy pressure distribution, respectively. Scan area = 6 mm × 6 mm, scan increments $dx = dy = 0.2$ mm, photodiode aperture = 0.2 mm. (Beard, P.C., Two-dimensional ultrasound receive array using an angle-tuned Fabry–Perot polymer film sensor for transducer field characterization and transmission ultrasound imaging, *IEEE Trans. Ultrason. Ferroelectr. Freq. Control*, 52, 1002–1012 © 2005, IEEE.)

9.10 TIME-DELAY SPECTROMETRY

In pulse calibration using either a broadband hydrophone or a suitable reflector, the potential errors may include limited bandwidth in the emitted pulse and nonlinear distortion in the wave propagation. Time-delay spectrometry (TDS) is a spectroscopic measurement technique to determine complex transfer functions in reverberant surroundings (Heyser 1967, Koch 2003). It can be used to quantitatively calibrate the acoustic field of a transducer in the frequency domain and determine the sensitivity of the hydrophone under free-field conditions. Particular features of this technique are the continuous frequency response, the high signal-to-noise ratio, and the ease of comparative measurements.

The principle behind TDS is that the time delays of the propagation from transmitter to hydrophone are proportional to the shifts in instantaneous frequency. When the excitation signal is swept at a constant rate, the frequency shift is proportional to the delay. The received signal is the sum of the waves along the individual propagation paths. So, it is feasible to extract the direct signal (from the transmitter to the hydrophone) from the whole received sequence, virtually eliminating the effects of multiple transmission paths, standing waves, and other interferences due to reflected signals. In other words, selectivity in the time domain is proportional to that in the frequency domain at the sweep rate. In addition, the measurement system with a finite impulse response, which is the combined response of the transmitter and hydrophone, also produces a distribution of time delays, which results in a broad spectrum of the measured signal relative to that of excitation. However, the measurement of high-frequency or nonlinear, distorted waveforms is hindered by the nonideal transfer function of the hydrophone due to its finite aperture, limited signal-to-noise ratio, or imperfect manufacturing process (Figure 9.19).

The impulse response of the measurement system consisting of a pure delay τ is

$$h_d(t) = \delta(t - \tau) \qquad (9.73)$$

Then, the signal picked up by the hydrophone, $g(t)$, is

$$g(t) = e(t) * h_d(t) = e(t - \tau) \qquad (9.74)$$

where
  $e(t)$ is the excitation signal
  the symbol $*$ is the convolution

If $f_T(t)$ is the frequency of the transmitter at time t and $f_R(t)$ is the central frequency of the hydrophone, then

$$f_R(t) = f_T(t - \tau) = f_T(t) - \tau \frac{\partial f_T}{\partial t} = f_T(t) - \tau \cdot S \qquad (9.75)$$

where S is the frequency sweep rate under the assumption of an ideal flat frequency response within the frequency range of excitation and determines the frequency resolution obtainable. However, with appropriate choice of τ or the frequency offset $f_R - f_T$, very narrow receiving filters can be used, which allows its use in weak fields or with highly attenuating media. The offset frequency is equal to the product of S and the ultrasonic transit time $t_d = l/c$:

$$f_T - f_R = \frac{f_{stop} - f_{start}}{\Delta t} \cdot \frac{l}{c} = S \cdot t_d \qquad (9.76)$$

And, the spectral broadening is given by

$$\frac{f_{stop} - f_{start}}{\Delta t} \cdot t_x = S \cdot t_x \qquad (9.77)$$

where t_x is the excitation duration. The degree of spatial discrimination Δx is given by

$$\Delta x = c\Delta \tau = \frac{cB}{\partial f / \partial t} \qquad (9.78)$$

where c is the sound speed in the medium. Thus, the minimum spatial discrimination is obtained with the minimum Δf and the maximum S. The time ΔT required for a filter of width B to achieve the equilibrium output is given approximately by $B \cdot \Delta T \geq 1$, which produces the constraint

$$B^2 \geq S \qquad (9.79)$$

In order to remove the unwanted reflections, B should be made as small as possible. If Δx is sufficiently small, the shape of the filter used is relatively unimportant. A flat filter produces more blurring than a smooth one, such as a Hanning or a Butterworth filter.

For n discrete transmission paths, the impulse response is

$$h_d(t) = \sum_{i=1}^{n} \delta(t - \tau_i) \qquad (9.80)$$

So,

$$g(t) = e(t) * h_d(t) = \sum_{i=1}^{n} e(t - \tau_i) \qquad (9.81)$$

Therefore, in order to avoid overlap, the arrival time of the nearest reflection, t_r, should satisfy

$$t_r > t_d + t_x \qquad (9.82)$$

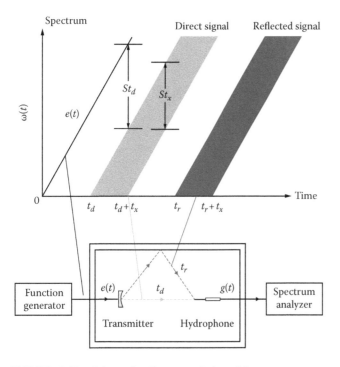

FIGURE 9.19 Schematic diagram of time-delay spectrometry (TDS) used for the calibration of an ultrasound transducer.

The width of the time window, t_x, is determined by S and B:

$$t_x = \frac{B}{S} \qquad (9.83)$$

If the frequency offset in the data analysis of received acoustic signal is set to B, the direct path between the transmitter and the hydrophone is selected and reverberation from the tank walls is filtered out.

Gated time-delay spectrometry (GTDS) measurement allows the phase-locked (i.e., coherent) detection of the hydrophone signals in the time domain. A narrow intermediate frequency (IF) filter can select the length of the acoustic signal by setting a gate and canceling out all unwanted and longer signals. Then, the filtered signal will be converted to the time domain by an inverse FFT. The FFT algorithm and the gating procedure are implemented by digital filtering techniques, so the technical effort of the GTDS measurement is minimal. GTDS is less expensive and suitable for a wide frequency range but limited to devices with a smooth frequency response. A decrease in the distance between the transducer and hydrophone improves the accuracy, and the use of a focusing transducer is preferred.

For heterodyne time-delay spectrometry (HTDS), a network analyzer with excitation signal shifted by a local oscillator frequency f_{LO} is connected to the transmitting port A and the hydrophone signal is connected to the receiving port B. A reference signal R is necessary for the phase-locked loop in the network analyzer to set the phase relation between ports A and B. Thus, the signals at port A have a frequency between $f_{LO} + f_{start}$ and $f_{LO} + f_{stop}$, while those at ports B and R are in the frequency range of $f_{stop} - f_{start}$. A double-conversion heterodyne system can also be implemented to determine the sweeping spectra of the received signal (Harris et al. 2004), which provides better selectivity than a single heterodyne scheme.

There are five parameters in TDS implementation: F, the frequency range of interests; Δt, the sweep time; l, the distance between the transmitter and hydrophone; B, the filter bandwidth; and $f_R - f_T$, the frequency offset. The frequency range of interest is determined by the application and constrained by the bandwidth of the other electronics in the system, for example, preamplifiers. The sweep time Δt is only limited by the stability of system, primarily by thermal effects on the operation of electronics. The sweep rate depends on the instrument. Usually, the slower the rate, the better the frequency resolution and the signal-to-noise ratio. The direct distance is determined by the field characteristics of the transmitter (the diffractive near-field region should be avoided). The resolution bandwidth is usually minimized to enhance signal-to-noise ratio and exclude reflections. However, the lowest Δf may not be used in order to accommodate small fluctuations in the propagation path. The frequency offset is set by the calibration requirement and limited only by the stability of the circuitry.

The bandwidth of the IF filter controls the sensitivity of the TDS system to temperature fluctuation. The fractional change of velocity consequent upon a temperature change is related to the fractional change of frequency offset by

$$\frac{\Delta(f_T - f_R)}{f_T - f_R} \approx \frac{\Delta c}{c} \qquad (9.84)$$

A change of 1°C changes the velocity by approximately 0.2% as well as the frequency offset. So, the measurement arrangement is relatively robust to temperature fluctuations. In addition, small velocity variations due to low concentrations of dissolved salts or gases have negligible influence on the measurement technique. A concentration of 10 mg/mL NaCl would have less effect than a 4°C temperature change.

The TDS approach has several advantages. It yields a continuous instead of discrete frequency response calibration in an easy and quick way. Virtual free-field conditions in a smaller measurement tank is available in most labs, eliminating influence of standing waves and reverberations. In comparison to other conventional techniques, such as gated burst transmission, TDS offers improved signal-to-noise ratio in the determination of transmitting and receiving frequency responses, directivity patterns, and effective aperture of ultrasound transducer. Although using an impulse source and Fourier transform can also illustrate the frequency response, TDS achieves much improved signal-to-noise ratio, which is quantitatively equal to the product of the sweep time and the frequency range. For impulse techniques, this product is unity. This approach is very convenient for magnitude, but not phase, information of acoustic pressure. The achieved measurement accuracy is in good agreement with that using independent calibration techniques. The applications include the rapid assessment of not only the reproducibility of batch-produced probes but also the absolute hydrophone frequency responses. The linearity and directionality of the transducer can also be easily checked. A key element in the successful implementation of TDS is the availability of a transmitter with adequate bandwidth. The choice of experimental parameters requires a little care because of their interdependence (Chivers 1986).

9.11 INFRARED THERMOGRAPHY

Infrared (IR) thermography is a promising alternative to hydrophone scanning for the determination of acoustic intensity. A short ultrasound beam of 0.1–0.3 s duration heats an absorbing thin sheet with known acoustic and thermal parameters, and the acoustic intensity is inferred from temperature measurements using an infrared camera (Shaw and Nunn 2010). Relatively low acoustic power of the ultrasound transducer provides the fundamentals of linear acoustic propagation and a temperature rise of less than 50°C to avoid damage of the absorber. Short-duration sonication can be applied to reduce the diffusion effects so that the temperature rise is proportional to the intensity. Advantages of this technique include its noninvasive nature and the ability to estimate the intensity in an entire plane simultaneously. Because time-consuming mechanical scanning is not required, assessment of 2D and 3D ultrasound beams is rapid and quantitative. Its drawbacks include the expensive IR camera with limited frame rate and complications due to the presence of the air interface residing between the IR camera and the heated surface, such as convection in the air and beam reflection at the air interface.

The IR method depends on the acoustic frequency and the interfering pattern in the absorber mostly due to reflections from the air interface. Reflections from the water/absorber interface can be ignored, but the total reflection from the absorber/air interface is included in modeling the intensity field in the absorber. In the absorber, the acoustic field consists of the primary and the reflected waves:

$$I(x, y, z) = I_w(x, y, z)[\exp(-2\alpha z') - \exp(-4\alpha l + 2\alpha z')] \quad (9.85)$$

The heat deposited in the layer is

$$q(x, y, z) = -\frac{\partial I(x, y, z)}{\partial z} = 2\alpha I_w(x, y, z)[\exp(-2\alpha z')$$

$$+ \exp(-4\alpha l + 2\alpha z')] \quad (9.86)$$

where
   α is the absorption coefficient of the layer
   z' is the propagation distance inside the layer
   $z' = 0$ is the water/absorber interface
   $z' = l$ is the absorber/air interface
   l is the thickness of the absorber

The heat transfer equation for the temperature rise is

$$\frac{\partial T}{\partial t} = \kappa \Delta T + \frac{q(x, y, z)}{\rho C_v} \quad (9.87)$$

where
   κ is the thermal diffusivity
   C_v is the specific heat capacity of the absorber

In the air, there is no heat deposition. The boundary condition at the absorber/air interface is $\partial T/\partial z = 0$ assuming no heat conduction to the air, while there is no temperature rise at the absorber/water interface, $T = 0$, if the water is thick enough. If diffusion is negligible, the energy deposition and temperature rise in the absorber are linearly related, and the distribution of temperature rise in the absorber's surface is

$$T = \frac{4\alpha \Delta t}{\rho C_v} I_w(x, y, F)\exp(-2\alpha l) \quad (9.88)$$

Then, free-field intensity can be determined as

$$I_w = \frac{\rho C_v}{4\alpha} \cdot \exp(2\alpha l) \cdot \frac{T}{\Delta t} \quad (9.89)$$

Because of diffusion, the maximum temperature inside the absorber induces heat movement outward and causes even some temperature rise in the water. Its influence should be

---

TABLE 9.2
International Standards for the Measurement of Medical Ultrasound Fields

IEC 60500 (1974–01)	IEC standard hydrophone
IEC 60565 (1977–01)	Calibration of hydrophones
IEC 60565A (1980–01)	First supplement
IEC/TR 60854 (1986–10)	Methods of measuring the performance of ultrasonic pulse-echo diagnostic equipment
IEC 60866 (1987–05)	Characteristics and calibration of hydrophones for operation in the frequency range of 0.5–15 MHz
IEC 61101 (1991–12)	The absolute calibration of hydrophones using the planar scanning technique in the frequency range of 0.5–15 MHz
IEC 61102 (1991–11)	Measurement and characterization of ultrasonic fields using hydrophones in the frequency range of 0.5–15 MHz
IEC 61102-am1 (1993–09)	Amendment 1—measurement and characterization of ultrasonic fields using hydrophones in the frequency range of 0.5–15 MHz
IEC 61266 (1994–12)	Ultrasonics—hand-held probe Doppler fetal heartbeat detectors—performance requirements and methods of measurement and reporting
IEC/TS 61390 (1996–07)	Ultrasonics—real-time pulse-echo systems—test procedures to determine performance specifications
IEC 61685 (2001–07)	Ultrasonics—flow measurement systems—flow test object
IEC 61689 (1996–08)	Ultrasonics—physiotherapy systems—performance requirements and methods of measurement in the frequency range of 0.5–5 MHz
IEC 61828 (2001–05)	Ultrasonics—focusing transducers—definitions and measurement methods for the transmitted fields
IEC 61846 (1998–04)	Ultrasonics—pressure pulse lithotripters—characteristics of fields
IEC 61847 (1998–01)	Ultrasonics—surgical systems—measurement and declaration of the basic output characteristics
IEC/TS 61895 (1999–10)	Ultrasonics—pulsed Doppler diagnostic systems—test procedures to determine performance
IEC 62092 (2001–08)	Ultrasonics—hydrophones—characteristics and calibration in the frequency range from 15 to 40 MHz
IEC 62359 (2005–04)	Ultrasonics—field characterization—test methods for the determination of thermal and mechanical indices related to medical diagnostic ultrasonic fields
IEC 60601-2-5 (2001)	Medical electrical equipment: particular requirements for safety of ultrasonic physiotherapy equipment
IEC 60601-2-37 (2004-10) Ed. 1.1	Medical electrical equipment: particular requirements for the safety of ultrasonic medical diagnostic monitoring equipment

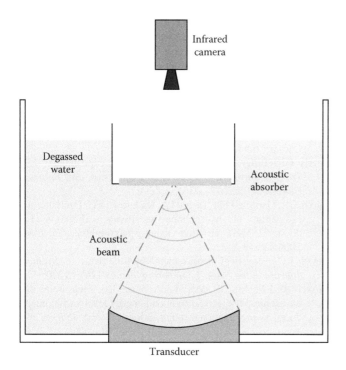

FIGURE 9.20 Schematic diagram of quantitative estimation of acoustic intensity using infrared thermography.

accounted for (Shaw et al. 2010). The axial and radial diffusion of heat is given by (Giridhar et al. 2012)

$$f_b(r, t) = \frac{r_b^2}{4\kappa} \int_0^{4\kappa t/r_b^2} \left[\frac{1 - \exp[-(kr_b)^2 u]}{1 + u} \right] \cdot \exp\left[\left(\frac{u}{1+u} \right) \frac{r^2}{r_b^2} \right] du$$

(9.90)

where
  k is the acoustic wavenumber
  r_b is the ultrasound beam radius at $z' = 0$

In practice, limits of $(kr_b)^2 \gg 1$ and $4\kappa t/r_b^2 \ll 1$ are satisfied, which correspond to the high-frequency limits and short heat diffusion in comparison to the ultrasound beam width.

Furthermore, using a three-layer model (water, absorber, and air) and considering the wave transmission and thermal exchange at each interface, the acoustic and thermal field in the absorber can be described with higher accuracy (Yu et al. 2013). The acoustic field and subsequent temperature distribution on the axis of absorber oscillate like the standing wave, but they decay significantly at the absorber/water interface (Table 9.2). Experimental results are compared with theoretical simulation results, and great similarities are found between them in the 3D space (Figures 9.20 and 9.21).

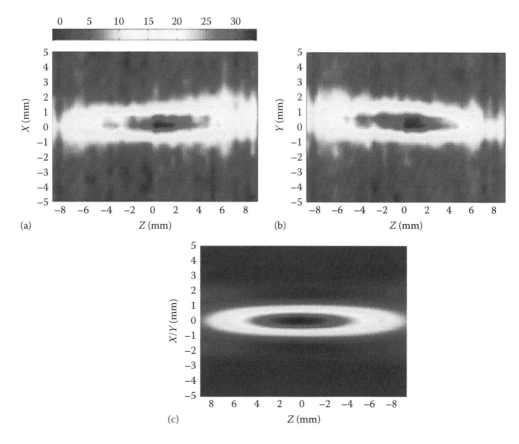

FIGURE 9.21 Comparison of acoustic intensity distribution derived by the proposed method at (a) xz and (b) yz planes in the IR measurement at a heating time of 200 ms with (c) the theoretical simulation in the free field. (From Yu, Y. et al., *Ultrasound Med. Biol.*, 39, 2021, 2013.)

REFERENCES

Acoustical Measurements, 1988.

AIUM. *Standard Methods for Measuring Performance of Pulse-Echo Ultrasound Imaging Equipment*, 1990.

AIUM/NEMA. *Acoustic Output Measurement Standard for Diagnostic Ultrasound Equipment*, 1998a.

AIUM/NEMA. *Standard for Real-Time Display of Thermal and Mechanical Acoustic Output Indices on Diagnostic Ultrasound Equipment*, 1998b.

Almqvist M, Holm A, Persson HW, Lindström K. Characterization of air-coupled ultrasound transducers in the frequency range 40 kHz–2 MHz using light diffraction tomography. *Ultrasonics* 2000;37:565–575.

Bamber JC, Phelps J. The effective directivity characteristic of a pulsed ultrasound transducer and its measurement by semi-automatic means. *Ultrasonics* 1977;15:169–174.

Barker LM, Hollenbach RE. Shock wave studies of PMMA, fused silica, and sapphire. *Journal of Applied Physics* 1970;41:4208–4226.

Beard PC. Two-dimensional ultrasound receive array using an angle-tuned Fabry–Perot polymer film sensor for transducer field characterization and transmission ultrasound imaging. *IEEE Transactions on Ultrasonics, Ferroelectrics, and Frequency Control* 2005;52:1002–1012.

Beard PC, Hurrell A, Mills TN. Characterisation of a polymer film optical fibre hydrophone for the measurement of ultrasound fields for use in the range 1–30 MHz: A comparison with PVDF needle and membrane hydrophones. *IEEE Transactions on Ultrasonics, Ferroelectrics, and Frequency Control* 2000;47:256–264.

Brendel K, Ludwig G. Calibration of ultrasonic standard probe transducers. *Acustica* 1976;36:203–208.

Chen JF, Zagzebski JA. Frequency dependence of backscatter coefficient versus scatterer volume fraction. *IEEE Transactions on Ultrasonics, Ferroelectrics, and Frequency Control* 1996;43:345–353.

Chivers RC. Time-delay spectrometry for ultrasonic transducer characterization. *Journal of Physics E: Scientific Instruments* 1986;1:834.

Clement GT, Hynynen K. Field characterization of therapeutic ultrasound phased arrays through forward and backward planar projection. *Journal of the Acoustical Society of America* 2000;108:441–446.

Davision L, Graham RA. Shock compression of solids. *Physics Reports* 1979;55:255–379.

Fisher GA. Transducer test system design. *Hewlett Packard Journal* 1983;34:24–25.

Gendreu P, Fink M, Royer D. Optical imaging of transient acoustic fields generated by piezocomposite transducers. *IEEE Transactions on Ultrasonics, Ferroelectrics, and Frequency Control* 1995;42:135–143.

Giridhar D, Robinson RA, Liu Y, Sliwa J, Zderic V, Myers MR. Quantitative estimation of ultrasound beam intensities using infrared thermography—Experimental validation. *Journal of the Acoustical Society of America* 2012;131:4283–4291.

Gloersen WB, Harris GR, Stewart HF, Lewin PA. A comparison of two calibration methods for ultrasonic hydrophones. *Ultrasound in Medicine and Biology* 1982;8:545–548.

Hanafy A, Zanelli CI, McAvoy BR. Quantitative real-time pulsed Schlieren imaging of ultrasonic waves. *IEEE Ultrasonics Symposium Proceedings*, 1991, pp. 1223–1227. Dec. 8–11, Orlando, FL.

Harris GR. A discussion of procedures for ultrasonic intensity and power calculations from miniature hydrophone measurements. *Ultrasound in Medicine and Biology* 1985;11:803–817.

Harris GR. Hydrophone measurements in diagnostic ultrasound fields. *IEEE Transactions on Ultrasonics, Ferroelectrics, and Frequency Control* 1988;35:87–101.

Harris GR. Medical ultrasound exposure measurements: Update on devices, methods, and problems. *IEEE Ultrasonics Symposium Proceedings*, 1999, pp. 1341–1352. Oct. 17–20, Caesars Tahoe, NV.

Harris GR. Progress in medical ultrasound exposimetry. *IEEE Transactions on Ultrasonics, Ferroelectrics, and Frequency Control* 2005;52:717–736.

Harris GR, Gammell PM, Lewin PA, Radulescu EG. Interlaboratory evaluation of hydrophone sensitivity calibration from 0.1 to 2 MHz via time delay spectrometry. *Ultrasonics* 2004;42:349–353.

Heyser RC. Acoustical measurements by time delay spectrometry. *Journal of the Audio Engineering Society* 1967;15:370.

Hinkelman LM, Liu DL, Metlay LA, Waag RC. Measurements of ultrasonic pulse arrival time and energy level variations produced by propagation through abdominal wall. *Journal of the Acoustical Society of America* 1994;95:530–541.

Hinkelman LM, Szabo TL, Waag RC. Measurements of ultrasonic pulse distortion produced by human chest wall. *Journal of the Acoustical Society of America* 1997;101:2365–2373.

Ide M, Ohdaira E. Measurement of diagnostic electronic linear arrays by miniature hydrophone scanning. *IEEE Transactions on Ultrasonics, Ferroelectrics, and Frequency Control* 1988;35:214–219.

IEC Standard 61828. *Ultrasonics: Focusing Transducers Definitions and Measurement Methods for the Transmitted Fields*, 2001.

IEC Standard 60601-2-37. *Medical Electrical Equipment, Part 2: Particular Requirements for the Safety of Ultrasonic Medical Diagnostic and Monitoring Equipment*, 2002.

IEEE Standard 790-1989. *IEEE Guide for Medical Ultrasound Field Parameter Measurements*, 1990.

Jansson TT, Mast TD, Waag RC. Measurements of differential scattering cross section using a ring transducer. *Journal of the Acoustical Society of America* 1998;103:3169–3179.

Klann M, Koch C. Measurement of spatial cross sections of ultrasound pressure fields by optical scanning means. *IEEE Transactions on Ultrasonics, Ferroelectrics, and Frequency Control* 2005;52:1546–1554.

Koch C. Amplitude and phase calibration of hydrophones by heterodyne and time-gated time-delay spectrometry. *IEEE Transactions on Ultrasonics, Ferroelectrics, and Frequency Control* 2003;50:344–348.

Koch C, Molkenstruck W, Reibold R. Shock-wave measurement using a calibrated interferometric fiber-tip sensor. *Ultrasound in Medicine and Biology* 1997;23:1259–1266.

LeDet EG, Zanelli CI. A novel, rapid method to measure the effective aperture of array elements. *IEEE Ultrasonics Symposium Proceedings*, 1999, pp. 1077–1080. Oct. 17–20, Caesars Tahoe, NV.

Lee CP, Wang TG. Acoustic radiation force. *Journal of the Acoustical Society of America* 1993;94:1099–1109.

Lewin PA. Miniature piezoelectric polymer ultrasonic hydrophone probes. *Ultrasonics* 1981;19:213–216.

Liu DL, Waag RC. Estimation and correction of ultrasonic wavefront distortion using pulse-echo data received in a two-dimensional aperture. *IEEE Transactions on Ultrasonics, Ferroelectrics, and Frequency Control* 1998;45:473–490.

Lloyd EA. *Ultrasonic Techniques in Biology and Medicine*. London, U.K.: Iliffe, 1964.

Lum P, Greenstein M, Grossman C, Szabo TL. High frequency membrane hydrophone. *IEEE Transactions on Ultrasonics, Ferroelectrics, and Frequency Control* 1996;43:536–543.

Madsen EL, Insana MF, Zagzebski JA. Method of data reduction for accurate determination of acoustic backscatter coefficients. *Journal of the Acoustical Society of America* 1984;76:913–923.

Margulis MA, Margulis IM. Calorimetric method for measurement of acoustic power absorbed in a volume of a liquid. *Ultrasonics Sonochemistry* 2003;10:343–345.

Menssen W, Molkenstruck W, R. R. Fiber optic sensor system. *IEEE Ultrasonics Symposium*, New York, 1991, pp. 347–350.

Miller HB. *Acoustical Measurements: Methods and Instrumentation*. Stroudsburg, PA: Hutchinson Ross Publishing, 1982.

Paltauf G, Schmidt-Kliber H. Optical method for two dimensional ultrasonic detection. *Applied Physics Letters* 1999;75:1048–1050.

Perkins MA. A versatile force balance for ultrasound power measurement. *Physics in Medicine and Biology* 1989;34:1635–1651.

Pitts TA, Sagers A, Greenleaf JF. Optical phase contrast measurement of ultrasonic fields. *IEEE Transactions on Ultrasonics, Ferroelectrics, and Frequency Control* 2001;48:1686–1694.

Preston RC, Bacon DR, Corbett SS, Harris GR, Lewin PA, McGregor JA, O'Brien WD, Szabo TL. Interlaboratory comparison of hydrophone calibrations. *IEEE Transactions on Ultrasonics, Ferroelectrics, and Frequency Control* 1988a;35:206–213.

Preston RC, Bacon DR, Smith RA. Calibration of medical ultrasonic equipment-procedures and accuracy assessment. *IEEE Transactions on Ultrasonics, Ferroelectrics, and Frequency Control* 1988b;35:110–121.

Preston RC, Lewin PA, Bjørøn L. Application of time-delay spectrometry for calibration of ultrasonic transducers. *IEEE Transactions on Ultrasonics, Ferroelectrics, and Frequency Control* 1988c;35:185–205.

Raman CV, Nath NS. The diffraction of light by high frequency ultrasonic waves. *Proceedings of the Indian Academy of Science II*, 2, 406–412, 1935.

Schafer ME, Lewin PA. A computerized system for measuring the acoustic output from diagnostic ultrasound equipment. *IEEE Transactions on Ultrasonics, Ferroelectrics, and Frequency Control* 1988;35:102–109.

Schafer ME, Lewin PA. Transducer characterization using the angular spectrum method. *Journal of the Acoustical Society of America* 1989;85:2202–2214.

Schneider B, Shung KK. Quantitative analysis of pulsed ultrasonic beam patterns using a schlieren system. *IEEE Transactions on Ultrasonics, Ferroelectrics, and Frequency Control* 1996;43:1181–1186.

Shaw A, Khokhlova V, Bobkova S, Gavrilov L, Hand J. Calibration of HIFU intensity fields measured using an infra-red camera. *Journal of Physics: Conference Series* 2010;279:012019.

Shaw A, Nunn J. The feasibility of an infrared system for real-time visualization and mapping of ultrasound fields. *Physics in Medicine and Biology* 2010;55:N321–N327.

Shaw A, Pay N, Preston RC, Bond A. A proposed standard thermal test object for medical ultrasound. *Ultrasound in Medicine and Biology* 1999;25:121–132.

Shekhawat GS, Dravid VP. Nanoscale imaging of buried structures via scanning near-field ultrasound holography. *Science* 2005;310:89–92.

Sherman GC. Integral transform formulation of diffraction theory. *Journal of the Optical Society of America* 1967;57:1490–1498.

Shombert DG, Harris GR. Use of miniature hydrophones to determine intensities typical of medical ultrasound devices. *IEEE Transactions on Ultrasonics, Ferroelectrics, and Frequency Control* 1986;UFFC-33:287–294.

Shombert DG, Smith SW, Harris GR. Angular response of miniature ultrasonic hydrophones. *Medical Physics* 1982;9:484–492.

Staudenraus J, Eisenmenger W. Fibre-optic probe hydrophone for ultrasonic and shock-wave measurements in water. *Ultrasonics* 1993;31:267–273.

Szabo TL, Melton Jr HE, Hempstead PS. Ultrasonic output measurements of multiple mode diagnostic ultrasound systems. *IEEE Transactions on Ultrasonics, Ferroelectrics, and Frequency Control* 1988;35:220–231.

Teo TJ, Reid JM. Angular spectrum decomposition: Improving the resolution. In: Berkout AJ, Ridder J, Van der Wal L, eds., *Acoustical Imaging*. New York: Plenum, 1985, pp. 143–154.

Thebald PD, Robinson SP, Thompson AD, Preston RC, Lepper PA, Wang Y. Technique for the calibration of hydrophones in the frequency range 10 to 600 kHz using a heterodyne interferometer and an acoustically compliant membrane. *Journal of the Acoustical Society of America* 2005;118:3110–3116.

Torr GR, Watmough DJ. A constant-flow calorimeter for the measurement of acoustic power at megahertz frequencies. *Physics in Medicine and Biology* 1977;22:444–450.

Waag RC, Campbell JA, Ridder J, Mesdag P. Cross-sectional measurements and extrapolations of ultrasonic fields. *IEEE Transactions on Sonics and Ultrasonics* 1985;32:26–35.

Wang Y, Tyrer J, Ping Z, Wang S. Measurement of focused ultrasonic fields using a scanning laser vibrometer. *Journal of the Acoustical Society of America* 2007;121:2621–2627.

Whittingham TA. The purpose and techniques of acoustic output measurement. In: Duck FA, Baker AC, Starritt HC, eds., *Ultrasound in Medicine*. Bristol, U.K.: Institute of Physics Publication, 1998, pp. 129–148.

Wu JR. Determination of velocity and attenuation of shear waves using ultrasonic spectroscopy. *Journal of the Acoustical Society of America* 1996;99:2871–2875.

Yadav HS, Murty DS, Veerma SN, Sinha KHC, Gupta BM, Chand D. Measurement of refractive index of water under high dynamic pressures. *Journal of Applied Physics* 1973;44:2197–2200.

Yu Y, Shen G, Zhou Y, Bai J, Chen Y. Quantitative assessment of acoustic intensity in the focused ultrasound field using hydrophone and infrared imaging. *Ultrasound in Medicine and Biology* 2013;39:2021–2033.

Zhang E, Beard P. Broadband ultrasound field mapping system using a wavelength tuned, optically scanned focused laser beam to address a Fabry Perot polymer film sensor. *IEEE Transactions on Ultrasonics, Ferroelectrics, and Frequency Control* 2006;53:1330–1338.

Ziskin MC, Lewin PA. *Ultrasonic Exposimetry*. Boca Rotan, FL: CRC Press, 2000.

Ziskin MC, Szabo TL. Impact of safety considerations on ultrasound equipment and design and use. *Clinical Diagnostic Ultrasound* 1993;28:151–160.

10 Ultrasonic Physiotherapy

10.1 INTRODUCTION

Ultrasound has been used widely and frequently in the clinical field of physiotherapy since World War II, such as an enhancer for healing and recovering process. But the nature of ultrasound practice has changed significantly over the past two decades. There are many reports on the positive effects of ultrasound therapy at low intensities, similar to those of laser therapy. Tissues, essentially the dense collagenous ones, absorb the acoustic energy in the ultrasound exposure for therapeutic purpose. However, such an effect is a little less in muscle, nerve, and tissues with significant edema (Watson 2006a, 2008). It is also used by a plenty of osteopaths, chiropractors, and sports therapists in their professional groups. Thus, ultrasound is involved in the treatment of a sprained ligaments, inflamed tendons and tendon sheaths, lacerations, soft tissue damage, scar tissue, varicose ulcers, amputations, neuromata, strained and torn muscles, inflamed and damaged joint capsules, fasciitis, delayed-onset muscle soreness, and the bone fractures. It is found that 93% of physiotherapists in Canada, 81% in Australia (93% in Brisbane), 64% in the northeastern United States, and 88% in the United Kingdom as well as in Denmark, Finland, New Zealand, and Switzerland use ultrasound daily (Robinson and Snyder-Mackler 1988, Lindsay et al. 1990, Pope et al. 1995, Robertson and Spurritt 1998). In the Netherlands, ultrasound is used in 17% of all episodes of care by the primary care physical therapists (Robebroeck et al. 1998). The number of annual ultrasound treatments performed in the United States is estimated to be 15 million (Naslund 2001).

In 1985, a survey was carried out among physiotherapists at all National Health Service (NHS) hospital departments and those from the Organization of Chartered Physiotherapists in Private Practice (OCPPP) in England and Wales regarding the use of therapeutic ultrasound (ter Haar et al. 1987). A total of 2420 physiotherapists at 204 NHS departments of which 2027 (84%) and all of 191 private practitioners had used ultrasound on patients at some time. NHS departments performed 95,400 physiotherapy treatments in the week of questionnaires, of which 18,861 (20%) involved ultrasound, whereas the corresponding numbers for private physiotherapists were 5,290 treatments and 2,857 (54%) ultrasound one. This represents over 980,000 and 148,000 treatments involving ultrasound per year in the NHS and private practice, respectively. Six hundred and eighty-three ultrasound machines of 13 different types and 287 machines in 18 models were in use in NHS departments and private practice, respectively. The most popular ultrasound devices have frequencies of 0.75, 1.0, 1.5, and 3.0 MHz. Of NHS departments, 16% only used pulsed mode and 12% used continuous mode, whereas the corresponding numbers of the private practitioners are 17% and 19%. The exposure mode (pulsed or continuous one) is chosen empirically in order to minimize the thermal effects. Intensities used vary from 0.1 to 3.0 W/cm^2 and the common use is in the range of 1.0–3.0 W/cm^2. Ultrasound has been used mainly for the treatment of acute soft tissue injuries, the relief of inflammation, and the relief of associated pain. It is also usually applied to treat lesions that are slow to heal (i.e., gravitational ulcers and pressure sores), but rarely to accelerate remodeling processes (i.e., scar tissue and fascia) and the bone repair as listed in Table 10.1. However, ultrasound is contraindicated in the situations with the possibility of spreading infection or disease as listed in Table 10.2. Despite the absence of conclusive evidence, it is also contraindicated for malignant disease to avoid stimulating metastasis as well as the pregnant uterus. Regions that undergo deep radiotherapy will be avoided for fear of tissue breakdown. Regions of lowered pain sensitivity should not be treated using ultrasonic burst at high intensity since damage might be caused unwittingly due to the absence of the normal warnings. Patients wearing cardiac pacemakers should be excluded because of the feasibility of disturbance by the radiofrequency emitted from the ultrasound device. Great caution is taken in treating any tissue containing metal implant because the significant reflection at the interface would increase the consequent acoustic pressure.

Another survey on the physical therapy was carried out in the Dutch primary health-care system among 17,201 patients, addressing reasons for referral, treatment goals (in terms of impairments and disabilities), and physical therapy interventions in 1998 (Robebroeck et al. 1998). Patients treated with ultrasound ($n = 3,959$) for recent onset of soft tissue injuries, mainly aiming to reduce pain and swelling, were compared with those references ($n = 13,242$) as listed in Table 10.3. As expected, ultrasound was used more commonly for injuries at the shoulder, elbow, knee, and ankle, but infrequently for improvement of joint motion (relative occurrence of 1.5% vs. 21.5% in the reference group). Treatment goals that cannot be pursued by means of ultrasound theoretically, such as the improvement of muscle strength, respiratory disorders, posture, and function and stabilization of the spine and other joints (relative occurrence varying from 0% to 0.9%), were chosen infrequently. In addition, the deep tissue (i.e., hip) seems inappropriate for ultrasound physiotherapy. Massage was used more frequently in the ultrasound group than in the control despite small differences between them, which may be due to the presence of ultrasound-induced micro-massage effect with similar deep transverse frictions (Robebroeck et al. 1998).

TABLE 10.1
Conditions Treated with Ultrasound

	NHS Departments	Private Practitioners
Strained tendons	95%	94%
Tenosynovitis	95%	94%
Torn ligaments	91%	87%
Scar tissue tension	90%	75%
Contusions and hematomata	89%	70%
Tendonitis	87%	94%
Torn muscles	84%	86%
Bursitis	80%	74%
Local edema	79%	68%
Dupuytren's syndrome	73%	51%
Torn joint capsules	70%	63%
Adhesive capsulitis (periarthritis)	68%	70%
Low back pain	65%	80%
Carpal tunnel syndrome	61%	52%
Chondromalacia patellae	58%	61%
Post-herpetic neuralgia	56%	45%
Gravitational ulcers	49%	25%
Muscles spasm	48%	55%
Osteoarthritis	47%	52%
Cervical spondylosis	44%	64%
Pressure sores	39%	17%
Amputation neuromata	39%	10%
Treatment soreness	38%	36%
Episiotomy scars	30%	9%
Rheumatoid arthritis	27%	21%
Lacerations	27%	14%
Sciatica	27%	45%
Bell's palsy	22%	17%
Neuritis	18%	21%
Neuralgia (trigeminal)	18%	16%
Ankylosing spondylitis	15%	23%
Bone fractures	11%	10%
Bone nonunions	4%	6%
Strained ligaments	0.5%	94%
Torn tendons	0%	68%

Source: Reprinted from *Ultrasound Med. Biol.*, 13, ter Haar, G., Dyson, M., and Oakley, E.M., The use of ultrasound by physiotherapist in Britain, 1985, 659–663, Copyright 1987, with permission from Elsevier.

10.2 SYSTEM STRUCTURE AND TREATMENT PROTOCOL

An ultrasonic physiotherapy system usually consists of three major elements: (1) power supply unit, including a signal generator that produces a sinusoidal signal at a single frequency and a power amplifier; (2) control unit that sets the output levels (the maximum temporal average effective acoustic intensity according to current regulations being ≤ 3 W/cm^2), the transmission mode (continuous or pulsed wave), and sonication duration and ensures adequate acoustic transmission to the target, such as using a pulse-echo detection between the

TABLE 10.2
Situations in Which Ultrasound Was Thought to Be Contraindicated

	NHS Department	Private Practitioners
Thrombophlebitis	94%	90%
Infectious conditions	78%	79%
Acute epsis of bone or soft tissue	89%	91%
Tumor		
Benign	63%	63%
Malignant	94%	88%
Deep x-ray treatment		
During treatment	87%	88%
Within 6 months	81%	74%
More than 6 months	22%	25%
Pregnancy	88%	63%
Cardiac pacemakers	75%	68%
Cardiac disease	44%	33%
Diabetes	21%	17%
Where eye may be irradiated	91%	89%
Where reproductive organs may be irradiated	73%	63%
Regions of lowered pain sensitivity	31%	24%
Epiphysial plate	55%	46%
Presence of internal fixation devices	54%	37%

Source: Reprinted from *Ultrasound Med. Biol.*, 13, ter Haar, G., Dyson, M., and Oakley, E.M., The use of ultrasound by physiotherapist in Britain, 1985, 659–663, Copyright 1987, with permission from Elsevier.

therapeutic pulses; (3) piezoelectric transducer, which usually has a large radiation area of several squared centimeters (smaller ones are also available for special cases). The choice of frequency in the ultrasound physiotherapy is usually 1 or 3 MHz. Lower frequencies are assumed to work well for the deep tissues, such as a hip joint or deep back ligament, while 3 MHz is recommended for more superficial lesions at depths of 1–2 cm. The acoustic intensity should be kept to the minimum value concomitant with effective outcome at the desired depth, and its usual setting is 0.5–1.0 W/cm^2. The continuous wave (CW) mode generates the heating, which is the most obvious ultrasound effect experienced by the patients, and may be a better choice for a small target (i.e., muscle lying beneath a layer of fatty tissue). In contrast, mechanical mechanisms dominate in the pulsed-wave (PW) mode.

Before the starting of ultrasound physiotherapy, the wave entry site (i.e., the skin) will be smeared uniformly with acoustic coupling gel or (i.e., hand or legs) be immersed in water for excellent transmission of acoustic energy into the target with minimal absorption, attenuation, or disturbance. The coupling medium should fill all available spaces. Poor coupling during ultrasound application would cause unexpected

TABLE 10.3
Treatment Goals Pursued with Ultrasound and with Reference Interventions

| Treatment Goal | Relative Occurrence (%) | | | | | |
| | Ultrasound ($n = 3,951$) | | Reference ($n = 13,089$) | | Kruskal–Wallis Analysis | |
	Mean	Std	Mean	Std	p	r
Pain reduction	66.4	45.8	26.2	30.9	<0.001	0.44
Reduction of swelling	1.3	34.6	2.0	10.2	<0.001	0.20
Recovery of range of motion	1.5	11.0	21.5	28.2	<0.001	−0.47
Regulation of muscle tone	3.3	16.8	19.3	2.9	<0.001	−0.35
Improvement of muscle strength	0.2	3.9	6.7	18.7	<0.001	−0.20
Reduction of respiratory problems	0	0	2.6	15.4	<0.001	−0.09
Improvement of posture	0.1	3.5	4.8	15.3	<0.001	−0.19
Improvement of function in spine and other joints	0.9	8.9	94	21.3	<0.001	−0.23
Improvement of stabilization in spine and other joints	0.1	1.9	2.2	10.7	<0.001	−0.12
Alleviation of other impairments	12.2	31.8	5.2	18.4	<0.001	−0.04

Source: Robebroeck, M.E. et al., *Phys. Ther.*, 78, 470, 1998.

heating of the transducer cap. So, special attention is paid when treating the uneven skin contours, such as the dorsal surface of the hand unless it is edematous. Although the acoustic transmission and absorption characteristics are different among popular ultrasound gels (Poltawski and Watson 2007), they had no clinically significant performance (Watson 2008). Various wound dressing materials, including alginates, foams, honey-impregnated dressings, hydrocolloids, hydrogel sheets, low-adherence dressings, vapor-permeable films, and odor-absorbent dressings, have a very wide variation in their transmission characteristics from excellent to zero. If excellent, there is no disturbance to apply ultrasound therapy in chronic wound management, which has clinical advantages since the removal of wound dressings takes several hours on each occasion. Pulse mode with a duty cycle of 20% or a pulse ratio of 1:4 is often applied in the treatment sessions that are on the order of minutes (a maximum of 10 min). In order to prevent skin heating and to prevent a stress pattern applied to cells, the transducers are slowly moved manually by the operator in overlapping circular or linear paths around

the region of interest. The slow rate is to ensure that sufficient effect to be induced in the tissue during long exposure while the overlapping pattern is to address the uneven distribution of ultrasound energy. In addition, after about a minute, air bubbles may be found on the patient's skin, which will affect the propagation of the subsequent ultrasound wave. The transducer should be aligned at right angles to the skin. If the incident angle of the ultrasound beam is 15° or less, the transmitted energy to the target may not be sufficient for effective outcomes.

The penetration depth and acoustic absorption are closely related to each other. It was first stated in 1948 that the higher the frequency, the greater the attenuation of ultrasound energy near the skin surface (Hüter 1948). So, the frequency in the range of 0.8–3 MHz is selected according to the desired penetration depth and the thermal and acoustical properties of the mammalian tissue, which is well known and generally accepted in ultrasound physiotherapy (McDiarmid et al. 1996, Ward and Robertson 1996, Young 1996, Demmink et al. 2003). Human studies show that skin with an area of 10.0 cm^2 must be sonicated for at least 10 min to raise the temperature of the underlying muscle to therapeutic levels (40°C). Effective heating occurred at depths of 0.8–1.6 cm using 3 MHz ultrasound at 1.0 W/cm^2 in CW and at depths of 2.5–5.0 cm using 1 MHz ultrasound at 1.5–2.0 W/cm^2 in CW. There is an opportunity to stretch tissue after thermal ultrasound for 5–10 min before the tissue cools. Higher intensities at 3 MHz, 1.5–2.0 W/cm^2, CW caused pain; meanwhile, no pain was found using the same intensity at 1 MHz. In addition, painful hot spots could also be experienced at lower intensities if a transducer with a beam non-uniformity ratio (BNR) greater than 8:1 is used.

The Chartered Society of Physiotherapy (CSP) recommends that the output of a device should be measured every week and the output power should be calibrated at least once a month. The Institute of Physical Sciences in Medicine (IPSM) guidelines suggest a simple calibration every week by the operators and a complete calibration every 3 months by a hospital physicist. It is noted that measuring the maximum output power only is not sufficient since the output pressure of most therapy machines is nonlinear. More than 30% error may exist at the lower level if the output at full power is correct. Therefore, the whole range of clinical settings should all be tested over. However, a complete calibration of output power, acoustic pressure, acoustic field, and pulse duration is time consuming. If either patients or physiotherapists find or feel any abnormalities of device, it is essential to have a simple, reliable, accurate, and quantitative testing, which is important for patient safety as well as consistent and effective treatment regimes.

However, ultrasound physiotherapeutic machines in Europe, North America, and Australia were found to have more than 20% or 30% errors in the output settings, which may lead to uniformly depressing outcomes (Stewart et al. 1974, Allen and Battye 1978, Repacholi and Benwell 1979, Snow 1982, Hekkenberg et al. 1986, Pye and Milford 1994). The acoustic output power, which is frequently measured by physiotherapists, physicists, engineers, and manufacturers in the calibration, should not deviate by more than ±20% of the indicated values or greater than 10% of the maximum value. Another important variable is the spatial average acoustic intensity, defined as the ratio of output acoustic power over the effective radiating area (ERA). However, the manufacturers usually do not measure the ERA but simply use the area of the piezoelectric disc to derive the spatial average acoustic intensity, leading to systematic errors in all intensity settings.

10.3 MECHANISMS AND BIOEFFECTS

There are two possible mechanisms that make ultrasound valuable in physiotherapy and introduce significant bioeffects, thermal, and mechanical effects (Patrick 1966, Dyson and Pound 1970).

Initially, the major effect at lower intensities is thought to be the thermal effect. The amount of the temperature rise is determined by not only the total absorbed acoustic energy that depends on the treatment duration, acoustic intensity, and frequency-dependent absorption coefficient but also the diffusivity of the circulation to the tissues. Protein-rich tissue rather than loose connective tissue has higher temperature elevation at the same sonicating conditions. In Europe, some health insurance companies have classified ultrasound therapy as a deep-heating modality as it can selectively increase the temperature of periarticular structures and areas at bone–muscle interfaces (Robebroeck et al. 1998). Physical therapy makes a distinction between the temperatures applied. A weak heating of 1°C–2°C would result in a 13% increase in the metabolic rate for each degree Celsius; a moderate heating of 2°C–4°C would reduce muscle spasms, pain, and chronic inflammation and promote blood flow; and heating over 4°C would decrease the viscoelastic properties of collagenous tissue (Castel 1993, Draper et al. 1995). The temperature in the tissue to 40°C–45°C for at least 5 min leads to vasodilatation (widening of blood vessels), increased blood perfusion to the sonicated region, reduction in muscle spasm, increased extensibility of collagen fibers, and a pro-inflammatory response. As a result, the metabolic rate is increased and the healing process is accelerated. At a high intensity, such thermal effects become more significant. The excessive heat in some cases may be due to sonication over bony surfaces where the subperiosteal blood supply is poor, or high energy being delivered to a small area. The thermal effect has been known since the beginning of ultrasound therapy, but the therapeutic effect produced by ultrasound is relatively inefficient when applied at common doses.

Tissue subject to the ultrasound wave with the alternating compressional and rarefractional pressure will oscillate about its equilibrium position in a small amplitude. In addition, the tissue also has net movement in the direction of the wave propagation, which is known as acoustic streaming. The term cavitation was first used by Sir John Thornycroft in the early twentieth century and defined as the formation and life of bubbles in liquids. Cavitation and mechanical reaction of the tissue to the acoustic pressure (i.e., micro-massage) seem more important in the treatment of soft tissue lesions than thermal

effects. The effective micro-massage at depth is of the greatest value in the absorption of the edema (free liquid) whenever the soft tissues are damaged by trauma, either a major accident (i.e., a fractured bone or a crushed limb) or a minor one (i.e., the skin cut or muscle fibers torn). Stable cavitation is beneficial to injured tissue, whereas transient one for tissue damage. However, collapse cavitation could also cause damage to the surrounding tissue due to extremely high local temperatures and high pressure waves (Wells 1977). Occurrence of inertial cavitation (IC) also leads to the formation of free radicals, and the derived compounds, such as hydrogen peroxide (H_2O_2), are toxic to tissues to extend the area of initial injury, thereby offsetting any potential ultrasound benefit (Al-Karmi et al. 1994). Bulk streaming is far less mechanically powerful than micro-streaming, which is formed as eddies of flow adjacent to an oscillating source. Micro-streaming is always associated with and secondary to cavitation, and it may alter cell membrane structure, function, and permeability to stimulate tissue repair (Schs 1988). The enhanced cell membrane and ion permeability could result in an increased Ca^{2+} influx into the cell and subsequent intracellular concentration (Mortimer and Dyson 1988). Calcium is one of the most important intracellular messengers for extracellular signals known so far (Berridge 1993). The mechanical effects of ultrasound were introduced in the 1960s to increase the efficiency and reduce the side effects in tissue regeneration (Dyson et al. 1968), soft tissue repair (Paul 1960, Dyson et al. 1976), angiogenesis in chronically ischemic tissues (Hogan et al. 1982a), protein synthesis (Webster et al. 1978), and bone repair.

Besides acoustic streaming and cavitation, the frequency resonance hypothesis is another possible explanation of ultrasound-induced molecular effects (Robertson and Baker 2001). The central premise is that the mechanical energy of the ultrasound pulse is absorbed by proteins and protein complexes, resulting in a transient conformational shift (modifying its three-dimensional structure) and alteration to signaling mechanisms within the cell either by inducing the structural conformation of an individual protein or disrupting the effector function of a multimolecular complex. Alternatively, the resonating and shear force of ultrasound may also result in the dissociation of functional multimolecular complexes or the release of a sequestered molecule by dislodging an inhibitor molecule from the multimolecular complex, which may be the reason for the decreased activity of the dimeric and tetrameric forms of creatine kinase in the sonication. Signal transduction pathways consist of a series of enzymatic proteins that are modulated by phosphate molecules, leading to distinct changes in conformation and the regulated enzymatic activity of the protein. So, the frequency resonance hypothesis could explain the increased enzymatic activity but no alteration to the activity of the enzymes creatine kinase, lactate dehydrogenase, hexokinase, and pyruvate kinase in sonothrombolysis. Ultrasound has been shown to modulate a number of proteins associated with inflammation and repair, such as interleukin [IL]-1, IL-2, IL-6, IL-8, interferon-γ, fibroblast growth factor-β, vascular endothelial growth factor, and collagen (Enwemeka et al. 1990a).

Ultrasound physiotherapy could induce certain biological effects, such as the remarkable increase in cell metabolism and membrane properties (i.e., lymphocyte adhesion, the release of histamine from mast cells, the activity of lysosomes, the release of growth factors from macrophages, angiogenesis, membrane permeability calcium flux, and the motility and proliferation of T-cells, osteoblasts, and fibroblasts), enhancement of the quantity and quality of collagen fibers laid down in wounds, acceleration of lymph flow, vasodilation, vasoconstriction, increased vascular permeability with extravasation of fluid into the area of injury and enhanced inflammatory response, the affected enzyme activity, the synthesis of proteins, and possibly gene regulation due to activated signal-transduction pathways (Fyfe and Chahl 1985, Young and Dyson 1990a). An increase in intracellular calcium in fibroblasts was found during sonication, suggesting that the disrupted cell membrane permits influx of calcium. However, the cells expel the calcium rapidly and return to a homeostatic state after ultrasound exposure. Cells employ calcium as a cofactor in regulating the activity of enzymes. Activation of calcium-sensitive signal-transduction pathways (i.e., protein kinase C and cyclic AMP) would result in gene activation and/or the modulation of RNA translation to a protein product. Total ionic conductance in cellular and paracellular pathways is greater in pulsed than in continuous-mode ultrasound (Dinno et al. 1989). Ultrasound interacts with one or more components of inflammation, resulting in earlier resolution of inflammation, acceleration of fibrinolysis, stimulation of macrophage-derived fibroblast mitogenic factor, heightened fibroblast recruitment, enhanced angiogenesis, increased matrix synthesis, more dense collagen fibrils, and increased tissue tensile strength *in vitro* (Speed 2001). There are also some alterations in the blood chemistry, such as a decrease in blood glucose and the liberation of H-substances, which can aggravate some conditions. The central nervous system (CNS) is more responsive to heat, whereas the autonomic nervous system responds to the rhythmical vibration. The pulsed ultrasound at a ratio of 1:5 has the optimum analgesia effect on the nerve roots, which is often most helpful in overcoming the "pain-spasm-more pain" cycle in the patients. Some of mechanical effects of physiotherapeutic ultrasound are listed in Table 10.4.

10.4 CLINICAL TRIALS

Ultrasound has been recognized as an appropriate method in physiotherapy to treat acute and chronic musculoskeletal disorders, heating deep structures (i.e., joints, muscle, and bone), accelerating the restoration of tissue regeneration (Dyson et al. 1968), increasing pain thresholds (Aleya et al. 1956), stimulating the growth of bone (Duarte 1983), and increasing the extensibility of tendon (Gersten 1955). Based on these findings, ultrasound therapy has been used in clinics to relieve pain and enhance immobility in acute periarticular disorders and osteoarthritis, but not in chronic periarticular inflammatory disorders.

TABLE 10.4

Cellular and Molecular Effects of Nonthermal Ultrasound

Increase in Protein or Cellular Function	Producing Cell Type	Effector Function
Interleukin-1β	Osteoblasts, monocytes	General inflammatory mediator
Interleukin-2	T cells	T-cell growth
Interleukin-8	Osteoblasts	Endothelial cell migration and proliferation
Vascular endothelial growth factor	Osteoblasts, monocytes	Endothelial cell migration and proliferation
Basic fibroblast growth factor	Osteoblasts	Endothelial cell migration and proliferation
Fibroblast growth factor	Monocytes	Fibroblast growth
Collagen	Osteoblasts, fibroblasts	Wound healing
Chloramphenicol acetyl transferase	HeLa, NIH/3T3, C1271	Gene expression of liposomal transfection
Increased proliferation	Fibroblasts	Enhanced wound healing
Increased proliferation	Osteoblasts	Enhanced wound healing
Lymphocyte adhesion	Endothelial cells	Enhanced lymphocyte trafficking
Vasodilation	Capillary, endothelium	Enhanced blood flow

Source: Johns, L.D., *J. Athlet. Train.*, 37, 293, 2002.

However, the clinical outcome is quite debating. In a review of 35 randomized trials of therapeutic ultrasound carried out between 1975 and 1999, only 10 studies were considered to have acceptable methods, and there was little evidence that ultrasound therapy is more effective to treat the case of pain and some musculoskeletal injuries and to promote soft tissue healing (Robertson and Baker 2001). Twenty-two clinical papers were selected from 293 papers published since 1950 in the investigation of assessing the ultrasound effect and comparing ultrasound treatment with sham, control, and untreated groups for musculoskeletal disorders. An analysis of the effect of proper randomization on the result was not possible because there is inadequate description of the methods used. It is concluded that ultrasound is usually used empirically in treatment of musculoskeletal disorders, but lacking firm evidence from well-designed and controlled studies (Gam and Johannsen 1995).

Most popular applications of ultrasound in physiotherapy are described in the following section with clinical observation, statistical analysis for its effectiveness, and discussion for discrepancies between groups and future direction.

10.4.1 PAIN RELIEF

There are a large number of patients with low back pain from sciatica to intervertebral disc, whose common feature is their dermatome distribution. Ultrasound can raise the threshold of pressure-induced pain, which provides different stimuli than transcutaneous electrical nerve stimulation (TENS), so it is one of the common modalities used by physiotherapists for pain relief (Partridge 1987, Holmes and Rudland 1991). However, it does not decrease the pain threshold through a systemic mechanism, or even by the spread of a local response unless the remote areas are innervated by the same nerve. In the ultrasound physiotherapy, there are some degree of analgesia immediately after the first session, which lasts for a few hours. After the subsequent sessions, this analgesia will gradually increase and last for a longer time, until the target is no more tender than the other scar. The session duration varies with the severity of the strain, and the whole treatment should be no more than 10 sessions, unless there seems a good medical reason for continuing. Several mechanisms have been postulated to explain the ultrasound-induced pain relief. The acoustic absorption by the nocioceptive fibers (A-delta and C) may increase in afferent input and inhibit the pain transmission at the cord level (Wall 1989). However, the nerve conduction velocity and alternations in the large and fastest fibers can only be measured, and increase with the sonication, which may be related to the temperature rise.

Twenty healthy volunteers (12 men and 8 women) ranging in age from 22 to 51 years were recruited at the University of Alberta (Mardiman et al. 1995). Pain threshold over the muscle belly of the wrist extensors 7 cm distal to the lateral epicondyle of the humerus, on a line drawn from the lateral epicondyle of the humerus to the dorsal radial tubercle was measured with a pressure dolorimeter at the pressure increase rate of about 1 kg/s. The pressure increase was terminated once the subjects felt discomfortable. At each test site, only one pain threshold was measured before and after the treatment. The subject was seated on a chair with the arm supported comfortably in full pronation and 90° elbow flexion. Continuous ultrasound (1.1 MHz, 5 cm² element, 3 W/cm²) was applied over the muscle bellies of the wrist extensors to an area twice the size of the ultrasound head for 5 min, and the transducer was moved around at the rate of about 3 cm/s over the radial nerve and the lateral cutaneous nerve of the forearm. A three-way analysis of variance (ANOVA) with repeated measures on all factors (i.e., arm, time, site) and Newman–Keuls *post hoc* analysis was used to reveal the experimental difference. Eleven subjects reported warmth on the treated arm. As listed in Table 10.5, there were significant differences in pain threshold between ultrasound-treated and control sites. However, experimental pain is an acute type rather than chronic pain (Wolff 1983) and does not elicit emotional and cognitive responses in the practice (Chapman 1983). Furthermore, pain threshold was not correlated with the level of clinical pain.

Shoulder pain is a common and debilitating complaint in the general population. Its prevalence in the United Kingdom is about 7%–34%. Most patients are treated conservatively

TABLE 10.5
Comparison of the Pain Thresholds (kg/cm²) in the Control and Ultrasound-Treated Arms

	Treated Site			Untreated Site		
	Pre-Test	Post-Test	Δ	Pre-Test	Post-Test	Δ
Treated arm	1.53 ± 0.078	1.93 ± 0.124	0.40*	1.95 ± 0.121	1.96 ± 0.128	0.01
Control arm	1.49 ± 0.071	1.56 ± 0.078	0.07	1.95 ± 0.129	1.94 ± 0.130	−0.01
Δ	0.04	0.37*		0.00	0.02	

* Significant difference ($p < 0.05$).

with nonsteroidal anti-inflammatory drugs (NSAIDs), corticosteroid injections, and physiotherapy. There are a wide range of modalities in the treatment of shoulder pain by physiotherapists, including mobilization, manipulation, acupuncture, electrotherapy, and exercise. A systematic review of randomized clinical trials for soft tissue disorders concluded that therapeutic ultrasound was not an effective treatment for shoulder pain (van der Heijden et al. 1997). However, it also highlighted the poor methodological quality of almost all trials evaluated. A total of 221 participants (mean age of 56 years) were recruited for a prospective double-blind randomized study (Ainsworth et al. 2007), 113 of them were randomized to the ultrasound (the average acoustic intensity of 0.5 W/cm²: 0.2–1.0 W/cm² and the average duration of 4.5 min: 3–7 min), and 108 to the placebo ultrasound group. In over 95% of treatments, pulsed ultrasound at a ratio of 1:4 was applied. To 46% of participants, 1 MHz ultrasound was delivered and 3 MHz given to 39%. There was 76% and 71% follow-ups at 6 weeks and 6 months, respectively. The mean (95% CI) reduction in Shoulder Disability Questionnaire (SDQ) scores at 6 weeks was 17 points (13–26) for ultrasound and 13 points (9–17) for placebo ultrasound ($p = 0.06$). There were no significant differences at the 5% level in averaged changes between groups at any time as shown in Figure 10.1. Therefore, it is concluded that ultrasound was not superior to placebo in the short-term management of shoulder pain (Ainsworth et al. 2007).

Therapeutic ultrasound is also used routinely by podiatrists and physiotherapists in their treatment of plantar fasciitis and plantar heel pain. Pain on the plantar aspect of the heel is usually diagnosed as plantar fasciitis. The etiology of plantar heel pain is believed to be associated with lower limb biomechanics, pronation producing tension on the soft tissues of the plantar surface or part of a systemic inflammatory condition. Nineteen patients experienced episodes of heel pain (seven bilateral) were recruited in a study (Crawford and Snaith 1996). Those with previous experience of ultrasound treatment, the presence of fluffy calcaneal spur on radiograph, generalized joint pain, non-specific urethritis, the use of pain control (i.e., analgesics, non-steroidal anti-inflammatory drugs, steroids, heel pads, or orthoses appliances), a diagnosis of a seropositive or seronegative arthropathye were excluded.

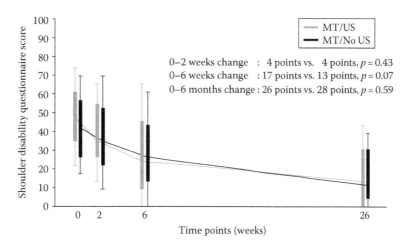

FIGURE 10.1 Median, inter-quartile ranges (IQRs) and 80% confidence interval (CI) of shoulder disability questionnaire (SDQ) scores. MT, manual therapy and home exercise programme; US, ultrasound; No US, placebo US. (Ainsworth, R., Dziedzic, K., Hiller, L., Daniels, J., Bruton, A., and Broadfield, J., A prospective double-blind placebo-controlled randomized trial of ultrasound in the physiotherapy treatment of shoulder pain, *Rheumatology*, 46, 815–820, 2007 by permission of Oxford University Press.)

Patients' feet were examined for the presence of a positive Helbing's sign (i.e., lateral bowing of the tendon Achillis), which is a soft tissue indicator of excessive (>4°) subtalar joint pronation. Eight minutes of therapeutic ultrasound at an intensity of 0.5 W/cm², a frequency of 3 MHz, and a pulse ratio of 1:4 was applied to the patients twice weekly for 4 weeks. The pain was assessed using a 10 cm horizontal visual analogue scale at the first and last visits. Both treated and placebo groups showed a reduction in pain with a mean improvement of 30% (from 6.7 to 4.5) in the treated group and of 25% (from 7.5 to 5.6) in the placebo group. There is no significant difference in improvement (Wilcoxon's signed-ranks test) between the treated and placebo groups, between pain duration and improvement (Fisher's exact test), and between the patient's age and improvement in either group. Complete resolution of pain did not occur in any episode, and four episodes had worsened pain during the treatment (two in each group). Pain neither improved nor worsened in two episodes (one in each group). So, it is concluded that the therapeutic ultrasound used in that study is no more effective than placebo in the treatment of plantar heel pain (see Figure 10.2).

10.4.2 WOUND HEALING

A wound is a type of injury in which skin is torn, cut, or punctured (an open wound) or where blunt force trauma causes a contusion (a closed wound). In pathology, it specifically refers to a sharp injury which damages the skin dermis. There are about 50 million patients globally suffering from hard-to-close wounds, including about 15% of diabetic foot ulcers in the United States that eventually required amputations. The total market for advanced wound care was $3.4 billion in 2010 and will further grow to $4.6 billion by 2016 at a compound annual growth rate (CAGR) of 4.9% (Figure 10.3). Healing of the injured tissue is a result of the movement, division, death of specific cells, and the synthesis of intracellular and extracellular materials. The principal cellular responses

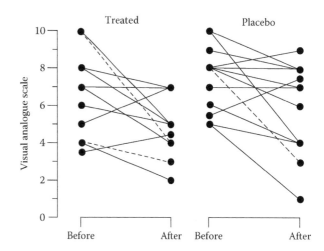

FIGURE 10.2 Pain scores of each patient before and after session in the ultrasound treatment and placebo groups (dash line: two episodes with identical pain scores). (From Crawford, F. and Snaith, M., *Ann. Rheum. Dis.*, 55, 265, 1996.)

immediately after the injury involve the interaction of platelets with thrombin and collagen, leading to local blood coagulation, mast cell degranulation, and the liberation of a number of chemical mediators for the inflammatory phase of tissue repair (Yurt 1981). After the coagulation process, various cellular infiltrates appear in the wound chronologically.

Leucocytes (white blood cells) are the first cells to arrive at the wound site, and then the edge of the injury is infiltrated with granulocytes and macrophages within a few hours. In a few days, the proliferative phase of wound repair begins, fibroblasts reaching the wound bed, attracting and stimulating the proliferation by mediators released from the macrophages. These fibroblasts gradually replace the majority of the leucocytes and increase the rate of collagen synthesis. By 3 or 4 days after injury, the presence of endothelial cells starts the neovascularization of the wound bed, which is essential for successful repair (Banda et al. 1982, Polverini and Leibovich

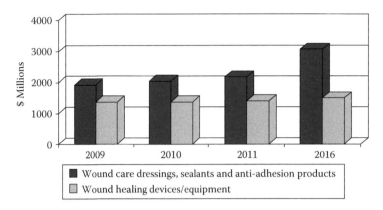

FIGURE 10.3   Total advanced wound care market by the selected segment from 2009 to 2016.

1984, Knighton et al. 1986). In mammalian skin, the complete healing process produces scar tissue. Scar tissue generally forms to cover on its superficial aspect by epidermal cells, but inferior to natural dermis, and can cause certain physical and psychological problems, even life threatening. Controlling the development of scar tissue and its replacement by tissue in order to resemble the dermis more closely are considerably important in clinics.

Ultrasonic stimulation may increase the rate of angiogenesis. Angiogenesis (the formation of new blood vessels) is vital in wound healing, re-establishing circulation at the injury site, and limiting ischemic necrosis. Immediately after injury, the ends of the severed blood vessels thrombose, and then capillary buds sprout from the functioning vessels in close proximity to the wound bed. The basement membrane of the capillaries adjacent to the wound bed breaks down under the influence of proteinases released from endothelial cells to convert plasminogen to plasmin, which causes proteolysis of laminin and fibronectin. Endothelial cells form pseudopodia that protrude through the gaps in the basement membrane and then migrate away from the parent vessel into the perivascular space. Capillary sprouts then vacuolize to create a central lumen and eventually connect to another sprout or capillary to form capillary loops, which later either disappear or develop into larger vessels. The vascularization rate is associated with the degree and cause of damage. Complete vascularization of a surgical incised wound is generally achieved in 6–7 days, while it takes 12–16 days for third-degree burns (Hughs and Dann 1941). Since capillary growth occurs during inflammation and formation of granulation tissue, and is accompanied by an increased accumulation of leucocytes (Branemark 1965), initiation of angiogenesis may be due to the release of some substances from these cells resident in the wound bed, such as lymphocytes which is supported by the observed disruption of granulocytic leucocytes (Sidky and Auerbach 1975). Macrophages also play a crucial role in generating angiogenic factors (Clark et al. 1976), inducing angiogenesis *in vivo* and stimulating endothelial cell proliferation *in vitro*. Furthermore, both mechanical factors (i.e., blood pressure, wall tension, wall stress, increased blood flow, increased hematocrit or stretch of vessels by growing tissue) and metabolic or chemical factors (i.e., cytokines, monokines,

and PDGF) are involved in the angiogenesis (Gospodarowicz et al. 1978). Under normal circumstances, the microvascular system stays in a dormant state with various control mechanisms preventing the growth of rampant capillary.

Ultrasound has been used for over 40 years to stimulate repair (Summer and Patrick 1964). The timing of ultrasound therapy in the wound healing is critical. The accelerated tissue repair by ultrasound mainly applies to the early phases of the healing process (both inflammatory and proliferative phases), and the application in the remodeling phase may even have adverse effects on the tissue strength. The use of ultrasound may assist in pain relief or in the alleviation of other inflammation symptoms, such as edema, and increase the extensibility of collagen, thus facilitating the stretching of scars or adhesions (Hudlicka and Tyler 1986).

The effect of 5 min daily therapeutic ultrasound ($f = 0.75$ or 3.0 MHz, $I_{SATA} = 0.1$ W/cm^2, pulse duration of 10 ms, pulse ratio of 1:4) on the angiogenesis in full-thickness excised lesion in the flank skin of adult rats was assessed quantitatively using microfocal x-ray. The wound bed and 1 cm of the adjacent normal skin were excised, fixed in neutral formalin for 72 h, and prepared for histological examination. The tissue was cut into 7 μm sections and stained with hematoxylin and eosin (H&E). The number of cells in middle area of the wound bed (i.e., midway between the dorsal and ventral margins and the superficial and deep margins of the wound bed) in each injured flank skin was counted. There were 30% and 20% more blood vessels in the granulation tissue of the wounds 5 days after ultrasound treatment at 0.75 and 3.0 MHz than in the control, respectively ($p < 0.05$) (Figure 10.4). Blood vessels grew into the wound from its edges and deep aspect. The number of polymorphonuclear leucocytes (polymorph) in the 0.75 MHz group was significantly lower than the other two groups ($p < 0.05$) (Young and Dyson 1990b). There are a significantly higher number of macrophages but less fibroblasts and shorter depth of granulation tissue in the control group than in the two ultrasound-treated ones ($p < 0.05$). The fibroblasts in the ultrasound-treated groups were aligned conducively for efficient wound contraction (i.e., approximately parallel to the wound surface) as shown in Figure 10.5, while those in the control were more random. The greater amount of granulation tissue by the ultrasound treatment is partially

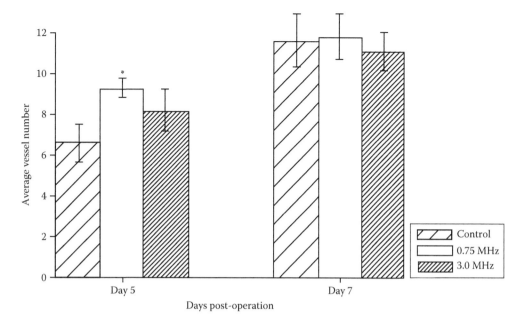

FIGURE 10.4 Comparison of the average number of blood vessels per m of wound bed in the ultrasound (0.75 and 3.0 MHz) treated and control groups*, significant difference ($p < 0.05$). (Reprinted from *Ultrasound Med. Biol.*, 16, Young, S.R. and Dyson, M., The effect of therapeutic ultrasound on angiogenesis, 261–269, Copyright 1990b, with permission from Elsevier.)

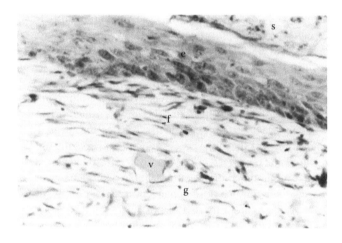

FIGURE 10.5 Specific alignment of the fibroblasts below the surface of the epidermis in the microscopic image of a wound 5 days after the ultrasound treatment (3.0 MHz, 0.1 W/cm² SATA). e, epidermis; f, fibroblast; g, granulation tissue; s, scab; v, blood vessel. (Reprinted from *Ultrasonics*, 28, Young, S. and Dyson, M., Effect of therapeutic ultrasound on the healing of full thickness excised skin lesions, 175–180, Copyright 1990a, with permission from Elsevier.)

due to the higher number of matrix-synthesizing fibroblasts, which suggests that these wounds had progressed further into the proliferative phase of repair than the control (Young and Dyson 1990a). This acceleration of repair may be due to stimulation of the resident inflammatory cell population (i.e., macrophages) to synthesize and release wound factors (i.e., IL-1, fibroblast growth factor, and tumor necrosis factor-alpha), which attract fibroblasts and endothelial cells into the wound area. However, by 7 days after injury, the significant difference in the degree of vascularization disappeared, and there are more polymorphs ($p < 0.05$ between the control and

3.0 MHz group, $p > 0.05$ between two ultrasound groups) and macrophages ($p < 0.05$ between the control and 0.75 MHz group) in the control group. The number of fibroblasts and the depth of granulation tissue were significantly higher in the two ultrasound-treated groups than in the control group ($p < 0.05$). Either the control group may simply have caught up with the ultrasound-treated groups by 7 days after injury, or the ultrasound-treated groups may have entered the remodeling phase of repair where the cell growth rate is expect to decrease. No tissue hypertrophy in the ultrasound groups suggests that the control mechanisms that limit the development of granulation tissue are not impaired by ultrasound. The main vessels were arranged predominantly parallel to each other and grew toward the wound surface in all groups. Branching was also observed. The low frequency (i.e., 0.75 MHz) seems to have a greater effect on angiogenesis than 3.0 MHz, which suggested a contribution from nonthermal ultrasound effects.

Scar tissue is inferior to normal tissue mechanically (elasticity and movement) and cosmetically. In addition, it has some functional problems even than the original injury, such as peritoneal adhesions, post-traumatic cerebral scarring, and esophageal stricture. Therefore, scar formation should be reduced. In addition, ultrasound can affect wound contraction and reduce all or part of a skin defect by the centripetal movement of the surrounding skin (Young and Dyson 1990a). Early decrease in the wound size in this manner could reduce the need for scar tissue production if it is accompanied by compensatory intussusceptive growth. Acceleration through the early phase of repair, which leads to the earlier beginning of the proliferative phase, may be due partially to ultrasound-induced cell membrane permeability changes in the wound bed (Young and Dyson 1990b). Calcium ion fluxes, in response to cell membrane permeability changes, act as chemical signals (second messengers) that control the enzymatic activity of the cell and stimulate the synthesis of specific proteins and their secretion.

Sonophoresis or phonophoresis with an appropriate drug, such as dexamethasone phosphate (0.334%), can reduce the amount of collagen deposited in a wound nearly as well as a hydrocortisone injection. The value of this in the management of chronic inflammation in the soft tissue should be considered.

Wound debris, edema, poor blood supply, loss of skin, presence of infection, and eschar present in the upper extremity wounds with delayed closure. Both ulcers and upper extremity wounds can be extensive in the area and as deep as bone. Factors that delay healing, such as arterial and venous insufficiency, nutritional problems and immobility, are likely to be more serious in chronic ulcers. It was found beneficial using 5 min of ultrasound per 5 cm^2 (Nussbaum et al. 1994); however, this approach did not prove beneficial in another study even using similar dosage and pulse ratio (ter Riet et al. 1996). It suggests that factors other than dosage, pulse ratio, and duration affect the outcome of ultrasound treatment (Nussbaum 1998). Many studies show that there is a risk of inhibiting tissue repair at thermal dosages. Although inhibition was not evident in the leg ulcer study, the slow rate of healing does not support a thermal approach to wound healing. Optimal duration, area, and frequency of treatment need further clarification. Studies specific to open wounds in hand therapy are needed.

The process of tissue repair is a complex series of cascaded, chemically mediated events of producing scar tissue that constitutes an effective material to restore the continuity of the damaged tissue. The ultrasound effect varies according to its involvement in the tissue repair process. Immediately following injury and still during the tissue bleeding phase, it is generally inadvisable to apply ultrasound because there would be a clinical disadvantage if ultrasound has the capacity to enhance local blood flow. It suggests to start sonication to the wound as soon as active bleeding has stopped. In the inflammatory phase, a stimulative effect of ultrasound was found on the mast cells, platelets, white cells with phagocytic roles, and the macrophages (Maxwell 1992, Nussbaum 1997, ter Haar 1999). The sonication induces the degranulation of mast cells and releases arachidonic acid which itself is a precursor for the synthesis of inflammatory mediators, such as prostaglandins and leukotriene (Mortimer and Dyson 1988, Nussbaum 1997, Leung et al. 2004). The overall effect of therapeutic ultrasound is pro-inflammatory by increasing the activity of these inflammatory mediators rather than anti-inflammatory (El Hag et al. 1985, Hashish et al. 1988). The benefit of this action mode is not to increase the inflammatory response but rather to act as an inflammatory optimizer. The inflammatory response is essential to the effective tissue repair, and inhibition of these events would inhibit the following repair phases (Hashimoto et al. 2002, Watson 2006b) and the progress to the next proliferation phase. For tissues with an inflammatory reaction but no repair to be achieved, the benefit of ultrasound is to promote the normal resolution of the inflammatory

events. During the proliferative phase (scar production), ultrasound also has a stimulative effect on cellular up-regulation, pro-proliferative in the same way that it is pro-inflammatory (not changing the normal proliferative phase but maximizing its efficiency and producing the required scar tissue in an optimal fashion), though the primary active targets are now the fibroblasts, endothelial cells, and myofibroblasts (Dyson and Smalley 1983, Young and Dyson 1990a,b, Ramirez et al. 1997, Nussbaum 1998). Low-intensity pulsed ultrasound (LIPUS) increases the synthesis of protein (Harvey et al. 1975), fibroplasia, and collagen (Enwemeka 1989, Turner et al. 1989, Enwemeka et al. 1990a,b, Huys et al. 1993). Scar tissue is an essential component of the repair, and it is the best outcome for most musculoskeletal tissues. In many ways, tissue regeneration would be ideal but is not available in most musculoskeletal tissues. Functional scar tissue may be considered as the second best, but the therapeutic purpose is to promote the construction of the most efficient scar possible. During the remodeling phase, the generic scar generated in the initial stages is refined to adopt functional characteristics of the tissue (Watson 2006b). For example, a scar in ligament will not become ligament but behaves more like it (Vanable 1989). The ultrasound application appears to enhance the events normally associated with the remodeling process, which may last for a year or more and is an essential component of quality repair, to enhance the appropriate orientation of the newly formed collagen fibers and also to change the collagen profile from mainly type III to a more dominant type I, thus increasing tensile strength and enhancing scar mobility (Huys et al. 1993, Nussbaum 1998). If a tissue is repairing in a compromised or inhibited fashion, the ultrasound therapy at an appropriate dose will enhance this activity. If the tissue is healing normally, the sonication will speed the process and thus enable the tissue to reach its end point faster. Thus, optimal ultrasound parameters have a promotional effect on the whole healing cascade, and the outcome may be dose dependent (ter Haar 1999, Watson 2006a). There are a myriad of possible combinations of frequency, power, acoustic intensity, duty cycle, and treatment times, and some combinations are more effective than others as shown in the published data.

10.4.3 Burn Trauma

Ultrasound is used empirically in the management of burn injuries, inspired by the demonstrated benefit of ultrasound in the healing of nonthermal injuries and its effect on scar extensibility. In second- and third-degree burns, ultrasound can be applied throughout the collagen synthesis phase of healing and continued long after the wounds have closed in order to improve the quality and mobility of the maturing scar. Burned areas may include the anterior aspect of the wrist, elbow, and axilla. Eighty-seven percent of individual sessions resulted in increased joint mobility. The effect appeared to be transient as four of eight joints in the ultrasound group (1 MHz, 1.0 W/cm^2, 10 min followed by stretching on alternate day for 2 weeks) and two of six joints in the placebo group lost range over 2 weeks (Ward et al. 1994).

10.4.4 Tendinosis

When muscles and bones move, stresses are exerted on the tendons and ligaments that are attached to them. Some damage on a cellular scale in the muscles and tendons is generated when the motion is in new ways or out of capacity. If the increase in demand is made gradually, muscle and tendon tissues will usually heal, build in strength, and adapt to new loads. Tendinosis (also known as chronic tendinitis, chronic tendinopathy, or chronic tendon injury) is an accumulation over time of microtears in the connective tissue in and around the tendon that do not heal properly in the wrist, forearm, elbow, shoulder, knee, and heel, leading to reduced tensile strength, thus increasing the chance of tendon rupture. Tendinosis is often misdiagnosed as tendinitis due to the limited understanding of tendinopathies by the medical community. Classical characteristics of tendinosis include degenerative changes in the collagenous matrix, hypercellularity, hypervascularity, and a lack of overt inflammatory cells. The degenerative process is poorly understood but may be due to failure of the internal tendon cells to repair and remodel the extracellular matrix after injury. There are tremendous variations in matrix composition, alteration of collagen fiber type distribution (i.e., a relative increase in type III collagen over type I one), and fibrocartilaginous change in some tendon lesions (i.e., fibrovascular proliferation and the focal expression of type II collagen) found from extensive studies of normal and degenerated human tendons. After injury, an increase in matrix turnover is necessary to remove damaged matrix and to remodel scar tissue.

Because tendons injuries are slower to heal than muscles, it often requires patience and careful rehabilitation with physical therapy. For tendinosis, treatment intends to reduce inflammation and promote healing in order to reduce pain and weakness in the tendon (Fulcher et al. 1998). A reduction of disability is expected after the pain reduction in the healing phase when graded exercises begin. Partial tears heal by the rapid production of disorganized type III collagen, which is weaker than the normal tendon. Recurrence of injury in the damaged tendon is common. Ultrasound improved the extensibility of mature collagen by promoting fibers to remodel, leading to greater elasticity without loss of strength. These *in vitro* studies indicate that both ultrasound and laser can reduce the inflammation, stimulate the tissue repair, and promote healing (Saunders 2003). Similar biostimulatory effects produced by both modalities suggest that their use in the treatment of tendinosis (Low and Reed 2000b).

Thirty-six patients with supraspinatus tendinosis referred by their general practitioner or consultant for physiotherapy (17 women and 19 men, age ranging from 37 to 76 years) were randomly divided into three groups: control, laser-, and ultrasound-treated groups (Saunders 2003). A 50 mW, 820 nm infrared laser with a dose of 30 J/cm² was used because of its penetration properties. The ultrasound parameters used were the acoustic intensity of 1.5 W/cm², an element area of 9.6 cm², a driving frequency of 1 MHz frequency, and a pulse ratio of 1:4. Nine sessions were

performed on the patients three times a week for 3 weeks. There was a significantly greater improvement in the muscle force, which was measured using a Penny and Giles myometer, produced by the laser group compared with the others ($p < 0.01$). Although the ultrasound group seemed to have an improvement compared with the control, but with no significance ($p = 0.6$). More than 90% of patients in the laser group reported a pain reduction using the 100 mm pain analogue scale, compared with 58% in the ultrasound group and 50% in the control group. A 24 pain diary was used to measure any changes in function and disability due to the shoulder pain. There were improvement in both the laser and ultrasound groups compared with the control, but with different significance ($p < 0.05$ and $p > 0.05$, respectively). There was a greater reduction in tenderness in the laser group ($p < 0.01$), but not for ultrasound ($p = 0.08$) as shown in Figure 10.6. These data suggest that the laser therapy improves the symptoms of supraspinatus tendinosis and the improvement of the symptoms by ultrasound was not significantly different from the control group that received advice only. So, laser therapy should be the physiotherapy choice for supraspinatus tendinosis rather than ultrasound.

Increment of intensity may make a difference to the effectiveness on tendinosis since deeper tissues need more energy to achieve biostimulation (Vasseljen 1992, Vecchio et al. 1993). A review of the biophysical effects of ultrasound illustrated that there was insufficient evidence and scientific foundation for its clinical use (Baker et al. 2001). Pulsed ultrasound was preferred for soft tissue repair (Young 1996). Lower frequencies (i.e., 1 MHz) assume to have higher penetration. However, it is also found that once the temperature of tissue has been increased to 40°C–45°C, the tissues can stretch up to 10 min. Increases in the tensile strength, energy absorption, mobility, improved collagen fibril alignment, reduction in inflammatory infiltrate and scar tissue in tendons have already been demonstrated in some investigations

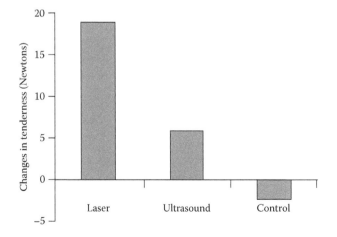

FIGURE 10.6 Comparison of changes in minimal force (N) required to produce tenderness after treatment of laser, ultrasound, and control. (Reprinted from *Physiotherapy*, 89, Saunders, L., Laser versus ultrasound in the treatment of supraspinatus tendinosis, 365–373, Copyright 2003, with permission from Elsevier.)

(Enwemeka 1989, Gan et al. 1995), but not the others (Roberts et al. 1982, Turner et al. 1989). However, attention should be paid to extrapolate these results to human tendon lesions because of biological differences between different types of collagen in tendon.

The rate of healing of open wounds can be conveniently measured by wound tracing. When the injury involves deep tissues such as tendon and ligament, healing has to be measured by outcomes such as swelling, tenderness, pain, range of motion, function, and muscle force. In animal studies invasive measures such as inflammatory infiltrate, collagen deposition, and tensile strength of excised tissue can also be used. But these are not applicable for human studies.

Ultrasound treatment has been evaluated during the early inflammatory and proliferative phases and the later remodeling phase of tendon healing. Low-intensity ultrasound for early treatment promotes fibroplasia and collagen synthesis, and increasing tensile strength of the tendon should prevent the complication of re-rupture (Enwemeka 1989, Turner et al. 1989, Enwemeka et al. 1990a, Huys et al. 1993). However, later ultrasound treatment during the phase of scar maturation may promote better alignment of collagen fibrils and prevent formation of adhesions, which should enhance functional outcome (Stevenson et al. 1986, Huys et al. 1993). There is no accepted protocol for ultrasound treatment for tendon repair. Benefits might occur from daily treatment initiated on day 1 or day 7 after surgery and continued for 10 days. An adverse effect could occur, however, if daily treatment is continued for more than 10 days (Roberts et al. 1982). Treatment initiated at day 7 and carried out three times weekly for 5 weeks appears safe but has not proved beneficial. Tensile strength is enhanced by early treatment but is unchanged by treatment during the remodeling phase. The effect of ultrasound on tendon repair needs to be further investigated, and human studies are urgently needed.

There were 400 randomized trials of the use of physiotherapy in various musculoskeletal disorders, such as lateral epicondylitis, periarthrosis humeroscapularis, shoulder bursitis, tendinitis of the shoulder and elbow, ankle distorsion (sprain), osteoarthritis of the knee, low back pain, myofascial pain, traumatized perineum, and breast pain (Beckerman et al. 1993). Although some evidence seems to support its use in elbow disorders, no evidence suggests its effectiveness (van der Heijden et al. 1997, Green et al. 1998). No evidence of benefit for ultrasound in soft tissue lesions may be due to the performance itself, but poor clinical study design or technical factors may also play a role. Significant heterogeneity of study populations presenting various disorders together with shoulder pain prevents to make a clear conclusion. It is possible that ultrasound is only effective in the early stages after tissue injury so that causative factors in the etiology of the lesion and its chronicity should also be defined in the studies. The non-inflammatory, degenerative nature of soft tissue lesions, especially those affecting the tendon, and the presence of more complex underlying pathologies may also contribute to the ineffectiveness of ultrasound therapy.

10.4.5 OSTEOARTHRITIS

Osteoarthritis (OA) is the abnormalities in the joints due to their degradation in articular cartilage and subchondral bone. Symptoms include joint pain, tenderness, stiffness, locking, and sometimes an effusion. Hereditary, developmental, metabolic, and mechanical deficits may initiate processes leading to loss of cartilage. When there is less well protection of cartilage, bone surface may be exposed and damaged. Regional muscles may atrophy, and ligaments may become more lax as a result of decreased movement secondary to pain. Treatment generally involves a combination of exercise, lifestyle modification, and analgesics. Joint replacement surgery can improve the quality of life if the pain becomes worsen and debilitating. OA is the most common form of arthritis and the leading cause of chronic disability of about 1.9 million people in Australia, 8 million people in the United Kingdom, and 27 million people in the United States, accounting for 25% of visits to primary care physician and half of NSAID prescriptions (Figure 10.7) (van Manen et al. 2012). Although NSAIDs are widely used to treat the pain and stiffness associated with knee OA, the high incidence of serious upper gastrointestinal side effects with NSAIDs and the increased risk for hospitalization can limit their use (Scheiman 1996). There are about 250 million patients with the knee OA worldwide (3.6% of the population), occurring more frequently in women than in men (Vos et al. 2012). It is estimated that 80% of the population have radiographic evidence of OA by age 65, although only 60% of them have symptoms. In 2004, OA causes moderate to severe disability in 43.4 million people (WHO 2008). In the United States, an aggregate cost of $14.8 billion ($15,400 per stay) is related to OA, which was the second-most expensive condition in hospital stays in 2011.

Physiotherapy, including short-wave diathermy, TENS, hot packs, and exercise, is a non-pharmacological intervention for knee OA recommended by the American College of Rheumatology and the European League Against Rheumatism. Ultrasound has recently been recommended as a therapeutic intervention for hand OA by the European League Against Rheumatism (EULAR) Standing Committee for International Clinical Studies Including Therapeutics (ESCISIT). However, such recommendations are based on expert opinion alone, and there is no evidence from well-designed studies. Although several studies demonstrated that ultrasound decreased pain in osteoarthritis or acute periarticular inflammatory conditions, most of them are uncontrolled or lack of blind evaluation. Many studies that have reported improvement in pain with ultrasound did not include appropriate analysis. Therefore, varying results could be due to differences in either the treatment parameters or the natural history of the condition.

In a randomized, placebo-controlled, double-blind study performed at the Physical Therapy and Rehabilitation Department of Osmangazi University Hospital (Tascioglu et al. 2010), 82 idiopathic knee OA patients (56 women and 26 men, aged 54–70 years) with symptom for at least 3 years and Grade II or III bilateral knee OA confirmed radiologically according to the Kellgren–Lawrence grading system were

FIGURE 10.7 Age-standardized disability-adjusted life year (DALY) rates from osteoarthritis by country (per 100,000 inhabitants).

randomly divided into three groups: placebo group, 5 min continuous ultrasound group, and 5 min pulsed ultrasound group with a pulse ratio of 1:4. Ultrasonic waves have a frequency of 1 MHz and an intensity of 2 W/cm^2 generated from a 5 cm diameter transducer. All treatments were applied daily for 5 days a week for 2 weeks. The primary outcome was knee pain on movement, which was assessed using a visual analogue scale (VAS) consisting of a 10 cm horizontal line, with anchor points of 0 (no pain) and 10 (maximum pain). Secondary outcomes were the Western Ontario and McMaster Universities Osteoarthritis Index (WOMAC) score, active range of knee flexion, and 20 m walking time. Comparison of the baseline and post-treatment results is presented in Table 10.6. The reductions in pain and WOMAC scores were significantly higher in patients treated with pulsed ultrasound than those in the placebo group. Although there were significant improvements in total WOMAC scores and walking speed in both the continuous and pulsed ultrasound groups in comparison to the baseline, only the patients in the pulsed ultrasound group had significant improvement over placebo group. No superiority was observed for continuous ultrasound over placebo, and no systemic or local side effects were reported during or after the therapy.

Improvement in pain was also observed in the placebo group, which is known as the placebo effect and may be attributed to the attention, interest, and concern displayed by the physician or physiotherapist and to patients' expectations of the outcome. Despite its widespread use, the efficacy of therapeutic ultrasound in knee OA has only been subjected to limited study and the results of these studies are somewhat conflicting. Continuous ultrasound at a frequency of 1 MHz and an acoustic intensity of 1 W/cm^2 for 5 min to the knee OA for 10 sessions is effective as ibuprofen phonophoresis in pain scores, knee range of movement, walking time, and WOMAC scores (Kozanolu et al. 2003, Ozgonenel 2009). Pulsed ultrasound could enhance the therapeutic effects of isokinetic strengthening excise for treating the knee OA patients with peri-articular soft tissue pain (Huang et al. 2005a,b). In comparison to hot pack, short-wave diathermy, ultrasound, and TENS, it is found that using therapeutic ultrasound before isokinetic exercises leads to augmented exercise performance, reduced pain, and improved function (Cetin et al. 2008). In a Cochrane study

TABLE 10.6

Comparison of the Baseline and Post-Treatment Results in the Patients with Primary Knee Osteoarthritis

	Placebo US (n = 27)	Continuous US (n = 27)	Pulsed US (n = 28)
Visual analogue scale scores			
Baseline	7.26 ± 1.46	6.67 ± 1.41	6.89 ± 1.39
Post-treatment	6.67 ± 1.78*	5.22 ± 1.70*	5.25 ± 1.90***
Total Western Ontario and McMaster Universities OA index scores			
Baseline	45.81 ± 12.28	44.56 ± 16.15	43.43 ± 8.26
Post-treatment	44.33 ± 12.78*	43.44 ± 16.48*	35.61 ± 8.73***
Active range of knee flexion			
Right knee			
Baseline	119.70 ± 4.50	121.48 ± 4.93	122.00 ± 6.03
Post-treatment	119.74 ± 4.32	121.67 ± 4.90	122.00 ± 5.43
Left knee			
Baseline	119.48 ± 5.75	120.67 ± 5.52	121.39 ± 5.12
Post-treatment	120.44 ± 6.23	120.85 ± 5.40	121.93 ± 5.02
20 m walking time			
Baseline	23.44 ± 2.24	23.07 ± 2.96	22.57 ± 2.08
Post-treatment	23.19 ± 2.54	22.85 ± 2.99*	20.00 ± 1.94**

Source: Tascioglu, F. et al., *J. Int. Med. Res.*, 38, 1233, 2010.
* $p < 0.05$, ** $p < 0.01$, *** $p < 0.001$ for baseline versus post-treatment (paired samples t-test).

by searching randomized or quasi-randomized controlled trials in CENTRAL, CINAHL, EMBASE, MEDLINE, and PEDro until July 23, 2009 that comparison of therapeutic ultrasound with a sham intervention or no intervention in knee OA patients revealed five small size trials in a total of 341 patients. The quality of methodology and reporting was poor, and a high degree of heterogeneity among the trials was revealed for function (88%). There was an effect in favor of ultrasound, which corresponded to a difference in pain scores between ultrasound and control of −1.2 cm on a 10 cm VAS. For function, a difference of −1.3 units was found on a standardized WOMAC score. Thus, therapeutic ultrasound may be beneficial for the knee OA, although the magnitude of the effects on pain relief and function is uncertain due to the low quality of the evidence (Rutjes et al. 2010).

However, in a randomized controlled study in acceptable methodological quality, both ultrasound and sham exercises showed significant improvement in pain and range of movement, but with no differences between them. In a recent Cochrane review on the use of therapeutic ultrasound for knee OA, it is concluded that ultrasound therapy seemed to have no benefit over placebo or short-wave diathermy (Rutjes et al. 2010). The dosage of ultrasound used in previous studies varies considerably, and little guidance of the optimal dosage is available in the literature. Similarly, the applied acoustic intensity, the size of the treatment area, and the exposure duration also varied between each other. Further research into the effects of therapeutic ultrasound parameters is in a great need to develop the treatment guidelines (Falconer et al. 1990).

10.4.6 Bone Fracture Healing

A bone fracture is a medical condition in which there is a break in the continuity of the bone, and it can be the result of high force impact or stress, or trivial injury as a result of certain medical conditions that weaken the bones, such as osteoporosis, bone cancer, or osteogenesis imperfecta, where the fracture is then properly termed a pathologic fracture (Marshall and Browner 2012). Treatment of bone fractures is broadly classified as surgical or conservative, the latter basically referring to any non-surgical procedure, such as pain management, immobilization, or other non-surgical stabilization. It is estimated that there are over 5.6 million fractures costing $17 billion per year in the United States, of which 5%–10% demonstrate delayed healing or nonunion (Einhor 1995). By 2025, annual fractures and costs are projected to increase by 50% and $25 billion, respectively.

In an effort to reduce the substantial associated disability and socioeconomic costs, various interventions have been proposed (Duarte 1983, Hadjiargyrou et al. 1998). Initially, fracture sites have been considered an absolute contraindication or ambivalence for the use of therapeutic ultrasound by many textbooks in physiotherapy since it was thought that ultrasound could cause bone resorption and damage epiphyseal plates. Low-intensity ultrasound has emerged as an effective modality in the treatment of fracture repair, for both fresh fractures and those experiencing inhibited repair. Continuous ultrasound (1.0 W/cm^2) appears to be harmful in earlier animal studies, while pulsed one (30 mW/cm^2) seems to promote accelerated healing, although the exact mechanism of bone repair is not clearly known. Recent controlled trials in humans have yielded conflicting results, so the use of ultrasound therapy is not generally involved in the current management of fractures or nonunion. In order to clarify the effect of LIPUS for fracture healing, a systematic review and meta-analysis of randomized controlled trials found in CINAHL (1982 to December 2000), MEDLINE (1966 to December 2000), Health-STAR (1975 to December 2000), and EMBASE (1983 to December 2000) was carried out (Busse et al. 2002). Inclusion criteria are skeletally mature patients of either sex with one or more fractures; blinding of both the patient and the assessor(s) as to fracture healing; administration of low-intensity pulsed ultrasound treatments to at least one of the treatment groups; and assessment of time to fracture healing, as determined radiographically by bridging of three or four cortices. Daily 20 min sessions of 1.5 MHz ($\pm5\%$) ultrasound with a pulse duration of 200 μs ($\pm10\%$), a pulse repetition rate of 1 kHz ($\pm10\%$), and a spatial average acoustic intensity of 30 mW/cm^2 ($\pm30\%$) were applied in all six selected studies. The reduction in the healing time for tibial and radial shaft fractures ($n = 111$) was 41% among smokers ($p < 0.006$) and 26% among nonsmokers ($p < 0.05$), and 51% among smokers ($p < 0.003$) and by 34% among nonsmokers ($p < 0.0001$) for distal radial fractures. Despite a similar trend toward a reduced incidence of tibial delayed union, there was no significant difference except for smokers within the previous 10 years. The weighted average effect size was 6.41 (95% confidence interval 1.01–11.81), which converts to a mean difference in a healing time of 64 days between the treatment and control groups as listed in Table 10.7. Therefore, LIPUS may significantly reduce the time to the healing of nonoperatively treated fractures, yield substantial cost savings, and decrease in disability associated with delayed union and nonunion of fractures. But no additional benefit of ultrasound was found for the use of intramedullary nails, which may be due to osteoblastic effect by reaming of fractures. Although baseline healing time differs by bone size and site of fracture, the process of fracture healing is consistent across all fractured bones, and the effect of ultrasound versus placebo on the time to fracture healing is similar. However, there are large discrepancies between the intensity set by the machines and the actual value emitted from the transducer. Further larger clinical trials are required to determine the optimal role of ultrasound therapy in fracture healing.

TABLE 10.7
Summary of the Trials Included in the Meta-Analysis

Location	Sample Size		Fracture		Healing Time (*d*)		Effect Size	Quality Score
	Treated	Control	Open	Closed	Treated	Control		
Tibial shaft	33	34	3	64	114 ± 7.5	12 ± 15.8	5.41	5
Distal radius	30	31	0	61	61 ± 3	98 ± 5	8.82	5
Scaphoid	15	15	N/A	N/A	43 ± 11	62 ± 19	1.20	4

Source: Busse, J.W. et al., *Can. Med. Assoc. J.*, 166, 437, 2002.

10.4.7 SPRAINED ANKLE

Inversion injuries of the lateral ligament of the ankle are a common cause for attendance at the Accident and Emergency Department and are responsible for a large amount of short-term absence from work and sport. Treatments include immobilization in plaster of Paris, cast-bracing, physiotherapy, and no care at all. It suggests that early physiotherapy was the best treatment, but the role of ultrasound has yet to be clarified. A total of 110 patients (12–65 years) with injury to the lateral ligament of the ankle within the previous 48 h were included and divided into two groups undergoing physiotherapy and combination of physiotherapy and ultrasound treatments, respectively (Williamson et al. 1986). There was no significant difference between the scores for any criterion at any stage, and patients in both groups reached their end points at the same rate. It implies that ultrasound treatment does not hasten recovery after lateral ligament sprain of the ankle and a physiotherapy regimen of early exercise, ice packs, and walking does not require the addition of ultrasound.

The optimal ultrasound parameters (i.e., the acoustic pressure, wave mode, the frequency, and the treatment duration) have not been reliably established by comparative measurement. In a prospective randomized trial, the relative efficacy of high- (3 MHz) and low-frequency ultrasound (45 kHz) in the treatment of unilateral acute ankle sprains was compared. There was a statistically significant improvement in length of stride, symmetry of swing phase duration, cadence, and walking velocity in the group treated with low-frequency ultrasound (longwave), which has a more widely divergent shape and a very short near field so that high uniformity in the acoustic field (Figure 10.8) (Bradnock et al. 1996). Thus, difference between therapeutic modalities is important factor in the investigation of effective ultrasound therapy.

10.4.8 DELAYED-ONSET MUSCLE SORENESS

Exercise-induced muscle soreness is a common occurrence in athletics. Delayed-onset muscle soreness (DOMS) is easily induced by relatively intense, slow, eccentric muscle actions, such as for athletes starting a new season, a new training program, or a weightlifting program involving unaccustomed concentric and eccentric work. These eccentric actions produce micro-injury to the active muscle fibers, exhibiting muscular soreness, loss of joint range of motion, swelling, and decreased force production (Clarkson et al. 1992). Clinical signs of DOMS include increases in plasma enzymes, muscular fiber degeneration, and the protein degradation. DOMS usually has a gradual onset within 24 h post-exercise, peaks at 24–72 h, and then declines. It is temporary, repairable damage to muscle (symptoms gradually subsiding within 10 days), and its degree is influenced by increased muscle pressure causing swelling within the exercised muscular area.

Thirty-six college-age females (age = 21.5 ± 2.0 years; height = 164.5 ± 6.2 cm; weight = 57.5 ± 6.5 kg) volunteered for participation in the study done at Brigham Young

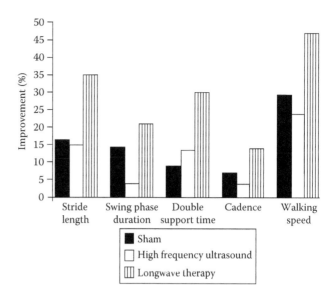

FIGURE 10.8 Comparison of improvement in the stride length, swing phase duration ration (injured or uninjured), double support time, cadence and walking speed before and after treatment with sham ultrasound, high-frequency ultrasound and longwave therapy. (Reprinted from *Physiotherapy*, 82, Bradnock, B., Law, H.T., and Roscoe, K., A quantitative comparative assessment of the immediate response to high frequency ultrasound and low frequency ultrasound ('longwave therapy') in the treatment of acute ankle sprains, 78–84, Copyright 1996, with permission from Elsevier.)

University (Stay et al. 1998). To exercise and induce DOMS, participants performed concentric and eccentric dumbbell curls. The elbow flexors were isolated by stabilizing the arm on a preacher curl bench. Post-exercise measurements were taken to assess upper-arm swelling, perceived soreness in the elbow flexors, relaxed-elbow extension angle, and elbow-flexion strength immediately post-exercise and 24, 48, 72, and 96 h post exercise. The soreness is rated on the scale: 0 (none), 1 (very slight), 2 (mild), 3 (moderate), 4 (sever), and 5 (extreme). After the post-exercise assessments, participants received either pulsed ultrasound treatment (7 min at 20% duty cycle, 1 MHz, 1.0 W/cm²) or a sham treatment to an area of 10 cm × 5 cm on the anterior surface of the upper arm and elbow joint. A 3 × 6 factorial repeated-measures analysis of variance was used to test for significant differences ($p < 0.05$). Tukey's tests were used for all post hoc comparison. There were no significant interactions between time and group ($F_{10,58} = 0.67$, $p = 0.75$). Neither were there significant differences in soreness among groups ($F_{2,33} = 0.49$, $p = 0.62$) as listed in Table 10.8. Similar conclusion for upper-arm circumference, relaxed-elbow extension angle, and 1-repetition maximum (1RM) elbow-flexion strength. Pulsed ultrasound did not significantly diminish the effects of DOMS on soreness perception, swelling, relaxed-elbow extension angle, and strength. A reduction in pain and tenderness and increased muscle strength in delayed-onset muscle soreness (DOMS) have been reported by some group, but not confirmed by the others (Hasson et al. 1990, Craig et al. 1999).

TABLE 10.8
Mean Perceived Soreness Level of Elbow Flexors by Treatment Group and Time

Group	Pretest	Time Post-Exercise				
		Immediately	24 h	48 h	72 h	96 h
Twice daily	0.0 ± 0.0	0.9 ± 0.8	2.7 ± 0.5	2.7 ± 0.9	1.4 ± 0.9	0.5 ± 0.5
Once daily	0.0 ± 0.0	1.2 ± 0.9	2.5 ± 0.8	2.6 ± 0.8	1.2 ± 0.8	0.2 ± 0.5
Placebo	0.0 ± 0.0	1.5 ± 1.1	3.0 ± 1.0	2.5 ± 0.9	1.5 ± 0.9	0.2 ± 0.5
Time mean	0.0 ± 0.0	1.2 ± 1.0[a]	2.7 ± 0.8[a]	2.6 ± 0.8[a]	1.4 ± 0.9[a]	0.3 ± 0.5

Source: Stay, J.C. et al., *J. Athlet. Train.*, 33, 341, 1998.

[a] Immediately, 24, 48, and 72 h post-exercise means were significantly different from pretest mean.

The percentage deviations from baseline for isometric contraction, maximum-extension torque, and knee extension work were significantly less at 48 h for subjects who received pulsed ultrasound compared with placebo treatment and control subjects, which suggests that pulsed ultrasound accelerates restoration of normal muscle performance and was effective in decreasing DOMS (Hasson et al. 1990). Ice massage, ice massage with exercise, and exercise alone have not significantly prevented or reduced soreness, strength, or range-of-motion losses associated with DOMS (Isabell et al. 1992). A 20 min ice pack application followed by pulsed ultrasound (7 min at 20% duty cycle, 1 MHz, 1.0 W/cm^2) applied immediately post-exercise and once daily for 3 days may be more effective than ice alone for the prevention and treatment of DOMS (Mickey et al. 1996). However, when massage, microcurrent electrical stimulation, upper body ergometry, and a post-exercise resting control group were compared, treatments applied immediately and 24 h post-exercise did not prevent soreness or strength loss (Weber et al. 1994). In addition, thermal CW ultrasound at an intensity of 1.5 W/cm^2 had a negative effect on DOMS, aggravating pain, and muscle stiffness (Ciccone et al. 1991).

10.4.9 TENNIS ELBOW SYNDROME

The tennis elbow syndrome or lateral epicondylitis is a disorder of the common extensor origin of the arm and usually caused by excessive quick, repetitive movements of the wrist and forearm in work and sports. These movements may rupture the proximal attachment of the long extensor muscles and cause local inflammation and pain. Cervical root irritation, shoulder problems, local bursitis or radiohumeral synovitis, and posterior interosseus nerve entrapment are implicated in epicondylalgia and adverse neural tension also contribute to the epicondylar pain (Yaxley and Gwendolen 1993). Acute cases will improve with anti-inflammatory drugs or local corticosteroid injections. In contrast, chronic cases are often difficult to treat. Outcome from conservative treatment methods is variable. Local steroid injections are widely used in chronic cases and more effective than ultrasound and placebo ultrasound therapy in a clinical trial (Binder and Hazelman 1983). Continuous and pulsed ultrasound treatments have produced conflicting results

(Binder et al. 1985, Lundeberg et al. 1988), and pulsed ultrasound was no better than placebo (Haker and Lundeberg 1991).

Thirty-nine patients (14 men and 25 women, 31–53 years) with clinically diagnosed chronic unilateral epicondylitis from the Oulu University Hospital were randomized into two treatment groups (Pienimäki et al. 1996), of which 20 were treated with progressive slow, repetitive wrist and forearm stretching, muscle conditioning and occupational exercises and 19 with pulsed ultrasound (frequency of 1 MHz, radiated area of 5 cm^2, 0.3–0.7 W/cm^2, pulse ratio of 1:5, pulse duration of 2 ms, sonication of 10–15 min, two to three times a week, 6–8 weeks for the treatment). The principal exclusion criteria were cubital OA, carpal or radial tunnel syndrome, rheumatoid arthritis, severe cervical spondylosis or cervical radicular syndrome, painful shoulder or rotator cuff tendinitis, previous arm fractures, and no clinical signs of the tennis elbow syndrome. All included patients had a positive Mill's test, local tenderness on palpation over the lateral epicondyle, and resisted wrist and/or middle finger extension produced typical pain at the origin on the lateral epicondyle. In the follow-up, pain at rest and under strain had decreased, sleep disturbance was alleviated, and subjective ability to work increased in the exercise group more significantly than in the ultrasound group ($p < 0.05$). The isokinetic torque of wrist flexion increased by 45% in the exercise group but declined by 4% in the ultrasound group ($p = 0.0002$). Maximum isometric grip strength increased 12% in the exercise group and remained unchanged in the ultrasound group ($p = 0.05$). So, progressive exercise is more effective than ultrasound in treating chronic lateral epicondylitis, reducing pain, and improving patients' ability to work (Pienimäki et al. 1996). It may correct the ill-effects of prolonged immobilization, counter patients' fear of using the forearm and hands, and help them to return to work (Table 10.9).

Slow and repetitive exercises for strengthening the soft tissues of the upper limb seem to be beneficial in treating chronic strain injuries. As the etiology of strain injuries (i.e., the tennis elbow syndrome) may be repeated, rapid movements and slow progressive strengthening exercises may allow tissue healing. Early mobilization has good effects on the tensile strength of connective tissue scars in acute muscle injury (Kannus et al. 1992). Pulsed ultrasound (1 MHz, 0.5 W/cm^2) was ineffective as a sole approach in

TABLE 10.9

Positive Clinical Manual Tests in Both Treatment Group before and after Treatment

	Before Treatment	After Treatment	Chane	Z Value	p Value
Palpation, lateral epicondyle					
EG	20	13	−7	−2.4	0.02
UG	19	17	−2	−1.3	0.18[a]
Resisted wrist extension					
EG	20	12	−8	−2.5	0.01
UG	18	7	−1	−0.53	0.59[a]
Resisted middle finger extension					
EG	19	11	−8	−2.2	0.02
UG	17	15	−2	−1.34	0.18[a]
Mill's test					
EG	20	9	−11	−2.9	0.003
UG	19	16	−3	−1.6	0.11[a]

Source: Reprinted from *Physiotherapy*, 82, Pienimäki, T.T., Tarvainen, T.K., Siira, P.T., and Vanharanta, H., Progressive strengthening and stretching exercises and ultrasound for chronic lateral epicondylitis, 522–530, Copyright 1996, with permission from Elsevier.

EG, exercise group; UG, ultrasound group.

[a] No significance tested with non-parametric Wilcoxon's test for matched pairs.

treating chronic epicondylar pain, although it has beneficial effects on protein synthesis.

10.4.10 CARPAL TUNNEL SYNDROME

Carpal tunnel syndrome is the entrapment mononeuropathy caused by compression of the median nerves at the wrist (Gelberman et al. 1998). Its symptoms include hand weakness, pain, numbness, or tingling in the hand, especially in the thumb, index, and middle fingers (Simovic and Weinberg 2000), and it becomes worse at night, often waking the patient. Standard treatments include splints, local injection of corticosteroids, and surgical decompression. The benefit of non-surgical treatment seems to be limited, although not all patients respond to surgery. Meanwhile, the efficacy of most conservative treatment options for carpal tunnel syndrome is still little known (Gerritsen et al. 2002). Stimulation of nerve regeneration and conduction by low-level laser therapy and ultrasound therapy proves that they may facilitate recovery from nerve compression. Mechanisms suggested as underlying therapeutic effects with low-level laser therapy included increased adenosine triphosphate (ATP) production by the mitochondria and increased cellular oxygen consumption, increased serotonin and endorphins, anti-inflammatory effects, and improved blood circulation (Bakhtiary and Rashidy-Pour 2004). Satisfying short to medium term effects of ultrasound treatment in patient with mild to moderate idiopathic carpal tunnel syndrome were found (Ebenbichler et al. 1998).

Fifty patients with clinically diagnosed carpal tunnel syndrome (40 in both wrists and 10 in the right wrist) referred to the rehabilitation clinic of the Semnan Medical Sciences University were included using criteria of positive Phalen's test, positive Tinnel's test, and standard electrophysiological criteria such as prolongation of nerve conduction velocity (i.e., motor latency > 4 ms or sensory latency > 3.5 ms). Patients were excluded if they had secondary entrapment neuropathies, electroneurographic and clinical signs of axonal degeneration of the median nerve, or had been treated with ultrasound or low-level laser therapy for the syndrome, or had required regular analgesic or anti-inflammatory drugs, or had a history of steroid injection into the carpal tunnel, thyroid disease, diabetes, or systemic peripheral neuropathy. They were randomly assigned to two treatment groups: low-level laser (9 J, 830 nm infrared laser at five points) and ultrasound (1 MHz, 1.0 W/cm², pulse ratio of 1:4, 15 min per session) treatments (Bakhtiary and Rashidy-Pour 2004). The carpal tunnel was treated daily five times a week for 3 weeks. Outcome measures consisted of pain assessment by means of a 10 cm VAS; pinch strength measured with a standard dynamometer between the tips of the thumb and the little finger; and hand grip strength measured with a handheld dynamometer. Improvement was significantly more pronounced in the ultrasound group than in low-level laser therapy group for motor latency (mean difference 0.8 m/s, 95% CI: 0.6–1.0), motor action potential amplitude (2.0 mV, 95% CI: 0.9–3.1), finger pinch strength (6.7 N, 95% CI: 5.0–8.2), and pain relief (3.1 points on a 10-point scale, 95% CI: 2.5–3.7). Effects were sustained in the follow-up period as listed in Table 10.10. Therefore, ultrasound treatment seems more effective than laser therapy for the treatment of carpal tunnel syndrome.

The differences between therapeutic effects on latencies and amplitudes may be because there is myelin involvement in most patients with carpal tunnel syndrome, but axonal involvement is not always seen in mild or moderate diseases. Conservative approaches seem better over surgery in patients with mild or moderate symptom. Short-term effects of steroid injections into the carpal tunnel can modestly or completely relieve pain in up to 92% of the patients despite great variations in the long-term recurrence rates (Gonzales and Bylak 2001). The value of this treatment has been limited by potential adverse effects to nerves and tendons with repeated injections. Wearing wrist splints at night seems suitable only when symptoms are mainly nocturnal (Burk et al. 1994). Ultrasound therapy has similar effectiveness as steroid injection or wrist splinting, but with no complications or limits. It could elicit anti-inflammatory and tissue stimulation effects and accelerate the healing process in damaged tissues, leading to pain relief, increased grip strength, and changed electrophysiological parameters toward normal values better than laser therapy in patients with mild to moderate carpal tunnel syndrome diagnosis.

10.4.11 OSSIFICATION

Ossification (or osteogenesis) is the process of laying down new bone material by osteoblasts cells, which is synonymous with bone tissue formation. There are two processes resulting in the formation of normal and healthy bone tissue:

TABLE 10.10

Mean Changes of Pain, Force Measures, and Recorded Electrophysiological Parameters at the End of Therapy and 4 Weeks Later

	End of Therapy			Four Weeks Follow-Up		
	Ultrasound	Laser	*p*	Ultrasound	Laser	*p*
Pain (VAS; 10 cm scale)	−5.6 ± 1.5	−2.4 ± 1.2	<0.001	−6.3 ± 1.6	−2.0 ± 1.3	<0.001
Handgrip strength (N)	36.6 ± 19.1	19.4 ± 15.3	0.008	39.3 ± 21.5	21.2 ± 18.4	<0.001
Finger pinch (N)	9.1 ± 4.1	2.6 ± 1.0	<0.001	9.9 ± 5.5	2.9 ± 1.5	<0.001
Motor distal latency (ms)	−1.0 ± 0.6	−0.3 ± 0.3	<0.001	−1.1 ± 0.5	−0.2 ± 0.2	<0.001
CMAP amplitude (mV)	3.0 ± 1.6	1.0 ± 2.9	<0.001	3.6 ± 1.5	1.1 ± 2.9	<0.001
Thumb sensory latency (ms)	−0.7 ± 0.5	−0.2 ± 0.7	<0.001	−0.7 ± 0.5	−0.2 ± 0.6	0.003
Thumb SAP amplitude (µV)	9.5 ± 7.3	4.5 ± 7.6	0.004	10.1 ± 6.9	4.4 ± 7.4	<0.001
Index sensory latency (ms)	−0.8 ± 1.0	0.1 ± 1.2	0.003	−0.8 ± 1.0	0.1 ± 1.1	0.004
Index SAP amplitude (µV)	16.1 ± 16.4	7.0 ± 14.2	0.007	6.8 ± 15.2	6.5 ± 11.9	0.003

Source: Bakhtiary, A.H. and Rashidy-Pour, A., *Aust. J. Physiother.*, 50, 147, 2004.

intramembranous ossification is the direct laying down of bone into the primitive connective tissue (mesenchyme), while endochondral ossification involves cartilage as a precursor. In the case of bone, ultrasound may cause damage to the periosteal cells. On the other hand, bone repair can be stimulated by ultrasound to enhance fracture healing. It is generally assumed that sonication of these discs may influence normal bone development.

Paired metatarsal long bone rudiments (II–IV) were dissected from 16- and 17-day-old fetal mice along with metatarsus V to facilitate orientation of the triplets (MT). The tissue between the metatarsae was not disrupted to avoid damage to the developing bone rudiments. After 24 h of culture, MT were treated with 1 MHz ultrasound in pulsed (pulse duration of 2 ms; spatial-average temporal-peak acoustic intensity of 0.1, 0.33, 0.49, and 0.77 W/cm^2; pulse repetition frequency of 100 Hz) or continuous mode (spatial-average temporal-average acoustic intensity of 0.1 and 0.5 W/cm^2) for 5 min or 1 min (Wiltink et al. 1995). The length of MT and calcification zone were determined daily using a measuring ocular, and resorption was evaluated by the appearance of translucent spots in the mineralized zone. After 1 week of culture, the metatarsal long bone rudiments were fixed and paraffin sections were prepared or histological evaluation and for measurement of the relative contribution to the various cartilage zone in the total bone length. In contrast to CW ultrasound exposure, PW ultrasound resulted in significantly increased longitudinal growth after 4 days of culture ($p < 0.01$) (Figure 10.9). The fraction of the proliferative cartilage zone at day 7 as compared to the total length of the MT was 0.56 ± 0.03 ($n = 5$) in the treated MT group and 0.47 ± 0.03 ($n = 5$) in the paired control MT. There was no difference in the length of the hypertrophic cartilage zone between the treated and control MT ($n = 5$), indicating that the accelerated MT growth was mainly due to increase in growth of the proliferative cartilage zone. Histology revealed a significantly increased length of the proliferative zone ($p < 0.02$), whereas the length of the hypertrophic cartilage zone was unaltered. This may

suggest that the proliferation of the cartilage cells is stimulated without influence on cell differentiation.

There are few clinical trials to assess the effectiveness of ultrasound in ossification (Holmes and Rudland 1991). Mechanisms at the tissue level is also unclear for the consequently physiological effects. One of the contraindications to treatment is sonication of the epiphyseal discs of growing long bones. The cartilage proliferation seemed to be stimulated by stable cavitation rather than inertial cavitation by PW ultrasound without affecting the hypertrophic cartilage zone, which suggests the stimulated proliferation of cartilage cells without influence on their differentiation, corroborating with the stimulation of cartilage proliferation during fracture healing. In addition, a correlation was found between the elevation

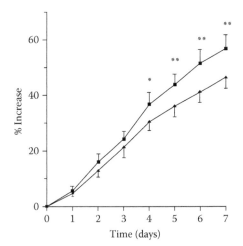

FIGURE 10.9 Comparison of percentage increase in length of 16-day-old metatarsal long bone rudiments (MT) treated by ■: pulsed ultrasound group at intensity of 0.77 W/cm^2, and ◆: sham (*, $p < 0.05$, **, $p < 0.01$). (Reprinted from *Ultrasound Med. Biol.*, 21, Wiltink, A., Nijweide, P.J., Oosterbaan, W.A., Hekkenberg, R.T., and Helders, P.J.M., Effect of therapeutic ultrasound on endochondral ossification, 121–127, Copyright 1995, with permission from Elsevier.)

of the cytosolic calcium concentration in preosseous chondrocytes and increase in cell proliferation of resting chondrocytes, whereas in hypertrophic cells no such correlation was found.

10.4.12 NERVE REPAIR

Low-intensity ultrasound enhances tissue repair and that thermal ultrasound (1.0 W/cm^2, CW) delays recovery of motor nerve (Hong et al. 1986). Ultrasound treatment applied over the distribution of peripheral nerves may be beneficial in relieving symptoms of reflex sympathetic dystrophy (Portwood et al. 1987). When applied to neuromas following amputation, it may relieve the pain, leading to improved prosthetic use (Uygur and Sener 1995). Neuromas in the hand often lead to decreased prehension and grip; therefore, it is prudent to use a therapy technique that may reduce sensitivity.

10.4.13 HAND

Chronic swelling, contractures, and adhesions in the hand interfere with function. Exudate in the hand, particularly around the metacarpophalangeal and interphalangeal joints, quickly progresses to fibrosis, which restricts movement and causes pain. Extensibility of tissue can be increased by raising its temperature to 40°C–45°C, enough to cause dissolution of the cross-links within and between protein molecules. A combination of thermal ultrasound and stretch improves human tissue extensibility. Lengthening has been demonstrated in the ligament of normal knees (Reed and Ashikaga 1997) and in burn scar contracture of the finger (Baryza 1996). When bone is superficial, as in the hand and wrist, ultrasound at 1 MHz causes tendons adjacent to bone to heat more quickly than skin. Subjective skin warmth works as a good indicator that dense tissue adjacent to bone has become heated.

Metal implants and joint replacement prostheses are not a contraindication to the use of thermal ultrasound. A temperature rise in the vicinity of metal is not different than near bone. Pain from periosteal heating should not be confused with pain from heating of major peripheral nerves, which in the wrist or forearm typically produce a pain that radiates into the hand. The movement rate of the transducer can be increased over the nerve to reduce discomfort. Decreased sensation in the hand is a contraindication to the effect of thermal ultrasound. Chronic inflammation commonly leads to painful restriction of movement due to fibrosis and formation of adhesions. A number of studies are on the effect of pulsed ultrasound in chronic inflammation with benefits of increased grip strength and pain-free wrist extension. However, evidence supporting the use of ultrasound in chronic inflammatory conditions is equivocal. Studies on tendinitis (Downing and Weinstein 1986, Lundeberg et al. 1988, Holdsworth and Anderson 1993) and osteoarthritis have showed either no benefit or a low success rate in improving pain and mobility (Stratford et al. 1989).

Support for ultrasound therapy of the hand to accelerate healing and to improve scar extensibility is largely from studies that do not directly involve the hand. LIPUS seems the most appropriate to promote healing of open wounds, to resolve acute and subacute inflammation, and to enhance repair in tendon, nerve, and bone. Treatment in the proliferative stage of healing leads to improved recovery of function. The use of thermal ultrasound and stretch to improve mobility in the hand needs to be investigated. Ultrasound should be combined with other appropriate intervention, such as rest, compression, elevation, controlled movement, stretch, and exercise (Nussbaum 1998).

10.5 DISCUSSION

There is controversy about the effect of therapeutic ultrasound in the treatment of pain relief, a range of musculoskeletal injuries, and soft tissue healing in comparison to placebo ultrasound and control, which may be due to some confounding factors, including technical variables, the complexity and variety of underlying pathologies in soft tissue lesions, methodological limitations of clinical studies, and true lack of effect. The dosages used in all published studies varied considerably without discernable reason (Robertson and Baker 2001). Early clinical trials attempting to figure out the effectiveness of ultrasound therapy were methodologically flawed, such as lack of control groups, standardized treatment, assessment criteria, and statistical analysis of the results (Holmes and Rudland 1991, Nussbaum 1997). Because only 22 of the 293 articles were methodologically adequate, any contribution of ultrasound to the treatment outcomes was not evident on the basis of the controls (Gam and Johannsen 1995). However, most methodologically adequate studies ($n = 13$) lacked evidence of either meaningful outcomes or statistically significant differences (van der Windt et al. 1999). Randomized controlled trials (RCTs) are widely recognized as the best way of comparing the effectiveness of different treatments (Altman 1996); however, as other research methods, it can also be biased and have deficiencies if performed inappropriately. When methodologically flawed trials were excluded, there were few RCTs providing clinical evidence for the efficacy of therapeutic ultrasound, from which few conclusions can be drawn with positive evidence due to heterogeneity and omission of important details (Robertson and Baker 2001).

In order to objectively and scientifically evaluate the value of ultrasound application in physiotherapy, the following items should be considered. Adequate controls, including placebo treatment and randomized group allocation, are believed to present if subjects were randomly allocated to an active ultrasound and a placebo ultrasound treatment groups, with both of them receiving apparently identical treatments. This methodological design is crucial to investigate an intervention that is alleged to placebo effect (Hashish et al. 1986). A true control group, which received neither active nor placebo ultrasound, should be considered. This allowed the contribution of ultrasound to be distinguished from other components of multiple interventions, which is the aim of randomized controlled trials. Including both placebo ultrasound and the standard treatment groups makes it possible to distinguish the placebo component in the outcome. Adequate experimental blinding, including blinding of the assessor, the subjects and the users of the ultrasound equipment (i.e. therapists), is complex and important but insufficient in most reported studies. However, ensuring that both subjects and therapists are unaware of the settings of an ultrasound machine (double blinding) is difficult. Adequate description and technical calibration of treatment variables should be included in the published paper for replication purpose. Discrepancies between the dosage displayed on the machine and the actually emitted to the patient could affect outcome and may be a reason for the ineffectiveness of current ultrasound therapy (Dyson 1987). As the transducer keeps moving during the treatment to avoid hot spots, the area and ultrasound energy delivered to the target cannot be determined precisely (Williams 1983, Hoogland 1989). There should be at least one meaningful outcome measure with validity for the treated target. For example, pain, swelling, and function are used in soft tissue injuries, whereas the lesion in the wound-healing studies can be evaluated more closely by wound tracing. Adequacy of sample size for trials is necessary to evaluate the treatment effect. To attain an 80% probability of detecting a therapeutic effect ($\alpha = 0.05$), a minimum of 26 subjects per group are required for a two-group study if a large difference in outcome is expected using a parametric statistic analysis (Portney and Watkins 1993). If a nonparametric statistic is used, up to 20% more subjects are needed if a large treatment effect is to be identified. Acceptable statistical analysis of results should be performed. Although robust, a Student's t-test is dependent on a level of homogeneity of variance as other parametric tests (Portney and Watkins 1993). This is especially important when small numbers of subjects are involved, in which a non-parametric analysis should be used.

In the few methodologically adequate studies, treatment was provided for a wide range of problems so that few conclusions can be drawn. In addition, those studies with significant findings have not been replicated by various researchers in a different facilities using the same procedure. Ultrasound dosages varied considerably in the studies, and those with significant outcomes seem to have a higher total energy output. However, there is little scientific basis for choosing the dosage in clinical practice, and no clear range for effectiveness of therapeutic ultrasound. Unless *in vitro* promising effects are not only consistent with healing but also sufficient to alter a relevant outcome positively, they do not justify the clinical use of ultrasound. This doubt can be dissolved until methodologically adequate studies can demonstrate a better and consistent outcome than those treated with placebo ultrasound in clinics (Robertson and Baker 2001).

Physical therapists intended to overlook the tenuous nature of the scientific basis for the use of ultrasound therapy. Some of their knowledge is not based on scientific fact as found in the literature, but empirically and being passed between users. Thus, the training in ultrasound use for more clinical trials and a more basic understanding of its potential as a therapeutic agent are in a great need. In addition, physiotherapists should be more vigilant about the necessity of ultrasound

calibration for the safety and efficacy concerns (ter Haar et al. 1987). There is currently insufficient biophysical evidence as a scientific foundation for the clinical use of ultrasound therapy since a large extent of investigation is *in vitro* with relatively little evidence that these changes occur *in vivo*, and further extrapolation to the clinical situation is therefore questionable. In an *in vivo* condition, any change in the extracellular fluid initiates a protective reaction to minimize the effect on cells, tissues, and organs, which may be at least partially responsible for the discrepancy between the *in vitro* results and the findings of a small number of high-quality randomized controlled trials. Cavitation, the major mechanism for cell damage *in vitro*, has less propensity of occurrence *in vivo* at therapeutic intensities. Furthermore, there are also some differences between animal results and human outcome. For example, the benefits of ultrasound in the healing of superficial wounds in rats, guinea pigs, and rabbits seem to exceed those in human tissue, which may be attributed partially to the relatively condense and tight human skin.

Sonication of normal muscle resulted in vasoconstriction and decreased blood flow (Hogan et al. 1982a,b), but no change in blood flow was found using pulsed ultrasound at the same duration and intensity (Rubin et al. 1990). In another study, increased muscle blood flow occurred at high ultrasound intensities with intolerance due to pain caused by excessive heating (Paul and Imig 1955), although no change was found at low intensity values (Bickford and Duff 1953). The possibility of ultrasound-induced angiogenesis in other situations except wound healing is controversial (Rubin et al. 1990, Hanahan and Folkman 1996). Although there are *in vitro* data of stimulation of fibroblast proliferation with ultrasound (Ramirez et al. 1997), good evidence is not found *in vivo*. Damage to fibroblasts treated with 1 and 3 MHz ultrasound therapy was found in *in vivo* studies, such as changes in the plasma membrane (ter Haar 1999) and in intracellular organelles (i.e., lysosomes and mitochondria) (Williams 1983). Mechanical trauma can cause not only mast cell degranulation but also increased passive cell membrane permeability (McCance and Huether 1998). Despite membrane and intracellular changes *in vitro*, the outcomes of treating soft tissue injuries in animals with ultrasound are also contradictory (Ramirez et al. 1997). Increased collagen deposition was found after pulsed ultrasound treatment (0.1–0.3 W/cm^2, 1 MHz) of wounded pigs (Byl et al. 1992), but no alteration in healing following sonication (0.2 W/cm^2, 3 MHz) of repaired cockerel tendon, which has a similar degree of collagen cross-linkage to that found in human tendon (Turner et al. 1989). Increased cellular activity due to heating is more difficult to address, and the type of cell affected is usually not specified. By far, the most significant difficulty with the concept of increased cellular or enzymatic activity is the implication that this process will accelerate healing. Unfortunately, there is no evidence to connect these events (Hashish et al. 1988, Lundeberg et al. 1988, Grant et al. 1989, Nykanen 1995, ter Riet et al. 1996).

The discrepancy between *in vitro* results and *in vivo* studies may also be due to the blood circulation on heat dissipation (Reed and Ashikaga 1997). Homeostatic mechanisms intend to counteract the temperature rise in the tissues, but cannot quickly reverse the thermal effect (Vander et al. 1998). Dissipation by blood perfusion is highly variable and difficult to estimate but is known poor in fatty tissue and tendon and confined to the skin at clinically acceptable doses (Johnson et al. 1976, Rowell 1986). Therefore, temperature evaluation in the tissue during heating primarily depends on both the extent of conduction into the surrounding tissues and dissipation by blood perfusion (ter Haar 1987, Barnett et al. 1997).

Differences in the duration and the number of treatments, the area of sonication, and the frequency of ultrasound, the acoustic intensity, and pulse ratio account for the variation in outcomes. It is premature to abandon the ultrasound therapy because of the current lack of clinical evidence for its effectiveness. Reliable methods of characterizing the output and measuring the performance of ultrasound physiotherapy can ensure the delivery of a standard ultrasound dosage. There is a need for therapists to prove the efficacy of different ultrasound dosages across the therapeutic range. Currently favored dosages for ultrasound treatment are difficult to justify. Adequate randomized double-blind placebo-controlled clinical studies are required for the use of ultrasound therapy at specific doses in soft tissue lesions (Crawford and Snaith 1996, Speed 2001). Until sufficient clinical studies are available to identify a dose–response relationship, guidelines for clinical physiotherapists can be established (Robertson 2002). The discrepancies between ultrasound dosage recommendations are evident in current textbooks used in the courses of entry-level physical therapy: 0.10–0.25 W/cm^2 for treating acute and post-traumatic lesions with an average duration of 5 min (Low and Reed 2000a), 0.50–3.0 W/cm^2 with a frequency of 1 or 3 MHz for 3–10 min for pain (Cameron 1999), for chronic lesions and scar tissue, 0.25–1.0 W/cm^2 with a longer time than for acute lesions (Low and Reed 2000a), and 1.0–2.50 W/cm^2 with a frequency of 1 or 3 MHz for 5–10 min for soft tissue shortening (Cameron 1999). If the treated area has a relatively superficially located bone, the percentage of ultrasound energy reflected will be higher than that deeper bone, which would effectively increase the dosage in the intervening area. Some studies with nonsignificant findings used a similar intensity as those in which ultrasound produced a significant outcome even if treating a similar problem. There are three possible reasons: (1) some studies with nonsignificant findings did not have enough statistical power (either using too few subjects to identify an effect or including too heterogenous patients) to identify an effect of ultrasound, (2) therapeutic ultrasound was not used at the optimal stage or dosage for the particular problem studied, and (3) the effects of therapeutic ultrasound may not be sufficiently large or predictable to be reliably identified.

Although the absorption coefficient increases linearly with frequency and the distance rate of energy deposition decreases with frequency at a certain intensity (Wells 1977), it is oversimplified to draw the conclusion that it is an advantage to use 1 MHz ultrasound for heating tissues at 2.5–5 cm in depth and 3 MHz one to heat tissues at <2.5 cm in depth (Gann 1991). The thermography shows that the geometries and properties of the tissue, especially the bone in the direction of acoustic beam, have a great influence on the heating depths. The high absorption in bone and its low specific heat and thermal conduction tend to equalize the heating depth. The thermal parameters, such as the intensity distribution, the thermal conductivity of the tissue, and the treatment time, play a role in the effective heating depth (Demmink et al. 2003).

Therapeutic stimulation of cells and molecules within cells by ultrasound is involved in the inflammatory and healing processes. In a view point of cellular biology, a strong argument can be made that, if a stimulus reaches a critical threshold (i.e., depolarization thresholds), the cell will respond, regardless of whether the cell is *in vitro* or *in vivo* (i.e., insulin, histamine, aspirin). Importantly, cells in tissue culture usually respond to nanogram or microgram per milliliter quantities of a stimulus. However, due to pharmocokinetics (i.e., administration, absorption, distribution, and elimination of a drug), higher concentrations and multiple doses per day are normally required to achieve clinical efficacy. The identification and scientific understanding of the mechanisms proposed by the frequency resonance hypothesis and ultrasound-induced mechanical effects can be elucidated and provide insight into a comprehensive strategy for the clinical indications of therapeutic ultrasound at various application (Johns 2002).

Concern has been raised for the accuracy of clinical ultrasound machines. In an evaluation of machines used in NHS practice in Scotland, it was found that almost 70% of machine outputs differed from the expectation by more than 30%, which means that a significant proportion of machines in use are not delivering what the therapists intend to do (Pye and Milford 1994). So, more accurate tests and calibration procedures should be proposed. Weekly checks using simple acoustic radiation balance is helpful to evaluate its performance, but a more comprehensive and rigorous quality assurance system is almost certainly required.

Infection risk is a topical issue in current practice, and the possibility of passing on infective microorganisms as a result of health-care events is a growing concern. A study was conducted in Australia, and microbiological cultures were obtained from 44 transducers and 43 gels in hospitals, private practices, and other health-care facilities (Schabrun et al. 2006). Twenty-seven percent of transducer heads were contaminated with fairly low levels of contamination across the sample, and the majority of organisms isolately found in normal skin and environmental flora. In contrast, 28% of gels were heavily contaminated with opportunistic and potentially pathogenic organisms, including *Stenotrophomonas maltophilia*, *Staphylococcus aureus*, *Acinetobacter baumannii*, and *Rhodotorula mucilaginosa*. No multi-resistant organisms were identified. Cleaning the treatment head with 70% alcohol significantly reduced the level of contamination on transducer heads. However, this clinically norm approach does not deal with the microorganism colonies in the gel bottles sampled, and re-usable gel containers are likely to be a bigger problem.

REFERENCES

Ainsworth R, Dziedzic K, Hiller L, Daniels J, Bruton A, Broadfield J. A prospective double blind placebo-controlled randomized trial of ultrasound in the physiotherapy treatment of shoulder pain. *Rheumatology* 2007;46:815–820.

Al-Karmi A, Dinno M, Stoltz D, Crum L, Matthews J. Calcium and the effects of ultrasound on frog skin. *Ultrasound in Medicine and Biology* 1994;20:73–81.

Aleya WS, Rose DL, Shires EB. Effect of ultrasound on threshold of vibration perception in a peripheral nerve. *Archives of Physical Medicine and Rehabilitation* 1956;37:265–267.

Allen KGR, Battye CK. Performance of ultrasonic therapy instruments. *Physiotherapy* 1978;64:174–179.

Altman DG. Better reporting of randomised controlled trials: The CONSORT statement. *British Medical Journal* 1996;313:570–571.

Baker KG, Robertson VJ, Duck FA. A review of therapeutic ultrasound: Biophysical effects. *Physical Therapy* 2001;81:1351–1358.

Bakhtiary AH, Rashidy-Pour A. Ultrasound and laser therapy in the treatment of carpal tunnel syndrome. *Australian Journal of Physiotherapy* 2004;50:147–151.

Banda MJ, Knighton DR, Hunt TK, Werb Z. Isolation of a nonmitogenic angiogenesis factor from wound fluid. *Proceedings of the National Academy of Sciences of the United States of America* 1982;79:7773–7777.

Barnett SB, Rott H-D, ter Haar GR, Ziskin MC, Maeda K. The sensitivity of biological tissue to ultrasound. *Ultrasound in Medicine and Biology* 1997;23:805–812.

Baryza M. Ultrasound in the treatment of post burn skin graft contracture: A single case study. *Physical Therapy* 1996;76:S54.

Beckerman H, Bouter LM, van der Heijden GJ, de Bie RA, Koes BW. Efficacy of physiotherapy for musculoskeletal disorders: What can we learn from research? *British Journal of General Practice* 1993;43:73–77.

Berridge MJ. Inositol trisphosphate and calcium signalling. *Nature* 1993;361:315–325.

Bickford RH, Duff RS. Influences of ultrasonic irradiation on temperature and blood flow in the human skeletal muscle. *Circulation Research* 1953;1:534–538.

Binder AI, Hazelman BL. Lateral humeral epicondylitis—Natural history and the effect of conservative therapy. *British Journal of Rheumatology* 1983;22:73–76.

Binder AI, Hodge G, Greenwood AM, Hazelman BL, Page Thomas DP. Is therapeutic ultrasound effective in treating soft tissue lesions? *British Medical Journal* 1985;290:512–514.

Bradnock B, Law HT, Roscoe K. A quantitative comparative assessment of the immediate response to high frequency ultrasound and low frequency ultrasound ('longwave therapy') in the treatment of acute ankle sprains. *Physiotherapy* 1996;82:78–84.

Branemark PO. Capillary form and function: The microcirculation of granulation tissue. *Bibliotheca Anatomica* 1965;4:9.

Burk DT, McHarl Burke M, Stewart GW, Cambre A. Splinting in carpal tunnel syndrome: In search of the optimal angle. *Archives of Physical Medicine and Rehabilitation* 1994;73:1241–1244.

Busse JW, Bhandari M, Kulkarni AV, Tunks E. The effect of low-intensity pulsed ultrasound therapy on time to fracture healing: A meta-analysis. *Canadian Medical Association Journal* 2002;166:437–441.

Byl NN, McKenzie AL, West JM, Whitney JD, Hunt TK, Scheuenstuhl HA. Low-dose ultrasound effects on wound healing: A controlled study with Yucatan pigs. *Archives of Physical Medicine and Rehabilitation* 1992;73:656–664.

Cameron MH. *Physical Agents in Rehabilitation*. Philadelphia, PA: WB Saunders, 1999.

Castel JC. Therapeutic ultrasound. *Rehabilitation and Therapy Products Review* 1993: Jan/Feb 22–32.

Cetin N, Aytar A, Atalay A, Akman MN. Comparing hot pack, short-wave diathermy, ultrasound, and TENS on isokinetic strength, pain, and functional status of women with osteo-arthritic knees: A single-blind, randomized, controlled trial. *American Journal of Physical Medicine and Rehabilitation* 2008;87:443–451.

Chapman CR. On the relationship of human laboratory and clinical pain research. In: Melzack R, ed., *Pain Measurement and Assessment*. New York: Raven Press, 1983, pp. 243–249.

Ciccone C, Leggin B, Callamaro J. Effects of ultrasound and trolamine salicylate phonophoresis on delayed-onset muscle soreness. *Physical Therapy* 1991;71:666.

Clark RA, Stone RD, Leung DYK, Silver I, Hunt TK. Role of macrophages in wound healing. *Surgical Forum* 1976;27:16–18.

Clarkson PM, Nosaka K, Braun B. Muscle function after exercise-induced muscle damage and rapid adaptation. *Medicine and Science in Sports and Exercise* 1992;24:512–520.

Craig JA, Bradley J, Walsh DM, Baxter GD, Allen JM. Delayed onset muscle soreness: Lack of effect of therapeutic ultrasound in humans. *Archives of Physical Medicine and Rehabilitation* 1999;80:318–323.

Crawford F, Snaith M. How effective is therapeutic ultrasound in the treatment of heel pain? *Annals of the Rheumatic Diseases* 1996;55:265–267.

Demmink JH, Helders PJM, Hobaek H, Enwemeka C. The variation of heating depth with therapeutic ultrasound frequency in physiotherapy. *Ultrasound in Medicine and Biology* 2003;29:113–118.

Dinno MA, Dyson M, Young SR, Mortimer AJ, Hart J, Crum LA. The significance of membrane changes in the safe and effective use of therapeutic and diagnostic ultrasound. *Physics in Medicine and Biology* 1989;34:1543.

Downing D, Weinstein A. Ultrasound therapy of subacromial bursitis: A double blind trial. *Physical Therapy* 1986;66: 194–199.

Draper DO, Castel JC, Castel D. Rate of temperature increase in human muscle during 1 MHz and 3 MHz continuous ultrasound. *Journal of Orthopaedic and Sports Physical Therapy*. 1995;22:142–150.

Duarte LR. The stimulation of bone growth by ultrasound. *Archives of Orthopaedic and Trauma Surgery* 1983;101:153–159.

Dyson M. Mechanisms involved in therapeutic ultrasound. *Physiotherapy* 1987;73:116–120.

Dyson M, Franks C, Suckling J. Stimulation of healing varicose ulcers by ultrasound. *Ultrasonics* 1976;14:232–236.

Dyson M, Pond JB. The effect of pulsed ultrasound on tissue regeneration. Physiotherapy 1970;56(4):136–142.

Dyson M, Pond JB, Joseph J, Warwick R. The stimulation of tissue regeneration by means of ultrasound. *Clinical Science* 1968;35:273–285.

Dyson M, Smalley D. Effects of ultrasound on wound contraction. In: Millner R, Rosenfeld E, Cobet U, eds., *Ultrasound Interactions in Biology & Medicine*. New York: Plenum Press, 1983, pp. 151–158.

Ebenbichler GR, Resch KL, Nicolakis P, Wiesinger GF, Uhl F, Ghanem A, Fialka V. Ultrasound treatment for treating the carpal tunnel syndrome: Randomised "sham" controlled trial. *British Medical Journal* 1998;316:731–735.

Einhor TA. Enhancement of fracture-healing. *Journal of Bone and Joint Surgery. American Volume* 1995;77:940–956.

El Hag M, Coghlan K, Christmas P, Harvey W, Harris M. The anti-inflammatory effect of dexamethazone and therapeutic ultrasound in oral surgery. *British Journal of Oral and Maxillofacial Surgery* 1985;23:17–23.

Enwemeka CS. The effects of therapeutic ultrasound on tendon healing. A biomechanical study. *American Journal of Physical Medicine and Rehabilitation* 1989;68:283–287.

Enwemeka CS, Rodriguez O, Mendosa S. The biomechanical effects of low-intensity ultrasound on healing tendons. *Ultrasound in Medicine and Biology* 1990a;16:801–807.

Enwemeka CS, Rodriguez O, Mendosa S. Inflammation, cellularity, and fibrillogenesis in regenerating tendon: Implications for tendon rehabilitation. *Physical Therapy* 1990b;69:816–825.

Falconer J, Hayes K, Chang R. Therapeutic ultrasound in the treatment of musculoskeletal conditions. *Arthritis Care and Research* 1990;3:85–91.

Fulcher SM, Kiefhaber TR, Stern PJ. Upper-extremity tendinitis and overuse syndromes in the athlete. *Clinics in Sports Medicine* 1998;17:433–448.

Fyfe MC, Chahl LA. The effect of single or repeated applications of "therapeutic" ultrasound on plasma extravasation during silver nitrate induced inflammation of the rat hindpaw ankle joint in vivo. *Ultrasound in Medicine and Biology* 1985;11:273–283.

Gam AN, Johannsen F. Ultrasound therapy in musculoskeletal disorders: A meta-analysis. *Pain* 1995;63:85–91.

Gan BS, Huys S, Sherebrin MH, Scilley CG. The effects of ultrasound treatment on flexor tendon healing in the chicken limb. *Journal of Hand Surgery (Edinburgh, Scotland)* 1995;20:809–814.

Gann N. Ultrasound: Current concepts. *Clinical Management* 1991;11:64–69.

Gelberman RH, Rydevik BL, Pess GM, Szabo RM, Lundeberg G. Carpal tunnel syndrome. *Orthopedic Clinics of North America* 1998;19:115–124.

Gerritsen AA, de Krom MC, Struijs MA, Scholten RJ, de Vet HC, Bouter LM. Conservative treatment options for carpal tunnel syndrome: A systematic review of randomized controlled trials. *Journal of Neurology* 2002;294:772–780.

Gersten JW. Effect of ultrasound on tendon extensibility. *American Journal of Physical Medicine and Rehabilitation* 1955;34:362–369.

Gonzales MH, Bylak J. Steroid injection and splinting in treatment of carpal tunnel syndrome. *Orthopedics* 2001;24:479–481.

Gospodarowicz D, Brown KD, Birdwell CR, Zetter BR. Control of proliferation of human vascular endothelial cells. *Journal of Cell Biology* 1978;77:774–788.

Grant A, Sleep J, McIntosh J, Ashurst H. Ultrasound and pulsed electromagnetic energy treatment for perineal trauma: A randomized placebo-controlled trial. *British Journal of Obstetrics and Gynaecology* 1989;96:434–439.

Green S, Buchbinder R, Glazier R, Forbes A. Systematic review of randomised controlled trials of interventions for painful shoulder: Selection criteria, outcome assessment and efficacy. *British Medical Journal* 1998;316:354–360.

Hadjiargyrou M, McLeod K, Ryaby JP, Rubin C. Enhancement of fracture healing by low intensity ultrasound. *Clinical Orthopaedics and Related Research* 1998;355:s216–s229.

Haker E, Lundeberg T. Pulsed ultrasound treatment in lateral epicondylitis. *Scandinavian Journal of Rehabilitation Medicine* 1991;23:115–118.

Hanahan D, Folkman J. Patterns and emerging mechanisms of the angiogenic switch during tumorigenesis. *Cell* 1996;86:353–364.

Harvey W, Dyson M, Pond JB, Grahame R. The stimulation of protein synthesis in human fibroblasts by therapeutic ultrasound. *Rheumatology and Rehabilitation* 1975;14:237.

Hashimoto I, Nakanishi H, Shono Y, Toda M, Tsuda H, Arase S. Angiostatic effects of corticosteroid on wound healing of the rabbit ear. *Journal of Medical Investigation* 2002;49:61–66.

Hashish I, Hai HK, Harvey W, Feinmann C, Harris M. Reduction of postoperative pain and swelling by ultrasound treatment: A placebo effect. *Pain* 1988;33:303–311.

Hashish I, Harvey W, Harris M. Anti-inflammatory effects of ultrasound therapy: Evidence for a major placebo effect. *British Journal of Rheumatology* 1986;25:77–81.

Hasson S, Mundorf R, Barnes W, Williams J, Fujii M. Effect of pulsed ultrasound versus placebo on muscle soreness perception and muscular performance. *Scandinavian Journal of Rehabilitation Medicine* 1990;22:199–205.

Hekkenberg RT, Oosterbaan WA, van Beekum WT. Evaluation of ultrasound therapy devices. *Physiotherapy* 1986;72:390–395.

Hogan RDB, Burke KM, Franklin TD. The effect of ultrasound on the microvascular hemodynamics in skeletal muscle: Effects during ischemia. *Microvascular Research* 1982a;23:370.

Hogan RDB, Franklin TD, Fry FJ. The effect of ultrasound on microvascular hemodynamics in skeletal muscle: Effect on arterioles. *Ultrasound in Medicine and Biology* 1982b;8:45–55.

Holdsworth L, Anderson D. Effectiveness of ultrasound used with a hydrocortisone coupling medium or epicondylitis clasp to treat lateral epicondylitis: Pilot study. *Physiotherapy* 1993;79:19–25.

Holmes MAM, Rudland JR. Clinical trials of ultrasound treatment in soft tissue injury: A review and critique. *Physiotherapy Theory and Practice* 1991;7:163–175.

Hong C, Liu H, Yu J. Ultrasound thermotherapy: Its effect on the recovery of nerve conduction in experimental compression neuropathy. *Archives of Physical Medicine and Rehabilitation* 1986;67:618.

Hoogland R. *Ultrasound Therapy.* Delft, the Netherlands: Enraf-Nonius, 1989.

Huang M-H, Lin Y-S, Lee C-L, Yang R-C. Use of ultrasound to increase effectiveness of isokinetic exercise for knee osteoarthritis. *Archives of Physical Medicine and Rehabilitation* 2005a;86:1545–1551.

Huang M-H, Yang R-C, Lee C-L, Chen T-W, Wang M-C. Preliminary results of integrated therapy for patients with knee osteoarthritis. *Arthritis Care and Research* 2005b;53:812–820.

Hudlicka O, Tyler KR. *The Growth of the Vascular System.* London, U.K.: Academic Press, 1986.

Hughes, AFW, Dann L. Vascular regeneration in experimental wounds and burns. British Journal of Experimental Pathology 1941;22(1):9–14.

Hüter T. Messung der Ultraschallabsorption in tierischen Geweben und ihre Abhängigkeit von der Frequenz. *Naturwissenschaften* 1948;53:285–287.

Huys S, Ban BS, Sherebrin MH, Scilley CG. Comparison of effects of early and late ultrasound treatment on tendon healing in the chicken limb. *Journal of Hand Therapy* 1993;6:58–59.

Isabell WK, Durrant E, Myrer W, Anderson S. The effects of ice massage, ice massage with exercise, and exercise on the prevention and treatment of delayed-onset muscle soreness. *Journal of Athletic Training* 1992;27:208–217.

Johns LD. Nonthermal effects of therapeutic ultrasound: The frequency resonance hypothesis. *Journal of Athletic Training* 2002;37:293–299.

Johnson JM, Brengelmann GL, Rowell LB. Interactions between local and reflex influences on human forearm skin blood flow. *Journal of Applied Physiology* 1976;41:826–831.

Kannus P, Jozsa L, Renström P, Järvinen M, Kvist M, Lehto M, Oja P, Vuori I. The effects of training, immobilisation and remobilisation on musculoskeletal tissue. *Scandinavian Journal of Rehabilitation Medicine* 1992;2:100–118.

Knighton DR, Ciresi KF, Fiegel VD, Austen LL, Butler CL. Classification and treatment of chronic nonhealing wounds. Successful treatment with autologous platelet-derived wound healing factors (PDWHF). *Annals of Surgery* 1986;204:323–330.

Kozanolu E, Basaran S, Guzel R, Guler-Uysal F. Short term efficacy of ibuprofen phonophoresis versus continuous ultrasound therapy in knee osteoarthritis. *Swiss Medical Weekly* 2003;133:333–338.

Leung MC, Ng GY, Yip KK. Effect of ultrasound on acute inflammation of transected medial collateral ligaments. *Archives of Physical Medicine and Rehabilitation* 2004;85:963–966.

Lindsay DM, Dearness J, Ricardson C et al. A survey of electromodality usage in private physiotherapy practices. *Australian Journal of Physiotherapy* 1990;36:249–256.

Low J, Reed A. *Electrotherapy Explained: Principles and Practice.* Oxford, U.K.: Butterworth-Heinemann, 2000.

Lundeberg T, Abrahamsson P, Haker E. A comparative study of continuous ultrasound, placebo ultrasound and rest in epicondylalgia. *Scandinavian Journal of Rehabilitation Medicine* 1988;20:99–101.

Mardiman S, Wessel J, Fisher B. The effect of ultrasound on the mechanical pain threshold of healthy subjects. *Physiotherapy* 1995;81:718–723.

Marshall ST, Browner BD. Emergency care of musculoskeletal injuries. In: Townsend Jr CM, ed., *Sabiston Textbook of Surgery: The Biological Basis of Modern Surgical Practice.* Amsterdam, the Netherlands: Elsevier, 2012, pp. 480–520.

Maxwell L. Therapeutic ultrasound: Its effects on the cellular & molecular mechanisms of inflammation and repair. *Physiotherapy* 1992;78:421–426.

McCance KL, Huether SE. *Pathophysiology: The Biological Basis for Disease in Adults and in Children.* St. Louis, MO: Mosby, 1998.

McDiarmid T, Ziskin MC, Michlovitz SL. Therapeutic ultrasound. In: Michlovitz SL, ed., *Thermal Agents in Rehabilitation.* Philadelphia, PA: F.A. Davis, 1996, pp. 168–212.

Mickey CA, Bernier JN, Perrin DH. Ice and ice with nonthermal ultrasound effects on delayed-onset muscle soreness. *Journal of Athletic Training* 1996;31:S-19.

Mortimer AJ, Dyson M. The effect of therapeutic ultrasound on calcium uptake in fibroblasts. *Ultrasound in Medicine and Biology* 1988;14:499–506.

Naslund J. Modes of sensory stimulation: Clinical trials and physiological aspects. *Physiotherapy* 2001;87:413–423.

Nussbaum E. The influence of ultrasound on healing tissues. *Journal of Hand Therapy* 1998;11:140–147.

Nussbaum E, Biemann I, Mustard B. Comparison of ultrasound/ultraviolet-C and laser for treatment of pressure ulcers with spinal cord injury. *Physical Therapy* 1994;74:812–823.

Nussbaum EL. Ultrasound: To heat or not to heat—That is the question. *Physical Therapy Reviews* 1997;2:59–72.

Nykanen M. Pulsed ultrasound treatment of the painful shoulder: A randomized, double-blind, placebo-controlled study. *Scandinavian Journal of Rehabilitation Medicine* 1995;27:105–108.

Ozgonenel L, Aytekin E, Dumusoglu G. A double-blind trial of clinical effects of therapeutic ultrasound in knee osteoarthritis. *Ultrasound in Medicine & Biology* 2009;35(1):44–49.

Partridge CJ. Evaluation of the efficacy of ultrasound. *Physiotherapy* 1987;73:166–168.

Patrick MK. Ultrasound in physiotherapy. *Ultrasonics* 1966;4:10–14.

Paul J. Cell and tissue culture. American Journal of the Medical Sciences 1960;239(2):258.

Paul WC, Imig CJ. Temperature and blood flow studies after ultrasonic irradiation. *American Journal of Physical Medicine and Rehabilitation* 1955;34:370–375.

Pienimäki TT, Tarvainen TK, Siira PT, Vanharanta H. Progressive strengthening and stretching exercises and ultrasound for chronic lateral epicondylitis. *Physiotherapy* 1996;82: 522–530.

Poltawski L, Watson T. Relative transmissivity of ultrasound coupling agents commonly used by therapists in the UK. *Ultrasound in Medicine and Biology* 2007;33:120–128.

Polverini PJ, Leibovich JS. Induction of neovascularisation in vivo and endothelial proliferation in vitro by tumor associated marcrophages. *Laboratory Investigation* 1984;51:635–642.

Pope GD, Mockett SP, Wright JP. A survey of electrotherapeutic modalities: Ownership and use in the National Health Service in England. *Physiotherapy* 1995;81:82–91.

Portney L, Watkins M. *Foundations of Clinical Research.* East Norwalk, CT: Appleton & Lange, 1993.

Portwood M, Lieberman J, Taylor R. Ultrasound treatment of reflex sympathetic dystrophy. *Archives of Physical Medicine and Rehabilitation* 1987;68:116–118.

Pye SD, Milford C. The performance of ultrasound therapy machines in Lothian Region, 1992. *Ultrasound in Medicine and Biology* 1994;20:347–359.

Ramirez A, Schwane JA, McFarland C, Starcher BC. The effect of ultrasound on collagen synthesis and fibroblast proliferation in vitro. *Medicine and Science in Sports and Exercise* 1997;29:326–332.

Reed B, Ashikaga T. The effects of heating with ultrasound on knee joint displacement. *Journal of Orthopaedic and Sports Physical Therapy* 1997;26:131–137.

Repacholi MH, Benwell DA. Using surveys of ultrasound therapy devices to draft performance standards. *Health Physics* 1979;36:679–686.

Robebroeck ME, Dekker J, Oostendorp RAB. The use of therapeutic ultrasound by physical therapists in Dutch primary health care. *Physical Therapy* 1998;78:470–478.

Roberts M, Rutherford JH, Harris D. The effect of ultrasound on flexor tendon repairs in the rabbit. *Hand* 1982;14:17–20.

Robertson VJ. Dosage and treatment response in randomized clinical trials of therapeutic ultrasound. *Physical Therapy in Sport* 2002;3:124–133.

Robertson VJ, Baker KG. A review of therapeutic ultrasound: Effectiveness studies. *Physical Therapy* 2001;81:1339–1350.

Robertson VJ, Spurritt D. Electrophysical agents: Implications of their availability and use in undergraduate clinical placements. *Physiotherapy* 1998;84:335–344.

Robinson AJ, Snyder-Mackler L. Clinical application of electrotherapeutic modalities. *Physical Therapy* 1988;68: 1235–1238.

Rowell LB. *Human Circulation: Regulation during Physical Stress.* New York: Oxford University Press, 1986.

Rubin MJ, Etchison MR, Condra KA, Franklin TD, Snoddy AM. Acute effects of ultrasound on skeletal muscle oxygen tension, blood flow, and capillary density. *Ultrasound in Medicine and Biology* 1990;16:271–277.

Rutjes AWS, Nüesch E, Sterchi R, Jüni P. Therapeutic ultrasound for osteoarthritis of the knee or hip. *Cochrane Database of Systematic Reviews* 2010;1:CD003132.

Saunders L. Laser versus ultrasound in the treatment of supraspinatus tendinosis. *Physiotherapy* 2003;89:365–373.

Schabrun S, Chipchase L, Rickard H. Are therapeutic ultrasound units a potential vector for nosocomial infection? *Physiotherapy Research International* 2006;11:61–71.

Scheiman JM. NSAIDs, gastrointestinal injury, and cytoprotection. *Gastroenterology Clinics of North America* 1996;25: 279–298.

Schs F. Mechanical transduction in biological systems. *CRC Critical Reviews in Biomedical Engineering* 1988;16:141–169.

Sidky YA, Auerbach R. Lymphocyte-induced angiogenesis: A quantitative and sensitive assay of the graft-vs-host reaction. *Journal of Experimental Medicine* 1975;141:1084–1100.

Simovic D, Weinberg DH. Carpal tunnel syndrome. *Archives of Neurology* 2000;57:754–755.

Snow CJ. Ultrasound therapy units in Manitoba and Northwestern Ontario: Performance evaluation. *Physiotherapy Canada* 1982;34:185–189.

Speed CA. Therapeutic ultrasound in soft tissue lesions. *Rheumatology* 2001;40:1331–1336.

Stay JC, Ricard MD, Draper DO, Schulthies SS, Durrant E. Pulsed ultrasound fails to diminish delayed-onset muscle soreness symptoms. *Journal of Athletic Training* 1998;33: 341–346.

Stevenson J, Pang C, Lindsay W, Zuker R. Functional, mechanical, and biochemical assessment of ultrasound therapy on tendon healing in the chicken toe. *Plastic and Reconstructive Surgery* 1986;77:965–970.

Stewart HF, Harris GR, Herman BA, Robinson RA, Haran ME, McCall GR, Carless G, Rees D. Survey of the use and performance of ultrasonic therapy equipment in Pinellas County, Florida. *Physical Therapy* 1974;54:707–715.

Stratford P, Levy D, Gauldie S, Miseferi D, Levy K. The evaluation of phonophoresis and friction massage as treatments for extensor carpi radialis tendinitis: A randomized controlled trial. *Physiotherapy Canada* 1989;41:93–99.

Summer W, Patrick MK. *Ultrasonic Therapy: A Textbook for Physiotherapists.* Amsterdam, the Netherlands: Elsevier Publishing Company, 1964.

Tascioglu F, Kuzgun S, Armagan O, Ogutler G. Short-term effectiveness of ultrasound therapy in knee osteoarthritis. *Journal of International Medical Research* 2010;38:1233–1242.

ter Haar G. Basic physics of therapeutic ultrasound. *Physiotherapy* 1987;73:110–113.

ter Haar G. Therapeutic ultrasound. *European Journal of Ultrasound* 1999;9:3–9.

ter Haar G, Dyson M, Oakley EM. The use of ultrasound by physiotherapist in Britain, 1985. *Ultrasound in Medicine and Biology* 1987;13:659–663.

ter Riet G, Kessels A, Knipschild P. A randomized clinical trial of ultrasound in the treatment of pressure ulcers. *Physical Therapy* 1996;76:1301–1311.

Turner SM, Powell ES, Ng CS. The effect of ultrasound on the healing of repaired cockerel tendon: Is collagen cross-linkage a factor? *Journal of Hand Surgery (Edinburgh, Scotland)* 1989;14:428–433.

Uygur F, Sener G. Application of ultrasound in neuromas: Experience with seven below-knee stumps. *Physiotherapy* 1995;81:758–761.

van der Heijden GJMG, van der Windt DAWM, de Winter AF. Physiotherapy for patients with shoulder disorders: A systematic review of randomised controlled clinical trials. *British Medical Journal* 1997;315:25–30.

van der Windt DAWM, van der Heijden GJMG, van der Berg SG, ter Riet G, de Winter AF, Bouter LM. Ultrasound therapy for musculoskeletal disorders: A systematic review. *Pain* 1999;81:257–271.

van Manen MD, Nace J, Mont MA. Management of primary knee osteoarthritis and indications for total knee arthroplasty for general practitioners. *Journal of the American Osteopathic Association* 2012;112:709–715.

Vanable J. Integumentary potentials and wound healing. In: Borgens R, ed., *Electric Fields in Vertebrate Repair.* New York: Alan Liss, 1989, pp. 171–224.

Vander A, Sherman J, Luciano D. *Human Physiology: The Mechanism of Body Function.* Boston, MA: McGraw-Hill, 1998.

Vasseljen O. Low-level laser versus traditional physiotherapy in the treatment of tennis elbow. *Physiotherapy* 1992;78: 329–334.

Vecchio P, Cave M, King V, Adebajo AO, Smith M, Hazleman BL. A double-blind study of the effectiveness of low level laser treatment of rotator cuff tendinitis. *British Journal of Rheumatology* 1993;32:740–742.

Vos T, Flaxman AD, Naghavi M, Lozano R, Michaud C, Ezzati M, Shibuya K et al. Years lived with disability (YLDs) for 1160 sequelae of 289 diseases and injuries 1990–2010: A systematic analysis for the Global Burden of Disease Study 2010. *Lancet* 2012;380:2163–2196.

Wall PD. Introduction. In: Wall PD, Melzack R, eds., *Textbook of Pain*, 2nd edn. New York: Churchill Livingstone, 1989, pp. 1–18.

Ward AR, Robertson VJ. Comparison of heating of nonliving soft tissue produced by 45 kHz and 1 MHz frequency ultrasound machines. *Journal of Orthopaedic and Sports Physical Therapy* 1996;23:258–266.

Ward R, Hayes-Lundy C, Reddy R, Brockway C, Mills P, Saffle J. Evaluation of topical therapeutic ultrasound to improve response to physical therapy and lessen scar contracture after burn injury. *Journal of Burn Care Rehabilitation* 1994;15:74–77.

Watson T. Electrotherapy and tissue repair. *Sportex-Medicine* 2006a;29:7–13.

Watson T. Tissue repair: The current state of the art. *Sportex-Medicine* 2006b;28:8–12.

Watson T. Ultrasound in contemporary physiotherapy practice. *Ultrasonics* 2008;48:321–329.

Weber MD, Servedio FJ, Woodall WR. The effects of three modalities on delayed-onset muscle soreness. *Journal of Orthopaedic and Sports Physical Therapy* 1994;20:236–242.

Webster DF, Pond JB, Dyson M, Harvey W. The role of cavitation in the *in vitro* stimulation of protein synthesis in human fibroblasts by ultrasound. *Ultrasound in Medicine and Biology* 1978;4:343–351.

Wells PNT. *Biomedical Ultrasonics.* London, U.K.: Academic Press, 1977.

Williams AR. *Ultrasound: Biological Effects and Potential Hazards.* London, U.K.: Academic Press, 1983.

Williamson JB, George TK, Simpson DC, Hannah B, Bradbury E. Ultrasound in the treatment of ankle sprains. *Injury* 1986;17:176–178.

Wiltink A, Nijweide PJ, Oosterbaan WA, Hekkenberg RT, Helders PJM. Effect of therapeutic ultrasound on endochondral ossification. *Ultrasound in Medicine and Biology* 1995;21:121–127.

Wolff BB. Laboratory methods of pain measurement. In: Melzack R, ed., *Pain Measurement and Assessment.* New York: Raven Press, 1983, pp. 7–13.

World Health Organization. *The Global Burden of Disease: 2004 Update*, 2008.

Yaxley GA, Gwendolen AJ. Adverse tension in the neural system. A preliminary study of tennis elbow. *Australian Physiotherapy Journal* 1993;39:15–22.

Young S. Ultrasonic therapy. In: Kitchen S, Bazin S, eds., *Clayton's Electrotherapy*. Philadelphia, PA: W.B. Saunders, 1996, pp. 243–267.

Young S, Dyson M. Effect of therapeutic ultrasound on the healing of full thickness excised skin lesions. *Ultrasonics* 1990a;28:175–180.

Young SR, Dyson M. The effect of therapeutic ultrasound on angiogenesis. *Ultrasound in Medicine and Biology* 1990b;16:261–269.

Yurt RW. Role of the mast cell in trauma. In: Dineen P, Hildick-Smith G, eds., *The Surgical Wound*. Philadelphia, PA: Lea and Febiger, 1981, pp. 37–62.

11 Hyperthermia

11.1 RATIONALE OF HYPERTHERMIA

There is plenty of evidence showing that *in vivo* cancer cells are more sensitive to heat at temperatures between 42°C and 45°C than normal ones (Crile 1963, Cavaliere et al. 1967, Dickson and Shah 1972, Miller et al. 1977, Overgaard 1977). However, such difference in heat sensitivity disappears at temperatures above 45°C. Subsequently, heat begins to indiscriminately damage both normal and cancer cells. In contrast, heat at low-temperature hyperthermia (between 37°C and 41.5°C) enhances cell growth as well as the growth, proliferation, and metastases of tumors (Dickson and Ellis 1976, Dickson 1977). Therefore, a narrow temperature range of 42°C–45°C is usually set as the operation parameter for effective and safe hyperthermia while keeping surrounding normal tissues at physiologically acceptable levels. Temperatures above 42.5°C for 30–60 min have a cytotoxic effect both *in vitro* and *in vivo*. Because of the variations in tumor response to heating, disappearance of some tumors requires a temperature in the range of 47°C–50°C for 30 min (Dickson 1977, Magin and Johnson 1979). In the treatment of disease localized to one or more circumscribed regions, local hyperthermia of the involved volume of tissue is preferable to whole-body hyperthermia as the local temperatures can be raised to higher levels without the toxicity and hazards associated with the latter (Short and Tumer 1980). Penetration depth of the heating beam is an important consideration in hyperthermia systems. Another critical factor is the noninvasiveness of the technique. The potential of metastasis caused by delivering heat invasively might increase with disruption of blood vessels and stress applied to the tumor. Hyperthermia is also used in the treatment of arthritis, asthma, multiple sclerosis, and infectious diseases such as syphilis and gonorrhea (Short and Tumer 1980). The opportunity for visualization of the tumor volume in the operating room (e.g., sonography and x-ray) makes the use of a contact, sterilizable, and superficial applicator capable of various depths of penetration.

Most of the heat-producing methods can be classified into three major modalities: hot fluid perfusion (Cavaliere et al. 1967, Stehlin et al. 1975), immersion, or irrigation (Cockett et al. 1967, Kim et al. 1977); ohmic heating produced by electrical currents generated from radio frequency (RF) sources and by electrical waves generated from microwave sources (LeVeen et al. 1976, Hornback et al. 1977, Kim et al. 1977, Kato et al. 1985); and mechanical friction caused by an ultrasound wave vibrating the molecules. There are different kinds of interaction between electromagnetic (EM) fields and biological systems resulting in heat production. The friction associated with the rotation of polar atoms and molecules causes temperature rise in a time-varying EM field. Another

kind of interaction is oscillation of free electrons and ions with immobile atoms and molecules of the tissues. At microwave frequencies (300 MHz to 30 GHz), the rotation of water molecules dominates all interactions. Therefore, water containing tissues (e.g., skin and muscle) are usually good microwave absorbers (Cheung et al. 1981).

The thermal dose during a heating process is calculated as (Sapareto and Dewey 1984)

$$t_{43} = \sum_t R^{[43-T(t)]}\Delta t \qquad (11.1)$$

where

t_{43} is the equivalent time at 43°C
$T(t)$ is the average temperature during time Δt
$R = 0.5$ for temperatures above 43°C and 0.25 otherwise

For obtaining the desired accuracy, Δt is chosen to be smaller than 0.2% of the total heating time. The cumulative equivalent minutes (CEM) t_{43} of 120 min has been accepted widely as a prognostic indicator of therapeutic effect in different tissues. Changes in pH, nutrients, and blood flow may have only minimal influence on the determination of the thermal dose. The clinical use of the equivalent thermal dose depends on the validity of the linear Arrhenius relationship between the activation energy and the break temperature. Normal tissue has a predictable threshold of thermal dose, beyond which a tremendous increase in tissue damage will occur. However, there are great variations on the CEM threshold for tumors because of their different histologies. In a clinical trial involving 25 patients treated by microwave hyperthermia, 140 CEM was predictive of a 50% tumor response and 210 CEM was predictive of a 100% tumor response ($p = 0.003$). The degree of tumor necrosis is a function of the thermal dose (Vargas et al. 2004).

11.2 EFFECT OF THERMAL CONDUCTION AND BLOOD PERFUSION

However, the temperature distribution in the target volume during a hyperthermia treatment is not uniform as expected; undesirably high temperatures in some regions lead to thermal injury, whereas too low temperatures in some other regions have no therapeutic effect. Thermal conduction and blood perfusion are the two mechanisms for biological heat transfer and would smooth out the temperature rise. However, thermal conductivity of tumor has not yet been measured directly. Blood perfusion in the normal tissues is, somehow, governed by vasomotor reflexes, which would increase the blood perfusion and decrease the temperature if it exceeds 42°C for a few minutes (Guy et al. 1974). The temperature elevation measured in low-perfusion case (the blood supply to the kidney being cut off) was four to six times higher than in a kidney with normal perfusion, but the maximum temperatures were still at the edges of the kidney (Hynynen et al. 1987a).

Blood flow in the tissues has a significant effect on the thermal energy distribution, and there are great variations in the blood flow in tumors. The rapidly growing periphery is usually well perfused, whereas the central part could be necrotic. But the thickness of the well-perfused periphery remains almost unchanged with the growth of tumor. Subsequently, the blood flow per unit volume of the tumor decreases gradually (Tannock 1970, Folkman and Contran 1976, Endrich et al. 1979). The blood flow ratio between surrounding normal tissues and the malignant tumor is in the range of 3:1–28:1, and such difference leads to the therapeutically useful temperature rise in the tumor more than that in the healthy tissue with the same heat input. The newly formed blood vessels inside the malignant tumor are usually small and consist of irregular channels lined by endothelium or naked tumor cells (Willis 1967), which cannot expand in response to heat (Song 1978) or even can induce vasoconstriction with the use of vasodilating drugs (Kruuv et al. 1967). Decreased blood flow in tumors can reduce the diffusion and consequent heat-sink phenomenon. Because more heat can be removed from a region with higher blood flow, depositing energy evenly throughout the tumor would heat the central region more and the tumor margin less, due to conduction and blood perfusion. So, a temperature difference of as much as 10°C between normal and malignant tissues can be achieved. Decreased blood flow in the tumor is the most important factor during hyperthermia and can restrict the temperature rise exactly to the malignant tumor. This allows high temperatures over 45°C to the center of the tumors without significant damage and lower temperatures at the proliferating margins and in the surrounding tissue. On the other hand, a uniform temperature distribution is difficult to achieve because of the lack of homogeneity inside the whole tumor (Hynynen et al. 1981).

11.3 ULTRASOUND HYPERTHERMIA TECHNOLOGY

The key event in the application of ultrasound for the treatment of cancer was first mentioned in 1933 but with no specific effect on Ehrlich's carcinoma (Szent-Györgyi 1933). Afterward, enthusiasm for this approach has bloomed after the success in human skin metastases in 1944 (Horvath 1944). However, a conference in Erlangen in 1944 concluded that its use should not be continued because of the potential adverse clinical results (Kremkau 1979). Because ultrasound has been used to heat nonmalignant tissue for physiotherapeutic purposes (Lehmann 1965), the interest revived for the possible use of hyperthermia in cancer therapy. Noninvasive, deep-heating methods to induce hyperthermia can be generated

using planar ultrasound, focused ultrasound (Beard et al. 1982, Ocheltree et al. 1984, Ibbinni and Cain 1990), scanned focused ultrasound (Lele 1983, Hynynen et al. 1987a), and ultrasound arrays (Fessenden et al. 1984). Positive tumor responses up to 12 cm in depth and cure of sarcomas and carcinomas in mice were found after ultrasonically induced hyperthermia with scanned and focused ultrasound (Marmor et al. 1977, Lele 1983, 1984, 1986, Sapareto and Dewey 1984). Water can serve as both a coupling medium to transfer the ultrasound energy from the transducers to the tissue target and a coolant (the temperature is set as low as patient comfort will allow, 15°C–20°C) for the skin surface and the ultrasound transducer.

Ultrasound has several advantages in comparison to other hyperthermia modalities. Tumors absorb acoustic energy better than normal tissue compared to EM energy. The acoustic impedance of the body fluids is similar to that of the soft tissue, but with much lower absorption. Thus, there is no possibility of excessive heating or significant reflections at the interfaces. No substantial hazard in ultrasound hyperthermy for tumors or tissues was found close to organs containing or surrounded by fluids, such as the cerebral ventricles, spinal cord, eye, heart, aorta, stomach, gall bladder, pelvis of the kidney, urinary bladder, and fetal sac. Ultrasound hyperthermia is found to have little effect on tumor metastasis because the integrity of blood vessels is not compromised and the tumor is not disrupted mechanically. Since only 1%–2% of the beam is reflected at the interface of fat and muscle, the reflection coefficient remains almost constant if the incident angle is less than ~40°. However, complete reflection occurs at ~63.5°. Unlike RF or microwaves, ultrasound does not perturb the temperature probe significantly. Diffraction spreading of the ultrasound beam is insignificant, and geometric focusing is apparent over appreciable distances, unlike the spreading and stray radiations for microwave fields. Its depth of penetration and relatively easy generation under favorable circumstances for localized energy accumulation allow deeper heating than microwaves. A variety of piezoelectric ceramic materials are available for producing ultrasound beams of 10 cm or more in diameter. In addition, relatively uniform high-intensity fields are already achievable. No special RF-shielded room is required. Excellent focusing is possible because the wavelengths are much smaller than the source size. Both imaging guidance and thermometry are available for ultrasound hyperthermia.

However, the presence of bone or air cavities (e.g., in lung and bowel) is a big concern for the acoustic wave propagation. At the interface of soft tissue and bone or gas, ~33%–39% and 100% of the incident energy is reflected backward at normal incidence, respectively. Therefore, the gas interfaces can reflect the beams into unexpected locations and sometimes even cause a focal spot resulting in patient discomfort. Reflection at the interface of bone and the subsequent temperature rise are significant, so ultrasound is not suitable for lung or brain cancer and also not recommended for deep heating of extremities (Cheung and Neyzari 1984). Hot spots in front of bones were also found when using a weakly focused, low-frequency single ultrasound beam in hyperthermia (Marmor et al. 1979,

Hynynen and DeYoung 1988). They can be reduced to an acceptable level if the intensities at the bone surface are about 10%–50% of that in the target volume, which depends on the tumor perfusion rate. It is to be noted that the intended higher temperature rise in the bone is for the treatment of osteogenic sarcoma. At lower frequencies, the beams can propagate long distances in tissues. Plastic bags filled with water need to be placed on the skin at the wave exit site to allow the beam to propagate out of the body. In contrast, because the air cavities in the aero-digestive tracts (e.g., oral and nasal cavities, respiratory and gastrointestinal tract) are convex toward soft tissues, refocusing of the ultrasound energy in the soft tissues is negligible. Another concern is the difference in the speed of sound in tissues, which can change the transmission angle and thus alter the direction of the beam propagation. The required acoustic window size at the skin site is a function of depth, frequency, and target size. The optimal frequency for deep-heating seems to be between 0.5 and 1 MHz, and in practice the highest possible frequency should be selected to minimize the hot spots at bone surfaces behind the tumor.

Insonation with plane wave ultrasound results in spatially nonuniform hyperthermia, which is characterized by the existence of a small, almost punctate, region of maximum temperature rise, whose depth is dependent on the acoustic and heat transfer properties of the medium and cannot be altered by spatial manipulation of the planar transducer. Furthermore, the resultant nonuniformity of temperature distributions within the tumor renders impossible any precise correlation of the temperature and duration of hyperthermia with consequent effects on tumors. The distance Z_m from the piston to the last maximum is given by $Z_m \approx 0.75a^2/\lambda$ with $a \geq \lambda$, where a is the radius of the transducer, and λ is the ultrasonic wavelength in the medium (Zemanek 1971). The operating ultrasound frequency is chosen to have much more absorbed power at depth than near the skin surface. Frequency modulation can reduce the possibility of stationary hot spots induced by standing waves or phase reinforcement. Thus, a single nonfocused transducer cannot have the desired power deposition at depth and is not optimal even for therapy of superficial lesions (Lele and Parker 1982).

11.3.1 OVERLAPPING TRANSDUCERS

One of the solutions to produce preferential high heating in a deep tumor is the superimposition of two or more planar beams at the target through different surface entries in a converging geometry to form a larger, more spread-out focus, even though each single beam is attenuated significantly along its propagation path (Fessenden et al. 1984). Consequently, the effects on tissues outside the treatment volume can be minimized. The total power absorbed is approximately n times that from a single element, where n is the number of elements. The phase should be set carefully while superimposing the acoustic energy. The longest wavelength (or the lowest driving frequency) that can be utilized is determined not only by the lateral (or radial) dimension of the tumor but also by its axial length or thickness, and should be approximately one-fifth of that dimension. However, because of some overlap of the beams in the pre-focal region during multiple sonications, hot spots can be produced resulting in tissue damage. A large number of treatments (30%) were limited by pain, although 24 of the 57 clinical treatments achieved therapeutic temperatures (Fessenden et al. 1984). Furthermore, when multiple tilted beams are used, the large incident angles can cause significant distortion in the heating field (Figure 11.1).

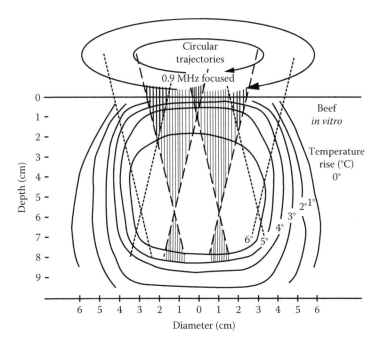

FIGURE 11.1 Temperature distributions produced by a focused ultrasound transducer in circular trajectories. (From Lele, P.P. and Parker, K.J., *Br. J. Cancer*, 45, 108, 1982.)

11.3.2 Focused Transducer

In the first clinical ultrasound hyperthermia using nonfocused beams, pain was a common problem (Marmor et al. 1979, Corry et al. 1982, Fessenden et al. 1984). To cause a selectively effective heating in a deep-lying tumor without undesirable temperature rise in the overlying skin or mucosa, a better solution is to use focused ultrasound and to distribute the energy within the tumor around its periphery (Lele 1979, 1980) (Figure 11.2). Focused ultrasound is a noninvasive, safe, and may be the best modality to produce well-controlled thermal accumulation localized to deep-seated tumors. It combines superior penetration depth with good beaming and focusing ability and, therefore, is an excellent tool for the treatment of smaller lesions that can be reached through intervening soft tissues in the acoustic path. Focusing of the energy also enables control of the depth and permits generation of localized hyperthermia even in irregularly shaped tumors without subjecting any significant volumes of surrounding normal tissues to unnecessary heat stress, which is not possible with plane wave sources and is specially important if the tumor is located close to a bone or in the lungs or intestines. Furthermore, the angle of incidence and the amplitude of excitation of the transducer can be controlled at every point in its trajectory, and thus the safety of critical target areas and surrounding tissues can be ensured. The accuracy of the temperature reached in the predetermined volume can be only ±0.1°C. At the selected level, hyperthermia can be sustained in the tumor for a long duration (i.e., 30–60 min) with no influence to the overlying skin or other adjacent or subjacent tissues. The ultrasonic focus is moved around the tumor periphery in order to achieve uniform temperature distributions. In addition, pain was never a problem when a sharply focused and higher-frequency ultrasound beam was mechanically scanned around the tumor (Lele 1983).

A significant advantage of focused ultrasound and scanned fields over the other hyperthermia techniques is that the shape and size of the treated volume can be accurately controlled under computer controlled scanning. To avoid higher temperatures both in front of and behind the focal area when scanning a single beam oriented normal to the surface, which depends on the *f*-number of the transducer, the frequency, the diameter of the scanning path, and the focal depth (Moros et al. 1990), the transducer can be tilted and rotated in such a manner that the beam is not overlapping with the scan axis (Lele 1983). Using available steered and focused ultrasound techniques, tumors (whether situated superficially or at depth) could be heated to a uniform and controllable temperature without undesirable temperature elevation in surrounding normal tissues. The small size of the focal region enables heating of tumors even when located near ultrasound-reflective targets (Lele and Parker 1982) (Figure 11.3).

11.3.3 Phased-Array Transducer

In order to heat larger tumors, a multielement transducer with independent power inputs allows variable power output over the heated area to compensate for variations in the cooling by blood flow and thermal conduction as well as to adapt to the geometry of the tumors (Underwood et al. 1987). In order to compensate for the conduction effects at

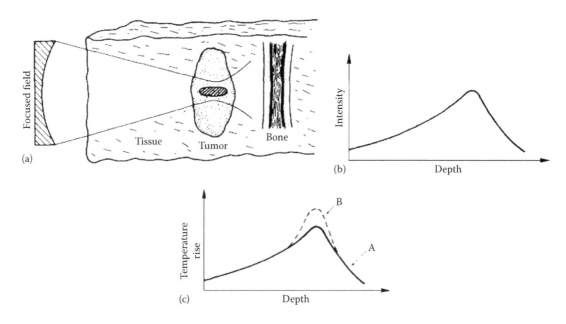

FIGURE 11.2 (a) Schematic diagram of ultrasound hyperthermia in the treatment of deep-seated tumor, and the corresponding (b) intensity and (c) temperature distribution patterns in a focused ultrasound field. In (A) the acoustic and thermal properties of the tumor are the same as those of the tissues. In (B) The acoustic absorption coefficient of the tumor is higher than that of tissues or/and the heat diffusivity of the tumor is lower than that of tissues. (With kind permission from Springer Science + Business Media: *Radiat. Environ. Biophys.*, Induction of deep, local hyperthermia by ultrasound and electromagnetic fields: Problems and choices, 17, 1980, 205–217, Lele P.P.)

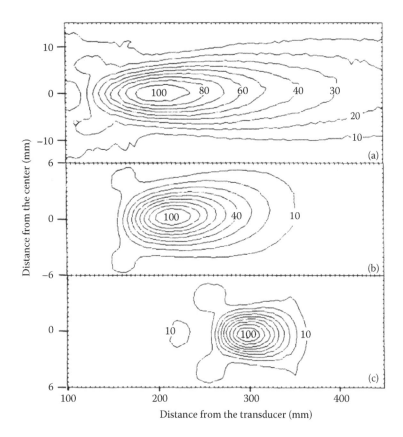

FIGURE 11.3 Acoustic pressure amplitude distribution measured in the axial plane of three different focused transducers in water (a) $f = 0.5$ MHz, $D = 70$ mm, $R = 350$ mm, (b) $f = 1.0$ MHz, $D = 70$ mm, $R = 250$ mm, and (c) $f = 1.0$ MHz, $D = 130$ mm, $R = 340$ mm. (With kind permission from Springer Science + Business Media: *Methods. Ext. Hypertherm. Heating*, Biophysics and technology of ultrasound hyperthermia, 1990, 61–115, Hynynen, K., in: Gautherie, M., ed., Springer-Verlag Berlin, Heidelberg, Germany.)

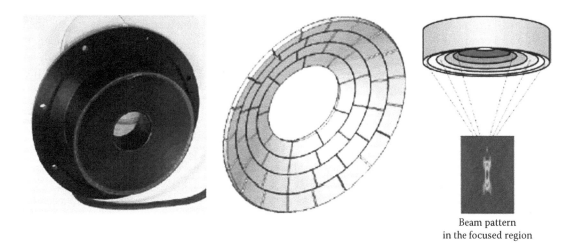

Beam pattern
in the focused region

FIGURE 11.4 Example of a phased-array concave transducer with 56 channels and an annular array with focused beam pattern.

the outer edge of the heated field, more power can be applied to the outer ring of concentric planar transducer for obtaining uniform temperature distributions (Munro et al. 1982). The disadvantages of electrical focusing by the phased-array design are the limited shifting capabilities, complexity in the electronics and control, and high cost in manufacture and maintenance (Figure 11.4).

11.3.4 INTERSTITIAL TRANSDUCER

However, hyperthermia effects have not been illustrated in large (several centimeters in diameter) (Perez et al. 1989) or deep tumors or those located behind an air cavity or bone because of the low heating efficiency (Corry et al. 1988, Kapp et al. 1988, Shimm et al. 1988a). One of the solutions is to deliver energy directly to a target via implanted small sources.

There are three types of interstitial hyperthermia techniques in use: RF current systems, microwave antennae, and hot sources (Strohbehn 1987, Gautherie 1990). All of them suffer from the lack of control over the three-dimensional power deposition or poor energy penetration. The interstitial (or intraoperative) ultrasound applicator has easy assemblies for rapid insertion of different elements, an expandable latex membrane filled with degassed water to conform to the tumor surface and to achieve good acoustic coupling, a completely immersible housing with watertight RF connection, and inlet/outlet ports for circulating water adjacent to the RF cable. The fragile ceramic cylindrical elements should be handled with care to avoid breakage. Flexible and bare-ended microthermocouples are placed superficially or implanted into the tissue. Most of them have a right-angle bend and are parallel to the direction of ultrasound propagation to minimize viscous heating. Interstitial applicators have several features compared to the extracorporeal design. First, the ultrasound field is emitted equally throughout the applicator length, with maxima at both ends of the cylinder, which is more desirable than a uniform power over the whole length because it compensates for the energy loss by conduction at the ends of the implant. The catheter cooling would limit excessive temperatures next to the applicators and produce a wider and more uniform temperature distribution. The cool tip and the nonuniform energy deposition pattern along the antenna length are inherent problems of the microwave probes (Satoh and Stauffer 1988). Second, the ultrasound field is independent of the insertion depth, which affects some microwave fields. Third, large catheter spacing (up to 20 mm) could be used, whereas a spacing of 10–12 mm is required for the hot source techniques (Stea et al. 1990). Fourth, each applicator can be made of several elements, which are independently driven, thus allowing three-dimensional control over the whole tumor volume. Tumors with higher perfusion rates could also be treated by reducing the distance between the catheters. Finally, the treatment system is simple, inexpensive, portable, and easy to use. Altogether, interstitial ultrasound sources are potentially the most promising devices of generating therapeutic temperatures through standard interstitial radiation therapy catheters (Hynynen 1992). In interstitial ultrasound catheter design, there is not much difference between the frequencies of 6 and 10 MHz (Figure 11.5). However, the frequency of 20 MHz had a significantly lower penetration. Although a lower frequency of 4 MHz could have some benefit, a larger outer diameter makes it difficult to circulate cooling fluid evenly and a smaller inner diameter would make it almost impossible to pass wires into multiple element arrays.

The nonuniformity in the produced temperature distribution is due to thermal conduction away from the acoustical focal region, variation of acoustic absorption coefficients of different tissues, acoustic impedance mismatches at tissue interface, differences in local blood flow, and the response of blood circulation to the heating (Marmor et al. 1979). The temperature elevation due to multiple sonications in the near field is strongly dependent on the perfusion rate, as illustrated in Figure 11.6. The steady state of temperature

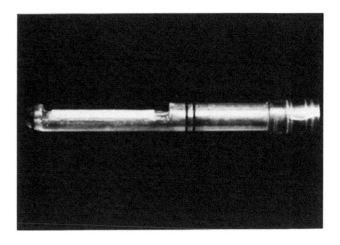

FIGURE 11.5 Photograph of the eight-element transrectal ultrasound array. (Reprinted from *Int. J. Radiat. Oncol., Biol., Phys.*, 26, Fosmire, H., Hynynen, K., Drach, G.W., Stea, B., Swift, P., and Cassady, J.R., Feasibility and toxicity of transrectal ultrasound hyperthermia in the treatment of locally advanced adenocarcinoma of the prostate, 253–259, Copyright 1993, with permission from Elsevier.)

change occurs at about 200 s for the rate of 10 kg/m³/s, where the corresponding value is 1500 s for the low perfusion rate. Subsequently, the thermal dose decreases with the perfusion rate (24.4 min for rate of 0.5 kg/m³/s and 0.15 min for 10 kg/m³/s).

The temperature induced by ultrasound hyperthermia in a dog's thigh rose with time until reaching the steady state. The profile at 3 cm depth had higher temperatures than that at 1 cm, and the experimental temperature profile could be simulated using the established model (Figure 11.7). In tissues with uniform perfusion (i.e., thigh), the temperature distributions could be relatively uniform in the sonicated region. The maximum temperature could be reached at depths up to 10 cm (limited by the size of the dogs used). More weakly focused transducers allow the use of simple scanning mechanics and control algorithms, but more sharply focused transducers can achieve better control over the temperature in the target with greater depth, large variations in blood perfusion in the tumor, and a well-perfused tissue layer lying under a poorly perfused one (e.g., kidney under the muscle) (Hynynen et al. 1987a).

Although uniform temperature distributions can be obtained inside the target volume with a steep fall-off outside the tumor (Parker and Lele 1980, Lele and Parker 1982, Lele 1983, Dickinson 1984, Roemer et al. 1984, Hynynen et al. 1986, 1987a), the important operating parameters, such as scanning pattern and speed, transducer geometry, ultrasound parameters, and blood perfusion pattern and flow rate, have significant effects on the temperature fluctuations and the consequent thermal dose delivered to the tumor (Hynynen et al. 1986). The magnitude of the temperature fluctuations is linearly related to the blood perfusion and scan time, while the average temperature remains constant for all scans at a given blood flow. Significant temperature fluctuations are not present even either in the high blood flow cases for fast

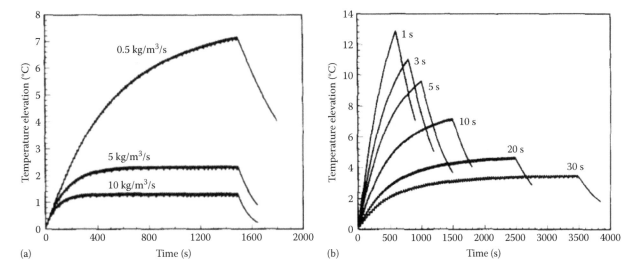

FIGURE 11.6 Temperature versus time profiles for (a) 0.5, 5, and 10 kg/m³/s perfusion rates with pulse delay of 10 s and (b) different delays between pulses with perfusion of 0.5 kg/m³/s. $f = 1$ MHz, $R = 10$ cm, $D = 10$ cm, $P = 12.4$ W, $T = 60$ min, pulse = 5 s (100 pulses), depth = 2 cm. (Reprinted from *Ultrasound Med. Biol.*, 19, Damianou, C. and Kynynen, K., Focal spacing and near-field heating during pulsed high temperature ultrasound therapy, 777–787, Copyright 1993, with permission from Elsevier.)

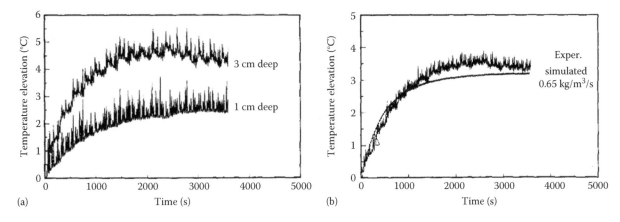

FIGURE 11.7 (a) Temporal dependent temperature elevation at two different depths for a transducer of $f = 1$ MHz, $R = 10$ cm, $D = 10$ cm, pulse = 7 s (132 pulses), delay = 20 s measured in the experiment. (b) Comparison of experimental (18 W) and simulated (11.1 W) temperature elevations at the depth of 2 cm with a 20-s delay. (Reprinted from *Ultrasound Med. Biol.*, 19, Damianou, C. and Kynynen, K., Focal spacing and near-field heating during pulsed high temperature ultrasound therapy, 777–787, Copyright 1993, with permission from Elsevier.)

scanning speeds or in the low blood flow cases even for slow scanning speeds. The largest temperature fluctuations, which is a consequence of the heating and cooling during scanned ultrasound hyperthermia, are always located in the focal plane, where the highest acoustic intensity is not a matter of transducer geometry and driving frequency with dominance of the local heat transfer, and independent of the focal depth. To reduce the fluctuations due to blood perfusion inhomogeneities, the power output should be controlled spatially according to the measured temperatures. Uniform temperature distributions are achieved as long as the interval between the sonicated locations in the scan is equal to or less than the ultrasound beam diameter. During clinical hyperthermia, only the average temperatures are usually measured to calculate the thermal dose. However, in presence of fluctuations,

the calculated thermal dose may be significantly different (Moros et al. 1988).

In order to generate perfusion-insensitive hyperthermia and to avoid inertial cavitation, pulses with high power but short duration of sharply focused ultrasound beams were used (Britt et al. 1984, Davis and Lele 1989, Billard et al. 1990, Hunt et al. 1991, Hynynen 1991, Dorr and Hynynen 1992). Subsequently, the vascular perfusion to the heated volume would not have any effect on the resulting temperature rise (ter Haar 1999). Shorter ultrasound pulses also have a smaller effect on the near-field heating than longer ones.

Acoustic propagation at frequency around 1 MHz in continuous mode is almost linear. However, at higher frequencies and with weakly focused fields, nonlinearity occurs. Both theoretical (Swindell 1985) and experimental (Hynynen et al.

1987a) studies showed that the increased absorption due to a higher power and the presence of higher harmonics can enhance the hyperthermia efficiency. The subsequent temperature increases more at the focus than in surroundings as a function of the intensity until the absorption in the overlying tissues starts to rise. Thus, nonlinear propagation can be utilized to improve the temperature distributions obtained during scanned focused ultrasound hyperthermia. However, care should be taken of the acoustic intensity threshold of transient or inertial cavitation in tissues.

11.4 THERMOMETRY

Thermometry is a key concern in the thermal treatment. Multiple thermocouples should be placed in the tumor under the guidance of orthogonal x-ray, computer tomography (CT), or ultrasound imaging based on the treatment planning and clinical considerations. Controlling the power output based on the measured temperatures can significantly improve the achieved temperature distributions (Das and Lele 1984, Johnson et al. 1987) if all thermocouples are located at significantly different blood perfusion regions. However, it is very difficult to properly heat tissue volumes close to large blood vessels (diameter larger than 0.5 mm) even during scanned focused hyperthermia utilizing a multipoint control system. Patient movements should be eliminated since misalignment of only a few millimeters can cause the beam to hit bone (especially for a tumor close to it). If the thermocouples are not parallel to the imaging beam, they cannot be figured out in the sonography. In such a situation, the thermocouple locations and the geometrical information are used to direct the beams into the tumor. Overall, the invasive nature of thermocouple alignment severely limits its clinical application (Fessenden et al. 1984).

The Radiation Therapy Oncology Group (RTOG) has already established quality assurance guidelines for clinical trials using ultrasound hyperthermia. Thermal mapping instead of measurements only at a limited number of points is required to obtain profiles of the temperature across the tumor dimensions, including margins of normal tissue. More highly localized hot or cold spots require more compounded temperature measurements. Acceptable thermocouples include a polyurethane-sheathed single-sensor thermocouple in a polyurethane catheter, a fiber-optic probe in a steel needle, a single-sensor thermocouple in a steel needle, and a manganin–constantan multisensor thermocouple. Unacceptable types include fixed or static probes that do not provide profiles of the temperature across the tumor dimensions, copper–constantan multisensor thermocouples, and Teflon-sheathed thermocouples inserted into a Teflon catheter (Waterman et al. 1991).

However, there are some artifacts in the measurement of temperature from viscous heating caused by the relative motion of the probe and the surrounding medium, absorption heating caused by the difference in the acoustic absorption coefficient of the probe/catheter and the surrounding medium, and reflection and scattering of the acoustic wave from the probe once its size becomes comparable to the wavelength.

The magnitude of the artifact depends on a variety of factors, including the type of sheath (metal or plastic), the use of catheters, the material and diameter of catheters and thermocouples, the acoustic properties (i.e., absorption coefficient) of the local tissue, the acoustic intensity, the orientation of thermocouple with respect to the ultrasound propagating beam, and the local blood flow rate. The smallest artifact was found from a bare copper–constantan thermocouple in a diameter of ≤50 μm without any plastic insulation at the junction. Immediately after the sonication is turned on, the temperature measured by thermocouple increases much more rapidly than the tissue temperature, which is due to a viscous artifact, until the thermal equilibrium is established between the thermocouple and the surrounding tissue. Meanwhile, the artifactual heating is extended to a volume of tissue by thermal conduction, which could last for several minutes in the continuous sonication. Similarly, when the sonication is terminated, the measured temperature initially decays very rapidly in approximately 30 s, while the complete dissipation of the artifact takes considerable time. However, the temperature artifact is typically underestimated because its mechanism is not well understood (Waterman 1990). The indiscriminate use of catheters for ultrasound also results in temperature artifacts, but prudent selection of the catheter can give rise to no greater artifacts than needle thermocouples (Waterman and Leeper 1990). The key factors of a plastic catheter are its wall thickness and diameter. The maximum orientation-induced artifact is produced when the probe is perpendicular to the direction of propagation of the beam, but it vanishes when the probe is parallel to the direction of propagation (Fry and Fry 1954a,b, Hynynen and Edwards 1989).

Another disadvantage of thermocouples and thermistors is the inaccurate readings in spatially varying temperature fields due to the thermal conduction along the metal, which leads to the higher temperatures to be underestimated and the lower ones to be overestimated. The magnitude of this error is determined by the spatial variation of the temperature gradient, the diameters of the wire, the thermal conductivities of the metals, the number of wires in the bundle, the thickness of the insulating material and metallic sheath, and the contact of the probe, catheter, and tissue. This conduction error is estimated using idealized models (Dickinson 1985, Samulski et al. 1985):

$$\frac{d^2 \Delta T(x)}{dx^2} = \frac{\Delta T(x)}{L(x)^2} - \frac{d^2 T_t(x)}{dx^2} \tag{11.2}$$

where

$\Delta T(x) = T_p(x) - T_t(x)$ is the temperature difference between the thermally conductive probe and the unperturbed surround tissue

$L(x)$ is the error length or thermal smearing length

The error length is especially evident in copper–constantan thermocouples with a common constantan wire substrate. The conduction error will be greater when a sheathed thermocouple is inserted into a catheter as heat flows along the wires.

In order to reduce the conduction error, the wall thickness of both the sheath and catheter should be minimized. However, it is hard to estimate the actual conduction error in clinical data using the error length, and the reliability of existing models is not well established (Samulski et al. 1985). Furthermore, the use of a fluoroptic sensor in a steel needle can eliminate absorption heating and the conduction error (Samulski et al. 1990).

Overall, the general guidelines for reliable temperature measurement in the ultrasound hyperthermia are as follows: (1) the use of a thermometer with an error of $\pm 0.5°C$ or less, including the contribution of calibration, viscous and absorption heating, and thermal conduction. Thermal mapping must be carried out in all patients; (2) oblique insertion of the catheters to reduce the measurement artifacts after local anesthesia or sedation; (3) a step size of 0.5 cm for catheter tracks <5 cm and a maximum of 1.0 cm step size for those >5 cm; (4) no use of Teflon catheters and careful application of multisensory probes after testing; (5) accurate determination of the temperature artifact for each sensor; (6) calibration of all probes periodically using established standard operation protocols (SOP) (Shrivastava et al. 1988); (7) determination of the location of all catheters by an accurate and reliable imaging modality; and (8) avoiding small air gaps adjacent to the probe. It is noted that the measurement artifacts and conduction errors of thermometry are much greater in ultrasound hyperthermia than microwaves (Waterman et al. 1991).

In the absence of a noninvasive technique with sufficient resolution, the region of maximum temperature, generally restricted to a few millimeters, can be located only by careful and thorough scanning of an implanted thermometer in small steps within the volume of tissue, which may be inadvisable or impracticable in many patients because of the location of the tumor and patient movement. The thermal damage to the tissue could be measured using magnetic resonance imaging (MRI) techniques (Hynynen et al. 1993). Spatial temperature distributions can also be noninvasively measured using MRI thermometry based on the proton resonance frequency (PRF) shift (De Poorter et al. 1995, Ishihara et al. 1995, Denis de Senneville et al. 2005). To a first approximation, the relationship between water PRF and temperature is linear, independent of tissue composition, and unaffected by thermally induced tissue changes; and the dependence is 0.0094 ppm/K. Fast spoiled gradient-echo (FSPGR) images can be continuously acquired in a coronal plane through the acoustic focus for thermometry in the sonication. A major drawback to heating with a moving ultrasound transducer in an MRI is that, as the transducer moves along the scanning pathway, the susceptibility distribution in the bore changes, giving rise to periodic variations in the measured temperature, which would result in an overestimated thermal dose and excessive fluctuations in the output power. By averaging over consecutive six images, standard deviation of measured temperatures is reduced from 1°C to 0.4°C, similar to the measured noise level of static heating. However, this method may also reduce real peaks or troughs, resulting in an underestimation of the thermal dose (Figure 11.8).

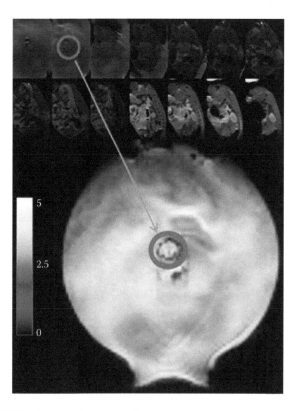

FIGURE 11.8 MR images of a Met-1 tumor prior to heating (top two rows) and temperature through the heating sequence (bottom). (From Fite, B.Z. et al., *PLoS One*, 7, e35509, 2012.)

Meanwhile, the two-dimensional temperature mapping can also be realized using noninvasive diagnostic ultrasound system based on the detection of shifts in echo location of backscattered ultrasonic bursts from a region of tissue undergoing thermal therapy. The echo shifts are due to both the local temperature dependence of sound speed and thermal expansion in the heated region, and in linear relationship to the tissue temperature rise. However, thermoacoustic lens effect would introduce ripples in the estimation. This approach has an accuracy of 0.5°C and a spatial resolution of 2 mm (Johnson et al. 1977, Ueno et al. 1990, Simon et al. 1998, Arthur et al. 2005, Rivens et al. 2007) (Figure 11.9).

Furthermore, the estimation of local tissue elasticity from shear wave imaging (SWI) can result in 2D temperature mapping. The propagation of shear wave with a duration of 100 µs at three depths was acquired at 17,000 frames/s. Tissue stiffness was found to decrease for temperatures up to 43°C, and there was a high correlation between SWI-based temperature estimation and stiffness variation maps ($r^2 = 0.91–0.97$). In comparison to echo-shift approach, SWI is much more robust to motion artifacts (Arnal et al. 2011).

11.5 TREATMENT PLANNING AND CONTROL

Owing to the highly localized nature of ultrasound hyperthermia and the complications created by bone and gas in the treatment field, the treatment geometry (i.e., the location of the target volume and the surrounding structures) has to be known exactly in advance. Ultrasonic beams need to be directed into the target volume with the same precision as radiation beams in order to cover the treatment volume properly without overheating bones or surrounding normal tissues. Therefore, 3D CT-based treatment planning of deep tumors is as important in ultrasound hyperthermia as it is in radiotherapy. For abdominal targets, sonography is used to identify bowel gas that may obstruct the acoustic beam and used to guide the thermocouple probes into desired locations. Hyperthermia treatment planning can be divided into two phases: optimization of the treatment geometry and the sonication parameters, and estimation of the temperature distribution (Lele and Goddard 1987). The anatomical information is obtained from sequential CT scans, and then the power deposition pattern and thermal field are calculated by utilizing the average acoustic attenuation and thermal values for different tissue layers, blood perfusion pattern, the transducer characteristics, and the scanning path. The scanning path and angle, together with the sonication parameters, are then modified iteratively to generate a uniform temperature distribution only in the whole tumor without hot spots in the bones or skin. In comparison, the perfusion values for different tissues and especially for tumors are difficult to obtain. Sometimes, only the best and worst cases can be estimated. Combination of current treatment planning capability and the ability to scan the focal zone around the desired treatment volume allows controlled and preferential power deposition

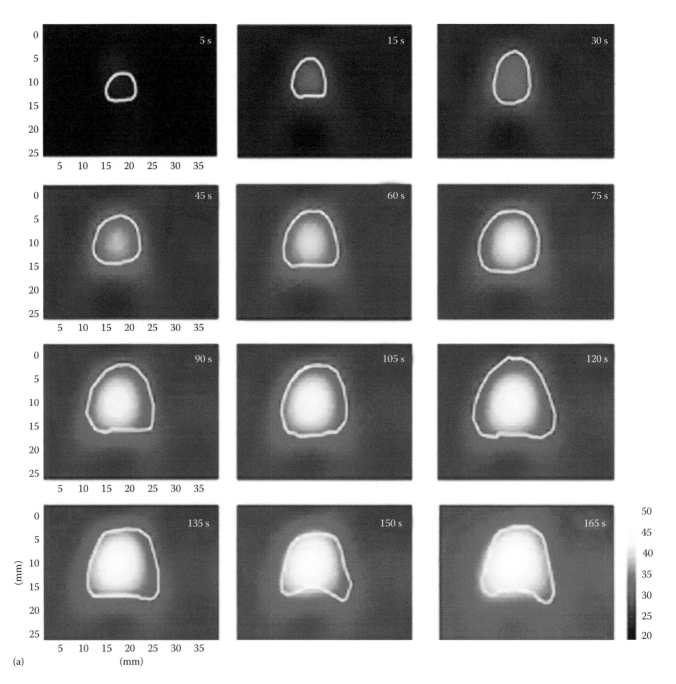

FIGURE 11.9   Comparison of the ultrasound temperature images based on (a) echo time-shift estimation.   (*Continued*)

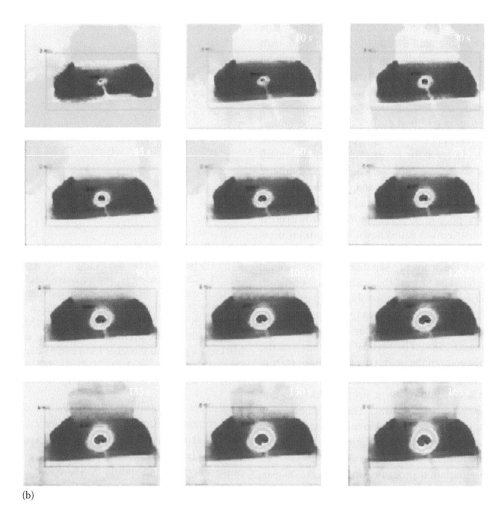

(b)

FIGURE 11.9 (*Continued*) Comparison of the ultrasound temperature images based on (b) infrared temperature maps in a liver sample acquired at the various ablation times. (From Geng, X., Zhou, Z., Li, Q., Wu, S., Wang, C.-Y., Liu, H.-L., Chuang, C.-C., and Tsui, P.-H., Comparison of ultrasound temperature imaging with infrared thermometry during radio frequency ablation, *Jap. J. Appl. Phys.*, 53, 047001, 2014.)

and, consequently, the temperature distribution in the desired volume (Hynynen et al. 1987a).

The inhomogenety in blood flows makes it challenging to have a uniform temperature distribution in the entire tumor. The necrotic center is easy to heat, but the well-perfused advancing front needs stronger sonication for the same hyperthermia level. Because of thermal conduction toward the surrounding tissue, the temperature in small tumors is lower than that in large ones even with the equal blood flow, absorption, and heat input. So, it is easy to heat the large tumors, which is in accordance with the finding that 43% of the tumors less than 5 cm in diameter and 73% of tumors larger than 5 cm can reach temperatures over 45°C (Storm et al. 1979). The blood flow in the tissues varies a lot: skin 0.8–28 mL/kg/s, muscle 0.5–5 mL/kg/s, fat 0.58–5 mL/kg/s, kidney 75 mL/kg/s, and liver 10 mL/kg/s. The heat transfer equations in inhomogeneously perfused tissues are not well formulated. However, a semiempirical model based on the experimental measurements in phantoms, animals, and patients and on some mathematical approximations may provide guidelines for the practicing hyperthermia

physiologist, whereas only a few temperature monitoring points are required to calibrate the model. The level of the hyperthermia and its distribution (i.e., the volume extent and uniformity) are governed by the pattern of heat generation and by heat diffusion from thermal conduction and blood flow. Therefore, it is not always possible to heat a large, well-perfused, and deep-lying tumor with even an ideal focused ultrasound system. The tissue density, specific heat, and thermal conduction are fairly well known for tissues and do not vary as the blood perfusion rate (Bowman 1982). However, due to variations in perfusion, cold spots are often observed. One way to minimize this effect is to monitor the temperature and adjust the power correspondingly to obtain a more uniform temperature distribution (Lele 1983). During the treatment, if patient discomfort is reported, the treatment plan (i.e., the scan location, size, depth, or beam angle, and power) will be adjusted for the therapeutic deposition pattern. The scanning pattern (as well as the temperature fluctuations) to obtain a uniform temperature distribution in the whole scanned volume depends on the tissue perfusion rate and focal diameter (Lele 1983, Hynynen et al. 1986).

If the normal tissue temperature maximum during the fluctuations is close to the pain threshold (44°C–45°C), then slow scanning can cause periodic pain during the treatment (Hynynen et al. 1987b), which is especially important if the beam is propagating to a bone or skin surface. In a tissue with low perfusion, a fairly large spacing between the scans can be used, and more energy is required at the edges of the heated region than at the center in order to compensate the increased thermal conduction effects at the edges. The optimal spacing between concentric scans is less than or equal to the diameter of the beam focus (Moros et al. 1988), and this scanning pattern gives a fairly uniform temperature elevation across the scanned volume irrespective of perfusion rates. However, if the perfusion rate distribution is unknown, the optimal heating can be provided by using a multipoint feedback controller, if required, where the scans are separated by the distance of the beam diameter. The size of the scan has a significant effect on the resulting temperature distribution, particularly for deep tumors. The greatest variation in temperature is at or close to the skin surface, which may be due to several reasons (ter Haar and Hopewell 1983). The vertical positioning of the thermocouple in the center of the field is of crucial importance. Tiny gas bubbles lying in the coupling medium between the transducer and the skin could result in strong reflections and thus produce localized surface heating. Thermal conduction of the transducer housing toward the skin takes place because of the acoustic mismatch between the piezoelectric material and the coupling medium or tissue.

A transient 3D simulation program was used to investigate the effects of scanning speed and pattern, blood circulation, transducer geometry, and ultrasound parameters on the temperature fluctuations during scanned ultrasound hyperthermia (Moros et al. 1988). It is shown that the largest temperature fluctuations are always located on the scanning path in the focal plane, increase linearly with scan times and as a weak exponential with the blood perfusion rate, and have essentially the same pattern and magnitudes regardless of the focal depth. Moreover, the smaller the focus of the beam, the larger the temperature fluctuations. To avoid them, scan times of 10 s or less were needed when single, 2 cm circular scans were simulated at practical blood flow values. Simulations showed that relatively uniform temperature distributions can be achieved at constant power for an outer scan diameter of up to 5 cm and blood perfusion values of up to 15 kg/m³/s as long as the spacing between the concentric scans was not greater than the beam width (Moros et al. 1988).

The maximum spacing between the pulses is a strong function of both the *f*-number and frequency. The stronger the focusing (i.e., an *f*-number of 1.0 or smaller), the smaller the near-field heating. Increasing the interval between pulses and the frequency, or decreasing the pulse duration and *f*-number, could minimize the near-field heating. For a transducer with an *f*-number of 1 and the driving frequency between 1 and 1.5 MHz, the near-field heating can be reduced to an acceptable level by using a pulse length of 5 s or shorter and by having an interval of at least 20 s between the pulses. If the tumor of size of 1 cm × 1 cm × 3 cm is treated using a 1 MHz

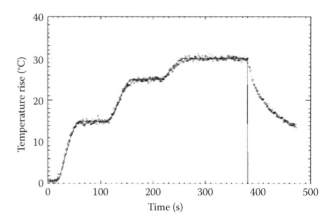

FIGURE 11.10 Temperature evolution of the focal point in a three-step FUS heating a fresh meat sample with a target temperature rise of 15°C, 25°C, and 30°C, respectively. *Solid line*: predefined target temperature profile; *circle*: measured temperature by MR thermometry in real-time. (Salomir, R., Vimeux, F.C., de Zwart, J.A., Grenier, N., and Moonen, C.T.W.: Hyperthermia by MR-guided focused ultrasound: Accurate temperature control based on fast MRI and a physical model of local energy deposition and heat conduction. *Magn. Reson. Med.*, 2000. 43. 342–347. Copyright Wiley-VCH Verlag GmbH & Co. KGaA. Reproduced with permission).

transducer of *f*-number of 1, and 1 s pulse (20 s interval), the total treatment duration is 256 s × 7 s × 21 s = 10.45 h using the radial and axial spacing of 0.625 and 4.25 mm, respectively. When the transducer *f*-number is increased to 2, the pulse interval time should be increased to 40 s to avoid the near-field heating. Subsequently, the treatment time becomes 64 (one layer) × 2 (two layers total) × 41 s = 1.45 h (Damianou and Kynynen 1993) (Figure 11.10).

11.6 ULTRASOUND HYPERTHERMIA-INDUCED EFFECTS

The biophysics of ultrasonic cell killing at the level of hyperthermia is not fully understood. The interaction between ultrasound and tumor can be categorized as thermal effects, cavitation, and direct effects (Hill 1968), and so the mechanism of ultrasonic action *in vivo* and *in vitro* is thought to be complex because of their synergy. The direct effect of ultrasound alone did not influence cell killing, but it enhanced the hyperthermic cell killing synergistically when both agents simultaneously acted on the cells (ter Haar et al. 1980, Kondo and Kano 1987). Shear stress arising from direct effects of ultrasound could enhance the heat-induced inactivation of cultured cells, which inflicts lethal damage at 44°C but not before or after ultrasound hyperthermia (Dunn 1985, Kondo and Kano 1987).

The phenomenon of vasodilation usually occurs in a normal blood vessel during a hyperthermia treatment. As a result, the blood flow is increased by a factor of 3–20 upon heating at 42°C–45°C because of a decrease in vascular resistance. Without the vasodilation effect, the temperature in the tissue could be about 50°C, whereas 45°C is maintained in the real practice. The temperature elevation induced by ultrasound hyperthermia in the muscle and the fat layers that have much less blood circulation and supply, even during full vasodilation, is higher than that in the kidney and the liver. However, the vessels in a tumor have a different reaction. At temperatures of 41°C–43°C, blood flow in the tumor either remains unchanged or increases slightly by a factor of less than 2. The newly formed tumor vessels seem vulnerable to heat so that the blood flow in most of the animal tumors decreases at 42°C–43°C. Consequently, the heat dissipation becomes much less in tumors during heating, as a result of vasoconstriction, and thereby results in a greater temperature rise for greater damage in tumor relative to normal tissues. In addition, the intrinsically acidic intratumor environment becomes more acidic upon heating and accentuates the thermal damage (Song et al. 1984). Avascular tissues (e.g., the lens oculi) also need special attention because of their low capacity of heat dissipation. Cataracts occur at intensity levels much lower than those used for hyperthermia.

Sonication of mammalian tissues with 0.75 MHz ultrasound at intensities above 100 mW/cm² can produce gas cavities primarily in subcutaneous tissues (ter Haar and Daniels 1981). These cavities would increase sound backscattering, the attenuation of the tissue, and, thus, superficial temperatures (ter Haar and Hopewell 1983). Cavitation occurs under O_2-, Ar-, or N_2-saturated conditions in the experiments of uracil sonolysis (McKee et al. 1977) and radical formation but can be suppressed under N_2O saturation. Subsequently, cell death and decrease of ultrasound-induced clonogenicity can be inhibited. In addition, N_2O also inhibits formation of double-strand breaks of DNA and liberation of iodine from potassium iodide at hyperthermic temperature as well as the other cavitation-induced chemical and biological effects (McKee et al. 1977, Henglein 1985, Kondo et al. 1986). However, N_2O does not act as a radical scavenger, as illustrated by physicochemical and biological analyses.

11.7 CLINICAL OUTCOME

In treatments of the brain, an individualized mould is used and a small part of the plastic membrane covering the water tank is bolused up to form a pathway for the ultrasound to propagate to the skin and through the opening in the skull bone into the tumor. Patients with deep pelvis, rectal, and vaginal tumor sit on a specially made chair that has an opening in the seat, on top of the bath. In the case of abdominal and pelvic tumors, and tumors in the chest wall and extremities, the patient lies on top of the water bath (Figure 11.11).

Each ultrasound hyperthermia treatment should be evaluated carefully for any damage outside of the heated volume. A second problem with ultrasound hyperthermia is pain. The first clinical patient treatments using nonfocused stationary transducers were often power-limited by pain to reach a therapeutic temperature level in the tumor. Significant pain associated with the ultrasound exposure was noted in almost 10 of 26 patients (Marmor et al. 1979). Better localization of the ultrasonic energy could overcome this problem. A series of 147 treatments with $T_{50} = 40.5°C$ and $T_{90} = 38.5°C$, which are the

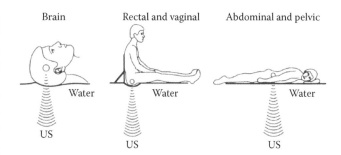

FIGURE 11.11 The different treatment positions used in the ultrasound hyperthermia system. (With kind permission from Springer Science + Business Media: *Methods of External Hyperthermic Heating*, Biophysics and technology of ultrasound hyperthermia, 1990, 61–115, Hynynen, K., in: Gautherie M, ed., Springer-Verlag Berlin, Heidelberg, Germany.)

temperatures that 90% and 50% of all measured temperatures are no less than the specific values, respectively, illustrated pain rather than temperature as the limiting factor (Samulski et al. 1990). The discomfort usually occurred when bones were located close behind the tumor (Marmor et al. 1979, Corry et al. 1982). So, accurate treatment planning with consideration of several other factors, such as tissue interfaces, gas, nonlinear propagation, the beam path available, beam distortion, the tissue geometry and its relationship with respect to the temperature sensors, and the heating field, is required in order to avoid unintended heating and to obtain adequate temperature distribution in the target volume, which results in a fairly complicated treatment procedure compared with superficial microwave or regional heating and has probably been one of the factors discouraging the use of ultrasound hyperthermia.

Six planar ultrasound transducers with a diameter of 7 cm and a driving frequency of 350 kHz were mounted on a spherical shell section with a radius of curvature of 26 cm for near-field peak suppression, and a few percent modulation of the transducers could minimize the effects of standing waves. In the hyperthermia of realistic perfused tumors at depth, small wedges are used to rotate the transducer axes a few degrees away from the radial direction, yielding a waist (3–6 cm in diameter) rather than a single point where the central axes of individual beam are close to each other to produce an ideal ellipsoidal shape (Fessenden et al. 1984). Fifty-seven treatments (mostly the pelvis, and the arm, axilla, back, and inguinal regions) were performed on 15 patients using this system. Six or more thermocouples were implanted into each tumor. During about 50% of all treatments, some degree of local discomfort, varying from mild to severe, was found among patients, which was often related to the power applied. For 20% of the treatments, radicular pain suggesting nerve stimulation was experienced. A little adjustments of the wave entry angle could reduce or eliminate them in many cases. Two small superficial blisters of little clinical significance and one small burn extending into subdermal fat were found (Fessenden et al. 1984). In fact, a randomized hyperthermia study recently reported by the RTOG confirms that even superficial tumors are difficult to heat effectively, particularly

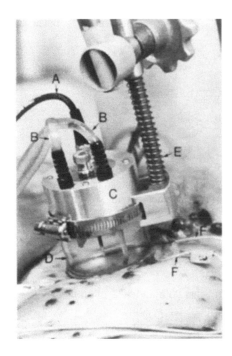

FIGURE 11.12 Clinical use of a 4-cm ultrasound transducer to superficial tumor, A: excitation input, B: inflow and outflow of circulating water, C: ultrasound transducer, D: circulating water cuff, E: spring loaded holder, F: thermocouples. (From Marmor, J.B. et al., *Cancer*, 43, 188, 1979.)

when larger than 3 cm in size (Perez et al. 1989). Ultrasound offers the unique property of deep soft tissue penetration at small wavelengths (millimeter range), thereby allowing the potential for subsequent energy focusing and scanning to conform closely to target dimensions.

Hyperthermia resulted in a significant rate of objective tumor regressions in superficial neoplasms despite nonuniform heating. Because of small wavelengths of ultrasound, energy focusing and scanning can conform closely to target dimensions. Twenty-five patients who had recurrent or

metastatic tumor in superficial locations refractory to conventional treatment modalities were treated by ultrasound hyperthermia. Objective tumor responses were seen with hyperthermia alone, although the effects may be transitory due to the short follow-up time and partial responses in most cases (Marmor et al. 1979) (Figures 11.12 and 11.13). Because local control is a critical concern for patients with inoperable or advanced head and neck cancers, safe, effective, and well-tolerated ultrasound hyperthermia would be a choice in clinical treatment either alone or in conjunction with surgery or radiation (Marmor et al. 1979). The high-frequency beams would allow even superficial tumors overlying a bone to be heated.

Nineteen patients with colorectal ($n = 3$), portahepatus ($n = 1$), pancreas ($n = 7$), liver ($n = 1$), pelvis ($n = 3$), sacrum ($n = 2$), and abdomen ($n = 2$) cancers were heated by 60 min intraoperative ultrasound in the operating room during surgical resection. Temperatures were measured at a total of 133 fixed locations in the tumor volume (Ryan et al. 1992). The average maximum temperature was 46.6°C with $T_{50} = 42.9$°C and $T_{90} = 39.2$°C. Although the corresponding values in the intraoperative hyperthermia were greater than in the clinical series, the single-element applicator was severely limited in performance even while using adequate power. In addition, some tumor dimensions exceeded the capability of applicator (i.e., 10 cm in a circle). It seems that a multielement intraoperative applicator would be a better way to improve the performance of ultrasound hyperthermia (Cain and Umemura 1986, Ryan et al. 1991) in the situation of constrained access to the treatment site (Ryan et al. 1992).

Brain tumors are good candidates for local therapy because of the absence of an existing therapy. However, the skull bone has large attenuation for the acoustic beam energy required for hyperthermia (Fry 1965). Clinical trials showed that at least some brain tumors can be repeatedly treated and that good therapeutic temperatures can be reached in the tumor without heating the surrounding brain tissues (Guthkelch et al. 1991).

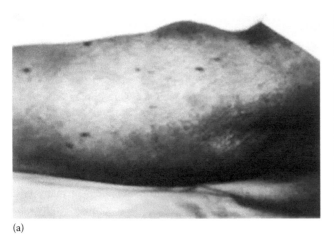

(a)

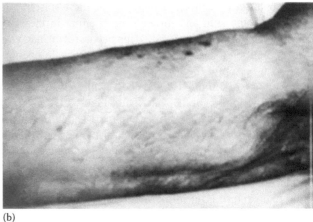

(b)

FIGURE 11.13 (a) View of the left arm in a patient with mycosis fungoides and (b) complete regression of the nodule after six ultrasound treatments at 44°C. (From Marmor, J.B. et al., *Cancer*, 43, 188, 1979.)

11.8 COMBINATION WITH RADIATION

Strong evidence from cell culture, spontaneous animal tumors (Dewhirst et al. 1982), and superficial recurrent or metastatic human neoplasms through both retrospective and prospective randomized trials has shown that properly applied hyperthermia can significantly enhance the effects of radio and chemotherapy by increasing the regression of local tumor (Dewey et al. 1980, Hahn 1982, Arcangeli et al. 1985, Overgaard and Overgaard 1987, Herman et al. 1988, Valdagni et al. 1988, Perez et al. 1989), which means that hyperthermia interacts synergistically (Dewey et al. 1977) probably by interfering with the repair of sublethal damage (BenHur et al. 1974). This combination with radiation and some chemotherapeutic agents has the potential to enhance cytotoxic responses and seems more effective in destroying malignant tissue (Crile 1963, Overgaard and Overgaard 1972, Hahn et al. 1975, Prakash et al. 1980, Dewey 1984, Overgaard 1989). Synergism between the two modalities occurs even at temperatures as low as 41°C (Harisiadis et al. 1978). A number of clinical studies also support the enhancement of local control when heat is added to irradiation (Emami and Perez 1984, Arcangeli et al. 1987, Overgaard 1989). However, similar responses have not yet been shown in larger (several centimeters in diameter) (Perez et al. 1989) or deeper tumors (Shimm et al. 1988b), which may be due to the inadequate thermal exposure throughout the target (Corry et al. 1988, Kapp et al. 1988, Shimm et al. 1988a).

International experience with localized deep microwave hyperthermia to the prostate (Yerushalmi et al. 1982, Shani et al. 1990) has no significant complications. Standard radiotherapy to the prostate gland and periprostatic tissues (i.e., the seminal vesicles) was delivered using a four-field approach with cerrobend blocking and 1.8–2 Gy daily for 5 days a week over a 7-week course with a total dose of 67–70 Gy based on the minimum tumor volume. Cylindrical PZT 4 crystals were cut to four to eight half-cylindrical elements, either driven individually or two neighboring elements driven from the same power source, and then fixed to a plastic support. Once the applicator was in the desired location, the water bolus was inflated to fill the rectal cavity and the thermostated water circulating system was turned on. On average, it took 25 ± 12 min to reach the treatment temperature (>42.5°C) and then to maintain for 30 min. Two to three thermocouple probes containing seven sensors each were placed in the prostate to measure temperatures throughout the gland with the depth from the rectal wall ranging from 5 to 25 mm. Fourteen patients (age range: 53–86 years, mean: 72) with American Urological Society stage C_2 or D_1 adenocarcinoma of the prostate were treated with transrectal ultrasound hyperthermia (TRUSH) concurrently with standard radiotherapy to the prostate. Twenty-two heat treatments were delivered to 14 patients; 8 patients received two TRUSH procedures, each separated by 1 week. Three patients had well-differentiated, six patients had moderately differentiated, and five patients had poorly differentiated adenocarcinoma of the prostate. Treatments have been well tolerated (good in 17/22, fair in 3/22, and treatment

limiting in 2/22) with a few complications secondary to position intolerance and pain that was felt to reflect nerve pain radiating down the anterior thighs and penis and disappeared with angular adjustment of the TRUSH applicator. There has been one episode of hypotension related to narcotic administration and three episodes of rapidly resolving pain during hyperthermia treatment. Mild hematuria occurred in 5/22 and moderate hematuria in 2/22 transperineal thermometer catheter placements. Mild hematuria, treated with hydration and observation, occurred in 5/22 and moderate hematuria, treated with catheterization, hydration, and observation, in 2/22 transperineal thermometer probe placements. Treatment has been well tolerated, and complications were related to transperineal thermometer probe placement and the difficulty of lying still on one side for over an hour (treatment setups required approximately 1/2 h for thermometry probe placement, 1/2 h for simulation, and 10 min for TRUSH applicator placement) (Fosmire et al. 1993).

A scanned focused ultrasound (SFUS) system was designed to heat deep pelvic and abdominal tumors with a fixed transducer geometry and frequency, slow scanning speed, fixed power output, and simple concentric octagonal scanning patterns. It was used to heat 66 tumors at various anatomical locations in 52 patients for a total of 160 treatments. The time-averaged temperatures over 30 min in the best treatment of each tumor were 44.0°C ± 2.4°C (mean ± std) and 39.6°C ± 1.5°C at the location of the highest and the lowest sensors, respectively. After extracting bad candidates for statistical analysis, 64% of the sensors reached a temperature over 42.5°C, and the highest and the lowest temperatures achieved were 45.9°C ± 2.3°C and 40.7°C ± 1.4°C, respectively. Because the acoustic power was kept same during the scanning, the outside tumor region received twice as much energy as the inner region, which could compensate the thermal conduction losses and thus improve the temperature distribution in low-perfusion cases (Hynynen et al. 1989). The maximum practical diameter of the target that SFUS can treat is 60 mm due to its scanning speed; larger tumors have to be treated in portions. Within 15–30 min following hyperthermia that was given once a week, the patients received radiotherapy in daily increments of 1.5–3.0 Gy to a maximum of 30–60 Gy for 1–4 weeks, the precise dose depending on the patient tolerance and other clinical considerations. Sharp periodic pain accompanying the scanning cycle at specific locations reflected pain felt from the arm or leg outside the treatment region (eight patients), and the hot feeling in the sonicated volume was experienced in 128 treatments. Some sharp periodic pains also occurred at low measured peak temperatures, which might be due to the presence of bone or gas in the acoustic path. Occasionally, there was no discomfort to patients with tumor temperatures reaching 46°C or higher. However, tumors on the chest wall or head and neck that were close to a bone or a skin/air interface were difficult to heat with SFUS, and only a part of them could be elevated to therapeutic temperatures because of patient discomfort. In addition, placing water balloons on the distal skin surfaces could reduce skin heating (Hynynen et al. 1990).

Selective heating of irregularly shaped tumors at depth can now be accomplished through focusing and controlled scanning of the energy deposition patterns by ultrasound. Concurrent radiotherapy was delivered (range 1000–7640 cGy, mean 4320 cGy) to 67 SFUS hyperthermia patients; 4 received concomitant chemotherapy. A 62% overall response rate was observed, with 22% of treated tumors demonstrating a complete response (defined as complete disappearance of the treated tumor) and 40% exhibiting a partial response (defined as ≥50% reduction in tumor volume). Dramatic local pain reduction was achieved in 42% of the tumors treated. The acute tolerance of SFUS hyperthermia was quite good, and chronic toxicities (persistent skin blisters/burns) were identified in two patients. The versatility of the SFUS system is demonstrated, as well as its future potential for improving control of advanced locoregional malignancies treated with curative intent (Harari et al. 1991).

Between October 1986 and May 1990, 220 SFUS hyperthermia treatments were carried out on 87 tumors (the most common being recurrent colorectal and breast adenocarcinomas, metastatic squamous cell carcinomas, and metastatic melanomas) in 71 cancer patients in the Department of Radiation Oncology, the University of Arizona Cancer Center, in whom conventional therapies had been exhausted, were considered nonbeneficial, too large or too deep for electromagnetic devices, or too inaccessible for interstitial approaches. The majority of patients presented with advanced locoregional failure or symptomatic metastatic disease. Multisensor manganin–constantan thermocouple probes were used for interstitial temperature measurements. The assessment of tumor dimensions and location was accomplished by a combination of CT scan, palpation, and ultrasound imaging of the tumor. Deep pelvic and abdominal tumors often required CT-guided probe insertion to ensure accurate positioning. A concentric and octagonal scanning pattern for ultrasound heating was then directed into the target volume using the most accessible entry path. The outermost scan was generally repeated twice before proceeding to the innermost part to compensate for thermal conduction at the tumor periphery. The goal was to achieve minimum intratumoral temperatures of 42.5°C for 30 min using the maximum power tolerable for patients and to deliver a maximum of four weekly treatments per patient. Manual adjustments in the scanning pattern, location of ultrasound focus, and power output were made during hyperthermia treatments in response to continuous assessment of the patient's discomfort. Radiation therapy was administered daily within 10–30 min from the termination of hyperthermia with a range of 150–400 cGy. Complete response (CR) was defined as disappearance of all measurable tumor as determined by inspection, palpation, or radiographic study. Partial response (PR) was defined as ≥50% reduction in the original tumor volume for at least two consecutive evaluations. No response (NR) was defined as <50% volume reduction, stable tumor, or progression during treatment and thereafter. Pain response was assessed subjectively by the patient for the pain reduction and/or the cessation of analgesic medication for pain control. Eighteen of 83 tumors (22%)

achieved CR, 33 tumors (40%) exhibited PR, and 32 tumors (38%) demonstrated NR. Significant pain palliation was achieved in 35 of the 83 tumors (42%), and 10 of them (12%) showed a stable tumor following therapy. About 45% of the ultrasound scanning did not encompass the whole tumor volume, and there was a strong relationship between CR and the entire volume scanning (38% vs. 9%). Thirteen of the 18 CRs (72%) had virtually complete scanning. Fifty-three of the 71 (75%) patients had tumors at depths more than 3 cm from the skin, with several ones extending to more than 10 cm in depth, considered as the limit of effective scanning. About 34% of the sensors had time-averaged temperatures ≥42.5°C for fully scanned tumors and 45% in partially scanned tumors, which was due to the excessive tumor volume. Although the insertion of interstitial thermometry probes was not comfortable, it did not prevent the initiation of treatment and withdrawal from therapy. Approximately 75% of all SFUS treatments were characterized by the transient ultrasound- or thermal-related pain syndromes. However, adjustment of power, scanning pattern, or beam entry angle can relieve the pain. Superficial skin blisters or burns are the most common subacute toxicities and observed in 8 of the 83 tumors following SFUS hyperthermia (Harari et al. 1991).

11.9 HYPERTHERMIA-MEDIATED DRUG RELEASE

Hyperthermia has been shown to enhance extravasation of drug carriers in solid tumors and overcome barriers to drug delivery by the tumor microenvironment (Kong and Dewhirst 1999). The use of temperature-sensitive drug carriers (e.g., liposomes) would trigger the rapid and local release of high concentrations of an active anticancer drug (Yatvin et al. 1978, Kong and Dewhirst 1999, Needham and Dewhirst 2001, Ponce et al. 2006, Tashjian et al. 2008). This localized therapy depends on the stability and contents of released drug carriers as well as the ability to accurately maintain temperature elevations only in targeted regions with minimal influence in the surrounding tissue. Recently, pulsed high-intensity focused ultrasound was used to release a drug from thermosensitive liposomes (Dromi et al. 2007, O'Neill and Li 2008, Patel et al. 2008) in a feedback control loop by MRI thermometry, which has been applied for tumor coagulation using intracavitary (Smith et al. 2001, Chopra et al. 2008) and externally (Palussiere et al. 2003) focused ultrasound (FUS) treatment. MRI thermometry-controlled FUS beam was scanned in a circular trajectory in a diameter of 10–15 mm to achieve temporally and spatially uniform heating in thigh to 43°C for 20–30 min. Lyso-thermosensitive liposomal doxorubicin was infused intravenously during hyperthermia. Unabsorbed liposomes were flushed from the vasculature by saline perfusion 2 h later, and tissue samples were harvested from sonicated and unsonicated regions. The fluorescence intensity of the homogenized tissue samples was used to determine the concentration of doxorubicin. It was found that higher concentrations of doxorubicin were present in the heated thigh muscle than in the normal muscle (8.3 vs. 0.5 ng/mg, mean per-animal difference = 7.8 ng/mg, $p < 0.05$ in Wilcoxon's matched pairs signed rank test). In the region of thermal coagulation, doxorubicin concentrations were 9.7 times higher than that in the contralateral thigh (3.7–18.8 times), but there was no statistical significance in pairwise difference.

Hyperthermia increased both the rate and uniformity of 100 nm liposome extravasation in angiogenic tumor vessels (Huang et al. 1994, Ning et al. 1994, Gaber et al. 1996, Wu et al. 1997, Kong et al. 2000, 2001) and was thought to cause morphological changes in tumor endothelial cells, increasing the occurrence and size of pores in the vessel wall, enabling and increasing extravasation of normally impermeable particles (Kong et al. 2000, 2001). Without saline perfusion, intact liposomes would continue to circulate with a plasma half-life of approximately 1.75 h (Poon and Borys 2009). Because a small fraction of the body is heated, free doxorubicin released in the vasculature of the target region during heating will have a minimal impact on systemic doxorubicin concentrations, even when compared to that leaking from the liposomes at body temperature in a small amount (Gasselhuber et al. 2010). Free doxorubicin in the vasculature is rapidly taken up by tissue with an initial half-life of 8 min and then followed by slow elimination with a terminal half-life of 30 h (Greene et al. 1983). However, tissue coagulation may change the mechanism for drug accumulation. In some cases, hyperthermia may allow triggered release in healthy vessels prior to vascular shutdown. In other cases, high temperatures may cause liposome extravasation due to vessel damage, or rapid vessel collapse preventing liposomes from reaching the target tissue. Coagulation could also prevent the removal of residual liposomes from damaged vessels by saline perfusion, exaggerating measurements of tissue drug concentration, which would vary spatially within a heated region. Rapid triggered release of large concentrations of the drug in the tumor vasculature and perivascular space may lead to the subsequent diffusion of free doxorubicin into deeper cell layers for greater antitumor effect. Altogether, MRI-controlled FUS hyperthermia can enhance local drug delivery with temperature-sensitive drug carriers (Staruch et al. 2011).

The transport of solutes between the tissue and the blood within the central nervous system (CNS) differs significantly from that in other organs. In contrast to the systemic circulation, brain microvessels form the blood–brain barrier (BBB), an endothelial barrier that is characterized by its limited permeability to water-soluble or nonhydrophobic solutes (Rapoport et al. 1979). Moreover, recent identifications of multidrug-resistant (MDR) efflux transporters in the BBB such as P-glycoprotein (P-gp) (Cordon-Cardo et al. 1989) and MDR resistance–associated protein (MRP) (Seetharaman et al. 1998) have illustrated the dynamics of the BBB as a restrictive barrier to both hydrophilic and hydrophobic molecules. Hyperthermia has been shown to increase BBB permeability of compounds (Hahn and Strande 1976), but at high temperature (43°C) for a long duration (Shivers and Wijsman 1998), which is unfortunately associated with substantial BBB disruption, increased accumulation of unwanted molecules in the brain, and undesirable disturbances of brain function.

Ultrasound-induced mild hyperthermia (USHT, 0.4 W/cm^2 at 41°C) can enhance drug absorption in BBB endothelial cells and produce a significant and comparable increase in hydrophobic (R123 and [^{14}C]-antipyrine) but not hydrophilic molecules ([^{14}C]-sucrose and 2-[^{3}H]-deoxy-D-glucose) accumulation. The enhanced permeability is reversible and size dependent. However, USHT did not affect P-gp activity or the activity of glucose transporters (Cho et al. 2002). Local pharmacokinetics of active drugs are determined by the equilibration between the blood and the brain. Mild USHT enhances passive diffusion of hydrophobic drugs and allows them to bypass efflux transporters. So, it is an efficient, noninvasive, and reversible novel method for enhancing drug penetration through the BBB. Peak intensities on the order of 1 W/cm^2 are known to produce acoustic streaming and bioeffects, even for short pulse durations (Duck et al. 1998). However, this effect is most likely not relevant for the BBB *in vivo* because of the suppression in the solid tissue. USHT can mediate enhanced cellular accumulation of P-gp substrates in bovine brain microvessel endothelial cells (BBMECs); however, the mechanism is still unknown. USHT failed to affect sucrose accumulation in cells, suggesting that the integrity of the cell membrane following USHT was intact and that cells were viable. In contrast, cellular accumulation of both R123 (log partition coefficient of 0.53) and antipyrine (log partition coefficient of 0.4) increased immediately after USHT. These observations are consistent with the hypothesis that USHT may produce subtle effects on the cellular membrane such that it affects the permeability of hydrophobic molecules rather than hydrophilic ones. The activity of several membrane-bound proteins is reduced by hyperthermia (Lepock et al. 1983, Bats and Mackillop 1985), probably through irreversible protein transitions that are found to occur in membranes (Lepock et al. 1983). Subtle membrane effects produced by USHT may also lead to impairment of P-gp activity. Consistently, mild USHT did not significantly affect the activity of glucose transporters (Cho et al. 2002) (Figures 11.14 and 11.15).

11.10 FUTURE WORK

The current ultrasound hyperthermia systems are far from optimal. The treatment planning and control programs need to be improved to the same level as that of radiotherapy. In most hyperthermia cases, the simplified model is sufficiently good to predict the treatment fields (Wang et al. 1991). A more complicated and improved numerical model for calculating the ultrasonic power deposition in layered media is needed, taking into account the ultrasound wave reflection and refraction at the tissue interfaces, although in most cases the soft tissue interfaces can be ignored. However, in some instances, the acoustic focus may be shifted several millimeters off axis in layered media by a sharply focused beam at a high power but short duration or because of the entry angle of the beam being larger than 60° (Davis and Lele 1989, Billard et al. 1990). Acoustic pressure distribution, both the amplitude and location, will be affected, introducing significant uncertainty in the thermal exposure. Furthermore, the presence of a shear

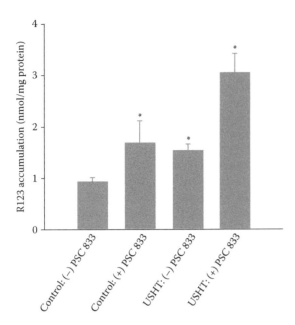

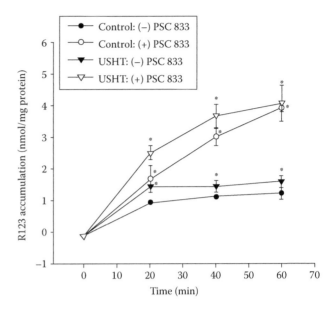

FIGURE 11.14 Comparison of P-gp activity expressed by BBMECs between control (37°C) and ultrasound hyperthermia treatment (0.4 W/cm² at 41°C for 20 min) in the presence or absence of 1 μM PSC 833. (With kind permission from Springer Science + Business Media: *Pharm. Res.*, Ultrasound-induced mild hyperthermia as a novel approach to increase drug uptake in brain microvessel endothelial cells, 19, 2002, 1123–1129, Cho, C-W., Liu, Y., Cobb, W.N., Henthorn, T.K., Lillehei, K., Christians, U., and Ng, K.-Y.). *$p < 0.05$, compared with untreated control (i.e., control less PSC 833).

FIGURE 11.15 Accumulation of R123 in BBMECs subjected to control (37°C) or ultrasound hyperthermia (0.4 W/cm² at 41°C for 20 min) in the presence or absence of PSC 833 (1 μM). (With kind permission from Springer Science + Business Media: *Pharm. Res.*, Ultrasound-induced mild hyperthermia as a novel approach to increase drug uptake in brain microvessel endothelial cells, 19, 2002, 1123–1129, Cho, C-W., Liu, Y., Cobb, W.N., Henthorn, T.K., Lillehei, K., Christians, U., and Ng, K.-Y.). *$p < 0.05$, compared with untreated control (i.e., control less PSC 833).

wave would cause more energy to be reflected at the boundary and more shift of the beam. The other practical situations should also be included in the simulation for appropriate treatment planning, such as apodization across the transducer surface, properties of sheath and catheter for acoustic wave propagation and temperature elevation, the perfusion pattern in the target and intervening tissue, and difference of sound speed in the tissue (e.g., the skin) (Fan and Hynynen 1992). The optimal treatment parameters, such as temperature, exposure time, treatment volume, the number of treatments, and interval between treatments, are not yet known.

Improvement of the ultrasound hyperthermia system may include the following aspects: (1) A multipoint feedback control can be implemented to set the power output at different focal point locations according to the temperature measurement (Lin et al. 1990). (2) Variable frequency would provide better control over penetration depth and allow tumors overlying or close to bones to be heated. (3) Real-time adaptation of scanning pattern and transducer orientation would allow better energy distribution throughout the tumor (especially in the larger depth) to improve outcomes of well-perfused tumors and reduce the associated pain. (4) New treatment strategies are required to reduce the treatment duration for larger tumors without inducing large temperature fluctuations (Hynynen et al. 1986, Moros et al. 1988). A combination of mechanical and electrical scanning and focusing can shorten the treatment duration. (5) Better

control over surface temperature would improve temperatures achieved in superficial tumors. (6) Accurate and reliable thermography is highly desired to control the power delivery in real time throughout the tumor and surrounding normal tissues for automatic power and temperature control since the calculation is time consuming and not ready for reliable practice (Lele 1980). The development of special, reliable, and nonperturbing temperature-sensing probes is desirable too. (7) If a tumor is close to a body cavity, an intracavitary ultrasonic applicator with multiple elements appears to offer good control over the power deposition, and the system is easy to operate. (8) A better patient positioning system would enhance comfortability of treatment, increase alignment accuracy, and reduce treatment duration (Hynynen et al. 1990).

REFERENCES

Arcangeli G, Arcangeli G, Guerra A, Lovisolo G, Cividalli A, Marino C, Mauro F. Tumor response to heat and radiation: Prognostic variables on the treatment of neck node metastases from head and neck cancer. *International Journal of Hyperthermia* 1985;1:207–217.

Arcangeli G, Benasso M, Cividalli A, Lovisolo GA, Mauro F. Radiotherapy and hyperthermia: Analysis of clinical results and identification of prognostic variables. *Cancer* 1987;60:950–956.

Arnal B, Permot M, Tanter M. Monitoring of thermal therapy based on shear modulus changes: I. Shear wave thermometry. *IEEE Transactions on Ultrasonics, Ferroelectrics, and Frequency Control* 2011;58:369–378.

Arthur RM, Straube WL, Trobaugh JW, Moros EG. Non-invasive estimation of hyperthermia temperatures with ultrasound. *International Journal of Hyperthermia* 2005;21:589–600.

Bats DA, Mackillop WJ. The effect of hyperthermia on the sodium-potassium pump in Chinese hamster ovary cells. *Radiation Research* 1985;103:441–451.

Beard RE, Magin RL, Frizzell LA, Cain CA. An annular focus ultrasonic lens for local hyperthermia treatment of small tumors. *Ultrasound in Medicine and Biology* 1982;8:177–184.

BenHur E, Elkind MM, Bronk BB. Thermally enhanced radioresponse of cultured Chinese hamster cells: Inhibition of repair of sublethal damage and enhancement of lethal damage. *Radiation Research* 1974;58:38–51.

Billard B, Hynynen K, Roemer RB. Effect of physical parameters on high temperature ultrasound hyperthermia. *Ultrasound in Medicine and Biology* 1990;16:409–420.

Bowman FH. Heat transfer mechanisms and thermal dosimetry. *National Cancer Institute Monograph* 1982;61:437–445.

Britt H, Pounds W, Lyons E. Feasibility of treating malignant brain tumors with focussed ultrasound. *Progress in Experimental Tumor Research* 1984;28:232–245.

Cain CA, Umemura SI. Concentric-ring and sector-vortex phased-array applicators for ultrasound hyperthermia. *IEEE Transactions on Microwave Theory and Techniques* 1986;MTT-345:542–551.

Cavaliere R, Ciocatto EC, Giovanella BC, Heidelberger C, Johnson RO, Margotti M, Mondovi B, Moricca G, Rossi-Fanelli A. Selective heat sensitivity of cancer cells: Biochemical and clinical studies. *Cancer* 1967;20:1351–1381.

Cheung AY, Golding WM, Samaras GM. Direct contact applicators for microwave hyperthermia. *Journal of Microwave Power* 1981;16:151–159.

Cheung AY, Neyzari A. Deep local hyperthermia for cancer therapy: External electromagnetic and ultrasound techniques. *Cancer Research* 1984;44:4736s–4744s.

Cho C-W, Liu Y, Cobb WN, Henthorn TK, Lillehei K, Christians U, Ng K-Y. Ultrasound-induced mild hyperthermia as a novel approach to increase drug uptake in brain microvessel endothelial cells. *Pharmaceutical Research* 2002;19:1123–1129.

Chopra R, Baker N, Choy V, Boyes A, Tang K, Bradwell D, Bronskill MJ. MRI-compatible transurethral ultrasound system for the treatment of localized prostate cancer using rotational control. *Medical Physics* 2008;35:134–135.

Cockett AT, Kazmin M, Nakamura R, Fingerhut A, Stein JJ. Enhancement of regional bladder megavoltage irradiation in bladder cancer using local bladder hyperthermia. *Journal of Urology* 1967;97:1034–1039.

Cordon-Cardo C, O'Brien JP, Casals D, Rittman-Grauer L, Biedler JL, Melamed MR, Bertino JR. Multidrug-resistance gene (P-glycoprotein) is expressed by endothelial cells at blood-brain barrier sites. *Proceedings of the National Academy of Sciences of the United States of America* 1989;86:695–698.

Corry PM, Barlogie B, Tilchen EJ, Armour EP. Ultrasound induced hyperthermia for the treatment of human superficial tumors. *International Journal of Radiation Oncology, Biology, Physics* 1982;8:1225–1229.

Corry PM, Jabboury K, Kong JS, Armour EP, McCraw FJ, LeDuc T. Evaluation of equipment for hyperthermia treatment of cancer. *International Journal of Hyperthermia* 1988;4:53–74.

Crile G. The effects of heat and radiation on cancers implanted on the feet of mice. *Cancer Research* 1963;23:372–380.

Damianou C, Kynynen K. Focal spacing and near-field heating during pulsed high temperature ultrasound therapy. *Ultrasound in Medicine and Biology* 1993;19:777–787.

Das H, Lele PP. Design of a power modulator for control of tumor temperature. In: Overgaard J, ed., *Hyperthermic Oncology*. London, U.K.: Taylor & Francis Group, 1984, pp. 707–714.

Davis B, Lele PP. A theoretical study of rapid hyperthermia by scanned focused ultrasound. *The Winter Annual Meeting of the American Society of Mechanical Engineering*, 1989, pp. 51–58, San Francisco, CA, Dec. 10–15.

De Poorter J, De Wagter C, De Deene Y, Thomsen C, Ståhlberg F, Achten E. Noninvasive MRI thermometry with the proton resonance frequency (PRF) method: *In vivo* results in human muscle. *Magnetic Resonance in Medicine* 1995;33:74–81.

Denis de Senneville B, Quesson B, Moonen CT. Magnetic resonance temperature imaging. *International Journal of Hyperthermia* 2005;21:515–531.

Dewey SC, Hopwood LE, Sapareto SA, Gerweck LE. Cellular responses to combinations of hyperthermia and radiation. *Radiology* 1977;123:463–474.

Dewey WC. Interaction of heat with radiation and chemotherapy. *Cancer Research* 1984;44:4714–4720.

Dewey WC, Freeman ML, Raaphorst GP, Clark EP, Wong RSL, Highfield DP, Spiro IJ, Tamasovic SP, Denman DL, Coss RA. Cell biology of hyperthermia and radiation. In: Meyn RE, Withers HR, eds., *Radiation Biology in Cancer Research*. New York: Raven Press, 1980, pp. 589–621.

Dewhirst MW, Moon T, Carlin D. Analysis of tumor volume and thermal dosimetric effects on tumor response to heat, radiation and heat plus radiation: Results of phase III randomized clinical trial in pet animals. In: Nussbaum GH, ed., *Physical Aspects of Hyperthermia*. American Association of Physicists in Medicine, 1982, pp. 495–510.

Dickinson RB. An ultrasound system for local hyperthermia using scanned focused transducer. *IEEE Transactions on Biomedical Engineering* 1984;BME-31:120–125.

Dickinson RJ. Thermal conduction errors of manganin-constantan thermocouple arrays. *Physics in Medicine and Biology* 1985;30:445–453.

Dickson JA. The effects of hyperthermia in animal tumor systems. Recent Results. *Cancer Research* 1977;59:43–111.

Dickson JA, Ellis HA. The influence of tumour volume and the degree of heating on the response of solid Yoshida sarcoma to hyperthermia (40–42°C). *Cancer Research* 1976;36:1188–1195.

Dickson JA, Shah DM. Effects of hyperthermia (42°C) on the biochemistry and growth of a malignant cell line. *European Journal of Cancer* 1972;8:561–571.

Dorr L, Hynynen K. The effects of tissue heterogeneities and large blood vessels on the thermal exposure induced by short high power ultrasound pulses. *International Journal of Hyperthermia* 1992;8:45–59.

Dromi S, Frenkel V, Luk A, Traughber B, Angstadt M, Bur M, Poff J, Xie J, Libutti SK, Li KCP, Wood BJ. Pulsed-high intensity focused ultrasound and low temperature-sensitive liposomes for enhanced targeted drug delivery and antitumor effect. *Clinical Cancer Research* 2007;13:2722–2727.

Duck F, Baker A, Starritt H. *Ultrasound in Medicine*. Philadelphia, PA: Institute of Physics Publishing, 1998.

Dunn F. Cellular inactivation by heat and shear. *Radiation and Environmental Biophysics* 1985;24:131–139.

Emami B, Perez CA. Hyperthermia in the treatment of cancer. *Applied Radiology* 1984;79:150–161.

Endrich B, Reinhold HS, Gross JF, Intaglietta M. Tissue perfusion inhomogeneity during early tumour growth in rats. *Journal of the National Cancer Institute* 1979;62:387–395.

Fan X, Hynynen KH. The effect of wave reflection and refraction at soft tissue interfaces during ultrasound hyperthermia treatments. *Journal of the Acoustical Society of America* 1992;91:1727–1736.

Fessenden P, Lee ER, Anderson TL, Strohbehn JW, Meyer JL, Samulski TV, Marmor JB. Experience with a multitransducer ultrasound system for localized hyperthermia of deep tissues. *IEEE Transactions on Biomedical Engineering* 1984;31:126–135.

Fite BZ, Liu Y, Kruse DE, Caskey CF, Walton JH, Lai C-Y, Mahakian LM, Larrat B, Dumont E, Ferrara KW. Magnetic resonance thermometry at 7T for real-time monitoring and correction of ultrasound induced mild hyperthermia. *PLoS One* 2012;7:e35509.

Folkman J, Contran R. Relation of vascular proliferation of tumour growth. *International Review of Experimental Pathology* 1976;16:207–248.

Fosmire H, Hynynen K, Drach GW, Stea B, Swift P, Cassady JR. Feasibility and toxicity of transrectal ultrasound hyperthermia in the treatment of locally advanced adenocarcinoma of the prostate. *International Journal of Radiation Oncology, Biology, Physics* 1993;26:253–259.

Fry FJ. Recent developments in ultrasound at biophysical research laboratory and their application to basic problems in biology and medicine. In: Kelly E, ed., *Ultrasound Energy.* Urbana, IL: University of Illinois Press, 1965, pp. 202–228.

Fry WJ, Fry RB. Determination of absolute sound levels and acoustic absorption coefficients by thermocouple probes-experiment. *Journal of the Acoustical Society of America* 1954a;26:311–317.

Fry WJ, Fry RB. Determination of absolute sound levels and acoustic absorption coefficients by thermocouple probes-theory. *Journal of the Acoustical Society of America* 1954b;26:294–310.

Gaber MH, Wu NZ, Hong K, Huang SK, Dewhirst MW, Papahadjopoulos D. Thermosensitive liposomes: Extravasation and release of contents in tumor microvascular networks. *International Journal of Radiation Oncology, Biology, Physics* 1996;36:1177–1187.

Gasselhuber A, Dreher MR, Negussie A, Wood BJ, Rattay F, Haemmerich D. Mathematical spatio-temporal model of drug delivery from low temperature sensitive liposomes during radiofrequency tumour ablation. *International Journal of Hyperthermia* 2010;26:499–513.

Gautherie M. *Interstitial, Endocavitary and Perfusion Hyperthermia: Methods and Clinical Trials. Clinical Thermotherapy.* Berlin, Germany: Springer-Verlag, 1990.

Geng X, Zhou Z, Li Q, Wu S, Wang C-Y, Liu H-L, Chuang C-C, Tsui P-H. Comparison of ultrasound temperature imaging with infrared thermometry during radio frequency ablation. *Japanese Journal of Applied Physics* 2014;53:047001.

Greene RF, Collins JM, Jenkins JF, Speyer JL, Myers CE. Plasma pharmacokinetics of adriamycin and adriamycinol—Implications of the design of in vitro experiments and treatment protocols. *Cancer Research* 1983;43:3417–3421.

Guthkelch AN, Carter LP, Cassady JR, Hynynen K, Iacono RP, Johnson PC, Obbens EAMT, Roemer RB, Seeger JF, Shimm DS, Steal B. Treatment of malignant brain tumors with focused ultrasound hyperthermia and radiation: Results of a phase I trial. *Journal of Neuro-Oncology* 1991;10:271–284.

Guy AW, Lehmann JF, Stonebridge JB. Therapeutic applications of electromagnetic power. *Proceedings of the IEEE* 1974;62:55–75.

Hahn GM. *Hyperthermia and Cancer.* New York: Plenum Press, 1982.

Hahn GM, Braun J, Har-Kedar I, Thermochemotherapy: Synergism between hyperthermia (42-43 degrees) and adriamycin (of bleomycin) in mammalian cell inactivation, *Proceedings of the National Academy of Sciences of the United States of America*, 1975;72:937–940.

Hahn GM, Strande DP. Cytotoxic effects of hyperthermia and adriamycin on Chinese hamster cells. *Journal of the National Cancer Institute* 1976;57:1063–1067.

Harari PM, Hynynen KH, Roemer RB, Anhalt DP, Shimm DS, Stea B, Cassady JR. Development of scanned focussed ultrasound hyperthermia: Clinical response evaluation. *International Journal of Radiation Oncology, Biology, Physics* 1991;21:831–840.

Harisiadis L, Sung D, Kessaris N, Hall EJ. Hyperthermia and low dose irradiation. *Radiology* 1978;129:195–198.

Henglein A. Sonolysis of carbon dioxide, nitrous oxide and methane in aqueous solution. *Zeitschrift für Naturforschung* 1985;40b:100–107.

Herman TS, Teicher BA, Jochelson M, Clark J, Svensson G, Coleman CN. Rationale for use of local hyperthermia with radiation therapy and selected anticancer drugs in locally advanced human malignancies. *International Journal of Hyperthermia* 1988;4:143–158.

Hill CR. The possibility of hazard in medical and industrial applications of ultrasound. *British Journal of Radiology* 1968;47:561–569.

Hornback NB, Shupe RE, Shidnia H, Joe BT, Sayoc E, Marshall C. Preliminary clinical results of combined 433 Megahertz microwave therapy and radiation therapy in patients with advanced cancer. *Cancer* 1977;40:2854–2863.

Horvath J. Ultraschallwirkung beim menschlichen Sarkom. *Strahlentherapie* 1944;75:119–125.

Huang SK, Stauffer PR, Hong K, Guo JW, Phillips TL, Huang A, Papahadjopoulos D. Liposomes and hyperthermia in mice: Increased tumor uptake and therapeutic efficacy of doxorubicin in sterically stabilized liposomes. *Cancer Research* 1994;54:2186–2191.

Hunt JW, Lalonde R, Ginsberg H, Urchuk S, Worthington A. Rapid heating: Critical theoretical assessment of thermal gradients found in hyperthermia treatments. *International Journal of Hyperthermia* 1991;7:703–718.

Hynynen K. Biophysics and technology of ultrasound hyperthermia. In: Gautherie M, ed., *Methods of External Hyperthermic Heating.* Heidelberg, Germany: Springer-Verlag Berlin, 1990, pp. 61–115.

Hynynen K. The threshold for thermally significant cavitation in dog's thigh muscle *in vivo. Ultrasound in Medicine and Biology* 1991;17:157–169.

Hynynen K. The feasibility of interstitial ultrasound hyperthermia. *Medical Physics* 1992;19:979–987.

Hynynen K, Darkazanli A, Unger E, Schenck J. MRI-guided non-invasive ultrasound surgery. *Medical Physics* 1993;20:107–115.

Hynynen K, DeYoung D. Temperature elevation at muscle-bone interface during scanned, focussed ultrasound hyperthermia. *International Journal of Hyperthermia* 1988;4:267–279.

Hynynen K, DeYoung D, Kundrat M, Moros E. The effect of blood perfusion rate on the temperature distribution induced by multiple, scanned and focussed ultrasonic beams in dogs' kidneys *in vivo. International Journal of Hyperthermia* 1989;5:485–497.

Hynynen K, Edwards DK. Temperature measurements during ultrasound hyperthermia. *Medical Physics* 1989;16:618–626.

Hynynen K, Roemer RB, Anhalt DP, Johnson C, Xu ZX, Swindell W, Cetas T. A scanned, focused, multiple transducer ultrasonic system for localized hyperthermia treatments. *International Journal of Hyperthermia* 1987a;3:21–35.

Hynynen K, Roemer RB, Moros E, Johnson C, Anhalt D. The effect of scanning speed on temperature and equivalent thermal exposure distributions during ultrasound hyperthermia in vivo. *IEEE Transactions on Microwave Theory and Techniques* 1986;34:552–559.

Hynynen K, Shimm D, Anhalt D, Stea B, Sykes H, Cassady JR, Roemer RB. Temperature distributions during clinical scanned, focused ultrasound hyperthermia treatments. *International Journal of Hyperthermia* 1990;6:891–908.

Hynynen K, Shimm D, Roemer RB, Anhalt D, Cassady JR. Temperature distributions during clinical ultrasound hyperthermia. *The 9th Annual Conference of IEEE Engineering in Medicine and Biology*, New York, 1987b, pp. 1644–1645.

Hynynen K, Watmough DJ, Mallard JR. The effects of some physical factors on the production of hyperthermia by ultrasound in neoplastic tissues. *Radiation and Environmental Biophysics* 1981;19:215–226.

Ibbinni MS, Cain CA. The concentric-ring array for ultrasound hyperthermia: Combined mechanical and electrical scanning. *International Journal of Hyperthermia* 1990;6:401–419.

Ishihara Y, Calderon A, Watanabe H, Okamoto K, Suzuki Y, Kuroda K, Suzuki Y. A precise and fast temperature mapping using water proton chemical shift. *Magnetic Resonance in Medicine* 1995;34:515–531.

Johnson C, Kress R, Roemer RB, Hynynen K. Multipoint feedback control system for scanned, focussed ultrasound hyperthermia. *35th Annual Meeting of Radiation Research Society*, Atlanta, GA, 1987, p. 12.

Johnson SA, Christensen DA, Johnson CC, Greenleaf JF, Rajagopalan B. Non-intrusive measurement of microwave and ultrasound-induced hyperthermia by acoustic temperature tomography. *IEEE Ultrasonics Symposium*, 1977, pp. 977–982, Oct. 26–28, Phoenix, AZ.

Kapp DS, Fessenden P, Samulski TV, Bagshaw MA, Cox RS, Lee ER, Lohrbach AW, Meyer JL, Prionas SD. Phase I evaluation of equipment for hyperthermic treatment of cancer. *International Journal of Hyperthermia* 1988;4:75–115.

Kato H, Hiraka M, Nakajima T, Ishida T. Deep-heating characteristics of an RF capacitive heating device. *International Journal of Hyperthermia* 1985;1:15–28.

Kim JH, Hahn EW, Tokita N, Nisce LZ. Local tumor hyperthermia in combination with radiation therapy. I. Malignant cutaneous lesions. *Cancer* 1977;40:161–169.

Kondo T, Kano E. Enhancement of hyperthermic cell killing by non-thermal effect of ultrasound. *International Journal of Radiation Biology* 1987;51:157–166.

Kondo T, Kuwabara M, Sato F, Kano E. Influence of dissolved gases on chemical and biological effect of ultrasound. *Ultrasound in Medicine and Biology* 1986;12:151–155.

Kong G, Braun RD, Dewhirst MW. Hyperthermia enables tumor-specific nanoparticle delivery: Effect of particle size. *Cancer Research* 2000;60:4440–4445.

Kong G, Braun RD, Dewhirst MW. Characterization of the effect of hyperthermia on nanoparticle extravasation from tumor vasculature. *Cancer Research* 2001;61:3027–3032.

Kong G, Dewhirst MW. Hyperthermia and liposomes. *International Journal of Hyperthermia* 1999;15:345–370.

Kremkau FW. Cancer therapy with ultrasound: An historical review. *Journal of Clinical Ultrasound* 1979;7:287–300.

Kruuv JA, Inch WR, McCredie JA. Blood flow and oxygenation of tumour in mice. II. Effect of vasodilator drugs. *Cancer* 1967;20:60–65.

Lehmann JF. *Ultrasound Therapy. Therapeutic Heat and Cold.* New Heaven, CT: Licht, 1965, pp. 321–386.

Lele PP. A strategy for localised chemotherapy of tumours using ultrasonic hyperthermia. *Ultrasound in Medicine and Biology* 1979;5:95–97.

Lele PP. Induction of deep, local hyperthermia by ultrasound and electromagnetic fields: Problems and choices. *Radiation and Environmental Biophysics* 1980;17:205–217.

Lele PP. Physical aspects and clinical studies with ultrasound hyperthermia. In: Storm FC, ed., *Hyperthermia in Cancer Therapy.* Boston, MA: Hall Medical, 1983, pp. 333–367.

Lele PP. Ultrasound: Is it the modality of choice for controlled, localized heating of deep tumors? In: Overgaard J, ed., *Hyperthermic Oncology.* London, U.K.: Taylor & Francis Group, 1984, pp. 129–154.

Lele PP. Rationale, technique and clinical results with scanned focussed ultrasound (SIMFU) systems. *8th Annual Conference of IEEE Engineering in Medicine and Biology Society*, New York, 1986, 1435–1440.

Lele PP, Goddard J. Optimizing insonication parameters in therapy planning for deep heating by SIMFU. *9th Annual Conference of IEEE Engineering in Medicine and Biology Society*, Boston, MA, 1987, pp. 1650–1651.

Lele PP, Parker KJ. Temperature distributions in tissues during local hyperthermia by stationary or steered beams of unfocused or focused ultrasound. *British Journal of Cancer* 1982;45:108–121.

Lepock JR, Cheng KH, Al-Qysi H, Kruuv JA. Thermotropic lipid and protein transitions in Chinese hamster lung cell membranes: Relationship to hyperthermic cell killing. *Canadian Journal of Biochemistry and Cell Biology* 1983;61:421–427.

LeVeen H, Wapnick S, Piccone V, Falk G, Ahmed M. Tumor eradication by radiofrequency therapy: Response in 21 patients. *JAMA* 1976;235:2198–2200.

Lin WL, Roemer RB, Hynynen K. Theoretical and experimental evaluation of a temperature controller for scanned focussed ultrasound hyperthermia. *Medical Physics* 1990;17:615–625.

Magin RL, Johnson RK. Effects of local tumour hyperthermia on the growth of solid mouse tumours. *Cancer Research* 1979;39:4534–4539.

Marmor JB, Nagar C, Hahn GM. Tumor regression and immune recognition after localized ultrasound heating. *Radiation Research* 1977;70:633.

Marmor JB, Pounds D, Postic TB, Hahn GM. Treatment of superficial human neoplasms by local hyperthermia induced by ultrasound. *Cancer* 1979;43:188–197.

McKee JR, Christman CL, O'Brien WD, Wang SY. Effects of ultrasound on nucleic acid bases. *Biochemistry* 1977;16:4651–4654.

Miller RC, Connor WC, Heusinkveld RS, Boone MLM. Prospects for hyperthermia in human cancer therapy. Part I. Hyperthermic effects in man and spontaneous animal tumours. *Radiology* 1977;123:489–495.

Moros EG, Roemer RB, Hynynen K. Pre-focal plane high-temperature regions induced by scanning focused ultrasound beams. *International Journal of Hyperthermia* 1990;6:351–366.

Moros EG, Roemer RB, Hynynen KH. Simulations of scanned focused ultrasound hyperthermia: The effects of scanning speed and pattern on the temperature fluctuations at the focal depth. *IEEE Transactions on Ultrasonics, Ferroelectrics, and Frequency Control* 1988;35:552–560.

Munro P, Hill RP, Hunt JW. The development of improved ultrasound heaters suitable for superficial tissue heating. *Medical Physics* 1982;9:888–897.

Needham D, Dewhirst MW. The development and testing of a new temperature-sensitive drug delivery system for the treatment of solid tumors. *Advanced Drug Delivery Reviews* 2001;53:285–305.

Ning S, Macleod K, Abra RM, Huang AH, Hahn GM. Hyperthermia induces doxorubicin release from long-circulating liposomes and enhances their anti-tumor efficacy. *International Journal of Radiation Oncology, Biology, Physics* 1994;29:827–834.

O'Neill BE, Li KCP. Augmentation of targeted delivery with pulsed high intensity focused ultrasound. *International Journal of Hyperthermia* 2008;24:506–520.

Ocheltree KB, Benkeser PJ, Frizzell LA, Cain CA. An ultrasonic-phased array applicator for hyperthermia. *IEEE Transactions on Sonics and Ultrasonics* 1984;SU-31:526–531.

Overgaard J. Effect of hyperthermia in malignant cells in vivo. A review and a hypothesis. *Cancer* 1977;39:2637–2646.

Overgaard J. The current and potential role of hyperthermia in radiotherapy. *International Journal of Radiation Oncology, Biology, Physics* 1989;16:535–549.

Overgaard J, Overgaard M. Hyperthermia as an adjuvant to radiotherapy in the treatment of malignant melanoma. *International Journal of Hyperthermia* 1987;3:483–501.

Overgaard K, Overgaard J. Investigations on the possibility of a thermic tumor therapy. II. Action of a combined heat-Roentgen treatment. *European Journal of Cancer* 1972;8:573–575.

Palussiere J, Salomir R, Le Bail B, Fawaz R, Quesson B, Grenier N, Moonen CT. Feasibility of MR-guided focused ultrasound with real-time temperature mapping and continuous sonication for ablation of VX2 carcinoma in rabbit thigh. *Magnetic Resonance in Medicine* 2003;49:89–98.

Parker KJ, Lele PP. The effect of blood flow on temperature distributions during localized hyperthermia. *Annals of the New York Academy of Sciences* 1980;335:64–65.

Patel PR, Luk A, Durrani AK, Dromi S, Cuesta J, Angstadt M, Dreher M, Wood B, Frenkel V. In vitro and in vivo evaluations of increased effective beam width for heat deposition using a split focus high intensity ultrasound (HIFU) transducer. *International Journal of Hyperthermia* 2008;24:537–549.

Perez CA, Gillepsie B, Pojak T, Hornback NB, Emami B, Ruben B. Quality assurance problems in clinical hyperthermia and its impact on therapeutic outcome: A report by the Radiation Therapy Oncology Group. *International Journal of Radiation Oncology, Biology, Physics* 1989;16:551–558.

Ponce AM, Vujaskovic Z, Yuan F, Needham D, Dewhirst MW. Hyperthermia mediated liposomal drug delivery. *International Journal of Hyperthermia* 2006;22:205–213.

Poon RT, Borys N. Lyso-thermosensitive liposomal doxorubicin: A novel approach to enhance efficacy of thermal ablation of liver cancer. *Expert Opinion on Pharmacotherapy* 2009;10:333–343.

Prakash O, Fabbri M, Drocourt M, Escanye JM, Marchal C, Gaulard ML, Robert J. Hyperthermia induction and measurement using ultrasound. *IEEE Ultrasonics Symposium*, 1980, pp. 1063–1066. Nov. 5–7, Boston, MA.

Rapoport SI, Ohno K, Pettigrew KD. Drug entry into the brain. *Brain Research* 1979;172:354–359.

Rivens I, Shaw A, Civale J, Morris H. Treatment monitoring and thermometry for therapeutic focused ultrasound. *International Journal of Hyperthermia* 2007;23:121–139.

Roemer RB, Swindell W, Clegg S, Kress R. Simulation of focused, scanned ultrasonic heating of deep-seated tumors: The effect of blood perfusion. *IEEE Transactions on Sonics and Ultrasonics* 1984;SU-31:457–466.

Ryan TP, Colacchio TA, Douple EB, Strohbenhn JW, Coughlin CT. Techniques for intraoperative hyperthermia with ultrasound: The Dartmouth experience with 19 patients. *International Journal of Hyperthermia* 1992;8:407–421.

Ryan TP, Hartov A, Colacchio TA, Coughlin CT, Stafford JH, Hoopes PJ. Analysis and testing of a concentric ring applicator for ultrasound hyperthermia with clinical results. *International Journal of Hyperthermia* 1991;7:587–603.

Salomir R, Vimeux FC, de Zwart JA, Grenier N, Moonen CTW. Hyperthermia by MR-guided focused ultrasound: Accurate temperature control based on fast MRI and a physical model of local energy deposition and heat conduction. *Magnetic Resonance in Medicine* 2000;43:342–347.

Samulski TV, Grant WJ, Oleson JR, Leopold KA, Dewhirst MW, Vallario P, Blivin J. Clinical experience with a multi-element ultrasonic hyperthermia system: Analysis of treatment temperatures. *International Journal of Hyperthermia* 1990;6:891–908.

Samulski TV, Lyons BE, Britt RH. Temperature measurements in high thermal gradients: II. Analysis of conduction effects. *International Journal of Radiation Oncology, Biology, Physics* 1985;11:963–971.

Sapareto SA, Dewey W. Thermal dose determination in cancer therapy. *International Journal of Radiation Oncology, Biology, Physics* 1984;10:787–800.

Satoh T, Stauffer PR. Implantable helical coil microwave antenna for interstitial hyperthermia. *International Journal of Hyperthermia* 1988;4:497–512.

Seetharaman S, Maskell L, Scheper RJ, Barrand MA. Changes in multidrug transporter protein expression in endothelial cells cultured from isolated human brain microvessels. *International Journal of Clinical Pharmacology and Therapeutics* 1998;36:81–83.

Shani A, Ygaek D, Levy E, Katsnelson R, Singer D, Fishelovitz Y, Yerushalmi A. Combined radiotherapy and intracavitary hyperthermia for the treatments of prostatic carcinoma—A decade of experience. *Strahlentherapie und Onkologie* 1990;166:532.

Shimm D, Cetas TC, Oleson JR, Cassady JR, Sim DA. Clinical evaluation of hyperthermia equipment: The University of Arizona Institute Report for the NCI Hyperthermia equipment evaluation contract. *International Journal of Hyperthermia* 1988a;4:39–51.

Shimm DS, Cetas TC, Oleson JR, Gross ER, Buechler DN, Fletcher AM, Dean SE. Regional hyperthermia for deep-seated malignancies using the BSD annular array. *International Journal of Hyperthermia* 1988b;4:159–170.

Shivers RR, Wijsman JA. Blood-brain barrier permeability during hyperthermia. *Progress in Brain Research* 1998;115:413–424.

Short JG, Tumer PF. Physical hyperthermia and cancer therapy. *Proceedings of the IEEE* 1980;68:133–141.

Shrivastava P, Luk K, Oleson JR, Dewhirst MW, Pajak T, Paliwal B, Perez CA, Sapareto SA, Saylor T, Steeves R. Hyperthermia quality assurance guidelines. *International Journal of Radiation Oncology, Biology, Physics* 1988;16:571–587.

Simon C, VanBaren P, Ebbini ES. Two-dimensional temperature estimation using diagnostic ultrasound. *IEEE Transactions on Ultrasonics, Ferroelectrics, and Frequency Control* 1998;45:1088–1099.

Smith NB, Merrilees NK, Dahleh M, Hynynen K. Control system for an MRI compatible intracavitary ultrasound array for thermal treatment of prostate disease. *International Journal of Hyperthermia* 2001;17:271–282.

Song CW. Effect of hyperthermia on vascular functions of normal tissues and experimental tumours. *Journal of the National Cancer Institute* 1978;60:711–713.

Song CW, Lokshina A, Rhee JG, Patten M, Levitt SH. Implication of blood flow in hyperthermic treatment of tumors. *IEEE Transactions on Biomedical Engineering* 1984;BME-31:9–16.

Staruch R, Chopra R, Hynynen K. Localised drug release using MRI-controlled focused ultrasound hyperthermia. *International Journal of Hyperthermia* 2011;27:156–171.

Stea B, Cetas TC, Cassady JR, Guthkelch AN, Iacono RP, Lulu B, Lutz W, Obbens EAMT, Rossman K, Seeger JF, Shetter A, Shim DS. Interstitial thermoradiotherapy hyperthermia of brain tumors: Preliminary results of a phase I clinical trial. *International Journal of Radiation Oncology, Biology, Physics* 1990;19:1463–1471.

Stehlin JS, Giovanella BC, Ipolyi PD, Muenz LR, Anderson RF. Results of hyperthermic perfusion for melanoma of the extremities. *Surgery, Gynecology, and Obstetrics* 1975;140:339–348.

Storm FK, Harrison WH, Elliott RS, Morton DL. Normal tissue and solid tumour effects of hyperthermia in animal models and clinical trials. *Cancer Research* 1979;39:2245–2251.

Strohbehn JW. Interstitial techniques for hyperthermia. In: Field SB, Franconi C, eds., *Physics and Technology of Hyperthermia*. Boston, MA: Nijhoff, 1987, pp. 221–239.

Swindell W. A theoretical study of nonlinear effects with focussed ultrasound in tissues: An acoustic Bragg peak. *Ultrasound in Medicine and Biology* 1985;11:121–130.

Szent-Györgyi A. Chemical and biological effects of ultrasonic radiation. *Nature* 1933;131:278.

Tannock IF. Population kinetics of carcinoma cells, capillary endothelial cells and fibroblasts in a transplanted mouse mammary tumour. *Cancer Research* 1970;30:2470–2476.

Tashjian JA, Dewhirst MW, Needham D, Viglianti BL. Rationale for and measurement of liposomal drug delivery with hyperthermia using non-invasive imaging techniques. *International Journal of Hyperthermia* 2008;24:79–90.

ter Haar G. Therapeutic ultrasound. *European Journal of Ultrasound* 1999;9:3–9.

ter Haar G, Daniels S. Evidence for ultrasonically induced cavitation in vivo. *Physics in Medicine and Biology* 1981;26:1145–1149.

ter Haar G, Hopewell JW. The induction of hyperthermia by ultrasound: Its value and associated problems I. Single, static, plane transducer. *Physics in Medicine and Biology* 1983;28:889–896.

ter Haar G, Stratford IJ, Hill CR. Ultrasonic irradiation of mammalian cells in vitro at hyperthermic temperatures. *British Journal of Radiology* 1980;53:784–789.

Ueno S, Hashimoto M, Fukukita H, Yano T. Ultrasound thermometry in hyperthermia. *IEEE Ultrasonics Symposium*, Honolulu, HI, 1990, pp. 1645–1652.

Underwood HR, Burdette EC, Ocheltree KB, Magin RL. A multielement ultrasonic hyperthermia applicator with independent element control. *International Journal of Hyperthermia* 1987;3:257–267.

Valdagni R, Amichetti M, Pani G. Radical radiation alone versus radical radiation plus microwave hyperthermia for N_3 (TNM-UICC) neck nodes: A prospective randomized clinical trial. *International Journal of Radiation Oncology, Biology, Physics* 1988;15:13–24.

Vargas HI, Dooley WC, Gardner RA, Gonzalez KD, Venegas R, Heywang-Kobrunner SH, Fenn AJ. Focused microwave phased array thermotherapy for ablation of early-stage breast cancer: Results of thermal dose escalation. *Annals of Surgical Oncology* 2004;11:139–146.

Wang H, Ebbini E, Cain CA. Effect of phase errors on field patterns generated by an ultrasound phased array hyperthermia applicator. *IEEE Transactions on Ultrasonics, Ferroelectrics, and Frequency Control* 1991;38:521–531.

Waterman FM. Determination of the temperature artifact during ultrasound hyperthermia. *International Journal of Hyperthermia* 1990;6:131–142.

Waterman FM, Dewhirst MW, Fessenden P, Samulski TV, Stauffer P, Emami B, Corry P et al. RTOG quality assurance guidelines for clinical trials using hyperthermia administered by ultrasound. *International Journal of Radiation Oncology, Biology, Physics* 1991;20:1099–1107.

Waterman FM, Leeper JB. Temperature artifacts produced by thermocouples used in conjunction with 1 and 3 MHz ultrasound. *International Journal of Hyperthermia* 1990;6:383–399.

Willis RA. *Pathology of Tumours*, 4th edn. London, U.K.: Butterworths & Co, 1967, pp. 133–135.

Wu NZ, Braun RD, Gaber MH, Lin GM, Ong ET, Shan S, Paphadjopoulos D, Dewhirst MW. Simultaneous measurement of liposome extravasation and content release in tumors. *Microcirculation* 1997;4:83–101.

Yatvin MB, Weinstein JN, Dennis WH, Blumenthal R. Design of liposomes for enhanced local release of drugs by hyperthermia. *Science* 1978;202:1290–1293.

Yerushalmi A, Servadio C, Leib A, Fishelovitz Y, Rokowski E, Stein JA. Local hyperthermia for treatment of carcinoma of the prostate: A preliminary report. *Prostate* 1982;3:623–630.

Zemanek J. Beam behavior within the nearfield of a vibrating piston. *Journal of the Acoustical Society of America* 1971;49:181–191.

12 Shock Wave Lithotripsy and Treatment

12.1 KIDNEY STONE DISEASE AND SWL HISTORY

Urinary stone disease is one of the most common and painful disorders of the urinary tract and has plagued humankind for centuries. The oldest bladder stone was found in an Egyptian mummy that is dated to 4,800 BC. Once a crystal nucleation occurs in the collecting system of the kidney, it will grow into a stone depending on the urine conditions (i.e., pH or supersaturation of calcium salts) (Coe et al. 2005, Worecester et al. 2008). The high concentration of calcium in urine is due to intestinal hyperabsorption, reduced renal reabsorption of calcium, and calcium release from bone that may be associated with low bone mass. Nearly 80% of kidney stones are mainly calcium oxalate with some calcium phosphate, and 10% of them are mainly calcium phosphate with some calcium oxalate. A typical calcium oxalate stone has a core of apatite (Johrde and Cocks 1985) and contains an organic matrix of mucoproteins, mucopolysaccharides, inorganic material, and bound water (Cheng et al. 1985). Other popular stones are uric acid, struvite, and cystin. The composition of kidney stone is usually characterized using infrared spectroscopy or x-ray crystallography (Worecester et al. 2008). Urinary stones are typically classified by their location in the kidney, ureter, or bladder. Urolithiasis is one of the most prevalent urological disorders and affects 5%–15% of the population worldwide (Teichman 2004, Moe 2006). Among many others, Peter the Great, Issac Newton, Napoleon, Louis XIV, George IV, and Benjamin Franklin all suffered from this disease. Typical symptoms of acute renal colic are intermittent colicky flank pain that may radiate to the lower abdomen or groin, often associated with nausea and vomiting (Eskelinen et al. 1998). As the stone moves and the body tries to push it out, blood may appear in the urine. As the stone moves down the ureter closer to the bladder, the more need to urinate or a burning sensation during urination will occur. If fever and chills accompany any of these symptoms, an infection may be present. Lower urinary tract symptoms such as dysuria, urgency, and frequency occur when a stone enters the ureter. Comorbid diseases should also be identified since it may increase the risk of kidney stone formation. Other factors such as a personal or family history of kidney stones with previous treatments, anatomical abnormalities, and surgery of the urinary tract also contribute to the occurrence of calculi. The recurrence of kidney stone disease within 5 years of the first stone event at a rate of as high as 50% indicates its chronic nature (Sutherland et al. 1985). The health-care costs related to kidney stones has grown from $1.8 billion in 1995 (Clark et al. 1995) to $2.5 billion in 2005 in the United States (Pearle et al. 2005), despite a shift from inpatient to outpatient treatment.

Currently, around 13% of men and 7% of women in the United States will have the kidney stone in their lifetime, mostly in the age range of 40–70 years (Neuhausel 1987, Stamatelou et al. 2003, Pearle et al. 2005). In the United Kingdom, at least one renal stone will form in approximately 8% of male and 4% of female patients (Ajayi et al. 2007). Unenhanced helical computed tomography is the best radiographical test for diagnosing urolithiasis in patients with acute flank pain (Vieweg et al. 1998). Most kidney stones are visible on computed tomography, except for stones induced by certain drugs, such as indinavir (Figure 12.1 and Tables 12.1 and 12.2).

Up to 98% of small stones (<2–3 mm in diameter) may pass spontaneously through urination within 4 weeks of the onset of symptoms (Miller and Lingeman 2007). However, the corresponding rate of spontaneous passage decreases to less than 53% for larger stones (5–10 mm in diameter) without causing symptoms (Gettman and Segura 2005), and they can cause obstruction of the ureter. Initial stone location also affects the success rate of spontaneous stone passage, increasing from 48% for stones located in the proximal ureter to 79% for those at the vesicoureteric junction regardless of stone size (Gettman and Segura 2005). Lithotomy, operations to remove calculi via the perineum by means of a surgical incision like other surgery before the invention of anesthesia, was a fairly common procedure and performed since about 300 BC, with intense pain, using specially designed surgical instruments, such as dilators of the canal, forceps and tweezers, lithotomes (stone cutter) and cystotomes (bladder cutter), urethrotomes (for incisions of the urethra), and conductors (grooved probes used as guides for stone extraction). It was quite successfully performed by some practitioners in the seventeenth century. Lateral vesical stone lithotomy was invented in 1727, and its duration was about 1 min (an important feat before anesthesia). But the mortality rate of lithotomy was 24%, may be due to the absence of antibacterial drugs at that time. After the invention of litholapaxy in 1878 (Bigelow 1878), the mortality rate dropped to 2.4%. Open surgery for non-passable calculi in the upper urinary tract now accounts for less than 1% of stone procedures in the United States. Because of the recurrence of the stone disease, multiple sessions are involved, with high potential for complications and a long in-hospital recovery period (Lingeman et al. 1986b).

During World War II, engineers of Dornier GmbH (an aircraft manufacturer in Germany) found the pitting of metal surfaces after supersonic flights (Coombs et al. 2000). The effect of shock waves on biologic tissues *in vivo* was investigated between 1968 and 1971 (Brendel et al. 1987, Brummer et al. 1990). In 1974, a research grant for the "Application of Extracorporeal Shock Wave Lithotripsy" was approved by the German government. The first clinical trial of shock wave

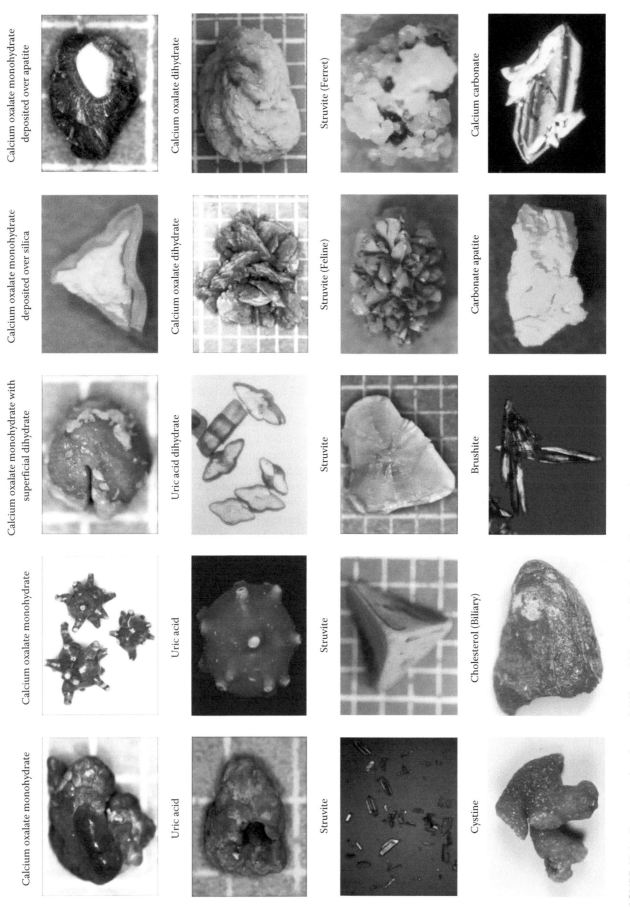

FIGURE 12.1 Representative photos of different kidney calculi taken from human patients.

TABLE 12.1
Hardness of Human Calculi

Calculi	Wet (MPa)	Dry (MPa)
Calcium oxalate monohydrate	650 ± 150	833 ± 178
Uric acid	300 ± 100	480 ± 100
Magnesium ammonium phosphate	190 ± 90	300 ± 60
Calcium carbonate (mixed bile pigment 14%)	1709	1730
Mixed bile pigments	28–44	50–166
Cholesterol	11–32	19–63

Source: *Ultrasonics*, 31, Coleman, A.J. and Saunders, J.E., A review of the physical properties and biological effects of the high amplitude acoustic fields used in extracorporeal lithotripsy, 80, Copyright 1992, with permission form Elsevier.

lithotripsy (SWL) on kidney stone and gallstone was performed in Munich, Germany, in 1980 and 1985, respectively (Ogden et al. 2001c, Thiel 2001). In 1983, the first commercial model, Dornier HM3, was released and installed in Stuttgart, Germany (Figure 12.2). The Food and Drug Administration (FDA) approved the clinical use of the HM3 in the United States in 1984, and since then extracorporeal SWL (ESWL) has gained rapid acceptance, revolutionizing the practice pattern of urologists on an outpatient basis in the treatment of urinary lithiasis, and becoming the front-line option worldwide for uncomplicated renal and proximal ureteral calculi because of its ease of use, noninvasive nature, high efficacy in treating solitary uncomplicated kidney and ureter stones, significantly lower morbidity than both surgical procedures for stone removal (pyelolithotomy/ureterolithotomy) and percutaneous nephrostolithotomy, and wide availability of lithotripters (Chaussy et al. 1984, Lingeman et al. 1986a, 1989, Chaussy and Fuchs 1989). In the treatment of urinary stones, ESWL has been used in ~85% of all cases after its introduction (Chaussy and Fuchs 1989). Stone-free rates for stones in the proximal and distal ureter are 83% and 85%,

respectively (Segura et al. 1997). Preliminary reports in children followed up to 36 months have found that renal growth does not appear to be affected, since no difference in growth rates was observed between treated and nontreated kidneys (Brown and Preminger 1988, Roth and Beckmann 1988, Krysiewicz 1992). Despite major concerns about potential hazards of SWL in pediatric patients, highly satisfactory results have been reported. The overall stone-free rates were initially 37%–52% and increased to 57%–100% in long-term follow-up after SWL (Brinkmann et al. 2001, Muslumanoglu et al. 2003). SWL monotherapy for staghorn calculi in children results in a stone-free rate of 73.3% after an average of two SWL sessions (Orsola et al. 1999). SWL also found its application in even complex cases, such as multiple stones, bilateral stones, stones in solitary kidneys, and staghorn calculi (Chaussy et al. 1980, 1982, Chaussy and Fuchs 1989). The motivation for shock wave (SW) propagation through the back or flank is to avoid bones (i.e., spine and ribs) and gas (e.g., in the intestines and lungs).

By 1989, clinical reports from Europe and the United States showed that approximately 85% of patients were treated by SWL without the need for open procedures (Chaussy and Fuchs 1989, Finlayson and Ackermann 1989, Kerbl et al. 2002). However, the use of lithotripsy has steadily declined to about 50% today, particularly due to the significant advance of minimally invasive ureteroscopic approaches and the lack of similar effectiveness of SWL as the first lithotripter (Pearle et al. 2005). The use of SWL by urologists in metropolitan hospitals is not very popular, which is attributed to their willingness to take up greater technical challenges, such as endoscopic procedures (Kijvaikai et al. 1997). The ureteroscope (URS) equipped with a camera is used to snare the small stone in the ureter or kidney or break it into passable tiny pieces through the urethra and bladder under general or local anesthesia during this procedure. A stent placed in the ureter can relieve swelling and promote healing. Percutaneous nephrolithotomy (PCNL) involves surgically removing a kidney stone using small telescopes and instruments inserted through a small incision in the back of the patient under general anesthesia.

TABLE 12.2
Fracture Toughness and Material Properties of Renal Calculi

Stone Composition (wt%)	Fracture Toughness (MPa m$^{1/2}$)	Density (kg/m^3)	Longitudinal Wave Speed (m/s)	Transverse Wave Speed (m/s)	Young's Modulus (GPa)	Shear Modulus (GPa)	Vickers Hardness (GPa)
Cystine (100)	High	1624 ± 73	4651 ± 138	2125 ± 9	20.065	7.333	0.238 ± 0.014
COM (100)	0.136 ± 0.021	2038 ± 34	4535 ± 58	2132 ± 25	25.162	9.264	1.046 ± 0.088
Brushite (95)/COM (5)	0.119 ± 0.030	2157 ± 16	3932 ± 134	1820 ± 22	19.486	7.145	0.727 ± 0.148
Uric acid (100)	0.090 ± 0.028	1546 ± 12	3471 ± 62	1464 ± 12	9.224	3.314	0.312 ± 0.044
CA (95)/COD (5)	0.057 ± 0.003	1732 ± 116	2724 ± 75	1313 ± 20	8.504	2.986	0.556 ± 0.170
MAPH (90)/CA (10)	0.056 ± 0.014	1587 ± 68	2798 ± 82	1634 ± 25	10.519	4.237	0.257 ± 0.080

Source: With kind permission from Springer Science+Business Media: *J. Mater. Sci. Lett.*, Characterization of fracture toughness of renal calculi using a microindentation technique, 12, 1993, 1461, Zhong, P., Chuong, C.J., and Preminger, G.M.

CA, carbonate apatite; COD, calcium oxalate dihydrate; COM, calcium oxalate monohydrate; MAPH, magnesium ammonium phosphate hydrogen.

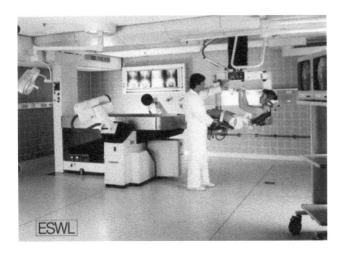

FIGURE 12.2 Donier HM3 lithotripter for the extracorporeal treatment of kidney stone.

12.2 DEVELOPMENT

Despite the great success of the Dornier HM3 lithotripter, there were a number of concerns about its design, such as the size and inconvenience due to water bath; the need for regional, general, or epidural anesthesia; the use of two expensive x-ray units; and the expensive disposable electrodes, which led Dornier and other manufacturers to make fundamental changes in the design. First-generation lithotripters did not allow treatment of mid- and lower ureteral stones in a satisfactory manner because of stone-targeting or positioning limitations. Meanwhile, a substantial body of evidence suggested that the HM3 lithotripter might be associated with significant renal trauma (Lingeman et al. 1990, Janetschek et al. 1997, Evan et al. 1998). Despite the lack of knowledge of SWL, manufacturers tried various ways to upgrade to second-generation devices. All lithotripter improvements are focused on four key features: (1) an energy source to generate the shock wave (SW), (2) a device to focus the SW at a certain location, (3) a coupling medium for SW propagation, and (4) a stone localization system.

The large beam size of the HM3 lithotripter in the focal region (≈15 × 90 mm) led to the hypothesis that a smaller focus might have higher stone comminution and less renal injuries. Thus, the aperture size was enlarged, which resulted in a broad acoustic field along the shock wave axis, resulting in the most dramatic difference in these new lithotripters of SWL operation under no or slight intravenous sedation. Many of these lithotripters have a focal width of around 5 mm (Cleveland and McAteer 2007). Although second-generation lithotripters were reported to have fewer side effects, unfortunately their stone fragmentation abilities also became less (Daniel and Burns 1990, Bierkens et al. 1992). Although the 1000 L water bath in the original Dornier HM3 lithotripter was theoretically ideal for coupling the SWs to the patient, the water processing plant required a dedicated facility. Modern lithotripters are dry-head devices, enclosing

the SW source and transmitting the SWs through a rubber or silicon membrane, through which the treatment head makes contact with the patient. Among the currently available models, the exception is the SLX from Storz Medical, which uses a partial water bath for coupling. However, such dry heads are less efficient than water baths (Chan et al. 1995) because defects of only 2% coverage reduced the stone fragmentation ability by 20%–40% (Pishchalnikov et al. 2006). Some second-generation lithotripters utilized sonography, which has the advantages of real-time imaging, absence of ionization, and the ability to image radiolucent calculi, to identify renal calculi with much lower costs than x-ray systems but with comparable outcomes (Marberger et al. 1988, Vallancien et al. 1988). However, most ureteral stones cannot be identified with ultrasound. Currently, all lithotripters in the United States have fluoroscopic imaging, usually in the form of a portable C-arm, which was introduced by Direx in the late 1980s (Servadio et al. 1988). Most second-generation lithotripters offer both ultrasound and x-ray localization capabilities. As the spark gap of the electrohydraulic (EH) lithotripter widens with the procedure of stone fragmentation, variations in the path of the arc discharge, SW amplitude, and location from pulse to pulse have increased gradually (Cleveland et al. 2000). Meanwhile, electrode cost (about $85 for the Dornier HM3 lithotripter, although small in comparison to $165 for the disposable Nitinol stone basket) was also a factor in the development of electromagnetic (EM) and piezoelectric (PE) shock wave generators in the late 1980s because they involved no disposables costs. The most successful second-generation lithotripter was the Siemens Lithostar, an EM type machine. In contrast, Richard Wolf never sold a lithotripter in the United States even with the approval for the Piezolith 2300 in 1988.

A hypothesis that increasing the peak pressures at the focus would improve clinical outcomes led to the third-generation lithotripters (Dornier Delta, Siemens Multiline, Storz Modulith). Unfortunately, these models cannot duplicate the success of Dornier HM3 and are also associated with more side effects (Kohrmann et al. 1995). Furthermore, SWL is uncomfortable for the patients irrespective of the machine used. Although a smaller focused beam size can have anesthesia-free SWL, most urologists in the United States prefer general anesthesia in order to minimize patient and renal motion. If not sedated, the patient will move, trying to get more comfortable. Thus, a totally anesthesia-free device has not been successful and accepted clinically (Cleveland and McAteer 2007). Two studies reported that clinical outcomes with third-generation lithotripters are improved with general anesthesia in comparison to intravenous sedation (Eichel et al. 2001, Sorensen et al. 2002). The technical feature of the third-generation lithotripter is the compact and modular design with improved imaging quality, which are largely practical concerns for user convenience and multifunctionality of the system rather than a rigorous understanding of the underlying mechanisms in SWL (Lingeman 1997, Lingeman et al. 2003a, Zhou et al. 2004). The acoustic characteristics of the SWs from different lithotripters are remarkably similar. Moreover, the evolution of lithotripter design thus far has overwhelmingly, and perhaps mistakenly, only considered the contribution of the compressive component of the lithotripter shock wave (LSW), almost neglecting that of the tensile component (Zhou et al. 2004).

Patients with the single-chamber pacemaker implanted in the abdomen and close to *F*2 should not undergo SWL (Asroff et al. 1993). Although several patients had abdominal aortic aneurysmal rupture after SWL (Neri et al. 2000), experimental and clinical data demonstrate that patients with aortic and renal aneurysms can be treated safely without complications (Deliveliotis et al. 1995). Major vein thrombosis has been reported after SWL, but very rare (Desmet et al. 1989, Brodmann et al. 1998). Sixty-two of 3,423 (1.81%) patients after SWL experienced a gastrointestinal (GI) injury complication (Makcr and Iayke 2004). Small bowel and colon perforation, ureterocolic fistula formation, GI anastomosis dehiscence, cecal ulcers, colon erythema, bruising and hematomas, bleeding per rectum, pancreatitis and peripancreatic hematoma and abscess formation, liver and spleen subcapsular hematomas, and ileus have been reported in case studies (Maker and Iayke 2004). But SWL does not cause severe permanent malfunctionalities on testicles and the ovary (Vieweg et al. 1992, Basar et al. 2004). Pregnancy is the only absolute contraindication because of the potential disruption on the fetus (Ohmori et al. 1994). Embryotoxic or teratogenic sequelae may not occur when SWs are focused outside the uterus (Deliveliotis et al. 2001). Until now, although various side effects have been reported, most are rare and do not hamper the effectiveness and application of SWL (Skolarikos et al. 2006).

Biliary stones or gallstones have a higher incidence than kidney stones and are classified into two types: cholesterol (usually yellow) and pigment gallstones (usually black) (Stranne et al. 1990). While most gallstones have a hardness of about one order of magnitude lower than kidney stones, they are more difficult to break *in vitro* (Stranne et al. 1990). Although SWL was used to treat gallstones in Europe, it never received acceptance in the United States. SWL was also used to break salivary stones, but infrequently (Ottaviani et al. 1997, Siddiqui 2002, McGurk et al. 2005). In biliary stone disease, it is used in 20% of cases, which are unsuitable for minimally invasive surgical procedures including laparoscopic cholecystectomy (Sauerbruch and Paumgartner 1991).

At the same time, endoscopic devices, such as EH nephrolithotripsy (EHL), and Ho–yttrium aluminum garnet (Ho:YAG) laser, liquid drop impact, and the use of explosives, lasers, and ultrasonically activated invasive probes, have also been developed to break the calculi (Miller 1983, 1985, Bailey et al. 1984, Field 1991). EHL can be combined with Ho:YAG laser via flexible ureteroscopes to treat renal calculi of size 2–4 cm mini-invasively with similar or superior results as PCNL. Treatment strategies were the aggressive treatment of infection, EHL stone debulking, Ho:YAG

laser lithotripsy to weaken hard stones structurally, manual piston irrigation to maintain visibility, and bladder drainage to maintain low intrarenal pressures. The complete stone-free rate was 92% with all residual fragments <4 mm with renal calculi (COM and/or apatite) 22–42 mm long (mean 33 mm) among 13 patients in the 3-month follow-up. All patients were rendered infection-free and symptom-free, and one was hospitalized overnight for the management of preexisting pulmonary problems and one rehospitalization for colic management but no other unplanned emergent care (Mariani 2004). Choldocholithiasis and intrahepatic duct stones pose a significant health risk. Although it is estimated that 20% of patients with choledocholithiasis, notably those with smaller stone (<3 mm), will pass their stones spontaneously, 80% of patients will require a therapeutic intervention (Bergdahl and Holmlund 1976). Standard endoscopic stone removal techniques consist of endoscopic biliary sphincterotomy followed by balloon or basket extraction of stones, with 80%–90% success rate with complications less than 10% (Sivak 1989, Shaw et al. 1993, Freeman et al. 1996). Mechanical lithotripsy (ML) as second-line treatment for choldocholithiasis and intrahepatic duct stones has a success rate of 88%–92% (Schneider et al. 1988, Shaw et al. 1993). Difficult biliary stone disease includes large stones >2 cm, stones above a narrow duct segment, impacted stones, and stones lodged in the cystic duct. EHL is also used in the management of difficult choledocholithiasis and large intrahepatic bile duct stones via peroral endoscopic choledochoscopy effectively and safely. Ninety-three patients who had endoscopic retrograde cholangiopancreatography (ERCP) and failed standard stone extraction techniques because of large stones (81 patients) or a narrow caliber bile duct (13 patients) underwent EHL and achieved 96% successful fragmentation (61 complete, 28 partial) and a final stone clearance of 90% in a mean follow-up of 26.2 months. Seventy-six percent of patients required one EHL session, 14% required two sessions, and 10% required three or more. All patients with successful stone fragmentation required post-EHL balloon or basket extraction of fragments. Only minor complications, such as cholangitis and/or jaundice (13 patients), mild hemobilia (1 patient), mild post-ERCP pancreatitis (1 patient), biliary leak (1 patient), and bradycardia (1 patient), were observed (Arya et al. 2004) (Figure 12.3). In a series of 50 patients, EHL under direct cholangioscopic vision resulted in 100% fragmentation rate and 92% final stone clearance rate (Bonnel et al. 1991). In another smaller series of 40 patients in Japan, no serious complications and only 18% with minor complications were reported following percutaneous transhepatic EHL under cholangioscopic control in the management of choledocholithiasis, hepatolithiasis, and cholecystolithiasis (Yoshimoto et al. 1989). Overall, EHL had a similar stone-free rate as SWL, but with fewer complications and requiring fewer sessions (Binmoeller et al. 1993, Adamek et al. 1995, 1996); 92%–94% patients successfully treated with SWL, required additional endoscopic removal of stone fragments.

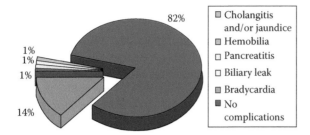

FIGURE 12.3 Complications of EHL. (With permission from Macmillan Publishers Ltd. *Am. J. Gastroenterol.*, Arya, N., Nelles, S.E., Haber, G.B., Kim, Y.-I., Kortan, P.K., Electrohydraulic litho-tripsy in 111 patients: A safe and effective therapy for difficult bile duct stones, 99, 2332, copyright 2004.)

12.3 LIMITATIONS IN RENAL INJURY AND STONE FRAGMENTATION

Although considered as a primary option for the treatment of virtually all stones, predisposing factors to SWL limitations are stone composition, size, location, and number, as well as renal morphology and shock wave rate and energy (Madbouly et al. 2002, Abdel-Khalek et al. 2004). Stone fragility during lithotripsy depends to a greater or lesser extent on the chemical composition (Dretler 1988), the crystalline structure, flaw size and distribution, microhardness (Gracewski et al. 1990) and elasticity (Johrde and Cocks 1985), stone volume and shape (Sauerbruch et al. 1991), as well as the gas content (Vakil and Everbach 1991). Stone-free rate for stones <2 and 2–3 cm is 91% and 50%–70%, respectively (Chaussy et al. 1984), and decreases further for staghorn stones (Lingeman et al. 1987). Proximal ureteral stones can be treated more successfully (65%–81%) than lower ones (58%–67%). SWL success rate varies, 62%–92% for small stones and 33%–84% for larger stones (Sheir et al. 2003, Lingeman et al. 2007). Some calculi, such as brushite, calcium oxalate monohydrate, and cysteine stones, are found more resistant to SWL (Dretler 1988, Zhong et al. 1993, Kim et al. 2007). Complete stone fragmentation requires more sessions or an ancillary procedure to remove residual fragments, and the maximum size of stone that can be treated is around 2.5 cm because of the limited discharge capacity of the ureter. The number of SWs was varied to complete *in vitro* comminution (passable through a sieve with 3 mm round or 2 mm square holes) for 195 human stones, representing six major stone types classified by their dominant mineral composition: 400 ± 333 g^{-1} ($n = 39$) for uric acid, 965 ± 900 g^{-1} ($n = 75$) for calcium oxalate monohydrate, 1134 ± 770 g^{-1} ($n = 21$) for hydroxyapatite, 1138 ± 746 g^{-1} ($n = 13$) for struvite, 1681 ± 1363 g^{-1} ($n = 23$) for brushite, and 5937 ± 6190 g^{-1} ($n = 24$) for cystine. The variation for these natural stones (83% ± 15%) was greater than that for artificial (i.e., gypsum-based) stones (17% ± 8%) (Williams et al. 2003). Aspects of renal anatomy, such as lower pole calyx, acute infundibulopelvic angle, and calyceal diverticula, could pose a barrier to the clearance of stone debris (Lingeman et al. 2007). Other anatomic features, including ureteropelvic

obstruction, calyceal diverticulum, and fusion anomalies such as horseshoe kidney, can also negatively affect the outcome (Lingeman et al. 2007). In a review of 309 patients who had 373 renal calculi with complete follow-up, the initial fragmentation rate was 94%. The stone-free rate for stones <10 mm was 77%, for 11–20 mm it was 69%, and for >20 mm it was 50%. The stone-free rate was 59% and 75% with 1 and 3 months of SWL, respectively. Additional procedures to render patients stone-free after SWL were needed in only 22 cases (7%). The stone-free rates for lower, upper, middle calyceal, and renal pelvic calculi were 74%, 70%, 78.5%, and 75%, respectively (Nomikos et al. 2007). The low success rates in the lower pole are due to the renal anatomy and gravity on the retention of fragments. The slower sound speed in fat (approximately 1450 m/s) results in the refraction of the SWs as they pass through fat (Duck 1990), which causes the focus to broaden and get shifted. Obese patients (high body mass index, BMI) have larger distance between the skin and stone, which makes stone fragmentation by SWL more difficult (Siener 2006); it even fails when the distance between the skin and stone is more than 9–10 cm (Pareek et al. 2005, Perks et al. 2008). Currently, around 30%–50% of patients need re-treatment (Tan et al. 2002, Papadoukakis et al. 2006), and some undergo more than three sessions (Tolley and Downey 1999, Auge and Preminger 2002). The presence of multiple stones leads to a higher recurrence after SWL (Chaussy et al. 1984, Evan and McAteer 1996, Sun et al. 1996, Moody et al. 2001).

PCNL is superior to SWL or URS in the treatment of large stones and staghorn calculi despite the higher complication rate and longer hospital stay (Preminger et al. 2005). SWL and PCL had stone-free rates of 63% and 100% for stones of 1 cm or smaller in the lower pole, respectively (Albala et al. 2001). For stones larger than 1 cm, the corresponding values were 21% and 91%. The 1997 American Urological Association (AUA) Guidelines preferred SWL for proximal ureteral stones less than 1 cm in size with a stone-free rate of 84%, while the median success rate of URS was only 56% (Segura et al. 1997). In contrast, the 2007 AUA–European Association of Urology (EAU) Ureteral Stone Guidelines Panel highlighted that the corresponding stone-free rates with SWL and URS were 90% and 80%, respectively, due to the technical advance in URS (Preminger et al. 2007). Overall, URS is generally preferred in the distal ureter for small ureteral stones, but either SWL or URS is acceptable in the proximal ureter (Wolf 2007). For larger stones, URS seems to have a higher stone-free rate and fewer recessions in comparison to SWL, but with a slightly higher complication rate and longer hospital stay (Preminger et al. 2007).

Initially, the resulting complications after SWL treatment were reported to be reversible and did not affect the clinical course of the patient. However, a substantial amount of evidence illustrates that these severe acute renal and extrarenal injuries have the potential for long-term adverse effects (Evan and McAteer 1996). SWs could rupture a broad range of blood vessels, and the renal papilla was particularly susceptible to SW-induced damage. Management of intraparenchymal hemorrhage, massive renal hematomas, edema, and acute tubular necrosis of the kidney is generally conservative, while the severe ones require transfusion or even nephrectomy. The renal trauma initiated an inflammatory response that could result in an immediate decrease in renal function of the treated kidney or scarring with permanent loss of functional renal volume. Tubular injury could occur, but usually secondary to vessel rupture. Such hematomas usually spontaneously resorb within 6 weeks but occasionally last for 6 months or more (Knapp et al. 1988, Lingeman et al. 1988). Extrarenal damage, such as intra-abdominal bleeding, splenic rupture, and hematomas of the liver and pancreas, also occurred. The lesion size was dose dependent and different from the dimensions of the focal zone (Willis et al. 2005). Even a brief exposure to SWs could initiate a vasoconstrictive response in the kidney (Willis et al. 2006, Handa et al. 2009a). Although glomerular and tubular functions were affected acutely, the kidney recovered in a short time. Furthermore, gastrointestinal, cardiovascular, genital, and reproductive systems may also be affected by SWL treatment.

Post-ESWL renal abnormalities are as much as 63%–85% of cases when evaluated by imaging methods, including intravenous urography (IVU), radionuclide renography, sonography, computed tomography (CT), and magnetic resonance imaging (MRI) or positron emission tomography (PET) imaging (Kaude et al. 1985, Baumgartner et al. 1987, Rubin et al. 1987, Lingeman et al. 1988). Clinically significant hemorrhage, causing ongoing morbidity or mortality, was estimated to be ~22% for open surgery versus 3%–6% for percutaneous procedures and <1% for SWL (Brown and Preminger 1988, Das et al. 1988, Roth and Beckmann 1988). Occasionally, the anuric renal failure following ESWL without obstruction or myoglobinuria but with unilateral nephrolithiasis and two normally functioning kidneys was found. Attention should be paid to older patients (Liguori et al. 2004). When CT or MRI is performed routinely after SWL, the hematoma rate was 20%–25% (Dhar et al. 2004) (Figure 12.4).

The renal trauma and vascular disruption associated with SWL may allow bacteria in urine to enter the bloodstream. Moreover, when the infected calculi are destroyed, bacteria are released into the urine and may be absorbed systemically (Moody et al. 2001). As a consequence, bacteriuria, bacteremia, clinical urinary tract infection, urosepsis, perinephric abscess formation, endocarditis, candida and Klebsiella endophthalmitis, candidal septicemia, tuberculosis, and, rarely, death could occur (Moody et al. 2001). Bacteriuria was found in 7.7%–23.5% of patients after SWL, including 7.7% of those without infection-related stones (Muller-Mattheis et al. 1991, Moody et al. 2001).

SWL can also cause trauma to thin-walled vessels in the kidneys and adjacent tissues (Evan and McAteer 1996), such as hemorrhage, release of cytokines/inflammatory cellular mediators, and tissue infiltration by inflammatory response cells, which may lead to short-term complications, formation of scar, and chronic renal injury. Renal complications can be classified into early effects on kidney anatomy with the introduction of hematuria and hematoma (i.e., thin-walled veins in the corticomedullary junction), and late complications that

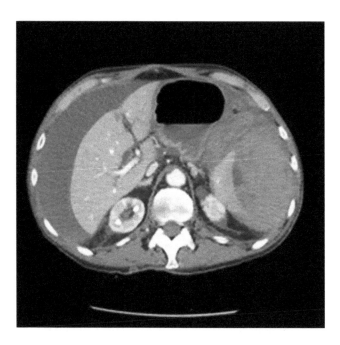

FIGURE 12.4 Axial CT scan image showing large splenic laceration with hematoma in the peritoneal cavity. (Doran, O. and Foley, B.: Acute complications following extracorporeal shock-wave lithotripsy for renal and ureteric calculi. *Emerg. Med. Australas.* 2008. 20. 106. Copyright Wiley-VCH Verlag GmbH & Co. KGaA. With permission.)

affect kidney function (Karlsen et al. 1991). SWL-induced acute renal damage may also result in severe injury to the nephron, microvasculature, and the surrounding interstitium (Delvecchio et al. 2003); renal tubules and vessels are more vulnerable than renal blood flow to SWs (Connors et al. 2000). These injuries may be the reason for SWL-induced chronic effects on renal function. Endothelial cell damage was found in mid-sized arteries, veins, and glomerular capillaries immediately after SWL (Karlsen et al. 1991, Recker et al. 1992). Symptomatic intrarenal, subcapsular, or perirenal fluid collections and hematomas are rare and occur in <1% of patients after SWL (Dhar et al. 2004). Treatment of the hematomas is conservative in most cases. Blood and urine markers, such as renin, creatinine, N-acetyl-β-D-glucosaminidase (NAG), β-galactosidase (BGAL), β-2-microglobulin (B2M), and proteinuria, return to almost normal levels within a few days (Neal et al. 1991, Recker et al. 1992, Perez-Blanco et al. 2001).

Inadequate spontaneous fragment passage is another concern of clinical complications of SWL. Although the complete elimination of stone fragments occurs in the first week for most patients, it may take up to several months, depending on stone size, location, quality of fragmentation, age and ambulation of the patient, hydration, and individual anatomy (Riehle 1986, Riehle et al. 1986). Obstruction with or without symptomatic colic may occur 4–6 weeks postESWL or longer.

Some patients, particularly the older ones and those have undergone multiple sessions, were found at increased risk of new-onset hypertension and diabetes mellitus (Newman et al. 1987, Abrahams et al. 1988, Lingeman et al. 1988, Williams et al. 1988, Ackaert and Schroder 1989, Begun et al. 1989, Neurburg et al. 1989, Rigatti et al. 1989, Janetschek et al. 1997, Frauscher et al. 1999, Krambeck et al. 2006), which may be associated with the complexity and severity of the stone disease (Parks et al. 2004). At 19 years of follow-up, SWL of 578 patients for renal and proximal ureteral stones was associated with the development of hypertension and diabetes mellitus, which is significantly higher than in a cohort of conservative treatment (Krambeck et al. 2006). Hypertension was more prevalent in the SWL group (OR = 1.47, 95% confidence interval [CI]: 1.03, 2.10, $p = 0.034$) and related to bilateral treatment ($p = 0.033$). In the SWL group, diabetes mellitus developed in 16.8% of patients. Patients treated with SWL were more likely to have diabetes mellitus than controls (OR = 3.23, 95% CI: 1.73–6.02, $p < 0.001$), which was related to the number of delivered SWs and intensity ($p = 0.005$ and 0.007).

There is no difference in the incidence of new-onset hypertension in the general population, which is ~6% (Jewett et al. 1998). Systolic hypertension did not increase after SWL, but diastolic hypertension did in a number of studies (Lingeman and Kulb 1987, Williams et al. 1988, Yokoyama et al. 1992). An increasing number of SWs correlate with more severe diastolic hypertension (Yokoyama et al. 1992). However, these preliminary surveys were of uncertain validity because of their retrospective nature, the lack of control groups, and a low response rate (only up to 33%). Patients over the age of 60 years may have a higher possibility of developing hypertension than the normal population, so they should be excluded from postoperative screening for hypertension in SWL. Randomized controlled trials did not reveal any evidence that SWL caused changes in blood pressure (Zanetti et al. 1992, Jewett et al. 1998, Elves et al. 2000). Furthermore, a recent study indicated that SWL might be responsible for a decrease in blood pressure, possibly due to the alteration in the intrarenal metabolism. Cardiac arrhythmias during SWL are not uncommon, with an incidence of 11%–59% (Zanetti et al. 1999). No correlation is demonstrated between ventricular premature contractions and patient age, gender, heart disease, stone size and location, presence of a ureteral catheter or nephrostomy tube, mode of anesthesia, the number of shock waves, or types of lithotripters (Zanetti et al. 1999). Of 148 urolithiasis patients who underwent SWL in 1984, 2 of the 21 patients who returned after 17–21 months had developed hypertension requiring treatment but became normotensive when given medication (Williams et al. 1988). In the other patients, there was a statistically significant increase in both systolic ($p = 0.0002$) and diastolic ($p = 0.015$) blood pressures. Although the pathogenesis is unknown, hypertension occurs immediately or after several weeks or months after SWL in about 8% of patients who also had a perirenal hematoma and a remarkable decrease in the percentage of effective renal plasma flow (ERPF) to the treated kidney. Quantitative radionuclide renography showed a slight but insignificant decrease in the ERPF of the treated kidney from a mean of 260–255 mL/mm, while the total ERPF to both kidneys increased slightly but not significantly from a mean of 520–540 mL/mm.

However, the percentage of total ERPF to the treated kidney declined from a mean of 50% before ESWL with statistical significance ($p = 0.048$). The mean diastolic blood pressure increased from 78 to 84 mmHg after SWL ($p = 0.015$), exceeding any age-related increase in blood pressure (systolic blood pressure rises at an average rate of 0.5–1.0 mmHg/year beyond the age of about 45 years, Dollery [1979]). There was a definite trend between the decreased ERPF to the treated kidney and increased systolic and diastolic blood pressures in 18 patients, but the correlation was not significant in this small series ($p = 0.12$ for the correlation with systolic blood pressure and $p = 0.22$ for diastolic blood pressure). In addition, no partial and total parenchymal obstructive patterns attributed to acute tubular necrosis and edema caused by hemorrhage were observed in the follow-up after SWL (Williams et al. 1988). In another study, a decrease in ERPF of more than 5% was found in 10 of 33 of the treated kidneys, and MRI showed renal trauma (edema or hemorrhage) in 24 of 38 of these treated kidneys (Kaude et al. 1985).

In the SWL group, diabetes developed in 16.8% of patients versus 6.6% in controls (Krambeck et al. 2006). In a population-based study in Taiwan, a total of 2,921 of 23,569 patients who were diagnosed with urinary calculi from 2001 to 2003 together with 6,171 of 70,707 matched enrollees as a comparison cohort were found to be suffering from diabetes mellitus during the 5-year follow-up. The stratified Cox proportional analysis showed that, after removing individuals who died during follow-up and adjusting for patient monthly income, geographic location, urbanization level, hypertension, hyperlipidemia, and obesity, the hazard of receiving a first diagnosis of diabetes during the 5-year follow-up was 1.32 times greater for patients with urinary calculi than for those in the comparison cohort (95% CI: 1.26–1.39, $p < 0.001$) (Chung et al. 2011).

However, the relationship is still controversial. In a large population-based cohort of kidney stone formers, there was no association between ESWL and the subsequent long-term risk of hypertension (Krambeck et al. 2011). In a survey of 772 renal stone patients and 505 ureteral stone patients treated with SWL, between 1984 and 1994, the rates of new onset of hypertension in the renal stone and ureteral stone groups were 22.8% and 20.0% in men and 23.1% and 20.5% in women, respectively; and the rates of new onset of diabetes mellitus (DM) in the renal stone and ureteral stone groups were 7.4% and 11.0% in men and 8.7% and 8.7% in women, respectively. SWL treatment was not a significant risk factor for new-onset hypertension and DM, as found by logistic regression analysis (Sato et al. 2008). After an average follow-up of 8.7 years for a total of 5287 urolithiasis from 1985 to 2008, 423 patients (8%) were treated with SWL, and new-onset DM had developed in 743 (12%). However, no association was evident between SWL and the development of DM before (hazard ratio 0.98, 95% CI: 0.76–1.26) or after (hazard ratio 0.92, 95% CI: 0.71–1.18) SWL, controlling for age, sex, and obesity (de Cógáin et al. 2012). Prevalence of diabetes and hypertension in 727 patients treated for urolithiasis 20 years ago between 1985 and 1989 was compared to those of the background population of British Columbia. In univariate analysis, SWL using HM3 (either unmodified or modified model) was not associated with increased DM or hypertension (Chew et al. 2012).

Fluoroscopy requires more space, has the risk of ionization to both the patient and operators, and is not useful for radiolucent calculi (Buchholz and van Rossum 2001). Sonography suffers from low resolution and acoustic shadowing effects and is also not effective for all stone types (Preminger 1989). Although the fluoroscopic and ultrasound imaging units equipped with lithotripters have been continuously improved for stone localization, effective and reliable determination of the treatment end point remains a problem. One of the reasons for overtreating stone patients is the inability of conventional imaging to show when the stone is broken to completion. Experienced operators seek softening of margins, loss of density, and movement of particles as signs of fragmentation; however, such features are mostly subjective and empirical, so are hard to be quantified for universal use (Leighton et al. 2008).

Despite of the development of the lithotripter and numerous available models, EH lithotripters are still considered the gold standard in terms of stone fragmentation efficacy, because of the superiority proven both experimentally and clinically (Lingeman et al. 2002). The clinical results demonstrate that the first commercial lithotripter (Dornier HM3) has not been surpassed by any other model (Kerbl et al. 2002), although no consensus study was done to support it (Tailly et al. 2008). The efficacy of three generations of lithotriptors, namely the original HM3, Lithostar® Plus (LSP), and the Modulith® SLX, in stone disintegration and dilatation of the pyelocaliceal system using identical protocol inclusion and follow-up criteria but with different modes of anesthesia was evaluated by abdominal x-ray and renal sonography 1 day and 3 months after treatment and compared with each other at a single prospective, randomized trial. This study indicated that the HM3 lithotriptor disintegrated caliceal and renal pelvic stones better than the LSP and SLX models, with fewer complications and re-treatments (Gerber et al. 2005) (Table 12.3). In another study, several popular

TABLE 12.3
Baseline Characteristics, Treatment Parameters, and Outcome in 234 Patients (264 Caliceal or Renal Pelvic Stones) Treated with HM3, LSP, or SLX

	HM3	p-Value (HM3 vs. LSP)	LSP	p-Value (LSP vs. SLX)	SLX	p-Value (HM3 vs. SLX)
No. stone	82		75		107	
No. patients (% men/women)	76 (66/10)		69 (46/23)		89 (58/31)	
Mean age ± std	47 ± 16	0.3	45 ± 15	0.06	49 ± 16	0.43
Mean stone vol ± std						
Renal pelvis	0.64 ± 0.35	0.55	0.7 ± 0.42	0.07	0.5 ± 0.36	0.15
Upper calix	0.36 ± 0.33	0.4	0.25 ± 0.26	0.79	0.27 ± 0.23	0.38
Mid calix	0.18 ± 0.13	0.95	0.19 ± 0.13	0.94	0.19 ± 0.13	0.86
Lower calix	0.34 ± 0.28	0.97	0.35 ± 0.26	0.79	0.37 ± 0.35	0.75
No. anesthesia (%)						
Sedoanalgesia	0		0		83 (93)	
Epidural	71 (94)		61 (88)		6 (7)	
Spinal	4 (5)		5 (8)		0	
General	1 (1)		3 (4)		0	
Mean fluoroscopy time (s)	53 ± 48	0.46	61 ± 72	<0.001	106 ± 50	<0.001
Mean no. applied shock waves	1702 ± 628	<0.001	2297 ± 430	<0.001	3349 ± 401	<0.001
Mean treatment time (min)	27 ± 11	<0.001	35 ± 11	0.009	39.6 ± 11	<0.001
Shock wave rate (no./min)	67 ± 19	0.52	69 ± 19	<0.001	84 ± 26	<0.001
Stone free on postoperative day 1 (%)	91	<0.001	65	0.015	48	<0.001
3–5 mm fragments (%)	7	0.006	21	0.06	35	<0.001
>6 mm fragments (%)	1	0.002	4	0.1	15	<0.001
No collecting system dilation (%)	85	0.05	73	0.69	71	0.02
Re-treatment (%)	4	0.05	13	<0.001	38	<0.001
Obstructive pyelonephritis (%)	1	0.02	8	0.4	5	0.12
Stone free 3 months later (%)	89		83		81	
<2 mm residual fragments (%)	10		10		9	
3–5 mm residual fragments (%)	1	0.52	5	0.02	9	0.005
>6 mm residual fragments (%)	0		2		2	

Source: J. Urol., 173, Gerber, R., Studer, U.E., and Danuser, H., Is newer always better? A comparative study of 3 lithotriptor generations, 2015, Copyright 2005, with permission from Elsevier.

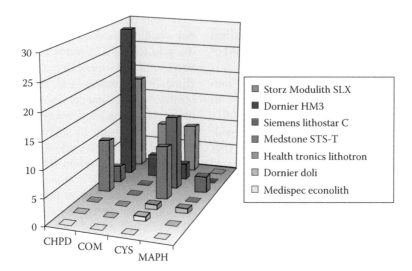

FIGURE 12.5 Comparison of the residual fragments after 2000 shocks using different lithotripter at the same center. CHPD, calcium hydrogen phosphate dihydrate; COM, calcium oxalate monohydrate; CYS, cystine; MAPH, magnesium ammonium phosphate hexahydrate. (From *J. Urol.*, 164, Teichman, J.M.H., Portis, A.J., Cecconi, P.P., Bub, W.L., Endicott, R.C., Denes, B., Pearle, M.S., and Clayman, R.V., In vitro comparison of shock wave lithotripsy machines, 1259–1264, Copyright 2000, with permission form Elsevier.)

lithotripters used in the United States were compared at the same center. At the FDA treatment limits, the residual fragment percentage of calcium hydrogen phosphate dihydrate, calcium oxalate monohydrate, and cystine and magnesium ammonium phosphate hexahydrate stones was 0% for the HM3, Modulith SLX, and Lithostar C, 10% for the STS-T, 3% for the LithoTron 160, 29% for the Doli and 18% for the Econolith ($p = 0.04$); 0% for the HM3, Modulith SLX, Lithostar C, STS-T, and LithoTron 160, 4% for the Doli, and 9% for the Econolith ($p = 0.15$); 1% for the HM3, 0% for the Modulith SLX, 1% for the Lithostar C, 10% for the STS-T, 14% for the LithoTron 160, 3% for the Doli, and 9% for the Econolith ($p = 0.44$); and 1% for the HM3, 0% for the Modulith SLX, 1% for the Lithostar C, 10% for the STS-T, 14% for the LithoTron 160, 3% for the Doli, and 9% for the Econolith ($p = 0.44$), respectively (Teichman et al. 2000). Dornier HM3 still has satisfactory performance, as shown in Figure 12.5. Contrary to the advance in ureteroscopy, modifications to the original lithotripters have focused on operator convenience and on optimizing the compressive wave component of the shock wave, neglecting the contribution of the tensile component of the waveform, which is based on the nascent and naïve understanding of SWL technology and the mechanisms of the effects of SWs, both good and bad. As a consequence, a completely safe and effective lithotripter has yet to be developed. Current lithotripters are less effective now than when they were first introduced, which is an embarrassment and significantly different from the other fields (Citron 1996, Madaan and Joyce 2007).

12.4 TECHNOLOGY AND MEASUREMENT

Current popular approaches of extracorporeal shock wave generation are electrohydraulic (EH), electromagnetic (EM), and piezoelectric (PE) type, although explosive method was initially used.

In EH sources, a high-voltage (10–30 kV) capacitor is discharged between two electrodes immersed in water to produce the rapid evaporation of water that generates an acoustic wave by expanding the surrounding water in the form of a spherically spreading shock front (Coleman et al. 1992, 1993). Collapse of such evaporation will generate another shock wave about 100 ms later, which also contributes to stone fragmentation, but not as significant as the first and primary one (Zhou et al. 2012). To focus such spherically diverging SWs, the electrode is placed in the first focus (F1) of an ellipsoidal reflector. The shock waves hit the metal reflector and get almost completely reflected and then are focused at the second focus of the ellipse (F2) (Figure 12.6). Some spreading shock waves travel directly from the electrode to the target, the so-called direct wave. Because of the short propagation distance, the direct wave arrives earlier than the primary focused one. To prevent the bubbles from drifting up from the electrode and accumulating along the beam path, which may collect against the skin or water balloon membrane and interfere with the propagation of subsequent SWs as bubble shields, the ellipsoidal reflector is tilted at 14° in the Dornier HM3 and the water in the bath is continuously degassed. However, due to cavitation erosion on the electrode tip, the gap becomes large

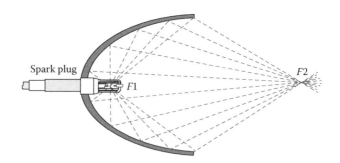

FIGURE 12.6 Schematic diagram of an electrohydraulic lithotripter.

with use, which leads to great variation in the shock wave production and in shift of the shock wave center. Meanwhile, misalignment by only 1 mm can lead to a significant loss in focusing (a reduction in measured peak pressures of the order of 50%) and a lengthening and broadening of the focal zone (Hunter et al. 1986).

The EM generator consists of a coil of wire placed against a thin metal membrane (Figure 12.7). The current pulse through the coil by discharge from a high-voltage capacitor induces a repulsive force on the metal membrane, which deflects and generates a pressure pulse in the water. The produced wave is focused by means of an acoustic lens for a planar membrane or a paraboloidal reflector for a cylindrical membrane (Rassweiler et al. 2005). EM sources have lifetimes in excess of 1 million shots with much less variation (Rassweiler et al. 2005). The high-intensity ultrasonic wave due to the displacement of the plate has a smooth waveform with no discontinuities but gradually forms a shock with the leading positive pressure sharpening through nonlinear effects as it propagates toward the focus (Ison 1987, Delius et al. 1989a, Folberth et al. 1992, Leighton 2007). In addition, the pressure waveforms at the focus produced by the EM source have a relatively small trailing positive pressure after the tensile wave, which has little impact on the stress inside the stone but affects the cavitation process.

PE shock wave lithotripters consist of an array of ceramic elements positioned on a spherically concave surface. When driven by the discharge of a high-voltage capacitor simultaneously, the ceramic array produces spherically converging waves toward its geometrical center (Figure 12.8) (Delius et al. 1989a). The PE source has a lifetime of up to 5 million shots (Rassweiler et al. 2005), but increased potential for mechanical damage and electrical breakdown on the

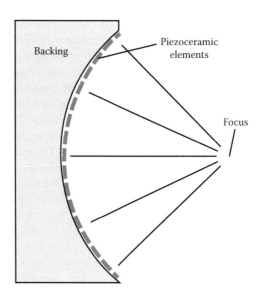

FIGURE 12.8 Schematic diagram of an electromagnetic lithotripter.

ceramic elements over time (Delius et al. 1989a). The spherically focusing PE lithotripters have a wide area of shock wave entry at the skin surface, which has the minimal patient discomfort but a very narrow focal size at $F2$ (Preminger 1989). Furthermore, the coda is much more significant in the pressure waveform at the focus, which is because the PE crystals need a couple of cycles to return to equilibrium after the beginning or termination of excitation. This may be reason for the poor performance of all PE lithotripters. In addition, each element could be driven individually in amplitude and phase as a phased-array transducer. However, the cost increases dramatically with the number of elements, so it is not popular in the lithotripter design.

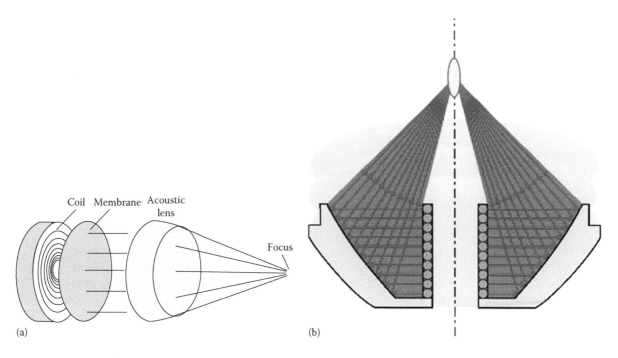

FIGURE 12.7 Schematic diagram of an electromagnetic lithotripter with planar membrane (a) and cylindrical membrane (b).

Polyvinylidene difluoride (PVDF) membrane hydrophone is the common device for the measurement of LSW. However, the adhesion between water and PVDF is not strong, and the tensile phase of the LSW can generate cavitation bubbles at the surface of the PVDF. As a result, the capability of the PVDF hydrophone to measure large tensile waves is limited because the formed bubbles cause pressure value close to 0. When the cavitation bubbles collapse, they can irreversibly damage the hydrophone. Pitting of the PVDF hydrophone was visible in a 2 cm diameter region around the sensitive element after exposure to SWL acoustic fields. Without standard and extensive calibration, the hydrophone cannot be used further. Recent development of the fiber-optic probe hydrophone (FOPH) can solve these questions because the adhesion between water and the optical fiber (silica) is very high and the optical fiber is more robust to damage from cavitation. Once the tip is broken, a new one can be prepared quickly with the advantage of self-calibration. But the main drawback with the FOPH is that its low sensitivity is not suitable for weak signal measurement.

In the measured waveform, p^+ and p^- are defined as the maximum peak positive and minimum negative pressures, respectively; t^+ and t^- are defined as the zero-crossing duration for positive and negative pressures, respectively. The most characteristic feature of all shock waves at the focus is the relatively rapid rise in pressure at the leading edge of the pulse. The rise time t_r is defined as the time needed for the pressure to rise from 10% to 90% of p^+, the fall time t_f is the time needed to decrease from 90% to 10% of p^+, and the pulse width t_w is the half-amplitude width of the initial positive pressure half cycle (Coleman and Saunders 1989). The distortion of the generated SWs may be classified as small for $t_r > 0.4t_w$, medium for $t_r > 0.4t_w > t_r > 30$ ns, and large for $t_r = 30$ ns. The pulse intensity integral (or energy flux, or energy density, or energy flux density) is

$$PII = \frac{1}{\rho c} \int_0^t p^2(t')dt' \tag{12.1}$$

The energy that passes through a specific area A can be calculated as

$$E = \iint PII\, dA \tag{12.2}$$

The acoustic intensity is determined by

$$I = \frac{PII}{\tau} \tag{12.3}$$

where τ is pulse duration, which is defined as 1.25 times the duration of the interval between t_5 and t_6 in the time integral of pulse intensity, which is 10% and 90% of the maximum integral value, respectively, according to AIUM/NEMA (1981) (Figure 12.9). The pressure distribution along two orthogonal axes (x and y) at the focal plane and along the lithotripter axis

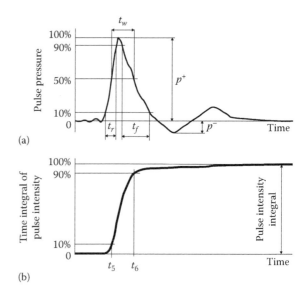

FIGURE 12.9 (a) Definition of the amplitude (p^+: peak positive pressure, p^-: peak negative pressure) and temporal (t_r: rise time, t_f: fall time, t_w: pulse width) parameters of the pressure pulse in the focus. (b) According to AIUM/NEMA (1981), pulse duration, τ, is defined as 1.25 times the duration of the interval between t_5 and t_6 in the time integral of pulse intensity. (From *Ultrasound Med. Biol.*, 21, Buizza, A., dell'Aquila, T., Giribona, P., and Spagno, C., The performance of different pressure pulse generators for extracorporeal lithotripsy: A comparison based on commercial lithotripters for kidney stones, 262, Copyright 1995, with permission from Elsevier.)

(z) is scanned, and then the acoustic peak pressure >50% of the focal peak pressure is used to determine the half-peak amplitude beam width and length (L_x, L_y, and L_z), where the focal area is defined as

$$A_f = \frac{\pi L_x L_y}{4} \tag{12.4}$$

The pulse acoustic energy in the focus is determined by

$$E = TA_f I_{SATA} \tag{12.5}$$

where

I_{SATA} is the spatial-average temporal-average acoustic intensity

T is the interval time between consecutive pulses

The energy has another approximate form

$$E = 0.7 PII \cdot A_f \tag{12.6}$$

The focal pressure gain (G) is defined as the ratio of the half-amplitude beam width at the aperture of the acoustic source and the half-amplitude width at the focus. Because of the particular importance of finite-amplitude effects on the shape of the acoustic pulse generated in ESWL, considerable distortion of the waveform from a sinusoidal variation is expected and observed.

From the measured pressure waveform, it is found that lithotripsy shock wave (LSW) does not have a dominant frequency, and its energy is spread over a very large frequency range (determined by the zero-crossing frequency of one complete cycle), mostly between 100 kHz and 1 MHz, which means that LSW-induced resonance is not the dominant reason for fragmentation of kidney stones. Regardless of the shock wave source, the pressure waveforms at the focus of all lithotripters have similar features but with considerable differences in the amplitude and beam size in the focal region.

The pressure pulses from different commercial extracorporeal lithotripters using three different ways (EH, EM, PE) to generate SWs were measured using the established characterization system and protocol and compared with each other (the total uncertainty in the absolute measurements of about ±36%). The lithotripters included in the survey were the Dornier HM3, Wolf Piezolith 2200 and 2300, Siemens Lithostar, Technomed Sonolith 2000 and 3000, and EDAP LT01. Peak negative pressures on most lithotripters were found to increase at a slower rate than peak positive pressures with the output voltage. Both the rise time and the pulse width decrease, and both p^+ and p^- distributions become narrower with the output; however, the rise time of 30 ns in the EH lithotripters seems consistent over the entire range of output settings. The peak positive and negative pressures at the maximum output are up to 114 and 10 MPa, respectively. The mean spatial-peak temporal-average intensity of the lithotripters is 5.0×10^2 W/m² delivered at a PRF of 1 Hz. The spatial-peak pulse-average intensity and the maximum acoustic energy vary from 6.6×10^7 to 1.24×10^9 W/m² and from 2.0×10^{-3} to 9.0×10^{-2} J, respectively (Coleman and Saunders 1989). The focus, the position of the spatial peak pressure, was found to shift by ~3 mm away from the source at the higher setting.

If E_{max} on the Dornier HM3 is set at 100%, the PE lithotripters have a value of ~5% and the EM lithotripter ~30%. The value of E_{max} on Sonolith 2000 and the Sonolith 3000 is ~20% and 4% of that on the Dornier HM3, respectively (Buizza et al. 1995). High values of E_{max} may provide peak pressures above the threshold for stone fragmentation over a larger beam area (A_f). EH generators have a large degree of waveform distortion over most of their operating settings, whereas PE generators give large waveform distortion only at the highest-output settings and give medium or small distortion at most settings. The EM generator gives medium waveform distortion over the lower operating range and large distortion at high-output settings. The degree of pain associated with the passage of the shock waves is generally less on machines that have large values of focal pressure gain.

Using a 25 μm PVDF membrane hydrophone, shock waves were measured inside a pig model. The pressure waveforms were very similar to those in water, but the amplitude of the peak pressure was reduced to about 70%. The focal region *in vivo* was 82 mm × 20 mm, which was larger than that *in vitro* (57 mm × 12 mm), and the broad lithotripter focus is due to both nonlinear effects and inhomogeneities in the tissue. The shock rise time was on the order of 100 ns, which was also larger than that in water and attributed to higher absorption in tissue (Cleveland et al. 1998) (Figure 12.10 and Tables 12.4 and 12.5).

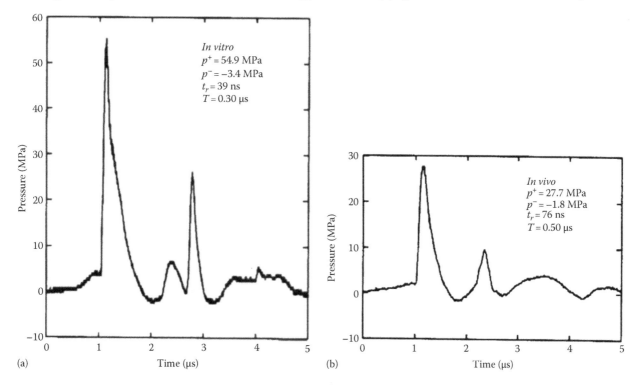

FIGURE 12.10 Comparison of representative waveforms measured (a) *in vitro* and (b) *in vivo*. The durations of the positive phase of the two waveforms are very close, the peak-positive pressure (p^+) *in vivo* is almost half the *in vitro* pressure, and the rise time has been doubled. (From *Ultrasound Med. Biol.*, 24, Cleveland, R.O., Lifshitz, D.A., Connors, B.A., Evan, A.P., Willis, L.R., and Crum, L.A., In vivo pressure measurements of lithotripsy shock waves in pigs, 298, Copyright 1998, with permission for Elsevier.)

TABLE 12.4
Comparison of Pressure Output from Different Shock Wave Lithotripter

Lithotripter	Source	p^+ (MPa)	p^- (MPa)	t_r (ns)	τ (ns)	$L_x \times L_y \times L_z$ (mm³)	I_{SPTA} (kW/m²)	I_{SPPA} (GW/m²)	E (mJ)	Variation (%)
Dornier MFL 5000	EH	65.2 ± 10.2	7.1 ± 0.6	166 ± 65	475 ± 79	4 × 3.1 × 13.6	1.01 ± 0.12	2.21 ± 0.57	6	16.4
Dornier MPL 9000	EH	62.3 ± 2.6	7.8 ± 1.1	129 ± 12	756 ± 191	6.8 × 7.9 × 58.8	0.89 ± 0.04	1.25 ± 0.25	24.5	8.2
Medas Lithoring Multi 1	EH	65.5 ± 15	7.9 ± 0.6	158 ± 107	522 ± 35	3.7 × 2.7 × 23.8	1.26 ± 0.17	2.93 ± 0.88	6.1	27.8
Technomed Sonolith 3000	EH	43.7 ± 10.9	5.8 ± 1.3	251 ± 188	861 ± 700	9.6 × 6.8 × 34.7	0.65 ± 0.11	1.14 ± 0.61	19.5	24.8
Dornier Compact	EM	53.2 ± 0.1	11.1 ± 0.1	91 ± 3	4440 ± 75	7.2 × 6.9 × 65	0.75 ± 0.005	0.17 ± 0.005	20.7	0.8
Siemens Lithostar Plus	EM	30.8 ± 0.4	5.3 ± 1.1	335 ± 62	2990 ± 1200	4 × 5.5 × 59.5	0.18 ± 0.01	0.09 ± 0.06	2.2	4.9
Storz Modulith	EM	84.5 ± 0.3	10.1 ± 0.1	183 ± 3	592 ± 6	3.1 × 3.8 × 29.5	1.52 ± 0.005	2.56 ± 0.03	10.4	0.4
EDAP LT01	PE	78.5 ± 1.4	1.9 ± 0.7	47 ± 1	5186 ± 105	2.3 × 2.5 × 20	1.36 ± 0.03	0.26 ± 0.01	N/A	2.9
EDAP LT02	PE	59.2 ± 0.6	7.5 ± 0.6	57 ± 2	5438 ± 268	5.2 × 5.2 × 32	0.52 ± 0.01	0.095 ± 0.005	N/A	2.1
Wolf Piezolith 2500	PE	30.8 ± 0.5	6.1 ± 0.04	181 ± 3	402 ± 5	3.4 × 3.2 × 14.3	0.19 ± 0.005	0.47 ± 0.01	1	1.7

Source: *Ultrasound Med. Biol.,* 21, Buizza, A., dell'Aquila, T., Giribona, P., and Spagno, C., The performance of different pressure pulse generators for extracorporeal lithotripsy: A comparison based on commercial lithotripters for kidney stones, 264, Copyright 1995, with permission from Elsevier.

TABLE 12.5

Variation Is for p^+ in a Sequence of 100 Shocks at Medium-Output Setting

	Output	p^+ (MPa)	std (%)	p^- (MPa)	std (%)	t_r (ns)	t_w (ns)	A_f	G	p^+ Skin (MPa)
Sonolith 2000	13.5 kV	21	20	3.6	16	30	380	3.8	4	10.5
Sonolith 3000	14–20 kV	52–78	50	4.0–9.5	60	30	300	0.07	37	3.5
Dornier HM3	15–25 kV	33–50	16–25	7.1–9.5	18	30	560	1.77	5	20
Piezolith 2200/2300	1–4	56–114	5	9.5–9.9	10	130–70	250–180	0.03	250	0.6
EDAP LT01	5%–100%	9–105	6	6.2–6.4	6	450–30	600–200	0.11	100	1.3
Lithostar	12.9–19 kV	26–44	2–3	2.8–5.0	7	120–30	340	0.24	19	6.4

Source: Ultrasound Med. Biol., 15, Coleman, A.J. and Saunders, J.E., A survey of the acoustic output of commercial extracorporeal shock wave lithotripters, 213–227, Copyright 1989, with permission for Elsevier.

The bubble wall will not hold its intact condition because of the significant pressure difference between the near-vacuum inside the bubble and the roughly atmospheric pressure in the surrounding fluid. The bubble collapse is very violent, and rectified gas diffusion occurs during acoustic cavitation (Crum and Hansen 1982, Coleman et al. 1992). The main collapse is followed by rebounds, after which the gas in the remnant bubble (size on the order of 10 μm) will slowly dissolve into the fluid over hundreds of milliseconds. However, the nature of cavitation makes it challenging to investigate the temporal and spatial characteristics of the bubble dynamics. Photography has a limited viewing depth and cannot adequately record bubble dynamics throughout the substantial volume of the acoustic field. Light from an illuminating laser light is scattered by the bubble and measured by a photodetector to monitor the bubble radius changes. But it has several limitations, such as the small sampled volume, the requirement of visual access at high magnification, and quantitative recovery of the actual size of only a single spherical bubble. Therefore, only qualitative information of bubble clouds or nonspherical bubbles induced by lithotripsy shock wave can be obtained. Detection of bubbles acoustically is useful within a living subject and can be classified into two modes: passive cavitation detection (PCD) and active cavitation detection (ACD) (Coakley 1971, Roy et al. 1990, Madanshetty et al. 1991). In PCD, many focused transducers are used to pick up the acoustic emissions from cavitation bubbles, which have a double-spike structure. In ACD, an ultrasound transducer is used to emit an acoustic pulse and receive the reflected or scattered signal from the bubbles using the same probe (like ultrasound imaging to show the location and lifespan of bubbles but not the exact size and number) or another one (like laser scattering method) (Coleman et al. 1993, Zhong et al. 1997, Cleveland et al. 2000).

A standard stone phantom was used to compare the comminution efficiency of EH, EM, and PE lithotripters at both low and high intensities. A certain number of SWS (i.e., 50, 100, 200, or 400) were focused either at the anterior or posterior surface of the stone. The stone volume loss was proportional to the number of SWs delivered for machines at the lower-intensity settings, but in a rapid form at higher intensities. The EH machine had the greatest stone loss, followed

by EM and PE lithotripters for both low- and high-intensity settings with the same number of SWs. Craters in the stone produced by the PE, EM, and EH machines were narrow and deep, in the shape of a right angle circular cone, and shallow and wide, respectively. At high intensities, PE and EH lithotripters produced more damage to the stone when their focus was at the anterior surface, while the focal point of the EM device should be at the posterior surface (Chuong et al. 1992). Plaster of Paris and a dental plaster were used as stone phantoms for struvite and calcium oxalate monohydrate, respectively (Heimbach et al. 2000, Liu and Zhong 2002).

With approval from the Institutional Animal Care and Use Committee, young female pigs were used as *in vivo* model and underwent induction of general endotracheal anesthesia. Surgical dissection through a standard midline laparotomy incision exposed the right kidney and the proximal portion of the right ureter. An artificial stone phantom was introduced into the lumen of the ureter and advanced into the renal pelvis in a retrograde fashion using a lithotomy forceps. A polyurethane internal ureteral stent was placed in the right ureter and renal pelvis. The ureterotomy was closed in one layer with a running suture. Radiopaque staples were placed in the renal fascia lateral, medial, and inferior to the lower pole of the kidney to facilitate fluoroscopic localization during SWL treatment. The midline incision was closed with a running suture. After the SWL, bilateral nephrectomies were performed through the previous midline incision, and the animals were euthanized humanely. The kidneys were bisected to expose the renal pelvis and the residual stone fragments in the collecting system. The residual fragments were collected, air-dried at room temperature for 24 h, and then filtered through a stack of standard sieves to determine the stone comminution efficiency (Figure 12.11).

The propensity of vascular injury produced by the lithotripter can be evaluated using a vessel phantom made of a single cellulose hollow fiber of 0.2 mm inner diameter with circulating degassed water seeded with a small amount of microbubbles (i.e., 0.1% contrast agent by volume) as cavitation nuclei (Figure 12.12). The vessel phantom was immersed in a testing chamber filled with a highly viscous fluid (i.e., castor oil) to suppress cavitation activity outside the vessel. In addition, a low pulse repetition rate (i.e., 0.1 Hz) of the LSWs

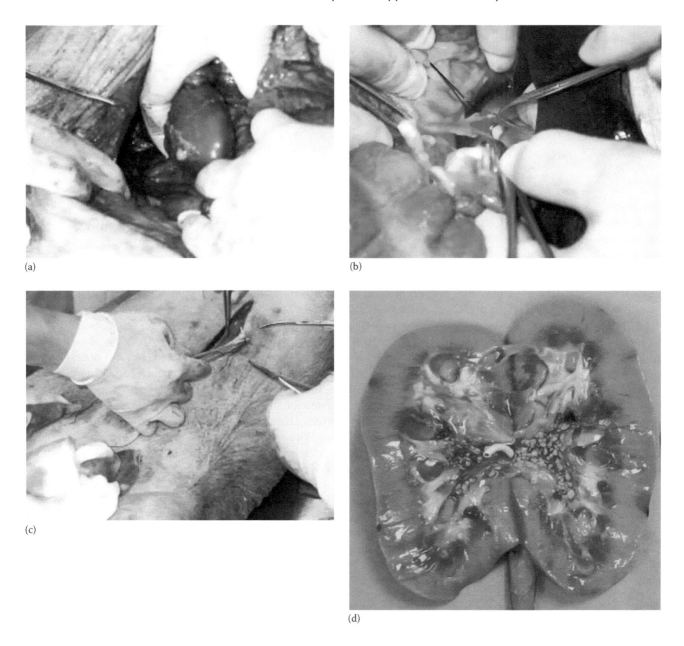

FIGURE 12.11 Procedure of *in vivo* stone fragmentation experiment: (a) mobilization of the kidney, (b) insertion of artificial stone into the kidney through the ureter, (c) closure of midline incision, and (d) biosection of the kidney of SWL to collect fragments.

was employed so that before each shock wave exposure any visible bubbles outside the vessel phantom could be removed to minimize the shear stress generated either by the scattering of LSW from bubbles in castor oil or by simultaneous bubble expansion both inside and outside the vessel phantom. Rupture of the vessel phantom could be easily identified since the circulating fluid would leak out, forming a liquid droplet in castor oil at the rupture site (Zhong et al. 2001, Zhong and Zhou 2001).

An agar gel phantom can also be used to evaluate the propensity and distribution of shock wave–induced renal injury. The center of the gel phantom is aligned with the focus of the lithotripter. After the passage of shock waves, the bubble nuclei that are naturally and homogeneously distributed inside the gel

phantom will expand and form visible air cavities, which can be recorded photographically and then analyzed quantitatively using an image processing toolbox (Figure 12.13).

In vivo renal injury could be evaluated using the animal protocol as described before. After the SWL, gross renal damage was documented with digital photography. The kidneys were then fixed in 4% formaldehyde, refrigerated at 4°C for 7 days, and sectioned in 3 mm slices. The slices were individually analyzed for injury using computerized volumetric analysis (trapezoid model in the first order of approximation), and the results were summed for the entire kidney as a whole. Hematoxylin and eosin-stained sections of the kidneys were qualitatively examined for shock wave–induced histologic changes (Figure 12.14).

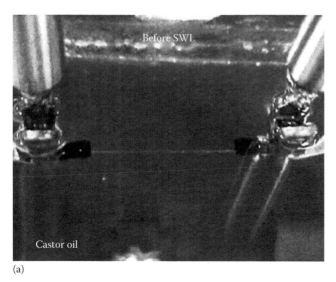
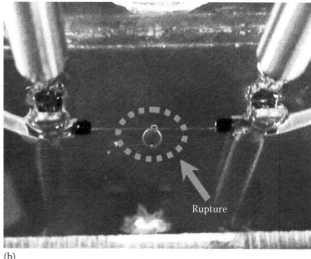

(a)   (b)

FIGURE 12.12 (a) Regenerated cellulose hollow fiber immersed in castor oil. (b) Formation of liquid droplet at the rupture site after SW exposure.

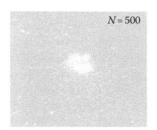
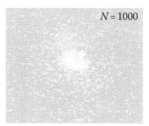
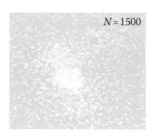

FIGURE 12.13 Dose dependence of cavitation-induced damage in gel phantom *in vitro* experiment.

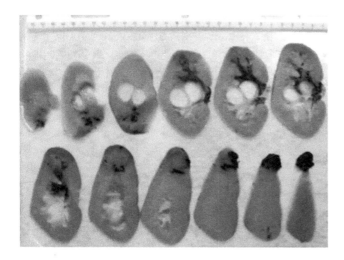

FIGURE 12.14 Histology photo of porcine kidney after SWL.

12.5 MECHANISMS OF SWL: STONE FRAGMENTATION AND RENAL INJURY

Although SWL is the preferred modality for most symptomatic upper urinary tract calculi, new lithotripters are not as effective and safe as Dornier HM3. Our understanding of the interaction between SWs and kidney stones and renal tissue for fragmentation and injury was nascent and naïve at the time of its introduction and initial success. Although continuous research for three decades have significantly increased the knowledge, the mechanisms are still controversial, which may result in poor of progress in SWL evolution. A fundamental understanding of the mechanisms of stone comminution and tissue injury is necessary for improving and developing SWL technologies, although multiple factors (i.e., beam size, total acoustic energy, and coupling condition, etc.) may also contribute to the reduced effectiveness of the newer generation models. The interaction between stress waves and cavitation and their synergy is a strategy for better SWL performance and needs further investigation. Because of its low duty cycle, a typical lithotripter pulse would generate less than 2°C temperature rise in tissue after 5000 SWs (Coleman and Saunders 1989). Only with the full understanding of SWL technology, can new lithotripter designs become mature and achieve a real improvement on the original Dornier HM3 machine.

The events involved in stone comminution are not completely and accurately known but may include compression fracture, spallation, squeezing, and acoustic cavitation. The imperfections in the calculi amplify stresses significantly (Leighton 1994, Leighton et al. 2008, Turangan et al. 2008) and cause the imperfections to grow into microcracks. Stone fragmentation is a sequence of dynamic fatigue (Coleman et al. 1987). Repetitive action of shock wave–induced stress (either a tensile or shear stress, or both) progressively weakens the stone, resulting in the nucleation, growth, and coalescence of microcracks and leading to large-scale failure in fracture and eventual fragmentation in brittle materials (Lokhandwalla and Sturtevant 2000) (Figure 12.15).

As the lithotripsy shock wave passes through the stone, it can generate direct stress on the stone and local tensile stress around the preexisting defects (such as grain boundaries, cavities, inclusions, and similar flaws) in the calculi (Lokhandwalla and Sturtevant 2000). Because of the complicated geometry of the stone surface and its internal structure, the compressive pulse of the lithotripsy shock wave will generate pressure gradients to tear and shear the crack (Chaussy et al. 1980).

During spallation (also known as the Hopkinson effect), the reflected compressive wave from the stone–fluid interface at the distal surface of the stone due to acoustic impedance

mismatch has an inverted waveform, which superposes to the incoming tensile stress of the shock wave, resulting in a large tensile stress for stone fragmentation because most solids are much weaker in tension than in compression (Chaussy 1982, Chaussy et al. 1982, Sass et al. 1991, Ding and Gracewski 1994, Cleveland and Sapozhnikov 2005). However, if the stone size is too small to induce such a wave superposition (the minimum length in the propagation direction is calculated as the product of the longitudinal wave speed in the stone and interval time between the peak positive and negative pressures), spallation will not occur (Xi and Zhong 2001). The negative pressure of the lithotripter shock wave exerts tensile stresses in a similar way as in the spall mechanism, and the large tensile stress near the distal end of the stone can fail the material. Tear and shear forces, as well as spallation alone, remain relevant in small focal sizes. The focal width may play a critical role in stone breakage, at least for initial fragmentation.

Circumferential quasistatic compression or squeezing by evanescent waves in the stone is of importance when the stone is smaller than the beam width of the lithotripter in the focal zone (Eisenmenger 2001). The velocity of the longitudinal wave within the stone is much higher than the sound speed in the surrounding fluid, leaving the thin waves in the fluid encircling and squeezing the stone (Eisenmenger 2001, Sapozhnikov et al. 2007). Cleavage surfaces either parallel or perpendicular to the wave propagation direction were found in the stone phantom after SWL treatment, which is in agreement with the squeezing mechanism. A quantitative model of binary fragmentation by quasistatic squeezing was established to predict the number of pulses required for cleavage into two parts, which was verified by the experimental results for spherical stone phantoms of 5, 12, and 15 mm diameter at a pulse pressure of 11 MPa (Eisenmenger 2001). For the squeezing to be effective, the focal width of the lithotripter must be higher than the stone dimension, but a steep shock front is not required (Figure 12.16).

Most kidney stones are brittle in nature and, therefore, can be reasonably approximated using linear elasticity up to the point of failure. The brittle solid absorbs only a small amount of energy prior to failure, while the ductile fracture involves extensive plastic deformation. However, the fracture behavior of materials is also dependent on their intrinsic properties (i.e., crystal structure, presence of flaws, impurities) and extrinsic loading conditions (i.e., strain rate). For most crystalline materials, there is competition between the tendency to deform plastically, with attendant work-hardening, and the tendency of the flaws to propagate and fragment the material. Microcracks would be distributed all around the main crack. Hence, the cohesive law is equivalent to a constitutive law for the microcracked material, applicable under static or quasistatic loading situations. If the fragmentation process is linear (i.e., a fragment forms after every N shocks), then the worst case of producing n fragments needs a total of nN shocks. However, the best case will be a logarithmic fragmentation process, that is, the number of fragments doubles after every N shocks, and the total number of shocks needed for

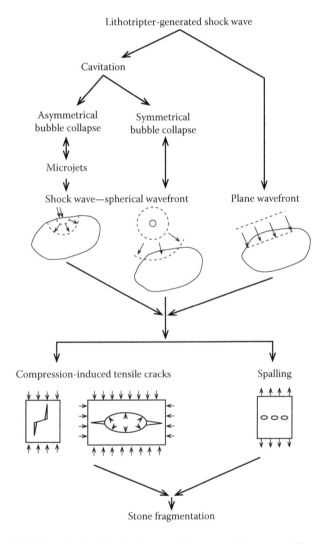

FIGURE 12.15 Mechanisms of stone fracture. (From Lokhandwalla, M. and Sturtevant, B., Fracture mechanics model of stone comminution in ESWL and implications for tissue damage, *Phys. Med. Biol.*, 45, 1924, 2000. With permission.)

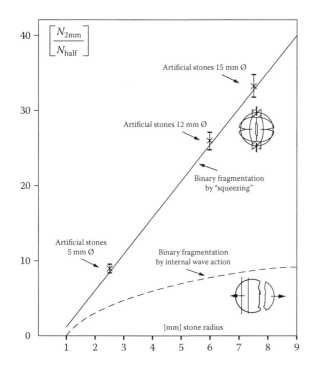

FIGURE 12.16 Comparison of dose relationship of stone fragmentation and the first cleavage, determined by the quasistatic squeezing model of binary fragmentation, and experimental results at a pulse pressure of 11 MPa, a pulse duration of 1.8 ms, and −26 dB focal width of 18 mm. (From *Ultrasound Med. Biol.*, 27, Eisenmenger, W., The mechanisms of stone fragmentation in ESWL, 688, Copyright 2001, with permission for Elsevier.)

n fragments is reduced to $N \cdot \log_2(n)$. A quasi-steady uniaxial tension test predicts a static fracture strength for renal calculi (Lokhandwalla and Sturtevant 2000).

Kidney stones are much weaker in shear similar to many materials, especially those composites, as the bonding strength between layers has a low ultimate shear stress (Chaussy et al. 1980, 1982, Dahake and Gracewki 1997, Xi and Zhong 2001). The shear waves are generated by two mechanisms: the passage of the shock wave in the fluid outside the stone as squeezing the stone (Eisenmenger 2001), and the interaction of the shear wave propagating from the stone surface toward the center with an internal wave (Cleveland and Sapozhnikov 2005, Sapozhnikov et al. 2007), which may contribute more to the maximum tensile stress in the stones (Eisenmenger 2001, Cleveland and Sapozhnikov 2005). A shear wave and an inverted diffracted compression wave are produced at the edges of the stone. Because of the higher speed of the longitudinal wave than of the shear wave, there is a conical wavefront of supersonic shear waves at the stone boundary. The maximum tensile stress occurs as a result of the interaction of the reflected longitudinal wave from the rear stone surface with a shear wave generated from the traveling wave in the fluid outside the lateral surface of the stone. The spatial extent of the shear waves is related to the spatial extent of the shock front. The shear wave-induced loads were significantly larger than the loads generated by the classic Hopkinson or spall effect. The leading compressional wave in the stone suffers more from diffraction due to the longer rise time, reducing by 30% as the rise time increases from 25 to 150 ns. As a result, the peak tensile stress is reduced dramatically. Therefore, in order to utilize the shear wave–induced high stresses inside kidney stones in stone comminution more effectively, lithotripters should have large focal widths and short rise times (Cleveland and Sapozhnikov 2005) (Figure 12.17).

The regions of both tensile and shear stress can be determined by the geometry of the stone (for the possibility of superfocusing effect) and its elastic properties (i.e., density, longitudinal wave speed, and shear wave speed) (Gracewski et al. 1993, Xi and Zhong 2001). After the initial fracture in a spherical stone, the irregular geometry of residual large fragments makes it more resistant to subsequent shock waves. Initial damage may be due to a combination of spallation (Gracewski et al. 1993, Xi and Zhong 2001) and quasi-static or dynamic squeeze (Eisenmenger 2001, Cleveland and Sapozhnikov 2005, Sapozhnikov et al. 2007) mechanism that generates an uneven fracture of the stone. Furthermore, acoustic pulse energy correlates closely with stone comminution (Granz and Hohler 1992, Delius et al. 1994b, Eisenmenger 2001).

The SW drives the bubble dynamics with the zero-order effect of gas diffusion, as described by the Gilmore–Akulichev formulation. A preexisting bubble in the size of 1–10 μm can expand to a peak radius of ~1400 μm in 250 μs and then collapse very violently, emitting far UV or soft x-ray photons. Gas diffusion does not appreciably mitigate the amplitude of the pressure wave radiated at the primary collapse but significantly reduces the collapse temperature. Diffusion also increases the initial bubble radius to 40 μm and extends the duration of ringing following the primary collapse. Bubble dynamics are sensitive to p^+, p^-, and t^- but not to t_r (Church 1989) (Figure 12.18).

Cavitation bubbles produced close to the stone by the tensile component of each LSW can pit and erode stone surfaces (Averkiou and Crum 1993). Bubbles expand, stabilize, and finally collapse violently, creating secondary shock waves and microjets a few hundred microseconds after the arrival of shock wave in the focal region (Loske et al. 2002). The speed of the liquid microjet due to the nonspherical involution of the bubble wall on collapse is as high as 100 m/s (Crum 1988, Delius et al. 1988), and the impact of the microjet can generate a blast wave in the vicinity of the bubble with an amplitude much greater than that of LSW pulse (Ball et al. 2000, Leighton 2004, Johnsen and Colonius 2008, Turangan et al. 2008). The strong pressure and flow following the microjet may significantly enhance stone erosion and the collapse of neighboring bubbles. It is found that a cloud of cavitation bubbles is formed in SWL and acts most strongly on the proximal surface of the stone rather than its distal and lateral surfaces (Delius et al. 1988a, Sass et al. 1991, Pishchalnikov et al. 2003). Cavitation damage grinds up small fragments that are ineffective in being fragmented by the stress wave (Zhu et al. 2002). Suppression of cavitation using highly viscous media significantly reduces the disintegrative shock wave efficacy (Delius et al. 1988a). Overpressure significantly reduces the bubble lifetimes in the

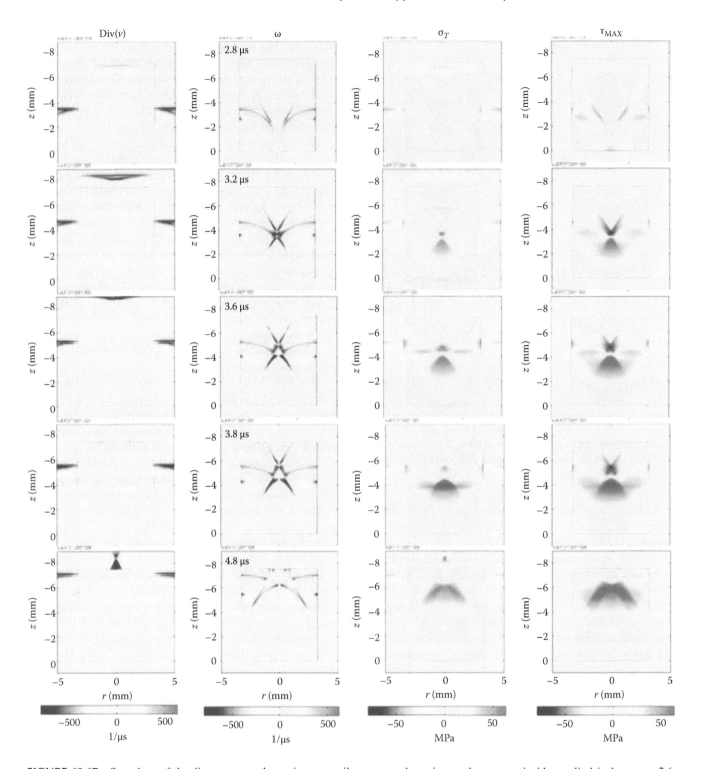

FIGURE 12.17 Snapshots of the divergence, curl, maximum tensile stress, and maximum shear stress inside a cylindrical stone at 2.6, 3.2, 3.6, 3.8, and 4.8 μs after the SW is incident on the stone. The shock wave is incident from beneath the stone. (With permission from Cleveland, R.O. and Sapozhnikov, O.A., Modeling elastic wave propagation in kidney stones with application to shock wave lithotripsy, *J. Acoust. Soc. Am.*, 118, 2670, Copyright 2005, American Institute of Physics.)

free field. At 1.5 bar static pressure, a denser bubble cluster was measured with high PRF (2–3 Hz) than low PRF (0.5–1 Hz). However, overpressure did not suppress cavitation of bubbles stabilized on a cracked surface. So, the use of overpressure and PRF in SWL may not reduce the cavitation erosion to stones (Sapozhnikov et al. 2002) (Figure 12.19).

The formation and subsequent collapse of the bubble cluster at the proximal face of the stone appear to involve the coalescence of numerous individual cavitation bubbles to form a large single bubble or cluster of bubbles that collapse with damaging force. As the proximal face becomes etched and pitted, these outward-facing defects would tend to capture

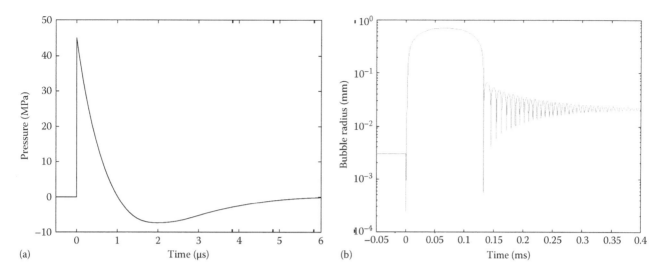

(a)   Time (μs)

(b)   Time (ms)

FIGURE 12.18 (a) Typical lithotripter shock wave with P_{max} = 45 MPa. (b) Bubble response to the shock wave, with the initial bubble radius of 3 μm.

FIGURE 12.19 High-speed images of the bubble cloud on the proximal surface of an artificial stone induced by a lithotripsy shock wave incident from below and arriving at 180 μs. During the collapse, the bubble cloud pinches in (arrows at 680 ms) and forms a microjet. (With permission from Pishchalnikov, Y.A., Williams, J.C., and McAteer, J.A., Bubble proliferation in the cavitation field of a shock wave lithotripter, *J. Acoust. Soc. Am.*, 130, E87–E93, Copyright 2011, American Institute of Physics.)

the cavitation nuclei (Bailey et al. 1999a). An existing crack might act as a focus for bubble cluster formation by providing a source of cavitation nuclei. A crack might physically stabilize cavitation bubbles (Crum 1979, Bailey et al. 1999b), keeping them from being swept away by fluid motion induced by bubble dynamics elsewhere on the stone. Bubble clusters are associated with fractures in stones, and cluster collapse along the line of a fracture may promote its growth (Pishchalnikov et al. 2003). In comparison, the cluster in the distal end appeared to be a solitary bubble that collapsed sooner than the proximal cluster (duration of distal cluster ~400 μs, proximal cluster ~600 μs) and caused very little damage to the stone. The size and shape of a stone will influence the bubble cluster behavior (Figure 12.20).

In an *in vitro* experiment, spherical plaster-of-Paris stone phantoms (D = 10 mm), either in degassed water or in castor oil to delineate the contribution of stress waves and cavitation to stone comminution, aligned to a Dornier HM3 lithotripter were exposed up to 500 shocks at 20 kV and a PRF of 1 Hz. In degassed water, there is a progressive increase in stone comminution efficiency (from 3% after 25 shocks to 66% after 500 shocks). However, the percentage of dischargable fragments in the castor oil became saturated quite soon, and the corresponding increase was 2%–11%. Therefore, although stress wave–induced fracture is important for the initial fragmentation of kidney stones, cavitation is necessary to produce fine passable fragments. Stress waves and cavitation work synergistically, rather than independently, to produce effective and successful stone comminution in SWL (Zhu et al. 2002) (Figure 12.21).

The complex shape, composition, the amount of hydration, the crystallization of minerals, as well as organic matrix material, the size and distribution of flaws, the microhardness, the surface features, the fracture strength, and the elasticity the of natural stones make it impossible to assess the susceptibility of a stone to SWL (Lokhandwalla and Sturtevant 2000, Cleveland et al. 2001, Eisenmenger 2001, Xi and Zhong 2001, Zhu et al. 2002, Cleveland and Sapozhnikov 2005). A compressive hoop stress applied to the stone surface is less pronounced for natural stones with more complex geometries (Cleveland and Sapozhnikov 2005). In addition, the cavitation bubbles created by a lithotripter shock wave need a long time (more than 1 s) to dissipate (Church 1989). If shock waves are delivered at a rate faster than 1 Hz, a large number of remnant bubbles will form in a small volume of the lithotripter field and shield the subsequent shock waves, which is known as the bubble shielding effect (Leighton 1995, 2007, Leighton et al. 2004). It shows that p^+ is not significantly reduced at fast rates, but the negative pressure is, which is similar to an acoustic diode (Pishchalnikov et al. 2005).

The mechanisms of disintegrating renal and urinary calculi (common bile duct stones, bladder stones, kidney stones) by a shock wave from an intracorporeal probe that is advanced to the stone through the working channel of an

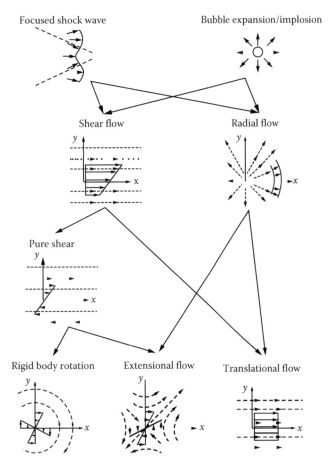

FIGURE 12.20 Kinematics of flow (velocity profiles in solid arrows and flow field in dashed arrows) induced by focused shock wave and bubble motion. (From Lokhandwalla, M., McAteer, J.A., Williams, J.C., and Sturtevant, B., Mechanical hemolysis in shock wave lithotripsy (SWL): II. In vitro cell lysis due to shear, *Phys. Med. Biol.*, 46, 1247, 2001. With permission.)

endoscope to generate a controlled and very fast electric discharge are similar to those of extracorporeal type, as shown in Figure 12.22. The expanding spark plasma and later the collapse of a cavitation bubble create sharply rising shock waves that disintegrate the stone in seconds. The Olympus LUS-1 and LUS-2, Circon-ACMI USL-2000, Karl Storz Calcuson, and Richard Wolf model 2271.004 at the maximum power settings were compared with each other for the time of complete Ultracal-30 cylindrical stone penetration. All probes had outer diameters of 3.4 mm except for the Circon-ACMI unit (3.8 mm). Efficiencies of the LUS-2 and USL-2000 units were essentially equivalent, and all others were significantly less efficient (Kuo et al. 2003). However, the physical characteristics of these devices were not measured to understand the performance discrepancies.

A clinical dose of shock waves leads to injury in most all subjects (McAteer and Evan 2008). Renal injury from SWL is primarily a vascular lesion, such as hematuria (the rupture of blood vessels, although in most cases resolving naturally over a day or two or several weeks at most [Lingeman et al. 2006]), hemorrhages, thrombi, and arrhythmias (Evan et al. 1998) accompanied by a reduction in blood flow secondary to an SWL-induced vasoconstrictive response (Willis et al. 1999). Venules in the medulla are usually damaged, followed by rupture of cortical arterioles. However, SWL-induced injury is not always accompanied by subcapsular bleeding. Reduction of renal functionality, infections, alterations to the autonomous neural system, and the release of cell mediators and hormones were also found (Lingeman et al. 1990, 2003, Seitz et al. 1991, Coleman and Saunders 1992, Madbouly et al. 2002, Skolarikos et al. 2006, Knoll and Wendt-Nordahl 2008). In addition, there were 1.8% cases of gastrointestinal injury, including colonic perforation or duodenal erosions (Maker and Iayke 2004). Multiple sessions of SWL

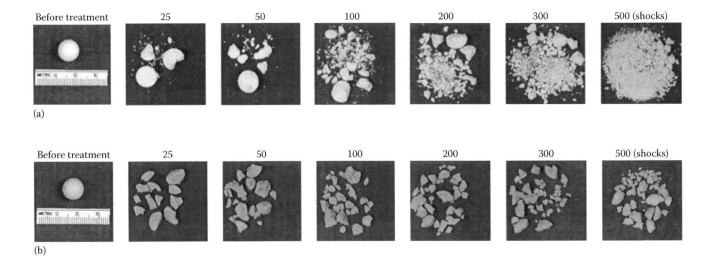

FIGURE 12.21 Dose-dependent fragments of plaster-of-Paris stone phantoms treated in water (a) and castor oil (b) using an HM3 lithotripter at 20 kV. (From *Ultrasound Med. Biol.*, 28, Zhu, S.L., Cocks, F.H., Preminger, G.M., and Zhong, P., The role of stress waves and cavitation in stone comminution in shock wave lithotripsy, 664–665, Copyright 2002, with permission for Elsevier.)

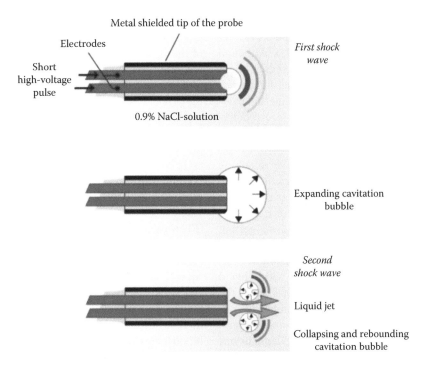

FIGURE 12.22 Schematic diagram of the working principle of the endoscopic lithotripter.

may increase the risk of the transition from calcium oxalate stones to stones of calcium phosphate and brushite (Parks et al. 2004).

The sharp shock front will lead to nonuniform straining of the tissue and to the generation of shear forces. It is well known that tissue structures are sensitive to shear stress, and the shear-induced damage in SWL is due to the tissue distortion by LSW (Sturtevant 1996, Lokhandwalla and Sturtevant 2000, 2001, Lokhandwalla et al. 2001) (Figure 12.23). However, individual

FIGURE 12.23 Anatomical structure of the renal injury in kidney after SWL.

shocks (~0.1%) do not produce sufficient shear, but the cumulative shear of the many shocks before tissue relaxation to its unstrained state between shocks is damaging. The net shear deformation from a typical LSW is estimated using a separate dynamic shock simulation. The simulation suggests that the larger interstitial volume (~40%) near the papilla tip gives the tissue a relaxation time comparable to the delivery rates (~1 Hz), thus allowing the shear to accumulate. If the interstitial volume is smaller (~20%), the tissue relaxes completely before the arrival of next LSW. Decrease of renal injury using lower delivery rates and broader focal zones suggested by the simulation is consistent with clinical observations (Freund et al. 2007). Tear and shear forces with the microstructure of the tissue were reduced (Evan et al. 2007a, Freund et al. 2007).

Shock-induced acoustic streaming causes the rupture of cell membrane, even in the absence of cavitation. Expansion and collapse of a bubble outside the cell and at a sufficiently large distance from it can generate a radial flow field and consequently result in significant cell damage. A shear flow due to the velocity gradient along the shock front and a radial flow due to curvature of the wavefront also occur. Thus, the LSW-induced flow field can be decomposed into a translational flow, a rigid body rotation, and an extensional flow. Only the extensional flow causes deformation of an elementary fluid particle; the other two only displace and rotate the fluid particle without any deformation. So, a uniform translational flow follows the planar shock wave with uniform strength and causes no cell lysis. Pores smaller than 1–10 nm are energetically unfavorable, and the membrane heals itself (Litster 1975). However, larger pores grow unstably, leading to membrane rupture. In the poration regime, cell deformation is not sufficient for lysis but will enhance transport across the cell membrane (Lokhandwalla and Sturtevant 2001).

Cavitation forces or shear stresses are the major mechanisms for SWL-induced tissue injury (Howard and Sturtevant 1997, Zhong et al. 2001, Rassweiler et al. 2011). Although cavitation bubbles have a desirable effect in the region of the stone, their formation in tissue and blood vessel lumen has unwanted consequences. Evidence suggests that inertial cavitation plays an important role in the renal injury incurred during SWL, and significant injury typically occurs only after a sufficient dose of shock waves (Leighton et al. 1995, Zhong et al. 2001). Cavitation occurring within blood vessels is responsible for SW-induced hemorrhage (Zhong et al. 2001). When cavitation is suppressed by applying overpressure in excess of the amplitude of the tensile wave of LSW, cell injury is significantly reduced but not eliminated, suggesting that cavitation and shear stress are both mechanisms of tissue damage in SWL (Williams et al. 1999, Sapozhnikov et al. 2002). Cavitation is more likely to induce injury within blood vessels than within the surrounding tissue, which is due to the constrained effect of the surrounding tissue on bubble dynamics and subsequently an incomplete growth-and-collapse cycle (Figure 12.24). When cavitation bubbles collapse asymmetrically in a blood vessel, high-velocity microjets of fluid focused to a small spot will be formed. Bubbles may rupture vessel walls by greater stresses during the explosive growth phase of the bubble cycle and pushing the vessel outward. This is consistent with clinical observation, and the capillaries are more vulnerable to that damage due to their small size. The rapid and large intraluminal bubble expansion causes a significant dilation of the vessel wall, leading to consistent rupture of the hollow fibers (inner diameter of 200 μm) after less than 20 LSWs in an XL-1 lithotripter. The rupture is dose dependent, and it varies with the spatial location of the vessel

phantom in the lithotripter field. Furthermore, when the large intraluminal bubble expansion was suppressed by inversion of the lithotripter pressure waveform, no rupture of the hollow fiber was found after 100 LSWs. Theoretical calculation of SWL-induced bubble dynamics in blood confirms that the propensity of vascular injury due to intraluminal bubble expansion increases with the tensile pressure of the lithotripter shock wave, and with the reduction in the inner diameter of the vessel (Zhong et al. 2001).

In an *ex vivo* experiment, bubble dynamics induced by high-intensity focused ultrasound (HIFU) burst in rat mesentery was observed using high-speed photography to provide insight into the mechanics of bubble–vessel interactions (Chen et al. 2011). Liquid jets were directed away from the nearest vessel wall, and invagination exceeded dilation, which seem sensitive to the mechanical properties of biological tissues. Evidence from lithotripsy supports the contention that cell lysis is induced by stresses in the cells resulting from bubble oscillations, since photographic evidence suggests that there is little significant translational motion of bubbles in the lithotripter field (Figure 12.25).

While cavitation occurred almost immediately in the urine, the first unambiguous cavitation was significant and observed *in vivo* only after about 1000 shock waves. Thus, the initial vessel rupture with minimal or no cavitation activity present results in pooled blood to provide nuclei for extensive and spreading cavitation to produce significant damage (Shao et al. 2003). Using finite-volume simulation methods, the strong re-entrant jet that forms upon collapse of a small bubble is found to have the spreading injury after multiple shocks but limited penetration into the tissue (using a viscous fluid with comparable viscosity).

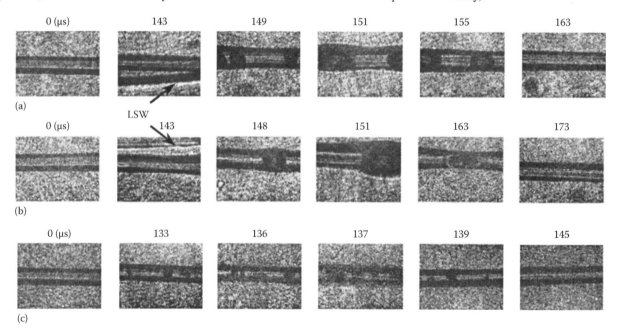

FIGURE 12.24 Representative high-speed images of bubble dynamics in a 200 mm hollow fiber. (a) Lithotripter shock wave (LSW) at 16 kV, (b) LSW at 24 kV, and (c) the inverted LSW at 24 kV. The number above each image frame indicates the time delay in ms after the spike discharge. (From *Ultrasound Med. Biol.*, 27, Zhong, P., Zhou, Y., and Zhu, S.L., Dynamics of bubble oscillation in constrained media and mechanisms of vessel rupture in SWL, 125, Copyright 2001, with permission from Elsevier.)

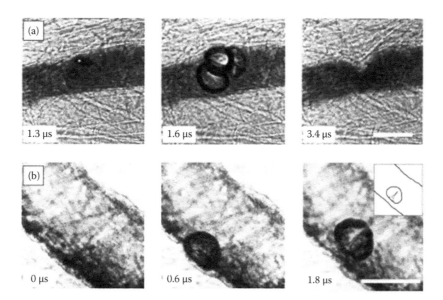

FIGURE 12.25 Representative high-speed photographs of bubble–vessel interactions with the behavior of (a) dilation and subsequent invagination by a group of bubbles and (b) a liquid jet directed away from this wall at different the time after arrival of the ultrasound pulse. Scale bars represent 50 μm. (With permission from Chen, H., Kreider, W., Brayman, A.A., Bailey, M.R., and Matula, T.J., Blood vessel deformations on microsecond time scales by ultrasonic cavitation, *Phys. Rev. Lett.*, 106, 034301, 2011. Copyright 2011 by the American Physical Society.)

Microjetting is a manifestation of the instability of the collapsing cavity and is enhanced by the presence of a rigid wall close to the imploding cavity (Blake and Gibson 1987) or by pressure gradients. However, renal tissues do not have rigid boundaries, and there are no pressure gradients at the focus during the collapse phase of the cavities because cavities usually collapse about 300 μs after the shock wave has passed the focus (Cleveland et al. 2000). Hence, microjetting may not be a dominant mechanism.

It is found that the highest expected values for the elasticity of the membrane and surrounding tissue are insufficient to suppress bubble growth using the Rayleigh–Plesset equation to simulate the bubble dynamics with respect to lithotripsy shock wave and the Voigt viscoelastic model for the elastic membrane and exterior tissue. The reduced confinement of a cylindrical vessel should not alter this conclusion. However, the higher viscosities are arguably more relevant to the deformations caused by growing bubbles and suppress bubble expansion significantly to resist injury (Freund 2008).

The degree of renal alteration (i.e., intrarenal hematomas, a cord-like scar at the corticomedullary junction, and a wedge-shaped necrosis of papillary apex) and damage to small and mid-sized veins and arteries depend on the number of shock waves administered (Pishchalnikov et al. 2003). Computed tomography (CT) has shown wedge-shaped enhancements of contrast medium in the shock wave path, and histologic examinations within the first 24 h and 1 day after treatment, respectively, showed increased alterations in the 17 kidneys of beagle dogs that received 2000, 3000, or 4000 shock waves with the Wolf Piezolith 2200 machine. In addition, the size of SWL-induced lesion and tubular injury increased significantly in size with the discharge voltage from 12 to 24 kV, while changes in renal hemodynamics were already maximal at the lowest discharge voltage, which means frank injury to tubules and vessels is insensitive to renal blood flow (Connors et al. 2000). Bilateral glomerular filtration rate (GFR), renal plasma flow (RPF), and *para*-aminohippurate (PAH, a measure of tubular function) extraction were measured 1 h before and 1 and 4 h after SWL in all treated and sham 6–8-weeks-old pigs, and the kidneys were dissected for morphometric analysis. PAH extraction was not significantly affected at 12 kV, transiently reduced at 18 kV, and reduced during SWL at 24 kV. GFR and RPF were significantly and similarly reduced 1 h after treatment and then returned to the baseline 3 h later. Furthermore, the injury induced by SWL affected a larger fraction of small kidneys than large ones, and the renal vasoconstriction induced by SWL was greatest in small kidneys in an experiment using 6- and 10-week-old pigs (Willis et al. 1999). SWL-induced lesions occupied a significantly greater volume of the small kidneys than in the large kidneys (6.1 ± 1.7 vs. 1.5 ± 0.2 vol%). RPF was significantly reduced by SWL but was more significant in small kidneys. RPF was also significantly reduced in the contralateral kidneys, but only at 1 h after SWL. The reduction in GFR seems independent of the kidney size.

The low-voltage shock waves caused vasoconstriction of the entire kidney, which protected the treated segment of the kidney from hemorrhagic injury. This same vasoconstrictive effect may be the reason for reduced renal injury seen in animals exposed to increasing energy doses (voltage stepping) (Willis et al. 2006). SWL to one kidney generally impaired the renal hemodynamics in the contralateral kidney (Connors et al. 2000). SWL exerts an acute change in renal hemodynamics (i.e., vasoconstriction) that occurs away from the focal region, as measured by a transient reduction in both GFR and RPF. Not only was there a decrease in GFR and RPF of the kidneys exposed to SWL, but also RPF (but not GFR) decreased in contralateral, untreated renal units of minipigs (Willis et al. 1996, 1997). Prolonged vasoconstriction may result in tissue ischemia and permanent renal damage. Moreover, it appears that the hemodynamic changes not only exist but are significant enough to produce ischemia-reperfusion injury. Although previous studies have suggested that the hemodynamic effects are transient in nature in the normal kidneys, patients with baseline renal dysfunction may be at significant risk for permanent renal damage (Willis et al. 1996, 1997).

SWL causes renal damage directly through cellular injury and indirectly through vascular injury and resulting in ischemia, which gives rise to oxygen free-radical compounds. Oxygen free radical is believed to contribute to parenchymal damage via lipid peroxidation and the disruption of cellular membranes, and its toxicity is attributed to its ability to initiate lipid peroxidation of cellular membranes. Oxygen free radicals may be an integral element in SWL-induced injury to small vessels and renal parenchyma as well as SWL-induced renal vasoconstriction through an indirect mechanism. Tissue ischemia caused by vascular injury can create abnormally high levels of free radicals and oxidants produced by the normal cellular metabolism as reperfusion occurs (Sarica et al. 1996). The toxicity of free radicals is attributed to their ability to initiate lipid peroxidation of cellular membranes (Cohen et al. 1998). Intrarenal hemorrhage from vascular injury may compound the problem since the iron of red blood cells can catalyze the production of free radicals (Agil et al. 1995). Using microdialysis probes inserted into the renal parenchyma at the right upper and lower poles and left lower pole of juvenile female swine, conjugated dienes, a measure of lipid peroxidation and free-radical activity signifying renal cellular damage, were quantitatively assessed. A dose-related increase in conjugated diene ratio levels from the right lower pole (SWL site) and to a lesser degree from the ipsilateral right upper pole was observed, which were significantly different from sham group measurements or the contralateral kidney of the treatment group. The elevation at the treated site was significantly greater than that of the upper pole of the ipsilateral kidney. The increase in free-radical activity at a remote site suggests detrimental global effects of SWL and may be due to vasoconstriction throughout the treated kidney with resulting ischemia reperfusion injury (Delvecchio et al. 2003). The right upper or left lower pole demonstrated the biochemically detrimental cellular effects for gross and histological damage to glomeruli, renal capillaries, and renal tubules. Nitric oxide (NO) production by a nonenzymatic pathway has already been shown with short-time kinetics (min), which illustrates the biochemical responsible for these therapeutic effects (Gotte et al. 2002). This regenerative pathway seems to involve free-radical production and revascularization events taking place in the SW-treated area (Seidl and Steinback 1994). NO exerts a potent and immediate vasodilatory action and modulates the subsequent angiogenesis (Palmer et al. 1987, Moncada and Higgs 1993). After a total of 10,000 SWs delivered to 10 juvenile swine, conjugated diene levels in the collected dialysate fluid in the treated right kidney increased almost 100-fold compared to that in the contralateral one; however, there was no increase in those premedicated with allopurinol during SWL, confirming SWL-induced free-radical activity as well as the antioxidant and protective nature of allopurinol (Munver et al. 2002).

Renal trauma caused by SWL may cause hypertension as the result of a perirenal hematoma via the well-known Page kidney effect: trauma → perirenal hemorrhage → fibrosis → compression of renal parenchyma → increased interstitial pressure → decreased renal perfusion → renin release → generation of angiotensin II → hypertension (Peterson and Finlayson 1986). Up to 18 months after SWL, decreased renal plasma flow may result either from increased interstitial pressure caused by fibrosis due to intrarenal hemorrhage or pressure from a perirenal fibrotic process. The late renal changes and hypertension may not have been due to the trauma of SWL but to some other process unrelated to renal lithiasis, such as unilateral renal artery stenosis (Williams et al. 1988).

The plasma concentrations of urinary nitrite, a stable metabolite of nitric oxide (NO), and adrenomedullin (AM) of 20 patients with renal pelvic or caliceal stones ≤2 cm before and 24 h after anesthesia-free SWL were measured by the Griess reaction and high-performance liquid chromatography (HPLC), respectively, and then compared with those of 10 control patients without any urological symptoms. High-energy SWT caused a statistically significant increase in both NO and AM (from 29.9 ± 7.6 to 39.02 ± 8.45 μmol/L and from 20.51 ± 3.0 pmol/mL to 32.54 ± 4.3 pmol/mL, respectively), which suggested that SWL can stimulate the NO–cGMP signaling pathway to increase NO production in the kidney and that the increased levels of NO and AM secretion during renal parenchymal ischemia may be protective enough for SWL-induced renal trauma (Sarica et al. 2003).

12.6 TECHNICAL IMPROVEMENT

Exploiting innovation in a developed technology is usually based on the potential benefits and risks in such an industry for the investment. The collaboration between health services or research laboratories (i.e., in academia, government, or industry) and manufacturers depends on the requirement of market and regulatory environments (Citron 1996). There are multiple factors that may have limited the improvement in SWL: the initial design had great success and excellent performance, so there were few easy paths for improvement; there is no complete understanding on the mechanisms of stone fragmentation and tissue injury, which is arguably necessary in order to improve current designs; the high acceptance of SWL and saturation of market reduced the incentive to innovate (Leighton and Cleveland 2009). According to clinical experience and mechanism understanding, new technologies to significantly increase the efficacy and safety of SWL have begun to address the following issues: improve acoustic coupling, enhance stone targeting, determine the endpoint of SWL, and minimize the occurrence of residual stone fragments (Rassweiler et al. 2011).

Even without the development of new-generation litho-tripters, a number of strategies of SWL delivery have been proposed to enhance the performance of current machines. Pretreatment of patients by antioxidants might protect the renal parenchyma against free-radical injury, as well as acute and/or chronic SWL-induced renal injury (Ogiste et al. 2003). Chemolytic pretreatment increases stone fragility. Calcium channel blockers, steroids, and alpha blockers may improve spontaneous passage of ureteral stones. Stenting before SWL seemed to reduce complications caused by residual stone fragments, especially for a large stone, but not decrease the incidence of steinstrasse after SWL of small to mid-sized stones (Preminger et al. 1989, Bierkens et al. 1991). Overall steinstrasse occurs in 1%–4% of patients who undergo SWL (Madbouly et al. 2002). The rate increases in 5%–10% of patients with large stones (>2 cm^2) (Bierkens et al. 1991) and in up to 40% of patients with partial or complete staghorn cal-culi (Wirth et al. 1992). The European Association of Urology (2010) recommends pre-SWL stenting for renal stones with a diameter of >20 mm (approximately 300 mm^2), and a Double-J stent to reduce obstructive and infective complica-tions after the use of SWL (Türk et al. 2010). Stents are associ-ated with significant symptoms of discomfort, such as urinary frequency, urgency, dysuria, and hematuria (Shen et al. 2011). A systematic meta-analysis of 876 patients illustrated that the use of Double-J stenting before SWL did not benefit stone-free rate, incidence of steinstrasse, hematuria, nausea and vomiting, fever, urinary tract infection, pain and analgesia, and auxiliary treatment but induced more lower urinary tract symptoms (Shen et al. 2011). Recognition of SWL limitations, correction of preexisting renal or systemic disease, treatment of urinary tract infection, and use of prophylactic antibiotics have been used in clinics to reduce the side effects. Decrease in the number of shock waves, pretreatment using low pres-sure pulses, the schedule for targeting checks, appropriate management of delivery rate, and energy protocol are new treatment strategies without modifying lithotripter configura-tions. The current standard is to start at low or moderate out-put levels with as few SWs as possible in order to minimize acute and chronic renal injuries and at a slow pulse repetition rate (60 SWs/min or slower) in order to enhance stone frag-mentation and reduce tissue damage.

The actions involved in the varying PRF are different for stone fragmentation and tissue injury. In stone fragmentation, cavitation bubbles collapse at the stone surface to generate fine fragments that can be discharged spontaneously after SWL treatment (Zhu et al. 2002, Cleveland and McAteer 2007). At a high PRF, remanent microbubbles after bubble collapse persist between shock waves, which serve as nuclei to form new bubbles in the following pulse, and subsequently enhance the bubble cavitation (Pishchalnikov et al. 2008). However, the presence of these small bubbles reduces the amplitude of the coming LSWs, alters the bubble dynamics, and affects the delivered energy to the stone for effective fragmentation, col-lectively known as the bubble shielding effect (Pishchalnikov et al. 2005). In contrast, bubbles that form within the renal vessel would damage the vessel wall in either the expansion or

collapse stage (Zhong et al. 2001). Stress might also accumu-late if the PRF is faster than the relaxation time of the kidney tissue. Recent studies in a pig model indicate that a low PRF can significantly reduce renal injury and improve the stone fragmentation (Delius et al. 1988b, Koga et al. 1996, Evan et al. 2007a, Connors et al. 2009). Critical meta-analyses have concluded that treatment at a 1 Hz PRF was more effective than that at 2 Hz (Semins et al. 2008). Damage is dramati-cally reduced at low rates of shock wave delivery. Despite these sound evidences, few urologists are willing to adopt this simple and effective strategy in practice because lower-ing the PRF increases the procedure duration, particularly at a high-volume stone center that has a dozen or more cases a day. Furthermore, juvenile pigs treated at 24 kV continu-ously up to 2000 shocks showed a mean gross lesion volume of approximately 4%, whereas the corresponding value was around 0.5% or less using a pause protocol that should be fea-sible in most clinical settings.

The vasoconstrictive effect induced by shock waves to the kidney has been well documented in animal experiments (Connors et al. 2003). The initial 100 pulses at a low setting induce a vasoconstrictive response in the kidney, which pro-tects it from the subsequent LSW exposure, as has been con-firmed in recent Doppler ultrasound measurements (Handa et al. 2009a). This practical strategy may allow the patient the opportunity to acclimatize to the procedure (Lingeman et al. 2006, McAteer et al. 2008).

The output voltage of lithotripsy is usually set constant throughout the treatment. In vitro studies have found that dif-ferent strategies on the setting could lead to different frag-mentation patterns (Zhou et al. 2004). Although the initial rate of stone fragmentation produced by the power increase strategy was low compared to that of the other two (consistent output and power decrease), a gradual increase in the output voltage was found to help in maintaining a relatively constant rate of fragmentation throughout the entire course of SWL treatment by enhancing bubble cavitation and introducing a strong synergy between stress waves and cavitation in the later stage of SWL. In contrast, the other strategies had high initial stone fragmentation, which decreased significantly after about 500–1,000 shocks (Zhou et al. 2004). Its improved performance in stone breakage was further confirmed in both *in vivo* with surgically implanted stones in pigs and in patients (Maloney et al. 2006, Demirci et al. 2007). Stepwise treat-ment might also contribute to low hematoma rates (Mobley et al. 1993). A subsequent clinical trial of 50 patients con-firmed greater success for stepwise power ramping in com-parison with conventional protocols using a constant power (Demirci et al. 2007).

Cavitation contributes to both desired stone comminu-tion and undesired tissue damage, so appropriately control-ling it can lead to improved SWL outcome (Sheir et al. 2005). In EH lithotripters, the greatest cavitation occurs not at the geometric focus but at a proximal site of 1–3 cm. *In vitro* stone comminution, hemolysis, and free-radical production were assessed along the focal axis, and pig kidneys treated with SWL *in vivo* were sectioned to determine the extent of hemorrhagic injury along the focal axis. At *F*2, The weight loss of gypsum stones after 200 SWs *in vitro* at 18 kV placed at *F*2 and *F*2 – 2 cm was 11.3% ± 1.1% and 16.1% ± 4.2%, respectively. Hemolysis of 10% hematocrit blood was similar at *F*2 – 2 cm (14.7% ± 2.3%) and *F*2 (15.2% ± 3%). Hydroxyl radical production in iodine solution after 1500 SWs at 20 kV was greatest at *F*2 – 2 cm (0.384 ± 0.035 mM). Stone comminution may be achieved more rapidly without greater tissue damage by a simple pre-focal alignment (Sokolov et al. 2002).

EH lithotripters have both a direct and a focused SW. The direct one propagates as a spherically diverging wave, arrives the focal region earlier, and causes the bubble nuclei to grow to a big size to shield the coming focused SWs. A baffle in the electrode can block the propagation of the direct wave and significantly reduce the induced bubble growth. As a result, the therapeutic efficacy may be improved (Matula et al. 2005).

Using a pressure-release reflector (made out of polyurethane foam, which has acoustic impedance lower than that of water), an inverted lithotripsy shock wave could be produced with a leading tensile wave followed immediately by a much stronger compressive wave. Although the pressure amplitude, pulse duration, and acoustic energy were similar to the standard pulse, the change of wave component sequence led to significant suppression of cavitation (Bailey et al. 1998, 1999). Red blood cell lysis *in vitro* was significantly lower with the pressure release reflector than with a rigid reflector. Furthermore, the lower pole of the right kidney of 6-week-old anesthetized pigs was treated with 2000 SWs at 24 kV using an unmodified Dornier HM3 lithotriptor or with a pressure release reflector insert. SWs from the rigid reflector induced a characteristic morphological lesion and functional changes, such as bilateral reduction in renal plasma flow and unilateral reduction in the GFR and *para*-aminohippurate extraction. In comparison, those from the

pressure-release reflector only had hemorrhage of vasa recta vessels near the tips of renal papillae, and the only change in kidney function was a decrease in the GFR at the 1 and 4 h periods (Evan et al. 2002). However, the pressure-release reflector could not achieve satisfactory stone comminution efficiency (Loske and Prieto 2002). Polyurethane foam is not robust to SW damage and cannot be used for a long time (Figures 12.26 and 12.27).

To reduce the potential of vascular injury without compromising the stone comminution capability of a Dornier HM3 lithotripter, *in situ* pulse superposition was developed to suppress intraluminal bubble expansion. A thin-shell ellipsoidal reflector insert was designed and fabricated to fit snugly into the original reflector of an HM3 lithotripter. The inner surface of the reflector insert shared the same first focus with the original HM3 reflector but had its second focus located 5 mm proximal to the aperture. The original lithotripter shock wave was partitioned into a leading lithotripter pulse from the reflector insert and an ensuing second compressive wave from the remaining HM3 reflector 4 μs later to be imposed on the tensile component of the primary pulse. There were no significant changes on the pulse amplitude and beam width of HM3 at 20 kV and upgraded reflector at 22 kV. After 2000 shocks, stone comminution efficiencies produced were comparable. Using the original HM3 reflector, about 30 shocks were required to break a standard vessel phantom made of cellulose hollow fiber (i.d. = 0.2 mm) at the focus while no damage occurred in the whole focal region after 200 shocks. Therefore, the upgraded reflector could significantly reduce the propensity for vessel rupture in SWL while maintaining satisfactory stone comminution (Zhong and Zhou 2001, Zhou and Zhong 2003), which was also confirmed in the animal experiments (stone comminution efficiency increased from 87.6% ± 1.8% to 91.6% ± 8.8%, but the volume of renal injury

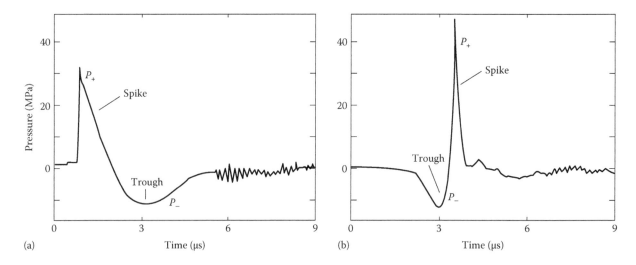

FIGURE 12.26 Pressure waveforms produced by (a) a rigid reflector and (b) a pressure-release reflector at the focus at a charging voltage of 18 kV. Although pulse durations P_+ and P_- are comparable, the sequences of pressure component are different (spike-then trough waveform of the rigid reflector and the trough-then-spike waveform of the pressure-release reflector). (With permission from Bailey, M.R., Blackstock, D.T., Cleveland, R.O., and Crum, L.A., Comparison of electrohydraulic lithotripters with rigid and pressure-release ellipsoidal reflectors. I. Acoustic fields, *J. Acoust. Soc. Am.*, 104, 2520, Copyright 1998, American Institute of Physics.)

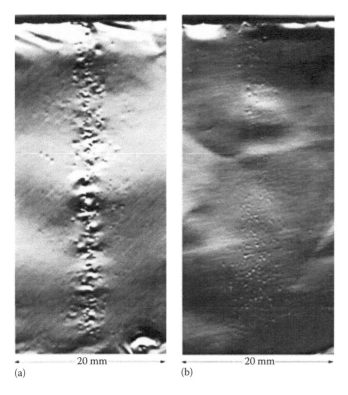

(a)   20 mm

(b)   20 mm

FIGURE 12.27 Comparison of bubble pit on an aluminum foil by 10 pulses produced by (a) a rigid reflector and (b) a pressure-release reflector. (With permission from Bailey, M.R., Blackstock, D.T., Cleveland, R.O., and Crum, L.A., Comparison of electrohydraulic lithotripters with rigid and pressure-release ellipsoidal reflectors. II. Cavitation fields, *J. Acoust. Soc. Am.*, 106, 1154, Copyright 1999, American Institute of Physics.)

decreased from 1.69% ± 0.35% to 0.92% ± 0.49%) (Zhou 2011) (Figures 12.28 and 12.29).

The electroconductive lithotripter (e.g., Sonolith 4000) uses a highly conductive solution to enclose the electrode so that the discharge energy is more precise and consistent between the anode and the cathode. *In vitro* experiments showed a linear relation between the output voltage and the

produced pressure with reduced variation, improved energy transfer to the stone, a linear increase of stone fragmentation with output voltage without a saturation effect, and much longer lifetime of the electrode. In 142 evaluable treatments with a 3-month follow-up, the overall stone-free rate of the Sonolith 4000 lithotripter was 82%, and the retreatment rate in stone-free patients was 10%. For stones equal to or less than 10 mm, the 3-month stone-free rate, re-treatment rate, secondary procedure rate, and efficiency quotient were 85%, 5%, 0%, and 81%, respectively. For stones between 11 and 20 mm, these values were 83%, 4%, 2%, and 78%, respectively. These clinical results confirm the improvements in efficacy observed *in vitro* with very satisfactory tolerance (Flam et al. 1994).

Delivery of tandem shock waves from two sources at an appropriate interval can control the bubble cavitation and subsequently improve SWL efficacy and safety, so it is an area of active investigation with promise. Many devices have been developed using this concept. Stone comminution efficiency was significantly enhanced at a delay of about 400 and 250 μs between two piezoelectric generators to rectangular and spherical stone phantoms, respectively (Loske et al. 2002), but without more tissue damage on rabbits (Loske et al. 2005). A related concept is to deliver shock waves from separate sources aligned to the same focal point at appropriate triggers (Sokolov et al. 2001, 2003). Dual-head lithotripters have already been applied to patients. Despite the inherent complexity associated with twin generators and coupling and twice the number of shocks delivered, initial reports suggest that it is safe and effective with minimal morphological changes to the renal parenchyma compared with the standard design (Sheir et al. 2008, Handa et al. 2009b). No gross lesions of the surrounding organs, subcapsular hemorrhage, or parenchymal damage was found at the outer surfaces of the kidneys undergoing twin-head SWL after 3,000 shocks. Microscopically, the parenchymal changes were minimal. In contrast, a single-head instrument revealed large subcapsular hematomas at

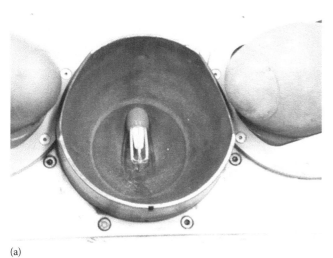

(a)

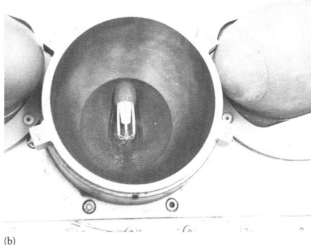

(b)

FIGURE 12.28 Photographs of (a) the original HM3 reflector and (b) the upgraded reflector with a single piece of reflector inserted.

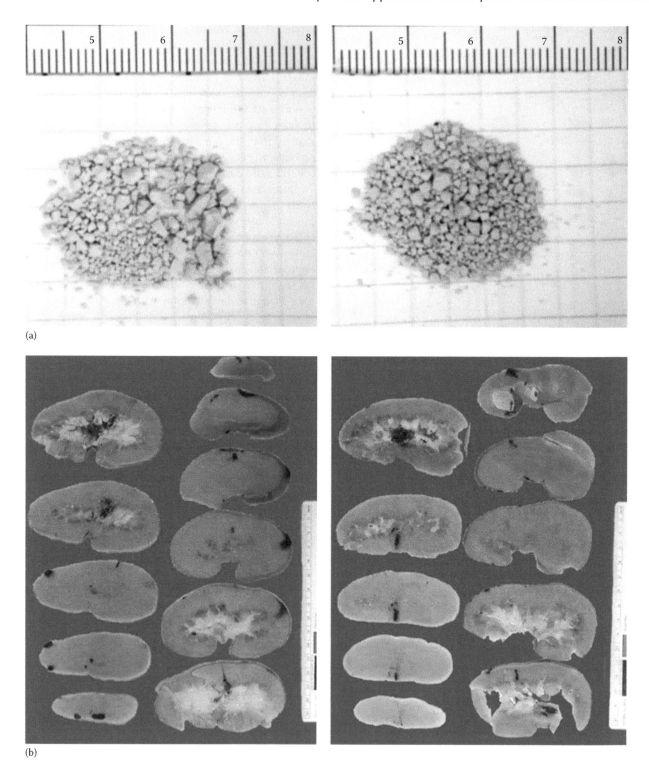

(a)

(b)

FIGURE 12.29 Photographs of (a) the stone fragments and (b) injuries on the kidney after 2000 shocks using original HM3 reflector at 20 kV (left) and the upgraded reflector at 24 kV (right).

both anterior and posterior surfaces and on coronal section extending into the parenchyma. However, dual-head lithotripters are not currently in wide use.

To control the collapse of cavitation bubbles induced during SWL, a piezoelectric annular array (PEAA) shock-wave generator was fabricated and combined with an EH lithotripter. By changing the delay time, the PEAA-generated shock wave could be used to interact with cavitation bubbles induced by the EH source at various stages of their oscillation. A maximum increment of 60%–80% in stone fragmentation could be achieved when the shock wave–bubble interaction occurred during the collapsing phase of the bubbles, with intensified collapse of the bubbles near the surface of the stone, strong secondary shock-wave emission, and increased stress concentration at the impact site of the solid boundary, as observed using high-speed imaging (Xi and Zhong 2000).

If stones are larger than the width of the focal zone, the energy deposited into the stone would be low (Cleveland 2008). Based on the mechanism of binary fragmentation by squeezing, a wide-focus, low-pressure lithotripter was proposed and proved to produce better stone comminution than its counterpart with a narrow focus and high pressure (Qin et al. 2010). The first clinical EM model (XX-ES, Xixin Medical, China) with a wide focal zone of 18 mm and a low acoustic pressure of 17 MPa was evaluated in seven hospitals in China for a total of 297 patients with an average of 1532 pulses and with no pain medication. The stone-free rate was 86% after a 3-month follow-up (Eisenmenger et al. 2002). One thousand and five hundred SWs produced from XX-ES at a slow rate of 27 SW/min or from a Dornier HM3 EH lithotripter at an output voltage of 18 kV and a delivery rate of 30 SW/min were delivered to the left kidneys of anaesthetized pigs. The XX-ES-treated kidneys showed no significant change in renal hemodynamic function and no detectable tissue injury, while those treated with the HM3 had a modest decline from baseline ($\approx$20%) in both GFR and RPF and two pigs had 0.1% functional renal volume focal injury localized to the renal papillae. The number of SWs to break gypsum model stones to completion using XX-ES and the HM3 was 634 ± 42 and 831 ± 43, respectively ($p < 0.01$). Altogether, using a wide focal zone lithotripter operated at low pressure and slow rate improved stone fragmentation and reduced tissue injury simultaneously (Evan et al. 2007b).

During SWL, stone motion due to respiratory motion can be 50 mm or more with each breath, which is much larger than the 4–15 mm beam width of most clinical lithotripters, and is usually not on the lithotripter axis. Depending on respiratory rate, the length of excursion, the focal width of lithotripter, and SW delivering rate, 50% or more of the shots may miss the target (Cleveland et al. 2004, Cleveland and McAteer 2007). In an *in vitro* study, gypsum cement (6.5 × 7.5 mm) was connected to a motorized positioner to simulate ventilatory motion up to 48 mm (±24 mm about the focus) and exposed to 400 SWs from a Storz Modulith SLX lithotripter at different energy levels and firing rates. Stone fragmentation efficiency was reduced significantly for a motion of ≥10 mm, and three-quarters of the shockwaves missed the stone for motion >20 mm. Therefore, ventilatory gating or stone tracking is necessary for successful SWL treatment (Cleveland et al. 2004). Immobilization by high-frequency ventilatory respiration anesthesia was clinically effective, but it was too invasive. Systems with respiratory belts and shock wave triggering (i.e., Lithostar) have been abandoned because of increased treatment time. Spectral Doppler ultrasound can be used to monitor SWL *in vitro*. If a shock wave hits a stone phantom, a high peak followed by a decaying signal, whose duration from 30 to 150 ms is dependent on shock wave energy, stone size, gas concentration in the water, and the degree of fragmentation, shows up in the Doppler spectra. Thus, hits were reliably distinguished from misses, which was also confirmed in the clinical treatments (Bohris et al. 2003). Another solution

is using sophisticated ultrasound imagers and tracking algorithms to actively and synchronously steer the beam of a PE lithotripter with the moving stone (Thomas and Fink 1996, Orkisz et al. 1998, Chang et al. 2001, Bohris et al. 2003). Although tracking can improve the hit rate by about 50%, it is not currently in clinical use. Another reason of patient motion is due to the pain in SW exposure, and sedation or anesthesia could minimize such influence. One hundred patients treated on the modified HM3 machine with intravenous sedation at low voltage (12–15 kV) had similar results as those at standard settings (Newman et al. 1989).

The Hounsefield units (HU) of renal stones on pretreatment noncontrasted CT could predict stone-free rates of SWL. Some resistant stones, such as calcium oxalate monohydrate or cystine stones, may be identified, and such patients will subsequently undergo endoscopic modalities (Weizer et al. 2007). Body mass index (BMI) and HU density of urinary calculi may predict the stone-free rate after SWL. The radiographs of 64 patients treated with a DoliS lithotripter from March 2000 to April 2004 with lower pole kidney stones of 0.5–1.5 cm were reviewed. The average skin-to-stone distance (SSD) was calculated by distances from the center of the stone to the skin at angles of 0°, 45°, and 90°. The mean SSD was 8.12 ± 1.74 cm for the stone-free group versus 11.53 ± 1.89 cm for the residual stone group ($p < 0.01$) 6 weeks after SWL. Logistic regression analysis revealed only SSD rather than BMI and HU to be a significant predictor of outcome (OR = 0.32, 95% CI: 0.29–0.35, $p < 0.01$) (Pareek et al. 2005). Dual x-ray absorptiometry is the standard for measuring stone mineral content (SMC), from which the fragility of the stone may be predicted. In a total of 102 patients with a solitary renal and upper ureteral stone of size 16.68 ± 7 mm (5–30 mm) and SMC of 0.63 ± 0.83 g (0.01–5.54 g) treated with SWL, logistic regression (LR) analysis showed that SMC affects the fragmentation outcome and 95% of the stones would not fragment within 3000 shock waves if SMC is more than 1.27 g (Mandhani et al. 2003). In addition, an artificial neural network (ANN) model can improve the prediction of stone-free status compared to an LR model. Between February 1989 and December 1998, 984 patients (780 men and 204 women of age 40.85 ± 10.33 years) with ureteral stones were treated. An LR model was constructed and ANN was trained on 688 randomly selected patients to predict stone-free status at 3 months. The LR and ANN models illustrated a sensitivity of 100% and 77.9%, a specificity of 0.0% and 75%, a positive predictive value of 93.2% and 97.2%, an overall accuracy of 93.2% and 77.7%, and an average classification rate of 50% and 76.5%, respectively. The neural network showed a higher ability to predict those who failed to respond to ESWL (Gomha et al. 2004). Real-time monitoring of the effectiveness of SWL, including passive acoustic monitoring and improved active ultrasonic diagnostics, may benefit the treatment (Sorensen et al. 2008). Resonant scattering of the SWs could differentiate intact and fractured stones *in vitro*. The reflected SWs are followed by pressure fluctuations that correspond to reverberations within the stone, which are due to vibrational resonance of

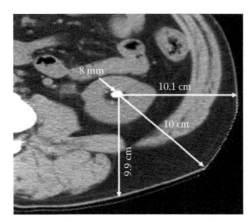

FIGURE 12.30 Three measurements (0°, 45°, and 90°) from center of stone to skin used to determine SSD in SF patient with an 8 mm lower pole kidney stone. (From *Urology*, 66, Pareek, G., Hedican, S.P., Lee, F.T., and Nakada, S.Y., Shock wave lithotripsy success determined by skin-to-stone distance on computed tomography, 943, Copyright 2005, with permission form Elsevier.)

the stone after SW excitation but not cavitation. Identification of fracture was possible through spectrum analysis (i.e., higher energy in distinct bands), which was in a good agreement with numerical simulation with linear elasticity theory (Owen et al. 2006). Therefore, the operator can prefer to terminate the treatment when the stone has already completely fragmented or is resistant to SWL in order to limit the morbidity associated with SWs (Figure 12.30).

After SWL, most of fine particles can be discharged through ureter with urine naturally, but larger fragments (on the order of 1 mm to several mm) are slow to wash out or can be retained and may obstruct the ureter, thus leading to complications such as acute renal pain, hydronephrosis, infection, and renal failure (Salem et al. 2010). Incomplete fragmentation, residual stone fragments, steinstrasse, and obstruction are among the problems urologists confront when SWL fails to completely fragment the stone treated. The increased number of residual fragments, but not their location, decreased the probability of stone clearance (Osman et al. 2005). Growth of residual fragments <4 mm has been found in 21%–59% patients after SWL (Sun et al. 1996, Osman et al. 2005). Forty-three percent of these patients with residual calculi had a symptomatic episode or needed an intervention after a mean 26-month follow-up (Streem et al. 1996). They can be symptomatic or act as a nidus for stone formation. Feasibility of repositioning kidney stones using acoustic radiation force by delivering short bursts of focused ultrasound from the imaging probe was investigated by implanting artificial and human stones into a kidney-mimicking phantom that simulated a lower pole and collecting system. The stone motion speed was estimated to be on the order of 1 cm/s. Such an approach is promising to facilitate spontaneous clearance of small kidney stones and residual stone fragments (Shah et al. 2010).

Overall, more improvements or novel lithotripsy technologies will be developed based on the further understanding of the working mechanisms. The success of translating them with proven concept or performance in animal experiments

from academia to clinical practice depends on multidisciplinary collaborations between researchers, urologists, and lithotripter manufacturers (Zhong 2007).

12.7 SHOCK WAVE THERAPY

The application of shock waves to the loosening of cement in the revision of total hip was thought to be feasible since the introduction of SWL to urology and found an osteoblastic response pattern to numerous orthopedic disorders (Brummer et al. 1990, Delius 1994, Delius et al. 1998, Coombs et al. 2000, Thiel 2001). In 1986, the influence of SWs wave on bone was firstly investigated (Haupt 1997). The first shock wave therapy (SWT) for nonunion fracture was done in Bochum, Germany, in 1988. In the same year, SWT was applied for nonunions and delayed unions with a success rate of 85% despite the poor control (Haupt et al. 1992, Haupt 1997, Coombs et al. 2000). The application of SWT in certain orthopedic and musculoskeletal disorders dates back to about 20 years, and the success rate in nonunion of long bone fracture, calcifying tendonitis of the shoulder, medial and lateral epicondylitis of the elbow and proximal plantar fasciitis, patellar tendinitis (athletes with jumper's knee), and Achilles tendinitis, ranged from 65% to 91% with low and negligible complications (Sukul et al. 1993, Haupt 1997, Delius et al. 1998, Siebert and Buch 1998, Coombs et al. 2000). Recently, SWT was extended to the treatment of avascular necrosis of femoral head, osteochondritis dissecans, and non-calcifying tendonitis of the shoulder (Lussenhop et al. 1998, Siebert and Buch 1998, Ludwig et al. 2001, Ogden et al. 2001c, Thiel 2001). Overall, SWT is a novel, convenient, and cost-effective therapeutic modality without the need of surgery and surgical risks as well as pain. The relatively large amplitudes and low frequencies in SWT make it more suitable for generating transient cavitation than most other forms of medical ultrasound in mammalian tissue (Coleman and Saunders 1992). The use of extracorporeal SWT has gained significant acceptance in Europe, especially Germany, Austria, and Italy, as well as in Taiwan. In 2000, the FDA approved SWT for chronic plantar fasciitis and in 2003 for chronic lateral epicondylitis (Henney 2000, Rompe 2003). Many other clinical trials are going on, including lateral epicondylitis of the elbow, calcific tendinitis of the shoulder, and nonunion of fracture (Delius et al. 1998, Siebert and Buch 1998, Coombs et al. 2000, Ogden et al. 2001b). SWT also demonstrated its effects on nerve conduction, with reduction of pain in chronic degenerative pathologies, and potential bactericide action against *Staphylococcus aureus* (Ohtori et al. 2001, Gerdesmeyer et al. 2005, Takahashi et al. 2006). SWT had shown macroscopic effects but also caused interstitial and extracellular responses microscopically (Coombs et al. 2000, Ogden et al. 2001b). However, there is no general acceptance of the SWT for the treatment of fracture nonunions in humans because the clinical results were not confirmative with varied healing rates of 41%–100%. The rapid application has not been accompanied by well-conceived and promulgated studies, which has led to some skepticism regarding the actual efficacy of SWT for musculoskeletal disorders (Figure 12.31).

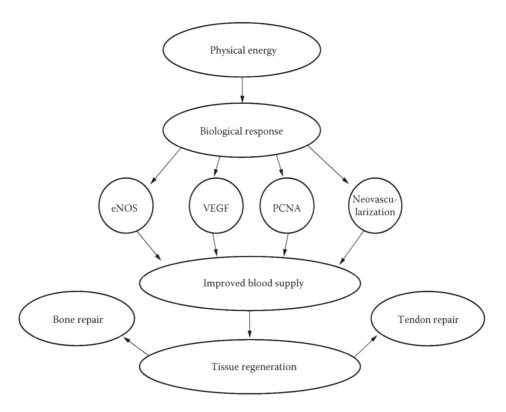

FIGURE 12.31 Mechanism of shock wave therapy, involving a cascade of interactions between physical energy and biologic responses. (From Wang, C.J., *Chang Gung Med. J.*, 26, 225, 2003.)

SWT in orthopedics (orthotripsy) is not used to disintegrate tissues, but rather to induce neovascularization, particularly elaboration of callus (Valchanou and Michailow 1991, Hirachi et al. 1999). The effect of SWT for nonunion and delayed union of long bone fractures has been studied, and the success rate ranged from 50% to 90% (Graff et al. 1987, Burger et al. 1991, 1992, Valchanou and Michailow 1991, Schleberger and Senge 1992, Augat et al. 1995, Vogel et al. 1997, Heller and Niethard 1998, Ikeda et al. 1999, Rompe et al. 2001, Schaden et al. 2001). The success rate of bony union associated with complete resolution of pain and functional recovery was 49% at initial 3-month follow-up of 55 patients, 82.4% at 6-month follow-up of 34 patients, and 88% at 9–12 month follow-up in 22 patients. SWT has shown effectiveness in promoting of bone growth, in promoting bone marrow stromal cell differentiation toward osteoprogenitor associated with induction of TGF-β1 and activation of Ras, and in inducing membrane hyperpolarization. An optimal dose of 0.16 mJ/mm^2 with 500 pulses induced rapid membrane hyperpolarization in 5 min, activation of Ras in 30 min, and cell proliferation in 2 days. The induced osteogenesis was illustrated by the increase in bone alkaline phosphatase activity in 6 days and osteocalcin mRNA expression in 12 days. Transfection of bone marrow stromal cells with a dominant negative Ras mutant (Asn-17 ras^H) abrogated the SW enhancement of osteogenic transcription factor (CBFA1) activation, osteocalcin mRNA expression, and bone nodule formations in human bone marrow stromal cells. The evidence of bone marrow stromal cell differentiation toward osteogenic lineage via membrane hyperpolarization, followed by Ras activation and specific osteogenic transcription factor CBFA1 expression, confirms the possibility of noninvasive physical agents in the treatment of fracture, osteoporosis, and osteopenic disorders (Wang et al. 2001b). The maximum stimulation of osteogenesis occurs at the interface of cortical and cancellous bones, while the tensile waves cause cavitation and osteocyte death, followed by osteoblast migration and new bone formation (Church 1989, May et al. 1990, Vogel et al. 1997, Coombs et al. 2000). SWs cause micro-fracture or micro-trauma and hematoma formation, which eventually lead to osteoblastic activities, increased callus formation, and bone healing (Church 1989, Delius 1994, Delius et al. 1995, 1998, Coombs et al. 2000, Ogden et al. 2001a) (Figure 12.32).

Osteoblast cultures isolated from cancellous bone fragments were treated with 500 pulses at 0.06, 0.18, 0.36, and 0.50 mJ/mm^2, and their proliferation, alkaline phosphatase activity, and mineralization were analyzed 24 and 96 h after SWT. Human osteoblasts (hOB) showed a dose-dependent increase in cell proliferation from 68.7% (at 0.06 mJ/mm^2, $p = 0.002$) up to 81.6% (at 0.5 mJ/mm^2, $p = 0.001$) after 24 h, which persisted until 96 h, and the peak levels of response in the numbers of alkaline phosphatase-positive hOB were between 0.18 and 0.5 mJ/mm^2. Mineralization was significantly higher in all SWT groups. SWT induced the upregulation of genes in physiologic processes, cell homeostasis, and bone formation as well as those in skeletal development and osteoblast differentiation (i.e., PTHrP, prostaglandin E2-receptor EP3, BMP-2 inducible kinase, chordin, cartilage oligomeric matrix protein, matrillin) (Raisz and Kream 1983), which are catalysis (25.7%), binding activity (73.2%), and physiologic processes

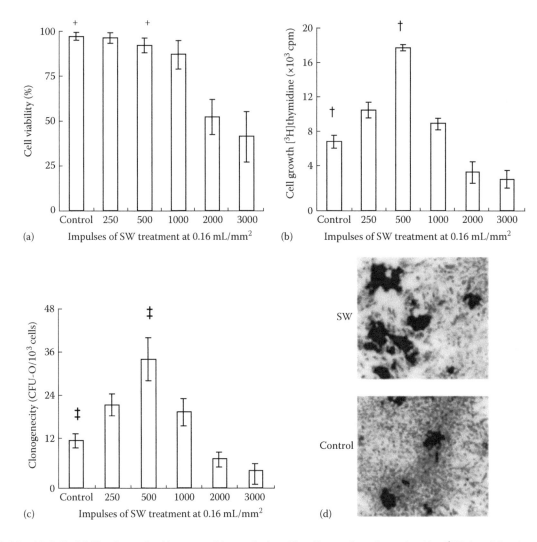

FIGURE 12.32 (a) Cell viability determined by trypan blue exclusion, (b) cell growth as determined by [³H] thymidine incorporation, (c) the plating efficiency of CFU-O formations as determined by colonies showing more than 32 aggregated cells after inoculated with 1×10^3 cell/well for 12-day culture after 250, 500, or 1000 shock waves at 0.16 mJ/mm² (*$p = 0.64$, †$p < 0.001$, and ‡$p < 0.001$), and (d) comparison of the tissue after the optimal dose of 500 pulses with the control. (From *Biochem. Biophys. Res. Commun.*, 287, Wang, F.-S., Wang, C.-J., Huang, H.-J., Chung, H., Chen, R.-F., and Yang, K.D., Physical shock wave mediates membrane hyperpolarization and Ras activation for osteogenesis in human bone marrow stromal cells, 648–655, Copyright 2001, with permission for Elsevier.)

(80.4%) with more extracellular, collagen, and proteinaceous extracellular matrix genes than those in intracellular space and nucleus (Hofmann et al. 2008). Despite an initial cytotoxic effect, SWT induced a delayed stimulatory response in cell proliferation and differentiation in human periosteal cells, which are critical for fracture healing in a dose-dependent manner. However, the expression of the typical marker OC and TGF-β1 expression, which are typical markers of osteoblasis, cell proliferation, and osteoblasts maturation in human cells, did not change as those in mice and rat BMSC and MG-63 osteoblasts (Wang et al. 2002a, Meury et al. 2006).

There was evidence of new bone formation, *de novo*, and against existing trabeculae (Johannes et al. 1994), local increase in bone mineral content, overgrowth of immature rabbit bone (Saisu et al. 1999), focal bone augmentation (in the osteoporotic femoral neck or radius), or the stimulation of longitudinal bone growth in a congenitally or posttraumatically shortened long bone.

SWT could disrupt the cement–bone and cement–prosthesis interfaces to allow easy extraction of the prosthesis and the cement mantle during revision (Karpman et al. 1987, Weinstein et al. 1988, May et al. 1990, Schreuers et al. 1991, Braun et al. 1992, Lewis 1992), which has a potential use to the bone surrounding an unstable (clinically symptomatic or painful) implant. Disrupting the calcific intra-tendinous deposit encourages resorption.

SWT enhanced callus formation and induced cortical bone formation in acute fractures in dogs at 12 weeks, which seemed to be time dependent (Haupt et al. 1992, Johannes et al. 1994). However, only one study showed a negative result (Forriol et al. 1994). High-energy SWT produces a significantly higher bone mass including BMD (bone mineral density), callus size, ash and calcium contents, and better bone strength, including peak load, peak stress, and modulus of elasticity, than the control group after fractures of the femurs in rabbits. However, the effects of low-energy SWT were less prevailing, with comparable results to the control. Superoxide mediates SW-induced ERK-dependent osteogenic transcription factor (CBFA-1) and mesenchymal cell differentiation toward osteoprogenitors (Wang et al. 2003b). SWs promote bone marrow stromal cell growth and differentiation toward osteoprogenitors associated with TGF-β1 and VEGF induction and enhance fracture healing and biomechanical strength (Wang et al. 2001b). Altered expression of bone morphogenetic protein in SWs promoted healing of fracture defect (Wang et al. 2003b).

The Achilles tendon near the insertion to bone of the right limb of 50 New Zealand white rabbits with weight 2.5–3.5 kg underwent SWT, and biopsies of the tendon–bone junction were performed at 0, 1, 4, 8, and 12 weeks. SWT produced an early release of a significantly higher number of neo-vessels and angiogenesis-related markers, including endothelial nitric oxide synthase (eNOS), vessel endothelial growth factor (VEGF), and proliferating cell nuclear antigen (PCNA) than the sham group, which lead to improvement of blood supply and tissue regeneration (Wang et al. 2001a). eNOS and VEGF levels began to rise as early as the first week and remained high for 8 weeks, then declined at 12 weeks, whereas the increases of PCNA and neo-vessels began at 4 weeks and persisted for 12 weeks. Therefore, the ingrowth of neovascularization associated with early release of angiogenesis-related markers plays a role to improve blood supply and tissue regeneration in SWT at the Achilles tendon–bone junction, which improves blood supply and increases cell proliferation and eventual tissue regeneration to repair tendon or bone tissues, alleviate pain, and initiate repairs of the chronically inflamed tissues by tissue regeneration (Wang et al. 2003a). Application of an energy flux density of 0.08 or 0.28 mJ/mm^2 caused only minor changes to the Achilles tendon. In contrast, the application of SWs at 0.60 mJ/mm^2 caused the formation of paratendinous fluid and swelling of the tendon. Histologic assessment showed fibrinoid necrosis and infiltration of inflammatory cells. In bone and contiguous tissues, the focal microinjury causes tissue changes and responses that concentrate autologous platelet-derived growth factors conducive to establishing more appropriate target tissue healing. SWs may affect lysosomes and mitochondria, interfering with metabolic activity (i.e., phosphate turnover, elaboration of extracellular matrix components in the osteoblast) within the cell (Ogden et al. 2001a). SWT relieves pain due to insertional tendinopathy by provoking a painful level of hyperstimulation analgesia (Figure 12.33).

Plantar heel pain (plantar fasciitis) can be debilitating, often with severe limitations on activity. Typically, patients present with pain in the plantar aspect of the heel while walking, particularly after rest. SWT is a safe and effective modality in the treatment of patients with proximal plantar fasciitis with a success rate ranging from 34% to 88%. The precise nature of the condition is poorly understood but may be due to an enthesitis at the attachment of the plantar fascia to the plantar medial tubercle of the calcaneum. There is considerable controversy regarding the effectiveness of SWT in the management of plantar heel pain. No adverse events were reported in two

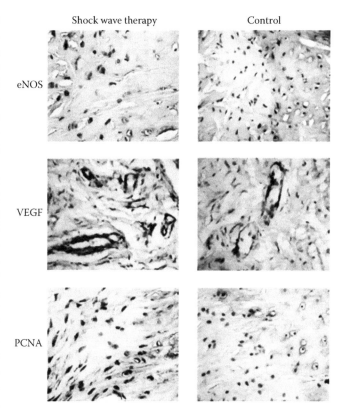

Shock wave therapy   Control

eNOS

VEGF

PCNA

FIGURE 12.33 Tissue biopsies from the tendon–bone junction stained with mouse anti-human eNOS, VEGF, and PCNA antibodies, respectively, and followed by HRP-conjugated goat anti-mouse antibody staining. (From Wang, C.J., Wang, F.S., Yang, K.D., Weng, L.H., Hsu, C.C., Huang, C.S., and Yang, L.C.: Shock wave therapy induces neovascularization at the tendon-bone junction. A study in rabbits. *J. Orthop. Res.* 2003. 21. 986. Copyright Wiley-VCH Verlag GmbH & Co. KGaA. With permission.)

trials (Rompe et al. 1996b, Speed et al. 2002). Pain for 1 week, a sensation of heat and numbness, and bruising were reported in one trial, while one patient in the placebo complained of a burning sensation in the heel and ankle (Buchbinder et al. 2002). Thirty-eight procedure-related complications (18 in the treatment arm), including mild neurological symptoms (numbness, tingling), were reported (Ogden et al. 2001a). Skin reddening, pain and local swelling, (less frequent) complaints of dizziness, sleep disturbance hematoma, nausea, and hair loss were the usual nonserious effects. The lack of convergent findings from randomized SWT trials for plantar heel pain has resulted in clinical uncertainty about its effectiveness. A meta-analysis of six randomized controlled trials ($n = 897$) from 1966 until September 2004 showed statistically significant results in favor of SWT for the treatment of plantar heel pain, but the effect size was very small, with a difference of 0.42 (95% CI: 0.02–0.83) representing less than 0.5 cm on a 10 cm visual analog scale (VAS) score of the morning (first-step) pain assessed at around 12 weeks (Thomson et al. 2005). This systematic review does not support the use of SWT for plantar heel pain in clinical practice.

Calcific tendonitis of the rotator cuff (RC) is a well-known source of shoulder pain, the most prevalent and costly work-related musculoskeletal disorder (Apfel 1981), and its overall incidence varies between 2.5% and 20% depending on both clinical criteria and radiographic technique. The disease is usually self-limiting, but the natural course is variable (Bosworth 1941, Harmon 1958, Uhthoff and Loehr 1998). Clinically, it is important to distinguish calcific tendonitis from an RC tear as a source of shoulder pain (Jim et al. 1993). The treatment of patients with calcific tendonitis typically is conservative, including use of subacromial cortisone injections, physical therapy, and systemic nonsteroidal anti-inflammatory drugs, despite their limited efficacy (Green et al. 1998). Failure of conservative treatment requires surgical intervention of the deposits, either with an open procedure or endoscopically, to relieve symptoms (Gazielly et al. 1997, Rochwerger et al. 1999). Recently, SWT has been applied as an alternative to surgery in those patients recalcitrant to traditional conservative treatment on an outpatient basis. The success rate of SWT on patients with calcific tendinitis of the shoulder ranged from 47% to 70%. High-energy SWT provides effective long-term improvement in pain, disability, motion, and power in patients with chronic calcific RC tendonitis when the shock waves are focused on the calcific deposit, while low-energy SWT does not provide effective short-term improvement, although this conclusion is based on only one high-quality study that was underpowered (Harniman et al. 2004). However, there is no universal agreement on the effective energy densities, usually considering energy density less than 0.2 mJ/mm^2 as low energy and between 0.2 and 0.4 mJ/mm^2 as high energy (Braun et al. 1992). Low-energy SWT ($<0.2 \text{ mJ/mm}^2$) seemed less effective for calcific RC tendonitis (Rompe et al. 1998, Loew et al. 1999). Trials have also examined the effect of SWT on various tendinopathies, including plantar fasciitis (Brendel et al. 1987, Brummer et al. 1990, 1992, Brazier et al. 1992), epicondylitis (Crum 1979, Brummer et al. 1989, Church 1989, Child et al. 1990, Burger et al. 1991, 1992, Coleman and Saunders 1992, Chaussy et al. 1997, Coombs et al. 2000), calcific RC tendonitis (Delius et al. 1987, 1994, 1995, 1997, Delius and Brendel 1988, Braun et al. 1992, Deam and Scott 1993, Delius 1994, Bao et al. 1997), and noncalcific RC tendonitis (Delius and Brendel 1988, Delius et al. 1988, 1990, Braun et al. 1992). Some researches emphasize the use of high-energy SWT to disintegrate the calcific deposit (Bao et al. 1997), whereas others emphasize the long-lasting hyperstimulation analgesic effect (Dear and Field 1988, Delius et al. 1989, 1995). SWs can decrease pain by increasing neovascularization and improving tissue regeneration (Delius et al. 1989) and disrupt the tendon and stimulate healing by causing alternations in cell permeability (Loew et al. 1995). In the high-energy SWT group, 20 and 16 patients had moderate and severe pain, respectively, among which 8 required intravenous analgesics during intervention. In the low-energy SWT group, the corresponding number was 22, 5, and 2. The number of patients with petechiae, bleeding, hematoma, or erythema after treatment with high-energy SWT, low-energy SWT, and sham group was 36, 32, and 8, respectively. No clinically significant adverse effects, including neurological disorders, tendon rupture, infection, bone edema, aseptic necrosis, or muscle ematoma, were found. Some patients complained petechial bruising,

subcutaneous hematoma, or skin reddening immediately after treatment, but all had resolved by 3 months. A complete disintegration and disappearance of the deposit was seen in 60% of patients 6 months after receiving high-energy SWT, which was threefold greater than sham treatment. Long-term observations 4 years after high-energy SWT found neither tendon lesions nor other adverse effects (Daecke et al. 2002). SWT is less expensive than surgery for the treatment of calcific tendonitis of the shoulder and more effective for calcifying tendonopathy than for impingement syndromes that do not involve any calcified masses (Haake et al. 2001, Schmitt et al. 2001, Speed et al. 2002). A double-blind, randomized, placebo-controlled trial was conducted between February 1997 and March 2001 among 144 patients recruited from primary care physicians, orthopedic surgeons, and sports physicians in seven orthopedic departments in Germany and Austria using high-energy SWT, low-energy SWT, or sham treatment (two sessions ~2 weeks apart and followed by physical therapy). Both high-energy and low-energy SWTs resulted in significant improvement in the 6-month mean (95% CI) Constant and Murley Scale (CMS) score compared with sham treatment 6 months after the intervention (high-energy SWT: 31.0 [26.7–35.3] points; low-energy SWT: 15.0 [10.2–19.8] points; sham treatment: 6.6 [1.4–11.8] points; $p < 0.001$ for both comparisons). Patients who received high-energy SWT also had significant 6-month CMS improvements compared with those who received low-energy SWT ($p < 0.001$). Both high-energy SWT and low-energy SWT had a beneficial effect on shoulder function, as well as on self-rated pain, and diminished size of calcifications, compared with placebo, but high-energy SWT seemed superior (Gerdesmeyer et al. 2003). There is moderate evidence from 11 nonrandomized trials that high-energy SWT is effective in treating chronic calcific RC tendonitis when SWs are focused at the calcified deposit, whereas low-energy SWT is not effective, although this conclusion is based on only one high-quality study, which was underpowered (Harniman et al. 2004). Noncalcific RC tendonitis tended to be treated with lower intensity SWT, which did not usually require anesthesia. Altogether, there was a correlation between functional improvement and the elimination of calcium deposit, with few recurrence of calcium deposit 2 years after SWT and MRI not showing any lasting damage to bone or soft tissue.

A total of 272 patients of lateral epicondylitis (tennis elbow) from 15 centers were randomly allocated to SWT (3×2000 pulses, $0.04–0.22 \text{ mJ/mm}^2$ under local anesthesia) or placebo. Transitory reddening of the skin (21.1%), pain (4.8%), and small hematomas (3.0%) were found in the SWT group (Haake et al. 2002). Other potential side effects include the misdirection of shock waves to neurovascular structures, which can result in nerve or vascular damage (Wang et al. 2002b). The success rate of SWT on patients with lateral epicondylitis of the elbow ranged from 48% to 73% (Rompe et al. 1996a, 1996b, Heller and Niethard 1998, Krischek et al. 1999, Ko et al. 2001, Ogden et al. 2001c, Wang and Chen 2002).

SWT showed favorable results in athletes with jumper's knee (patellar tendonitis). Preliminary results of SWT in 10 athletic patients with the diagnosis of primary jumper's knee and ACL-graft-related complications showed 80% in pain improvement (Sukul et al. 1993, Siebert and Buch 1997, Wang 2003).

SWT was applied to patients with early stages (I–III) of avascular necrosis of the femoral head (AVNFH) in adults (Ficat 1985, Burger et al. 1991, Dellian et al. 1993, Steinberg et al. 1995, Ludwig et al. 2001), with significant reduction of bone marrow edema despite insignificant change in the lesion size found in MRI. There was high success rate of SWT in patients with osteochondritis dessicans in the knee and the ankle (Sukul et al. 1993, Siebert and Buch 1998), and the effect seemed to correlate with the age of the patients and the size of the lesion.

Chronic ulcers are complex wounds that do not heal spontaneously and usually have multiple causative local or systemic factors associated with the nonhealing process (Lazarus et al. 1994, Nwomeh et al. 1998, Broughton et al. 2006) and require specialized medical and nursing care. As the average age of the population increases, their occurrence is becoming a major social and economic problem in the health-care system worldwide (Ferriera et al. 2006). Current treatment approaches are mechanical/surgical debridement followed by skin grafting or flap coverage, use of conventional or advanced dressings, medical or surgical correction of inadequate blood supply, and the use of recent wound-healing adjuvants. Recent studies have shown the effectiveness of SWT in stimulating growth factors, inducing angiogenesis and healing of fractures and injuries, in the treatment of chronic wounds in the lower extremities safely and cost-effectively (Saggini et al. 2008). Thirty-two patients with chronic posttraumatic venous and diabetic ulcers, unresponsive to conservative or advanced dressing treatments, were treated using SWT. Sixteen wounds healed completely within six sessions; decreased exudates, increased percentage of granulation tissue, and decreased wound size were found in the nonhealed ones. All patients reported significant decrease of pain. Between September 2005 and August 2006,

16 patients (18–85 years, 12 chronic venous ulcers and 4 chronic diabetic ulcers) with chronic ulcers of more than 3 months with a mean surface area of 5.29 cm^2, a granulation tissue percentage of 53.4%, a fibrin/necrotic tissue percentage of 46.6%, and pain in numeric box scale (NBS) of 6.65 that were unresponsive to conservative or advanced dressing treatments or mechanical debridement underwent SWT (100 pulses at 0.037 mJ/mm^2 on every cm^2 of the wound at a frequency of 4 Hz). Exclusion criteria were arrhythmias, presence of pacemaker, coagulopathies, tumors, pregnancy, presence of growth cartilage, local acute inflammation, exposed bone, and wound size area <1 cm^2 or more than 10 × 20 cm. The average session time was 16.84 (12–24) min. Eleven healed completely (69%), among which eight wounds healed after four sessions and the other three cases after six sessions, while there was a considerable improvement in the wound bed blood supply of nonhealed ulcers. All patients had a 1–3-point decrease of pain at self-assessment NBS (Figure 12.34).

SWs have been delivered in *in vitro* and *in vivo* studies for the tumor treatment (Russo et al. 1986, Randazzo et al. 1988, Bräuner et al. 1989, Brummer et al. 1989, 1992, Oosterhof et al. 1991, 1996, Dellian et al. 1993, Prat et al. 1994, Weiss et al. 1994, Wörle et al. 1994). Combining them with biological response modifiers, such as tumor necrosis factor alpha, has led to complete tumor regression in bone xenograft models, which may be due to vascular damage (Dellian et al. 1993). Genetic manipulation also has explored the potential benefits of extracorporeal SWs (Bao et al. 1997, Lauer et al. 1997). SWL-induced vein injury includes the induction of stress fibers and intercellular gaps to the complete detachment of endothelial cells and the damage of the basement membrane. An increased number of stress fibers may indicate an increased vessel wall permeability. This might result in the enhanced effects in tumor therapy when combining shock waves with drugs (Seidl et al. 1993).

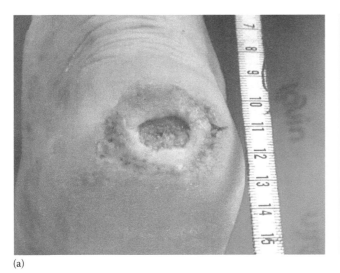
(a)

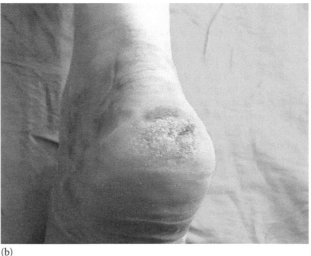
(b)

FIGURE 12.34 (a) A 5-month post-traumatic calcaneal ulcer (2.1 × 1.2 cm) on the left foot of a male patient (58 years) and (b) complete closure of the wound after six sessions (12 weeks) of SWT. (From *Ultrasound Med. Biol.*, 34, Saggini, R., Figus, A., Troccola, A., Cocco, V., Saggini, A., and Scuderi, N., Extracorporeal shock wave therapy for management of chronic ulcers in the lower extremities, 1263, Copyright 2008, with permission from Elsevier.)

REFERENCES

Abdel-Khalek M, Sheir KZ, Mokhtar AA, Eraky I, Kenawy M, Bazeed M. Predication of success rate after extracorporeal shock-wave lithotripsy of renal stones—A multivariate analysis model. *Scandinavian Journal of Urology and Nephrology* 2004;38:161–167.

Abrahams C, Lipson S, Ross L. Pathologic changes in the kidneys and other organs of dogs undergoing extracorporeal shock wave lithotripsy with a tubeless lithotripter. *Journal of Urology* 1988;140:391–394.

Ackaert KSJW, Schroder FH. Effects of extracorporeal shock wave lithotripsy (ESWL) on renal tissue. *Urological Research* 1989;17:3–7.

Adamek HE, Buttmann A, Wessbecher R, Kohler B, Riemann JF. Clinical comparison of extracorporeal piezoelectric lithotripsy (EPL) and intracorporeal electrohydraulic lithotripsy (EHL) in difficult bile duct stone. A prospective randomized trial. *Digestive Diseases and Sciences* 1995;40:1185–1192.

Adamek HE, Maier M, Jakobs R, Wessbecher FR, Neuhauser T, Riemann JF. Management of retained bile duct stones: A prospective open trial comparing extracorporeal and intracorporeal lithotripsy. *Gastrointestinal Endoscopy* 1996;44:40–47.

Agil A, Fuller CJ, Jialal I. Susceptibility of plasma to ferrous iron/hydrogen peroxide-mediated oxidation: Demonstration of a possible Fenton reaction. *Clinical Chemistry* 1995;41:220–225.

AIUM/NEMA. Safety standard for diagnostic ultrasound equipment, 1981.

Ajayi L, Jaeger P, Robertson W, Unwin R. Renal stone disease. *Medicine* 2007;35:415–419.

Albala DM, Assimos DG, Clayman RV, Denstedt JD, Grasso M, Gutierrez-Aceves J, Kahn RI et al. Lower pole I: A prospective randomized trial of extracorporeal shock wave lithotripsy and percutaneous nephrostolithotomy for lower pole nephrolithiasis-initial results. *Journal of Urology* 2001;166:2072–2080.

Apfel RE. Acoustic cavitation. In: Edmonds P, ed., *Methods of Experimental Physics*. New York: Academic Press, 1981, pp. 355–411.

Arya N, Nelles SE, Haber GB, Kim Y-I, Kortan PK. Electrohydraulic lithotripsy in 111 patients: A safe and effective therapy for difficult bile duct stones. *The American Journal of Gastroenterology* 2004;99:2330–2334.

Asroff SW, Kingston TE, Stein BS. Extracorporeal shock wave lithotripsy in patient with cardiac pacemaker in an abdominal location: Case report and review of the literature. *Journal of Endourology* 1993;7:189–192.

Augat P, Claes L, Sugar G. In vivo effect of shock waves on the healing of fractures bone. *Clinical Biomechanics* 1995;10:374–378.

Auge B, Preminger GM. Update on shockwave lithotripsy technology. *Current Opinion in Urology* 2002;12:287–290.

Averkiou MA, Crum LA. Cavitation: Its role in stone comminution and renal injury. In: Lingeman JE, Preminger GM, eds., *Topics in Clinical Urology: New Development in the Management of Urolithiasis*. New York: Igaku-Shoin, 1993, pp. 3–20.

Bailey A, Westcott A, Reynolds SE, Miller RA, Wickham JEA. Explosive nephrolithopaxy: Reality or fiction? *Urology* 1984;23:67–71.

Bailey MR, Blackstock DT, Cleveland RO, Crum LA. Comparison of electrohydraulic lithotripters with rigid and pressure-release ellipsoidal reflectors. I. Acoustic fields. *Journal of the Acoustical Society of America* 1998;104:2517–2524.

Bailey MR, Blackstock DT, Cleveland RO, Crum LA. Comparison of electrohydraulic lithotripters with rigid and pressure-release ellipsoidal reflectors. II. Cavitation fields. *Journal of the Acoustical Society of America* 1999a;106:1149–1160.

Bailey MR, Cleveland RO, Sapozhnikov OA, McAteer JA, Williams JC, Crum LA. Effect of increased ambient pressure on lithotripsy-induced cavitation in bulk fluid and at solid surfaces. *Journal of the Acoustical Society of America* 1999b;105:1267–1270.

Ball GJ, Howell BP, Leighton TG. Shock-induced collapse of a cylindrical air cavity in water: A free-Lagrange simulation. *Shock Waves* 2000;10:265–276.

Bao S, Thrall BD, Miller DL. Transfection of a reporter plasmid into cultured cells by sonoporation in vitro. *Ultrasound in Medicine & Biology* 1997;23:953–959.

Basar MM, Samli MM, Erbil M, Ozergin O, Basar R, Atan A. Early effects of extracorporeal shock-wave lithotripsy exposure on testicular sperm morphology. *Scandinavian Journal of Urology and Nephrology* 2004;38:38–41.

Baumgartner BR, Dickey KW, Ambrose SS, Walton KN, Nelson RC, Bernardino ME. Kidney changes after extracorporeal shock wave lithotripsy: Appearance on MR imaging. *Radiology* 1987;163:531–534.

Begun FP, Lawson RK, Kearns CM, Tieu TM. Electrohydraulic shock wave induced renal injury. *Journal of Urology* 1989;142:155–159.

Bergdahl L, Holmlund DEW. Retained bile duct stones. *Acta Chirurgica Scandinavica* 1976;142:145–149.

Bierkens AF, Hendrikx AJ, deKort VJ. Efficiency of second generation lithotripters: A multicenter comparative study of 2,206 extracorporeal shock wave lithotripsy treatments with the Siemens Lithostar, Dornier HM4, Wolf Piezolith 2300, Direct Tripter X-1, and Breakstone lithotripters. *Journal of Urology* 1992;148:1052.

Bierkens AF, Hendrikx AJ, Lemmens WA, Debruyne FM. Extracorporeal shock-wave lithotripsy for large renal calculi: The role of ureteral stents. A randomized trial. *Journal of Urology* 1991;145:699–702.

Bigelow HJ. *Litholapaxy or Rapid Lithotrity with Evacuation*. Boston, MA: A. Williams and Company, 1878, p. 29.

Binmoeller KF, Bruckner M, Thonke F, Soehendra N. Treatment of difficult bile duct stones using mechanical, electrohydraulic and extracorporeal shock wave lithotripsy. *Endoscopy* 1993;25:201–206.

Blake JR, Gibson DC. Cavitation bubbles near boundaries. *Annual Review of Fluid Mechanics* 1987;19:99–123.

Bohris C, Bayer T, Lechner C. Hit/miss monitoring of ESWL by spectral Doppler ultrasound. *Ultrasound in Medicine & Biology* 2003;29:705–712.

Bonnel DH, Liguory CE, Cornud FE, Lefebvre JF. Common bile duct and intrahepatic stones: Results of transhepatic electrohydraulic lithotripsy in 50 patients. *Radiology* 1991;180:345–348.

Bosworth B. Calcium deposits in the shoulder and subacromial bursitis: A survey of 12111 shoulders. *JAMA* 1941;116:2477–2489.

Braun W, Claes A, Rüter A, Paschke D. Effects of extracorporeal shock waves on the stability of the interface between bone and polymethylmethacrylate: An in vitro study on human femoral segments. *Clinical Biomechanics* 1992;7:47–54.

Bräuner T, Brümmer F, Hülser DF. Histopathology of shock wave treated tumor cells suspensions and multicell tumor spheroids. *Ultrasound in Medicine & Biology* 1989;15:451–460.

Brazier JE, Harper R, Jones NM, O'Cathain A, Thomas KJ, Usherwood T, Westlake L. Validating the SF-36 health survey questionnaire: New outcome measure for primary care. *British Medical Journal* 1992;305:160–164.

Brendel W, Delius M, Goetz A. Effect of shock waves on the microvasculature. *Progress in Applied Microcirculation* 1987;12:41–50.

Brinkmann OA, Griehl A, Kuwertz-Broking E, Bulla M, Hertle L. Extracorporeal shock wave lithotripsy in children. Efficacy, complications and long-term follow-up. *European Urology* 2001;39:591–597.

Brodmann M, Ramschak H, Schreiber F, Stark G, Pabst E, Pilger E. Venous thrombosis after extracorporeal shockwave lithotripsy in a patient with heterozygous APC-resistance. *Thrombosis and Haemostasis* 1998;80:861.

Broughton G, Janis JE, Attinger CE. The basic science of wound healing. *Plastic and Reconstructive Surgery* 2006;117:12S–34S.

Brown RD, Preminger GM. Changing surgical aspects of urinary stone disease. *Surgical Clinics of North America* 1988;68:1085–1104.

Brummer F, Brauner T, Hulser DF. Biological effects of shock waves. *World Journal of Urology* 1990;8:224–232.

Brummer F, Brenner J, Brauner T, Hulser DF. Effect of shock waves on suspended and immobilized L1210 cells. *Ultrasound in Medicine & Biology* 1989;15:229–239.

Brummer F, Suhr D, Hulser DF. Sensitivity of normal and malignant cells to shock waves. *Stone Disease* 1992;4:243–248.

Buchbinder R, Ptasznik R, Gordon J, Buchannan J, Prabaharan V, Forbes A. Ultrasound guided extracorporeal shockwave therapy for plantar fasciitis: A randomized controlled trial. *Journal of the American Medical Association* 2002;288:1364–1372.

Buchholz NP, van Rossum M. Shock wave lithotripsy treatment of radiolucent ureteric calculi with the help of contrast medium. *European Urology* 2001;39:200–203.

Buizza A, dell'Aquila T, Giribona P, Spagno C. The performance of different pressure pulse generators for extracorporeal lithotripsy: A comparison based on commercial lithotripters for kidney stones. *Ultrasound in Medicine & Biology* 1995;21:259–272.

Burger RA, Witzsch U, Haist J. Extracorporeal shock wave therapy of pseudo-arthrosis and aseptic osteonecrosis. *Journal of Endourology* 1991;5:48–50.

Burger RA, Witzsch U, Haist J. Extracorporeal shockwave therapy of pseudoarthrosis. *Journal of Urology* 1992;147:260–263.

Chan SI, Stothers I, Rowley A, Perler Z, Taylor W, Sullivan LD. A prospective trial comparing the efficacy and complications of the modified Dornier HM3 and MFL 5000 lithotriptors for solitary renal calculi. *Journal of Urology* 1995;153:1794–1797.

Chang CC, Liang SM, Pu YR, Chen CH, Chen I, Chen TS, Kuo CL, Yu FM, Chu ZF. *In vitro* study of ultrasound based real-time tracking of renal stones for shock wave lithotripsy: Part 1. *Journal of Urology* 2001;166:28–32.

Chaussy C. *Extracorporeal Shock Wave Lithotripsy: New Aspects in the Treatment of Kidney Stone Disease*. Basel, Switzerland: Karager, 1982.

Chaussy C, Brendel W, Schmiedt E. Extracorporeally induced destruction of kidney stones by shock waves. *Lancet* 1980;2:1265–1268.

Chaussy C, Eisenberger F, Jocham D. *High Energy Shock Waves in Medicine*. Stuttgart, Germany: Thieme, 1997.

Chaussy C, Fuchs GJ. Current state and future developments of noninvasive treatment of human urinary stones with extracorporeal shock wave lithotripsy. *Journal of Endourology* 1989;41:782–789.

Chaussy C, Schmiedt E, Jocham D, Brendel W, Forssmann B, Walther V. First clinical experience with extracorporeally induced destruction of kidney stones by shock waves. *Journal of Urology* 1982;127:417–420.

Chaussy C, Schuller J, Schmiedt E, Brandl H, Jocham D, Liedl B. Extracorporeal shock-wave lithotripsy (ESWL) for treatment of urolithiasis. *Urology* 1984;23:59–66.

Chen H, Kreider W, Brayman AA, Bailey MR, Matula TJ. Blood vessel deformations on microsecond time scales by ultrasonic cavitation. *Physical Review Letters* 2011;106:034301.

Cheng PT, Reid AD, Pritzker KP. Ultrastructural studies of crystal-organic matrix relations in renal stones. *Scanning Electron Microscope* 1985;1:201–207.

Chew BH, Zavaglia B, Sutton C, Masson RK, Chan SH, Hamidizadeh R, Lee JK et al. Twenty-year prevalence of diabetes mellitus and hypertension in patients receiving shock-wave lithotripsy for urolithiasis. *BJU International* 2012;109:444–449.

Child SZ, Hartman C, Schery LA, Carstensen EL. Lung damage from exposure to pulsed ultrasound. *Ultrasound in Medicine & Biology* 1990;16:817–825.

Chung S-D, Chen Y-K, Lin H-C. Increased risk of diabetes in patients with urinary calculi: A 5-year followup study. *Journal of Urology* 2011;186:1888–1893.

Chuong CJ, Zhong P, Preminger GM. A comparison of stone damage caused by different modes of shock wave generation. *Journal of Urology* 1992;148:200–205.

Church CC. A theoretical study of cavitation generated by an extracorporeal shock wave lithotripter. *Journal of the Acoustical Society of America* 1989;86:215–227.

Citron P. Medical devices: Factors adversely affecting innovation. *Journal of Biomedical Materials Research* 1996;32:1–2.

Clark JY, Thompson IM, Optenberg SA. Economic impact of urolithiasis in the United States. *Journal of Urology* 1995;154:2020–2024.

Cleveland RO. The advantage of a broad focal zone in SWL. In: Evan AP, Lingeman JE, McAteer JA, Williams JC, eds., *Second International Urolithiasis Research Symposium*. Melville, New York: American Institute of Physics, 2008, pp. 219–225.

Cleveland RO, Anglade R, Babayan RK. Effect of stone motion on *in vitro* comminution efficiency of Storz Modulith SLX. *Journal of Endourology* 2004;18:629–633.

Cleveland RO, Bailey MR, Fineberg NS, Hartenbaum B, Lokhandwalla M, McAteer JA, Sturtevant B. Design and characterization of a research electrohydraulic lithotripter patterned after the Dornier HM3. *Review of Scientific Instruments* 2000a;71:2514–2525.

Cleveland RO, Lifshitz DA, Connors BA, Evan AP, Willis LR, Crum LA. In vivo pressure measurements of lithotripsy shock waves in pigs. *Ultrasound in Medicine & Biology* 1998;24:293–306.

Cleveland RO, McAteer JA. The physics of shock wave lithotripsy. In: Smith AD, Badlani GH, Bagley DH, Clayman RV, eds., *Smith's Textbook of Endourology*. Hamilton, Ontario, Canada: BC Decker, 2007, pp. 317–332.

Cleveland RO, McAteer JA, Muller R. Time-lapse nondestructive assessment of shock wave damage to kidney stones *in vitro* using micro-computed tomography. *Journal of the Acoustical Society of America* 2001;110:1733.

Cleveland RO, Sapozhnikov OA. Modeling elastic wave propagation in kidney stones with application to shock wave lithotripsy. *Journal of the Acoustical Society of America* 2005;118:2667–2676.

Cleveland RO, Sapozhnikov OA, Bailey MR, Crum LA. A dual passive cavitation detector for localized detection of lithotripsy-induced cavitation *in vitro*. *Journal of the Acoustical Society of America* 2000b;107:1745–1758.

Coakley W. Acoustical detection of single cavitation events in a focused field in water at 1 MHz. *Journal of the Acoustical Society of America* 1971;49:792–801.

Coe FL, Evan AP, Worcester EM. Kidney stone disease. *Journal of Clinical Investigation* 2005;115:2598–2608.

Cohen TD, Durrani AF, Brown SA, Ferraro R, Preminger GM. Lipid peroxidation induced by shockwave lithotripsy. *Journal of Endourology* 1998;12:229–232.

Coleman AJ, Choi MJ, Saunders JE, Leighton TG. Acoustic emission and sonoluminescence due to cavitation at the beam focus of an electrohydraulic shock wave lithotripter. *Ultrasound in Medicine & Biology* 1992;18:267–281.

Coleman AJ, Saunders JE. A survey of the acoustic output of commercial extracorporeal shock wave lithotripters. *Ultrasound in Medicine & Biology* 1989;15:213–227.

Coleman AJ, Saunders JE. A review of the physical properties and biological effects of the high amplitude acoustic fields used in extracorporeal lithotripsy. *Ultrasonics* 1992;31:75–89.

Coleman AJ, Saunders JE, Crum LA, Dyson M. Acoustic cavitation generated by an extra corporeal shockwave lithotripter. *Ultrasound in Medicine & Biology* 1987;13:69–76.

Coleman AJ, Whitlock M, Leighton TG, Saunders JE. The spatial distribution of cavitation induced acoustic emission, sonoluminescence and cell lysis in the field of a shock wave lithotripter. *Physics in Medicine and Biology* 1993;38:1545–1560.

Connors BA, Evan AP, Blomgren PM, Handa RK, Willis LR, Gao S, McAteer JA, Lingeman JE. Shock wave lithotripsy at 60 SWs per minute reduces renal injury in the porcine model. *BJU International* 2009;104:1004–1008.

Connors BA, Evan AP, Willis LR, Blomgren PM, Lingeman JE, Fineberg NS. The effect of discharge voltage on renal injury and impairment caused by lithotripsy in the pig. *Journal of the American Society of Nephrology* 2000;11:310–318.

Connors BA, Evan AP, Willis LR, Simon JR, Fineberg NS, Lifshitz DA, Shalhav AL, Paterson RF, Kuo RL, Lingeman JE. Renal nerves mediate changes in contralateral renal blood flow after extracorporeal shockwave lithotripsy. *Nephron Physiology* 2003;95:67–75.

Coombs R, Schaden W, Zhou SSH. *Musculoskeletal Shockwave Therapy*. London, U.K.: Greenwich Medical Media, 2000.

Crum LA. Tensile strength of water. *Nature* 1979;278:148–149.

Crum LA. Cavitation microjets as a contributory mechanism for renal calculi disintegration in ESWL. *Journal of Urology* 1988;140:1587–1590.

Crum LA, Hansen GM. Generalized equations for rectified diffusion. *Journal of the Acoustical Society of America* 1982;72:1586–1592.

Daecke W, Kusnierczak D, Loew M. Long-term effects of extracorporeal shockwave therapy in chronic calcific tendinitis of the shoulder. *Journal of Shoulder and Elbow Surgery* 2002;11:476–480.

Dahake G, Gracewki SM. Finite difference predictions of P-SV wave propagation inside submerged solids. II. Effect of geometry. *Journal of the Acoustical Society of America* 1997;102:138–145.

Daniel MD, Burns JR. Renal function immediately after piezoelectric extracorporeal lithotripsy. *Journal of Urology* 1990;144:10–12.

Das G, Dick J, Bailey MJ, Fletcher MS, Birch B, Coptcoat MJ, Webb DR, Kellett MJ, Whitfield HN, Wickham JEA. 1500 cases of renal and ureteric calculi treated in an integrated stone centre. *British Journal of Urology* 1988;62:301–305.

de Cógáin M, Krambeck AE, Rule AD, Li X, Bergstralh EJ, Gettman MT, Lieske JC. Shock wave lithotripsy and diabetes mellitus: A population-based cohort study. *Urology* 2012;79:298–302.

Deam RK, Scott DA. Neurological damage resulting from extracorporeal shock wave lithotripsy when air is used to locate the epidural space. *Anaesthesia and Intensive Care* 1993;21:455–457.

Dear JP, Field JE. A study of the collapse of arrays of cavities. *Journal of Fluid Mechanics* 1988;190:409–425.

Delius M. Medical applications and bioeffects of extracorporeal shock waves. *Shock Waves* 1994;4:55–72.

Delius M, Brendel W. A model of extracorporeal shock wave action: Tandem action of shock waves. *Ultrasound in Medicine & Biology* 1988;14:515–518.

Delius M, Brendel W, Heine G. A mechanism of gallstone destruction by extracorporeal shock waves. *Naturwissenschaften* 1988a;75:200–201.

Delius M, Draenert K, Al Diek Y, Draenert Y. Biological effect of shockwave: In vivo effect of high-energy pulses on rabbit bone. *Ultrasound in Medicine & Biology* 1995;21:1219–1225.

Delius M, Draenert K, Draenert Y. Effects of extracorporeal shock waves on bone: A review of shock wave experiments and the mechanism of shock wave action. In: Siebert W, Buch M, eds., *Extracorporeal Shock Waves in Orthopaedics*. Berlin, Germany: Springer Verlag, 1997, pp. 91–107.

Delius M, Draenert K, Draenert Y, Börner M. Effects of extracorporeal shock waves on bone: A review of shock wave experiments and the mechanism of shock wave action. In: Siebert W, Buch M, eds., *Extracorporeal Shock Waves in Orthopaedics*, Heidelberg, Germany, Springer Science & Business Media, 1998, pp. 91–107.

Delius M, Enders G, Heine G. Biologic effects of shock waves: Lung hemorrhage by shock waves in dogs: Pressure dependence. *Ultrasound in Medicine & Biology* 1987;13:61–67.

Delius M, Hoffmann E, Steinbeck G, Conzen P. Biological effects of shock waves: Induction of arrhythmia in piglet hearts. *Ultrasound in Medicine & Biology* 1994a;20:279–285.

Delius M, Jordan M, Eizenhoefer H, Marlinghaus E, Heine G, Liebich HG, Brendel W. Biological effects of shock waves: Kidney haemorrhage by shock waves in dogs—Administration rate dependence. *Ultrasound in Medicine & Biology* 1988b;14:689–694.

Delius M, Jordan M, Liebich HG, Brendel W. Biological effects of shock waves: Effect of shock waves on the liver and gallbladder wall of dogs: Administration rate dependence. *Ultrasound in Medicine & Biology* 1990;16:459–466.

Delius M, Muller M, Vogel A, Brendel W. Extracorporeal shock waves: Properties and principles of generation. In: Ferrucci JT, Delius, M, Burhenne HJ, eds., *Biliary Lithotripsy*, London, U.K.: Year Book Medical Publishers, 1989a, pp. 9–15.

Delius M, Ueberle F, Gambihler S. Destruction of gallstones and model stones by extracorporeal shock-waves. *Ultrasound in Medicine & Biology* 1994b;20:251–258.

Delius M, Weiss N, Gambihler S, Goetz A, Brendel W. Tumor therapy with shock waves requires modified lithotripter shock waves. *Naturwibenschaften* 1989b;76:573–574.

Deliveliotis C, Kostakopoulos A, Stavropoulos N, Karagiotis E, Kyriazis P, Dimopoulos C. Extracorporeal shock wave lithotripsy in 5 patients with aortic aneurysm. *Journal of Urology* 1995;154:1671–1672.

Deliveliotis CH, Argyropoulos B, Chrisofos M, Dimopoulos CA. Shockwave lithotripsy in unrecognized pregnancy: Interruption or continuation? *Journal of Endourology* 2001;15:787–788.

Dellian M, Walenta S, Gamarra F, Kuhnle GE, Mueller-Klieser W, Goetz AE. Ischemia and loss of ATP in tumors following treatment with focused high energy shock waves. *British Journal of Cancer* 1993;68:26–31.

Delvecchio FC, Auge BK, Munver R, Brown SA, Brizuela RM, Zhong P, Preminger GM. Shock wave lithotripsy causes ipsilateral renal injury remote from the focal point: The role of regional vasoconstriction. *Journal of Urology* 2003;169:1526–1529.

Demirci D, Sofikerim M, Yalçin E, Ekmekçioğlu O, Gülmez İ, Karacagil M. Comparison of conventional and step-wise shockwave lithotripsy in management of urinary calculi. *Journal of Endourology* 2007;21:1407–1410.

Desmet W, Baert L, Vandeursen H, Vermylen J. Iliac-vein thrombosis after extracorporeal shock-wave lithotripsy. *New England Journal of Medicine* 1989;321:907.

Dhar NB, Thornton J, Karafa MT, Streem SB. A multi-variate analysis of risk factors associated with subcapsular hematoma formation following electromagnetic shock wave lithotripsy. *Journal of Urology* 2004;172:2271–2274.

Ding Z, Gracewski SM. Response of constrained and unconstrained bubbles to lithotripter shock wave pulses. *Journal of the Acoustical Society of America* 1994;96:3636–3644.

Dollery CT. Arterial hypertension. In: Beeson PB, McDermott W, Wyngaarden JB, eds., *Cecil Textbook of Medicine.* Philadelphia, PA: Saunders, 1979, pp. 1199–1223.

Doran O, Foley B. Acute complications following extracorporeal shock-wave lithotripsy for renal and ureteric calculi. *Emergency Medicine Australasia* 2008;20:105–111.

Dretler SP. Stone fragility—A new therapeutic distinction. *Journal of Urology* 1988;139:1124–1127.

Duck FA. *Physical Properties of Tissue.* London, U.K.: Academic Press, 1990.

Eichel L, Batzold P, Erturk E. Operator experience and adequate anesthesia improve treatment outcome with third-generation lithotripters. *Journal of Endourology* 2001;15:671–673.

Eisenmenger W. The mechanisms of stone fragmentation in ESWL. *Ultrasound in Medicine & Biology* 2001;27:683–693.

Eisenmenger W, Du XX, Tang C, Zhao S, Wang Y, Rong F, Dai D, Guan M, Qi A. The first clinical results of "wide-focus and low-pressure" ESWL. *Ultrasound in Medicine & Biology* 2002;28:769–774.

Elves AW, Tilling K, Menezes P, Wills M, Rao PN, Feneley RC. Early observations of the effect of extracorporeal shock-wave lithotripsy on blood pressure: A prospective randomized control clinical trial. *BJU International* 2000;85:611–615.

Eskelinen M, Ilkonen J, Lipponen P. Usefulness of history-taking, physical examination and diagnostic scoring in acute renal colic. *European Urology* 1998;34:467–473.

Evan AP, McAteer JA. *Q*-effects of shock wave lithotripsy. In: Coe FL, Favus MJ, Pak CYC, Parks HJ, Preminger GM, eds., *Kidney Stones: Medical and Surgical Management.* Philadelphia, PA: Lippincott-Raven, 1996, pp. 549–570.

Evan AP, McAteer JA, Connors BA, Blomgren PM, Lingeman JE. Renal injury in SWL is significantly reduced by slowing the rate of shock wave delivery. *BJU International* 2007a;100:624–627.

Evan AP, McAteer JA, Connors BA, Pishchalnikov YA, Handa RK, Blomgren PM, Willis LR, Williams JC, Lingeman JE, Gao S. Independent assessment of a wide-focus, low-pressure electromagnetic lithotripter: Absence of renal bioeffects in the pig. *BJU International* 2007b;101:382–388.

Evan AP, Willis LR, Lingeman JE, McAteer JA. Renal trauma and the risk of long-term complications in shock wave lithotripsy. *Nephron* 1998;78:1–8.

Evan AP, Willis LR, McAteer JA, Bailey MR, Connors BA, Shao Y, Lingeman JE, Williams JC, Fineberg NS, Crum LA. Kidney damage and renal functional changes are minimized by waveform control that suppresses cavitation in shock wave lithotripsy. *Journal of Urology* 2002;168:1556–1562.

Ferriera MC, Tuma PJ, Carvalho VF, Kamamoto F. Complex wounds. *Clinics* 2006;61:571–578.

Ficat RF. Idiopathic bone necrosis of the femoral head: Early diagnosis and treatment. *Journal of Bone and Joint Surgery* 1985;67B:3–9.

Field JE. The physics of liquid impact, shock wave interactions with cavities, and the implications to shock wave lithotripsy. *Physics in Medicine and Biology* 1991;36:1475–1484.

Finlayson B, Ackermann D. Overview of surgical treatment of urolithiasis with special reference to lithotripsy. *Journal of Urology* 1989;141:778–779.

Flam TA, Bourlion M, Thiounn N, Saporta F, Chiche R, Dancer P, Zerbib M, Debré B. Electroconductive lithotripsy: Principles, experimental data, and first clinical results of the Sonolith 4000. *Journal of Endourology* 1994;8:249–255.

Folberth W, Kohler G, Rohwedder A, Matura E. Pressure distribution and energy flow in the focal region of two different electromagnetic shock wave sources. *Journal of Stone Disease* 1992;4:1–7.

Forriol F, Solchaga L, Moreno JL, Candell J. The effect of shockwave on mature and healing cortical bone. *International Orthopaedics* 1994;8:325–329.

Frauscher F, Höfle G, Janetschek G. Re: A randomized controlled trial to assess the incidence of new onset hypertension in patients after shock wave lithotripsy for asymptomatic renal calculi. *Journal of Urology* 1999;162:806.

Freeman ML, Nelson DB, Sherman S, Haber GB, Herman ME, Dorsher PJ, Moore JP et al. Complications of endoscopic biliary sphincterotomy. *New England Journal of Medicine* 1996;335:909–918.

Freund J, Colonius T, Evan AP. A cumulative shear mechanism for tissue damage initiation in shock-wave lithotripsy. *Ultrasound in Medicine & Biology* 2007;33:1495–1503.

Freund JB. Suppression of shock-bubble expansion due to tissue confinement with application to shock-wave lithotripsy. *Journal of the Acoustical Society of America* 2008;123:2867–2874.

Gazielly DF, Bruyere G, Gleyze P, Thomas T. Open acromioplasty with excision of calcium deposits and tendon suture. In: Gazielly DF, Gleyze P, Thomas T, eds., *The Cuff.* Paris, France: Elsevier, 1997, pp. 181–184.

Gerber R, Studer UE, Danuser H. Is newer always better? A comparative study of 3 lithotriptor generations. *Journal of Urology* 2005;173:2013–2016.

Gerdesmeyer L, von Eiff C, Horn C, Henne M, Roessner M, Diehl P, Gollwitzer H. Antibacterial effects of extracorporeal shock waves. *Ultrasound in Medicine & Biology* 2005;31:115–119.

Gerdesmeyer L, Wagenpfeil S, Haake M, Maier M, Loew M, Wörtler K, Lampe R, Seil R, Handle G, Gassel S, Rompe JD. Extracorporeal shock wave therapy for the treatment of chronic calcifying tendonitis of the rotator cuff: A randomized controlled trial. *JAMA* 2003;290:2573–2580.

Gettman MT, Segura JW. Management of ureteric stones: Issues and controversies. *BJU International* 2005;95:85–93.

Gomha MA, Sheir KZ, Showky S, Abdel-Khalek M, Mokhtar AA, Madbouly K. Can we improve the prediction of stone-free status after extracorporeal shock wave lithotripsy for ureteral stones? A neural network or statistical model. *Journal of Urology* 2004;172:176–179.

Gotte G, Amelio E, Russo S, Marlinghaus E, Musci G, Suzuki H. Short-time non-enzymatic nitric oxide synthesis from L-arginine and hydrogen peroxide induced by shock waves treatment. *FEBS Letters* 2002;520:153–155.

Gracewski SM, Burns SJ, Everbach EC, Vakil N. Static mechanical properties of gallstones and their relationship to lithotripsy. *Gastroenterology* 1990;98:A590.

Gracewski SM, Dahake G, Ding Z, Burns SJ, Everbach EC. Internal stress wave measurements in solids subjected to lithotripter pulses. *Journal of the Acoustical Society of America* 1993;94:652–661.

Graff J, Pastor J, Senge T. The effect of high energy shock waves on bony tissue: An experimental study. *Journal of Urology* 1987;137:278–281.

Granz B, Hohler G. What makes a shock wave efficient in lithotripsy? *Journal of Stone Disease* 1992;4:123–128.

Green S, Buchbinder R, Glazier R, Forbes A. Systematic review of randomised controlled trials of interventions of painful shoulder: Selection criteria, outcome assessment, and efficacy. *BMJ* 1998;316:354–360.

Haake M, Boddeker IR, Decker T, Buch M, Vogel M, Labek G, Maier M et al. Side-effects of extracorporeal shock wave therapy (ESWT) in the treatment of tennis elbow. *Archives of Orthopaedic and Trauma Surgery* 2002;122:222–228.

Haake M, Rautmann M, Wirth T. Assessment of the treatment costs of extracorporeal shock wave therapy vs surgical treatment for shoulder disease. *International Journal of Technology Assessment in Health Care* 2001;17:612–617.

Handa RK, Bailey MR, Paun M, Gao S, Connors BA, Willis LR, Evan AP. Pretreatment with low-energy shock waves induces renal vasoconstriction during standard shock wave lithotripsy (SWL): A treatment protocol known to reduce SWL-induced renal injury. *BJU International* 2009a;103:1270–1274.

Handa RK, McAteer JA, Evan AP, Connors BA, Pishchalnikov YA, Gao S. Assessment of renal injury with a clinical dual-head lithotripter delivering 240 shock waves per minute. *Journal of Urology* 2009b;181:884–889.

Harmon PH. Methods and results in the treatment of 2580 painful shoulders with special reference to calcific tendinitis and the frozen shoulder. *American Journal of Surgery* 1958;95:527–514.

Harniman E, Carette S, Kennedy C, Beaton D. Extracorporeal shock wave therapy for calcific and noncalcific tendonitis of the rotator cuff: A systematic review. *Journal of Hand Therapy* 2004;17:132–151.

Haupt G. Use of extracorporeal shock wave in the treatment of pseudoarthrosis, tendinopathy and other orthopaedic disease. *Journal of Urology* 1997;158:4–11.

Haupt G, Haupt A, Ekkemkamp A, Gerety B, Chvapil M. Influence of shock waves on fracture healing. *Urology* 1992;39:529–532.

Heimbach D, Munver R, Zhong P, Jacobs J, Hesse A, Muller SC, Preminger GM. Acoustic and mechanical properties of artificial stones in comparison to natural kidney stones. *Journal of Urology* 2000;164:537–544.

Heller KD, Niethard FU. Using extracorporeal shockwave therapy in orthopedics: A meta-analysis. *Zeitschrift für Orthopädie und ihre Grenzgebiete* 1998;36:390–401.

Henney JE. From the Food and Drug Administration: Shock wave for heel pain. *JAMA* 2000;284:2711.

Hirachi K, Minami A, Kato H. Osteogenic potential of the shock wave: Experimental study in rabbit model. *Transactions of the Orthopaedic Research Society* 1999;45:558.

Hofmann A, Ritz U, Hessmann MH, Alini M, Rommens PM, Rompe J-D. Extracorporeal shock wave-mediated changes in proliferation, differentiation, and gene expression of human osteoblasts. *Journal of Trauma* 2008;65:1402–1410.

Howard D, Sturtevant B. In vitro study of the mechanical effects of shock-wave lithotripsy. *Ultrasound in Medicine & Biology* 1997;23:1107–1122.

Hunter PT, Finlayson B, Hirko RJ, Voreck WC, Walker R, Walck S, Nasr M. Measurement of shock wave pressures used for lithotripsy. *Journal of Urology* 1986;136(3):733–738.

Ikeda K, Tomitz K, Takayama K. Application of extracorporeal shock wave on bone: Preliminary report. *Journal of Trauma* 1999;47:946–950.

Ison KT. Physical and technical introduction to lithotripsy. In: Coptcoat MJ, Miller RA, Wickham JEA, eds., *Lithotripsy II. Textbook of Second Generation Extracorporeal Lithotripsy*, London, U.K.: BDI Publishing, 1987, pp. 7–12.

Janetschek G, Frauscher F, Knapp R, Hofle G, Peschel R, Bartsch G. New onset hypertension after extracorporeal shock wave lithotripsy: Age-related incidence and prediction by resistive index. *Journal of Urology* 1997;158:346–351.

Jewett MA, Bombardier C, Logan AG, Psihramis KE, Wesley-James T, Mahoney JE, Luymes JJ, Ibanez D, Ryan MR, Honey JDA. A randomized controlled trial to assess the incidence of new onset hypertension in patients after shock wave lithotripsy for asymptomatic renal calculi. *Journal of Urology* 1998;160:1241–1243.

Jim YF, Hsu HC, Chang CY, Wu JJ, Chang T. Coexistence of calcific tendinitis and rotator cuff tear: An arthrographic study. *Skeletal Radiology* 1993;22:183–185.

Johannes EJ, Kaulesar Sukul DM, Matura E. High-energy shock-wave for treatment of nonunion. An experiment on dogs. *Journal of Surgical Research* 1994;57:246–252.

Johnsen E, Colonius T. Shock induced collapse of a gas bubble in shockwave lithotripsy. *Journal of the Acoustical Society of America* 2008;124:2011–2020.

Johrde LG, Cocks FH. Fracture strength studies of renal calculi. *Journal of Materials Science Letters* 1985;4:1264–1265.

Karlsen SJ, Smevik B, Hovig T. Acute morphological changes in canine kidneys after exposure to extracorporeal shock waves: A light and electron microscopic study. *Urological Research* 1991;19:105–115.

Karpman RR, Magee FP, Gruen TW, Mobley T, Peltier LF. The lithotriptor and its potential use in the revision of total hip arthroplasty. *Orthopedic Reviews* 1987;16:81–85.

Kaude JV, Williams CM, Millner MR, Scott KN, Finlayson B. Renal morphology and function immediately after extracorporeal shock-wave lithotripsy. *American Journal of Roentgenology* 1985;145:305–313.

Kerbl K, Rehman J, Landman J, Lee D, Sundaram C, Clayman RV. Current management of urolithiasis: Progress or regress? *Journal of Endourology* 2002;16:281–288.

Kijvaikai K, Haleblian GE, Preminger GM, De la Rosette J. Shock wave lithotripsy or ureteroscopy for the management of proximal ureteral calculi: An old discussion revisited. *Journal of Urology* 1997;178:1157–1163.

Kim SC, Burns EK, Lingeman JE, Paterson RF, McAteer JA, Williams JC. Cystine calculi: Correlation of CT-visible structure, CT number, and stone morphology with fragmentation by shock wave lithotripsy. *Urological Research* 2007;35:319–324.

Knapp PM, Kulb TB, Lingeman JE, Newman DM, Mosbaugh JH, Steeler RE. Extracorporeal shock wave lithotripsy-induced perirenal hematomas. *Journal of Urology* 1988;139:700–703.

Knoll T, Wendt-Nordahl G. Perspective on lithotripsy adverse effects. In: Evan AP, Lingeman JE, McAteer JA, Williams JC, eds., *Second International Urolithiasis Research Symposium*. Melville, New York: American Institute of Physics, 2008, pp. 278–282.

Ko JY, Chen HS, Chen LM. Treatment of lateral epicondylitis of the elbow with shock waves. *Clinical Orthopaedics* 2001;387:60–67.

Koga H, Matsuoka K, Noda S, Yamashita T. Cumulative renal damage in dogs by repeated treatment with extracorporeal shock waves. *International Journal of Urology* 1996;3:134–140.

Kohrmann KU, Rassweiler JJ, Manning M, Mohr G, Henkel TO, Junemann KP, Alken P. The clinical introduction of a third generation lithotriptor Modulith SL20. *Journal of Urology* 1995;153:1379–1383.

Krambeck AE, Gettman MT, Rohlinger AL, Lohse CM, Patterson DE, Segura JW. Diabetes mellitus and hypertension associated with shock wave lithotripsy of renal and proximal ureteral stones at 19 years of followup. *Journal of Urology* 2006;175:1742–1747.

Krambeck AE, Rule AD, Li X, Bergstralh EJ, Gettman MT, Lieske C. Shock wave lithotripsy is not predictive of hypertension among community stone formers at long-term followup. *Journal of Urology* 2011;185:164–169.

Krischek O, Hopf C, Nafe B, Rompe JD. Shock-wave therapy for tennis and golfer's elbow: 1 year follow-up. *Archives of Orthopaedic and Trauma Surgery* 1999;119:62–66.

Krysiewicz S. Complications of renal extracorporeal shock wave lithotripsy reviewed. *Urologic Radiology* 1992;13:139–145.

Kuo RL, Paterson RF, Siqueira TM, Evan AP, McAteer JA, Williams JC, Lingeman JE. In vitro assessment of ultrasonic lithotriptors. *Journal of Urology* 2003;170:1101–1104.

Lauer U, Bürgeit E, Squire Z, Messmer K, Hofschneider PH, Gregor M, Delius M. Shock wave permeabilization as new gene transfer method. *Gene Therapy* 1997;4:710–715.

Lazarus GS, Cooper DM, Knighton DR, Margolis DJ, Pecorano RE, Rodeheaver G, Robson MC. Definitions and guidelines for assessment of wounds and evaluation of healing. *Archives of Dermatology* 1994;130:489–493.

Leighton TG. *The Acoustic Bubble*. London, U.K.: Academic Press, 1994.

Leighton TG. Bubble population phenomena in acoustic cavitation. *Ultrasonics Sonochemistry* 1995;2:S123–S126.

Leighton TG. From seas to surgeries, from babbling brooks to baby scans: The acoustics of gas bubbles in liquids. *International Journal of Modern Physics B* 2004;18:3267–3314.

Leighton TG. What is ultrasound? *Progress in Biophysics & Molecular Biology* 2007;93:3–83.

Leighton TG, Cleveland RO. Lithotripsy. *Proceedings of the Institution of Mechanical Engineers, Part H: Journal of Engineering in Medicine* 2009;224:317–342.

Leighton TG, Fedele F, Coleman AJ, McCarthy C, Ryves S, Hurrell AM, De Stefano A, White PR. A passive acoustic device for real-time monitoring of the efficacy of shockwave lithotripsy treatment. *Ultrasound in Medicine & Biology* 2008;34:1651–1665.

Leighton TG, Meers SD, White PR. Propagation through nonlinear time-dependent bubble clouds, and the estimation of bubble populations from measured acoustic characteristics. *Proceedings of the Royal Society of London A* 2004;460:2521–2550.

Leighton TG, White PR, Marsden MA. Applications of one-dimensional bubbles to lithotripsy, and to diver response to low frequency sound. *Acta Acustica* 1995;3:517–529.

Lewis G. Effect of lithotriptor treatment in the fracture toughness of acrylic bone cement. *Biomaterials* 1992;13:225–229.

Liguori G, Trombetta C, Bucci S, Salame L, Savoldi S, Belgrano E. Reversible acute renal failure after unilateral extracorporeal shock-wave lithotripsy. *Urological Research* 2004;32:25–27.

Lingeman JE. Extracorporeal shock wave lithotripsy: Development, instrumentation, and current status. *Urolithiasis* 1997;24:185–211.

Lingeman JE, Coury TA, Newman DM, Kahnoski RJ, Mertz JH, Mosbaugh JH, Steele RE, Woods JR. Comparison of results and morbidity of percutaneous nephrostolithotomy and extracorporeal shock wave lithotripsy. *Journal of Urology* 1987;138:485–490.

Lingeman JE, Delius M, Evan AP, Gupta M, Sarica K, Strohmaier W, McAteer JA, Williams JC. Bioeffects and physical mechanisms of SW effects in SWL. In: Segura JW, Conort P, Khoury S, Pak CYC, Preminger GM, Tolley DA, eds., *Stone Disease: First International Consultation on Stone Disease*. Paris, France: Health Publications, 2003a, pp. 251–286.

Lingeman JE, Kim SC, Kuo RL, McAteer JA, Evan AP. Shockwave lithotripsy: Anecdotes and insights. *Journal of Endourology* 2003b;17:687–693.

Lingeman JE, Kulb TB. Hypertension following extracorporeal shock-wave lithotripsy. *Journal of Urology* 1987;137:142A.

Lingeman JE, Lifschitz D, Evan AP. Surgical management of urinary lithiasis. In: Wein AJ, Walsh P, eds., *Campbell's Urology*, 8th edn. Philadelphia, PA: Saunders, 2002, pp. 3361–3451.

Lingeman JE, Matlaga BR, Evan AP. Surgical management of urinary lithiasis. In: Walsh P, Retik A, Vaughan ED, Wein AJ, eds., *Campbell's Urology*. Philadelphia, PA: W.B. Saunders, 2006, pp. 1431–1507.

Lingeman JE, Matlaga BR, Evan AP. Surgical management of upper urinary tract calculi. In: Wein AJ, Kavoussi LR, Novick AC, Partin AW, Peters CA, eds., *Campbell-Walsh Urology*. Philadelphia, PA: W.B. Saunders, 2007, pp. 1431–1507.

Lingeman JE, McAteer JA, Kempson SA, Evan AP. Bioeffects of extracorporeal shock-wave lithotripsy. *Urologic Clinics of North America* 1988;15:507–514.

Lingeman JE, Newman DM, Mertz JH, Mosbaugh PG, Steele RE, Kahnoski RJ, Coury TA, Woods JR. Extracorporeal shock wave lithotripsy: The Methodist Hospital of Indiana experience. *Journal of Urology* 1986a;135:1134–1137.

Lingeman JE, Saywell R, Woods JR, Newman DM. Cost analysis of ESWL relative to other surgical and non-surgical treatment alternatives for urolithiasis. *Medical Care* 1986b;24:115.

Lingeman JE, Woods JR, Toth PD. Blood pressure changes following extracorporeal shock wave lithotripsy and other forms of treatment for nephrolithiasis. *JAMA* 1990;263:1789–1794.

Lingeman JE, Woods JR, Toth PD, Evan AP, McAteer JA. The role of lithotripsy and its side effects. *Journal of Urology* 1989;141:793–797.

Litster JD. Stability of lipid bilayers and red blood cell membranes. *Physics Letters* 1975;53:193–194.

Liu Y, Zhong P. BegoStone—A new stone phantom for shock wave lithotripsy research (L). *Journal of the Acoustical Society of America* 2002;112:1265–1268.

Loew M, Daecke W, Kusnierczak D, Rahmanzadeh M, Ewerbeck V. Shock-wave therapy is effective for chronic calcifying tendinitis of the shoulder. *Journal of Bone and Joint Surgery* 1999;81:863–867.

Loew M, Jurgowski W, Mau HC, Thomsen M. Treatment of calcifying tendinitis of rotator cuff by extracorporeal shock waves: A preliminary report. *Journal of Shoulder and Elbow Surgery* 1995;4:101–106.

Lokhandwalla M, McAteer JA, Williams JC, Sturtevant B. Mechanical haemolysis in shock wave lithotripsy (SWL): II. In vitro cell lysis due to shear. *Physics in Medicine and Biology* 2001;46:1245–1264.

Lokhandwalla M, Sturtevant B. Fracture mechanics model of stone comminution in ESWL and implications for tissue damage. *Physics in Medicine and Biology* 2000;45:1923–1940.

Lokhandwalla M, Sturtevant B. Mechanical haemolysis in shock wave lithotripsy (SWL): I. Analysis of cell deformation due to SWL flow-fields. *Physics in Medicine and Biology* 2001;46:413–437.

Loske AM, Fermández F, Zendejas H, Paredes MI, Castano-Tostado E. Design of a dual-pulse lithotripter to enhance stone comminution during extracorporeal shock wave lithotripsy. *Biomedizinische Technik* 2005;50:46–47.

Loske AM, Prieto FE. Pressure-release versus rigid reflector for extracorporeal shockwave lithotripsy. *Journal of Endourology* 2002;16:273–280.

Loske AM, Prieto FE, Fermández F, van Cauwelaert J. Tandem shock wave cavitation enhancement for extracorporeal lithotripsy. *Physics in Medicine and Biology* 2002;47:3945–3957.

Ludwig J, Lauber S, Lauber HJ, Dreisilker U, Hotzinger H. High-energy shock wave treatment of femoral head necrosis in adults. *Clinical Orthopaedics* 2001;387:119–125.

Lussenhop S, Seeman D, Hahn M, Meiss L. The influence of shock waves on epiphyseal growth plates: First results of an in vivo study with rabbits. In: Siebert W, Buch M, eds., *Extracorporeal Shock Waves in Orthopaedics*. Berlin, Germany: Springer Verlag, 1998, pp. 109–118.

Madaan S, Joyce AD. Limitations of extracorporeal shock wave lithotripsy. *Current Opinion in Urology* 2007;17:109–113.

Madanshetty SI, Roy RA, Apfel RE. Acoustic microcavitation: Its active and passive acoustic detection. *Journal of the Acoustical Society of America* 1991;90:1515–1526.

Madbouly K, Sheir KZ, Elsobky E, Eraky I, Kenawy M. Risk factors for the formation of a steinstrasse after extracorporeal shock wave lithotripsy: A statistical model. *Journal of Urology* 2002;167:1239–1242.

Maker V, Iayke J. Gastrointestinal injury secondary to extracorporeal shock wave lithotripsy: A review of the literature since its inception. *American College of Surgeons* 2004;198:128–135.

Maloney ME, Marguet CG, Zhou Y, Kang DE, Sung JC, Springhart WP, Madden J, Zhong P, Preminger GM. Progressive increase of lithotripter output produces better *in-vivo* stone comminution. *Journal of Endourology* 2006;20:603–606.

Mandhani A, Raghavendran M, Srivastava A, Kapoor R, Singh U, Kumar A, Bhandari M. Prediction of fragility of urinary calculi by dual X-ray absorptiometry. *Journal of Urology* 2003;170:1097–1100.

Marberger M, Türk C, Steinkogler I. Painless extracorporeal lithotripsy. *Journal of Urology* 1988;139:695–699.

Mariani AJ. Combined electrohydraulic and holmium:YAG laser ureteroscopic nephrolithotripsy for 20 to 40 mm renal calculi. *Journal of Urology* 2004;172:170–174.

Matula TJ, Hilmo PR, Bailey MR. A suppressor to prevent direct wave-induced cavitation in shock wave therapy device. *Journal of the Acoustical Society of America* 2005;118:178–185.

May TC, Krause WR, Preslar AJ, Smith MJ, Beaudoin AJ, Cardea JA. Use of high energy shock waves for bone cement removal. *Journal of Arthroplasty* 1990;10:19–27.

McAteer JA, Evan AP. The acute and long-term adverse effects of shock wave lithotripsy. *Seminars in Nephrology* 2008;28:200–213.

McAteer JA, Evan AP, Connors BA, Pishchalnikov YA, Williams JC. Treatment protocols to reduce injury and improve stone breakage in SWL. In: Evan AP, Lingeman JE, McAteer JA, Williams JC, eds., *Second International Urolithiasis Research Symposium*. Melville, New York: American Institute of Physics, 2008, pp. 243–247.

McGurk M, Escudier MP, Brown JE. Modern management of salivary calculi. *British Journal of Surgery* 2005;92:107–112.

Meury T, Verrier S, Alini M. Human endothelial cells inhibit BMSC differentiation into mature osteoblasts in vitro by interfering with osterix expression. *Journal of Cellular Biochemistry* 2006;98:992–1006.

Miller NL, Lingeman JE. Management of kidney stones. *BMJ* 2007;334:468–472.

Miller R. New techniques for the treatment and disruption of renal calculi. *Journal of Medical Engineering & Technology* 1983; 7:1–4.

Miller RA. Endoscopic application of shock wave technology for the destruction of renal calculi. *World Journal of Urology* 1985;3:36–40.

Mobley TB, Myers DA, Grine WB, Jenkins JM, Jordan WR. Low energy lithotripsy with the Lithostar: Treatment results with 19.962 renal and ureteral calculi. *Journal of Urology* 1993;149:1419–1424.

Moe OW. Kidney stones: Pathophysiology and medical management. *Lancet* 2006;367:333–344.

Moncada S, Higgs A. Mechanisms of disease: The L-arginine-nitric oxide pathway. *New England Journal of Medicine* 1993;329:2002–2012.

Moody JA, Evan AP, Lingeman JE. Extracorporeal shock wave lithotripsy. In: Weiss RM, George NJR, O'Reilly PH, eds., *Comprehensive Urology*. Mosby International Limited, 2001, pp. 623–636.

Muller-Mattheis VG, Schmale D, Seewald M, Rosin H, Ackermann R. Bacteremia during extracorporeal shock wave lithotripsy of renal calculi. *Journal of Urology* 1991;146:733–736.

Munver R, Delvecchio FC, Kuo RL, Brown SA, Zhong P, Preminger GM. In vivo assessment of free radical activity during shock wave lithotripsy using a microdialysis system: The renoprotective action of allopurinol. *Journal of Urology* 2002;167:327–334.

Muslumanoglu AY, Tefekli A, Sarilar O, Binbay M, Altunrende F, Ozkuvanci U. Extracorporeal shock wave lithotripsy as first line treatment alternative for urinary tract stones in children: A large scale retrospective analysis. *Journal of Urology* 2003;170:2405–2408.

Neal DE, Kaack Mb, Harmon EP, Puyau F, Morvant A, Reichardson E, Thomas R. Renin production after experimental extracorporeal shock wave lithotripsy: A primate model. *Journal of Urology* 1991;146:548–550.

Neri E, Capannini G, Diciolla F, Carone E, Tripodi A, Tucci E, Sassi C. Localized dissection an delayed rupture of the abdominal aorta after extracorporeal shock wave lithotripsy. *Journal of Vascular Surgery* 2000;31:1052–1055.

Neuhausel DJ. Lithotripsy: A survey. *Journal of Clinical Engineering* 1987;12:283–295.

Neurburg J, Daus HJ, Recker F, Bohndorf K, Bex A, Guenther R, Hofstaedter F. Effects of lithotripsy on rat kidney: Evaluation with MR imaging, histology, and electron microscopy. *Journal of Computer Assisted Tomography* 1989;13:82–89.

Newman DM, Lingeman JE, Mosbaugh PG, Steele RE, Knapp PM, Hutchinson CL. Extracorporeal shock wave lithotripsy using only intravenous analgesia with an unmodified Dornier HM3 lithotripter. In: Lingeman JE, Newman DM, eds., *Shock Wave Lithotripsy 2: Urinary and Biliary Lithotripsy*. New York: Plenum Press, 1989.

Newman RC, Hackett R, Senior D, Brock K, Feldman J, Sosnowski J, Finlayson B. Pathologic effects of ESWL on canine renal tissue. *Urology* 1987;29:194–200.

Nomikos MS, Sowter SJ, Tolley DA. Outcomes using a fourth-generation lithotripter: A new benchmark for comparison? *BJU International* 2007;100:1356–1360.

Nwomeh BC, Yager DR, Cohen IK. Physiology of the chronic wound. *Clinics in Plastic Surgery* 1998;25:341–356.

Ogden JA, Alvarez RG, Levitt R, Marlow M. Shock wave therapy (Orthotripsy®) in musculoskeletal disorders. *Clinical Orthopaedics and Related Research* 2001a;387:22–40.

Ogden JA, Alvarez RG, Levitt R, Marlow M. Shock wave therapy for chronic proximal plantar fasciitis. *Clinical Orthopaedics* 2001b;387:47–59.

Ogden JA, Toth-Kischkat A, Schultheiss R. Principles of shock wave therapy. *Clinical Orthopaedics* 2001c;387:8–17.

Ogiste JS, Nejat RJ, Rashid HH, Greene T, Gupta M. The role of mannitol in alleviating renal injury during extracorporeal shock wave lithotripsy. *Journal of Urology* 2003;169:875–877.

Ohmori K, Matsuda T, Horii Y, Yoshida O. Effects of shock waves on the mouse fetus. *Journal of Urology* 1994;151:255–258.

Ohtori S, Inoue G, Mannoji C, Saisu T, Takahashi K, Mitsuhashi S, Wada Y, Takahashi K, Yamagata M, Moriya H. Shock wave application to rat skin induces degeneration and reinnervation of sensory nerve fibres. *Neuroscience Letters* 2001;315:57–60.

Oosterhof G, Cornel EB, Smits GAHJ, Debruyne FMJ, Schalken JA. The influence of high energy shock waves on the development of metastases. *Ultrasound in Medicine & Biology* 1996;22:339–344.

Oosterhof G, Smits GA, de Ruyter A, Schalken JA, Debruyne FMJ. Effects of high energy shock waves combined with biological response modifiers in different human kidney cancer xenografts. *Ultrasound in Medicine & Biology* 1991;17:391–399.

Orkisz M, Farchtchian T, Saighi D, Bourlion M, Thiounn N, Gimenez G, Debre B, Flam TA. Image based renal stone tracking to improve efficacy in extracorporeal lithotripsy. *Journal of Urology* 1998;160:1237–1240.

Orsola A, Diaz I, Caffaratti J, Izquierdo F, Alberola J, Garat JM. Staghorn calculi in children: Treatment with mono-therapy extracorporeal shock wave lithotripsy. *Journal of Urology* 1999;162:1229–1233.

Osman MM, Alfano Y, Kamp S, Haecker A, Alken P, Michel MS, Knoll T. t-Year-follow-up of patients with clinically insignificant residual fragments after extracorporeal shockwave lithotripsy. *European Urology* 2005;47:860–864.

Ottaviani F, Capaccio P, Rivolta R, Cosmacini P, Pignataro L, Castagnone D. Salivary gland stones: US evaluation in shock wave lithotripsy. *Radiology* 1997;204:437–441.

Owen NR, Bailey MR, Crum LA. The use of resonant scattering to identify stone fracture in shock wave lithotripsy. *Journal of the Acoustical Society of America* 2006;121:EL41–EL47.

Palmer RMJ, Ferrige AG, Moncada S. Nitric oxide release accounts for the biological activity of endothelium-derived relaxing factor. *Nature* 1987;327:524–526.

Papadoukakis S, Stolzenburg J, Truss MC. Treatment strategies of ureteral stones. *EAU-EBU Update Series* 2006;4:184–190.

Pareek G, Hedican SP, Lee FT, Nakada SY. Shock wave lithotripsy success determined by skin-to-stone distance on computed tomography. *Urology* 2005;66:941–944.

Parks JH, Worcester EM, Coe FL, Evan AP, Lingeman JE. Clinical implications of abundant calcium phosphate in routinely analyzed kidney stones. *Kidney International* 2004;66:777–785.

Pearle MS, Calhoun EA, Curham GC. Urologic diseases in America project: Urolithiasis. *Journal of Urology* 2005;63:848–857.

Perez-Blanco FJ, Arrabal Martin M, Ocete martin C, Arias Puerta JJ, Garcia-Valdecasas Bernal J, Rodriguez Cuartero A, Zuluaga Gómez A. Urinary glycosaminoglycans after extracorporeal shock wave lithotripsy in patients with kidney lithiasis. *Archivos Españoles de Urología* 2001;54:875–883.

Perks AE, Schuler TD, Lee J, Ghiculete D, Chung D-G, D'A Honey RJ, Pace KT. Stone attenuation and skin-to-stone distance on computed tomography predicts for stone fragmentation by shock wave lithotripsy. *Urology* 2008;72:765–769.

Peterson JC, Finlayson B. Effects of ESWL on blood pressure. In: Gravenstein J, Peter K, eds., *Extracorporeal Shock Wave Lithotripsy for Renal Stone Disease*. Boston, MA: Butterworth, 1986, pp. 145–150.

Pishchalnikov YA, McAteer JA, Williams JC. Effect of firing rate on the performance of shock wave lithotriptors. *BJU International* 2008;102:1681–1686.

Pishchalnikov YA, Neucks JS, VonDerHaar RJ, Pishchalnikova IV, Williams JC, McAteer JA. Air pockets trapped during routine coupling in dry-head lithotripsy can significantly reduce the delivery of shock wave energy. *Journal of Urology* 2006;176:2706–2710.

Pishchalnikov YA, Sapozhnikov OA, Bailey MR, Pishchalnikova IV, Williams JC, McAteer JA. Cavitation selectively reduces the negative-pressure phase of lithotripter shock waves. *Acoustics Research Letters Online* 2005;6:280–286.

Pishchalnikov YA, Sapozhnikov OA, Bailey MR, Williams JC, Cleveland RO, Colonius T, Crum LA, Evan AP, McAteer JA. Cavitation bubble cluster activity in the breakage of kidney stones by lithotripter shockwaves. *Journal of Endourology* 2003;17:435–447.

Pishchalnikov YA, Williams JC, McAteer JA. Bubble proliferation in the cavitation field of a shock wave lithotripter. *Journal of the Acoustical Society of America* 2011;130:E87–E93.

Prat F, Sibille A, Luccioni C. Increased chemocytotoxicity to colon cancer cells by shock wave induced cavitation. *Gastroenterology* 1994;106:937–944.

Preminger GM. Sonographic piezoelectric lithotripsy: More bang for your buck. *Journal of Endourology* 1989;3:321–327.

Preminger GM, Assimos DG, Lingeman JE, Nakada SY, Pearle MS, Wolf JS. Chapter 1: AUA guideline on management of staghorn calculi: Diagnosis and treatment recommendations. *Journal of Urology* 2005;173:1991–2000.

Preminger GM, Kettelhut MC, Elkins SL, Seger J, Fetner CD. Ureteral stenting during extracorporeal shock wave lithotripsy: Help or hindrance. *Journal of Urology* 1989;142: 32–36.

Preminger GM, Tiselius HG, Assimos DG, Alken P, Buck C, Gallucci M, Knoll T et al. 2007. Guideline for the management of ureteral calculi. *Journal of Urology* 2007;178:2418–2434.

Qin J, Simmons WN, Sankin G, Zhong P. Effect of lithotripter focal width on stone comminution in shock wave lithotripsy. *Journal of the Acoustical Society of America* 2010;127: 2635–2645.

Raisz LG, Kream BE. Regulation of bone formation. *New England Journal of Medicine* 1983;309:29–35.

Randazzo RF, Chaussy C, Fuchs GJ, Bhuta SM, Lovrekovich H, deKernion JB. The in vitro and in vivo effects of extracorporeal shock waves on malignant cells. *Urological Research* 1988;16:419–426.

Rassweiler JJ, Knoll T, Köhrmann K-U, McAteer JA, Lingeman JE, Cleveland RO, Bailey MR, Chaussy C. Shock wave technology and application: An update. *European Urology* 2011;59:784–796.

Rassweiler JJ, Tailly GG, Chaussy C. Progress in lithotripter technology. *EAU Update Series* 2005;3:17–36.

Recker F, Hofmann W, Bex A, Tscholl R. Quantitative determination of urinary marker proteins: A model to detect intrarenal bioeffects after extracorporeal lithotripsy. *Journal of Urology* 1992;148:1000–1006.

Riehle RA. Extracorporeal shock wave lithotripsy. *Bulletin of the New York Academy of Medicine* 1986;62:291–314.

Riehle RA, Fair WR, Vaughan ED. Extracorporeal shock-wave lithotripsy for upper urinary tract calculi. *JAMA* 1986;255:2043–2048.

Rigatti P, Colombo R, Centemero A, Francesca F, Di Girolamo V, Montorsi F, Trabucchi E. Histological and ultrastructural evaluation of extracorporeal shock wave lithotripsy-induced acute renal lesions: Preliminary report. *European Urology* 1989;16:207–211.

Rochwerger A, Franceschi JP, Viton JM, Roux H, Mattei JP. Surgical management of calcific tendinitis of the shoulder: An analysis of 26 cases. *Clinical Rheumatology* 1999;18:313–316.

Rompe JD. Extracorporeal shock wave therapy for lateral epicondylitis—A double blind randomized controlled trial. *Journal of Orthopaedic Research* 2003;21:958–959.

Rompe JD, Burger RA, Hopf C, Eysel P. Shoulder function after extracorporeal shock wave therapy for calcific tendinitis. *Journal of Shoulder and Elbow Surgery* 1998;7:505–509.

Rompe JD, Hopf C, Kullmer K, Heine J, Burger RA. Analgesic effect of extracorporeal shock-wave therapy on chronic tennis elbow. *Journal of Bone and Joint Surgery* 1996a;78B: 233–237.

Rompe JD, Hopf C, Kullmer K, Heine J, Burger RA. Low-energy extracorporeal shock wave therapy for persistent tennis elbow. *International Orthopaedics* 1996b;20:23–27.

Rompe JD, Rosendahl T, Schollner C, Thesis C. High-energy extracorporeal shock wave treatment of nonunions. *Clinical Orthopaedics and Related Research* 2001;387:102–111.

Roth RA, Beckmann CF. Complications of extracorporeal shock-wave lithotripsy and percutaneous nephrolithotomy. *Urologic Clinics of North America* 1988;15:155–166.

Roy RA, Madanshetty SI, Apfel RE. An acoustic backscattering technique for the detection of transient cavitation produced by microsecond pulses of ultrasound. *Journal of the Acoustical Society of America* 1990;87:2451–2458.

Rubin JI, Arger PH, Pollack HM, Banner MP, Coleman BG, Mintz MC, VanArsdalen KN. Kidney changes after extracorporeal shock wave lithotripsy: CT evaluation. *Radiology* 1987;162:21–24.

Russo S, Stephenson RA, Mies C, Huryk R, Heston WD, Melamed MR, Fair WR. High energy shock waves suppress tumor growth in vitro and in vivo. *Urology* 1986;135:626–628.

Saggini R, Figus A, Troccola A, Cocco V, Saggini A, Scuderi N. Extracorporeal shock wave therapy for management of chronic ulcers in the lower extremities. *Ultrasound in Medicine & Biology* 2008;34:1261–1271.

Saisu T, Goto S, Wada Y, Takahashi K, Mitsuhashi S, Harada Y, Minami S, Moriya H, Kamegaya M. Irradiation of the extracorporeal shock wave to the immature long bone causes overgrowth and local increase in bone mineral content. *Transactions of the Orthopaedic Research Society* 1999; 45:34.

Salem S, Mehrsai A, Zartab H, Shahdadi N, Pourmand G. Complications and outcomes following extracorporeal shock wave lithotripsy: A prospective study of 3,241 patients. *Urological Research* 2010;38:135–142.

Sapozhnikov OA, Khokhlova VA, Bailey MR, Williams JC, McAteer JA, Cleveland RO, Crum LA. Effect of overpressure and pulse repetition frequency on cavitation in shock wave lithotripsy. *Journal of the Acoustical Society of America* 2002;112:1183–1195.

Sapozhnikov OA, Maxwell AD, MacConaghby B, Bailey MR. A mechanistic analysis of stone fracture in lithotripsy. *Journal of the Acoustical Society of America* 2007;121:1190–1202.

Sarica K, Balat A, Erbagci A, Cekmen M, Yurekli M, Yagci F. Effects of shock wave lithotripsy on plasma and urinary levels of nitrite and adrenomedullin. *Urological Research* 2003;31: 347–351.

Sarica K, Kosar A, Yaman O, Beduk Y, Durak I, Gogus O. Evaluation of ischemia after ESWL: Detection of free oxygen radical scavenger enzymes in renal parenchyma subjected to high-energy shock waves. *Urologia Internationalis* 1996;57:221.

Sass W, Braunlich M, Dreyer HP, Matura E, Folberth W, Priesmeyer HG, Seifert J. The mechanisms of stone disintegration by shock waves. *Ultrasound in Medicine & Biology* 1991;17: 239–243.

Sato Y, Tanda H, Kato S, Ohnishi S, Nakajima H, Nanbu A, Nitta T, Koroku M, Akagashi K, Hanzawa T. Shock wave lithotripsy for renal stones is not associated with hypertension and diabetes mellitus. *Urology* 2008;71:586–591.

Sauerbruch T, Neubrand M, Lobentanzer H. *In vitro* fragmentation of gallstones. In: Paumgartner G, Sauerbruch T, Sackmann M, Burhenne HH, eds., *Lithotripsy and Related Techniques for Gallstone Treatment*. St. Louis, MO: Mosby Year Book, 1991, pp. 7–13.

Sauerbruch T, Paumgartner G. Gallbladder stones: Management. *Lancet* 1991;338:1121–1124.

Schaden W, Fischer A, Sailler A. Extracorporeal shock wave therapy of nonunion or delayed osseous union. *Clinical Orthopaedics and Related Research* 2001;387:90–94.

Schleberger R, Senge T. Non-invasive treatment of long-bone pseudarthrosis by shock waves (ESWL). *Archives of Orthopaedic and Trauma Surgery* 1992;111:224–227.

Schmitt J, Haake M, Tosch A, Hildebrand R, Deike B, Griss P. Low energy extracorporeal shock wave therapy (ESWT) of supraspinatus tendinitis: A prospective, randomised study. *Journal of Bone & Joint Surgery (British)* 2001;83: 873–876.

Schneider MU, Matek W, Bauer R, Domschke W. Mechanical lithotripsy of bile duct stones in 209 patients-effect of technical advances. *Endoscopy* 1988;20:248–253.

Schreuers BW, Bierkens AF, Huiskes R. The effect of the extracorporeal shock wave lithotripter on bone cement. *Biomedical Materials Research* 1991;25:157–164.

Segura JW, Preminger GM, Assimos DG, Dretler SP, Kahn RI, Lingeman JE, Macaluso JN. Ureteral stones clinical guidelines panel summary report on the management of ureteral calculi. *Journal of Urology* 1997;58:1915–1921.

Seidl M, Steinbach P, Wörle K, Hofstädter F. Induction of stress fibres and intercellular gaps in human vascular endothelium by shock-waves. *Ultrasonics* 1993;32:397–400.

Seidl M, Steinback P. Induction of stress fibres and intercellular gaps in human vascular endothelium by shock-waves. *Ultrasonics* 1994;32:397–400.

Seitz G, Neisius D, Wernert N, Gebhardt T. Pathological-anatomical alterations of human kidneys following extracorporeal piezoelectric shock wave lithotripsy. *Journal of Endourology* 1991;5:17–20.

Semins MJ, Trock BJ, Matlaga BR. The effect of shock wave rate on the outcome of shock wave lithotripsy: A meta-analysis. *Journal of Urology* 2008;179:194–197.

Servadio C, Livne P, Winkler H. Extracorporeal shock wave lithotripsy using a new, compact and portable unit. *Journal of Urology* 1988;139:685–688.

Shah A, Owen NR, Lu W, Cunitz BW, Kaczkowski PJ, Harper JD, Bailey MR, Crum LA. Novel ultrasound method to reposition kidney stones. *Urological Research* 2010;38:491–495.

Shao Y, Connors BA, Evan AP, Willis LR, Lifshitz DA, Lingeman JE. Morphological changes induced in the pig kidney by extracorporeal shock wave lithotripsy: Nephron injury. *Anatomical Record Part A: Discoveries in Molecular, Cellular, and Evolutionary Biology* 2003;275:979–989.

Shaw MJ, Mackiev RD, Moore JP, Dorsher PJ, Freeman ML, Meirr PB, Potter T, Hutton SW, Vennes JA. Results of a multicenter trial using a mechanical lithotripter for the treatment of large bile duct stones. *The American Journal of Gastroenterology* 1993;88:730–733.

Sheir KZ, El-Diasty TA, Ismail AM. Evaluation of a synchronous twin-pulse technique for shock wave lithotripsy. *BJU International* 2005;95:389–393.

Sheir KZ, Elhalwagy SM, Abo-elghar ME, Ismail AM, Elsawy E, El-diasty TA, Dawaba ME, Eraky IA, El-Kenawy MR. Evaluation of a synchronous twin-pulse technique for shock wave lithotripsy: A prospective randomized study of effectiveness and safety in comparison to standard single-pulse technique. *BJU International* 2008;101:1420–1426.

Sheir KZ, Madouly K, Elsobky E. Prospective randomized comparative study of the effectiveness and safety of electrohydraulic and electromagnetic extracorporeal shock wave lithotriptors. *Journal of Urology* 2003;170:389–392.

Shen P, Jiang M, Yang J, Li X, Li Y, Wei W, Dai Y, Zeng H, Wang J. Use of ureteral stent in extracorporeal shock wave lithotripsy for upper urinary calculi: A systematic review and meta-analysis. *Journal of Urology* 2011;186:1328–1335.

Siddiqui SJ. Sialolithiasis: An unusually large submandibular salivary stone. *British Journal of Surgery* 2002;193:89–91.

Siebert W, Buch M. *Extracorporeal Shock Waves in Orthopaedics.* Berlin, Germany: Springer Verlag, 1998.

Siener R. Links impact of dietary habits on stone incidence. *Urological Research* 2006;34:131–133.

Sivak MV. Endoscopic management of bile duct stones. *The American Journal of Surgery* 1989;158:228–240.

Skolarikos A, Alivizatos G, de la Rosette J. Extracorporeal shock wave lithotripsy 25 years later: Complications and their prevention. *European Urology* 2006;50:981–990.

Sokolov DL, Bailey MR, Crum LA. Use of a dual-pulse lithotripter to generate a localized and intensified cavitation field. *Journal of the Acoustical Society of America* 2001;110: 1685–1695.

Sokolov DL, Bailey MR, Crum LA. Dual-pulse lithotripter accelerates stone fragmentation and reduces cell lysis *in vitro*. *Ultrasound in Medicine & Biology* 2003;29:1045–1052.

Sokolov DL, Bailey MR, Crum LA, Blomgren PM, Connors BA, Evan AP. Prefocal alignment improves stone comminution in shockwave lithotripsy. *Journal of Endourology* 2002;16:709–715.

Sorensen C, Chandhoke P, Moore M, Wolf C, Sarram A. Comparison of intravenous sedation versus general anesthesia on the efficacy of the Doli 50 lithotriptor. *Journal of Urology* 2002;168:35–37.

Sorensen MD, Teichman JMH, Bailey MR. A prototype ultrasound instrument to size stone fragments during ureteroscopy. In: Evan AP, Lingeman JE, McAteer JA, Williams JC, eds., *Second International Urolithiasis Research Symposium.* Melville, New York: American Institute of Physics, 2008, pp. 348–352.

Speed CA, Richards C, Nichols D, Burnet S, Wies JT, Humphreys H, Hazleman BL. Extracorporeal shock-wave therapy for tendonitis of the rotator cuff: A double-blind, randomised, controlled trial. *Journal of Bone & Joint Surgery (British)* 2002;84:509–512.

Stamatelou KK, Francis ME, Jones CA, Nyberg LM, Curhan GC. Time trends in reported prevalence of kidney stones in the United States. *Kidney International* 2003;63:1817–1823.

Steinberg ME, Hyken GD, Steinberg DR. A quantitative system for staging avascular necrosis. *Journal of Bone and Joint Surgery* 1995;77B:34–41.

Stranne SK, Cocks FH, Gettliffe R. Mechanical property studies of human gallstones. *Journal of Biomedical Materials Research* 1990;24:1049–1057.

Streem SB, Yost A, Mascha E. Clinical implications of clinically insignificant stone fragments after extracorporeal shock wave lithotripsy. *Journal of Urology* 1996;155:1186–1190.

Sturtevant B. Shock wave physics of lithotripters. In: Smith AD, Badlani GH, Bagley DH eds., *Smith's Textbook of Endourology,* St. Louis, MO: Quality Medical, 1996, pp. 529–552.

Sukul D, Johannes EJ, Pierik E. The effect of high energy shock waves focused on cortical bone: An in vitro study. *Journal of Surgical Research* 1993;54:46–51.

Sun BY, Lee YH, Jiaan BP, Chen KK, Chang LS, Chen KT. Recurrence rate and risk factors for urinary calculi after extracorporeal shock wave lithotripsy. *Journal of Urology* 1996;156:903–905.

Sutherland JW, Parks HH, Coe FL. Recurrence after a single renal stone in a community practice. *Mineral and Electrolyte Metabolism* 1985;11:267–269.

Tailly GG, Baert JA, Hente KR, Tailly TO. Twenty years of single center experience in ESWL 1987–2007: An evaluation of 3079 patients. *Journal of Endourology* 2008;22:2211–2222.

Takahashi N, Ohtori S, Saisu T, Moriya H, Wada Y. Second application of low-energy shock waves has a cumulative effect on free nerve endings. *Clinical Orthopaedics and Related Research* 2006;443:315–319.

Tan YM, Yip SK, Chong TW, Wong MY, Cheng C, Foo KT. Clinical experience and results of ESWL treatment of 3,093 urinary calculi with the Storz Modulith SL 20 lithotripter at the Singapore general hospital. *Scandinavian Journal of Urology and Nephrology* 2002;36:363–367.

Teichman JMH. Clinical practice. Acute renal colic from ureteral calculus. *New England Journal of Medicine* 2004;350:684–693.

Teichman JMH, Portis AJ, Cecconi PP, Bub WL, Endicott RC, Denes B, Pearle MS, Clayman RV. In vitro comparison of shock wave lithotripsy machines. *Journal of Urology* 2000;164:1259–1264.

Thiel M. Application of shock waves in medicine. *Clinical Orthopaedics and Related Research* 2001;387:18–21.

Thomas JL, Fink M. Time reversal focusing applied to lithotripsy. *Ultrasound Imaging* 1996;18:106–121.

Thomson CE, Crawford F, Murray GD. The effectiveness of extra corporeal shock wave therapy for plantar heel pain: A systematic review and meta-analysis. *BMC Musculoskeletal Disorders* 2005;6:19–29.

Tolley D, Downey P. Current advances in shock wave lithotripsy. *Current Opinion in Urology* 1999;9:319–323.

Turangan CK, Jamaluddin AR, Ball GJ, Leighton TG. Free-Lagrange simulations of the expansion and jetting collapse of air bubbles in water. *Journal of Fluid Mechanics* 2008;598:1–25.

Türk C, Knoll T, Petrik A. *Guidelines on Urolithiasis.* Arnhem, the Netherlands: European Association of Urology, 2010, p. 19.

Uhthoff HK, Loehr JF. Tendinosis calcarea of the rotator cuff. In: Rockwood CA, Matsen FA, eds., *The Shoulder.* Philadelphia, PA: Saunders, 1998, pp. 989–1008.

Vakil N, Everbach EC. Gas in gallstones: Quantitative determination and possible effects on fragmentation by shock waves. *Gastroenterology* 1991;101:1628–1634.

Valchanou VD, Michailow P. High-energy shock waves in the treatment of delayed and nonunion of fractures. *International Orthopaedics* 1991;151:181–184.

Vallancien G, Aviles J, Munoz R, Veillon B, Charton M, Brisset JM. Piezoelectric extracorporeal lithotripsy by ultrashort waves with the EDAP LT01 device. *Journal of Urology* 1988;139:689–694.

Vieweg J, Teh C, Freed K, Leder RA, Smith RH, Nelson RH, Preminger GM. Unenhanced helical computerized tomography for the evaluation of patients with acute flank pain. *Journal of Urology* 1998;160:679–684.

Vieweg J, Weber HM, Miller K, Hautmann R. Female fertility following extracorporeal shock wave lithotripsy of distal ureteral calculi. *Journal of Urology* 1992;148:1007–1010.

Vogel A, Hopf C, Eysel P, Rompe JD. Application of extracorporeal shockwaves in the treatment of pseudarthrosis of the lower extremity. Preliminary results. *Archives of Orthopaedic and Trauma Surgery* 1997;116:480–483.

Wang CJ. An overview of shock wave therapy in musculoskeletal disorders. *Chang Gung Medical Journal* 2003;26:220–232.

Wang CJ, Chen HS. Shockwave therapy for patients with epicondylitis of the elbow. A one to two year follow-up study. *American Journal of Sports Medicine* 2002;30:422–425.

Wang CJ, Huang HY, Pai CH. Shock wave enhanced neovascularization at the bone-tendon junction. A study in a dog model. *Journal of Foot and Ankle Surgery* 2001a;41:16–22.

Wang CJ, Huang HY, Yang KD, Wang FS, Wong MY. Pathomechanism of shock wave injuries on femoral artery, vein and nerve. An experimental study in dogs. *Injury* 2002a;33:439–446.

Wang CJ, Wang FS, Yang KD, Weng LH, Hsu CC, Huang CS, Yang LC. Shock wave therapy induces neovascularization at the tendon-bone junction. A study in rabbits. *Journal of Orthopaedic Research* 2003a;21:984–989.

Wang F-S, Wang C-J, Huang H-J, Chung H, Chen R-F, Yang KD. Physical shock wave mediates membrane hyperpolarization and Ras activation for osteogenesis in human bone marrow stromal cells. *Biochemical and Biophysical Research Communications* 2001b;287:648–655.

Wang FS, Yang KD, Chen RF, Wang CJ, Sheen-Chen SM. Extracorporeal shock wave promotes growth and differentiation of bone-marrow stromal cells towards osteoprogenitors associated with induction of TGF-beta1. *Journal of Bone & Joint Surgery (British)* 2002b;84:457–461.

Wang FS, Yang KD, Kuo YR, Wang CJ, Sheen-Chen SM, Huang HC, Chen YJ. Temporal and spatial expression of bone morphogenetic proteins in extracorporeal shock wave-promoted healing of segmental defect. *Bone* 2003b;32:387–396.

Weinstein JN, Oster DM, Park JB, Park SH, Loening S. The effects of extracorporeal shock wave lithotripter on the bone-cement interface in dogs. *Clinical Orthopaedics* 1988;235:261–267.

Weiss N, Delius M, Gambihler S. The in vivo effects of shock waves and cisplatin on cisplatin sensitive and resistant rodent tumors. *International Journal of Cancer* 1994;58:693–699.

Weizer AZ, Zhong P, Preminger GM. New concepts in shock wave lithotripsy. *Urologic Clinics of North America* 2007;34:375–382.

Williams CM, Kaude JV, Newman RC, Peterson JC, Thomas WC. Extracorporeal shock-wave lithotripsy: Long-term complications. *American Journal of Roentgenology* 1988;150:311–315.

Williams JC, Saw KC, Paterson RF, Hatt EK, McAteer JA, Lingeman JE. Variability of renal stone fragility in shock wave lithotripsy. *Urology* 2003;61:1092–1097.

Williams JC, Woodward JF, Stonehill MA, Evan AP, McAteer JA. Cell damage by lithotriptor shock waves at high pressure to preclude cavitation. *Ultrasound in Medicine & Biology* 1999;25:1445–1449.

Willis LR, Evan AP, Connors BA, Blomgren PM, Fineberg NS, Lingeman JE. Relationship between kidney size, renal injury, and renal impairment induced by shock wave lithotripsy. *Journal of the American Society of Nephrology* 1999;10:1753–1762.

Willis LR, Evan AP, Connors BA, Fineberg NS, Lingeman JE. Effects of extracorporeal shock wave lithotripsy to one kidney on bilateral glomerular filtration rate and PAH clearance in minipigs. *Journal of Urology* 1996;156:1502–1506.

Willis LR, Evan AP, Connors BA, Fineberg NS, Lingeman JE. Effects of SWL on glomerular filtration rate and renal plasma flow in uninephrectomized minipigs. *Journal of Urology* 1997;11:27–32.

Willis LR, Evan AP, Connors BA, Handa RK, Blomgren PM, Lingeman JE. Prevention of lithotripsy-induced renal injury by pre-treating kidneys with low-energy shock waves. *Journal of the American Society of Nephrology* 2006;17:663–673.

Willis LR, Evan AP, Connors BA, Shao Y, Blomgren PM, Pratt JH, Fineberg NS, Lingeman JE. Shockwave lithotripsy: Dose-related effects on renal structure, hemodynamics, and tubular function. *Journal of Endourology* 2005;19:90–101.

Wirth MP, Theiss M, Frohmuller HG. Primary extracorporeal shockwave lithotripsy of staghorn renal calculi. *Urologia Internationalis* 1992;48:71–75.

Wolf JS. Treatment selection and outcomes: Ureteral calculi. *Urologic Clinics of North America* 2007;34:421–430.

Worecester EM, Evan AP, Coe FL. Pathogenesis of stone disease. In: Evan AP, Lingeman JE, McAteer JA, Williams JC, eds., *Second International Urolithiasis Research Symposium.* New York: American Institute of Physics, 2008, pp. 3–13.

Wörle K, Steinbach P, Hofstädter F. The combined effects of high-energy shock waves and cytostatic drugs or cytokines on human bladder cancer cells. *British Journal of Cancer* 1994;69:58–65.

Xi X, Zhong P. Improvement of stone fragmentation during shockwave lithotripsy using a combined EH/PEAA shock-wave generator—In vitro experiments. *Ultrasound in Medicine & Biology* 2000;26:457–467.

Xi X, Zhong P. Dynamic photoelastic study of the transient stress field in solids during shock wave lithotripsy. *Journal of the Acoustical Society of America* 2001;109:1226–1239.

Yokoyama M, Shoji F, Yanagizawa R, Kanemura M, Kitahara K, Takahasi S, Kawai K, Oda H, Osaka M, Handa H. Blood pressure changes following extracorporeal shock-wave lithotripsy for urolithiasis. *Journal of Urology* 1992;147:553–557.

Yoshimoto H, Ikeda S, Tanaka M, Kuroda Y. Choledochoscopic electrohydraulic lithotripsy and lithotomy for stones in the common bile duct, intrahepatic ducts, and gallbladder. *Annals of Surgery* 1989;210:576–582.

Zanetti G, Montanari E, Trinchieri A, Guarneri A, Ceresoli A, Mazza L, Austoni E. Long-term follow up of blood pressure after extracorporeal shock wave lithotripsy. *Journal of Endourology* 1992;6:195–196.

Zanetti G, Ostini F, Montanari E, Russo R, Elena A, Trinchieri A, Pisani E. Cardiac dysrhythmias induced by extracorporeal shockwave lithotripsy. *Journal of Endourology* 1999;13:409–412.

Zhong P. Innovations in lithotripsy technology. In: Evan AP, Lingeman JE, Williams JC, eds., *First Annual International Urolithiasis Research Symposium.* American Institute of Physics, 2007, pp. 317–325.

Zhong P, Chuong CJ, Preminger GM. Characterization of fracture toughness of renal calculi using a microindentation technique. *Journal of Materials Science Letters* 1993;12:1460–1462.

Zhong P, Cioanta I, Cocks FH, Preminger GM. Inertial cavitation and associated acoustic emission produced during electrohydraulic shock wave lithotripsy. *Journal of the Acoustical Society of America* 1997;101:2940–2950.

Zhong P, Zhou Y, Zhu SL. Dynamics of bubble oscillation in constrained media and mechanisms of vessel rupture in SWL. *Ultrasound in Medicine & Biology* 2001;27:119–134.

Zhong P, Zhou YF. Suppression of large intraluminal bubble expansion in shock wave lithotripsy without compromising stone comminution: Methodology and *in vitro* experiments. *Journal of the Acoustical Society of America* 2001;110:3283–3291.

Zhou Y, Cocks FH, Preminger GM, Zhong P. The effect of treatment strategy on stone comminution efficiency in shock wave lithotripsy. *Journal of Urology* 2004a;172:349–354.

Zhou Y, Cocks FH, Preminger GM, Zhong P. Innovations in shock wave lithotripsy technology: Updates in experimental studies. *Journal of Urology* 2004b;172:1892–1898.

Zhou YF. *Optimization of Pressure Filed in Shock Wave Lithotripsy: Pressure Waveform, Distribution and Sequence.* Saarbrücken, Germany: VDM Verlag Dr. Müller, 2011.

Zhou YF, Qin J, Zhong P. Characteristics of the secondary bubble cluster produced by an electrohydraulic shock wave lithotripter. *Ultrasound in Medicine & Biology* 2012;38:601–610.

Zhou YF, Zhong P. Suppression of large intraluminal bubble expansion in shock wave lithotripsy without compromising stone comminution: Refinement of reflector geometry. *Journal of the Acoustical Society of America* 2003;113:586–597.

Zhu SL, Cocks FH, Preminger GM, Zhong P. The role of stress waves and cavitation in stone comminution in shock wave lithotripsy. *Ultrasound in Medicine & Biology* 2002;28:661–671.

13 High-Intensity Focused Ultrasound

13.1 INTRODUCTION

Cancer is a major disease that involves abnormal cell growth with the potential to invade or spread to other parts of the body and has been known worldwide for centuries. About a quarter of all deaths in the United States are due to cancer (Jemal et al. 2009). According to the Surveillance Epidemiology and End Results (SEER) Cancer Statistics Review, there were more than 570,000 cancer deaths and ~1.6 million new cases diagnosed in 2011 in the United States. It is expected that the demand for cancer treatment of $16.8 billion in the United States will grow 10% annually through 2009, with the increased spending driven by higher incidence of the disease and wider cancer screening (The Freedonia Group 2009), including those for various cancer types (i.e., cancers of the breast, digestive system, and genital system, leukemia, and lymphoma), treatment modalities (i.e., surgery, chemotherapy, radiation therapy, the use of biotechnological drugs, hormonal therapy, vaccines, nanotechnology, and stem cells), and treatment providers (i.e., hospitals, outpatient facilities, and health care at home). The conventional therapy modalities (open surgery, chemotherapy, and radiotherapy) have significant morbidity and mortality associated with them with long in-patient stays and recovery periods. The extensive investigations of cancer focus on significantly improving the therapeutic efficacy, simultaneously reducing the local, regional, and systemic side effects.

Ultrasound is not limited to diagnosis in clinics. It was recognized long ago that localized temperature rise can be used for cancer therapy. As a result, minimally invasive devices, such as radio frequency (RF), lasers, microwaves, and cryoablation equipment, have been introduced and applied in clinics in recent years. However, these heat sources have to be positioned at or near the tumor and cancer, where the target is no larger than 3–4 cm in diameter. The interaction of ultrasound with tissues and the subsequent physical and biological changes have been investigated for decades (Wood and Loomis 1927). High-intensity focused ultrasound (HIFU) has emerged as a completely noninvasive and extracorporeal modality for selectively destroying primary solid tumors and metastatic disease in a controlled manner. Almost complete necrosis of tumor can be achieved within the focal region, especially those in surgically difficult locations, without damage to the intervening tissue. HIFU has been applied clinically in neurosurgery, ophthalmology, urology, gynecology, and oncology (ter Haar et al. 1991, Visioli et al. 1999, Gelet et al. 2000) in more than 30,000 patients, mainly in Asia and Europe. Preliminary reports suggest that toxicity is reduced significantly in comparison to current ablation techniques because of its noninvasive nature (Dubinsky et al. 2008). Therefore, *TIME* magazine selected HIFU as one of the 50 most inspired ideas, innovations, and revolutions of 2011 (Brock-Abraham et al. 2011).

13.2 HISTORY OF HIFU

The first tissue destruction using extracorporeal focused ultrasound energy in medical therapy began in 1942 (Lynn et al. 1942a). Seminal works were carried out at the University of Illinois by the Fry brothers as a neurosurgical tool for the treatment of Parkinson's disease in humans (Lynn et al. 1942, Fry et al. 1955, Ballantine et al. 1960, Fry and Fry 1960). Early attempts to generate lesions in the brain through the intact skull bone using HIFU were unsuccessful (Lynn et al. 1942b, Lynn and Putnam 1944). After removing a section of skull bone, small lesions were found well circumscribed and "trackless" in the brain without damaging overlying or surrounding tissues. Although the symptoms of Parkinsonism were claimed to be eliminated in 50 patients in the 1950s (Fry and Fry 1960), this treatment did not gain further popular acceptance and progress, mostly due to the introduction of an effective drug, L-dopa, at the same time. The requirement of an acoustic window in the skull and the lack of imaging modalities limited the progress of this neurosurgical research. Afterward, malignancies located deep inside the body were tried. The first case of breast cancer treatment by HIFU was reported in 1961 (Hickey et al. 1961).

HIFU was also applied in ophthalmology by Fred Lizzi's team at the Riverside Research Institute, New York, United States, for the treatment of glaucoma and retinal detachment and for sealing traumatic capsular tears in the 1980s (Lizzi et al. 1984, Coleman et al. 1985, Rosecan et al. 1985, Silverman et al. 1991). The novel device built by them and known as SonoCare was the first HIFU device approved by the U.S. Food and Drug Administration (FDA) to treat glaucoma patients to reduce intraocular pressure. However, its widespread use was hindered by the parallel development of laser devices, which are much easier to use.

HIFU was given a new life in the 1990s with the technical developments in ultrasound transducers, modes of energy delivery, computer control, and real-time imaging. Modern sonographic devices are designed with electronic beamforming and dynamic beamsteering and focusing mechanisms in various shapes and sizes operating under a wide range of parameters, including frequency, depth of field, field of view, depth of penetration, and frame rates, assisted by beamforming and image-processing algorithms optimized for specific organs. Diagnostic ultrasound and magnetic resonance imaging (MRI) provide precise targeting and good treatment follow-up with anatomical and functional imaging, which paves the way to practical implementation of HIFU therapy with its full potential. The ability of targeting subcutaneous tissue and producing almost instantaneous cell death by coagulation necrosis in deep-seated tumors selectively has made it a successful candidate for direct and rapid treatment of tumors in oncology (ter Haar et al. 1991, Visioli et al. 1999, Gelet et al. 2000). In 1996, superficial bladder cancers in 20 patients were ablated using HIFU (Vallancien et al. 1996). Wide application of HIFU in clinics began since the successful treatment of a patient with osteosarcoma in Chongqing, China, in 1997. Over the past 15 years, more than 30,000 cases of uterine fibroids and cancers of the liver, breast, pancreas, bone, and kidney have been performed using HIFU with promising results (Zhou 2011). MRI-guided focused ultrasound (ExAblate, InSightec) has obtained approval by the FDA for treatment of uterine fibroids and pain palliation of bone metastases when radiation is not an option. In several centers worldwide, HIFU is now being used clinically to treat solid tumors (both malignant and benign) in the prostate, liver, breast, kidney, bone, and pancreas, as well as soft-tissue sarcoma. Current data are very encouraging.

13.3 ADVANTAGES OF HIFU

The 1990s witnessed an explosion in minimally invasive alternatives to open surgery for localized malignancy. Open surgery is associated with significant morbidity and mortality, and it causes suppression of a patient's immune system, which in turn can lead to the risk of perioperative metastatic tumor dissemination, postoperative pain, and lengthy recovery. Laparoscopic surgery, in which small incisions are made in the abdomen and flexible tubes are inserted to deploy surgical instruments, such as scissors, suture holders, hemostats, and high-resolution video cameras for visual guidance, might be more acceptable to patients, but it usually takes longer than open surgery with comparable operative morbidity and mortality. In emergency medicine, conservative nonoperative management of solid-organ injuries in trauma patients who are hemodynamically stable is becoming the standard of care. Other minimally invasive techniques use a range of energy-based methods for *in situ* tumor destruction, including RF ablation (RFA), laser ablation, cryoablation, and HIFU. In the minimally invasive operation, transfusion requirements and their associated infection risks can be reduced significantly.

HIFU technology has the following advantages in the clinical application, but not limited to these (ter Haar 2008, Zhang and Wang 2010). As a completely noninvasive modality, no incision or transfusion is required in tumor ablation. Thus, there is no risk of tumor seeding along a needle track, which has been reported after procedures such as percutaneous ethanol injection and RFA, and there is no risk of hemorrhage from visceral or vascular puncture, which can occur during any of the minimally invasive procedures. Ultrasound has a short wavelength and a relatively high speed of sound in the soft tissue deep within the body, which makes possible the focusing of ultrasound energy in a millimeter-size spot in real time with high spatial and temporal resolution. At appropriate intensities, temperature in the focal zone can be raised up to 65°C–100°C within seconds while maintaining that of the intervening tissue at physiologically safe levels, which leads to the generation

of coagulative necrosis without loss of blood and collateral damage (i.e., hemorrhage). Thermal lesion production is predictable according to the *in situ* intensity and duration of the sonication. If a number of individual lesions are generated in a matrix formation, a contiguous large-volume treatment of tissue with arbitrary size and shape can be achieved according to a predetermined plan. A broad range of tumors can be treated if an acoustic transmission window is available. Pain is minimized due to the minimally invasive or noninvasive procedure. In a clinical trial of 68 liver cancer patients treated at the Royal Marsden Hospital, London, HIFU treatment was well tolerated by fully conscious patients with no local anesthesia or sedation on an outpatient basis (ter Haar 2001b). The cost of the procedure is lower than that of traditional modalities with less complications, such as scars in the patient's skin, contamination and infection after surgery, and reduction in immune response (i.e., hair and weight loss) after radio- and chemotherapy, as well as faster recovery. In radiotherapy, irreversible damage is also produced in some critical normal surrounding tissues. Theoretically, the patients can undergo an unlimited number of sessions. In contrast, neurotoxic effects may limit the dose of a cytotoxic agent in chemotherapy. The systemic maintenance is low. Most importantly, HIFU offers an alternative for patients who do not have any other option available or have a recurrence of cancer post conventional therapy. Palliative HIFU treatment aims either toward symptom control or local tumor control, but it can also be seriously contemplated for patients with poor prognoses. Such noninvasive or minimally invasive therapies are attractive to patients and clinicians alike since they reduce the length of hospital stays, thus saving health-care cost. In summary, compared to other competing noninvasive or minimally invasive modalities such as radiotherapy, HIFU technology is considerably safer and simpler to implement and has become a viable method of thermal ablation (ter Haar 2001a, Zhou 2011). Compared to other minimally invasive therapeutic modalities, HIFU can generate necrosis in any tumor (i.e., size and shape) in a well-controlled and highly precise manner (ter Haar 2001b, Kennedy et al. 2003, Zhou 2011).

13.4 SYSTEM STRUCTURE AND OPERATION

There are three types of HIFU devices employed for clinical usage or research purpose: extracorporeal, transrectal, and interstitial (Figure 13.1). The structure of HIFU systems is dependent on specific features and specifications of the target. The parameters of an HIFU transducer (i.e., frequency, bandwidth, shape, and size) are selected based on optimum performance for penetration depth and resolution for the ablation. Extracorporeal transducers are used for organs that are acoustically accessible through the skin, whereas transrectal devices are used for the treatment of the prostate, and interstitial probes are being developed for the treatment of biliary duct and esophageal tumors. According to the imaging guidance methods, they can also be classified as ultrasound- or MRI-guided ones.

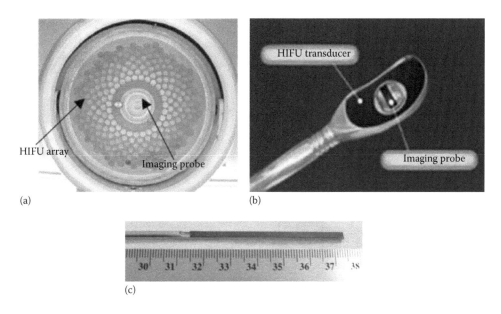

(a)   (b)

(c)

FIGURE 13.1 Photos of (a) an extracorporeal (FEP-BY02, Beijing Yuande Biomedical Engineering), (b) a transrectal (Ablatherm™, EDAP), and (c) interstitial (Guided Therapy Systems) HIFU transducer. (From *Ultrasound Med. Biol.*, 31, Makin, I.R.S., Mast, T.D., Faidi, W., Runk, M.M., Barthe, P.G., and Slayton, M.H., Miniaturized ultrasound arrays for interstitial ablation and imaging, 1541, Copyright 2005, with permission from Elsevier.)

13.4.1 EXTRACORPOREAL DEVICES

Extracorporeal HIFU devices have been used extensively for the treatment of various solid carcinomas in the liver, breast, thyroid, brain, or limbs and also in cosmetic surgery. Extracorporeal treatments are guided using either ultrasound (e.g., FEP-BY02 of Beijing Yuande Biomedical Engineering, China, and Model-JC of Chongqing HAIFU™ Co., China) or MRI (ExAblate 2000 of InSightec, Israel). A diagnostic ultrasound imaging probe is usually located at the center of a HIFU transducer. The HIFU transducer and the diagnostic ultrasound probe are mechanically coupled and mounted on a robotic arm with some freedom of movement (e.g., 5°). The location of the HIFU focus is displayed by superimposing a cross-marker on the B-mode ultrasound image, and the realignment continued to cover the desired tumor volume. Both pre- and post-HIFU treatment ultrasound images are used to monitor the effect of sonication on the targeted tissue. The high frame rate of sonography (usually 30 Hz, depending on the imaging depth and configuration) provides the valuable capability of real-time monitoring. Its Doppler image is also useful in detecting blood flow in the tumor and the formation of occlusion after ablation. However, the image quality is not satisfactory without registration with other diagnosis information (e.g., CT, MRI, PET). The tumor size may thus be underestimated (van Esser et al. 2007). Transcutaneous treatments require that there be an appropriate acoustic window on the entry site to propagate the focused ultrasound beam that is not affected by intervening gas or bone. In order to couple the ultrasound energy to the deep-seated target through the skin surface, a coupling gel, a water balloon, or another suitable liquid path is required. Thus, ultrasound-guided treatments (USgHIFU) can check the acoustic conditions in the HIFU propagation path using the same energy modality (i.e., the presence of air and ribs) and examine the changes in echogenicity in the B-mode image in real time. If the target cannot be well recognized in sonography before or during the HIFU ablation, the HIFU therapy will of course become ineffective in the target region and increase the potential of unintended side effects. Because of the large size of the water balloon, water cooling is not required, but the degassed water is suggested to be changed for every session.

HIFU beam can be generated by focusing the bursts from a flat transducer with an appropriate acoustic lens or a concave self-focusing single transducer, or arranging multiple piston transducers on a spherical bowl. The −6 dB beam size of extracorporeal HIFU system in its focal region is usually 1–3 mm in width and ~10 mm in length, depending on the geometrical size and driving frequency of the transducer. However, the detectable and treatable tumor/cancer in HIFU should be at least 1 cm in size. In order to achieve confluent regions of cell killing, the HIFU focus should be scanned throughout the whole volume. There are two ways of delivering acoustic energy during HIFU ablation. A single exposure is made with the transducer held stationary, and the transducer moved in discrete steps, either mechanically or electronically, to the next location with overlapped lesions. Many single coagulative necrotic dots form a line; several lines form a slice; and a few slices make a block. Such a strategy, instead of increasing the size of necrotic tissue at a single point, is the major difference between HIFU, microwave, and RF technology. With the progress of ablation, the HIFU transducer is moved from the deep target to the superficial tissue. The whole target is divided into slices of 5–10 mm thickness during treatment planning, and then each slice is completely ablated by scanning the HIFU beam point by point or in sweeps in each slice and then slice by slice toward the shallow region.

An alternative strategy is to move the HIFU focus of the active therapy transducer in predetermined trajectories (i.e., linear tracks or spirals) and moving velocities with continuous or high-duty-cycle exposure.

With developments in electrical control and ultrasound transducer fabrication, another focusing and scanning approach has become available, which uses the phased-array technology. Adjusting the amplitude and phase of each element individually allows more rapid electrical steering of the HIFU focus through the tissue and greater flexibility in the focal geometry. Tissue inhomogeneity in abdominal–pelvic or transcranial applications might cause focal beam distortion and might largely decrease the focusing ability in deep-seated tissues. Such a phase aberration can be corrected or compensated in the phased-array system. In addition, the generation of multiple foci is feasible. However, the largest electrical steering range may not be large enough to cover the whole target. So, mechanical scanning is also required. The performance of the phased-array transducer depends on the number of elements, which also increases the complexity and cost of the control system.

The combination of MRI and HIFU seemed highly advantageous in the mid-1990s (Cline et al. 1993, 1994, 1996) and was further developed (Damianou et al. 2004). MRI offers superb soft-tissue contrast, either without or with contrast agents so that excellent anatomical resolution and high sensitivity are provided for tumor detection, thereby allowing accurate planning of the tissue to be ablated as well as the assessment of tissue damage. The known coordinates can thus be used directly for image-guided therapeutic procedures. In order to use in the high magnetic fields of MRI, HIFU transducers must be specially designed for compatibility. The lead–zirconate–titanate (PZT) ceramic material, which was used frequently for most ultrasound transducers, causes magnetic field distortion. Currently, MRI-compatible transducers have already been fabricated and commercialized. In the planning stage, T2-weighted (T2w) images of turbo spin echo and native T1-weighted (T1w) 3D flash images with a high spatial resolution are taken to define the anatomical baseline and target volume. T2w sequences with an interval of 10 min during the ablation are used to monitor the development of edema, while the T1w perfusion is applied postoperatively with use of a paramagnetic contrast agent such as gadolinium. The capability of thermometry in MR guidance allows ablation monitoring, which is the reason for its approval by the FDA. However, MR-guided focused ultrasound (MRgFUS) is expensive, labor intensive, and superior in obese patients (limited to <113 kg for the gantry), but it has low temporal (1–4 s) and spatial (2 mm × 2 mm × 6 mm) resolution in some cases. Because of volumetric acquisitions, excellent contrast, and, above all, temperature imaging, MRI demonstrates major advantages in terms of monitoring HIFU ablation. Compared to sonography, enhanced MRI is more sensitive in the rapid assessment of HIFU-induced coagulative necrosis with no contrast enhancement in the ablation region and a thin peripheral rim surrounding it. Single-photon emission computed tomography (SPECT) with a radioisotope tracer can assess

tissue function of the treated tumor postoperatively. However, both MRI and SPECT are cost-intensive. The weight of the patient is limited to no more than ~115 kg for the MRI gantry. Lying in the closed MRI magnet for 1–2 h may introduce physical and psychological problems, especially in anxious or claustrophobic patients (Vaezy et al. 2001a). However, MRgFUS needs huge investment, which limits its availability to a few specialized centers. In contrast, mobility is one of the major advantages of the ultrasound-guided type used in very diverse settings, including outpatient clinics and field hospitals.

13.4.2 Transrectal Devices

The transrectal types have been developed for the treatment of benign and malignant prostate diseases by inserting per rectum. There are two commercially available devices: the Ablatherm™ (EDAP Technomed, France) and the Sonablate™ (Focus Surgery, USA). The Ablatherm device also employs a confocal design, in which a separate phased-array imaging probe is integrated into the HIFU transducer to provide high-frame-rate, high-resolution ultrasound images for monitoring and guidance of the prostate HIFU treatment. The Sonablate system has both imaging and treatment capabilities on the same ceramic substrate, which eliminates the need for aligning the imaging and HIFU transducers, provides absolute spatial accuracy for treatment at the focal zone, and prevents the HIFU beam from getting distorted as a result of the presence of a separate imaging transducer in the path of the HIFU beam. Both transverse and sagittal ultrasound images are available to monitor tissue changes and characterize the induced lesions quantitatively by acquiring the RF data pre-, intra-, and post-HIFU exposures for each treatment site. The implemented parameters, including backscattering power, tissue attenuation coefficient, and tissue speed of sound, are color coded and superimposed on the prostate tissue to illustrate the treatment progress. If the tissue changes are not sufficient during the first exposure, re-treatment is required. The HIFU system has two transducers with different focal lengths and geometries, the longer focal length for the deeper prostate tissue and the shorter one for the more superficial layers. Usually, degassed water is circulated during the HIFU treatment to couple the acoustic energy and avoid thermal damage to the interface tissue.

13.4.3 Interstitial Devices

HIFU technology also has attracted increasing interest for volume destruction in interstitial use by rotating the plane probe instead of the focused element (Makin et al. 2005). After completing the complete rotation, the probe can be repositioned under fluoroscopic or MRI guidance. It has been tried for biliary and esophageal tumors, and in bloodless partial nephrectomy in percutaneous and laparoscopic format.

The choice of the driving frequency of a therapeutic HIFU transducer involves a compromise between treatment depth and the heating rate. Frequencies near 1 MHz are used for most abdominal ablation, a low value of 0.5 MHz for deep treatments or those with large absorption in the propagation path (e.g., skull bone for transcranial application), and as high as 8 MHz for superficial treatments (e.g., prostatic application) (ter Haar and Coussios 2007). Usually, an extracorporeal device has a wide aperture, a long focal length, and high power output. Wide-aperture sources have the advantage of distributing the incident energy over a large skin area, thus reducing the acoustic intensity at the wave entry site and the consequent possibility of skin burn. Transrectal and interstitial sources operate at lower powers and higher frequencies because of the short distance between the source and the target. The integration of the imaging probe should be done appropriately to minimize interference with the HIFU beam, potential damage to the surface, mostly by the reflected HIFU bursts, and possible distortion in the HIFU beam propagation. The depth and field of view are chosen to visualize the entire tissue in both transverse and sagittal planes for treatment planning and monitoring with a satisfactory f-number. Meanwhile, the frame rate should be sufficiently high so that motion during the HIFU operation can be monitored and the corresponding compensation can be made to minimize such motion artifacts. In order to avoid the interference pattern (i.e., bright lines in sonography), diagnostic ultrasound has to be interleaved with the delivery of HIFU bursts. Backscattered RF data may be provided for a variety of quantitative treatment monitoring and control methodologies using digital signal processing.

In HIFU treatment, the first step is to differentiate and identify the malignant tissue from the normal surrounding tissue, from which 2D or 3D renderings can be sliced and viewed in sagittal, transverse, and coronal planes for an accurate description of the organ and peripheral anatomy. Accurate delineation of tumor margins is a key for successful ablation (Fornage et al. 1987, van der Ploeg et al. 2007). Then, the treatment plan is determined for the optimum HIFU dosage to be delivered and the coordinate of each site according to the exposure strategy, anatomical structure and properties, HIFU output energy, and field characteristics. Important tissues, such as blood vessels and nerves, should be avoided. The patient's position, either prone or supine, is determined by the tumor location and available acoustic window. For example, a prone position is used for breast cancer, as shown in Figure 13.2. As with other noninvasive treatments, control capabilities of system allow safe and effective tissue treatment. The imaging modality would provide quantitative feedback for sufficient but not excess energy dose to attain complete necrosis. After ablation, the generated lesion may be checked again by scanning or using a contrast agent. This working procedure would make a HIFU device efficacious and safe and render it easy to use in the clinical environment. The graphic user interface (GUI) of the HIFU system provides easy operation and the management of patients' database. Socioeconomic factors of the system design are also important, such as the equipment cost, the treatment procedure, and operator training, to determine the wider acceptability and successful usage of HIFU technology.

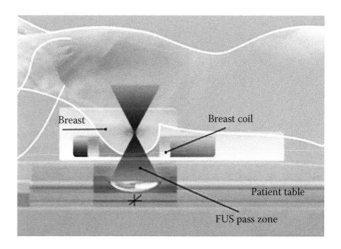

FIGURE 13.2 Schematic diagram of MRI-guided HIFU therapy for a breast cancer patient in prone position.

13.5 MECHANISM OF HIFU

The acoustic intensity of HIFU at the focal point is 1,000–20,000 W/cm² (several orders of magnitude greater than that of a clinical diagnostic ultrasound beam), with peak compressive pressure up to 70 MPa, peak rarefactional pressure up to 20 MPa, and a small focal region in the shape of cigar, which minimizes the potential for thermal damage to the intervening tissue. Absorption of the acoustic energy in the local tissue causes a rapid temperature rise to a cytotoxic level (more than 60°C) within seconds so that the tissue vasculature does not have a significant effect and leads to instantaneous and irreversible cell death via coagulative necrosis with a sharp demarcation (e.g., ~50 μm) between the treated and normal tissues in histological examinations (ter Haar et al. 1991), as shown in Figure 13.3. The temperature increase in HIFU ablation *in vivo* depends on local properties of the tissues that determine the energy absorption and the heat transfer induced by thermal conduction and blood perfusion, which vary significantly between different tissues and even within the same target. Coagulative necrosis caused by heat differs in microscopic scale and host response from the classic coagulative necrosis of ischemic type: heat coagulation favors the interaction of giant cells with chronic inflammation, whereas ischemic-type necrosis causes healing mainly with granulation tissue. However, lethal complications may develop if some vital blood vessels adjacent to the tumors are severely damaged. Large blood vessels are less vulnerable to HIFU damage than solid tumor tissues, which may be due to the blood flow dissipating the thermal energy from the vessel wall. So, HIFU is relatively safe in ablating solid tumors close to major blood vessels where surgical dissection is often contraindicated and hazardous. The blood flow in the vessels can act as a heat sink, reducing the rate of temperature rise. Overall, heating remains the predominant mode of HIFU ablation since the thermal mechanism is better understood and its effect is much easier to control than cavitation.

Thermal damage to tissue at high temperature can be predicted using an Arrhenius analysis or the Sapareto–Dewey iso

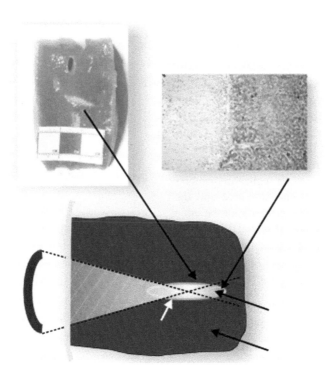

FIGURE 13.3 Schematic diagram of high-intensity focused ultrasound (HIFU) beam passing through overlying skin and other tissues without harming them and being focused to necrose a localized tumor region, which may lie deep within the body. There is a very sharp boundary between dead and live cells at this contour.

effect–thermal dose relationship, which demonstrated that the thermal damage to the tissue is approximately linearly dependent on the exposure time and exponentially on the temperature increase (Dewey 1994). The equivalent duration at 43°C (EM43°C or t_{43}) is calculated as

$$TD_{43°C}(t) = \int_{0}^{t} R^{43-T(t)} dt' \qquad (13.1)$$

where the value of the empirical constant R is given by

$$R = \begin{cases} 0.5 & T(t) > 43°C \\ 0.25 & 37°C > T(t) > 43°C \\ 0 & T(t) < 37°C \end{cases} \qquad (13.2)$$

Thermal doses of 120–240 min at 43°C irreversibly damage and coagulate critical cellular protein, tissue structural components, and the vasculature, leading to immediate tissue destruction, but the threshold varies with the tissue type in hyperthermia (Diederich 2005).

Heat generation as well as dissipation can be spatially heterogeneous and may be affected by temperature increases. Tissue coagulation may significantly modify heat conduction as well as energy absorption. As a consequence, heat losses and energy absorption are difficult to predict prior to the procedure. Therefore, the performance of HIFU therapeutic heating can be improved significantly with real-time feedback from the region of interest.

The second mechanism of lesion production is the mechanical effects associated with acoustic pulses at high intensities, including cavitation, micro-streaming, and radiation force. The mechanically driven bubbles cause damage to their surroundings, especially at resonance. In comparison to the thermal damage, which is relatively uniform and predictable in its spatial extent, cell killing due to acoustic cavitation, however, occurs only in close proximity to an oscillating bubble and is therefore more random in nature. Cavitation-damaged tissue contains holes or tears, but viable cells are still present. Therefore, cavitation itself may not be effective in destroying all cells. Gas cavity works as an effective enhancer of heat deposition, but higher concentration will lead to the change of lesion from a cigar shape to the shape of a tadpole with the head moving toward the source, which makes the control of ablation difficult and unpredictable despite enhanced therapy efficiency. If the temperature is close to 100°C, boiling may occur in the tissue. A vaporized cavity with complete tissue lysis may be produced by the motion of a large boiling bubble but without protein denaturation (Khokhlova and Hwang 2011). Cavitation is relatively unpredictable due to its randomness and can depend on many independent variables, such as pulse shape and length, frequency, acoustic intensity, and the presence of particles and voids in the medium. Specifically, inducing cavitation during exposures can improve the rate of heating through enhanced absorption. There is increasing interest, both in research and clinical applications, in using cavitation to improve HIFU treatment. Although the majority of the initial cell death is due to necrosis from thermal injury, HIFU can also induce apoptosis, where the cell

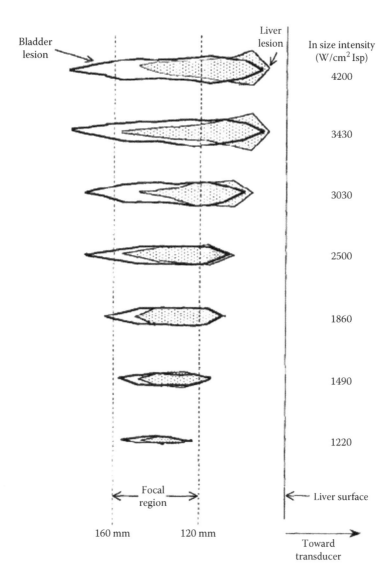

FIGURE 13.4 Comparison of the lesion size and location in the bovine liver (filled region) and the bladder (solid line) after a 2 s HIFU exposure at the frequency of 1.69 MHz and variable intensities of 1220–4200 W/cm². (From *Ultrasound Med. Biol.*, 22, Watkin, N.A., ter Haar, G.R., and Rivens, I., The intensity dependence of the site of maximal energy deposition in focused ultrasound surgery, 488, Copyright 1996, with permission from Elsevier.)

nucleus destroys itself with rapid DNA degradation by endonucleases (Sofuni et al. 2011). The velocity gradients associated with acoustic streaming may be quite high, and high shear stresses can cause transient damage to cell membranes, and viscous friction of the different layers of the fluid then results in temperature rise.

The outcomes of HIFU are a synergy of thermal effects, mechanical effects, and biological effects (Zhou 2011). Furthermore, it would be impossible to identify distinctly the thermal effect from the mechanical effect during HIFU ablation, since, in practice, the two effects occur simultaneously within a tissue (Figure 13.4). Therefore, the coagulative necrosis induced by HIFU can be considered as the result of biological effects from a combination of mechanical stresses and thermal damage to the tissue, though the latter is generally considered as the predominant mechanism of tissue ablation with HIFU. Granulation tissue, immature fibroblasts, inflammatory

cells, and new capillaries are found in the margin of the produced lesion. Small vessels (<2 mm in diameter) in the tumor, including branches of arteries and veins, are heavily destroyed, which is confirmed by the disappearance of endothelial cell nuclei, nondistinction of cellular margins, and disruption of junctions between individual cells. Scattered intravascular thrombi are often found in the destroyed vessels. As a result, there is reduced or no blood circulation in the HIFU-ablated tumor but a thin peripheral rim of contrast enhancement around the coagulative necrosis. Thermolysis in the capillary is not as effective as in larger vessels (Li et al. 2012).

Since HIFU is essentially an acoustic approach, any acoustic artifacts would also apply to HIFU, such as acoustic shadowing, reverberation, and refraction in sonography. Hence, the bones and lungs oppose the penetration of ultrasound, and some areas of the liver parenchyma adjacent to a rib may be difficult to reach with the focused beam. Sound waves are reflected from the gas in the bowel back toward the transducer, which may produce burns in the intervening tissue. Even a small amount of gas in the gastrointestinal tract can cause burns in the bowel wall anterior to the gas and in the abdominal wall musculature overlying the gas. Refraction artifacts can result in superficial energy deposition in the soft tissues adjacent to the target area.

13.6 HIFU MONITORING AND CONTROL

Ideally, ablation is preceded by precise 3D target definition and identification of nearby tissue that must be spared. The treatment should be performed while continuously monitoring the temperature of the targeted tissue as well as the adjacent tissue outside this region. A quantitatively accurate method of HIFU treatment monitoring and control in real time is an essential part of any image-guided HIFU treatment strategy. The goal is to monitor and control the temperature rise and thermal dose during treatment. Thermocouple-based and x-ray-based temperature measurements are not used in clinics because of their shortcomings (invasiveness and radiation, respectively). HIFU with MRI and ultrasound guidance have their advantages and disadvantages.

MR thermometry enables 3D thermal mapping and calculation of the thermal dose to achieve cytotoxic levels using specific pulse sequences and data processing and superposition on the anatomical image to allow real-time HIFU monitoring based on the relaxation time T_1, the diffusion coefficient (D), the water molecular apparent diffusion coefficient (ADC), or proton resonance frequency (PRF) of tissue water (Bohris et al. 1999). The use of temperature-sensitive contrast agents can also provide absolute temperature measurements. PRF is preferred at medium to high magnetic field strengths ($\geq$1 T) because of the excellent linearity and near-independence with respect to tissue type, together with good temperature sensitivity. In addition, motion artifacts can severely degrade the accuracy of MR temperature maps and must be carefully corrected. The PRF method typically employs rapid RF-spoiled gradient echo imaging. Gray levels on anatomical images are proportional to the MR signal magnitude, whereas its phase relates to the PRF and temperature variation:

$$\Delta\varphi = \gamma\alpha B_0 \Delta T T_E \qquad (13.3)$$

where
  $\Delta\phi$ is the phase difference
  ΔT is the temperature variation
  γ is the gyromagnetic ratio
  α is the temperature coefficient (~0.01 ppm/K)
  T_E is the echo time
  B_0 is the magnetic field

Fat suppression is necessary since their resonance frequencies do not depend on temperature. Therefore, in practice, lipid contribution to the MR signal can be conveniently suppressed in gradient echo imaging by frequency-selective slice excitation, selective fat saturation, or inversion recovery. The changes in PRF using a fast spoiled gradient-recalled echo sequence (SPGR) (Chung et al. 1996) is temperature dependent and found to be more reliable than T1w imaging (Hynynen et al. 1996). This approach at 1.5 T with segmented gradient-echo echo-planar imaging (GRE-EPI) sequences could minimize intra-scan motion effects and has been evaluated during liver tumor RFA with an accuracy of the measured temperature of 1.3°C ± 0.4°C at a frame rate of 0.6 s/image (Cernicanu et al. 2008). In addition, T1w or T2w fast spin-echo (FSE) has proven successful to image thermal lesions created by HIFU in rabbit liver *in vivo*, but the T1w method has higher contrast-to-noise ratio (CNR) with a repetition time (T_R) of 400–900 ms (Mylonas et al. 2010). Therefore, MRgFUS has the capability of closed-loop control on HIFU therapy, which is superior to sonography in obese patients (Yagel 2004). The accuracy of MR thermometry depends on several factors such as the signal-to-noise ratio (SNR), echo time (T_E), field strength, and artifacts. For optimal accuracy, T_E must be similar to T_2^* (a measure of the magnetic field homogeneity: for excellent field homogeneity, $T_2^* = T_2$). In comparison to visualization of anatomical structures, thermography requires rather low SNR values. Acceleration of image acquisition in order to increase temporal resolution usually implies a reduction in SNR. A robust unwrapping algorithm is required for accurate temperature computation individually for each voxel in order to correct for possible temporal discontinuities. As changes in phase variations between successive images are much smaller, temperature aliasing correction is performed on successive acquisitions. However, the temporal and spatial averaging effect leads to the underestimation of the temperature. The temperature measured in a single MRI voxel by water PRF shift attained a maximum value of only 73°C after 7 s of continuous HIFU exposure at the occurrence of boiling in a 4.7 T magnet, which agreed with the average temperature field in a theoretical simulation over the volume of the MRI voxel $0.3 \times 0.5 \times 2$ mm^3 (Khokhlova et al. 2009). In the case of HIFU ablation, the resolution should be better than the dimensions of the focal point. The costs involved in MRI-controlled treatments are relatively high, and its complexity makes portability impossible. In addition, MRI is extremely sensitive to RF interferences, magnetic susceptibility variations, motion artifacts, and displacement registration between scans. Magnetic susceptibility is spatially inhomogeneous (in particular near the intestines, stomach, esophagus, lungs, and heart) so that the temperature-dependent changes in magnetic susceptibility may not be negligible. The local magnetic susceptibility distribution changes during the motion. Therefore, organ displacement and deformation must be compensated in order to improve the precision, enable thermal dose computation (the history of the temperature corresponding to the same spatial localization), and preserve the validity of regions of interest at the beginning of the ablation intervention, but challenging because the phase variation is due to both temperature and movements, the contributions of which are hard to separate. Motion restraining devices can eliminate or reduce motion-related artifacts. For periodic motion artifacts, synchronization by triggering the MR pulse sequence based on electric signals provided by a pressure sensor ("respiratory gating") or cardiac electrode ("cardiac gating") is also used. Three-dimensional visualization of temperature evolution based on multi-slice images must be developed in order to accurately localize and monitor targeted regions designated for ablation and to avoid destruction of healthy neighboring tissue. Automatic and semiautomatic therapy controls require continuous assessment of the reliability of the available temperature maps. The regulation of temperature evolution at the focal point is based on temperature mapping and a physical model of local energy deposition and heat conduction, such as proportional–integral–derivative type of temperature control. Thus, temperature evolution at the focal point can be regulated automatically with a precision that is close to that of the temperature measurements. Furthermore, MRI may also be used to assess the efficacy of the therapy as well as the associated complications. Hemorrhage and edema can be confirmed by T_2^* and T_2 changes, respectively. However, many apoptotic and necrotic processes resulting from the ablation procedure may take up to 24 h to become clearly detectable (Figure 13.5).

In comparison, ultrasound imaging is more convenient and compatible, although it may be less than optimal (Qian et al. 2006). Ultrasound-based temperature imaging involves tracking the echo shifts in the backscattered ultrasound signals from the target undergoing ablation, which are due to the combination of the local temperature dependence of sound speed and thermal expansion in the heated region (Simon et al. 1998). These echo shifts estimated from the correlation of successive backscattered ultrasound frames are linearly proportional to the temperature rise between 37°C and 50°C in non-fatty tissue for both 2D and 3D assessments with the accuracy of 0.5°C and a spatial resolution of 2 mm

(Simon et al. 1998, Annand and Kaczkowski 2004, Anand et al. 2007, Daniels et al. 2007). A high-resolution spectral estimation method for tracking frequency shifts at two or more harmonic frequencies associated with temperature change was also available (Amini et al. 2005). However, during HIFU ablation with much higher temperature increases over shorter exposure times than conventional hyperthermia, a nonlinear dependence of sound speed and tissue expansion exists with temperature, especially after the irreversible changes in tissue. However, the accuracy and reliability of this technique become worse for temperature measurements in the focal region of HIFU systems, and an empirical relationship is used for clinical use (Qian et al. 2006).

Pulse-echo ultrasound imaging is based on the backscatter (or echo signals) from the tissue. The strength of the ultrasound backscattered signals depends on a range of acoustic and mechanical properties of the tissue. However, conventional B-mode imaging was found to be ineffective in discerning normal tissue and the thermally ablated region. The presence of hyperechoic (bright) regions in the focal region and its vicinity can persist after sonication in various HIFU experiments, which is due to acoustically generated (cavitation) or thermally generated (vapor) bubbles (Chan et al. 2002). Appearance of a hypoechoic area is found be to associated with the development of coagulation necrosis in the liver (Bush et al. 1993, Yang et al. 1993). It is to be noted that heating can lower the threshold of acoustical cavitation (Lafon et al. 2007). RF echo signals acquired from a region of interest were processed and analyzed using time domain, frequency domain, and time–frequency domain methods to estimate tissue properties, such as the coefficient of attenuation, parameter of nonlinearity (*B/A*), backscattering power,

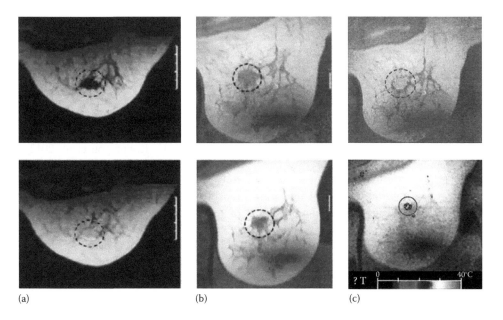

(a)   (b)   (c)

FIGURE 13.5 T1-weighted (T1w) MRI mammography of a breast cancer: (a) 2 days before HIFU ablation without (top) and after (bottom) i.v. contrast, showing strong tumor; (b) before treatment without i.v. contrast (top) and immediately after HIFU ablation with i.v. contrast (bottom), exhibiting a complete lack of contrast enhancement after HIFU ablation; and (c) T2w MRI treatment planning (top) and MRI thermometry during sonication (bottom). (From Zhou, Y., *J. Med. Imaging Health Inform.*, 3, 141, 2013.)

and speed of sound, both *ex vivo* and *in vivo*, toward HIFU monitoring. Among them, quantitative mapping of tissue attenuation is more exhaustive. In both *ex vivo* (e.g., liver, kidney, and muscle) and *in vivo* (e.g., prostate, liver, spleen, and abdominal wall) tissues, HIFU ablation caused approximately double the increase of the attenuation (Ribault et al. 1998, Zderic et al. 2004). As a result, the echogenicity decreased in the temperature range of 37°C–60°C (Gertner et al. 1998). However, the echogenicity did not drop in the prostate even with the increase of attenuation (Worthington et al. 2002). The acoustic nonlinearity (*B/A*) in tissue was quantified using both pulse echo and tomographic techniques since the thermal lesion was assumed to have a significantly different value (Law et al. 1985). However, not much study was reported for its application in the HIFU ablation. Measurement of tissue backscattering power has been investigated for real-time HIFU treatment monitoring and control. Changes in backscattering energy (CBE) was nearly monotonic after motion compensation, as proved both theoretically and experimentally (Arthur et al. 2003, 2005). Because of its relative simplicity and robustness, this technology has been successfully implemented in a commercial HIFU device (Sonablate 500, Focus Surgery, Inc.) with quantified backscattering power colorfully overlying on real-time sonography in order to provide robust real-time HIFU treatment monitoring and control and subsequently enhance the safety and efficacy of operation. However, because of the dependence on tissue inhomogeneity, calibration is challenging.

B-mode ultrasound imaging has a limitation in the measurements of the elastic moduli of soft tissue. Elastography was applied recently to detect the induced tissue strain by an externally applied controlled force that depends on the stiffness (characterized by its elastic or Young's modulus) of soft tissue (e.g., prostate and intravascular plaques) and the applied stress (force per unit area) (Ophir et al. 2002). The tissue displacement is measured by tracking the speckle in the ultrasound RF echo data, which is aligned parallel to the direction of the compression force using a cross-correlation technique, and images of tissue strain can be computed using a least-squares strain estimator by fitting a straight line to a number of displacement values or pixels. Because HIFU-induced thermal lesions are typically stiffer than normal tissue, several studies have been carried out for their visualization. Meanwhile, tissue disruption and inflammatory response could also be discerned from the elastogram. Despite encouraging *in vitro* and *ex vivo* results, there are still technical challenges, such as a feasible and accurate tissue compression mechanism and compensation for unwanted tissue motion due to breathing and cardiac cycle, to be overcome before *in vivo* applications (Righetti et al. 1999, Souchon et al. 2003, Thittai et al. 2011, Zhang et al. 2011). In comparison to gadolinium-enhanced T1w and T2w MRI, elastography generally underestimated the lesion volume with a statistically significant correlation of $\rho = 0.62$ ($p = 0.022$) (Curiel et al. 2005). The use of real-time sonoelastography to detect and estimate the volume of thermal lesions by HIFU was evaluated in porcine livers *in vivo*. The sonoelastographic measurements and

pathology showed good correlation in the area of the lesions ($r^2 = 0.9543$) as well as good gross volume (3.6% underestimate) (Zhang et al. 2008). In contrast, elastography-based thermography showed high ambiguity and low sensitivity in the coagulation temperatures.

Radiation force can also be produced remotely. Short (i.e., a few milliseconds) bursts of intensive ultrasound push small regions of tissue. Vibroacoustography is based on ultrasound-stimulated acoustic emission and a cyclical stress from the radiation force of a focused beam by modulating the acoustic power at a frequency of a few kilohertz (Fatemi and Greenleaf 1998). The tissue's mechanical response is then measured using a low-frequency hydrophone. Subsequently, variations in tissue stiffness could be visualized. This leads to the estimation of the beam's focal region in the tissue. It is also used to map HIFU-induced thermal lesions, which alter the emissions due to alternations in tissue absorption and stiffness (Alizad et al. 2004). Onset of coagulation necrosis is represented by a frequency shift to the higher side, and the re-emitted signal is sensitive to temperature (Konofagou et al. 2003). The therapeutic ultrasound transducer not only generates lesion but also induces tissue motion that is detected by a confocal diagnostic ultrasound transducer to characterize this motion and estimate tissue stiffness (Lizzi et al. 2003). This local and noninvasive strain generation approach does not need compression of large volumes of tissue as in conventional elastography and can avoid some artifacts. The contrast and SNRs in elastograms depend partially on the driving frequency, bandwidth of the imaging probe, and the axial resolution. Acoustic radiation force impulse (ARFI) imaging could visualize the size and boundaries of thermally and chemically induced lesions in soft tissues (Fahey et al. 2004). Its potential for monitoring and guiding HIFU treatment with acceptable contrast resolution and accuracy will be evaluated extensively (Bing et al. 2011). Harmonic motion imaging (HMI) is also a kind of radiation-force imaging approach, which induces a harmonic motion in the HIFU's focal zone by an oscillatory, internally applied radiation force to detect changes in localized stiffness. The oscillatory motion is estimated using cross-correlation of RF ultrasonic signals of the same location undergoing vibration. It has been recently used for imaging HIFU lesion, both *in vitro* and *in vivo* (Maleke et al. 2006, Hou et al. 2011, Konofagou et al. 2012). In addition, phase shift of tissue boiling in the HIFU ablation, a reverse lesion-to-background displacement contrast, could also be detected reliably in *ex vivo* studies (Hou et al. 2014). HMI was used to detect changes in stiffness in silicon phantoms and in an *in vivo* VX2 tumor implanted on the thighs of New Zealand rabbits. Inclusions as small as 4 mm and tumor as small as 10 mm in length and 4 mm in width were discerned from the surroundings (Curiel and Hynynen 2011).

A novel type of elastography, termed supersonic shear imaging or shear wave elastography (SWE), was developed recently to overcome the limitations associated with conventional ultrasound elastography. Localized radiation forces were produced remotely by focused ultrasound exposures and generated localized vibration for a transient shear wave

(Bercoff et al. 2004). Serial force generation at increasing depths was employed to generate a quasi-plane shear wave front that propagated throughout the whole region of interest. After generating this shear wave, ultrafast beamforming, at a very high frame rate (>2,000 and up to 17,000 frames/s) using plane-wave transmissions rather than line-per-line focused beam transmissions over a large field of view in graphical processing unit (GPU)–based platforms (Tanter and Fink 2014), was used to acquire successive RF data to calculate the shear wave speed quantitatively, which is associated with the shear modulus. *In vivo* studies have shown promising results in discerning malignant and benign abnormalities in the breast and liver (Tanter et al. 2008, Muller et al. 2009, Athanasiou et al. 2010). Stiffness of the right leg muscle of anesthetized rats was found to increase strongly with temperature between 38°C and 48.5°C as well as with the thermal dose during heating experiments in a waterbath. A thermal dose threshold was found at 202 min for an eightfold stiffness increase as a good indicator of thermal necrosis (Brosses et al. 2011). Changes in *ex vivo* bovine liver stiffness were observed only after 45°C. In contrast, between 25°C and 65°C, muscle stiffness varied, whose degree was associated with the thermal dose (Brosses et al. 2010). HIFU sonications were interleaved with fast imaging acquisitions, and elasticity and strain mapping was achieved every 3 s *in vivo* on sheep muscle. Tissue stiffness was found to increase up to four folds in the thermal lesion, and strain imaging elastograms showed strong shrinkages that blurred the temperature information and made the interpretation for accurate lesion characterization not easy. However, SWE provides a quantitative and reliable mapping of the thermal lesion, is robust to motion, and allows mapping temperature with the same method to predict the lesion growth (Bastien et al. 2011). Shear wave dispersion ultrasound vibrometry (SDUV) quantifies the shear elasticity and viscosity by evaluating the dispersion of shear wave propagation speed. SDUV measurements of *in vitro* prostate elasticity and viscosity are generally in agreement with the previously reported values (Mitri et al. 2011). Its application to monitor HIFU lesion needs more investigation.

When short-pulsed optical energy is deposited into a tissue, photoacoustic waves are produced as a result of the pressure rise from transient thermoelastic expansions, which can then be detected with a diagnostic imaging transducer in order to discover local tissue abnormalities (Wang 2008). The amplitude of photoacoustic (PA) signals is proportional to the absorbed energy for short pulses. Photoacoustic imaging (PAI) was used to monitor HIFU ablation in real time for accurate evaluation of thermal ablation outcomes through thermal dose calculations in the bovine kidney. The PA amplitude increases and tends to saturate after the coagulation, whose threshold is 240 TD_{43} minutes, and there is a linear relation between the PA signal amplitude changes and temperature (Cui and Yang 2011). In addition, PA waveforms obtained at different optical wavelengths and convolved with a continuous signal can be used in a time-reversing approach to precisely focus ultrasound energy on nonechogenic or moving targets (e.g., due to respiration) irrespective of the presence of aberrating layers

(e.g., an irregular layer of either intracoastal or intracranial fat tissue) (Funke et al. 2009). The target could be detected by PAI, and then a HIFU beam was automatically refocused with a time-reversal mirror to necrose the absorber (Prost et al. 2012). *In vivo* study needs higher contrast for guidance. Contrast PAI with intravascular injection of gold nanorods allowed the identification of a CT26 tumor subcutaneously inoculated on the hip of a BALB/c mouse and HIFU ablation under such guidance (Cui and Yang 2010).

The thermal properties of living tissue, such as thermal conductivity, capacity, and diffusivity, can be reconstructed noninvasively by solving the bioheat transfer equation and using quantitative spatial and temporal temperature distribution that is obtained by ultrasonic imaging or MRI (Sumi and Yanagimura 2007). However, additional work is required to translate from simulations and gel phantom experiments—to *in vivo* applications. Measurements of the time-varying acoustic strain and spatiotemporal variations in the beamformed RF echo shifts induced by the temperature-related sound speed changes are solved by the heat transfer equation to estimate the thermal diffusivity of the tissue in the heated zone. Larger temperature increases (i.e., >30°C) and *in vivo* testing are required to evaluate its performance and potential for clinical use (Anand et al. 2007, Anand and Kaczkowski 2008).

In summary, the detection of lesion formation and thermography are important features for ensuring the safety, efficacy, and efficiency in HIFU application. Current methods of MRI and sonography have both advantages and limitations. In order to meet the requirements for satisfactory clinical trials, the proposed modality should be invulnerable to motion artifacts, tissue deformation, and cavitation noise for accurate, reliable, and quantitative assessment. Spatial and temporal resolution and contrast, temperature range, as well as cost, operation ease, and maintenance should also be considered (Chen et al. 2010a). Therefore, technical improvement and development in the HIFU ablation are still being continuously carried out by various groups and companies.

13.7 CLINICAL APPLICATIONS

Cancer patients are usually classified using numerical subsets of the TNM components (T: the extent of the primary tumor, N: the absence or presence and extent of regional lymph node metastasis, and M: the absence or presence of distant metastasis) using the guideline of the Union International Contra Cancer 2002 (Sobin and Wittekind 2002) to reveal the clinical stage of the malignant disease, which was initially developed to stage bone sarcoma (Enneking et al. 1980). HIFU ablation for patients with early-stage cancer is curative, and a normal tissue margin is set to be about 1.5–2.0 cm. In contrast, it is palliative for those with advanced cancer, impeding tumor growth and improving the quality of life. The patient is kept under either epidural anesthesia, on 5 mg of diazepam as an oral sedative, or under general anesthesia. Altogether, HIFU can destroy tissue, kill tumor cells, restrain the malignant proliferation, reduce pain, and prevent metastasis.

13.7.1 Prostate Cancer

Transrectal HIFU treatment of prostate is one of the pilot investigations. Initial clinical trials for benign prostate hyperplasia (BPH) treatment were encouraging (Gelet et al. 1993, Sullivan et al. 1997), with increase in flow rate and decrease in post-void residual volume. However, the long-term results were disappointing (Madersbacher et al. 2000), with 43.8% of patients requiring a salvage transurethral resection of the prostate (TURP) within 4 years. Therefore, HIFU cannot replace the gold standard, TURP (Uchida et al. 1998).

Prostate cancer is the most common cancer in men, as seen by systemic prostate biopsy under ultrasound guidance, and 70% of diagnosed cases are still organ confined (Azzouz and De la Rosette 2006). Its estimated incidence was 110.5 per 100,000, and the mortality rate was 21.1 per 100,000, and it constituted 27.1% of all the reported cancer cases in men in the European Union in 2008, which may be due to an increasing life expectancy in the male population (Yancik 2005) and the use of serum prostate-specific antigen (PSA) assay for screening (Jemal et al. 2005). In contrast, treatment of prostate cancer presents different problems from those associated with BPH treatment (Madersbacher et al. 1995). Most patients with low-risk, intermediate-risk, and high-risk localized prostate cancers undergo encompass active surveillance, interstitial prostate brachytherapy, external-beam radiation therapy (EBRT) with hormonal control, and radical prostatectomy (RP) (Thompson et al. 2007, Heidenreich et al. 2010), all of which have seen significant technical developments during the last 10 years (Quinn and Babb 2002). The choice of an adequate therapy option depends on the tumor stage, the PSA value, Gleason score, patient's age, concomitant diseases, life expectancy, and patient preference. Some cannot tolerate EBRT because of existing comorbidities (Aus 2006). Currently, patients are more concerned with the post-treatment quality of life, such as the recovery time and the acute and chronic morbidity. Although RP was found to be superior for organ-confined prostate cancer in a prospective randomized trial due to a lower risk of cancer recurrence, cancer-related death, and improved survival (Iselin et al. 1999, Han et al. 2001), it is still associated with considerable operation-related morbidities (Rukstalis 2002, Augustin et al. 2003). The overall local relapse and disease progression (DP) after EBRT is about 30% 2 years after EBRT (Zelefsky et al. 2001, Pollack et al. 2002) and up to 60% within 10 years (Kuban et al. 2003). For patients with a rising PSA and metastasis diagnosed by CT, MRI with lymphotropic superparamagnetic nanoparticles (Harisinghani et al. 2003, Ghanem et al. 2005), or fluorocholine-18 PET/CT, this rate may increase to 60%–72% (Zagars et al. 1995, Zelefsky et al. 1998). Most of them will receive palliative androgen deprivation therapy (ADT) (Coen et al. 2002). For patients with an isolated local recurrence and a life expectancy over 5 years, a curative local therapy is still possible (Bianco et al. 2005, Ward et al. 2005). Untreated local recurrence will result in distant metastasis with a median survival of only 33 months (Lee et al. 1997).

Ablatherm (EDAP, Lyon, France) prototypes for use in clinical trials were available in 1993, the first success was achieved in 1995 as a single treatment module (Madersbacher et al. 1995), and the initial series were reported in 1996 (Gelet et al. 1996). Ablatherm Maxis™ (a 7.5 MHz ultrasound imaging probe and a 3 MHz HIFU transducer) was the first-generation HIFU device to be validated by the European Commission (EC) and commercialized in 2000 (Poissonnier et al. 2006). The most successful HIFU treatments have been those that have ablated the whole gland (Gelet et al. 1993, Sullivan et al. 1997). HIFU has a short learning curve (approximately 10 cases for a new operator/user with experience of ultrasound prostate imaging) with an established procedure. With experience, control rates for the treated tumor have risen from 50% at 8 months in the early days to 90% or more recently (Gelet et al. 1996, Chaussy and Thüroff 2003). Whole-gland versus focal treatment resulted in a reduced incidence of recurrent tumor. The HIFU-related morbidity is low, and the post-treatment management is simple. However, there is controversy about the use of HIFU for localized prostate cancer. The medical associations of France (Rebillard et al. 2008), Italy, and the United Kingdom (British Uro-Oncology Group [BUG] and British Association of Urological Surgeons [BAUS] Section of Oncology 2013) have approved HIFU for the primary and/or salvage treatment of prostate cancer, but the American Urology Association (Thompson et al. 2007), the National Comprehensive Cancer Network, the National Collaborating Centre for Cancer, the National Institute for Health and Clinical Excellence in the United Kingdom (National Institute for Health and Care Excellence 2014), and the German Association of Urology do not recommend it because of paucity of evidence concerning improved quality of life and long-term survival, lack of long-term follow-up data, and missing comparisons of HIFU with conventional therapy options. Since 2010, the European Association of Urology has recommended HIFU as an alternative option (Heidenreich et al. 2008).

Patients with a TURP history or a local recurrence after EBRT or surgery are suitable candidates for HIFU. Included patients for HIFU ablation are usually 60–75 years old with localized prostate cancer ($T_{1-2}N_{0-x}M_0$), a Gleason score of ≤ 7, a prostate-specific antigen (PSA) level of ≤ 15 ng/mL, and a prostate volume of ≤ 40 mL, who are not candidates for surgery or obese (Thüroff et al. 2003, Blana et al. 2004, Vallancien et al. 2004, Rebillard et al. 2005). Careful patient selection also depends on prognostic factors, such as pre-EBRT risk group, ADT use, and positive biopsy percentage. The disease-related risk level is classified as low risk = stage T_{1-2a} or PSA ≤ 10 ng/mL, and Gleason score ≤ 6; intermediate risk = T_{2b} or $10 <$ PSA ≤ 20 ng/mL or Gleason score = 7; and high risk = T_{2c} or PSA > 20 ng/mL or Gleason score ≥ 8. The initial HIFU limitation of prostate volume of no larger than 40 mL could be solved by debulking of the prostate by radiation or ADT for 3 months (reduction of prostate gland volume by up to 30%) (Ash et al. 2005). Exclusion criteria included a maximum anterior–posterior height greater than 40 mm, any focus of calcification, significant anorectal disease preventing probe insertion (i.e., previous hemorrhoidectomy), or carrying increased risk of fistulae (i.e., Crohn's disease). Patients received between one and three HIFU sessions (1 session: 5796%), and about 500 HIFU shots per session with a duration of about 1.5–2 h under general anesthesia if spinal anesthesia was not technically feasible or refused by the patient. Since the rectal wall is sensitive to temperature, transrectal HIFU system actively cools the rectal wall during treatment, continuously monitors the temperature of the rectal wall, and constantly measures the distance between the rectal wall and the prostate. A TURP performed just before a HIFU (now about 90%) seemed to reduce prolonged urinary retention (Chaussy and Thüroff 2003, Vallancien et al. 2004, Lee et al. 2006), and there was no significant difference in negative prostate biopsy rate between groups receiving HIFU alone or a combination of TURP and HIFU (Chaussy and Thüroff 2003). A bladder neck incision (BNI) was performed prior to HIFU to reduce the postoperative catheterization period and avoid bladder outlet obstruction. The evening after the HIFU session, the patient is returned to normal dietary intake, does not need any analgesic medication, and may be discharged the day after with a three-way catheter placed in the urethra for 3–5 days. The suprapubic catheter can be removed if the post-void residual urine is less than 50 mL. In the case of local recurrence after primary HIFU, the patient may still receive EBRT. HIFU results in short-term cancer control, as shown by a high percentage of negative biopsies, substantially decreased PSA levels, and promising medium-term biochemical disease-free survival (BDFS).

Extensive clinical trials have been carried out (Beerlage et al. 1999, Chaussy and Thüroff 2000, 2003, Gelet et al. 2000, Thüroff et al. 2003, Vallancien et al. 2004, Uchida et al. 2005, 2006, Aus 2006, Lee et al. 2006, Poissonnier et al. 2006). Although their follow-up and evaluation criteria are different, the performance of HIFU for prostate cancer can also be evaluated. Percentage of PSA nadir of ≤ 0.5 ng/mL was reported to be 55%–84%, 42%–84%, and 78% within 3 months, while it was 87% in the case of <1 ng/mL with a median follow-up of 12–24 months. The negative biopsy rate was 64%–93%, 80%–90%, and 64%–87% within 3 months, 51%–96% and 82%–94% within 1–2 years, and 87.2% within 4 years. The corresponding value is found to decrease with rising PSA nadir, with 89%, 54%, and 52% at a nadir of < 0.2, 0.21–1.00, and >1.00 ng/mL, respectively (Uchida et al. 2006a). BDFS was 71.5% and 54% with a PSA threshold of 0.4 and 0.2 ng/mL, respectively, at 5 years (Blana et al. 2004), 78% at 5 years, 59% at 6 years, and 98% at 8 years with an 8-year overall and cancer-specific survival rate of 83% and 98%, respectively. BDFS was better in patients with lower baseline PSA levels or lower PSA nadir after HIFU. The actuarial overall survival rate was 84%, 66%, 60%–70%, and 65%–66% at 5 years and 59% and 61% at 7 years. The success rates of HIFU were 85%–92.1%, 77%–86.4%, and 47%–82.1% in low-, intermediate- and high-risk groups, respectively.

Positive biopsy rates after EBRT were 25%–32%, and most of these patients were conveniently treated expectantly or with hormonal deprivation (Borghede et al. 1997, Crook

et al. 2000, Pollack et al. 2002). Salvage HIFU is used for local relapse and no metastasis after EBRT with acceptable morbidity in more intermediate-risk than low-risk patients (23.5% vs. 6.9%) as a curative approach (Murat et al. 2009). The prostate volume was 13 ± 9 mL at 3-month follow-up with a negative biopsy of 73% and the PSA nadir of 2.38 ± 6.22 ng/mL. At 3 years, the PFSR was 53% (35%–82%), 42% (28%–61%), and 25% (16%–38%) for low-, intermediate-, and high-risk patients, respectively, and the 5-year estimated overall survival was 84% (74%–94%). The poor results in the high-risk patients may be due to the unappreciated metastasis in the diagnosis such as transrectal color Doppler (Rouviere et al. 2006). Complications after salvage HIFU are much higher than after primary HIFU (Murat et al. 2009).

HIFU is well tolerated, with a low rate of complications. The common complications after HIFU ablation included stress urinary incontinence (5%–35%), which resolved spontaneously or after management with pelvic floor muscle exercise; epididymitis (7.6%); urinary impotence (55%–70%); infravesical obstruction; bladder outlet obstruction (BOO: 14.5%–30%); prolonged urinary obstruction/retention due to the scarring and delayed passage of necrotic debris (1%–14%), which required either a prolonged catheterization or, exceptionally, an endoscopic extraction of the necrotic debris (Blana et al. 2006); urethral/bladder neck stenosis or strictures (4%–30%); urinary infections (3.5%–23.8%), which required adapted antibiotics; erectile dysfunction (ED: 20%–77%); and retrograde ejaculation (1%–20%) (Blana et al. 2006). Repeated HIFU treatment is associated with much higher complication rates than a single session (Blana et al. 2004, Chaussy et al. 2005). Adverse events concerned the urinary tract (1%–58%), potency (1%–77%), the rectum (0%–15%), and pain (1%–6%) (Warmuth et al. 2010). The risk of ED might decrease in a nerve-sparing procedure, but this might increase the risk of cancer recurrence (Chaussy and Thüroff 2000, 2001, Poissonnier et al. 2006), which is not worse than that of RP (26%–100%), EBRT (8%–85%), and interstitial radiation (14%–61%) (Burnett et al. 2007). Two-thirds of the patients might expect to have sufficient erections 1 year following treatment. Urethral stricture was significantly lower in those with suprapubic catheter compared with urethral catheters (19.4% vs. 40.4%). The incidence of artificial urinary sphincter implantation was significantly reduced with the specific post-radiation parameters (20% vs. 6%). The most serious adverse effect found and reported was urethrorectal fistula (0%–0.7%, may be due to the rectal cooling system and thick rectal wall), which resolved after urinary catheter placement, fibrin glue injection, or surgery.

The natural history of prostate cancer does not allow the use of mortality as an outcome measure in most short- to medium-term therapy, so surrogates in the form of biochemical failure have emerged. The American Society for Therapeutic Radiology and Oncology (ASTRO) used the criterion of three successive PSA rises to define biochemical failure after radiotherapy (ASTRO 1997), although its optimal definition is unclear. The PSA nadir value by multivariate Cox regression analysis has been shown to be an independent and major predictor for HIFU success and biochemical recurrence (Lee et al. 2006, Poissonnier et al. 2006, Uchida et al. 2006b). Within the first 2 years after EBRT, PSA bounces and is not a reliable measure of DP, so prostate biopsies are not recommended because of delayed apoptotic cell death. Afterward, positive biopsy results are reliable and correlate with DP (Figures 13.6 and 13.7).

Although both the Ablatherm and Sonablate devices have been used in clinical trials, direct comparisons of their outcomes are difficult because of different follow-up, definitions of PSA end points and disease-free survival rates, and the limited number of patients using the latter device. Despite promising medium-term survival data, considering the biology and

natural course of prostate cancer, high-quality evidences on the efficacy and safety of HIFU in prostate are absent and there are some limitations in the current reports to assess patient-relevant outcomes, including incomplete outcome, limited evaluation of genitourinary outcomes, the poorly understood early post-treatment toxicity, the potential confounding due to the use of hormones for cytoreduction, the absence of formal exclusion/inclusion criteria, inevitably missing some key baseline data for analysis, and the absence of systematic evaluation. Therefore, randomized long-term controlled trials of good quality and sufficient size, comparing HIFU with conventional RP, EBRT, active surveillance, minimally invasive therapies, or other watchful waiting, are urgently required to further

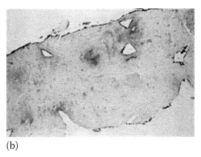

(a)   (b)

FIGURE 13.6   (a) Complete destruction of the glandular tissue due to coagulation necrosis lesion, which reaches the capsula and the periprostatic fat 48 h after HIFU treatment. (b) The necrotic prostatic tissue replaced by a fibrotic tissue, including the capsula, 3 months after HIFU therapy.

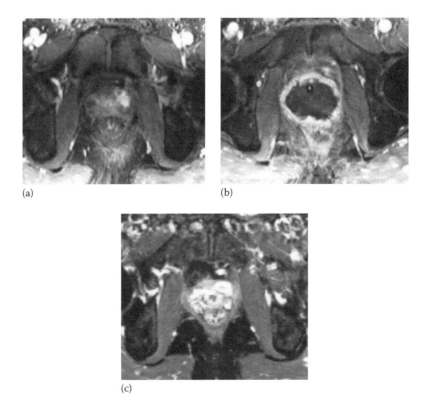

FIGURE 13.7   1.5 T dynamic contrast-enhanced MRI using gadolinium (a) prior to HIFU treatment, and (b) at 2 weeks and (c) 6 months after HIFU treatment. (With permission from Macmillan Publishers Ltd. *Br. J. Cancer,* Ahmed, H.U., Zacharakis, E., Dudderidge, T., Armitage, J.N., Scott, R., Calleary, J., Illing, R.O., Kirkham, A., Freeman, A., Ogden, C., Allen, C., and Emberton, M., High-intensity-focused ultrasound in the treatment of primary prostate cancer: The first UK series, 101, 24, copyright 2009.)

evaluate cancer-specific and overall survival rates before the indications for primary therapy can be expanded (Rebillard et al. 2008, Warmuth et al. 2010). Currently, the longest follow-up was for 8 years, whose cancer-specific survival rate was 98%. The management of early prostate cancer in elderly or debilitated population remains under debate. Although the recent NICE guidelines (Graham et al. 2008) have recommended HIFU only if patient data are added prospectively into clinical trials, national or international registries are also acceptable (NICE Guideline 2005, NICE 2008), which helps the timely diffusion of potentially successful technologies into clinical mainstream (Ahmed et al. 2009, Hou et al. 2009). In addition, a synergistic inhibitory effect of the HIFU and chemotherapy (e.g., docetaxel) could be useful for patients with high-risk prostate cancer.

13.7.2 BREAST TUMOR

Breast cancer is one of the most popular malignant diseases in women, with 411,000 annual deaths accounting for 14% of all female cancer deaths worldwide, and it is the fifth cancer mortality overall. There were an estimated 1.5 million new cases in 2002, mostly in the developed countries (361,000 in Europe and 230,000 in North America) with the average survival rate of 73% and 57% in developed and developing countries, respectively, which may be due to the routine screening programs in the West (Parkin et al. 2005). Radical mastectomy, that is, breast amputation with/without the pectoral muscle removal and requirement of skin graft (Punglia et al. 2007), has been the mainstream treatment for about 80 years (Halstedt 1907, Morrow and Harris 2000), but with a high local recurrence ratio due to an unsatisfactory surgical margin. Development of diagnostic techniques and the popularity of screening programs have resulted in a higher percentage of early stage detection of breast cancer (Veronesi et al. 2002). Since the 1970s, the gold standard treatment for patients with localized early-stage breast cancer desiring breast conservation has been lumpectomy (Fisher et al. 1989) followed by external radiation therapy (Fisher et al. 2002), possibly also in conjunction with interstitial laser coagulation, RF therapy, cryotherapy, interstitial radiotherapy, chemotherapy, and hormonal therapy (Hall-Craggs 2000). Despite the low morbidity rate, breast-conserving surgery also introduces certain complications, such as bleeding (2%–10%), infections (1%–20%), seroma formation (10%–80%), and chronic incisional pain (20%–30%) (Nystrom et al. 2002, Veronesi et al. 2002), as well as cosmetic worries. The long-term survival rate of both mastectomy and breast conservation surgery is similar (Fisher et al. 2002, Veronesi et al. 2002). Positive outcomes are found in 10%–53% of the patients because of large tumor size, young age, positive axillary lymph node, and extended intraductal component (Park et al. 2000, Singletary 2002).

There is increasing interest recently to use nonsurgical ablation as part of a breast conservation therapy, resulting in more psychological and cosmetic satisfaction. The breast is a good target for HIFU ablation because of the excellent acoustic window and easy immobilization. The first HIFU treatment on breast cancer was reported in 2001 (Hüber et al. 2001). The patient selection criteria were invasive breast cancer ($T_{1-2}N_{0-2}M_0$) proven in histology, single palpable tumors with size no larger than 6 cm, visible lesion boundaries in color Doppler ultrasound circumscribed with the distance from skin or rib cage and nipple of more than 0.5 and 2 cm, respectively, at least 18 years old, no breast implants, stable hematogenic parameters, no extensive intraductal components, no extensive calcifications in mammography, no multicentric disease and prior local therapies, chemotherapy, or hormonal therapy, and no history of active myocardial infarction in the last 6 months. General exclusion criteria were women who are pregnant or nursing, clinical evidence of brain metastases, subjects with tumors lying <5 mm from vital structures or either adjacent to the skin or the chest wall, concurrent antiarrhythmic disease with good prognostic factors, anticoagulant or immunosuppressive medication, ductal carcinoma *in situ* (DCIS), cancers with extensive intraductal component or lymphovascular invasion, tumors with very irregular margins, too large a size, scattered multiple foci or in proximity to the nipple, more than one focal breast lesion per quadrant, previous radiation or local thermal therapy, significant background illness or underlying medical condition (e.g., congestive heart failure, chronic obstructive pulmonary disease), metallic implants or other incompatibles with MRI (e.g., permanent implanted pacemakers), the inability to lie still for up to 150 min, and those who had previously documented severe intra-abdominal adhesions. Follow-up examinations on skin burns, local pain, discomfort, mammary edema, hemorrhage or infection, and fever were used to evaluate the HIFU-induced complications.

A HIFU-treated invasive grade 2 ductal carcinoma was partially necrotic, shown in the HIFU ablation as the white area with a hyperemic rim in the tissue and the lack of nuclear staining and mostly sublethal damage with chromatin clumping in histological slice (Figure 13.8) (Huber et al. 2001). Although slight changes were found on both unenhanced T1w and T2w images, the lack of contrast uptake in the treatment region (breast cancer and 1.5–2.0 cm margin) via the enhanced MRI was an indicator of coagulative necrosis in the postoperative MRI (3 days). Few strikes of enhancement were found in MRI with no mass contrast enhancement as the 1.8 cm invasive ductal carcinoma and dark void area at the site of it in a 44-year-old woman before MRgFUS (Figure 13.9). About 50% of the carcinoma and adjacent normal tissues showed thermal effects, and the remaining portion of the carcinoma appeared viable (Wu et al. 2003a). In the sonograph, a heterogeneous brightness increase, an obvious reduction in tumor size in both transverse and longitudinal dimensions, and an absence of blood flow in color Doppler were found after HIFU ablation, although its image quality and resolution were not as good as those of MRI (Wu et al. 2004d) (Figure 13.10). The disappearance of contrast uptake Tc-99m sestamibi in SPECT after the HIFU ablation in the treated lesion indicated a termination of tumor cell viability and a positive therapeutic outcome (see Figure 13.11) (Wu et al. 2005a).

Little severe (i.e., third-degree skin burn) and a few minor adverse effects (< 5%) were reported. The main reason for both

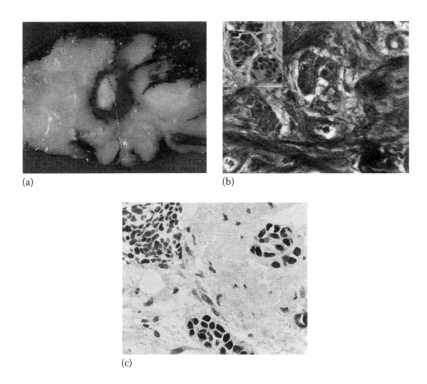

FIGURE 13.8 (a) Pathological slice of the HIFU ablation as the white area with a hyperemic rim. Microscopic specimen of (b) vital and degenerating tumor cells, and (c) two tumor cell clusters with p53 overexpression within necrotic cell debris 5 days after treatment HIFU ablation. Inset: H&E stain of tumor in the core biopsy before ablation. (From Zhou, Y., *J. Med. Imaging Health Inform.*, 3, 141, 2013.)

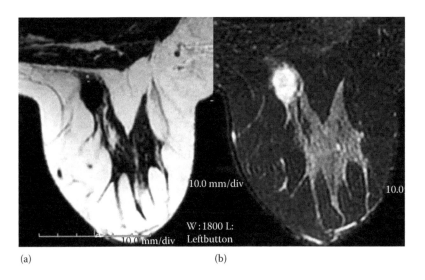

FIGURE 13.9 A 1.8 cm invasive ductal carcinoma in a 44-year-old woman, shown as an irregular enhancing mass in the upper outer quadrant of the right breast (a) in contrast-enhanced T1-weighted fat-saturated MR sagittal axial images. Three days after MRgFUS, (b) few strikes of enhancement and dark void area are seen in the sagittal image, which may be hyperemia due to reactive inflammation or residual tumor. (From Wu, F., *Br. J. Cancer*, 89, 2227, 2003.)

skin complications is the short distance between the breast lesion and the skin, resulting in therapeutic energy deposition. The skin at the wave entry site had no visible changes; no tumor bleeding or rupture of the large blood vessels or infection of the treated breast was found; and there were no local or systemic symptoms 3 months later. Five to ten percent of patients had 5–7 days of low-grade fever up to 38.5°C, and some had severe fever up to 39.5°C for 2–3 weeks postoperatively, which seems to correlate with the volume of destruction (Wu et al. 2004d). Despite the absence of anesthesia, no pain or discomfort was experienced by the patient during or after HIFU therapy except a mild pain (20%–30% within a week),

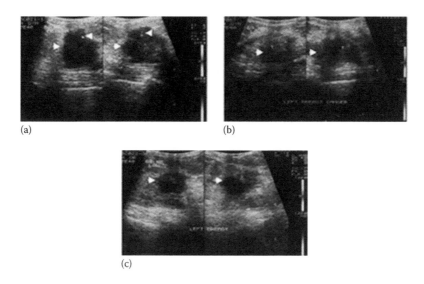

FIGURE 13.10 Sonography of a breast cancer (a) before and (b) after 6 months and (c) 12 months of HIFU ablation with an obvious reduction of tumor size and an absence of blood supply (arrowhead). (With kind permission from Springer Science+Business Media: *Breast Cancer Res. Treat.*, Extracorporeal high intensity focused ultrasound treatment for patients with breast cancer, 92, 2005, 54, Wu, F., Wang, Z.B., Zhu, H., Chen, W.Z., Zou, J.Z., Bai, J., Li, K.Q., Jin, C.B., Xie, F.L., and Su, H.B.)

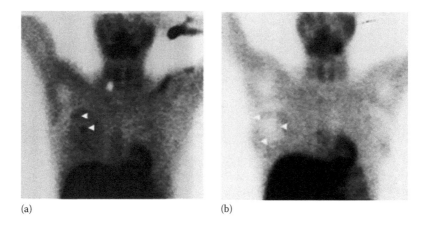

FIGURE 13.11 (a) Appearance and (b) disappearance of the radioisotope uptake of two breast lesions (arrows) before and 1 month after HIFU treatment with a margin of 1.5–2.0 cm in SPECT. (With kind permission from Springer Science+Business Media: *Breast Cancer Res. Treat.*, Extracorporeal high intensity focused ultrasound treatment for patients with breast cancer, 92, 2005, 55, Wu, F., Wang, Z.B., Zhu, H., Chen, W.Z., Zou, J.Z., Bai, J., Li, K.Q., Jin, C.B., Xie, F.L., and Su, H.B.)

warmth, and sensation of pressure in the treatment region (Huber et al. 2001), with a small percentage of them (5%–10%) needing 3–5 days of oral analgesics (Wu et al. 2003a). In total, 19 of 24 patients were free of neoplasia after one or two sessions of HIFU, as confirmed by breast biopsy, and remained free of metastasis (mean follow-up of 20.2 months) (Wu et al. 2005a). Forty-eight breast cancer patients had the coagulative necrosis shrunk by 20%–50% in volume at 6–12 months postoperatively. Half of the ablated tumor was reabsorbed within 1–2 years after the operation. All were alive at the last follow-up (1036 months), and all but one is disease-free (Wu et al. 2004a). After an average follow-up of 54.8 months, the 5-year disease-free survival was 95% in 22 patients with breast cancer (2–4.8 cm). Reduction in tumor size was obvious in the first 12 months (from 8.2% ± 6.1% at 3 months to 45.2% ± 22.1% at 12 months) but less remarkable after 3 years (from

80.3% ± 38.2% at 36 months to 90.4% ± 49.1% at 60 months) as listed in Table 13.1. In general, although MRI is more reliable for breast cancer delineation, the success rate of MRI-guided HIFU ablation is not as high as that under the guidance of sonography, which may be due to the larger ablation margin used. However, it took months to absorb the treated lesion. The anxiety regarding the persistence of a lump as the risk of recurrence, even in the absence of viable tumor cells by biopsy, is popular among patients and leads to the choice of lumpectomy after HIFU ablation (Wu et al. 2005a).

Homogeneous coagulative necrosis has characteristics of tumor cell distortion, pyknotic nuclei, shrink of nuclei, cell debris, and disappearance, but no cell apoptosis (Figure 13.12) (Wu et al. 2007b). There is also an obvious destruction of the non-neoplastic breast tissues and fat cells surrounding the necrosis, and granulation tissue with immature

TABLE 13.1

Percentage of Reduction in Tumor Volume

	3 Months	6 Months	12 Months	24 Months	36 Months	48 Months	60 Months
Number of patients	22	22	21	17	17	16	5
Average reduction	8.2 ± 6.1	26.7 ± 12.2	45.2 ± 22.1	72.3 ± 22.1	80.3 ± 38.2	87.3 ± 42.3	90.4 ± 49.1
Reduction range	2.5–18.7	19.2–49.1	25.2–70.1	50.2–82.3	58.1–90.6	70.3–92.1	80.5–96.3

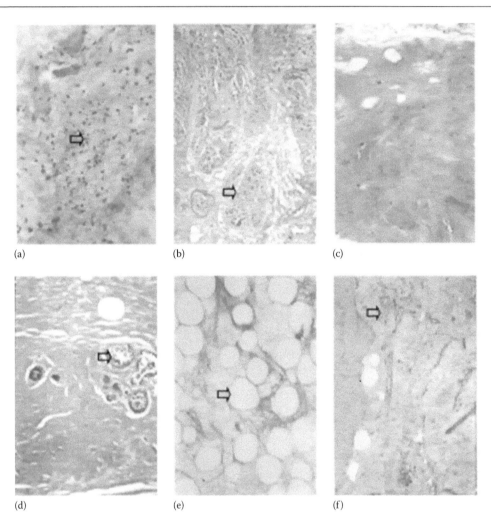

(a)   (b)   (c)

(d)   (e)   (f)

FIGURE 13.12 (a) Pyknotic nuclei of cancer cells, (b) disappearance of cancer cell nuclei, (c) denatured breast tissue, (d) coagulative necrosis of mammary gland tissue, (e) unviable fat cells, and (f) new-growth granulation tissue in the marginal region of coagulative necrosis 9 days after HIFU ablation in histological slice with H&E staining. (Wu, F., Wang, Z.-B., Cao, Y.-D., Zhu, X.-Q., Zhu, H., Chen, W.-Z., and Zou, J.-Z.: "Wide local ablation" of localized breast cancer using high intensity focused ultrasound. *J. Surg. Oncol.* 2007. 96. 134. Copyright Wiley-VCH Verlag GmbH & Co. KGaA. With permission.)

fibroblasts, inflammatory cells, and new capillaries in the margin (Wu et al. 2007b). Severely damaged tumor vessels illustrated the disappearance of endothelial cell nuclei, no distinction of cellular margins, and disrupted junctions between individual cells. Cellular discohesion, disruption of the smooth muscle, tunica media, and scattered intravascular thrombi were found in the treated vessels. Although cancer cells seemed normal in H&E staining after the ablation, the presence of some vacuoles in the cytoplasm, cell membrane disintegration, disappearance of the nuclear membrane, clumped chromatin at the periphery of the tumor nuclei, and no identification of organelle structures suggested irreversible cell death (Wu et al. 2006a). In the peripheral region, the treated breast cancer had an unrecognized, amorphous, electron-dense material. Cytological enzyme activity (i.e., NADH-diaphorase stain) was used to identify acute cell viability and intense cytoplasmic positivity (Wu et al. 2006b). VEGF expression in breast cancer might stimulate tumor growth, angiogenesis, and metastases. However, CA15-3 and VEGF positive expression was measured in only 52% and 30% patients receiving HIFU treatment, respectively (Wu et al. 2006b).

Furthermore, HIFU was also used for benign breast tumors. The first trial was carried out in 2001 with a success rate of 73% (Hynynen et al. 2001). The effectiveness of HIFU was evaluated 7 days after the session because some benign processes (e.g., edema, fibrosis, necrosis, and inflammation) had similar characteristics in T1w images as those of a malignant process during that period.

Overall, HIFU can reduce the breast cancer volume and increase the visualization of tumor (lesion) boundary for conservative surgery, which is critical for those at high surgical risks and elderly patients. For advanced-stage cancer with failure in control, HIFU ablation is a palliative approach to regress tumor growth and to relieve the pain due to tumor that is ineffective by antineoplastic and pharmacological modalities. The important issue is the lack of assessment to the margin status noninvasively to determine the treatment plan. So, contrast-enhanced MRI, instead of replacing histopathology, needs more investigation. However, for nonsurgical candidates with locally advanced breast cancer or metastases, the criteria of selecting patients are less important in the palliative HIFU ablation (Gianfelice et al. 2003).

There are great differences in the rate of complete breast cancer ablation using minimally invasive approaches, 76%–100% in RFA, 13%–76% in laser ablation, 0%–8% in microwave ablation, 36%–83% in cryoablation, and 20%–100% in HIFU ablation (Zhao and Wu 2010). Large randomized control studies are required to assess the long-term outcomes of these techniques compared to breast conservative surgery. RFA seems the most promising technique in terms of safety and complete ablation rates. Successful thermal ablation is a function of appropriate patient selection; breast lesions close to the skin and chest wall, as well as multiple lesions, should be excluded from thermal ablation. Determination of the status of the sentinel nodes in breast cancer is important in breast conserving therapy. As lymph drainage may be disturbed after local tumor ablation, it is desirable to perform sentinel node biopsy before the ablation. It is essential to perform such a relatively minor procedure that allows the detection of lymph node metastasis, which otherwise would not be amenable to this form of therapy.

A palpable firm lump would persist or even become larger for a certain time after HIFU ablation, although the lesion would become benign eventually. The fear of local recurrence will impact the patient's psychology, satisfaction, or cosmetic concern. As a result, some patients receive a modified radical mastectomy afterward. In addition, in the follow-up examination, it is hard to distinguish the firm lump and local recurrence postoperatively.

13.7.3 UTERINE FIBROIDS

Uterine fibroids (leiomyomas or myomas) are gonadal, steroid-dependent, benign smooth-muscle tumors and one of the most common female pelvic tumors with a prevalence of 25%–77% (Buttram and Reiter 1981, Stewart 2001). They are the most common anatomical cause of menorrhagia (abnormal uterine bleeding), dysmenorrhagia, dysmenorrhea,

dyspareunia, pelvic pain, the pressure effects on the urinary and gastrointestinal tracts for urinary frequency, and infertility in women of reproductive age (Vollenhoven et al. 1990). Uterine fibroids are more prevalent in Afro-Caribbean women with earlier onset than in Caucasian women (Stewart 2001). They are mostly located intramurally, but also subserosally (may be pedunculated), or lie in the submucosa. The fibroids can also occur in the fallopian tubes, broad ligament, or cervix. Although the risk of sarcomatous change within the leiomyomata is low (<1%), they are at risk of torsion, acute hemorrhage, and necrobiosis. The symptoms depend on the size, position, and number of fibroids and are usually classified into type 0: pedunculated fibroid without intramural extension, which can be easily excised at hysteroscopy; type I: sessile with an intramural extension less than 50%; and type II: sessile with an intramural extension of at least 50% (Wamsteker et al. 1993). However, the management of type I and type II uterine fibroids is more challenging. Though frequently asymptomatic, 10%–20% of women with fibroids need treatment, including hysterectomy, myomectomy, uterine artery embolization (UAE), and medical therapy (Buttram and Reiter 1981, Stewart 2001, Williams et al. 2006, Goodwin et al. 2008, Bradley 2009). Although hormonal treatment, such as gonadotrophin-releasing hormone agonists (GnRHa), can reduce the uterine size, after discontinuation it often grows to the previous size rapidly. Hysteroscopic myomectomy (hysterectomy accounts for more than 200,000 procedures in the United States each year), that is, resection of submucosal fibroids with deep intramural extension, requires multiple procedures for complete resection (subsequently, a long recovery period and high costs) and is associated with risks such as uterine perforation, vaginal vault prolapse, premature ovarian failure, fluid overload, incontinence, and ascending genitourinary infection, so that this aggressive option is unsuitable for women who prefer to preserve their uterus and fertility (Vercellini et al. 1997). Myomectomy involves substantial risks, such as hemorrhage, postoperative adhesion, and uterine rupture during pregnancy and labor (Bajekal and Li 2000). Therefore, an increasing number of patients do not undergo the invasive procedure for a benign and usually self-limiting condition. UAE has the advantage that it can potentially take care of all the fibroids in one procedure, while myomectomy will invariably leave some behind (Pinto et al. 2003, Lupattelli et al. 2005, Park et al. 2005). However, uterine necrosis and infection associated with UAE lead to emergent hysterectomy, ovarian failure, and vaginal dryness due to nontarget embolization or overembolization (Walker and Pelage 2002). All percutaneous ablation techniques require insertion of needle-like applicators. It may be difficult to find a safe puncture route for placement of applicators in deeply located fibroids.

MRgFUS was approved by the FDA in October 2004 after Phase I/II clinical trials in 1999. Clinical trials have been carried out across China (Wu et al. 2004a), the United States (Fennessy and Tempany 2006, LeBlang et al. 2010), Japan (Funaki et al. 2009), United Kingdom, Germany, and Israel. Most studies in Europe and North America are focused on

MRgFUS, while sonographically guided HIFU is much more popular in China. MRI with post-intravenous gadolinium is an optimal modality for characterizing fibroids and homogenous enhancement or areas of internal necrosis. The inclusion criteria for HIFU ablation were: child-bearing age (>18 years) with no desire for future fertility, type I or II uterine fibroids with no dominant fibroid greater than 10 cm in diameter, no other treatment before HIFU ablation, an 8-item symptom severity score (SSS) out of 21, and the Uterine Fibroid Quality of Life Questionnaire (UFS-QOL) of at least 40 points, and a normal cervical smear (Spies et al. 2002). The exclusion criteria were pregnancy, presence of bowel or scars from the previous abdominal surgery in the acoustic pathway, more than four multiple fibroids, other pelvic or uncontrolled systematic disease, inability to lie in the prone position, contraindication to contrast-enhanced MR or ultrasound (CEUS), inability to communicate with the operators during the procedure, and postmenopausal women.

Patients who underwent HIFU ablation were 18–59 years of age (mean ~40 years), dominant fibroid size varying in the range of 1.5–9.8 cm (mean ~5.5 cm, occasionally up to 13.5 cm), and a volume of 7.7–318.3 cm^3 (mean ~90 cm^3), with a treatment duration of 27–390 min (mean ~2 h). Before HIFU ablation, patients underwent careful bowel preparation including low-residue diet for 3 days, fasting for 12 h before treatment, administration of cathartics in the evening the day before treatment, and an enema in the early morning on the treatment day. An inflatable cushion of appropriate size filled with degassed water was used to compress and push away the bowel lying anterior to the uterus from the acoustic pathway for safe and feasible ablation. The temperature of the coupling medium (i.e., degassed water) could be set at 16°C to protect against accidental skin burns. To prevent irreversible thermal damage to the endometrium for potential risk of amenorrhea and infertility, intrauterine devices have to be removed before HIFU ablation. A catheter was inserted into the bladder to fill degassed, sterile water, cool the skin, and increase circulation in the subcutaneous tissues. HIFU treatment was performed under conscious sedation by using fentanyl and midazolam. The minimum applicable depth is restricted to 3.0 cm in practical application. MR images were used to determine the number of fibroids, location, size, treatment accessibility, contrast enhancement, and the presence of other pelvic diseases, during treatment planning. Special attention was paid to the neighboring organs, such as bladder, intestines, ovaries, and tubes. To avoid unintended thermal damage, there was at least 1 cm between the HIFU focus and the tumor margin. An average 70% (37.8%–92.4%) of the fibroid volume was treated. Pubic hair was removed, and lower abdominal skin was prepared by cleansing with an alcoholic solvent to remove oily deposits. Immediately after ablation, T1w MR images with contrast administration were acquired to visualize the treatment result, such as the nonperfused volume (NPV). If a low NPV was found in the treated fibroids, additional HIFU ablation was required. The transitional tissues (skin, fasciae, abdominal wall muscles) were thoroughly inspected for any signs of tissue damage. Pain was assessed according to the visual analog scale (VAS), and discomfort was classified as none, mild, moderate, and severe.

Sonography showed obvious increased hyperechogenicity (>90%) well confined to the treated fibroid region due to the generated microbubbles. These hyperechoic changes became gradually less evident and disappeared some days in the follow-up. The HIFU ablation indeed produced hemorrhagic necrosis in the treated uterine fibroids. The average fractional ablation, defined as the ratio of NPV to the original fibroid volume, immediately after HIFU treatment was 21.0%–97.0% (mean: ~75%) (Zhang et al. 2010). Decrease in the apparent diffuse coefficient (ADC) values is due to cell necrosis and loss of membrane integrity. Findings of blood supply change in the ultrasound color Doppler was also similar: 80.7% disappearance, 15.0% substantial decrease, and 4.3% no change (Ren et al. 2007). The nonperfused ablation ratio was 80% ± 12% in CEUS (Wang et al. 2012b), which seemed to be closely correlated with the volume change. Immediately after HIFU, the mean fibroid volume slightly enlarged (~10%) due to edema in interstitial tissues. During the follow-up, the ablated fibroids shrank significantly over time: 1.9%–60.0% (mean: ~30%) at 3 months, ~40% at 6 months, and ~45% at 24 months. The volume decrease in the uterus was 27.7% at 3 years. However, the volume shrinkage in ultrasound-guided HIFU was very high: 46.7%, 68.2%, 78.9%, and 90.1% on at 3, 6, 12, and 24 months follow-up at CEUS, respectively (Wang et al. 2012b), which may be due to the higher volume percentage treated in comparison to MRgFUS. The volume of menstrual bleeding decreased significantly at the first menstrual period after HIFU therapy ablation in 55.3% of patients, 71.1% and 89.5% of them at 3 and 6 months follow-up (Wang et al. 2012b). Vaginal expulsion of necrotic tissue was seen in 58% of patients, which disappeared after two to four menstrual cycles. HIFU treatments showed sustained symptomatic relief up to 3 years (Kim et al. 2011). Seventy-nine percent of patients had a 10-point reduction in SSS within the first 3 months, and the UFS-QOL score decreased to 16.0 ± 3.7 at 6 months follow-up and further to 11.0 ± 3.0 at 24 months after HIFU ablation (Wang et al. 2012b). At 3-year follow up, SSS and UFS-QOL had improved by 47.8 and 39.8, respectively. The reintervention rates were 14.0% for type 1/2 patients and 21.6% for type 3 patients at 24 months post treatment (Funaki et al. 2009).

In a Phase III clinical trial, only 51% patients had a 10-point reduction in SSS, and 28% of them underwent an alternative treatment by 12 months, which may be because only ~10% of the fibroid volume was treated in the FDA-approved study to maximize safety (Stewart et al. 2006). Pretreatment with GnRHa on patients with fibroids >10 cm had a beneficial effect and enhanced tissue response to HIFU (Smart et al. 2006). A novel volumetric ablation of moving a single focal spot outward in concentric circles improved both treatment efficacy and ablation homogeneity significantly (Köhler et al. 2009, Voogt et al. 2012). In addition, the use of microbubbles (e.g., SonoVue) may enhance the outcome of HIFU ablation as well as treatment assessment (Peng et al. 2012). The NPV of HIFU treatment with and without the use of SonoVue was 86.0% (28.8%–100.0%) and 83.0% (8.7%–100.0%), respectively.

The rate of massive gray-scale changes was higher, and the sonication time to achieve such changes was shorter with SonoVue.

Macroscopic and microscopic examinations showed that HIFU induced thermal ablation of a targeted adenomyosis with typical characteristics of coagulation necrosis under light microscopy (>80%), complete loss of cellular viability in nicotinamide adenine dinucleotide-diaphorase staining under electron microscopy (>95%), and the clear margin between the treated and untreated regions in 2,3,5-triphenyltetrazolium chloride staining (Yang et al. 2009). Epithelial cells showed pycnotic nuclei and shrunken cytoplasm. However, viable cells were still found. The endometrium in the target area showed diffuse, confluent hemorrhagic lesions surrounded by edematous swelling (Figures 13.13 and 13.14).

During the HIFU ablation, most patients felt minor abdominal pain, but as soon as the procedure stopped, the pain vanished. Minor adverse events after HIFU ablation included general discomfort (25%), pain at the treatment site (10%), abdominal tenderness, nausea, first-degree skin burns (skin redness or blisters), gross hematuria that persisted for approximately 1–3 days (14.2%), minimal vaginal bleeding lasting for about 1–3 days, and low-grade fevers up to 38.1°C that persisted for approximately 1–3 days (5.9%). Only a small percentage of patients (~10%) needed pain medication within 72 h of treatment and returned to normal activities after 2.3 ± 1.8 days (Hesley et al. 2008). No patients had amenorrhea after treatment. Nerve fiber damage (1.7%–3.4%), which was characterized by lower-extremity and buttock pain, was caused by HIFU in those patients who had large fibroids or whose fibroids were located deep in the pelvis or whose

sciatic nerve was in the far field of the ultrasound beam. However, nerve functions, including sensation and motion, recovered completely 2 months later with conservative management (Chapman and ter Haar 2007). Severe pelvic pain and complications, such as premature menopause and ovarian failure, were rare. Severe cramping abdominal pain 1 week after HIFU treatment was also found in some patients. If the navel was in the path of the ultrasound beam, thermal damage at the skin surface can occur as with scars because of the high acoustic absorption (Fruehauf et al. 2008, Ren et al. 2009).

The expulsion of necrotic tissue is a normal reaction after HIFU ablation of uterine fibroid, and patients should be informed to avoid unnecessary anxiety. However, the rate of complete expulsion seemed to be much lower (2.6%) (Wang et al. 2012b). If a large fibroid was expelled, it may obstruct the cervix and require hysteroscopic removal. Compared with other diagnostic modalities, contrast-enhanced MRI and CT are much better in the rapid assessment of the therapeutic response of a treated tumor. Because of their high cost, these examinations were performed only once a week after HIFU ablation. Doppler ultrasonography and CEUS were generally used for assessing vascularity changes in the treated fibroids, with similar results of residual unablated tumor and pathologic change in the treated lesions (Zhou et al. 2007). Treatment success (i.e., fibroid shrinkage, improved symptom relief, and fewer additional treatments) is closely dependent on the percentage of ablated fibroid tissue (Fennessy et al. 2007, Lenard et al. 2008, LeBlang et al. 2010). Reducing safety margins to sensitive structures (i.e., the bowel and the uterine serosa) is possible and has already proven to be safe

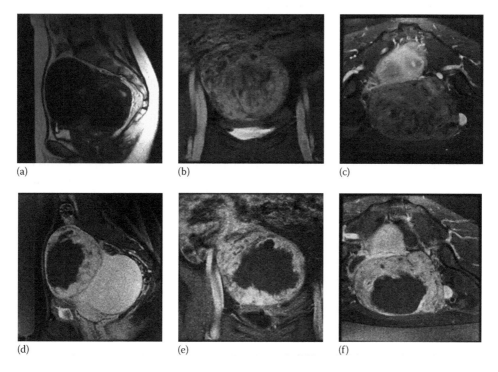

(a)   (b)   (c)

(d)   (e)   (f)

FIGURE 13.13 A uterine homogenous fibroid depicted in (a) sagittal T2 fast spin-echo, (b) coronal SPGR postgadolinium, and (c) axial SPGR postgadolinium obtained before treatment with slight heterogeneous enhancement, and a new large nonperfused area that is consistent with treatment-induced necrosis in the corresponding images immediately after treatment (d–f, respectively). (Courtesy of Chongqing Haifu Corp, Chongqing, China.)

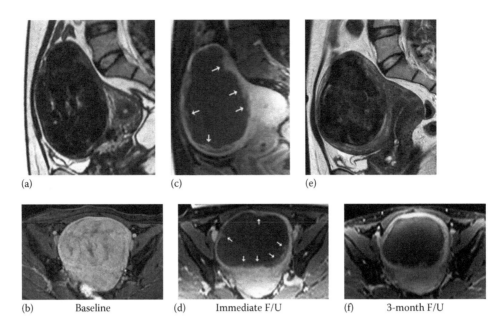

(a)	(c)	(e)
(b) Baseline	(d) Immediate F/U	(f) 3-month F/U

FIGURE 13.14   Baseline sagittal (a) T2-weighted and (b) contrast-enhanced T1-weighted MR image of a large intramural uterine fibroid, and axial image (b). The corresponding images (c) and (d) after MRgFUS, which demonstrated near-complete ablation with minimal residual contrast enhancement in the periphery (arrows) but with no complication. (e) and (f) Images at 3-month follow-up with obvious volume shrinkage of the fibroid.

(Morita et al. 2007, Zhang et al. 2010). However, the volume of treated tissue did not correlate well with the pretreatment plan and the MR-predicted value because bowel presented in the sonication pathway or the target was invisible (Chapman and ter Haar 2007). The large artery running across the fibroid with perfusion-mediated tissue cooling effects may reduce the acoustic energy deposited and HIFU treatment efficacy (Chen et al. 1991, Wu et al. 2004a). The consequent long treatment time required for ablation is especially critical for patients who feel abdominal pain during the treatment (Ren et al. 2007). Disruption of small mucosal blood vessels in the fibroids as well as glandular structures leads to diffuse interstitial extravasations of erythrocytes, which may enhance the cooling effect. In the endometrium, isolated hemorrhagic lesions were seen macroscopically with no signs of coagulative tissue necrosis. Till now, long-term results of HIFU ablation for uterine fibroids are scant to completely evaluate its outcome.

13.7.4 LIVER TUMOR

Hepatocellular carcinoma (HCC), both primary and metastatic type, is the fifth most common malignancy worldwide and increasing rapidly despite considerable advances in diagnostic modalities. Its prevalence is highest in Asia and Africa (20–60/100,000 persons in China) (Bosh et al. 2005), but its incidence in the Western world is also increasing (Taylor-Robinson et al. 1997, El-Serag and Mason 1999) due to the increased occurrence of hepatitis C (El-Serag and Mason 2000). Liver transplantation offers the only real hope for cure for small HCCs and cirrhosis, with a 5-year survival rate of 25%–30% (Pichlmayr et al. 1997). However, the shortage of donor livers is the greatest limitation for transplantation.

Surgery, as a curative option, is only for patients with single HCC without portal hypertension and preserved liver function (Llovet et al. 2003). Unfortunately, only 10%–20% of HCC lesions are resectable with a 5-year survival rate of 40% and an operative mortality of up to 5%. The reason for the majority of HCC patients not undergoing surgical resection or liver transplantation is either the advanced stage of the lesions or the presence of underlying cirrhosis. The survival of that subgroup is quite poor, less than 6 months (Livraghi et al. 1997). The poor outcome is mainly attributed to the multiple foci of origin of HCC, which results in low rates of success and a high risk of postoperative recurrence.

The other therapeutic modalities include systemic and regional chemotherapy, percutaneous ethanol injection (PEI), transcatheter arterial chemoembolization (TACE), cryoablation, immunotherapy, and ablative techniques (e.g., RFA, microwave coagulation, and laser-induced interstitial thermotherapy). They all offer potential local tumor control and occasionally achieve long-term disease-free survival. The favored alternative to surgery is combination chemotherapy, but it has an objective response rate of 20%–50% and a median survival rate of 12–18 months (Simmonds 2000). RFA and PEI are considered effective for relatively small and encapsulated HCC with the greatest dimension below 3 cm (Livraghi et al. 1997, Llovet et al. 2003). A tumor size of >2 cm was the only independent risk factor for local recurrence in RFA treatment (Murakami et al. 2007). TACE is commonly used for advanced and large-volume HCC (Bruix et al. 2004). However, it is almost impossible to achieve complete necrosis of large HCCs with embolization of the hepatic artery alone because of the blood supply of HCC by both artery and portal vein (Ikeda et al. 1991). HCC that was subjected to RFA after TACE for non-early HCC could achieve a relatively high complete local response. Because of the protective effect of portal venous blood flow on adjacent tumor cells from thermal destruction, local thermal therapies are unlikely to ablate tumors close to major liver vessels. So, the response is usually poor with no increased survival benefits for chemotherapy, immunotherapy and TACE. In addition, these localized procedures have disadvantages such as the increased risk of needle tract seeding, hemorrhage and bile leakage during liver puncture, the need for repeated performance over an extended period, and the inability to provide effective monitoring during treatment or to ensure complete tumor destruction.

For a multifocal and frequently recurrent malignancy, a noninvasive therapeutic modality that can selectively destroy multiple tumor nodules scattered throughout the liver without impairing liver function and allow repeated application when necessary is preferred. HIFU was approved for the treatment of HCC in China in 1999 (Wu et al. 1999). HCCs can be destroyed by producing a contiguous lesion lattice encompassing the tumor and appropriate margins of surrounding tissue determined by the control program. Inclusion criteria included a confirmed advanced HCC diagnosis, inoperable tumor, Karnofsky performance scale score of 70% or more, no Child-Pugh C cirrhosis, no history of hepatic encephalopathy, no ascites detected, no active infection, no more than 3–5 tumor nodules, the maximal tumor diameter of less than 8 cm, tumor involvement less than 70% volume of the liver, not in the end stage of liver disease (liver failure, hepato–renal syndrome), no severe concomitant diseases, no lesion behind the rib, detectable tumor in sonography, and no bowel adjacent to the tumor. Exclusion criteria were diffuse hepatocarcinoma, diffuse infiltrative lesions in more than one lobe, liver failure, hepato-renal syndrome, severe concomitant diseases, neoplastic thrombosis of the portal branches, prothrombin time elevated by < 50%, and platelet count less than 50,000.

The mean age of HCC patients who underwent HIFU ablation was about 55 years (range: 30–74 years), the tumor size about 9 cm (5.0–16.5 cm), and the treatment time usually limited to approximately 2 h. Since the ribs affect the propagation of ultrasound beams and deposition of acoustic energy, the efficacy of HIFU treatment may be decreased with a higher propensity of induced adverse effects when intra-abdominal tumors concealed by ribs are treated. Partial rib resection at the lower right chest wall, usually with posterior and lateral defects from rib 7 to 9, was performed under general anesthesia, and the range could be defined by sonography in the treatment position (larger than 6 cm × 6 cm in most cases). The lower part of the chest wall, close to the diaphragm, has little effect on respiratory movement. Although rib resection is an invasive technique, it offers those patients who have no other treatment option a possible solution. The ribs are susceptible to hematogenous metastases from cancers because of their remaining red marrow. So, rib resection may be helpful in preventing the metastasis. If the patients have large anterior and anterolateral chest wall defects, chest wall reconstruction may be carried out to limit paradoxical respiration. Ascites, if present, should be treated before HIFU treatment. If the tumor is in the dome of the liver, artificial

right pleural effusion may be induced before treatment. Skin preparation is a crucial step for success. Any hair, dirt, or topical creams on the skin surface overlying the target can cause acoustic deflection and result in accidental skin burn. So, the skin is cleaned by cotton soaked with alcohol, and gentle suction is performed in order to remove any dermal microbubbles. General anesthesia or epidural anesthesia after fasting overnight is required to prevent the patient from experiencing deep visceral-type pain and to ensure immobilization. Tumor positioning is facilitated either by breath holding by the anesthetist or by endobronchial intubation, the use of a dual lumen endotracheal tube, and selective ventilation of the contralateral lung, with the ipsilateral lung held in varying degrees of inflation. A nasogastric tube should be inserted for gastric decompression. The body is fixed in the desired position by cushioned straps. Continuous monitoring of the blood glucose before, during, and after treatment is necessary to avoid hypoglycemia, hyperosmolar non-ketotic diabetic coma, and ketotic acidosis. In MRgFUS, it is difficult to position the HCC patient, especially when tumors are located at the right lobe of the liver, due to the relatively small size of bore, and to monitor the target temperature variation during HIFU because of liver movement. HIFU ablation has also been used in the Western population who have more body fat, particularly in the abdominal wall (BMI of UK patients vs. Chinese population: 26.1 vs. 21.7 kg/m^2).

Standard follow-up was carried out according to the recommendation by the International Working Group on Image-guided Tumor Ablation, which categorizes study goals into technical success, technique effectiveness, patient morbidity, and oncological outcomes. The short-term therapeutic outcome was estimated with post-treatment complication rate, hospital mortality, improvement in tumor diagnosis including a decrease in tumor size, decrease in tumor blood supply, and change of imaging signals, suggesting coagulation necrosis in tumor areas, symptom relief including increase in Karnofsky performance scale (KPS) scores and/or decrease in NRS scores, and decrease of serum levels of AFP by more than 50%. Complete response (CR) was defined when all three items were reached, and partial response (PR) was defined as coagulation and symptom relief/AFP decrease. Long-term effectiveness was determined by the increase of the survival rates and improvement of QOL-LC scores during the follow-up periods.

Complete devascularization was achieved in 85.2% of small lesions (< 5 cm) and 45.8% of large lesions (>5 cm) at the first CT examination (Veltri et al. 2006). T1- and T2-weighted images of ablated region were of high signal and low signal, respectively. Enhanced MRI did not show enhanced signal, indicating the occurrence of coagulation necrosis, decrease or elimination of blood supply, and occasionally the obvious tumor shrinkage. Additional HIFU treatment was performed for the remaining incomplete necrosis, especially in liver cancer patients who had huge blocks or multifocal big nodules. The lesion reached its maximum size on the third day after the treatment and then was replaced by a thin fibrous scar after several months (Linke et al. 1973). Histology showed a homogenous dwell-delineated coagulation necrosis corresponding to the target volume in the depth of the liver. No viable tumor cells remained in the treated area (Chapelon et al. 1992). However, the size of the hyperechogenic area did not correlate well with the volume of the coagulation necrosis (Linke et al. 1973). After the second or third HIFU session, all the tumors achieved complete necrosis. In contrast to RFA and PEI, HIFU can achieve a complete necrosis of larger tumors, especially those close to the main hepatic blood vessels, diaphragm, or bile ducts, which are impossible for surgical resection. Those large tumors have a significant cooling effect during the ablation. A complete and a partial responses were achieved in 28.5%–68% and 60.3% of cases, while the values were only 0% and 16.7% in the control, respectively (Wu et al. 2004c, Li et al. 2007, Zhang et al. 2009). The great discrepancy of complete response is due to the evolution of HIFU system and physician experience with this new technology. The overall survival rates at 1, 2, 3, 4, and 5 years were 42.9%–100%, 30.9%–83.3%, 49.8%–69.4%, 31.8%–55.6%, and 31.8%–55.6%, respectively, which were significantly greater than those (3.4% and 0%, respectively) in the control group (Wu et al. 2004c, 2005, Li et al. 2007, Zhang et al. 2009, Zhu et al. 2009). The 1-, 2-, 3-, 4-, and 5-year cumulative local tumor progression rates were 0%, 0%, 16.7%, 33.3%, and 33.3%, respectively (Zhu et al. 2009). In another study, the overall recurrence rate was 61.9% (Ng et al. 2011). Both KPS and NRS scores improved after HIFU, but only

the KPS scores showed significant improvement. Symptom relief was observed 1–2 days after HIFU and became more obvious at 1 month, as seen by an obvious improvement in appetite, weight, and discomfort or pain in the liver region. The quality-of-life (QOL) score increased from 67.7 ± 5.9 to 83.1 ± 8.0 at 3 months after HIFU (69.0 ± 8.5 in the control group). The remission rate of symptoms was 86.6% (Li et al. 2004) (Figures 13.15 and 13.16).

HIFU has been combined with pre-procedural TACE widely in China in order to reduce the tumor size, blood supply, and the acoustic energy or dose required to achieve ablation. In a study, 89 patients were randomized into TACE ($n = 45$) and TACE–HIFU ($n = 44$) group. The 1-, 2-, 3- and 5-year overall survival rates for the TACE–HIFU and TACE groups were 72.7% vs. 47.2%, 50.0% vs. 16.7%, 31.8% vs. 2.8%, and 11.4% vs. 0%, respectively ($p < 0.01$). The corresponding 1-, 2-, 3- and 5-year disease-free survival rates were 34.1% vs. 13.9%, 18.2% vs. 5.6%, 9.1% vs. 0%, and 0% vs. 0% ($p < 0.01$) (Li et al. 2010b). The combination may remedy the limitation of each alone, eliminate residual cancer cells after TACE, and have synergistic effects. Furthermore, because HCCs are supplied almost entirely by the hepatic arteries, more iodized oil deposition in the inner tumor tissue after TACE is not only convenient to ultrasound orientation, but also changes the impedance and absorption coefficient for higher energy sediment in the focal zone to induce a high temperature in the coagulation.

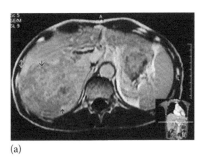
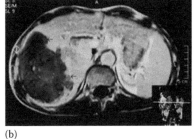

(a)   (b)

FIGURE 13.15 T1-weighted MRI showing (a) a hepatocellular carcinoma (18.2 cm × 13.8 cm × 9.2 cm) located in the right posterior lobe by arrow and (b) the tumor shrinkage and obvious necrosis and decreased blood supply after HIFU ablation. (Li, Y.-Y., Sha, W.-H., Zhou, Y.-J., and Nie, Y.-Q.: Short and long term efficacy of high intensity focused ultrasound therapy for advanced hepatocellular carcinoma. *J. Gastroenterol. Hepatol.* 2007. 22. 2152. Copyright Wiley-VCH Verlag GmbH & Co. KGaA. With permission.)

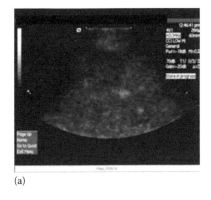
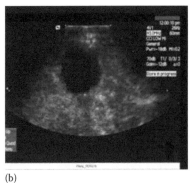

(a)   (b)

FIGURE 13.16 Microbubble contrast-enhanced sonography of a liver tumor (a) before and (b) after HIFU ablation. (From Leslie, T., *Br. J. Radiol.*, 85, 1369, 2012. With permission.)

Such improvements remain anecdotal, and further work would be needed (Kennedy et al. 2004).

Overall, HIFU treatment was well tolerated with acceptable side effects. Mild ascites disappeared after HIFU treatment, but 18.2% patients had a low-grade fever (temperature <38.5°C) that persisted for approximately 3–5 days. Mild or moderate and transient local pain (23.1%–83%) was controlled by injected or oral analgesics for 2–3 days. Severe pain required treatment with opioid analgesics in the immediate postoperative period (Li et al. 2004). Minor skin burns (2.5%) needed no therapy. Degree 1 skin burns (29.6%–39%) and blisters were treated with ice packs and topical aloe gels, and spontaneously subsided within 2 weeks; degree 2 skin burns (4.6%) recovered after 4 weeks of expectant treatment. There were no severe complications, such as hepatic functions failure, hemorrhagic accidents or damage to the bile ducts, during or following treatment. The severity of the HIFU-associated complications was related to the size of the targeted area and the acoustic dose delivered (Li et al. 2007). Autopsy revealed a perforation of diaphragm in 50% of the cases and a gastric perforation in 25% in HCC ablation when it was done close to the proximal surface of the liver (Yang et al. 1991). All these adverse events were local to the treatment site and self-limiting. One of the solutions was to establish artificial ascites by injecting normal saline solution or 0.2% hyaluronan water solution into the abdominal cavity or under the skin, which increased the distance between the skin surface and target and served as a heat sink to cool the overlying structures. It has been demonstrated that this method not only reduced the probability and extent of thermal damage to intervening structures but also had no adverse affect on the efficacy of HIFU ablation (Wu et al. 2008).

There was a transient and clinically insignificant drop in hemoglobin immediately after HIFU by 1.0 g/dL (0.9–3.2), which might have been due to dilution following i.v. fluid administration during anesthesia. A transient rise was also seen in the white blood cell by 1.71×10^9 L^{-1} (−8.10 to 9.94), and C-reactive protein (CRP) on day 2 by 9.8 mg/L (−52 to 70), which imply a mild, nonspecific systemic inflammatory response to the tumor ablation (Leslie et al. 2012). Two days after HIFU treatment, liver enzymes (alanine transaminase [ALT] and aspartate transaminase [AST]) rose slightly, but with no obvious statistical difference, and then fell to normal level after 2 weeks. ALT levels increased to two to three times the baseline during the first 3 days after therapy for 56.4% patients with large tumors (>5 cm in size) (Li et al. 2007). However, serum alpha fetoprotein (AFP) levels decreased significantly after HIFU treatment. In contrast, serum levels of CA19-9 decreased insignificantly, and a trend in improvement was observed afterwards.

In some HIFU treatments done in China, Lipiodol deposition via hepatic angiography is usually performed shortly before HIFU because the blood supply decreases after Lipiodol occlusion of tumor microvasculatures, resulting in reduced heat loss during ablation, and Lipiodol deposition in the tumor causes increased deposition of ultrasonic energy (Cheng et al. 1997). However, such a protocol may also have disadvantages. Nonspecific deposition of Lipiodol within the liver segment as well as tumor made the tumor margins poorly defined by sonography and affected the accuracy of tumor targeting, which is especially disastrous for vital vasculatures or the biliary system inside the target. Therefore, HIFU without Lipiodol deposition seems more effective than the combined strategy (Ng et al. 2011). However, MRI will not be influenced by the deposition of Lipiodol.

Large tumor size was the significant risk factor affecting the complete ablation rate and was a prognostic factor for local control. Liver cancer with 5–10 cm in the longest diameter has the rate of complete necrosis of 69.6% at the first MRI check, and the 1- and 2-year local recurrence rates after one session of RFA were 21.1% and 32.3%, respectively. For those large tumors located close to the liver surface, the 3-year local recurrence rate was over 50% (Zhu et al. 2009). Four variables (age, Child-Pugh classification, TNM stage, and portal vein tumor thrombosis) were found to be important risk factors of survival rates in univariate analysis. However, univariate analysis does not always reflect the actual significance. Multivariate analysis, considering multiple variables simultaneously, showed that only three variables (Child-Pugh classification, TNM stage, and portal vein tumor thrombosis) have independent prognostic value and that Child-Pugh scores are a particularly accurate indicator of hepatic damage (Li et al. 2010b).

A HIFU session of 2 h for a superficial 2–3 cm tumor may be as acceptable as surgical resection, but less favorable with other minimally invasive techniques (e.g., RFA). Ablating the whole large tumor in one session takes about 3–4 h, and more than 7 h in some difficult cases. It is impossible and not advisable to increase the HIFU intensity to shorten the therapeutic time, owing to the limitation of the HIFU device and for avoiding avoiding damage to overlying structures. Movement of liver during HIFU exposure could compromise treatment efficacy, and preventing this motion would be a further limitation of HIFU (Leslie et al. 2012). The damage of rib resections, which reduces the advantages of HIFU in comparison to other ablative therapies (Li et al. 2007), may be minimized by the advance of phased-array technology. However, no clinical trials have been reported and compared with the single-element concave transducer. For ethical reasons, it was impossible to perform a randomized study on HCC patients. The long-term survival rates are vulnerable to being confounded by many factors, such as a less active attitude toward the disease and poorer nutritional maintenance in the control group than in the HIFU group.

13.7.5 Renal Tumor

The last three decades have seen a dramatic change in the epidemiology of renal cancer as well as a genuine increase in the occurrence. The popular use of abdominal sonography and CT has had a significant impact on the diagnosis of asymptomatic patients with small renal masses (SRMs ≤ 4 cm) (Chow et al. 1999, Luciani et al. 2000). The detection rate of SRMs increased almost threefold while that of 4–7 and >7 cm was 50% and 26%, respectively (Hollingsworth et al. 2006). The incidence in the older patients (>65 years) has almost doubled (Luciani et al. 2000). However, renal cancer grows slowly with no immediate life threat (Bosniak 1995), and up to 14%–20% may actually be benign tumors (Silver et al. 1997, Pantuck et al. 2001, Remzi et al. 2006). Identification of SRMs as benign or malign is very difficult, and tumor size is not sufficient for distinguishing them (Remzi et al. 2006, Klatte et al. 2008). Synchronous metastatic disease occurred in only 5% of tumors of < 3 cm but not in those < 2.0 cm (Kunkle et al. 2007). Many of these lesions have a low tendency to grow.

The increasing incidence of small renal cell carcinoma (RCC) has led to a shift from radical nephrectomy (RN) to nephron-sparing surgery (NSS), which is the current standard for T_{1a} SRMs, especially for the young and healthy, given its favorable effect on renal function since RN is associated with an increased risk of chronic kidney disease and overtreatment (Clark et al. 2008, Ljungberg et al. 2009). NSS has excellent outcomes, with 5-year cancer-specific survival (CSS) rates up to 97% (Reddan et al. 2001). The choice of laparoscopic (L-NSS) or open (O-NSS) NSS depends on the surgeon's skills as well as on tumor features, and their total major urological complication rates were similar (O-NSS: 6.3% vs. L-NSS: 9.0%) in a meta-analysis (Novick et al. 2009). However, the perioperative mortality of surgery in octogenarians was as high as 15%–25% because of significant comorbidities (Porpiglia et al. 2008).

Because of the associated morbidity in surgery, minimally invasive ablative therapies are preferred. Cryotherapy and RFA have been applied clinically either laparoscopically or percutaneously under the guidance of high-resolution imaging techniques and become established techniques. Although their initial outcomes are encouraging, long-term studies are necessary to confirm their lasting efficacy. Both cryotherapy and RFA show reasonable rates of tumor control. A meta-analysis of medium-term data with more than 1300 treatments demonstrated rates of local tumor progression of 5.2% for cryotherapy and 12.9% for RFA, rates of distant metastases of 1% and 2.5%, and rates of major urological complication of 4.9% and 6.0% (Aron and Gill 2005, Kunkle and Uzzo 2008).

In contrast to cryoablation and RFA, HIFU does not require puncturing the tumor, thereby avoiding the high risk of hemorrhage or tumor spillage. Application of extracorporeal

HIFU for renal tumors is well tolerated with no serious perioperative complications. However, it did not destroy the tumor sufficiently to be considered as an alternative to surgical extirpation.

Currently, HIFU ablation of renal tumors in humans remains in the early stages of clinical trials. Clinical feasibility studies have been performed for both cure and palliation of RCC, and two-thirds of patients showed histologic evidence of lesions (Vallancien et al. 1993, Wu et al. 2003b). Hyperechoic changes on gray-scale sonography during HIFU ablation were not apparent despite the use of very high acoustic powers, which does not mean ablation failure (Marberger et al. 2005). Because hyperthermia injury can occur for coagulative necrosis, at least 12 days post-treatment is required to evaluate the outcome. The imaging in 4 out of 14 patients showed evidence of irregular enhancement suggestive of recurrent tumor. Follow-up showed continuing involution of the ablation zone, with a mean decrease of 12% and 30% in tumor size at 6 and 36 months, respectively. But histopathological examination revealed necrosis in only 15%–35% (Marberger et al. 2005, Hacker et al. 2006). Although there was a thin crescent-like rim of enhancement at the lesion periphery, central loss of enhancement in all lesions during the follow-up implied tumor destruction and is the hallmark of successful ablation (Goldberg et al. 2009). Embolization alone failed to control tumor growth but served both to reduce tumor perfusion shown in contrast-enhanced MRI and color Doppler sonography and to increase the acoustic absorption (Figure 13.17).

Despite large-volume ablations and the long treatment duration, side effects were low, including up to 20% fever and 5% skin burns. There was no evidence of significant morbidity associated with prolonged anesthesia as well as the risks of urinary leakage, hemorrhage, and tumor spillage. There was mild to moderate discomfort in most patients, which was managed with oral analgesia. All skin-related complications resolved completely without specific treatment. Some erythema of the skin was found in patients subjected to the highest doses.

Extracorporeal HIFU has faced major technical difficulties. Respiratory motion is a major problem in its clinical application. Respiratory motion tracking and correction will also allow tissue temperatures to increase more rapidly, thus reducing the sonication times and reducing the heating

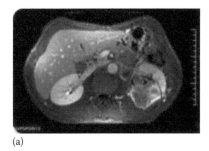
(a)

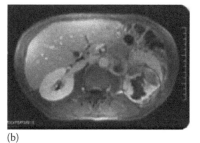
(b)

FIGURE 13.17 Contrast MR of (a) a large tumor in the left kidney and (b) 12 days following HIFU treatment. (Courtesy of Clinical HIFU Unit, Churchill Hospital, Oxford, U.K.)

of the prefocal tissues. Dual-lumen endotracheal intubation facilitated single-lung ventilation to minimize target movement during treatment. Even with it, precisely generating a small lesion in the predetermined location has proved virtually impossible. Structural heterogeneity is significant even in small tumors, yet is difficult to define prior to ablation, which leads to great attenuation at tissue interfaces (Ritchie et al. 2010), results in reduction of energy accumulation in the focal zone, and may be the main reason for the less consistent lesion production in comparison to that in normal renal parenchyma. Further investigations are under way for its efficacy of RRC treatment (Wu et al. 2003b).

While the extracorporeal approach shows unsatisfactory results, laparoscopic HIFU seems to be a promising alternative. Laparoscopic HIFU allows direct delivery of acoustic energy and can avoid the problems with respiratory movement and interphases to achieve a greater rate of tumor destruction as in extracorporeal approach but compromising the noninvasiveness (Klingler et al. 2008). Therefore, decreased morbidity, shorter hospitalization and convalescence, and preservation of renal function are expected. In a Phase I study, 3 of the 10 patients showed viable tumor cells after laparoscopic intracorporal HIFU ablation (Klingler et al. 2008). However, further prospective studies have to be carried out to determine the oncological outcome and to compare with open and laparoscopic surgery in SRMs (Klatte and Marberger 2009, Margreiter and Marberger 2010).

13.7.6 PANCREATIC CANCER

The pancreas is an essential gland organ in the digestive and endocrine system, producing hormones (e.g., insulin, glucagon, and somatostatin) into the bloodstream and secreting pancreatic juice to the small intestine or gut. Although pancreatic cancer is the 12th most common cancer in humans, its mortality ratio is as large as 98% (Jemal et al. 2011), the 4th leading cause of cancer death. Around the world, 338,000 new cases were diagnosed in 2012, and its prevalence is 4.1 per 100,000. About 55% of pancreatic cancer cases occur in more developed countries, such as Northern America and Europe, while Africa and Asia have the lowest incidence. The American Cancer Society had estimated that about 46,420 people (23,530 men and 22,890 women) would be diagnosed with pancreatic cancer and among them 39,590 people (20,170 men and 19,420 women) would die in the United States in 2014. Most of pancreatic cancers are diagnosed at the late stage with locally advanced (60%) and metastatic disease (20%), and only about 15%–20% of patients can undergo curative surgical resection (Parkin et al. 2006, Jemal et al. 2009). Gemcitabine is the gold standard drug for advanced pancreatic cancer; however, its clinical benefit response (CBR) is 12%–23.8%, and the median survival is prolonged only by 10 days. Erlotinib is the only targeted drug approved by the FDA. Chemotherapy, radiotherapy, and targeted drugs are rather ineffective for this malignancy (Nakakura and Yeo 2007). Adjuvant chemotherapy with 5-fluorouracil or the combination of 5-fluorouracil, leucovorin, irinotecan, and oxaliplatin (FOLFIRINOX) can increase the 5-year survival (about 10%–20%), but with significant toxicities (Hsueh 2011). In contrast, adjuvant chemoradiation is controversial with favoring practices. The median survival of pancreatic cancer patients is less than 3 months without therapy and less than 6–12 months with therapy. The overall 1-, 3-, and 5-year survival rates of pancreatic cancer patients is 16%, 5%, and 4%, respectively (Faivre et al. 1998). Most pancreatic cancer patients have severe abdominal pain, which is mainly due to the proximity of the pancreas to the duodenum, liver, stomach, jejunum, and transverse colon. The pain is usually dull and radiates to the waist, sometimes sharp and severe, and could be both neuropathic and inflammatory because of both tumor expansion and invasion of the celiac and mesenteric plexus. Although effective opioids for pain relief can enhance the quality of life for advanced pancreatic cancer patients, these analgesics have obvious adverse effects, such as vomiting, constipation, dysphoria, and respiratory depression.

HIFU has been used as a palliative approach for advanced pancreatic cancer since the late 1990s despite no approval by the FDA to date. Inclusion criteria usually are evidence of pancreatic cancer confirmed pathologically with either biopsy in initial laparotomy or sonography-guided fine-needle biopsy or diagnosed by computed tomography (CT) or positron emission tomography (PET)/CT and serum analysis; presence of inoperable pancreatic cancer on the basis of surgical consultation or refusal to undergo pancreaticoduodenectomy or other treatment; the minimum diameter of a solid tumor ($\geq$1.0 cm); a KPS score of at least 70%; adequate bone marrow (white blood cell count 42,500 mL^{-1}, platelet count 480,000 mL^{-1}, hemoglobin 48 g/mL^{-1}), renal (serum creatinine concentration <1.5 mg/dL^{-1}, blood urea nitrogen <20 mg%), and hepatic functions (serum transaminase level <2 times the upper normal range) except hyperbilirubinemia due to obstructive jaundice; and no palliative antitumor treatments performed in the previous 3 months. Exclusion criteria are the intolerance to HIFU treatment, radiotherapy or chemotherapy administered in the last 3 months, life expectancy <3 months, the tumor has invaded the duodenal wall, unstable hematogenic parameters, severe and active infection, and the patient had jaundice owing to biliary obstruction. Among all patients (3887, mean age of 60.8 years) found in the published papers, there were more men (1.7-fold more than women) with a little more cancer in the pancreas head with a mean size of 4.76 cm (2–11.9 cm) (Zhou 2014). HIFU is usually carried out as a daily procedure, with an average of 6.7 sessions on patients. Anesthesia may be used either to avoid the painful experience or to guarantee immobilization of the target (Wu et al. 2005c). If tumors are located in the pancreatic head, there is a substantial possibility of biliary obstruction or biliary duct damage caused by the thermal ablation. An endobiliary stent should be routinely placed before HIFU ablation.

Two hundred and fifty-one patients with advanced pancreatic cancer have been treated in China with reduction on tumor size, prolonged survival (mean: 12.5 months), but no pancreatitis (He et al. 2002). No skin burn, tumor hemorrhage, large blood vessel rupture, gastrointestinal perforation, or death within 1 month after HIFU therapy was observed. There was no evidence of dilatation of the bile or pancreatic duct, postinterventional pancreatitis, peritonitis, or jaundice during the follow-up period (He et al. 2002, Wu et al. 2005c). Obvious visceral pain that necessitated oral analgesic drugs resolved significantly in 84% of patients within 24–48 h after a single HIFU session, and the pain relief persisted during the follow-up period, which is similar to another initial non-randomized open-label human study (Wu et al. 2005c). Pain relief, including complete relief (CR) and partial relief (PR), was about 71.33% in the reported 1938 cases. The quality of life, such as appetite, sleeping, and mental status, improved in most cases, and the mean clinical benefit rate (CBR) in 508 cases was 71.06% (Zhou 2014). The mechanism of pain relief is not fully understood but hypothesized to the following mechanisms: the nerve fibers in the tumor are damaged or undergo apoptosis by the thermal effects, the targeted solar plexus may be inactivated to block the pain signal to be transferred to the brain, and the pressure on the nerve applied by

the tumor would be reduced due to tumor shrinkage. However, HIFU has less or no effect on the relief of obstructive pain. The average KPS increase by HIFU in the reported 290 cases was about 1.5-fold. In 806 cases, the median survival evaluated by the Kaplan–Meier method was 10.03 months (Zhou 2014) (Figure 13.18 and Table 13.2).

After clinical HIFU treatment, the serum amylase and urinary amylase levels are measured by a radioimmunometric assay as surrogate markers for traumatic pancreatitis. CA19-9, CA242, and CEA can be decreased by 49.41%, 34.93%, and 28.41%, respectively, which demonstrates the absence of pancreatitis (Table 13.3).

Most of the side effects associated with HIFU ablation are moderate and minor complications, such as first- and second-degree skin burns, edema, fever, and tumor warming, gastrointestinal (GI) dysfunction (e.g., abdominal distension and anorexia with slight nausea), thickening and swelling of the subcutaneous layer, and mild abdominal pain in the treated regions, which are inevitable and mostly associated with

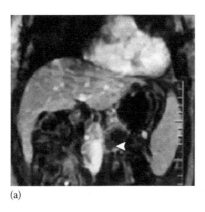 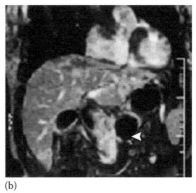

(a)   (b)

FIGURE 13.18 Dynamic contrast-enhanced gradient-echo T1-weighted MR images (a) before and (b) 2 weeks after HIFU ablation for advanced pancreatic cancer with a diameter of 4.5 cm without contrast enhancement in the treated lesion (arrowhead). (From Wu, F. et al., *Radiology*, 236, 1038, 2005. With permission.)

TABLE 13.2

Statistics of the Number of Sessions, Pain Relief, Clinical Benefit Rate, and Survival of Advanced Pancreatic Cancer Patients Undergoing HIFU Therapy or in Conjunction with Chemotherapy or Radiotherapy

	Session	Pain Relief	Complete Relief (CR)	Partial Relief (PR)	Clinical Beneficial Rate (CBR)	Survival (Months)
HIFU	6.7	71.33%	29.66%	39.83%	71.06%	10.03
	(n = 653)	(n = 1938)	(n = 1534)	(n = 1534)	(n = 508)	(n = 806)
HIFU + chemo	7.4	59.72%	8.35%	45.39%	74.76%	10.16
	(n = 471)	(n = 602)	(n = 395)	(n = 395)	(n = 353)	(n = 270)
Chemotherapy		31.50%	4.31%	23.22%	38.85%	7.4
		(n = 261)	(n = 100)	(n = 100)	(n = 222)	(n = 112)
HIFU + radio	5.2	65.91%	27.84%	38.07%	82.15%	15.55
	(n = 130)	(n = 176)	(n = 176)	(n = 176)	(n = 89)	(n = 101)
Radiotherapy		29.65%	3.76%	25.89%	60.36%	
		(n = 67)	(n = 67)	(n = 67)	(n = 95)	

Source: Zhou, Y.F., *Gastroenterol. Res. Pract.*, 2014, 205325, 2014.

ultrasound itself (Xiong et al. 2009). Third-degree skin burns found in the early application of HIFU were avoided after appropriate use of the coupling condition (Wu et al. 2004a). Two cases of mild pancreatitis were also found in the preliminary application (Wang and Sun 2002). A total of 15 cases of acute pancreatitis due to mechanical lysis and release of pancreatic enzymes were found and resolved usually within a week in the literature review (Zhou 2014). Although some patients had vertebral body and subcutaneous fat in MRI, all cases were asymptomatic with no need of further treatment (Jung et al. 2011). Most major complications, such as portal vein thrombosis (Orsi et al. 2010), delayed perforation of the pseudocyst (Xiong et al. 2009), transient upper GI bleeding, tumor-duodenal fistulas (Lee et al. 2011) could be avoided through careful operation.

The purpose of post-treatment diagnosis is to verify the generation of necrosis in the target and its size. CT can clearly demonstrate the tumor size and shape. But CT is insensitive to fat tissue, unreliable to assess the functionality of tumor's rim, and difficult to perform in hypovascular tumors. Contrast-enhanced CT (iodine), multiple-detector CT (MDCT), or MRI can assess necrosis by the absence of vascularity within the tumor but not its metabolic activity. Contrast-enhanced MRI works excellent in the rapid assessment of therapeutic response of ablated tumor, while PET or PET-CT is useful for staging pancreatic cancer. SPECT demonstrates the active metabolism of viable cancer cells. Because of malabsorption of tissue necrosis and slow atrophy of the cancer fibrosis by the scattered intravascular thrombi after HIFU ablation, the size of ablated tumors may not decrease significantly, but on the contrary may increase temporarily due to the edema on the edge. Thus, the diagnosed pancreatic tumor size is not a reliable benchmark to evaluate HIFU efficacy. The ablated tumor was replaced by a scar 10 months later, and there was no apparent mass lesion remaining (Sofuni et al. 2011). Furthermore, the inconsistently enhanced echogenicity in the target (0%–100%) is due to the induced cavitation or boiling bubbles and does not agree with the lesion size (Figures 13.19 and 13.20).

Hypovascularity of the tumor is changed to hypervascularity after chemotherapy, and then back to hypovascularity, which is more sensitive to heat shock by HIFU therapy, and the vasculature of large vessels through the tumor remains undamaged. HIFU increases the permeability of the vascular endothelial cells, which may be due to both intravascular cavitation and thermal effects, aids the distribution and penetration of the pharmacokinetics into the interstitial space of the tumor due to the acoustic radiation force, delays the drug clearance, and inhibits tumor cells to repair damage due to chemotherapy (Jang et al. 2010). Reduction in the vascularity through the tumor delays the drug clearance and increases the drug concentration. Hypovascularity leads to tumor cell damage and hypoxia, increases the cytotoxicity, and improves the sensitivity of radiotherapy. Since fibrosis produced after

TABLE 13.3
Comparison of the Serum Levels before and after HIFU Treatment

	CA19-9 (U/mL)		CA242 (U/mL)		CEA (ng/mL)	
	Pre-HIFU	Post-HIFU	Pre-HIFU	Post-HIFU	Pre-HIFU	Post-HIFU
Range	42.6 ± 8.6–583.8 ± 20.4	21.5 ± 6.6–305.7 ± 19.3	73.6 ± 41.7–114.4 ± 42.0	46.3 ± 13.4–85.2 ± 21.9	38.4 ± 12.4–53.8 ± 17.3	18.9 ± 33–33.9 ± 14.8
Change	49.41% (n = 701)		34.93% (n = 135)		28.41% (n = 114)	

Source: Zhou, Y.F., *Gastroenterol. Res. Pract.*, 2014, 205325, 2014.

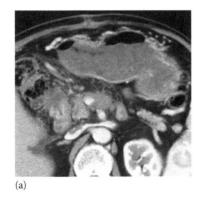
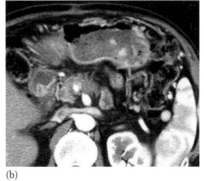

(a)   (b)

FIGURE 13.19 CT imaging shows no apparent change of pancreas (a) before and (b) after HIFU therapy. (With kind permission from Springer Science+Business Media: *J. Hepatobiliary Pancreat. Sci.*, The current potential of high-intensity focused ultrasound for pancreatic carcinoma, 18, 2011, 300, Sofuni, A., Moriyasu, F., Sano, T., Yamada, K., Itokawa, F., Tsuchiya, T., Tsuji, S., Kurihara, T., Ishii, K., and Itoi, T.)

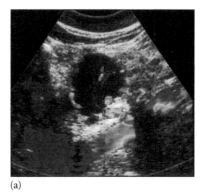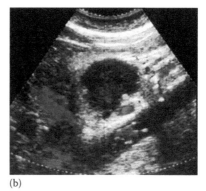

(a)   (b)

FIGURE 13.20 Sonography of pancreatic cancer (a) before and (b) after HIFU therapy showing the enhancement of echogenicity in the tumor but decrease in vascularity, an indicator of coagulative necrosis. (From Wang, K. et al., *Chin. J. Ultrasound Med.*, 22, 797, 2006.)

hyperthermia influences radiation effects, HIFU is carried out after or simultaneously with radiotherapy. The combination of HIFU and chemotherapy or radiotherapy can achieve a higher CBR and longer survival than the single modality. The side effects are associated with HIFU, chemotherapy, and radiotherapy themselves, but without enhancing the complications (Zhou 2014).

13.7.7 Bone Tumor

Bone tumor is about 0.8%–1.2% of body tumor, and its clinical symptoms include indolent bone pain, movement limitations, hypercalcemia, and a tendency for pathological fractures. Malignant bone tumors are usually secondary and often due to cancer metastasis, the third most common organ after the lungs and liver (i.e., in 90% of patients with metastatic breast cancer). Amputation has been used for the treatment of primary bone malignancies (Finn and Simon 1991). Recently, limb salvage with intact local function has gained more acceptance. Surgical removal and external beam radiotherapy are commonly applied to treat primary bone tumors and bone metastases, respectively. Surgical removal of the osteosarcomas may cause 40%–50% serious complications and disabilities when the tumors are located in the ilium, scapula, pubis, and limbs. Heat degradation of normal bone morphogenetic proteins can induce nonunion and resorption (Delloye 2003). Laser-induced interstitial thermotherapy (LITT), RFA, and microwave hyperthermia have been applied, with more frequent reunion of diseased bone segments, lower bone resorption, superior limb function, and substantial tumor control rates (Suk et al. 2002, Fan et al. 2003). To control and relieve bone pain, increase the quality of life, prevent complications, improve prognosis, and extend survival time are of importance for metastatic bone tumors, but are major challenges (Hoegler 1997). Radiotherapy provides pain relief but does not extend the patient survival time (Mereadante and Fulfaro 2007, Siegel and Pressey 2008). Systemic therapy (e.g., analgesics, chemotherapy, hormonal therapy, and biphosphonates) and thermotherapy (e.g., LITT and RFA) do not achieve pain relief (20%–30%) and long-lasting efficacy and frequently cause side effects or

need chronic administration (Mundy 2002, Roodman 2004, Callstrom et al. 2005, Yin et al. 2005).

Initially, HIFU was not considered as a suitable modality for bone tumors because of the strong reflection at the bone surface. However, malignant bone tumors, such as an aneurismal bone cyst, can destroy the integrity of cortical bone, which results in significant changes in the acoustic characteristics of the tumor-laden bone and the penetration of acoustic wave through it for diagnosis (Gomez et al. 1998). The high acoustic absorption and low thermal conductivity of bone cortex make it possible to use a relatively low level of ultrasound energy to achieve a localized necrosis of osteocytes without damaging the adjacent tissue (Smith et al. 2001). The treatable size of bone tumor increased with the absorption ratio of bone marrow to tumor, acoustic window of skin, and diameter of bone, but decreased with muscle depth and the specific absorption rate ratio (SARR) of the bone tumor to the surface skin, bone marrow, and bone (Lu et al. 2000). The optimal driving frequency depends on the tumor depth, ultrasound absorption of the bone marrow, and bone diameter, but is independent on the acoustic window area and SARR (Lu et al. 2000). Preliminary results demonstrated that HIFU together with chemotherapy in a limb salvage procedure is effective and well tolerated with fewer complications because of no surgical traumas and retainment of blood vessels for revascularization, repair of inactivated tumor bones, and no delay of postoperative chemotherapy to improve the prognosis (Liberman et al. 2009).

Recruited patients were those who had osteosarcoma histopathologically diagnosed by core biopsy at stage IIA and IIB without involvement of vital arteries and nerves, refused to undergo amputation and had a strong willingness for limb salvage surgery or was not a candidate for surgery, or were nonresponsive to multiple courses of chemotherapy. The exclusion criteria were primary bone malignancy of the spine or skull, a primary bone malignancy with a pathologic fracture, the distance between tumor and skin of <0.5 cm, joint deformity (bent in fixed position of less than 45°), invasion of tumor into the skin, the involvement of a joint or an artery or a nerve with tumor, or a KPS score no more than 70%. Patients underwent general or epidural anesthesia to improve immobilization and without allowing the feedback of experiencing pain and serious damage to sensitive structures, while conscious sedation was used to alleviate the ablative pain as in other studies. After suitable anesthesia, patients were carefully positioned—either prone, supine, or on side according to the tumor location—to couple the HIFU energy to the target. The power to be delivered was determined according to the tumor position, size, focal distance, and distance to the nearby nerves, blood vessels, and other organs. The margin from the edge of the tumor to the normal soft tissue and bone tissue was 1–2 and 3–5 cm, respectively. Skin protection was necessary for the treatment of the tibia, whose tumors often occurred subcutaneously. The accuracy of HIFU focusing was verified by a few low-energy sonications in order to achieve a mild increase of temperature exactly at the targeted volume. If found acceptable, the actual ablation was started at a full therapeutic power. Three to six cycles of neoadjuvant chemotherapy was given before HIFU, followed by four to six cycles of adjuvant chemotherapy after it. Chemotherapy not only inhibits growth of metastatic foci to prevent recurrence, but also reduces tumor size. Therefore, there may be a synergistic effect of the combined protocol (Li et al. 2010b). Cancers that are insensitive to chemotherapy, such as chondrosarcoma and malignant giant cell bone tumor, were treated with HIFU alone. Immediately after HIFU treatment, the adjacent joints were fixed by either plaster or splint support for 3–6 months. The mean age of patients was about 20 years (5–89 years), and the treatment duration varied (27.5–647.6 min), and mean number of treatment sessions was 1.35 (1–4 sessions). The biggest tumor, $20 \times 10 \times 8$ cm, needed two sessions.

After chemotherapy but before HIFU treatment, tumors showed low intensity on T1-weighted 1.5 T MRI, but a slightly higher intensity on T2-weighted signal. Meanwhile, bone imaging by PET-CT or ^{99m}Tc-methane-diphosphonate (^{99m}Tc-MDP) revealed abnormal radioactivity concentration. Four to six weeks after HIFU ablation, the tumors showed a slightly higher signal on T1-weighted MRI while a mixed signal on T2-weighted one, which was not enhanced with the injection of gadolinium–diethylenetriamine pentaacetic acid (Gd-DTPA). There was also a thin, regular intensification rim that clearly separated from the normal tissue (Chen et al. 2002b). And, PET-CT or ^{99m}Tc-MDP showed disappearance of radioactive uptake and a radioactive cold region, suggesting complete inactivation of the tumor foci and therapy success. There was no enhancement in color Doppler sonography, showing decrease or disappearance of blood supply. The presence of new-growth normal bone was determined with radiography and SPECT, which showed that the bone defects were repaired 1–6 months after treatment. The earliest neovascularization began 3 months after HIFU, and vascularization to the whole bone tumor needed more than a year. Furthermore, the levels of serum alkaline phosphatase (mean: 227.9, 78–412 U/L) and lactic acid dehydrogenase (mean: 263.3, 136–400 U/L) did not decrease 3 days after HIFU treatment but 1–2 months later.

On the basis of MRI or PET-CT, HIFU was effective. Among 13 patients with malignant bone tumors, the rate of complete response, partial response, and moderate response

was 46.2%, 38.4%, and 7.8%, respectively, with an overall response rate of 84.6%. For 12 patients with metastatic bone tumors, the corresponding value was 41.7%, 33.3%, and 8.3%, with an overall response rate of 75.0%. After HIFU treatment, 87.5% of the patients were completely relieved of pain, and all experienced significant relief. Physiological mechanisms of the resulting pain reduction are inhibition of signal transduction due to physical destruction of sensory nerves in the periosteum or corticalis, reduction of tumor volume with decompression and reduced stimulus for pain fibers, changes in the local microenvironment (e.g., reduction of osteoclastic and nerve-stimulating cytokines such as interleukins and TNF-α), and destruction of osteoclasts. The 1-, 2-, 3-, and 5-year survival rates were 100.0%, 84.6%, 69.2%, and 38.5%, respectively, for patients with primary bone tumors and 83.3%, 16.7%, 0%, and 0%, respectively, for patients with metastatic bone tumors (Li et al. 2010a). Of the 34 cases of stage IIb, 30 cases survived disease-free, 2 died of lung and brain metastases, and the other 2 had local recurrence after a mean of 17.6 months follow-up. Among 10 of the stage IIIb cases, 5 survived, 1 had local recurrence, and 5 died of lung metastases (Chen et al. 2002b, Chen and Zhou 2005). The lack of complete response may be due to insufficient ultrasound energy accumulation in the target by the adjacent anatomy (i.e., the presence of bowel in the acoustic pathway for tumors invading the pelvic cavity) and large tumor size (i.e., >10 cm in diameter in the proximal tibia, distal femur, proximal femur, and chest wall) as well as evolving physician experience with this new technology. Completion of the full treatment protocol is pivotal to survival (Chen et al. 2010b).

After HIFU ablation, all patients experienced mild local pain, which was controlled either with analgesics that were injected (20%) or administered orally (70%) for 2–3 days, or without medication (10%), which then disappeared within 3–4 days. In clinics, there were no significant changes in ECG, renal function, and blood electrolytes of patients before and after HIFU. Although ALP activity increased 3 days after HIFU, it returned to the same as or lower than the pretreatment level 1 week later. A low-grade fever (37.5°C–38.4°C) after HIFU ablation was normal (10%–20%) and persisted for 3–5 days. The symptoms of first-degree burns faded without intervention within 1–2 weeks, while those of second-degree burns healed without scars 4 weeks after treatment. Symptoms of venous skin engorgement seen in six patients before HIFU treatment disappeared in four patients, improved in one patient, and showed no obvious change in one patient. Local edema was observed in all patients immediately after the HIFU ablation, reached its peak the day after the ablation, and subsided within an average of 19 days (13–24 days). Peripheral nerve damage, shown as the lack of feeling in the affected limb, was observed in 10 patients. Two patients recovered completely, nine patients showed alleviated injury and regained function at different levels, and only one patient did not recover from sciatic nerve damage. It is to be noted that under ultrasound-guided HIFU ablation, nerves may not always be clearly detected. Bone fracture was observed in six (8%) patients, ligamentous laxity occurred in three (4%), and epiphysiolysis or secondary infection occurred in two (2%).

Interventional radiologists have used a combination of ablative techniques and vertebroplasty to treat vertebral body lesions, areas that cannot be treated with HIFU, with good success. However, till now, the reported cases are still quite limited. A large number of clinical trials are required to completely evaluate the HIFU performance (Figures 13.21 through 13.23).

HIFU is also used to successfully treat gynecological diseases such as white lesion of vulva and chronic cervicitis of more than 10,000 cases, showing both safety and efficacy. After 2 years of follow-up of 76 cases, 49 patients were successfully treated and 23 patients had improved symptoms (Li and Bian 2004). In comparison to laser therapy for symptomatic cervical ecotopy, ultrasound therapy has a better outcome and causes little damage to the normal tissue (Chen and Chen 2008).

13.7.8 Brain

For many years, the brain was considered as an elusive target for ultrasound therapy, and a cranial window (i.e., the temporal bone or partial skull open) is required for acoustic wave propagation (Lynn et al. 1942b). The large differences of acoustic properties (acoustic impedance and absorption coefficient) of water and skull bone result in severe distortion of the ultrasonic focus, significant reflection from the interface, and high attenuation in the acoustic wave transmission (Fry and Barger 1978, Hynynen and Jolesz 1998), which results in undesired heating of the scalp and skull before achieving sufficient heating of the underlying tissue (Sun and Hynynen 1998). Ultrasound at lower frequencies has less attenuation and distortion through the skull but suffers from poor focusing (Hynynen and Jolesz 1998). In contrast, higher frequencies result in tighter focusing and higher pressure amplitudes, but greater attenuation and focal distortion. The optimal transcranial ultrasound frequency for thermal treatments has been

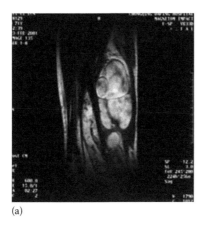
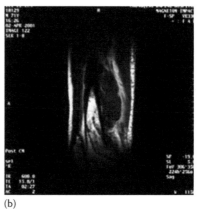

(a)   (b)

FIGURE 13.21   Contrast-enhanced magnetic resonance images (a) before and (b) 14 days after HIFU ablation in a 45-year-old patient with osteosarcoma at the upper right tibia.

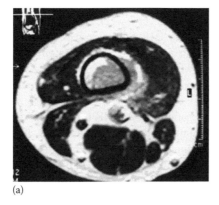
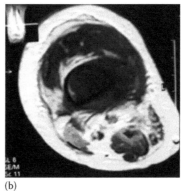

(a)   (b)

FIGURE 13.22   Dynamic contrast-enhanced gradient echo T1-weighted magnetic resonance images of a 9-year-old girl with osteosarcoma in the inferior segment of the right femoral bone (a) after neoadjuvant chemotherapy and 1 week before HIFU with abundant blood supply and tumor survival in the osteosarcoma lesion in a size of 42 × 46 × 65 mm, and (b) disappearance of contrast enhancement in the treated lesion 8 weeks after two HIFU treatments. (Li, C., Zhang, W., Fan, W., Huang, J., Zhang, F., and Wu, P.: Noninvasive treatment of malignant bone tumors using high-intensity focused ultrasound. *Cancer.* 2010. 116. 3938. Copyright Wiley-VCH Verlag GmbH & Co. KGaA. With permission.)

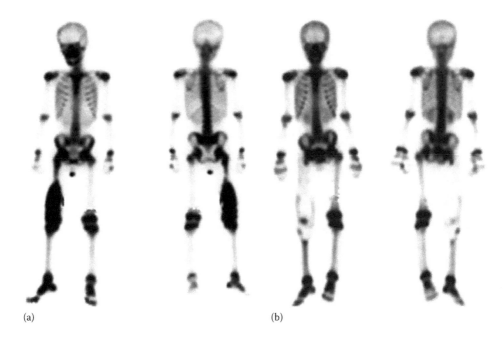

(a) (b)

FIGURE 13.23 Tc-99m methylene diphosphonate ECT image of a 10-year-old patient with a 46 cm long right femoral and tibial osteosarcoma (a) before HIFU with focal area of increased uptake, showing a local abnormality, and (b) 2 weeks after HIFU with the complete disappearance of the tracer uptake in the treated region. (Courtesy of Haifu Corp, Chongqing, China.)

determined to be at 600–700 kHz (Sun and Hynynen 1998, Hynynen and Clement 2007).

The use of a phased array allows correction of such limitations (i.e., phase aberration and defocusing) for the brain therapy. The element sizes required for steering are proportional to the wavelength, which means that higher frequencies require a larger number of smaller elements to populate an entire array to achieve adequate focusing and avoid grating lobes. However, increasing the number of elements (several hundreds) greatly increases the technical complexity since each element needs an independent driving signal, power amplifier, and matching circuitry (Sun and Hynynen 1998). In addition, the array elements have substantial increase of the electrical impedance due to their decreased width-to-thickness ratio (w/t). It was compensated using a multilayer lateral-mode coupling method instead of employing electrical matching circuits at a high cost of fabrication complexity and effort. For an n-layer lateral mode transducer, the electrical impedance is decreased by $(t/nw)^2$ times compared to a single-layer transducer driven in thickness mode (Song et al. 2012).

There are several strategies of correction that make completely noninvasive transcranial focused ultrasound possible (Kyriakou et al. 2014). The first is to derive the skull morphology (spatially varying geometry, thickness, and density) from 3D CT images, from which both the speed of sound and attenuation along each beam path can be estimated (Clement and Hynynen 2002, Aubry et al. 2003). Based on them, the phase aberration could be corrected using a conventional passive beamforming algorithm (Norton and Won 2000). Furthermore, the emissions from acoustically stimulated microbubbles could be mapped spatially and passively with sparse hemispherical imaging arrays after the distortion correction, which is particularly useful for nonthermal,

cavitation-mediated transcranial HIFU applications, such as blood–brain barrier disruption or sonothrombolysis where no real-time monitoring techniques currently exist (Jones et al. 2013). However, difficulties of image registration in practice have led to targeting inaccuracies. Not until the adoption of MRI and thermometry for the guidance of focused ultrasound did transcranial focused ultrasound neurosurgical interventions surgery become a clinical reality. With the possibility of image guidance and feedback, HIFU became a safe and viable treatment alternative in the brain treatment (Hynynen and McDannold 2006). In addition, 3D MR thermometry, which is achieved by a combination of k-space subsampling and a temporally constrained reconstruction method, monitoring of HIFU heating at both the focal spot and near-field tissue/bone interfaces provides a full characterization of the entire sonication field and subsequently thermal field. As a result, unintended near-field heating can also be assessed if a nonoptimal transducer designed for transcranial geometries is used *in vivo* to deliver therapeutic levels of HIFU (Odéen et al. 2014).

Secondly, the time reversal method, which is based on the reversibility of the linear wave equation (Fink 1992), can be used to precisely focus acoustic energy through skull (Aubry et al. 2003). The skull aberrations are corrected by reversely transmitting the signals picked up by an implantable hydrophone as a time reversal mirror. Subsequently, a sharp focus and several lesions are obtained through a human skull, demonstrating the accuracy and the steering capabilities of the phased-array system (Pernot et al. 2003). However, inserting a hydrophone into the neighborhood of the tumor reduces the significant advantage of HIFU application, namely noninvasiveness, and increases the complexity of the procedure (Thomas and Fink 1996, Pernot et al. 2007). In addition, the full wave propagation through the skull reconstructed using 3D CT images was simulated in a finite difference time domain (FDTD) to deduce the impulse response relating the targeted location and the ultrasound therapeutic array, thus providing a virtual time reversal mirror and adaptively correcting the distortions (Thomas and Fink 1996, Marquet et al. 2009). Furthermore, the use of amplitude correction allows either maintaining a spatially uniform focus or minimizing localized hotspots on the skull (White et al. 2005). In addition, the amplitude and phase of the driving pressure on each element could be determined from the temporal change in the monitored pressure of diverging acoustic waves emitted from virtual acoustic sources at the hot spot as well as the target point. This approach is an extension of the time reversal theory and can allow the phased-array HIFU system to be dynamically controllable. The primary peak is found to be strengthened through such a heuristic, iterative, and self-adaptive process, while the superficial secondary peaks are to be obliterated in the transcranial HIFU simulation using human body voxel data (Leduc et al. 2012).

Thirdly, an MRI scanner through a motion-sensitive sequence is used to measure the local displacements at the chosen focus by the acoustic radiation force. After the transmission of a set of spatially encoded ultrasonic waves, a noniterative inversion process is employed to accurately estimate the spatial–temporal aberration induced by the propagation medium and to maximize the acoustical intensity at the focus. So, strong distortions (up to 2π radians) could be recovered with acceptable errors (≤ 0.8 rad), negligible heat deposition, and limited acquisition time (Larrat et al. 2010b).

Finally, anisotropic, acoustic complementary metamaterial (CMM), which consists of unit cells formed by membranes and side branches with open ends, can be placed adjacent to the skull to restore acoustic fields distorted by aberrating layers. Numerical studies demonstrate that sound transmission can be enhanced, like virtually removing the skull in a noninvasive manner. The acoustic intensity at the focus is increased from 28% to 88% of the intensity in the free field (Shen et al. 2014). Although no physical device has been fabricated for testing, this technology has great potential in HIFU application.

The current clinical prototype (ExAblate 4000; InSightec, Haifa, Israel) has two operating frequencies of 230 and 650 kHz, with the higher frequency being used to treat patients with glioblastomas (McDannold et al. 2010), chronic neuropathic pain (Martin et al. 2009), and essential tremor (Chang et al. 2015). The accuracy of MRgFUS is 0.4 ± 1 mm along the right/left axis, 0.7 ± 1.2 mm along the dorsal/ventral axis, and 0.5 ± 2.4 mm in the rostral/caudal axis in the human cadaver model (Chauvet et al. 2013). Some studies have already been performed and reported recently for neurosurgical interventions for a variety of chronic, therapy-resistant neurological diseases in the thalamus, subthalamus, and basal ganglia (White 2006, Larrat et al. 2010a, Dervishi et al. 2013). Ongoing clinical trials on over 130 patients with glioblastomas (McDannold et al. 2010), neuropathic pain (Martin et al. 2009), essential tremor (Chang et al. 2015), Parkinson's

disease, and obsessive–compulsive disorder are very promising. Adjusting the ultrasound parameters allows not only ablation of pathological tissue, or silencing dysfunctional neuronal circuits, but also modulation of neural functions (Bauer et al. 2014). Essential tremor is a common movement disorder in adults (i.e., uncontrollably shaking hands), and HIFU destroys the ventralis intermedius nucleus of the thalamus for deep brain stimulation. Dramatic improvement in movement control and significant changes in brain mu rhythms were found after a single, same-day procedure. HIFU was performed as noninvasive central lateral thalamotomies in 12 patients with the ablation diameter of 3–4 mm and peak temperatures of $51°C$–$64°C$ (Jeanmonod et al. 2012). At 3 months, patients had a mean pain relief of 49% and six patients experienced immediate and persisting improvements. HIFU may reduce hyperexcitable activity in the motor cortex, resulting in normalized behavioral activity (Chang et al. 2015). Only one case of bleeding at the target and ischemia in thalamus was found.

The prevalence of stroke worldwide and the paucity of effective therapies have triggered interest in the use of transcranial ultrasound as an adjuvant to thrombolytic therapy (Bouchoux et al. 2012). HIFU can mechanically fragment a blood clot in the middle cerebral artery (MCA) within 30 s with no use of thrombolytics or microbubbles at an acoustic power of 300–450 W (Rosenschein et al. 1994, Maxwell et al. 2009, 2011, Burgess et al. 2012), which enables faster recanalization and provides a treatment option for those who are contraindicated to thrombolytics (~97%) (Rogers et al. 2011) (Figure 13.24).

For neurosurgical procedures in particular, focused ultrasound enables a same-day alternative to surgery, making potential risky procedures now viable. Finally, with the development of more advanced correction algorithms, phased arrays, and multichannel driving systems, devices will gain increased ability to precisely target locations within larger steerable volumes and with more power (Pajek and Hynynen 2012).

13.8 CLINICAL EXPERIENCE

Although HIFU has been finding its role in clinics as a noninvasive and effective therapeutic modality, particularly for those widespread or inoperable cancers, its technical development is still in its infancy (ter Haar 2008). The success of HIFU ablation depends on several factors, such as the ability to obtain an adequate and optimal acoustic window, the ability to precisely determine tumor size, the determination of 100% tumor cell killing, the ability to follow local recurrence, and patient selection. Despite being an overall safe and noninvasive therapeutic modality for cancer ablation, HIFU requires careful preoperative preparation as well as operative performance (Wang et al. 2013). Understanding the factors for complications, recruiting appropriate patients, preparing the preoperation carefully, selecting proper HIFU operation parameters, and paying attention to adjacent vital organs during the procedure are necessary steps for minimizing severe complications (Jung et al. 2011). Extensive

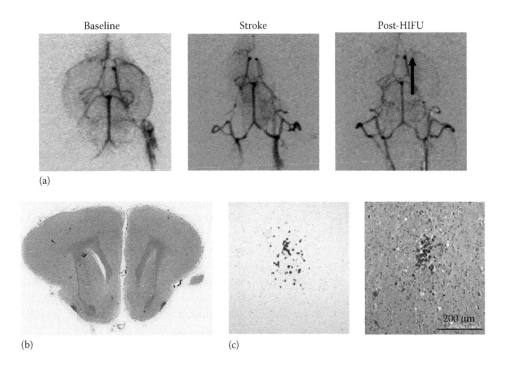

FIGURE 13.24 (a) Fifty percent of rabbit treated with 415 W power exhibiting recanalization in angiograms with no damage or leakage from the vessel. (b) H&E pathology showing no evidence of tissue damage or hemorrhage in the region of HIFU sonication or in the surrounding brain tissue. (c) Small fragments of the disrupted iron-loaded blood clots found in Prussian blue staining, and in the cortex, downstream of the MCA, in one animal. (From Burgess, A. et al., *PLoS ONE*, 7, e42311, 2012. With permission.)

research is going on in the area of focusing the HIFU pulses, the technique of gradual pulsed exposures to achieve a cumulative therapy result, improving imaging quality for accurate tumor determination and post-treatment evaluation, and developing real-time monitoring modality for lesion generation and temperature elevation. A treatment session lasting for 2 h for a superficial 2–3 cm tumor may be acceptable when compared to the alternative of surgical resection, but compares less favorably with other minimally invasive techniques. The longer treatment time may be justified on the grounds of lower morbidity and mortality than conventional surgery. Treatment time will be reduced with the development of HIFU technology, experience, and in combination with methods to reduce tumor perfusion, such as trans-arterial embolization (Wu et al. 2005a).

Nonsurgical ablation relies on imaging quality for an accurate determination of the tumor extent, which is why MRI guidance for HIFU initially became more rapidly accepted clinically than sonographically guided HIFU (Yagel 2004). Although MRI is more accurate than mammography or sonography in size assessment, it cannot exclude small amounts of residual invasive cancer. Three-dimensional sonography is likely to better delineate a volume of tissue to be treated than just a single plane or orthogonal planes, which provide valuable information of the treatment planning, monitoring, and post-treatment evaluation of HIFU. Definition of the tumor size and shape, especially the tumor margin, must be identified correctly in cancer resection or ablation. However, both MRI and sonography may underestimate tumor size (Boetes et al. 1995). The lack of a pathologic method to examine tumor

margins directly is still a major argument against this thermal approach because even MRI cannot delineate adequately the microscopic extent of tumor cells as a histological examination. The ideal tumor-free margin size of at least 10 mm is used for most microscopic cancerous tissues (Singletary 2002). However, the minimum value is always in debate. In a clinical histology study, the margin between HIFU-induced coagulation in the tumor and its surrounding normal tissue was 15–22 mm (mean 18.07 ± 5.8 mm). Pretreatment of HIFU on the margin of resectable tumor may also be good for a better surgical outcome.

There are a limited number of acoustic windows because bone, air, and gas affect the propagation of ultrasound beams into the targets beyond them. Small bubbles in the coupling media and local anesthesia cause acoustic scattering and thus limit the power delivery. Thus, it is recommended to place the local anesthesia beyond the lesion. The skin at the wave entry site should be shaved to avoid the trapping of bubbles, degassed with a vacuum cup aspiration device, and degreased with 95% alcohol. Artificial pleural effusion may be placed if necessary to ensure the acoustic window. Proper positioning is selected by observing the acoustic path in sonography. The gas in the stomach and colon should be evacuated by careful bowel preparation, such as providing liquid food and no milk for 3 days, fasting for 12 h before treatment, an enema in the early morning on the day of treatment, insertion of a urinary catheter (catharsis), and intraoperative bladder pressure. Drinking degassed water can remove the bowel gas quickly, but it has a short effective duration. Medicines may be more helpful, such as oral administration of a quick-solution gastroenter-ultrasound developer. Applying slight pressure to the abdomen, such as with a soft water balloon or a gel pad, also helps to compress the bowel loops and push them out of the acoustic pathway. Placement of water cushion may push the bowel loops. Patients with extensive scars or scars lying in the path of the acoustic beam should be excluded because scar tissue absorbs ultrasound strongly and may result in a skin burn. Nonextensive abdominal scar should not be traversed by angling the beam path. In addition, the ultrasonic beam distortion through inhomogeneous tissues and multiple interfaces or blockage by ribs could be corrected using certain algorithms and the phased-array structure. A distance of at least 1 cm between the tumor and the skin (to avoid superficial burn), and between the tumor and the chest wall (to avoid thermal accumulation in the underlying ribs and lung), is required. If a tumor is located more than 10 cm beneath the skin, the total attenuation would reduce the possibility of successful ablation.

In cancer ablation, accurate spatial control of the delivered acoustic energy is mandatory. Geometrical information from diagnostic imaging and from the positioning system must be co-registered to provide the reference of the HIFU transducer. Because of acoustic refraction through different heterogeneous interfaces, the actual acoustic focal point does not always coincide with the geometric one. Additionally, acoustic cavitation and boiling occurring in tissues induced by high-intensity ultrasound can also affect ultrasound propagation. Therefore, an examination prior to the ablation is preferred, usually by observing the induced bubble or acoustic interference pattern at low-power sonication. Respiratory motion during the treatment spreads the acoustic energy over a large area in the target than expected, and may result in incomplete tumor coagulation and damage to adjacent tissues. Although a variety of methods, such as anxiolytic, antispasmodic, analgesic, sedation, and local anesthesia, can be applied to reduce or minimize it, such a motion cannot be completely eliminated. If it is too serious during operation, general anesthesia with endotracheal intubation and mechanical ventilation should be applied to allow provisional suspension of breath with controlled pulmonary inflation as well as reduction of pain and discomfort associated with HIFU ablation. Real-time tracking the physiological or accidental motion of the patient during the HIFU exposure is critical in accurate control of the generated thermal dose according to the planning and would allow rapid focus shifting in synchronization with the target position. Motion artifacts due to respiration and heartbeat can be minimized during either end expiration or inspiration. However, the compensation of the tumor location is more challenging. An alternative solution is electrically steering the focus by the phased array in order to keep the exposed target consistently. Ideally, the tissue inhomogeneity and attenuation can be compensated using the phased-array design for accurate beamforming. The combination of all these factors makes use of the uniform and isotropic medium model as a tool to evaluate the focusing effects of focused ultrasound in biological tissues invalid.

Determination of the appropriate treatment planning with uniform lethal dose delivered over the entire target volume taking into consideration the physiological information (i.e., perfusion) and boundary conditions is preferred. Otherwise, active feedback control of the temperature evolution and online adjustment of the acoustical power is necessary. In order to estimate the thermal dose, the acoustic output of the device, the acoustic and biological characteristics of the tumor, and the attenuation along the ultrasound pathway (primarily abdominal wall and viscera) are required (Khokhlova and Hwang 2011). One of the major factors that limit the wide application of HIFU is the absence of ultrasound-based thermometry and the low frame rate and resolution of MRI-based HIFU. For clinical application, the intensity should be selected properly to destroy the tissue reliably and minimize side effects (i.e., heating of neighboring tissues, cavitation, and gas bubble formation) simultaneously (Fruehauf et al. 2008). Each session should be within 1 h, with all lesions covering the whole tumor volume, and multiple sessions may be necessary for satisfactory long-term outcome. A large-aperture transducer can decrease the acoustic intensity at the body surface and reduce the propensity of skin burn because of a wide convergent angle. Operators must monitor the changes in images of adjacent vital organs, such as the myocardium, diaphragm, and bowel loops. Detection of the complications as early as possible allows appropriate and immediate management.

Post-treatment assessment of the coagulation necrosis is usually performed with a contrast agent to assess the local perfusion within the tumor and the neighboring tissues. The best time for MRI follow-up may be approximately 1 week after ablation. Ultrasound contrast agent (microbubble)-enhanced sonography is also used to evaluate the HIFU outcome (Kennedy 2005). Another method currently being examined is the use of PET to assess changes in metabolic activity after HIFU treatment. Most protocols involve HIFU ablation followed by surgical resection or multiple large core-needle biopsies (LCNBs) for pathological tissue examination. Applying both H&E and NADH histochemical staining is reliable in evaluating the therapeutic effects of HIFU. Metastases in regional axillary lymph nodes of breast cancer patients, which are determined by the minimally invasive sentinel lymph node biopsy (SLNB), are an important prognostic factor (Krag et al. 1998, Veronesi et al. 2003). HIFU might affect the accuracy of SLNB by obstructing or changing the anatomy of breast lymphatics or lymph drainage. Use of a contrast agent and an isotope in the diagnosis of the sentinel node can improve the overall success rate to 91%, which is comparable with that of SLNB (Krag et al. 1998, Veronesi et al. 2003). If residual tumor or tumor recurrence in the lesion is found, the affected area will have to be re-treated.

A controversial question is the possibility of increased risk of metastases after the tumor has been disrupted by HIFU. A higher rate (17%–44%) of metastases was observed after HIFU ablation than that in the control group by Fry and Johnson. However, this conclusion was not confirmed in other studies. For example, a lower rate of lung metastases in the HIFU-treated group (Yang et al. 1991) and metastases in 16%

of the treated rats with prostatic cancer as compared to 28% in the control group were found (Chapelon et al. 1992). It was found that one of the most important biological consequences of HIFU treatment is the creation of a large amount of tumor antigens in the form of necrotic cells and the local release of a diverse array of endogenous danger signals from those damaged tumor cells, which has the potential to stimulate an antitumor immune response (Wu et al. 2004b, Hu et al. 2008). However, little is known about how such significant HIFU-induced changes in the tumor microenvironment may influence the host's antitumor immune response.

Despite the large number of clinical cases of HIFU on a variety of cancers with promise, large-scale randomized controlled trials at multiple centers with long-term follow-up have not been carried out to date to confirm these findings or to determine whether HIFU can improve the overall survival by inducing local tumor response with or without chemotherapy, radiotherapy, or targeted drugs (Yuan et al. 2012). Experience in China may not be applicable to the Western countries. Standard criteria are also required to evaluate both the short- and long-term efficiency and efficacy of HIFU. A standardized dose of HIFU, chemotherapy, or radiotherapy has not been established, so the current use is mostly empirical. An effective combination of treatment modalities is currently under investigation. Not until these technical and clinical issues have been resolved, can HIFU be considered as a candidate for conventional treatment for popular acceptance and application.

13.9 HIFU-RELATED THERAPEUTIC METHODS

Besides the thermal ablation, HIFU technology can also be used in treating cancers in the following ways.

13.9.1 IMMUNE RESPONSE

The host immune system protects abnormal cell growth, and lymphocyte (i.e., CD3+ T lymphocyte in the peripheral circulation) mediated immunity prevents the growth of primary tumors and subsequent metastases (Whiteside and Heberman 2003). Tumors can escape the immune surveillance by interfering with the migration of dentritic cells (DCs) and by failing to provide the necessary activation signals as well as by secreting active factors that inhibit the differentiation and functions of DCs, which makes breast cancer weakly immunogenic and, therefore, a poor candidate for immunotherapy (Allan et al. 2004). The tumor-infiltrating DCs are neither mature nor activated (Coventry et al. 2002). Thus, the immune system in most cancer patients fails to control the initial cancer growth and prevent local recurrence and metastasis after conventional therapies due to poor tumor antigen processing and severe impairment of the phenotypic and functional DCs by tumor-released, immune-suppressive cytokines (Gabrilovich et al. 1997, Satthaporn et al. 2004). DCs can infiltrate primary cancer, but with considerable variation in their number and distribution among patients (Lewko et al. 2000, Treilleux et al. 2004). Selective recognition and

destruction of tumor cells by the host immune system play an important role in antitumor immunity, which requires expression of tumor antigens. It has been demonstrated that the function of lymphocyte-mediated cellular immunity is suppressed after both surgical procedures (Vallejo et al. 2003, van Sandick et al. 2003) and local RFA (Toivanen et al. 1984).

Antigen-presenting cells (APCs) play a critical role in immune responses and can infiltrate local tumors and present tumor antigens to naive T lymphocytes. Two factors are essential for APCs to initiate an efficient antitumor response: the activation signals delivered directly or indirectly by tumor cells, and tumor-infiltrating APCs activated locally. Activating signals can induce the progression of infiltrating APCs from an immature to a mature stage. Afterward, APCs increase the expression of co-stimulatory molecules (i.e., CD80 and CD86) and become efficient in a process of cross-priming T cells (Lutz and Schuler 2002). However, the absence or blockade of these co-stimulatory molecules impairs tumor antigen-specific immune responses, indicating the requirement of APC activation in antitumor immunity (Pinzon-Charry et al. 2005). The transfer of tumor antigens from APCs to T lymphocytes is critical in initiating lymphocyte-mediated immunity. Macrophages, B lymphocytes, and DCs are the most potent APCs in the uptake, processing, and presentation of tumor antigens.

Recently, it was found that HIFU-induced immune response, suppression of the activity of tumor, and downregulation of tumor markers may be an effective approach of cancer treatment. HIFU ablation might modify the structure of tumor tissue and subsequently upregulate the host antitumor immunogenicity to control micro-metastasis and generate tumor resistance (Wu et al. 2006b), although the molecular mechanisms behind this phenomenon and the long-term therapeutic benefit remain unknown (Den Brok et al. 2004, Wu et al. 2004b, Hu et al. 2005, 2008, Xu et al. 2009a). It is conceivable that HIFU ablation may induce large amounts of tumor debris with and without typical characteristics of thermal damage, namely enhanced inflammation, and then a variety of cryptic antigens in the peripheral coagulation zone might be released and reabsorbed *in situ* after mechanical cell damage, thus stimulating an immunogenic response with the formation of tumor-specific T cells (Hansler et al. 2002), which may prevent both local and systemic relapse, ultimately enhance the immune function of tumor-bearing patients, and improve their prognosis (Wissniowski et al. 2003). Heat shock proteins (HSPs), whose tumor peptide complex is taken up by APCs and presented HSPs directly to tumor-specific T cells with high efficiency and enhanced tumor immunogenicity (Pockley 2003, Todryk et al. 2003), were upregulated (Madersbacher et al. 1998, Kramer et al. 2004, Hu et al. 2005, Wu et al. 2007a). In addition, HIFU can denature the secretion of immunosuppressive protein constituents originating from tumor cells (Zhou et al. 2008) and unfold the proteins from the native state to a more random state of lower organization (Lepock et al. 1993), which leads to either loss or preservation of antigenic determinants due to both thermal effects and cavitation. A similar phenomenon was also found in the upregulation of the HSPs by hyperthermia.

Animal experiments show that HSP70 and CD25+ cells in rabbit VX2 bone tumors significantly increased up to 3 weeks after HIFU treatment (Si et al. 2003), which may facilitate tumor-specific antigen presentation to T lymphocytes, stimulate the proliferation of T lymphocytes, and enhance the antitumor immune response. In a clinical trial, significant increases in natural killer (NK) cell activity, population of CD4+ lymphocytes, and the ratio of CD4+/CD8+ (T-cell helper/suppressor), as well as greater concentrations of dendritic cells, macrophages, and B lymphocytes, were found 7–10 days after HIFU ablation in peripheral blood (Figure 13.25) (Wu et al. 2004b), which suggests elimination or significant reduction of potential circulating tumor cells in those patients. Proliferating cell nuclear antigen (PCNA), cell adhesion molecule CD44v6, and matrix metalloproteinase-9 (MMP-9), which are molecular indicators of malignancy and nuclear immune reaction of breast cancer cell in proliferation, invasion, and metastasis, respectively, were stained by the biotin–streptavidin peroxidase in immune histochemistry, whereas the stromal and inflammatory tumor cells were negative. Positive cytoplasmic expressions of PCNA, *erbB2* mRNA, CD44v6, and MMP-9 were found in 44%, 36%, 56%, and 60% of 25 breast cancer patients in the control group, respectively, but no staining was observed in the HIFU ablation region (Wu et al. 2003a). The ablated region had the cellular structure of thermal fixation and large amounts of positively stained tumor cells with pyknotic nuclei (Figure 13.26). In contrast, typical characteristics of normal tissue were illustrated in the peripheral region. Because of unique tumor antigens by random mutations of cancer cells in each

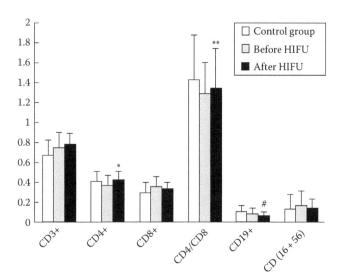

FIGURE 13.25 Effect of HIFU ablation on lymphocyte and subsets in cancer patients (compared with the preoperative group: *$p < 0.05$, **$p < 0.01$; compared with the control group: #$p < 0.05$). (From *Ultrasound Med. Biol.*, 30, Wu, F., Wang, Z.-B., Lu, P., Xu, Z.-L., Chen, W.-Z., Zhu, H., and Jin, C.-B., Activated anti-tumor immunity in cancer patients after high intensity focused ultrasound ablation, 1220, Copyright 2004, with permission from Elsevier.)

patient, personalized HSP vaccination does not require the identification of the unique antigens. Aseptic inflammation induced by pancreatic necrosis in HIFU ablation leads to the local accumulation of IL-1 and IL-2 (Schueller et al. 2003, Hu et al. 2007) (Table 13.4).

In the control group, APCs were observed in all tumor specimens, although with a high occurring frequency in the periphery than in the center of the breast cancers. S-100+ cells were distributed sparsely, and their cyoplasmic processes extended between the cancer cells (Figure 13.27). However, the infiltration of diffusely scattered CD68+ and CD20+ cells was more heterogeneous (Xu et al. 2009a). Almost no immune-stained cells were identified in the center of the tumors because of the death of both APCs and cancer cells, as well as no immediate wound healing after HIFU ablation. However, APCs were observed in the granulation tissue along the margins of ablation together with immature fibroblasts, new capillaries, and other inflammatory cells. Among them, CD68+ and CD20+ cells were usually in small clusters, whereas S-100+ cells had a scattered distribution. The number of CD68+ and CD20+ cells was higher than that of S-100+ cells without appreciable difference of the infiltration among them (Xu et al. 2009a). There was a statistical increase in the number of the APCs after HIFU ablation, compared to S-100+, CD68+, and CD20+ cells between the control and HIFU groups, respectively.

In summary, the immunologic abnormality and the lack of a host antitumor response on primary cancer could be reversed after the HIFU ablation (Yang et al. 1992, Rosberger et al. 1994, Wu et al. 2004b), may be due to the activation of T-lymphocyte-mediated antitumor immunity. The factors and mechanism involved in this phenomenon are very complicated and not fully understood. Acoustic radiation and bubble cavitation in the high-intensity acoustic field may also alter the tumor structures and impair the function of tumor-specific T cells (Burov and Dmitrieva 2002). It is shown that, *in vitro*, DCs and macrophages exposed to the supernatants of HIFU-treated tumor cells had a greater expression of co-stimulatory molecules with more IL-12 and TNF-α from DCs and macrophages ($p < 0.01$), respectively (Hu et al. 2005). It suggests that HIFU ablation may not only cause significant infiltration of APCs in the treated tumor but also induce the tumor infiltrating DCs from an immature to a mature stage. Therefore, the HIFU-induced immune response will have to be investigated further as an alternative modality in treating cancers.

13.9.2 Vessel Occlusion

Malignant tumors have several inherent characteristics in the histological structure and circulatory function of microvasculature that are invulnerable to hyperthermia damage, such as thin-walled lumen for easy compression, lack of perivascular smooth muscle with much less vasomotor control, large inter-capillary distance, increased permeability, arteriovenous shunts, heterogeneous blood flow with consequent foci of relative hypoxia, and acidosis. In addition to the coagulation production in HIFU ablation for an irreversible cell death,

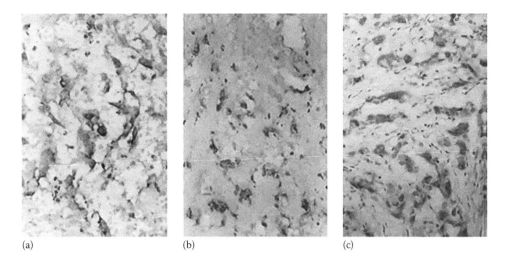

(a)   (b)   (c)

FIGURE 13.26 Positive expression of heat shock protein 70 (HSP-70) with (a) nuclear disappearance, (b) nuclear disruption in the peripheral region of ablated cancer debris, and (c) that in central treated region with pyknotic nuclei after HIFU treatment (streptavidin peroxidase immune-histochemical staining, 400×). (With kind permission from Springer Science+Business Media: *Ann. Surg. Oncol.*, Expression of tumor antigens and heat shock protein 70 in breast cancer cells after high intensity focused ultrasound ablation, 14, 2007, 1240, Wu, F., Wang, Z.-B., Cao, Y.-D., Zhou, Q., Zhang, Y., Xu, Z.-L., and Zhu, X.-Q.)

TABLE 13.4
Statistical Summary of Immune Factors before and after HIFU Ablation in Pancreatic Cancer Patients

	Pre-HIFU	Post-HIFU	Increase (%)
CD3+ ($n = 141$)	37.39 ± 11.78–60.3 ± 5.9	51.8 ± 6.4–59.6 ± 6.7	112.94
CD4+ ($n = 93$)	24.19 ± 7.02–32.6 ± 5.4	28 ± 10–34.7 ± 5.3	108.89
CD4+/CD8+ ($n = 93$)	0.9 ± 0.3–1.1 ± 0.1	1.09 ± 0.53–1.4 ± 0.1	125.90
NK ($n = 28$)	20.54 ± 9.1–21 ± 9	25 ± 13–25.52 ± 11.9	121.80

Source: Zhou, Y.F., *Gastroenterol. Res. Pract.*, 2014, 205325, 2014.

targeted vascular occlusion is another treatment strategy for some cancers with identifiable blood supply to cause a deprivation of nutrition and oxygen for the neoplastic cells in the clinical setting, similar to embolization (Dorr and Hynynen 1992, Chen et al. 1993, Susani et al. 1993, Fowlkes et al. 1994, Vaezy et al. 1999b). Two strategies, namely thermal and mechanical effects, have been utilized.

Blood flow occlusion of deep femoral arteries in left thighs of Sprague–Dawley rats was accomplished by HIFU at an intensity of 4,300 W/cm^2. Histologic studies have demonstrated that exposure to HIFU at 2,750 and 4,300 W/cm^2 leads to vacuolar degeneration and destruction of elastic fibers of the tunica media of the artery. In contrast, vascular contraction without tissue degeneration occurred at low intensity. Although these phenomena appeared to be mainly due to thermal effects, mechanical effects might have some role, particularly on vascular contraction. Vascular occlusion varied with ultrasound frequency and exposure. Arterial blood flow occlusion was achieved with 10 kW/cm^2 at 3 MHz, but not with 800 W/cm^2 at 1 MHz. Although immediate occlusion of blood vessels occurred with HIFU, long-term data indicated a trend of

vessel recanalization and return to pretreatment diameters. The renal artery branches of rabbits (diameter about 0.6 mm) were occluded by HIFU (Si et al. 2003). Complete cessation of blood flow was observed by color Doppler imaging and MRI, and lack of perfusion was also observed in the renal cortex in the contrast-enhanced image (Figure 13.28) (Wu et al. 2002). If the blood supply to the tumor is poor, sonography is not sensitive in detecting such a destructive response. Intravenous administration of ultrasound contrast agents (UCA, microbubbles) can improve the evaluation accuracy of vascularity. Postmortem histologic evaluation showed an infracted tissue volume corresponding to the wedge shape seen in the ultrasound and MRI images, with no damage to the surrounding soft tissue (Delon-Martin et al. 1995). HIFU may offer a noninvasive, nonsurgical technique to effectively obliterate blood flow.

The violent collapse of a bubble can exert strong mechanical forces to nearby structures, resulting in irreversible cell membrane damage and possibly cell death. The vascular endothelial surface can be damaged by exposing a vessel to HIFU pulses in the presence of circulating UCA (Hwang et al. 2005, 2006). If sufficiently treated, a nonocclusive

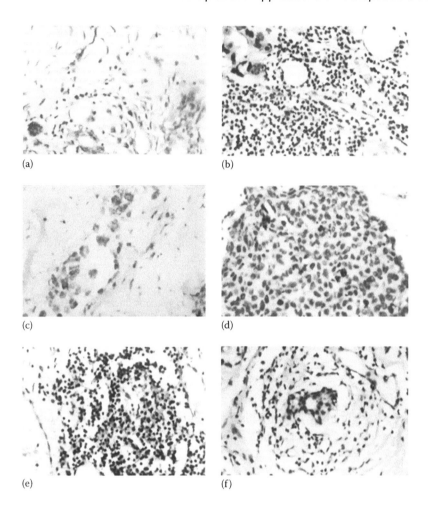

FIGURE 13.27   Positive expression of tumor-infiltrating (a) S-100 cells, (c) CD68 cells in a small cluster, (e) and CD20 cells within a cluster of lymphocytes at the margin of HIFU ablation, and the corresponding ones (b), (d), and (f) within breast cancer cells in the control group (streptavidin-peroxidase immune-histochemical staining, 200×). (From *Ultrasound Med. Biol.*, 35, Xu, Z.-L., Zhu, X.-Q., Lu, P., Zhou, Q., Zhang, J., and Wu, F., Activation of tumor-infiltrating antigen presenting cells by high intensity focused ultrasound ablation of human breast cancer, 53, Copyright 2009, with permission from Elsevier.)

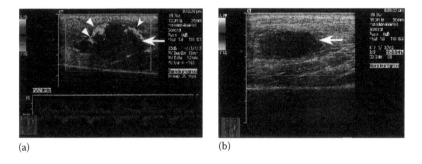

FIGURE 13.28   (a) Hypoechoic mass with plenty of blood flow in the tumor before HIFU and (b) no flow in the treatment region 2 weeks after HIFU for a thigh soft tissue sarcoma color Doppler ultrasound image. (From *Ultrasound Med. Biol.*, 28, Wu, F., Chen, W.-Z., Bai, J., Zou, J.-Z., Wang, Z.-L., Zhu, H., and Wang, Z.-B., Tumor vessel destruction resulting from high-intensity focused ultrasound in patients with solid malignancies, 541, Copyright 2002, with permission from Elsevier.)

fibrin thrombus can be formed along the luminal surface of the vessel. This nidus of fibrin clot is anchored to the damaged endothelial surface. A segment of a rabbit auricular vein was treated *in vivo* with low-duty-cycle, high peak-rarefaction pressure (i.e., 9 MPa) HIFU pulses in the presence of intravenously administered circulating microbubbles, followed by fibrinogen injection, which resulted in the formation of an acute occlusive intravascular thrombus (Hwang et al. 2010). Clot propagation was enhanced by injecting fibrinogen into the vessel after cavitation-induced damage to the endothelial surface. However, these vascular occlusions were not durable over a 14-day period, likely because of the inherent fibrinolytic system. Fibrinolysis is inhibited by plasminogen activator inhibitor-1, which in turn is upregulated in the setting of local inflammation caused by the effects of interleukin-1. By eliciting a local inflammatory response with a pro-inflammatory agent, clot stability may be promoted by a downregulation or suppression of the fibrinolytic system. When injecting a pro-inflammatory agent (ethanol, cyanoacrylate, or morrhuate sodium) with fibrinogen, long-term complete vascular occlusions could occur (Zhou et al. 2011a).

Furthermore, transcranial vascular occlusion is also possible at intensity ranges of 1,690–8,800 W/cm², duration < 15 s, and 0.68–3.3 MHz frequency. A threshold frequency–intensity product of 8,250 MHz W/cm² was needed for vascular occlusion with a sensitivity of 70% and a specificity of 86%. There are some complications such as skin burns, hemorrhage, and damage to the surrounding structures (Serrone et al. 2012).

Digital subtraction angiography (DSA) indicated a remarkable decrease in hypervascularity and almost disappearance of tumor vascularity and stain in the patients following HIFU (Figure 13.29). HIFU induced homogeneous coagulative necrosis of the tumor tissue as well as of small arteries and veins in the tumor. Vascular elasticity and collagen

fibrin in Victoria blue and Ponceau's histochemical staining were found to collapse and disrupt significantly in the treatment region, which indicated severe destruction of the HIFU ablation on the tumor's vascular wall (Figure 13.30). The damaged endothelial walls might accumulate collagen fibrin, activate platelets, and eventually cause the formation of thrombosis. Significant vessel rupture, vascular disruption, and vessel occlusion have been demonstrated in various HIFU experimental conditions (Yang et al. 1991, 1992, Chen et al. 1993, Susani et al. 1993, Delon-Martin et al. 1995, Vaezy et al. 1999a). The vascular destruction effects depended on the HIFU operation parameters (i.e., frequency, intensity, duty cycle, and duration) and the vessel characteristics (i.e., size, elasticity, and blood flow).

13.9.3 HEMOSTASIS

Vascular injury may result from both trauma and invasive medical procedures, and extensive hemorrhage may lead to a life-or-death situation. Trauma due to accidents, assault, terrorism, and war accounts for enormous loss of life and social productivity, and is the fifth cause of death, especially for all age groups under 35 years. Each year, about 2.6 million

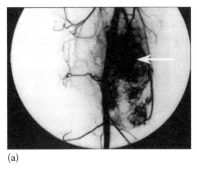
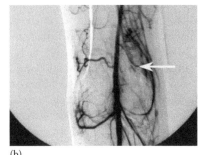

(a)　　　(b)

FIGURE 13.29 (a) Persistence of tumor vascularity and capillary stain within distal femur osteosarcoma, and (b) complete disappearance of tumor vascularity and stain within the tumor 3 months after HIFU ablation in DSA. (From *Ultrasound Med. Biol.*, 28, Wu, F., Chen, W.-Z., Bai, J., Zou, J.-Z., Wang, Z.-L., Zhu, H., and Wang, Z.-B., Tumor vessel destruction resulting from high-intensity focused ultrasound in patients with solid malignancies, 541, Copyright 2002, with permission from Elsevier.)

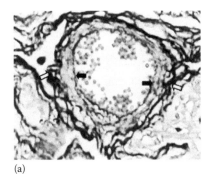
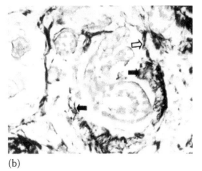

(a)　　　(b)

FIGURE 13.30 (a) Normal structure of elasticity fibrin (solid arrow) and collagen fibrin (open arrow) within the vascular wall in the untreated breast, and (b) significant collapse and disruption of them in the HIFU-treated one (Victoria blue and Ponceau's histochemical staining, 400×). (From *Ultrasound Med. Biol.*, 28, Wu, F., Chen, W.-Z., Bai, J., Zou, J.-Z., Wang, Z.-L., Zhu, H., and Wang, Z.-B., Tumor vessel destruction resulting from high-intensity focused ultrasound in patients with solid malignancies, 541, Copyright 2002, with permission from Elsevier.)

people are hospitalized in the United States as a result of acute injury, and the total life-time cost associated with both fatal and nonfatal injuries was over $260 billion. The liver is one of the most commonly injured intraperitoneal solid organs, and severe injury (grade IV and V) is the most common cause (>50%) of hemorrhagic death after abdominal trauma. The spleen is the most commonly injured organ in blunt abdominal trauma, with splenic injury accounting for 9% of deaths. The management of splenic trauma has changed over the last three decades toward splenic preservation partially because of the recognition of the critical role spleen plays in immunologic host defense and the small but definite risk of post-splenectomy sepsis. Low-grade splenic injuries are usually managed nonoperatively with great success, while more severe injuries more likely require surgical intervention. In 97% of trauma deaths (49% blunt force and 48% penetrating wounds in abdominal and pelvic injuries), hemorrhage is a common factor. In major pelvis fractures, the foremost clinical concern in initial management is the threat of exsanguinations from injuries to vessels (internal iliac, iliofemoral, and sacral) that are in close proximity to the bony pelvis. Thus, control of blood flow is essential in medicine for a variety of applications.

Rapid attainment of hemostasis (i.e., the simple application of pressure on a wound to surgically treat an internal bleeding site) is the primary treatment to prevent hemorrhagic shock following an intra-abdominal injury. However, there is increasing trend toward minimally invasive and noninvasive therapies, such as laparoscopic surgery. Operative intervention (e.g., temporary control of vascular inflow to facilitate definitive treatment with direct suture and ligation of bleeding vessels, angiographic embolization of bleeding vessels, perihepatic packing, resectional debridement, and partial and total organ resection) is the primary treatment for life-threatening intra-abdominal hemorrhage, and the employed technique depends upon the site and source (major vascular structure or parenchyma) of bleeding as well as the physiological status of the patient (shock, hypothermia, coagulopathy, etc.). In addition, hemostatic agents, including FDA-approved materials (oxidize cellulose, microfibrillar collagen powder, fibrin sealant, fibrin sealant dressing, fibrin glue, and topical thrombin) and some investigational ones (neutral microfibrillar microdispersed oxidized cellulose, cationic propygallate, gelatin foam containing multiple active ingredients, microcaps of aluminum sulfate, collagen-based product with human fibrinogen and thrombin, algae-derived polysaccharide polymer, and chitosan), and energy-based methods (i.e., electrocautery, argon beam coagulator, microwaves, and lasers) have been developed to achieve hemostasis intraoperatively. Most of them can provide effective hemostasis of arteries and veins up to ~3 mm in diameter (Ligasure, a newer electrocautery device, can seal vessels up to 7 mm in diameter). The tissue coagulation depth is 2 mm and 2 cm in laser and microwave and some electrocautery devices that involve placing needles deep into the tissue, respectively, which is their principal drawback.

In recent years, HIFU has been proposed as a noninvasive method for hemorrhage control because of its ability to treat deep-seated tissues without affecting the intervening and surrounding tissues (Vaezy et al. 1998, 1999, 2001). The first applications of HIFU in hemorrhage control were carried out in the late 1990s. The concept has been proven using handheld solid-cone HIFU devices with a fixed focus at an electrical power of up to 160 W under visual guidance, for simplicity. A safe and effective hemostasis of low-grade injuries in liver, spleen, lungs, and blood vessels (up to 3 mm in diameter) in anesthetized rabbit and pig hemorrhage models to blood vessels both surgically exposed and approximately 1 cm below the surface of the tissue was demonstrated. Furthermore, HIFU-induced hemostasis is especially useful for the occult injuries that are difficult to access with traditional surgical approaches, including the pelvis, the retroperitoneum, and deep parenchyma and posterior surfaces of the liver, and image guidance is required to locate the site of bleeding. In order to develop real-time image-guided (ultrasound B-mode imaging and color Doppler imaging) HIFU hemostasis systems, HIFU devices have to be integrated with real-time ultrasound imaging, which shows a hyperechoic region in the sonography and offers a valuable targeting aid in treatment guidance. Successful treatment of inaccessible bleeding depends on accurate and timely detection and localization of bleeding sites. Pulsed Doppler, contrast-enhanced B-mode ultrasound imaging, and vibrometry have shown promise in bleeding detection and localization. Slower bleeding (i.e., oozing from parenchyma or from venous injury) is more challenging than detecting fast bleeding. Acoustic streaming has been shown to be effective in detecting pooled blood and differentiating between nonclotted blood and clotted blood or tissue.

Acoustic hemostasis of various vascular injuries, including punctures and longitudinal and cross-sectional lacerations in arteries and veins, was achieved with HIFU faster (no more than a minute) than conventional suture ligation. There was a direct relationship between the overall HIFU dose and treatment outcome. In order to evaluate its performance for blunt abdominal trauma, hepatic injury (grade III) and occasionally concomitant splenic injury were induced by the impact of a nail gun driver discharge onto an aluminum disk placed on the abdominal wall of anesthetized juvenile domestic pigs. Complete hemostasis was achieved within 15 ± 6 min using HIFU treatment at a frequency of 5.7 MHz and intensities of over 2000 W/cm^2. In addition, incisions (lengths of 2–5 cm and depths of 3–10 mm, which resulted in both parenchymal hemorrhage and air leakage) were produced in pig lungs *in vivo*, and complete hemostasis and pneumostasis were achieved in 6 min on average using an intraoperative HIFU device (frequency of 5.7 MHz, acoustic power of 65 W) with no air leaks, which was better than or comparable to the control (the amount of blood in the chest cavity at 4 h: 40 ± 14 vs. 98 ± 22 mL in the stapling). The functionality of lung tissue in the injured lobe was preserved.

One potential advantage of using HIFU for acoustic hemostasis is that the energy could be delivered to deep tissue targets where hemorrhage would be occurring, allowing cauterization at depths of parenchymal tissues, or in difficult-to-access regions, while sparing intervening tissues

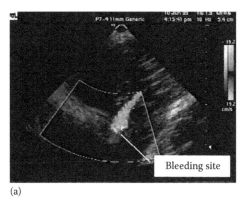
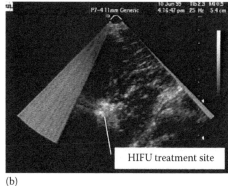

(a)   (b)

FIGURE 13.31 (a) Identification and (b) HIFU treatment of a bleeding pelvic vessel in a pig shown in sonography. (Vaezy, S., Martin, R., and Crum, L.A.: Therapeutic ultrasound, part II: High intensity focused ultrasound: A method of hemostasis. *Echocardiography.* 2001. 18. 313. Copyright Wiley-VCH Verlag GmbH & Co. KGaA. With permission.)

from harmful exposure. Furthermore, HIFU exposures work through unique physical and biological mechanisms, which are expected to provide additional advantages over other hemostasis methods. Examples of these mechanisms include: mechanical emulsification of tissue to exposure tissue factors that promote coagulation, the thermal energy deposition rate that becomes enhanced when vapor bubbles are formed due to boiling, and the formation of a HIFU-specific coagulum seal, all leading to effective hemostasis even under coagulopathic conditions (Figure 13.31).

If the diameter of injured vessels is >3 mm, a high rate of bleeding acts as a heat sink, which counteracts the energy deposition. By increasing the HIFU power, occlusion of vessels in a diameter of 5 mm, and up to 1 cm with manual pressure on the vessel during treatment, is possible. Hemostasis of a 20 cm^2 liver resection required 4.7 min of dual HIFU application with clamping (reduction of blood flow and heat sink effect); whereas a 15 cm^2 area treated without compression required 14 min. Blood vessels of up to 10 mm in diameter in large solid organs were occluded (liver: 6.0 ± 1.5 min, kidneys: 2.8 ± 0.6 min, and spleen: 3.6 ± 1.1 min) with an average width of 3 cm extended through the whole thickness (up to 4 cm) due to blood coagulation and vascular collapse with no occurrence of bleeding for up to 4 h at a frequency of 3.3 MHz and an *in situ* intensity of 9,000 W/cm^2. ALT and AST levels increased immediately post treatment to as high as 285% up to day 3 and returned to normal levels by day 7. Hematocrit and white blood cell counts showed no significant changes. Tissue necrosis and deposition of the blood/tissue homogenate coagulum occurred acutely at the injury site, and was reabsorbed and replaced by fibrous tissue by day 28. Scarring as well as tissue regeneration was found in a gross and histological examination up to 60 days post treatment. Leukocytes and fibroblasts were activated, migrating to the treatment area to produce collagen, elastin, and some proteoglycans as early as 14 days post treatment.

Puncture wounds (2–3 mm in diameter) were produced *in vivo* using a catheter in the swine femoral artery in a diameter of 5–6 mm and at a depth of 1–1.5 cm and were completely sealed in 79% and 33% with HIFU (3.2 MHz, *in situ*

intensity of 2900 W/cm^2) and sham treatments, respectively. The blood loss was significantly lower in the HIFU treatment group as compared to sham and control groups (HIFU: 19 ± 14 mL, sham: 81 ± 62 mL, control: 157 ± 13 mL). Both femoral artery and vein were patent after the HIFU treatment, with some narrowing at the ablation site. No complication (rebleeding, hematoma, or aneurysm) was observed secondary to HIFU treatment of femoral artery punctures.

The physical mechanisms involved in the HIFU hemostasis are both thermal and mechanical, leading to a variety of biological effects (i.e., coagulative necrosis, coagulum, and thrombus formation at the wound site). Mechanical disruption of tissue can lead to the release of tissue factors, enhancing coagulation, tissue fusion via collagen and elastin remodeling, and fibrin plug formation. The formation of coagulum (i.e., a blood/tissue homogenate) is an effective hemostatic seal over the injury, and the incision was plugged with a mixture of destroyed erythrocytes, neutrophils, lymphocytes, blood cells, and a few recognizable epithelial cells. Most cells in the mixture were destroyed, but some along the edge of the incision appeared to be viable. In addition to strong heating effects, acoustic cavitation was also involved in the production of the coagulum. The coagulum was initially full of microbubbles, produced by boiling and/or acoustic cavitation, which disappeared quickly as the coagulum cooled. There is a direct relationship between cavitation and hemolysis. The presence of the UCAs resulted in a 37% reduction in the normalized hemostasis time, as well as faster (by 60%) formation of the coagulum seal over the incision (Table 13.5), which shows the potential benefit of UCAs in improving the efficacy of HIFU hemostasis may be due to the combined effects of rapid tissue heating and boiling of the fluids (thermal effects) and the activity of bubbles (cavitation) in accelerating the formation of coagulum and achieving hemostasis.

Preclinical studies to date have shown that HIFU is a promising technology for effective hemostasis in trauma management especially in circumstances such as high-grade or inaccessible injuries. In addition to the role of HIFU as an adjunctive therapy in the operating room, the greatest potential for this technology is related to its ability to control bleeding in nonoperative settings. Control of intracavitary hemorrhage (abdominal, pelvic, and possibly thoracic) in the field or in the emergency room in an extracorporeal format is a unique potential application deserving future developments.

13.9.4 HISTOTRIPSY

Histotripsy, which has similar structure as HIFU but working at a little lower frequency (i.e., 750 kHz), has been developed to achieve mechanical fractionation of tissue structure instead of thermal ablation using a number of short, high-intensity ultrasound pulses (i.e., 20 μs duration, 1 kHz pulse repetition rate, and 18 MPa rarefactional pressure) (Parsons et al. 2006). At a fluid/tissue interface or inside a soft tissue, histotripsy results in localized tissue erosion in a liquefied core

with very sharply demarcated boundaries (Xu et al. 2005). Histology shows that the treated tissue could be fragmented to subcellular level surrounded by an almost imperceptibly narrow margin of cellular injury (Figure 13.32). Imaging-guided histotripsy has vast clinical applications, for example, as noninvasive surgery tool for tumor/cancer. Compared to noninvasive thermal therapy, histotripsy has some technical advantages: (1) Microbubbles are produced at the ultrasound focus, which are shown as bright spots on ultrasound imaging and provide real-time feedback for the therapeutic process. (2) The lesions appear darker on ultrasound imaging post treatment. (3) The lesions can be produced in a very controlled and precise manner. Disintegrating tissue rather than causing necrosis may aid reabsorption into the body, which cannot be accomplished with thermal HIFU.

TABLE 13.5

Normalized Hemostasis Times and Times to Coagulum Formation for Control and HIFU + UCA Treatments

Sampling Time	% Hemolysis
Pre-HIFU application	0 ± 0
HIFU: Pre-coagulum formation	0.79 ± 0.26
HIFU: Early coagulum formation	10.88 ± 8.58
HIFU: Peak coagulum formation	30.95 ± 14.38

Source: *Ultrasonics*, 45, Zderic, V., Brayman, A.A., Sharar, S.R., Crum, L.A., and Vaezy, S., Microbubble-enhanced hemorrhage control using high intensity focused ultrasound, 116, Copyright 2006, with permission from Elsevier.

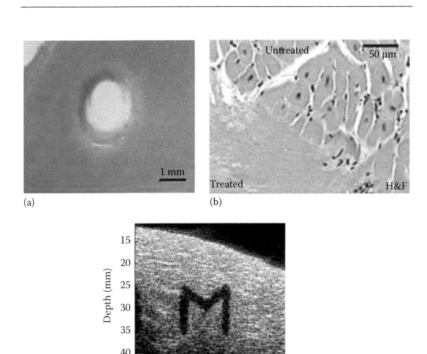

FIGURE 13.32 Representative (a) image and (b) histology of tissue erosion after histotripsy, and (c) "M" shaped lesion generated by histotripsy shown in ultrasound imaging.

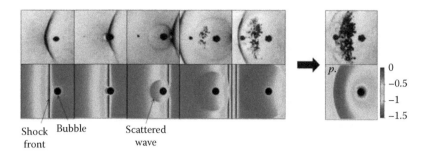

Shock   Bubble   Scattered
front   wave

FIGURE 13.33 Comparison of photographs (top) and simulations (bottom) of a shock scattering from a bubble. The constructive interference of the scattering wave and the rarefaction phase of an incident wave creates a large tensile pressure for cavitation. (From Maxwell, A. et al., *Acoust. Today*, 8, 29, 2012.)

The mechanism of histotripsy is acoustic cavitation. Energetic microbubble activities fragment and subdivide tissue, resulting in cellular destruction (Xu et al. 2007). Free bubbles are generally not stable by themselves and tend to dissolve as a result of the surface tension of the liquid. However, they are stabilized at the site of crevices in solid particles or inside macromolecule shells (Harvey et al. 1944, Yount 1979). Hundreds or thousands of histotripsy pulses can suddenly form a bubble cloud, which effectively can reflect and scatter the following pulses (Xu et al. 2005) (Figure 13.33). The result is that the cloud emerges from a single bubble, expands and collapses repeatedly, and moves toward the transducer. If the pulses are not delivered frequently, the bubble cloud will dissolve, and another cloud may form in the next activation process, which is referred to "cavitation memory" (Xu et al. 2009b, Wang et al. 2012a). Since the cavitation cloud is spatially confined to the focal volume, erosion occurs only in that region, which is a distinct advantage with no risk of overtreatment as can happen in thermal therapy.

Furthermore, another similar technology, boiling histotripsy, was also developed using the same driving frequency as HIFU. When shocks from in the pressure wave form at the focus of an acoustic beam, the heating rate could be more than 20 times higher than that from harmonic waves, which accelerates the thermal effects in HIFU, with subsequent temperature elevation (Bessonova et al. 2009). Consequently, ultrasound-induced heating is much more localized (to 0.1 mm radius) than acoustic intensity, and boiling occurs within several milliseconds with almost negligible heat diffusion (Canney et al. 2010). Contrary to bubble cavitation, which is a stochastic phenomenon and may occur in a much larger volume within the focus, shock heating and boiling are highly predictable and occur locally at the focus (Khokhlova et al. 2011). Histological studies have confirmed that the mechanical erosion is caused by the exploding boiling bubble and its further interaction with shocks with negligible thermal effect if the shock fronts are higher than 40 MPa at the focus, boiling initiation is in several milliseconds, pulse duration takes a little longer than the time to boil, and the repetition rate of the pulsing scheme is slow enough to avoid the thermal accumulation (Wang et al. 2012a). The major mechanisms may be acoustic atomization and the formation of a miniature fountain from the tissue into the boiling bubble. Both cavitation

clouds and shock-induced boiling are inherently self-limited to the focal volume (Figure 13.34).

Current clinical targets include kidney, breast cancer, prostate cancer, several cardiac applications (perforation of the arterial septum for congenital heart disease), benign prostatic hyperplasia (BPH), and breast fibroadenomas (Roberts et al. 2006, Xu et al. 2006, Lake et al. 2008). Ten or hundred transcutaneous pulses produced scattered damage on the normal rabbit kidneys characterized by focal hemorrhage and small areas of cellular injury in the targeted volume, and 1,000 or 10,000 pulses demonstrated complete destruction of the targeted volume (Roberts et al. 2006). A contiguous area of finely disrupted tissue was observed containing no recognizable cells or cellular components at day 0. Along the boundary of architectural disruption, a border several tubules wide contained cells that were not visibly disrupted but appeared damaged (pyknotic nuclei). Afterward, an inflammatory response developed in association with a steadily decreasing area of cellular and architectural disruption. By day 60, only a small fibrous scar persisted adjacent to a wedge of tubular dilation and fibrosis underlying a surface-contour defect, which showed the reabsorption of the resultant acellular material (Hall et al. 2007).

13.9.5 DENERVATION

There has been active investigation of ultrasound-inhibited nerve conduction for pain management, local anesthesia, and treatment of spasticity for nearly 50 years (Ballantine et al. 1960, Foley et al. 2004, 2007, 2008). Ultrasound irradiation has been used to treat neuroma, painful amputation stump neuromas, and phantom limbs (Tepperberg and Marjey 1953, Young and Henneman 1961). Sonication of peripheral nerves 1–3 mm proximal to the neuroma resulted in complete relief of pain without loss of sensation in 7 of 10 neuromas and partial relief in the other three cases. It was also recognized that focal and temporary nerve conduction inhibition has the potential of modulating the central nervous system functions (Fry et al. 1958), such as suppressing the transmission along peripheral nerves and neural pathways in the brain (Fry et al. 1958, Young and Henneman 1961, Bachtold et al. 1998). The nerve action potential decreased during sonication, correlated with temperature elevation in the nerve, and could recover

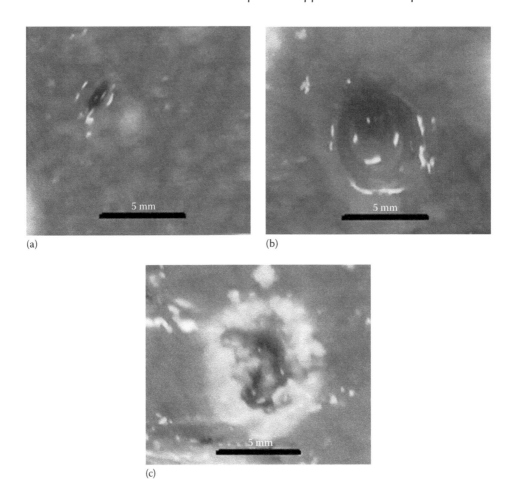

FIGURE 13.34 Comparison of the lesion produced by HIFU using different parameters: (a) 250 W 100% DC, (b) 1000 W 25% DC, and (c) 2000 W 12.5% DC.

either completely, partially, or not at all, the extent of inhibition and degree of recovery depending on the ultrasound dose. Sonication of sensory fibers is used for pain management, while that of motor nerves is employed for alleviating the symptoms of spasticity (Ballantine et al. 1960, Foley et al. 2004, 2007, 2008). Inhibition of brain function is effected by blocking the conduction of the fibers or fiber tracts of the white matter or by inhibiting the function of neurons at the cortex or at subcortical nuclei. HIFU has better localization and targeting, and a smaller focal area than the effective volume of transcranial magnetic stimulation (TMS), with temperature-sensitive imaging on MRI (Hynynen et al. 1997). Both TMS and HIFU penetrate deep into the central structures of the brain.

The mechanisms behind the nerve conductance blocks, either temporarily or permanently, have been attributed to ultrasound-induced temperature elevation (Lele 1963, Colucci et al. 2009), with no evidence of any mechanical effect of ultrasound (Halle et al. 1981, Moore et al. 2000) (Figure 13.35). The smallest and unmyelinated fibers are the most sensitive to ultrasound, while Alpha fibers with the thickest myelin sheath are the least sensitive, which is consistent with the clinical findings in sclerosis patients. Partial or total suppression of the visual evoked potential (VEP) in both

the optic nerve and visual cortex was obtained by sonications of the optic nerve in the area of its junction with the lateral geniculate nucleus, but an increase in the VEP often preceded its suppression. HIFU exposures of an *in vitro* frog nerve could temporarily block nerve conduction, which resulted in the action potential returning within 4 min, but up to 90 min for full recovery (Wang et al. 1999). At higher-power and/or repeated exposures, the action potential could be completely eliminated. The transient conduction block may be due to temperature-induced changes in the kinetics of ion channels (Xu and Pollock 1994). The temperature increase results in Na^+ inactivation and K^+ activation with larger impact on the inactivation, which causes the repolarization to occur before there is sufficient sodium current to generate a propagating pulse. In contrast, the main effect of focused ultrasound in stimulating neural structures is mechanical force, either acoustic radiation force or shock waves by the collapsing bubbles, that could produce a change in membrane potential. The burst sonications produced a small initial increase and then a reduction in the action potential that was proportional to the time-averaged power. An activation of stretch-sensitive ionic channels through conformational changes due to membrane tension by the mechanical force was responsible for these effects.

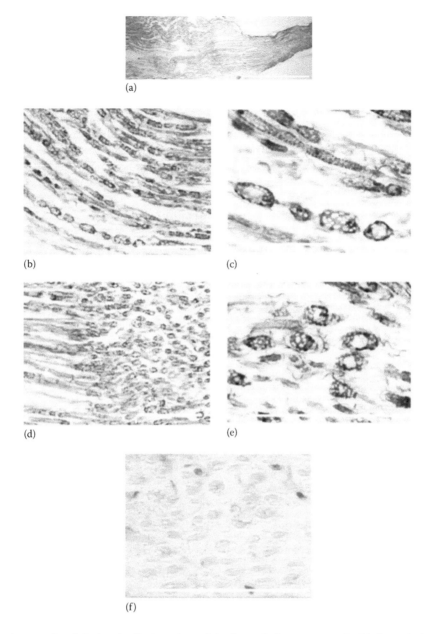

FIGURE 13.35 Microphotographs of the longitudinal sections of the frog sciatic nerve treated with focused ultrasound, showing (a) the irreversibly damaged nerve (the treated nerve segment appearing swollen and torn apart), (b) and (c) the swollen and vacuolated degenerated axis cylinders, (d)–(f) the complete dissociation of some fibers into free spheres, and (f) the fragmentation of the myelin sheaths into pale barely visible globules. Bars: (a) = 1 mm; (b)–(f) = 100 µm. (a)–(e): silver impregnation and (f): H&E staining. (From *Ultrasound Med. Biol.*, 35, Colucci, V., Strichartz, G., Jolesz, F., Vykhodtseva, N., and Hynynen, K., Focused ultrasound effects on nerve action potential *in vitro*, 1743, Copyright 2009, with permission from Elsevier.)

13.10 CONCLUSIONS

The innovation in modern medicine is progressing toward techniques that maximize the therapeutic effects and minimize intervention to the patient and length of hospital stay simultaneously. There are five interconnected stages related to the development and assessment of surgical innovations: innovation, development, exploration, assessment, and long-term study. Noninvasive thermal ablation therapies provide a minimally invasive approach to cancer therapy and are gaining rapid clinical acceptance. Image-guided HIFU therapy has great potential as a noninvasive treatment modality, such as for tumor ablation, hemostasis, neurosurgery, and cosmetic surgery. The well-established sonography provides a robust and economical solution for HIFU guidance and monitoring. The analysis of RF backscattered signal could monitor tissue viability and changes in real time during HIFU procedures with great promise. However, microscopic levels of pathologies, such as calcifications found in breast tumors and other multifocal prostate cancer, cannot be resolved in sonography mainly due to its relatively low contrast resolution, sensitivity, and specificity. Elastography, which involves quantitative ultrasound imaging for tissue characterization, shows promise to identify these occult lesions at their onset and then to treat them. It is anticipated that continuous advancements in both ultrasound imaging and HIFU technology will result in further clinical acceptance of ultrasound-guided HIFU as a noninvasive therapy modality in clinics. MRgFUS offers several distinct advantages in the guidance of thermal ablation. To date, MRI has better performance in accurate targeting (i.e., the lesion's number and location), delineating (i.e., the biological characteristics and natural history of the tumor), and thermal monitoring during the procedure as well as detecting residual disease postoperatively. MR temperature mapping allows enhanced control and safety, and MRI itself can evaluate the therapeutic efficacy, which makes it more attractive for a wide range of clinical applications. Further technical improvements are also necessary, such as accelerated temperature mapping, motion correction, feedback coupling to the HIFU device, and monitoring and control of volumetric heating techniques.

HIFU has been increasingly used for a variety of clinical applications in Asia and Europe. The overall complication rate will decrease as the surgeon's experience increases and with the evolution of HIFU technology (Coelho et al. 2010, Novara et al. 2010). This noninvasive technology has shown better acceptance by patients, with the consideration of psychology and cosmetics, than conventional therapies. For example, multiple Phase I and II studies have already proven the feasibility and safety of MRI- and US-guided HIFU ablation of small (< 2 cm) solitary breast cancer with a success rate of 20%–100% depending on system configuration, guidance technique, ablation protocol, and patient selection criteria. HIFU may also serve as a salvage method to eradicate small primary, residual, or locally recurring tumors that persist after the completion of breast conservation surgery as well as the treatment of breast fibroadenoma, a common benign tumor in young women, in outpatient settings. However, these studies are mostly preliminary, but with encouraging results. Further studies will be necessary before the widespread use of HIFU can be recommended because of scant evidence favoring this fairly new technique. Long-term medical benefit and perhaps the beneficial economic impact of this therapy for oncology are of clinical and social importance and will be an area of continuing interest. As its profile is raised and the technique becomes more widely available, it should be possible to coordinate large prospective and random clinical trials worldwide that will be necessary to develop the evidence base for the efficacy of HIFU, whether alone or in combination, to compare with conventional modalities, and to formulate a procedural guideline.

However, HIFU is still in its infancy, and there remain several important technical and clinical questions to be addressed. Ideally, HIFU should give almost the same results as surgical excision. With improved imaging, advances in transducer technology and energy delivery techniques, and better understanding of HIFU-induced bioeffects, it is highly possible that the versatility of HIFU will increase and its range of applicability will expand. Real-time imaging and treatment monitoring are the subjects of ongoing theoretical research and are likely to enable improvement in clinical outcome and to bring about a reduction in treatment duration. Co-registration of ultrasound with cross-sectional MRI/CT images should also provide additional safety and improve the user-friendliness of ultrasound-guided devices. The lack of pathological examination on tumor margin directly and accurately is a major obstacle for the thermal ablation approach. Extensive molecular biological research is going on to provide the basis for any possible immunological activation that remains to be elucidated.

REFERENCES

Ahmed HU, Zacharakis E, Dudderidge T, Armitage JN, Scott R, Calleary J, Illing RO, Kirkham A, Freeman A, Ogden C, Allen C, Emberton M. High-intensity-focused ultrasound in the treatment of primary prostate cancer: The first UK series. *British Journal of Cancer* 2009;101:19–26.

Alizad A, Wold LE, Greenleaf JF. Imaging mass lesions by vibro-acoustography. Modeling and experiments. *IEEE Transactions on Medical Imaging* 2004;23:1087–1093.

Allan CP, Turtle CJ, Mainwaring PN, Pyke C, Hart DN. The immune response to breast cancer, and the case for DC immunotherapy. *Cytotherapy* 2004;6:154–163.

Amini AN, Ebbini ES, Georgiou TT. Noninvasive estimation of tissue temperature via high-resolution spectral analysis techniques. *IEEE Transactions on Biomedical Engineering* 2005;52:221–228.

Anand A, Kaczkowski PJ. Noninvasive measurement of local thermal diffusivity using backscattered ultrasound and focused ultrasound heating. *Ultrasound in Medicine and Biology* 2008;34:1449–1464.

Anand A, Savery D, Hall C. Three-dimensional spatial and temporal temperature imaging in gel phantoms using backscattered ultrasound. *IEEE Transactions on Ultrasonics, Ferroelectrics and Frequency Control* 2007;54:23–31.

Annand A, Kaczkowski PJ. Monitoring formation of high intensity focused ultrasound (HIFU) induced lesions using backscattered ultrasound. *Acoustical Research Letters Online* 2004;5:88–94.

Aron M, Gill IS. Renal tumor ablation. *Current Opinion in Urology* 2005;15:298–305.

Arthur RM, Straube WL, Starman JD, Moros EG. Noninvasive temperature estimation based on the energy of backscattered ultrasound. *Medical Physics* 2003;30:1021–1029.

Arthur RM, Trobaugh J, Straube WL, Moros EG. Temperature dependence of ultrasonic backscattered energy in motion compensated images. *IEEE Transactions on Ultrasonics, Ferroelectrics and Frequency Control* 2005;52:1644–1652.

Ash D, Al-Qaisieh B, Bottomley D, Carey B, Joseph J. The impact of hormone therapy on post-implant dosimetry and outcome following iodine-125 implant monotherapy for localized prostate cancer. *Radiotherapy and Oncology* 2005;75:303–306.

ASTRO ASfTRaOCP. Consensus statement: Guidelines for PSA following radiation therapy. *International Journal of Radiation Oncology, Biology, Physics* 1997;37:1035–1041.

Athanasiou A, Tardivon A, Tanter M, Sigal-Zafrani B, Bercoff J, Deffieux T, Gennisson J-L, Fink M, Neuenschwander S. Breast lesions: Quantitative elastography with supersonic shear imaging-preliminary results. *Radiology* 2010;256:297–303.

Aubry JF, Tanter M, Pernot M, Thomas J-L, Fink M. Experimental demonstration of noninvasive transskull adaptive focusing based on prior computed tomography scans. *Journal of the Acoustical Society of America* 2003;113:84–93.

Augustin H, Hammerer P, Graefen M, Palissaar J, Noldus J, Fernandez S, Huland H. Intraoperative and perioperative morbidity of contemporary radical retropubic prostatectomy in a consecutive series of 1243 patients: Results of a single center between 1999 and 2002. *European Urology* 2003;43:113–118.

Aus G. Current status of HIFU and cryotherapy in prostate cancer—A review. *European Urology* 2006;50:927–934.

Azzouz H, De la Rosette JJMC. HIFU: Local treatment of prostate cancer. *EAU-EBU Update Series* 2006;4:62–70.

Bachtold MR, Rinaldi PC, Jones JP, Reines F, Price LR. Focused ultrasound modifications of neural circuit activity in a mammalian brain. *Ultrasound in Medicine and Biology* 1998;24:557–565.

Bajekal N, Li TC. Fibroids, infertility and pregnancy wastage. *Human Reproduction Update* 2000;6:614–620.

Ballantine HTJ, Bell E, Nanlapaz J. Progress and problems in the neurological applications of focused ultrasound. *Journal of Neurosurgery* 1960;17:858–876.

Bastien A, Pernot M, Tanter M. Monitoring of thermal therapy based on shear modulus changes: II. Shear wave imaging of thermal lesions. *IEEE Transactions on Ultrasonics, Ferroelectrics and Frequency Control* 2011;58:1603–1611.

Bauer R, Martin E, Haegele-Link S, Kaeqi G, von Specht M, Werner B. Noninvasive functional neurosurgery using transcranial MR imaging-guided focused ultrasound. *Parkinsonism & Related Disorders* 2014;20:S197–S199.

Beerlage HP, Thuroff S, Debruyne FMJ, Chaussy C, de la Rosette JJMC. Transrectal high-intensity focused ultrasound using the Ablatherm device in the treatment of localized prostate carcinoma. *Urology* 1999;54:273–277.

Bercoff J, Tanter M, Fink M. Supersonic shear imaging: A new technique for soft tissue elasticity mapping. *IEEE Transactions on Ultrasonics, Ferroelectrics and Frequency Control* 2004;51:396–409.

Bessonova OV, Khokhlova VA, Bailey MR, Canney MS, Crum LA. Focusing of high power ultrasound beams and limiting values of shock wave parameters. *Acoustical Physics* 2009;55:463–473.

Bianco FJ, Scardino PT, Stephenson AJ, Di Blasio CJ, Fearn PA, Eastham JA. Long-term oncologic results of salvage radical prostatectomy for locally recurrent prostate cancer after radiotherapy. *International Journal of Radiation Oncology, Biology, Physics* 2005;62:448–453.

Bing KF, Rouze NC, Palmeri ML, Rotemberg VM, Nightingale KR. Combined ultrasonic thermal ablation with interleaved ARFI imaging monitoring using a single diagnostic curvilinear array: A feasibility study. *Ultrasonic Imaging* 2011;33:217–232.

Blana A, Rogenhofer S, Ganzer R, Wild PJ, Wieland WF, Walter B. Morbidity associated with repeated transrectal high-intensity focused ultrasound treatment of localized prostate cancer. *World Journal of Urology* 2006;24:585–590.

Blana A, Walter B, Rogenhofer S, Wieland WF. High-intensity focused ultrasound for the treatment of localized prostate cancer: 5-year experience. *Urology* 2004;63:297–300.

Boetes C, Mus RD, Holland R, Barentsz JO, Strijk SP, Wobbes T, Hendriks JH, Ruys SH. Breast tumours: Comparative accuracy of MR imaging relative to mammography and US for demonstrating extent. *Radiology* 1995;197:743–747.

Bohris C, Schreiber WG, Jenne J, Simiantonakis I, Rastert R, Zabel H-J, Huber P, Bader R, Brix G. Quantitative MR temperature monitoring of high-intensity focused ultrasound therapy. *Magnetic Resonance Imaging* 1999;17:603–610.

Borghede G, Aldenborg F, Wurzinger E, Johansson KA, Hedelin H. Analysis of the local control in lymph-node staged localized prostate cancer treated by external beam radiotherapy, assessed by digital rectal examination, serum prostate-specific antigen and biopsy. *British Journal of Urology* 1997;80:247–255.

Bosh FX, Ribes J, Charies R, Diaz M. Epidemiology of hepatocellular carcinoma. *Clinical Liver Disease* 2005;9:191–211.

Bosniak MA. Observation of small incidentally detected renal masses. *Seminars in Urologic Oncology* 1995;13:267–272.

Bouchoux G, Bader KB, Korhagen J, Raymond JL, Shivashankar R, Abruzzo TA, Holland CK. Experimental validation of a finite-difference model for the prediction of transcranial ultrasound fields based on CT images. *Physics in Medicine and Biology* 2012;57:8005–8022.

Bradley LD. Uterine fibroid embolization: A viable alternative to hysterectomy. *American Journal of Obstetrics and Gynecology* 2009;201:127–135.

British Uro-Oncology Group (BUG), British Association of Urological Surgeons (BAUS) Section of Oncology. Multi-disciplinary term (MDT) guidance for managing prostate cancer, 2013.

Brock-Abraham C, Carbone N, Dodds E, Kluger J, Park A, Rawlings N, Suddath C, Sun FF, Thompson M, Walsh B, Webley K. The 50 best inventions. *Time* 2011;178:55–82.

Bruix J, Sala M, Llovet JM. Chemoembolization for hepatocellular carcinoma. *Gastroenterology* 2004;127:S179–S188.

Burgess A, Huang Y, Waspe AC, Ganguly M, Goertz DE, Hynynen K. High-intensity focused ultrasound (HIFU) for dissolution of clots in a rabbit model of embolic stroke. *PLoS ONE* 2012;7:e42311.

Burnett AL, Aus G, Canby-Hagino ED et al. Erectile function outcome reporting after clinically localized prostate cancer treatment. *Journal of Urology* 2007;178:597–601.

Burov VA, Dmitrieva NP. Nonlinear ultrasound: Breakdown of microscopic biological structures and nonthermal impact on a malignant tumor. *Doklady Biochemistry and Biophysics* 2002;383:101–104.

Bush NL, Rivens IH, ter Haar GR, Bamber JC. Acoustic properties of lesions generated with an ultrasound therapy system. *Ultrasound in Medicine and Biology* 1993;19:789–801.

Buttram VCJ, Reiter RC. Uterine leiomyomata: Etiology, symptomatology, and management. *Fertility and Sterility* 1981;36:433–445.

Callstrom MR, Charboneau JW, Goetz MP, Rubin J, Atwell TD, Farrell MA, Welch TJ, Maus TP. Image-guided ablation of painful metastatic bone tumors: A new and effective approach to a difficult problem. *Skeletal Radiology* 2005;35:1–15.

Canney MS, Khokhlova VA, Bessonova OV, Bailey MR, Crum LA. Shock-induced heating and millisecond boiling in gels and tissue due to high intensity focused ultrasound. *Ultrasound in Medicine and Biology* 2010;36:250–267.

Cernicanu A, Lepetit-Coiffe M, Roland J, Becker CD, Terraz S. Validation of fast MR thermometry at 1.5 T with gradient-echo echo planar imaging sequences: Phantom and clinical feasibility studies. *NMR in Biomedicine* 2008;21:849–858.

Chan AH, Fujimoto VY, Moore DE, Martin RW, Vaezy S. An image-guided high intensity focused ultrasound device for uterine fibroids treatment. *Medical Physics* 2002;29:2611–2620.

Chang JW, Min B-K, Kim B-S, Chang WS, Lee Y-H. Neurophysiologic correlates of sonication treatment in patients with essential tremor. *Ultrasound in Medicine and Biology* 2015;41:124–131.

Chapelon JY, Prat F, Sibille A, About El Fadil F, Henry L, Theilliere Y, Cathignol D. Extracorporeal selective focused destruction of hepatic tumours by high intensity ultrasound in rabbits bearing VX-2 carcinoma. *Minimally Invasive Therapy* 1992;1:287–293.

Chapman A, ter Haar G. Thermal ablation of uterine fibroids using MR-guided focused ultrasound-A truly non-invasive treatment modality. *European Radiology* 2007;17:2505–2511.

Chaussy C, Thuroff S. Results and side effects of high-intensity focused ultrasound in localized prostate cancer. *Journal of Endourology* 2001;15:437–440.

Chaussy C, Thüroff S. High-intensity focused ultrasound in prostate cancer: Results after 3 years. *Molecular Urology* 2000;4:179–182.

Chaussy C, Thüroff S. The status of high-intensity focused ultrasound in the treatment of localized prostate cancer and the impact of a combined resection. *Current Urology Reports* 2003;4:248–252.

Chaussy C, Thüroff S, Rebillard X, Gelet A. Technology insight: High intensity focused ultrasound for urologic cancers. *Nature Clinical Practice Urology* 2005;2:191–198.

Chauvet D, Marsac L, Pernot M, Boch AL, Guillevin R, Salameh N, Souris L, Darrasse L, Fink M, Tanter M, Aubry JF. Targeting accuracy of transcranial magnetic resonance-guided high-intensity focused ultrasound brain therapy: A fresh cadaver model: Laboratory investigation. *Journal of Neurosurgery* 2013;118:1046–1052.

Chen J, Chen W. A comparison between ultrasound therapy and laser therapy for symptomatic cervical ecotopy. *Ultrasound in Medicine and Biology* 2008;34:1770–1774.

Chen L, Rivens I, ter Haar GR et al. Histological changes in rat liver tumours treated with high-intensity focused ultrasound. *Ultrasound in Medicine and Biology* 1993;19:67–74.

Chen L, ter Haar G, Hill CR, Dworkin M, Carnochan P, Young H, Bensted JPM. Effect of blood perfusion on the ablation of liver parenchyma with high-intensity focused ultrasound. *Physics in Medicine and Biology* 1991;36:1661–1673.

Chen W, Zhou K. High-intensity focused ultrasound ablation: A new strategy to manage primary bone tumors. *Current Opinion in Orthopedics* 2005;16:494–500.

Chen W, Zhu H, Zhang L, Li K, Su H, Jin C, Zhou K, Bai J, Wu F, Wang Z. Primary bone malignancy: Effective treatment with high-intensity focused ultrasound ablation. *Radiology* 2010a;255:967–978.

Chen W-H, Sanghvi NT, Carlson RF, Uchida T. Real-time tissue change monitoring on the Sonablate 500 during high-intensity focused ultrasound (HIFU) treatment of prostate cancer. In: *10th International Symposium on Therapeutic Ultrasound.* Tokyo, Japan: AIP, 2010b, pp. 391–396.

Chen WZ, Wang ZB, Wu F, Bai J, Li KQ, Xie FL, Wang ZL. High intensity focused ultrasound alone for malignant solid tumours. *Chinese Journal of Oncology* 2002a;24:278–281.

Chen WZ, Wang ZB, Wu F, Zhu H, Zou JZ, Bai J, Li KQ, Xie FL. High intensity focused ultrasound in the treatment of primary malignant bone tumour. *Chinese Journal of Oncology* 2002b;24:612–615.

Cheng SQ, Zhou XD, Tang ZY, Yu Y, Bao S-S, Qian D-C. Iodized oil enhanced the thermal effect of high-intensity focused ultrasound on ablating experimental liver cancer. *Journal of Cancer Research and Clinical Oncology* 1997;123:639–644.

Chow W, Devesa S, Warren J, Fraumeni JJ. Rising incidence of renal cell cancer in the United States. *JAMA* 1999;281:1628–1631.

Chung AH, Hynynen K, Colucci V, Oshio K, Cline HE, Jolesz FA. Optimization of spoiled gradient-echo phase imaging for in vivo localization of caused ultrasound beam. *Magnetic Resonance in Medicine* 1996;36:745–752.

Clark A, Breau R, Morash C, Fergusson D, Doucette S, Cagiannos I. Preservation of renal function following partial or radical nephrectomy using 24-hour creatinine clearance. *European Urology* 2008;54:143–152.

Clement GT, Hynynen K. Correlation of ultrasound phase with physical skull properties. *Ultrasound in Medicine and Biology* 2002;28:617–624.

Cline HE, Hynynen K, Hardy CJ, Watkins RD, Schenck JF, Jolesz FA. MR temperature mapping of focused ultrasound surgery. *Magnetic Resonance in Medicine* 1994;31:628–636.

Cline HE, Hynynen K, Schneider E, Hardy CJ, Maier SE, Watkins RD, Jolesz FA. Simultaneous magnetic resonance phase and magnitude temperature maps in muscles. *Magnetic Resonance in Medicine* 1996;35:309–315.

Cline HE, Schenck JF, Watkins RD, Hynynen K, Jolesz FA. Magnetic resonance guided thermal surgery. *Magnetic Resonance in Medicine* 1993;30:98–106.

Coelho RF, Palmer KJ, Rocco B, Moniz RR, Chauhan S, Orvieto MA, Coughlin G, Patel VR. Early complication rates in a single-surgeon series of 2500 robotic-assisted radical prostatectomies: Report applying a standardized grading system. *European Urology* 2010;57:945–952.

Coen JJ, Zietman AL, Thakral H, Shipley WU. Radical radiation for localized prostate cancer: Local persistence of disease results in a late wave of metastases. *Journal of Clinical Oncology* 2002;20:199–204.

Coleman DJ, Lizzi FL, Driller J, Rosado AL, Chang S, Iwamoto T, Rosenthal D. Therapeutic ultrasound in the treatment of glaucoma. I. Experimental model. *Ophthalmology* 1985;92:339–346.

Colucci V, Strichartz G, Jolesz F, Vykhodtseva N, Hynynen K. Focused ultrasound effects on nerve action potential *in vitro*. *Ultrasound in Medicine and Biology* 2009;35:1737–1747.

Coventry BJ, Lee PL, Gibbs D, Hart DN. Dendritic cell density and activation status in human breast cancer-CD1a, CMRF-44, CMRF-56 and CD-83 expression. *British Journal of Cancer* 2002;86:546–551.

Crook J, Malone S, Perry G, Bahadur Y, Robertson S, Abdolell M. Postradiotherapy prostate biopsied: What do they really mean? Results for 498 patients. *International Journal of Radiation Oncology, Biology, Physics* 2000;48:355–367.

Cui H, Yang X. In vivo imaging and treatment of solid tumor using integrated photoacoustic imaging and high intensity focused ultrasound system. *Medical Physics* 2010;37:4777–4781.

Cui H, Yang X. Real-time monitoring of high-intensity focused ultrasound ablations with photoacoustic technique: An *in vitro* study. *Medical Physics* 2011;38:5345–5350.

Curiel L, Hynynen K. Localized harmonic motion imaging for focused ultrasound surgery targeting. *Ultrasound in Medicine and Biology* 2011;37:1230–1239.

Curiel L, Souchon R, Rouvière O, Gelet A, Chapelon JY. Elastography for the follow-up of high-intensity focused ultrasound prostate cancer treatment: Initial comparison with MRI. *Ultrasound in Medicine and Biology* 2005;31:1461–1468.

Damianou C, Pavlou M, Velev O, Kyriakou K, Trimikliniotis M. High intensity focused ultrasound ablation of kidney guided by MRI. *Ultrasound in Medicine and Biology* 2004;30:397–404.

Daniels MJ, Varghese T, Madsen EL, Zagzebski JA. Non-invasive ultrasound-based temperature imaging for monitoring radio-frequency heating-phantom results. *Physics in Medicine and Biology* 2007;52:4827–4843.

Delloye C. How to improve the incorporation of massive allografts? *La Chirurgia degli Organi di Movimento* 2003;88:335–343.

Delon-Martin C, Vogt C, Chignier E, Guers C, Chapelon JY, Cathignol D. Venous thrombosis generation by means of high-intensity focused ultrasound. *Ultrasound in Medicine and Biology* 1995;21:113–119.

Den Brok MH, Sutmuller RP, van der Voort R, Bennink EJ, Figdor CG, Ruers TJ, Adema GJ. In situ tumor ablation creates an antigen source for the generation of antitumor immunity. *Cancer Research* 2004;64:4024–4029.

Dervishi E, Larrat B, Pernot M, Adam C, Marie Y, Fink M, Delattre JY, Boch AL, Tanter M, Aubry JF. Transcranial high intensity focused ultrasound therapy guided by 7 TESLA MRI in a rat brain tumour model: A feasibility study. *International Journal of Hyperthermia* 2013;29:598–608.

Dewey WC. Arrhenius relationships from the molecule and cell to the clinic. *International Journal of Hyperthermia* 1994;10:457–483.

Diederich CJ. Thermal ablation and high-temperature thermal therapy: Overview of technology and clinical implementation. *International Journal of Hyperthermia* 2005;21:745–753.

Dorr LN, Hynynen K. The effect of tissue heterogeneties and large blood vessels on the thermal exposure induced by short high power ultrasound pulses. *International Journal of Hyperthermia* 1992;8:45–59.

Dubinsky TJ, Cuevas C, Dighe MK, Kolokythas O, Hwang JH. High-intensity focused ultrasound: Current potential and oncologic applications. *Ultrasound Imaging* 2008;190:191–199.

El-Serag HB, Mason AC. Rising incidence of hepatocellular carcinoma in the United States. *New England Journal of Medicine* 1999;340:745–750.

El-Serag HB, Mason AC. Risk factors for the rising rates of primary liver cancer in the United States. *Archives of Internal Medicine* 2000;160:3227–3230.

Enneking WF, Spanier SS, Goodman MA. A system for the surgical staging of musculoskeletal sarcoma. *Clinical Orthopaedics* 1980;153:106–120.

Fahey BJ, Nightingale KR, Stutz D, Trahey GE. Acoustic radiation force impulse imaging of thermally- and chemically-induced lesions in soft tissues: Preliminary ex vivo results. *Ultrasound in Medicine and Biology* 2004;30:321–328.

Faivre J, Forman D, Estève J, Obradovic M, Sant M. Survival of patients with primary liver cancer, pancreatic cancer and biliary tract cancer in Europe. EUROCARE Working group. *European Journal of Cancer* 1998;34:2184–2190.

Fan QY, Ma BA, Zhou YF, Zhang MH, Hao XB. Bone tumors of the extremities or pelvis treated by microwave-induced hyperthermia. *Clinical Orthopaedics and Related Research* 2003;406:165–175.

Fatemi M, Greenleaf JF. Ultrasound-stimulated vibroacoustic spectrography. *Science* 1998;280:82–85.

Fennessy FM, Tempany CM. A review of magnetic resonance imaging-guided focused ultrasound surgery of uterine fibroids. *Topics in Magnetic Resonance Imaging* 2006;17:173–179.

Fennessy FM, Tempany CM, McDannold NJ et al. Uterine leiomyomas: MR imaging-guided focused ultrasound surgery-results of different treatment protocols. *Radiology* 2007;243:885–893.

Fink M. Time reversal of ultrasonic fields. I. Basic principles. *IEEE Transactions on Ultrasonics, Ferroelectrics and Frequency Control* 1992;39:555–566.

Finn HA, Simon MA. Limb-salvage surgery in the treatment of osteosarcoma in skeletally immature individuals. *Clinical Orthopaedics and Related Research* 1991;262:108–118.

Fisher B, Anderson S, Bryant J, Margolese RG, Deutsch M, Fisher ER, Jeong JH, Wolmark N. Twenty-year follow-up of a randomized trial comparing total mastectomy, lumpectomy, and lumpectomy plus irradiation for the treatment of invasive breast cancer. *New England Journal of Medicine* 2002;347:1233–1241.

Fisher B, Bredmond C, Poinsson R, Margolese R, Wolmark N, Wickerham L, Fisher E, Deutsch M, Caplan R, Pilch Y. Eight year results of a randomized trial comparing total mastectomy and lumpectomy with or without irradiation in the treatment of breast cancer. *New England Journal of Medicine* 1989;320:822–828.

Foley JL, Little JW, Starr FL, Frantz C, Vaezy S. Image-guided HIFU neurolysis of peripheral nerves to treat spasticity and pain. *Ultrasound in Medicine and Biology* 2004;30:1199–1207.

Foley JL, Little JW, Vaezy S. Image-guided high-intensity focused ultrasound for conduction block of peripheral nerves. *Annals of Biomedical Engineering* 2007;35:109–119.

Foley JL, Little JW, Vaezy S. Effects of high-intensity focused ultrasound on nerve conduction. *Muscle Nerve* 2008;37:241–250.

Fornage BD, Toubas O, Morel M. Clinical, mammographic, and sonographic determination of preoperative breast cancer size. *Cancer* 1987;60:765–771.

Fowlkes JB, Ivey JA, Gardner EA, Rubin JM, Carson PL. New acoustical approaches to perfusion and other vascular dynamics. *Journal of the Acoustical Society of America* 1994;95:2855.

Freedonia Group. Cancer therapies to 2009—Market research, market share, market size, sales, demand forecast, market leaders, company profiles, industry trends, 2009.

Fruehauf JH, Back W, Eiermann A, Lang M-C, Pessel M, Marlinghaus E, Melchert F, Volz-Köster S, Volz J. High-intensity focused ultrasound for the targeted destruction of uterine tissues: Experiences from a pilot study using a mobile HIFU unit. *Archives of Gynecology and Obstetrics* 2008;277:143–150.

Fry FJ, Ades HW, Fry WJ. Production of reversible changes in the central nervous system by ultrasound. *Science* 1958;127:83–84.

Fry FJ, Barger JE. Acoustical properties of the human skull. *Journal of the Acoustical Society of America* 1978;63:1576–1590.

Fry W, Barnard J, Fry F, Krumins R, Brennan J. Ultrasonic lesions in the mammalian central nervous system with ultrasound. *Science* 1955;122:517–518.

Fry WJ, Fry FJ. Fundamental neurological research and human neurosurgery using intense ultrasound. *IRE Transactions on Medical Electronics* 1960;ME-7:166–181.

Funaki K, Fukunishi H, Sawada K. Clinical outcomes of magnetic resonance-guided focused ultrasound surgery for uterine myomas: 24-month follow-up. *Ultrasound in Obstetrics and Gynecology* 2009;34:584–589.

Funke AR, Aubry J-F, Fink M, Boccara A-C, Bossy E. Photoacoustic guidance of high intensity focused ultrasound with selective optical contrasts and time-reversal. *Applied Physics Letters* 2009;94:054102.

Gabrilovich DI, Corak J, Ciernik IF, Kavanaugh D, Carbone DP. Dendritic cell density and activation status in human breast cancer-CD1a, CMRF-44, CMRF-56 and CD83 expression. *British Journal of Cancer* 1997;86:546–551.

Gelet A, Chapelon JY, Bouvier R, Rouvière O, Lasne Y, Lyonnet D, Dubernard JM. Transrectal high-intensity focused ultrasound: Minimally invasive therapy of localized prostate cancer. *Journal of Endourology* 2000;14:519–528.

Gelet A, Chapelon JY, Bouvier R, Souchon R, Pangaud C, Abdelrahim AF, Cathignol D, Dubernard JM. Treatment of prostate cancer with transrectal focused ultrasound: Early clinical experience. *European Urology* 1996;29:174–183.

Gelet A, Chapelon JY, Margonari J, Theilliere Y, Gorry F, Souchon R, Bouvier R. High-intensity focused ultrasound experimentation on human benign prostatic hypertrophy. *European Urology* 1993;23:44–47.

Gertner MR, Worthington AE, Wilson BC, Sherar MD. Ultrasound imaging of thermal therapy in in vitro liver. *Ultrasound in Medicine and Biology* 1998;24:1023–1032.

Ghanem N, Uhl M, Brink I, Schafer O, Kelly T, Moser E, Langer M. Diagnostic value of MRI in comparison to scintigraphy, PET, MS-CT, and PET/CT or the detection of metastases of bone. *European Journal of Radiology* 2005;55:41–55.

Gianfelice D, Khait A, Boulanger Y, Amara M, Beblidia A. Feasibility of magnetic resonance imaging-guided focused ultrasound surgery as an adjunct to Tamoxifen therapy in high-risk surgical patients with breast carcinoma. *Journal of Vascular and Interventional Radiology* 2003;14:1275–1282.

Goldberg SN, Grassi CJ, Cardella JF et al. Image-guided tumor ablation: Standardization of terminology and reporting criteria. *Journal of Vascular and Interventional Radiology* 2009;20:S377–S390.

Gomez J, Pinar A, Vallcanera A, Moreno A, Cortina H. Sonographic findings in aneurysmal bone cyst in children: Correlation with computed tomography findings. *Journal of Clinical Ultrasound* 1998;26:59–64.

Goodwin SC, Spies JB, Worthington-Kirsch R, Peterson E, Pron G, Li S, Myers ER. Uterine artery embolization for treatment of leiomyomata: Long-term outcomes from the FIBROID Registry. *Obstetrics & Gynecology* 2008;111:22–33.

Graham J, Baker M, Macbeth F, Titshall V, Group GD. Diagnosis and treatment of prostate cancer: Summary of NICE guidance. *BMJ* 2008;336:610–612.

Hacker A, Michel MS, Marlinghaus E, Kohrmann KU, Alken P. Extracorporeally induced ablation of renal tissue by high-intensity focused ultrasound. *BJU International* 2006;97:779–785.

Hall TL, Kieran K, Ives K, Fowlkes JB, Cain CA, Roberts WW. Histotripsy of rabbit renal tissue in vivo: Temporal histologic trends. *Journal of Endourology* 2007;21:1159–1166.

Hall-Craggs M. Interventional MRI of the breast: Minimally invasive therapy. *European Radiology* 2000;10:59–62.

Halle JS, Scoville CR, Greathouse DG. Ultrasound's effect on the conduction latency of the superficial radial nerve in man. *Physical Therapy* 1981;61:345–350.

Halstedt WS. The results of radical operations for the cure of carcinoma of the breast. *Annals of Surgery* 1907;66:1–9.

Han M, Partin AW, Pound CR, Epstein JI, Walsh PC. Long-term biochemical disease-free and cancer-specific survival following anatomic radical retropubic prostatectomy. The 15-year Johns Hopkins experience. *Urologic Clinics of North America* 2001;28:555–565.

Hansler J, Neureiter D, Strobel D et al. Cellular and vascular reactions in the liver to radio-frequency thermo-ablation with wet needle applicators. Study on juvenile domestic pigs. *European Surgical Research* 2002;34:357–363.

Harisinghani MG, Barentsz JO, Hahn PF, Deserno WM, Tabatabaei S, van de Kaa CH, de la Rosette JJMC, Weissleder R. Noninvasive detection of clinically occult lymph-node metastases in prostate cancer. *New England Journal of Medicine* 2003;348:2491–2499.

Harvey EN, Barnes DK, McElroy WD, Whiteley AH, Pease DC, Cooper KW. Bubble formation in animals. I. Physical factors. *Journal of Cellular and Comparative Physiology* 1944;24:1–22.

He SX, Wang GM, Niu SG, Yao B, Wang XJ. The noninvasive treatment of 251 cases of advanced pancreatic cancer with focused ultrasound surgery. In: Andrew MA, Crum LA, Vaezy S, eds., *Second International Symposium on Therapeutic Ultrasound*, Seattle, WA, 2002, pp. 51–56.

Heidenreich A, Aus G, Bolla M, Joniau S, Matveev VB, Schmid HP, Zattoni F. EAU guidelines on prostate cancer. *European Urology* 2008;53:68–80.

Heidenreich A, Richter S, Thuer D, Pfister D. Prognostic parameters, complications, and oncologic and functional outcome of salvage radical prostatectomy for locally recurrent prostate cancer after 21st-century radiotherapy. *European Urology* 2010;57(3):437–445.

Hesley GK, Gorny KR, Henrichsen TL, Woodrum DA, Brown DL. A clinical review of focused ultrasound ablation with magnetic resonance guidance: An option for treating uterine fibroids. *Ultrasound Quarterly* 2008;24:131–139.

Hickey RC, Fry WJ, Meyers R, Fry FJ, Bradbury JT. Human pituitary irradiation with focused ultrasound: An initial report on effect in advanced breast cancer. *Archives of Surgery* 1961;83:620–633.

Hoegler D. Radiotherapy for palliation of symptoms in incurable cancer. *Current Problems in Cancer* 1997;21:129–183.

Hollingsworth JM, Miller DC, Daignault S, Hollenbeck BK. Rising incidence of small renal masses: A need to reassess treatment effect. *Journal of the National Cancer Institute* 2006;98:1331–1334.

Hou AH, Sullivan KF, Crawford ED. Targeted focal therapy for prostate cancer: A review. *Current Opinion in Urology* 2009;19:283–289.

Hou GY, Luo J, Marguet F, Maleke C, Vappou J, Konofagou EE. Performance assessment of HIFU lesion detection by harmonic motion imaging for focused ultrasound (HMIFU): A 3-D finite-element-based framework with experimental validation. *Ultrasound in Medicine and Biology* 2011;37:2013–2027.

Hou GY, Marquet F, Wang S, Konofagou EE. Multi-parametric monitoring and assessment of high-intensity focused ultrasound (HIFU) boiling by harmonic motion imaging for focused ultrasound (HMIFU): An *ex vivo* feasibility study. *Physics in Medicine and Biology* 2014;59:1121–1145.

Hsueh C-T. Pancreatic cancer: Current standards, research updates and future directions. *Journal of Gastrointestinal Oncology* 2011;2:123–125.

Hu Z, Yang X, Liu Y, Sankin GN, Pua EC, Morse MA, Lyerly HK, Clay TM, Zhong P. Investigation of HIFU-induced anti-tumor immunity in a murine tumor model. *Journal of Translational Medicine* 2007;5:34.

Hu ZL, Yang XY, Liu Y, Morse MA, Lyerly HK, Clay TM, Zhong P. Release of endogenous danger signals from HIFU treated tumor cells and their stimulatory effects on APCs. *Biochemical and Biophysical Research Communications* 2005;335:124–131.

Hu ZL, Yang XY, Liu YB, Sankin GN, Pua EC, Morse MA, Lyerly HK, Clay TM, Zhong P. Investigation of HIFU-induced anti-tumor immunity in a murine tumor model. *Journal of Translational Medicine* 2008;5:34–44.

Hüber PE, Jenne JW, Rastert R, Simiantonakis I, Sinn H-P, Strittmatter H-J, von Fournier D, Wannenmacher MF, Debus J. A new noninvasive approach in breast cancer therapy using magnetic resonance imaging-guided focused ultrasound surgery. *Cancer Research* 2001;61:8441–8447.

Hwang JH, Brayman AA, Reidy MA, Matula TJ, Kimmey MB, Crum LA. Vascular effects induced by combined 1-MHz ultrasound and microbubble contrast agent treatments in vivo. *Ultrasound in Medicine and Biology* 2005;31:553–564.

Hwang JH, Tu J, Brayman AA, Crum LA. Correlation between inertial cavitation dose and endothelial cell damage *in vivo*. *Ultrasound in Medicine and Biology* 2006;32:1611–1619.

Hwang JH, Zhou Y, Warren C, Brayman AA, Crum LA. Targeted venous occlusion using pulsed high-intensity focused ultrasound. *IEEE Transactions on Biomedical Engineering* 2010;57:37–40.

Hynynen K, Clement GT. Clinical applications of focused ultrasound-the brain. *International Journal of Hyperthermia* 2007;23:193–202.

Hynynen K, Freund WR, Cline HE, Chung AH, Watkins RD, Vetro JP, Jolesz FA. A clinical, noninvasive, MR imaging-monitored ultrasound surgery method. *Radiographics* 1996;16:185–195.

Hynynen K, Jolesz FA. Demonstration of potential noninvasive ultrasound brain therapy through an intact skull. *Ultrasound in Medicine and Biology* 1998;24:275–283.

Hynynen K, McDannold N. MRI-guided focused ultrasound for local tissue ablation and other image-guided interventions. In: Wu J, Nyborg WL, eds., *Emerging Therapeutic Ultrasound*. Singapore: Singapore World Scientific Publishing, 2006, p. 167.

Hynynen K, Pomeroy O, Smith DN, Huber PE, McDannold NJ, Kettenbach J, Baum J, Singer S, Joles FA. MR imaging-guided focused ultrasound surgery of fibroadenomas in the breast: A feasibility study. *Radiology* 2001;219:176–185.

Hynynen K, Vykhodtseva N, Chung AH, Sorentino V, Colucci V, Jolesz FA. Thermal effects of focused ultrasound on the brain: Determination with MR imaging. *Radiology* 1997;204:247–253.

Ikeda K, Kumada H, Saitoh S, Arase Y, Chayama K. Effect of repeated transcatheter arterial embolization on the survival time in patients with hepatocellular carcinoma. *Cancer* 1991;68:2150–2154.

Iselin CE, Robertson JE, Paulson DF. Radical perineal prostatectomy: Oncological outcome during a 20-year period. *Journal of Urology* 1999;161:163–168.

Jang HJ, Lee JY, Lee DH, Kim WH, Hwang JH. Current and future clinical applications of high-intensity focused ultrasound (HIFU) for pancreatic cancer. *Gut and Liver* 2010;4:S57–S61.

Jeanmonod D, Werner B, Morel A, Michels L, Zadicario E, Schiff G, Martin E. Transcranial magnetic resonance imaging-guided focused ultrasound: Noninvasive central lateral thalamotomy for chronic neuropathic pain. *Neurosurgical Focus* 2012;32:E1.

Jemal A, Bray F, Center MM, Ferlay J, Ward E, Forman D. Global cancer statistics. *CA: A Cancer Journal for Clinicians* 2011;61:69–90.

Jemal A, Siegel R, Ward E, Hao Y, Xu J, Thun M. Cancer statistics, 2009. *CA: A Cancer Journal for Clinicians* 2009;59:225–249.

Jemal A, Ward E, Wu X, Martin HJ, McLaughlin CC, Thun MJ. Geographic patterns of prostate cancer mortality and variations in access to medical care in the United States. *Cancer Epidemiology, Biomarkers & Prevention* 2005;14:590–595.

Jones RM, O'Reilly MA, Hynynen K. Transcranial passive acoustic mapping with hemispherical sparse arrays using CT-based skull-specific aberration corrections: A simulation study. *Physics in Medicine and Biology* 2013;58:4981–5005.

Jung SE, Cho SH, Jang JH, Han J-Y. High-intensity focused ultrasound ablation in hepatic and pancreatic cancer: Complications. *Abdominal Imaging* 2011;36:185–195.

Kennedy JE. High-intensity focused ultrasound in the treatment of solid tumors. *Nature Reviews: Cancer* 2005;5:321–327.

Kennedy JE, ter Haar GR, Cranston D. High intensity focused ultrasound: Surgery of the future? *British Journal of Radiology* 2003;76:590–599.

Kennedy JE, Wu F, ter Haar GR, Gleeson FV, Phillips RR, Middleton MR, Cranston DW. High-intensity focused ultrasound for the treatment of liver tumours. *Ultrasonics* 2004;42:931–935.

Khokhlova TD, Canney MS, Khokhlova VA, Sapozhnikov OA, Crum LA, Bailey MR. Controlled tissue emulsification produced by high intensity focused ultrasound shock. *Journal of the Acoustical Society of America* 2011;130:3498–3510.

Khokhlova TD, Canney MS, Lee D, Marro KI, Crum LA, Khokhlova VA, Bailey MR. Magnetic resonance imaging of boiling induced by high intensity focused ultrasound. *Journal of the Acoustical Society of America* 2009;125:2420–2431.

Khokhlova TD, Hwang JH. HIFU for palliative treatment of pancreatic cancer. *Journal of Gastrointestinal Oncology* 2011;2:175–184.

Kim HS, Baik JH, Pham LD, Jacobs MA. MR-guided high-intensity focused ultrasound treatment for symptomatic uterine leiomyomata. *Academic Radiology* 2011;18:970–976.

Klatte T, Marberger M. High-intensity focused ultrasound for the treatment of renal masses: Current status and future potential. *Current Opinion in Urology* 2009;19:188–191.

Klatte T, Patard JJ, de Martino M et al. Tumor size does not predict risk of metastatic disease or prognosis of small renal cell carcinomas. *Journal of Urology* 2008;179:1719–1726.

Klingler HS, Susani M, Seip R, Mauermann J, Sanghvi NT, Marberger M. A novel approach to energy ablative therapy of small renal tumours: Laparoscopic high-intensity focused ultrasound. *European Urology* 2008;53:810–816.

Köhler MO, Mougenot C, Quesson B, Enholm J, Le Bail B, Laurent C, Moonen CTW, Ehnholm G. Volumetric HIFU ablation under 3D guidance of rapid MRI thermometry. *Medical Physics* 2009;36:3521–3535.

Konofagou EE, Maleke C, Jonathan V. Harmonic motion imaging (HMI) for tumor imaging and treatment monitoring. *Current Medical Imaging Reviews* 2012;8:16–26.

Konofagou EE, Thierman J, Hynynen K. The use of ultrasound-stimulated acoustic emission in the monitoring of modulus changes with temperature. *Ultrasonics* 2003;41:337–345.

Krag D, Weaver D, Ashikaga T et al. The sentinel node in breast cancer-a multicentre validation study. *New England Journal of Medicine* 1998;339:941–946.

Kramer G, Steiner GE, Grobl M, Hrachowitz K, Reithmayr F, Paucz L, Newman M, Madersbacher S, Gruber D, Susani M, Marberger M. Response to sublethal heat treatment of prostatic tumor cells and of prostatic tumor infiltrating T-cells. *Prostate* 2004;58:109–120.

Kuban DA, Thames HD, Levy LB et al. Long-term multi-institutional analysis of stage T1-T2 prostate cancer treated with radiotherapy in the PSA era. *International Journal of Radiation Oncology, Biology, Physics* 2003;57:915–928.

Kunkle DA, Crispen PL, Li TC, Uzzo RG. Tumor size predicts synchronous metastatic renal cell carcinoma: Implications for surveillance of small renal masses. *Journal of Urology* 2007;177:1692–1696.

Kunkle DA, Uzzo RG. Cryoablation or radiofrequency ablation of the small renal mass: A meta-analysis. *Cancer* 2008;113:2671–2680.

Kyriakou A, Neufeld E, Werner B, Paulides MM, Szekely G, Kuster N. A review of numerical and experimental compensation techniques for skull-induced phase aberrations in transcranial focused ultrasound. *International Journal of Hyperthermia* 2014;30:36–46.

Lafon C, Bouchoux G, Souchon R, Chapelon JY. Monitoring and follow up of HIFU lesions by ultrasound. In: *Biomedical Imaging: From Nano to Macro*, 2007, pp. 1068–1071, April 12–15, Arlington, VA.

Lake AM, Hall TL, Kieran K, Fowlkes JB, Cain CA, Roberts WW. Histotripsy: Minimally invasive technology for prostatic tissue ablation in an in vivo canine model. *Urology* 2008;72:682–686.

Larrat B, Pernot M, Aubry JF, Dervishi E, Rinkus R, Seilhean D, Marie Y, Boch AL, Fink M, Tanter M. MR-guided transcranial brain HIFU in small animal models. *Physics in Medicine and Biology* 2010a;55:365–388.

Larrat B, Pernot M, Montaldo G, Fink M, Tanter M. MR-guided adaptive focusing of ultrasound. *IEEE Transactions on Ultrasonics, Ferroelectrics and Frequency Control* 2010b;57:1734–1747.

Law WK, Frizzell LA, Dunn F. Determination of the nonlinearity parameter B/A of biological media. *Ultrasound in Medicine and Biology* 1985;11:307–318.

LeBlang SD, Hoctor K, Steinberg FL. Leiomyoma shrinkage after MRI-guided focused ultrasound treatment: Report of 80 patients. *American Journal of Roentgenology* 2010;194:274–280.

Leduc N, Okita K, Sugiyama K, Takagi S, Matsumoto Y. Focus control in HIFU therapy assisted by time-reversal simulation with an iterative procedure for hot spot elimination. *Journal of Biomechanical Science and Engineering* 2012;7:43–56.

Lee HM, Hong JH, Choi HY. High-intensity focused ultrasound therapy for clinically localized prostate cancer. *Prostate Cancer and Prostatic Diseases* 2006;9:439–443.

Lee JY, Choi BI, Ryu JK, Kim Y-T, Hwang JH, Kim SH, Han JK. Concurrent chemotherapy and pulsed high-intensity focused ultrasound therapy for the treatment of unresectable pancreatic cancer: Initial experiences. *Korean Journal of Radiology* 2011;12:176–186.

Lee WR, Hanks GE, Hanlon A. Increasing prostate specific antigen profile following definitive radiotherapy for localized prostate cancer: Clinical observations. *Journal of Clinical Oncology* 1997;15:220–238.

Lele PP. Effects of focused ultrasonic radiation on peripheral nerve, with observations on local heating. *Experimental Neurology* 1963;8:47–83.

Lenard ZM, McDannold NJ, Fennessy FM, Stewart EA, Jolesz FA, Hynynen K, Tempany CMC. Uterine leiomyomas: MR imaging-guided focused ultrasound surgery-imaging predictors of success. *Radiology* 2008;249:187–194.

Lepock JR, Frey HE, Ritchie KP. Protein denaturation in intact hepatocytes and isolated cellular organelles during heat shock. *Journal of Cell Biology* 1993;122:1267–1276.

Leslie T, Ritchie R, Illing RO, ter Haar G, Phillips RR, Middleton MR, Wu F, Cranston DW. High-intensity focused ultrasound treatment of liver tumours: Post-treatment MRI correlates well with intra-operative estimates of treatment volume. *British Journal of Radiology* 2012;85:1363–1370.

Lewko B, Zoltowska A, Stepinski J, Roszkiewicz A, Moszkowska G. Dendritic and cancer cells in the breast tumors—An immunohistochemical study: Short communication. *Medical Science Monitor* 2000;6:892–895.

Li C, Bian D. Focused ultrasound therapy of vulvar dystrophies: A feasibility study. *Obstetrics & Gynecology* 2004;9:915–921.

Li C, Zhang W, Fan W, Huang J, Zhang F, Wu P. Noninvasive treatment of malignant bone tumors using high-intensity focused ultrasound. *Cancer* 2010a;116:3934–3942.

Li C, Zhang W, Zhang R, Zhang L, Wu P, Zhang F. Therapeutic effects and prognostic factors in high-intensity focused ultrasound combined with chemoembolisation for larger hepatocellular carcinoma. *European Journal of Cancer* 2010b;46:2513–2521.

Li CX, Xu GL, Jiang ZY, Li JJ, Luo GY, Shan HB, Zhang R, Li Y. Analysis of clinical effect of high-intensity focused ultrasound on liver cancer. *World Journal of Gastroenterology* 2004;10:2201–2204.

Li P-Z, Zhu S-H, He W, Zhu L-Y, Liu S-P, Liu Y, Wang G-H, Ye F. High-intensity focused ultrasound treatment for patients with unresectable pancreatic cancer. *Hepatobiliary & Pancreatic Diseases International* 2012;11:655–660.

Li Y-Y, Sha W-H, Zhou Y-J, Nie Y-Q. Short and long term efficacy of high intensity focused ultrasound therapy for advanced hepatocellular carcinoma. *Journal of Gastroenterology and Hepatology* 2007;22:2148–2154.

Liberman B, Gianfelice D, Inbar Y et al. Pain palliation in patients with bone metastases using MR-guided focused ultrasound surgery: A multicenter study. *Annals of Surgical Oncology* 2009;16:140–146.

Linke CA, Carstensen EL, Frizzell LA, Elbadawi A, Fridd CW. Localized tissue destruction by high-intensity focused ultrasound. *Archives of Surgery* 1973;107:887–891.

Livraghi T, Makuuchi M, Buscarini L. *Diagnosis and Treatment of Hepatocellular Carcinoma*. London, U.K.: Greenwich Medical Media, 1997.

Lizzi FL, Coleman DJ, Driller J, Ostromogilsky M, Chang S, Greenall P. Ultrasonic hyperthermia for ophthalmic surgery. *IEEE Transactions on Sonics and Ultrasonics* 1984;SU-31:473–480.

Lizzi FL, Deng CX, Ketterling JA, Alam SK, Mikaelian S, Kalisz A. Radiation-force technique to monitor lesions during ultrasonic therapy. *Ultrasound in Medicine and Biology* 2003;29:1593–1605.

Ljungberg B, Hanbury DC, Kuczyk MA, Merseburger AS, Mulders PFA, Patard JJ, Sinescu IC. EAU guidelines on renal cell carcinoma: The 2010 update. *European Urology* 2010;58(3):398–406.

Llovet JM, Burroughs A, Bruix J. Hepatocellular carcinoma. *Lancet* 2003;362:1907–1917.

Lu BY, Yang RS, Lin WL, Cheng KS, Wang CY, Kuo TS. Theoretical study of convergent ultrasound hyperthermia for treating bone tumours. *Medical Engineering and Physics* 2000;22:253–263.

Luciani LG, Cestari R, Tallarigo C. Incidental renal cell carcinoma-age and stage characterization and clinical implications: Study of 1092 patients (1982–97). *Urology* 2000;56:58–62.

Lupattelli T, Basile A, Garaci FG, Simonetti G. Percutaneous uterine embolisation for the treatment of symptomatic fibroids: Current status. *European Journal of Radiology* 2005;54:136–137.

Lutz MB, Schuler G. Immature, semi-mature and fully mature dendritic cells: Which signals induce tolerance or immunity? *Trends in Immunology* 2002;23:445–449.

Lynn JG, Putnam TJ. Histological and cerebral lesions produced by focused ultrasound. *American Journal of Pathology* 1944;20:637–649.

Lynn JG, Zwemer RL, Chick AJ, Gen J. The biological application of focused ultrasonic waves. *Science* 1942a;96:119–120.

Lynn JG, Zwemer RL, Chick AJ, Miller AF. A new method for the generation and use of focused ultrasound in experimental biology. *Journal of General Physiology* 1942b;26:179–193.

Madersbacher S, Grobl M, Kramer G, Dirnhofer S, Steiner GE, Marberger M. Regulation of heat shock protein 27 expression of prostatic cells in response to heat treatment. *Prostate* 1998;37:174–181.

Madersbacher S, Pedevilla M, Vingers L, Susani M, Marberger M. Effect of high intensity focused ultrasound on human prostate cancer in-vivo. *Cancer Research* 1995;55:3346–3351.

Madersbacher S, Schatzl G, Djavan B, Stulnig T, Marberger M. Long-term outcome of transrectal high-intensity focused ultrasound therapy for benign prostatic hyperplasia. *European Urology* 2000;37:687–694.

Makin IRS, Mast TD, Faidi W, Runk MM, Barthe PG, Slayton MH. Miniaturized ultrasound arrays for interstitial ablation and imaging. *Ultrasound in Medicine and Biology* 2005;31:1539–1550.

Maleke C, Pernot M, Konofagou EE. Single-element focused ultrasound transducer method for harmonic motion imaging. *Ultrasonic Imaging* 2006;28:144–158.

Marberger M, Schatzl G, Cranston D, Kennedy JE. Extracorporeal ablation of renal tumours with high-intensity focused ultrasound. *BJU International* 2005;95:52–55.

Margreiter M, Marberger M. Focal therapy and imaging in prostate and kidney cancer: High-intensity focused ultrasound ablation of small renal tumors. *Journal of Endourology* 2010;24:745–748.

Marquet F, Pernot M, Aubry JF, Montaldo G, Marsac L, Tanter M, Fink M. Non-invasive transcranial ultrasound therapy based on a 3D CT scan: Protocol validation and *in vitro* results. *Physics in Medicine and Biology* 2009;54:2597–2613.

Martin E, Jeanmonod D, Morel A, Zadicario E, Werner B. High-intensity focused ultrasound for noninvasive functional neurosurgery. *Annals of Neurology* 2009;66:858–861.

Maxwell A, Sapozhnikov O, Bailey M, Crum L, Xu Z, Fowlkes B, Cain C, Khokhlova V. Disintegration of tissue using high intensity focused ultrasound: Two approaches that utilize shock waves. *Acoustics Today* 2012;8:24–37.

Maxwell AD, Cain CA, Duryea, Yuan L, Gurm HS, Xu Z. Noninvasive thrombolysis using pulsed ultrasound cavitation therapy-histotripsy. *Ultrasound in Medicine and Biology* 2009;35:1982–1994.

Maxwell AD, Owens G, Gurm HS, Ives K, Myers Jr DD, Xu Z. Noninvasive treatment of deep venous thrombosis using pulsed ultrasound cavitation therapy (histotripsy) in a porcine model. *Journal of Vascular and Interventional Radiology* 2011;22:369–377.

McDannold N, Clement GT, Black PM, Jolesz FA, Hynynen K. Transcranial magnetic resonance imaging-guided focused ultrasound surgery of brain tumors: Initial findings in 3 patients. *Neurosurgery* 2010;66:323–332.

Mereadante S, Fulfaro F. Management of painful bone metastases. *Current Opinion in Oncology* 2007;19:308–314.

Mitri FG, Urban MW, Fatemi M, Greenleaf JF. Shear wave dispersion ultrasonic vibrometry for measuring prostate shear stiffness and viscosity: An in vitro pilot study. *IEEE Transactions on Biomedical Engineering* 2011;58:235–242.

Moore JH, Gieck JH, Saliba EN, Perrin DH, Ball DW, McCue FC. The biophysical effects of ultrasound on median nerve distal latencies. *Electromyography and Clinical Neurophysiology* 2000;40:169–180.

Morita Y, Ito N, Hikida H, Takeuchi S, Nakamura K, Ohashi H. Non-invasive magnetic resonance imaging-guided focused ultrasound treatment for uterine fibroids—Early experience. *European Journal of Obstetrics & Gynecology and Reproductive Biology* 2007;139(2):199–203.

Morrow M, Harris JR. Local management of invasive breast cancer. In: Harris JR, Lippmann ME, Morrow M, Osborne LK, eds., *Diseases of the Breast*. Philadelphia, PA: Lippincott, 2000, pp. 515–560.

Muller M, Gennisson J-L, Deffieux T, Tanter M, Fink M. Quantitative viscoelasticity mapping of human liver using supersonic shear imaging: Preliminary *in vivo* feasibility study. *Ultrasound in Medicine and Biology* 2009;35:219–229.

Mundy GR. Metastasis to bone: Causes, consequences and therapeutic opportunities. *Nature Reviews Cancer* 2002;2:584–593.

Murakami T, Ishimaru H, Sakamoto I, Uetani M, Matsuoka Y, Daikodu M, Honda S, Koshiishi T, Fujimoto T. Percutaneous radiofrequency ablation and transcatheter arterial chemoembolization for hypervascular hepatocellular carcinoma: Rate and risk factors for local recurrence. *Cardiovascular and Interventional Radiology* 2007;30:696–704.

Murat F-J, Poissonnier L, Rabilloud M, Belot A, Bouvier R, Rouviere O, Chapelon J-Y, Gelet A. Mid-term results demonstrate salvage high-intensity focused ultrasound (HIFU) as an effective and acceptably morbid salvage treatment option for locally radiorecurrent prostate cancer. *European Urology* 2009;55:640–649.

Mylonas N, Ioannides K, Hadjisavvas V, Iosif D, Kyriacou PA, Damianou C. Evaluation of fast spin echo MRI sequence for an MRI guided high intensity focused ultrasound system for in vivo rabbit liver ablation. *Journal of Biomedical Science and Engineering* 2010;3(3):241–246.

Nakakura EK, Yeo CJ. Periampullary and pancreatic cancer. In: Blumgart LH, ed., *Surgery of the Liver, Biliary Tract, and Pancreas*. Philadelphia, PA: Saunders, 2007, pp. 849–857.

Ng KKC, Poon RTP, Chan SC, Chok KSH, Cheung TT, Tung H, Chu F, Tso WK, Yu WC, Lo CM, Fan ST. High-intensity focused ultrasound for hepatocellular carcinoma—A single-center experience. *Annals of Surgery* 2011;253:981–987.

NICE Guideline. High-intensity focused ultrasound for prostate cancer, 2005. http://www.nice.org.uk

NICE Implementation Advice for Guidelines in Managing Prostate Cancer, 2008. http://www.nice.org.uk/nicemedia/pdf/CG58ImplementationAdvice.doc

Norton SJ, Won JJ. Time exposure acoustics. *IEEE Transactions on Geoscience and Remote Sensing* 2000;38:1337–1343.

Novara G, Ficarra V, D'Elia C, Secco S, Cavalleri S, Artibani W. Prospective evaluation with standardised criteria for postoperative complications after robotic-assisted laparoscopic radical prostatectomy. *European Urology* 2010;57:363–370.

Novick AC, Campbell SC, Belldegrun AS. Guideline for management of the clinical stage 1 renal mass, 2009. American Urological Association, 1–76.

Nystrom L, Andersson I, Bjurstam N, Frisell J, Nordenskjold B, Rutqvist LE. Long-term effects of mammography screening: Updated overview of the Swedish randomized trial. *Lancet* 2002;359:909–919.

Odéen H, de Bever J, Almquist S, Farrer A, Todd N, Payne A, Snell JW, Christensen DA, Parker DL. Treatment envelope evaluation in transcranial magnetic resonance-guided focused ultrasound utilizing 3D MR thermometry. *Journal of Therapeutic Ultrasound* 2014;2:19–29.

Ophir J, Alam SK, Garra BS, Kallel F, Konofagou EE, Krouskop TA, Berritt CRB, Righetti R, Souchon R, Srinivasan S, Varghese T. Elastography: Imaging the elastic properties of soft tissues with ultrasound. *Journal of Medical Ultrasonics* 2002;29:155–171.

Orsi F, Zhang L, Arnone P, Orgera G, Bonomo G, Vigna PD, Monfardini L, Zhou K, Chen W, Wang Z, Veronesi U. High-intensity focused ultrasound ablation: Effective and safe therapy for solid tumors in difficult locations. *American Journal of Roentgenology* 2010;195:245–252.

Pajek D, Hynynen K. Applications of transcranial focused ultrasound surgery. *Acoustics Today* 2012;8:8–14.

Pantuck AJ, Zisman A, Belldegrun AS. The changing natural history of renal cell carcinoma. *Journal of Urology* 2001;166:1611–1623.

Park CC, Mitsumori M, Nixon A, Recht A, Connolly J, Gelman R, Silver B, Hetelekidis S, Abner A, Harris JR, Schnitt SJ. Outcome at 8 years after breast-conserving surgery and radiation therapy for invasive breast cancer: Influence of margin status and systemic therapy on local recurrence. *Journal of Clinical Oncology* 2000;18:1668–1675.

Park HR, Kim MD, Kim NK, Yoon S-W, Park WK, Lee MH. Uterine restoration after repeated sloughing of fibroids or vaginal expulsion following uterine artery embolization. *European Radiology* 2005;15:1850–1854.

Parkin D, Pisani P, Ferlay J. Estimates of the worldwide incidence of eighteen major cancers in 1985. *International Journal of Cancer* 2006;54:594–606.

Parkin DM, Bray F, Ferlay J, Pisani P. Global cancer statistics, 2002. *CA—A Cancer Journal for Clinicians* 2005;55:74–108.

Parsons JE, Cain CA, Abrams GD, Fowlkes JB. Pulsed cavitational ultrasound therapy for controlled tissue homogenization. *Ultrasound in Medicine and Biology* 2006;32:115–129.

Peng S, Xiong Y, Li K, He M, Deng Y, Chen L, Zou M, Chen W, Wang Z, He J, Zhang L. Clinical utility of a micro-bubble-enhancing contrast ("SonoVue") in treatment of uterine fibroids with high intensity focused ultrasound: A retrospective study. *European Journal of Radiology* 2012;81:3832–3838.

Pernot M, Aubry JF, Tanter M, Boch AL, Marquet F, Kujas M, Seilhean D, Fink M. In vivo transcranial brain surgery with an ultrasonic time reversal mirror. *Journal of Neurosurgery* 2007;106:1061–1066.

Pernot M, Aubry JF, Tanter M, Thomas JL, Fink M. High power transcranial beam steering for ultrasonic brain therapy. *Physics in Medicine and Biology* 2003;48:2577–2589.

Pichlmayr R, Weimann A, Tusch G, Schlitt HJ. Indications and role of liver transplantation for malignant tumors. *Oncologist* 1997;2:164–170.

Pinto I, Chimeno P, Romo A, Paul L, Haya J, de la Cal M, Bajo J. Uterine fibroids: Uterine artery embolization versus abdominal hysterectomy for treatment—A prospective, randomized, and controlled clinical trial. *Radiology* 2003;226:425–431.

Pinzon-Charry A, Maxwell T, Lopez JA. Dendritic cell dysfunction in cancer: A mechanism for immunosuppression. *Immunology & Cell Biology* 2005;83:451–461.

Pockley AG. Heat shock proteins as regulators of the immune response. *Lancet* 2003;362:469–476.

Poissonnier L, Chapelon J-Y, Rouvière O, Curiel L, Bouvier R, Martin X, Dubernard JM, Gelet A. Control of prostate cancer by transrectal HIFU in 227 patients. *European Urology* 2006;51(2), 381–387.

Pollack A, Zagars GK, Antolak JA, Kuban DA, Rosen II. Prostate biopsy status and PSA nadir level as early surrogates for treatment failure: Analysis of a prostate cancer randomized radiation dose escalation trial. *International Journal of Radiation Oncology, Biology, Physics* 2002;54:677–685.

Porpiglia F, Volpe A, Billia M, Scarpa R. Laparoscopic versus open partial nephrectomy: Analysis of the current literature. *European Urology* 2008;53:732–742.

Prost A, Funke AR, Tanter M, Aubry JF, Bossy E. Photoacoustic-guided ultrasound therapy with a dual-mode ultrasound array. *Journal of Biomedical Optics* 2012;17:061205.

Prostate cancer: Diagnosis and treatment: National Institute for Health and Care Excellence, 2014. http://www.nice.org.uk/guidance/cg175.

Punglia RS, Morrow M, Winer EP, Harris JR. Local therapy and survival in breast cancer. *New England Journal of Medicine* 2007;356:2399–2405.

Qian ZW, Xiong L, Yu J, Shao D, Zhu H, Wu X. Noninvasive thermometer for HIFU and its scaling. *Ultrasonics* 2006;44:e31–e35.

Quinn M, Babb P. Patterns and trends in prostate cancer incidence, survival, prevalence and mortality, part I: International comparisons. *BJU International* 2002;90:162–173.

Rebillard X, Gelet A, Darvin JL, Soulie M, Prapotnich D, Cathelineau X, Rozet F, Vallancien G. Transrectal high-intensity focused ultrasound in the treatment of localized prostate cancer. *Journal of Endourology* 2005;19:693–701.

Rebillard X, Soulie M, Chartier-Kastler E, Davin J-L, Mignard J-P, Moreau J-L, Coulange C. High-intensity focused ultrasound in prostate cancer: A systematic literature review of the French Association of Urology. *BJU International* 2008;101:1205–1213.

Reddan DN, Raj GV, Polascik TJ. Management of small renal tumors: An overview. *American Journal of Medicine* 2001;110:558–562.

Remzi M, Özsoy M, Klingler H, Susani M, Waldert M, Seitz C, Schmidbauer J, Marberger M. Are small renal tumors harmless? Analysis of histopathology features according to tumor size in tumors 4 cm or less in diameter. *Journal of Urology* 2006;176:1–4.

Ren XL, Zhou XD, Yan RL, Liu D, Zhang J, He GB, Han ZH, Zheng MJ, Yu M. Sonographically guided extracorporeal ablation of uterine fibroids with high-intensity focused ultrasound: Midterm results. *Journal of Ultrasound in Medicine* 2009;28:100–103.

Ren XL, Zhou XD, Zhang J, He GB, Han ZH, Zheng MJ, Li L, Yu M, Wang L. Extracorporeal ablation of uterine fibroids with high-intensity focused ultrasound. *Journal of Ultrasound in Medicine* 2007;26:201–212.

Ribault M, Chapelon JY, Cathignol D, Gelet A. Differential attenuation imaging for the characterization of high intensity focused ultrasound lesions. *Ultrasonic Imaging* 1998;20:160–177.

Righetti R, Kallel F, Stafford RJ, Price RE, Krouskop TA, Hazle JD, Ophir J. Elastographic characterization of HIFU-induced lesions in canine livers. *Ultrasound in Medicine and Biology* 1999;25:1099–1113.

Ritchie RW, Leslie T, Phillips R, Wu F, Illing R, ter Haar G, Protheroe A, Cranston D. Extracoporeal high intensity focused ultrasound for renal tumours: A 3-year follow-up. *BJU International* 2010;106:1004–1009.

Roberts WW, Hall TL, Ives K, Wolf Jr JS, Fowlkes JB, Cain CA. Pulsed cavitational ultrasound: A noninvasive technology for controlled tissue ablation (histotripsy) in the rabbit kidney. *Journal of Urology* 2006;175:734–738.

Rogers VL, Go AS, Lloyd-Jones DM et al. Heart disease and stroke statistics-2011 update: A report from the American Heart Association. *Circulation* 2011;123:e18–e209.

Roodman GD. Mechanisms of bone metastasis. *New England Journal of Medicine* 2004;350:1655–1664.

Rosberger DF, Coleman DJ, Silverman R, Woods S, Rondeau M, Cunningham-Rundles S. Immunomodulation in choroidal melanoma: Reversal of inverted CD4/CD8 ratios following treatment with ultrasonic hyperthermia. *Biotechnology Therapeutics* 1994;5:59–68.

Rosecan LR, Iwamoto T, Rosado A, Lizzi FL, Coleman DJ. Therapeutic ultrasound in the treatment of retinal detachment: Clinical observations and light and electron microscopy. *Retina* 1985;5:115–122.

Rosenschein U, Frimerman A, Laniado S, Miller HI. Study of the mechanism of ultrasound angioplasty from human thrombi and bovine aorta. *American Journal of Cardiology* 1994;74:1263–1266.

Rouviere O, Mege-Lechevallier F, Chapelon JY, Gelet A, Bouvier R, Boutitie F, Lyonnet D. Evaluation of color Doppler in guiding prostate biopsy after HIFU ablation. *European Urology* 2006;50:490–497.

Rukstalis DB. Treatment options after failure of radiation therapy—A review. *Reviews in Urology* 2002;4:12–17.

Sapin-de Brosses E, Gennisson J-L, Pernot M, Fink M, Tanter M. Temperature dependence of the shear modulus of soft tissues assessed by ultrasound. *Physics in Medicine and Biology* 2010;55:1701–1718.

Sapin-de Brosses E, Pernot M, Tanter M. The link between tissue elasticity and thermal dose *in vivo*. *Physics in Medicine and Biology* 2011;56:7755–7765.

Satthaporn S, Robins A, Vassanasiri W, El-Sheemy M, Jibril JA, Clark D, Valerio D, Eremin O. Dendritic cells are dysfunctional in patients with operable breast cancer. *Cancer Immunology, Immunotherapy* 2004;53:510–518.

Schueller G, Stift A, Friedl J, Dubsky P, Bachleitner-Hofmann T, Benkoe T, Jakesz R, Gnant M. Hyperthermia improves cellular immune response to human hepatocellular carcinoma subsequent to co-culture with tumor lysate pulsed dendritic cells. *International Journal of Oncology* 2003;22:1397–1402.

Serrone J, Kocaeli H, Mast TD, Burgess MT, Zuccarello M. The potential applications of high-intensity focused ultrasound (HIFU) in vascular neurosurgery. *Journal of Clinical Neuroscience* 2012;19:214–221.

Shen C, Xu J, Fang NX, Jing Y. Anisotropic complementary acoustic metamaterial for canceling out aberrating layers. *Physical Review X* 2014;4:041033.

Si HP, Xiang LK, Wang Z, Li YY, Wang ZB. Immune changes in bone neoplasm rabbits transplanted with VX2 before and after high intensity focused ultrasound therapy. *Chinese Journal of Experimental Surgery* 2003;20:823–824.

Siegel HJ, Pressey JG. Current concepts on the surgical and medical management of osteosarcoma. *Expert Review of Anticancer Therapy* 2008;8:1257–1269.

Silver DA, Morash C, Brenner P, Campbell S, Russo P. Pathologic findings at the time of nephrectomy for renal mass. *Annals of Surgical Oncology* 1997;4:570–574.

Silverman RH, Vogelsang B, Rondeau MJ, Coleman DJ. Therapeutic ultrasound for the treatment of glaucoma. *American Journal of Ophthalmology* 1991;111:327–337.

Simmonds PC. Palliative chemotherapy for advanced colorectal cancer: Systematic review and meta-analysis. *BMJ* 2000;321:531–535.

Simon C, VanBaren P, Ebbini ES. Two-dimensional temperature estimation using diagnostic ultrasound. *IEEE Transactions on Ultrasonics, Ferroelectrics and Frequency Control* 1998;45:1088–1099.

Singletary SE. Surgical margins in patients with early-stage breast cancer treated with breast conservation therapy. *American Journal of Surgery* 2002;184:383–393.

Smart OC, Hindley JT, Regan L, Gedroyc WMW. Magnetic resonance guided focused ultrasound surgery of uterine fibroids-The tissue effects of GnRH agonists pre-treatment. *European Journal of Radiology* 2006;59:163–167.

Smith NB, Temkin JM, Shapiro F, Hynynen K. Thermal effects of focused ultrasound energy on bone tissue. *Ultrasound in Medicine and Biology* 2001;27:1427–1433.

Sobin LH, Wittekind C. *TNM Classification of Malignant Tumours*. New York: Wiley-Liss, 2002.

Sofuni A, Moriyasu F, Sano T, Yamada K, Itokawa F, Tsuchiya T, Tsuji S, Kurihara T, Ishii K, Itoi T. The current potential of high-intensity focused ultrasound for pancreatic carcinoma. *Journal of Hepatobiliary and Pancreatic Sciences* 2011;18:295–303.

Song J, Lucht B, Hynynen K. Large improvement of the electrical impedance of imaging and high-intensity focused ultrasound (HIFU) phased arrays using multilayer piezoelectric ceramics coupled in lateral mode. *IEEE Transactions on Ultrasonics, Ferroelectrics and Frequency Control* 2012;59:1584–1595.

Souchon R, Rouvière O, Gelet A, Detti V, Srinivasan S, Ophir J, Chapelon JY. Visualization of HIFU lesions using elastography of the human prostate *in vivo*: Preliminary results. *Ultrasound in Medicine and Biology* 2003;29:1007–1015.

Spies JB, Coyne K, Guaou GN, Boyle D, Skyrnarz-Murphy K, Gonzalves SM. The UFS-QOL, a new disease-specific symptom and health-related quality of life questionnaire for leiomyomata. *Obstetrics & Gynecology* 2002;99:290–300.

Stewart EA. Uterine fibroids. *Lancet* 2001;357:293–298.

Stewart EA, Rabinovici J, Tempany CM, Inbar Y, Regan L, Gostout BS, Hesley GK, Kim HS, Hengst S, Gedroyc WMW. Clinical outcomes of focused ultrasound surgery for the treatment of uterine fibroids. *Fertility and Sterility* 2006;85:22–29.

Suk KS, Shin HH, Hahn SB. Limb salvage using original low heat-treated tumor-bearing bone. *Clinical Orthopaedics and Related Research* 2002;397:385–393.

Sullivan LD, McLoughlin MG, Goldenberg LG, Gleave ME, Marich KW. Early experience with high-intensity focused ultrasound for the treatment of begin prostatic hypertrophy. *British Journal of Urology* 1997;79:172–176.

Sumi C, Yanagimura H. Reconstruction of thermal property distributions of tissue phantoms from temperature measurements-thermal conductivity, thermal capacity and thermal diffusivity. *Physics in Medicine and Biology* 2007;52:2845–2863.

Sun J, Hynynen K. Focusing of therapeutic ultrasound through a human skull: A numerical study. *Journal of the Acoustical Society of America* 1998;104:1705–1715.

Susani M, Madersbacher S, Kratzik C, Vingers L, Marberger M. Morphology of tissue destruction induced by focused ultrasound. *European Urology* 1993;23:8–11.

Tanter M, Bercoff J, Athanasiou A, Deffieux T, Gennisson J-L, Montaldo G, Muller M, Tardivon A, Fink M. Quantitative assessment of breast lesion viscoelasticity: Initial clinical results using supersonic shear imaging. *Ultrasound in Medicine and Biology* 2008;34:1373–1386.

Tanter M, Fink M. Ultrafast imaging in biomedical ultrasound. *IEEE Transactions on Ultrasonics, Ferroelectrics and Frequency Control* 2014;61:102–119.

Taylor-Robinson SD, Foster GR, Arora S, Hargreaves S, Thomas HC. Increase in primary liver cancer in the UK, 1979–94. *Lancet* 1997;350:1142–1143.

Tepperberg I, Marjey E. Ultrasound therapy of painful postoperative neurofibromas. *American Journal of Physical Medicine* 1953;32:27–30.

ter Haar G. Harnessing the interaction of ultrasound with tissue for therapeutic benefit: High-intensity focused ultrasound. *Ultrasound in Obstetrics & Gynecology* 2008;32:601–604.

ter Haar G, Coussios C. High intensity focused ultrasound: Physical principles and devices. *International Journal of Hyperthermia* 2007;23:89–104.

ter Haar G, Rivens IH, Chen L, Riddler S. High intensity focused ultrasound for the treatment of rat tumours. *Physics in Medicine and Biology* 1991;36:1495–1501.

ter Haar GR. Acoustic surgery. *Physics Today* 2001a;54:29–34.

ter Haar GR. High intensity ultrasound. *Seminars in Laparoscopic Surgery* 2001b;8:77–89.

Thittai AK, Galaz B, Ophir J. Visualization of HIFU-induced lesion boundaries by axial-shear strain elastography: A feasibility study. *Ultrasound in Medicine and Biology* 2011;37:426–433.

Thomas JL, Fink M. Ultrasonic beam focusing through tissue inhomogeneities with a time reversal mirror: Application to transskull therapy. *IEEE Transactions on Ultrasonics, Ferroelectrics and Frequency Control* 1996;43:1122–1129.

Thompson I, Thrasher JB, Aus G, Burnett AL, Canby-Hagino ED, Cookson, MS, D'Amico AV, Dmochowski RR, Eton DT, Forman JD, Gldenberg SL, Hernandez J, Higano CS, Kraus SF, Moul JW, Tangen CM. Guideline for the management of clinically localized prostate cancer: 2007 update. *Journal of Urology* 2007;177(6):2106–2131.

Thüroff S, Chaussy C, Vallacien G, Wieland W, Kiel HJ, Le Duc A, Desgrandchamps F, De La Rosette JJMCH, Gelet A. High-intensity focused ultrasound and localized prostate cancer: Efficacy results from the European Multicentric Study. *Journal of Endourology* 2003;17:673–677.

Todryk SM, Michael J, Goughy MJ, Pockley AG. Facets of heat shock protein 70 show immunotherapeutic potential. *Immunology* 2003;110:1–9.

Toivanen A, Granberg I, Nordman E. Lymphocyte subpopulations in patients with breast cancer after postoperative radiotherapy. *Cancer* 1984;54:2919–2923.

Treilleux I, Blay JY, Bendriss-Vermare N, Ray-Coquard I, Bachelot T, Guastalla JP, Bremond A, Goddard S, Pin JJ, Barthelemy-Dubois C, Lebecque S. Dentritic cell infiltration and prognosis of early stage breast cancer. *Clinical Cancer Research* 2004;10:7466–7474.

Uchida T, Baba S, Irie A et al. Transrectal high-intensity focused ultrasound in the treatment of localized prostate cancer: A multicenter study. *Acta Urologica Japonica* 2005;51:651–658.

Uchida T, Illing RO, Cathcart PJ, Emberton M. To what extent does the prostate-specific antigen nadir predict subsequent treatment failure after transrectal high-intensity focused ultrasound therapy for presumed localized prostate cancer. *BJU International* 2006a;98:537–539.

Uchida T, Muramoto M, Kyunou H, Iwamura M, Egawa S, Koshiba K. Clinical outcome of high-intensity focused ultrasound for treating benign prostatic hyperplasia: Preliminary report. *Urology* 1998;52:66–71.

Uchida T, Ohkusa H, Nagata Y, Hyodo T, Satoh T, Irie A. Treatment of localized prostate cancer using high-intensity focused ultrasound. *BJU International* 2006b;97:56–61.

Uchida T, Ohkusa H, Yamashita H, Shoji S, Nagata Y, Hyodo T, Satoh T. Five years experience of transrectal high-intensity focused ultrasound using the sonablate device in the treatment of localized prostate cancer. *International Journal of Urology* 2006c;13:228–233.

Vaezy S, Martin R, Crum LA. Therapeutic ultrasound, part II: High intensity focused ultrasound: A method of hemostasis. *Echocardiography* 2001a;18:309–315.

Vaezy S, Martin R, Kaczkowski P, Keilman G, Carter S, Caps M, Mourad P, Crum L. Occlusion of blood vessels using high-intensity focused ultrasound. *Journal of the Acoustical Society of America* 1998;103:2867–2867.

Vaezy S, Martin RW, Kaczkowski P, Keilman G, Goldman B, Yaziji H, Carter S, Caps M, Crum L. Use of high-intensity focused ultrasound to control bleeding. *Journal of Vascular Surgery* 1999a;29:533–542.

Vaezy S, Martin RW, Mourad P, Crum LA. Hemostasis using high intensity focused ultrasound. *European Journal of Ultrasound* 1999b;9:79–87.

Vaezy S, Shi X, Martin RW, Chi E, Nelson PI, Bailey MR, Crum LA. Real-time visualization of high-intensity focused ultrasound treatment using ultrasound imaging. *Ultrasound in Medicine and Biology* 2001b;27:33–42.

Vallancien G, Chartier-Kastler E, Harouni M, Chopin D, Bougaran J. Focused extracorporeal pyrotherapy: Experimental study and feasibility in man. *Seminars in Urology* 1993;11:7–9.

Vallancien G, Harouni M, Guillonneau B, Veillon B, Bougaran J. Ablation of superficial bladder tumours with focused extracorporeal pyrotherapy. *Urology* 1996;47:204–207.

Vallancien G, Prapotnich D, Cathelineau X, Baumert H, Rozet F. Transrectal focused ultrasound combined with transurethral resection of the prostate for the treatment of localized prostate cancer: Feasibility study. *Journal of Urology* 2004;171:2265–2267.

Vallejo R, Hord ED, Barna SA, Santiago-Palma J, Ahmed S. Perioperative immunosuppression in cancer patients. *The Journal of Environmental Pathology, Toxicology and Oncology* 2003;22:139–146.

van der Ploeg IM, van Esser S, van den Bosch MAAJ, Mali WPM, van Diest PJ, Borel Rinkes IH, van Hillegersberg R. Radiofrequency ablation for breast cancer: A review of the literature. *European Journal of Surgical Oncology* 2007;33:673–677.

van Esser S, Veldhuis WB, van Hillegersberg R, van Diest PJ, Stapper G, El-Ouamari M, Borel Rinkes IH, Mali WPM, van den Bosch MAAJ. Accuracy of contrast-enhanced breast ultrasound for pre-operative tumour size assessment in patients diagnosed with invasive ductal carcinoma of the breast. *Cancer Imaging* 2007;7:63–68.

van Sandick JW, Gisbertz SS, ten Berge IJ et al. Immune responses and prediction of major infection in patients undergoing transhiatal or transthoracic esophagectomy for cancer. *Annals of Surgery* 2003;237:35–43.

Veltri A, Moretto P, Doriguzzi A, Pagano E, Carrara G, Gandini G. Radiofrequency thermal ablation (RFA) after transarterial chemoembolization (TACE) as a combined therapy for unresectable non-early hepatocellular carcinoma (HCC). *European Radiology* 2006;16:661–669.

Vercellini P, Cortesi I, Oldani S, Moschetta M, De Giorgi O, Giorgio Crosignani P. The role of transvaginal ultrasonography and outpatient diagnostic hysteroscopy in the evaluation of patients with menorrhagia. *Human Reproduction Update* 1997;12:1768–1771.

Veronesi U, Cascinelli N, Mariani L et al. Twenty-year follow-up of a randomized study comparing breast conserving surgery with radical mastectomy for early breast cancer. *New England Journal of Medicine* 2002;347:1227–1232.

Veronesi U, Paganelli G, Viale G et al. A randomized comparison of sentinel-node biopsy with routine axillary dissection in breast cancer. *New England Journal of Medicine* 2003;349:546–553.

Visioli AG, Rivens IH, ter Haar GR, Horwich A, Huddart RA, Moskovic E, Padhani A, Glees J. Preliminary results of a phase I dose escalation clinical trial using focused ultrasound in the treatment of localised tumours. *European Journal of Ultrasound* 1999;9:11–18.

Vollenhoven BL, Lawrence AS, Healy DL. Uterine fibroids: A clinical review. *British Journal of Obstetrics and Gynaecology* 1990;97:285–298.

Voogt MJ, Trillaud H, Kim YS et al. Volumetric feedback ablation of uterine fibroids using magnetic resonance-guided high intensity focused ultrasound therapy. *European Radiology* 2012;22:411–417.

Walker WJ, Pelage JP. Uterine artery embolisation for symptomatic fibroids: Clinical results in 400 women with imaging follow up. *BJOG* 2002;109:1262–1272.

Wamsteker K, Emanuel MH, de Kruif JH. Transcervical hysteroscopic resection of submucous fibroids for abnormal uterine bleeding: Results regarding the degree of intramural extension. *Obstetrics & Gynecology* 1993;82:736–740.

Wang AK, Raynor EM, Blum AS, Rutkove SB. Heat sensitivity of sensory fibers in carpal tunnel syndrome. *Muscle Nerve* 1999;22:37–42.

Wang K, Liu L, Meng Z, Chen W, Zhou Z. High intensity focused ultrasound for treatment of patients with pancreatic cancer. *Chinese Journal of Ultrasound in Medicine* 2006;22:796–798.

Wang K, Zhu H, Meng Z, Chen Z, Lin J, Shen Y, Gao H. Safety evaluation of high-intensity focused ultrasound in patients with pancreatic cancer. *Oncology Research and Treatment* 2013;36:88–92.

Wang LV. Prospects of photoacoustic tomography. *Medical Physics* 2008;35:5758–5767.

Wang T-Y, Xu Z, Hall TL, Fowlkes JB, Cain CA. An efficient treatment strategy for histotripsy by removing cavitation memory. *Ultrasound in Medicine and Biology* 2012a;38:753–766.

Wang W, Wang Y, Wang T, Wang J, Wang L, Tang J. Safety and efficacy of US-guided high-intensity focused ultrasound for treatment of submucosal fibroids. *European Radiology* 2012b;22:2553–2558.

Wang X, Sun JZ. Preliminary study of high intensity focused ultrasound in treating patients with advanced pancreatic carcinoma. *Chinese Journal of General Surgery* 2002;17:654–655.

Wang Y-N, Khokhlova TD, Bailey MR, Hwang JH, Khokhlova VA. Histological and biochemical analysis of mechanical and thermal bioeffects in boiling histotripsy lesions induced by high intensity focused ultrasound. *Ultrasound in Medicine and Biology* 2012c;39:424–438.

Ward JF, Sebo TJ, Blute ML, Zincke H. Salvage surgery for radiorecurrent prostate cancer: Contemporary outcomes. *Journal of Urology* 2005;173:1156–1160.

Warmuth M, Johansson T, Mad P. Systematic review of the efficacy and safety of high-intensity focussed ultrasound for the primary and salvage treatment of prostate cancer. *European Urology* 2010;58:803–815.

Watkin N, Morris S, Rivens I, Woodhouse C. A feasibility study for the non-invasive treatment of superficial bladder tumours with focused ultrasound. *British Journal of Urology* 1996a;78:715–721.

Watkin NA, ter Haar GR, Rivens I. The intensity dependence of the site of maximal energy deposition in focused ultrasound surgery. *Ultrasound in Medicine and Biology* 1996b;22(4):483–491.

White J, Clement GT, Hynynen K. Transcranial ultrasound focus reconstruction with phase and amplitude correction. *IEEE Transactions on Ultrasonics, Ferroelectrics and Frequency Control* 2005;52:1518–1522.

White PJ. Transcranial focused ultrasound surgery. *Topics in Magnetic Resonance Imaging* 2006;17:165–172.

Whiteside TL, Heberman RB. Effectors of immunity and rationale for immunotherapy. In: Kufe DW, Pollock RE, Weichselbaum RR et al., eds., *Cancer Medicine*. Hamilton, Ontario, Canada: BC Decker, 2003, pp. 221–228.

Williams VS, Jones G, Mauskopf J, Spalding J, Duchane J. Uterine fibroids: A review of health-related quality of life assessment. *Journal of Womens Health* 2006;15:818–829.

Wissniowski TT, Hunsler J, Neureiter D et al. Activation of tumor-specific T lymphocytes by radio-frequency ablation of the VX2 hepatoma in rabbits. *Cancer Research* 2003;63:6496–6500.

Wood RW, Loomis AL. The physical and biological effects of high frequency sound waves of great intensity. *Philosophical Magazine* 1927;4:417–436.

Worthington AE, Trachtenberg J, Shearar MD. Ultrasound properties of human prostate tissue during heating. *Ultrasound in Medicine and Biology* 2002;28:1311–1318.

Wu CC, Chen WS, Ho MC, Huang KW, Chen CN, Yen JY, Lee PH. Minimizing abdominal wall damage during high-intensity focused ultrasound ablation by inducing artificial ascites. *Journal of the Acoustical Society of America* 2008;124:674–679.

Wu F, Chen W-Z, Bai J, Zou J-Z, Wang Z-L, Zhu H, Wang Z-B. Tumor vessel destruction resulting from high-intensity focused ultrasound in patients with solid malignancies. *Ultrasound in Medicine and Biology* 2002;28:535–542.

Wu F, Chen W, Bai J. A preliminary study of high intensity focused ultrasound in the treatment of hepatocellular carcinoma. *Chinese Journal of Ultrasound* 1999;8:213–216.

Wu F, Wang Z, Cao Y, Zhou Q, Zhang Y, Xu Z, Zhu X. Expression of tumor antigens and heat-shock protein 70 in breast cells after high-intensity focused ultrasound ablation. *Annals of Surgical Oncology* 2006b;14:1237–1242.

Wu F, Wang Z-B, Cao Y-D, Chen W-Z, Bai J, Zou J-Z, Zhu H. A randomised clinical trial of high-intensity focused ultrasound ablation for the treatment of patients with localised breast cancer. *British Journal of Cancer* 2003a;89:2227–2233.

Wu F, Wang Z-B, Cao Y-D, Xu Z-L, Zhou Q, Zhu H, Chen W-Z. Heat fixation of cancer cells ablated with high-intensity focused ultrasound in patients with breast cancer. *American Journal of Surgery* 2006a;192:179–184.

Wu F, Wang Z-B, Cao Y-D, Zhou Q, Zhang Y, Xu Z-L, Zhu X-Q. Expression of tumor antigens and heat shock protein 70 in breast cancer cells after high intensity focused ultrasound ablation. *Annals of Surgical Oncology* 2007a;14:1237–1242.

Wu F, Wang Z-B, Cao Y-D, Zhu X-Q, Zhu H, Chen W-Z, Zou J-Z. "Wide local ablation" of localized breast cancer using high intensity focused ultrasound. *Journal of Surgical Oncology* 2007b;96:130–136.

Wu F, Wang Z-B, Chen W-Z et al. Extracorporeal high intensity focused ultrasound ablation in the treatment of 1038 patients with solid carcinomas in China: An overview. *Ultrasonics Sonochemistry* 2004a;11:149–154.

Wu F, Wang Z-B, Lu P, Xu Z-L, Chen W-Z, Zhu H, Jin C-B. Activated anti-tumor immunity in cancer patients after high intensity focused ultrasound ablation. *Ultrasound in Medicine and Biology* 2004b;30:1217–1222.

Wu F, Wang Z-B, Zhu H, Chen W-Z, Zou J-Z, Bai J, Li K-Q, Jin C-B, Xie F-L, Su H-B. Extracorporeal high intensity focused ultrasound treatment for patients with breast cancer. *Breast Cancer Research and Treatment* 2005a;92:51–60.

Wu F, Wang ZB, Chen WZ, Bai J, Zhu H, Qiao TY. Preliminary experience using high intensity focused ultrasound for the treatment of patients with advanced stage renal malignancy. *Journal of Urology* 2003b;170:2237–2240.

Wu F, Wang ZB, Chen WZ, Zhu H, Bai J, Zou JZ, Li KQ, Jin CB, Xie FL, Su HB. Extracorporeal high intensity focused ultrasound ablation in the treatment of patients with large hepatocellular carcinoma. *Annals of Surgical Oncology* 2004c;11:1061–1069.

Wu F, Wang ZB, Chen WZ, Zou J-Z, Bai J, Zhu H, Li K-Q, Xie F-L, Jin C-B, Su H-B, Gao G-W. Extracorporeal focused ultrasound surgery for treatment of human solid carcinomas: Early Chinese clinical experience. *Ultrasound in Medicine and Biology* 2004d;30:245–260.

Wu F, Wang ZB, Chen WZ, Zou JZ, Bai J, Zhu H, Li KQ, Jin CB, Xie FL, Su HB. Advanced hepatocellular carcinoma: Treatment with high-intensity focused ultrasound ablation combined with transcatheter arterial embolization. *Radiology* 2005b;235:659–667.

Wu F, Wang ZB, Zhu H, Chen WZ, Zou JZ, Bai J, Li KQ, Jin CB, Xie FL, Su HB. Feasibility of US-guided high-intensity focused ultrasound treatment in patients with advanced pancreatic cancer: Initial experience. *Radiology* 2005c;236:1034–1040.

Xiong LL, Hwang JH, Huang XB, Yao SS, He CJ, Ge XH, Ge HY, Wang XF. Early clinical experience using high intensity focused ultrasound for palliation of inoperable pancreatic cancer. *Journal of the Pancreas* 2009;10:123–129.

Xu D, Pollock M. Experimental nerve thermal injury. *Brain* 1994;117:375–384.

Xu Z, Fan Z, Hall TL, Winterroth F, Fowlkes JB, Cain CA. Size measurement of tissue debris particles generated from pulsed ultrasound cavitational therapy-histotripsy. *Ultrasound in Medicine and Biology* 2009a;35:245–255.

Xu Z, Fowlkes JB, Cain CA. A new strategy to enhance cavitational tissue erosion by using a high intensity initiating sequence. *IEEE Transactions on Ultrasonics, Ferroelectrics and Frequency Control* 2006;53:1412–1424.

Xu Z, Fowlkes JB, Cain CA. Optical and acoustic monitoring of bubble cloud dynamics at a tissue-fluid interface in ultrasound tissue erosion. *Journal of the Acoustical Society of America* 2007;121:2421–2430.

Xu Z, Fowlkes JB, Rothman ED, Cain CA. Controlled ultrasound tissue erosion: The role of dynamic interaction between insonation and microbubble activity. *Journal of the Acoustical Society of America* 2005;117:424–435.

Xu Z-L, Zhu X-Q, Lu P, Zhou Q, Zhang J, Wu F. Activation of tumor-infiltrating antigen presenting cells by high intensity focused ultrasound ablation of human breast cancer. *Ultrasound in Medicine and Biology* 2009b;35:50–57.

Yagel S. High-intensity focused ultrasound: A revolution in non-invasive ultrasound treatment? *Ultrasound in Obstetrics & Gynecology* 2004;23:216–217.

Yancik R. Population aging and cancer: A cross-national concern. *Cancer Journal* 2005;11:437–441.

Yang R, Griffith SL, Rescorla FJ et al. Feasibility of using high intensity focused ultrasound for the treatment of unresectable retroperitoneal malignancies. *Journal of Ultrasound in Medicine* 1992;11:S37.

Yang R, Reilly CR, Rescorla FJ, Faught PR, Sanghvi NT, Fry FJ, Franklin TD, Lumeng L, Grosfeld JL. High-intensity focused ultrasound in the treatment of experimental liver cancer. *Archives of Surgery* 1991;126:1002–1010.

Yang RS, Kopecky KK, Rescorla FJ, Galliani CA, Wu EX, Grosfeld JL. Sonographic and computed tomography characteristics of liver ablation lesions induced by high-intensity focused ultrasound. *Investigative Radiology* 1993;28:796–801.

Yang Z, Cao Y-D, Hu L-N, Wang Z. Feasibility of laparoscopic high-intensity focused ultrasound treatment for patients with uterine localized adenomyosis. *Fertility and Sterility* 2009;91:2338–2343.

Yin JJ, Pollock CB, Kelly K. Mechanisms of cancer metastasis to the bone. *Cell Research* 2005;15:57–62.

Young RR, Henneman E. Functional effects of focused ultrasound on mammalian nerves. *Science* 1961;134:1521–1522.

Yount DE. Skins of varying permeability: A stabilization mechanism for gas cavitation nuclei. *Journal of the Acoustical Society of America* 1979;65:1429–1439.

Yuan Y, Shen H, Hu X-Y, Gu F-Y, Li M-D, Zhong X. Multidisciplinary treatment with chemotherapy, targeted drug, and high-intensity focused ultrasound in advanced pancreatic carcinoma. *Medical Oncology* 2012;29:957–961.

Zagars GK, Pollack A, Von Eschenbach AC. Prostate cancer and radiotherapy-The message conveyed by serum prostate specific antigen. *International Journal of Radiation Oncology, Biology, Physics* 1995;33:23–35.

Zderic V, Brayman AA, Sharar SR, Crum LA, Vaezy S. Microbubble-enhanced hemorrhage control using high intensity focused ultrasound. *Ultrasonics* 2006;45:113–120.

Zderic V, Keshavarzi A, Andrew MA, Vaezy S, Martin RW. Attenuation of porcine tissues *in vivo* after high-intensity ultrasound treatment. *Ultrasound in Medicine and Biology* 2004;30:61–66.

Zelefsky MJ, Fuks Z, Hunt M, Lee HJ, Lombardi D, Ling CC, Reuter VE, Venkatraman ES, Leibel SA. High-dose radiation delivered by intensity modulated conformal radiotherapy improves the outcome of localized prostate cancer. *Journal of Urology* 2001;66:876–881.

Zelefsky MJ, Leibel SA, Gaudin PB, Kutcher GJ, Fleshner NE, Venkatraman ES, Reuter VE, Fair WR, Ling CC, Fuks Z. Dose escalation with the three-dimensional conformal radiotherapy affects the outcome in prostate cancer. *International Journal of Radiation Oncology, Biology, Physics* 1998;41:491–500.

Zhang D, Zhang S, Wan M, Wang S. A fast tissue stiffness-dependent elastography for HIFU-induced lesions inspection. *Ultrasonics* 2011;51:857–869.

Zhang L, Chen W-Z, Liu Y-J, Hu X-Y, Zhou K, Chen L, Peng S, Zhu H, Zou H-L, Bai J, Wang Z-B. Feasibility of magnetic resonance imaging-guided high intensity focused ultrasound therapy for ablating uterine fibroids in patients with bowel lies anterior to uterus. *European Journal of Radiology* 2010;73:396–403.

Zhang L, Wang Z. High-intensity focused ultrasound tumor ablation: Review of ten years of clinical experience. *Frontiers of Medicine in China* 2010;4:294–302.

Zhang L, Zhu H, Jin C, Zhou K, Li K, Su H, Cheng W, Bai J, Wang Z. High-intensity focused ultrasound (HIFU): Effective and safe therapy for hepatocellular carcinoma adjacent to major hepatic veins. *European Radiology* 2009;19:437–445.

Zhang M, Castaneda B, Christensen J, Saad W, Bylund K, Hoyt K, Strang JG, Rubens DJ, Parker KJ. Real-time sonoelastography of hepatic thermal lesions in a swine model. *Medical Physics* 2008;35:4132–4141.

Zhao Z, Wu F. Minimally-invasive thermal ablation of early-stage breast cancer: A systemic review. *European Journal of Surgical Oncology* 2010;36:1149–1155.

Zhou Q, Zhu X-Q, Zhang J, Xu Z-L, Lu P, Wu F. Changes in circulating immunosuppressive cytokine levels of cancer patients after high intensity focused ultrasound treatment. *Ultrasound in Medicine and Biology* 2008;34:81–87.

Zhou XD, Ren XL, Zhang J, He GB, Zheng MJ, Tian X-f, Zhu T, Zhang M, Wang L, Luo W. Therapeutic response assessment of high intensity focused ultrasound therapy for uterine fibroid: Utility of contrast-enhanced ultrasonography. *European Journal of Radiology* 2007;62:289–294.

Zhou Y. Noninvasive treatment of breast cancer using high-intensity focused ultrasound. *Journal of Medical Imaging and Health Informatics* 2013;3:141–156.

Zhou Y, Zia J, Warren C, Starr FL, Brayman AA, Crum LA, Hwang JH. Targeted long-term venous occlusion using pulsed high-intensity focused ultrasound combined with a pro-inflammatory agent. *Ultrasound in Medicine and Biology* 2011a;37:1653–1658.

Zhou YF. High intensity focused ultrasound in clinical tumor ablation. *World Journal of Clinical Oncology* 2011;2:8–27.

Zhou YF. High-intensity focused ultrasound treatment for advanced pancreatic cancer. *Gastroenterology Research and Practice* 2014;2014:205325.

Zhou YF, Kargl SG, Hwang JH. The effect of the scanning pathway in high-intensity focused ultrasound therapy on lesion production. *Ultrasound in Medicine and Biology* 2011b;37:1457–1468.

Zhu H, Zhou K, Zhang L et al. High intensity focused ultrasound (HIFU) therapy for local treatment of hepatocellular carcinoma: Role of partial rib resection. *European Journal of Radiology* 2009;72:160–166.

14 Ultrasound-Mediated Drug Delivery/Gene Transfection

14.1 INTRODUCTION

Lack of control on the primary tumor, especially of the cervix, colon, ovaries, pancreas, and brain, is a critical factor for mortality (Howlader et al. 2011). Despite the extensive research and substantial progress in the development of anticancer agents in recent decades, their application in the treatment of solid tumors is not satisfactory. The overdose of chemo- and radiotherapeutic agents during clinical therapy usually destroys normal cells/tissues, and results in a variety of undesirable side effects, including cardiotoxicity, immune suppression, and nephrotoxicity (Awada and Piccart 2000, Floyd et al. 2005). There are three physiological barriers for the efficient and safe transportation of the most promising macromolecular agents (i.e., monoclonal antibodies, cytokines, antisense oligonucleotides, and genes): blood vessel walls (vasculature occupying 1%–10% of the tumor volume), collagen-rich interstitial space that accounts for a large volume of tumors, and the cancer cell membrane due to the heterogeneity of antigen and receptor expression for affinity-targeted delivery of drugs (Curti 1996). The typical diffusion time required for macromolecules to cross a distance of 200 μm (average distance between tumor capillaries) is longer than their average half-life time (Jain 2005). Uneven and slowed blood flow within tumors due to abnormal (aberrant branching and tortuosity), heterogeneous, and inefficient distribution of a vasculature (Jain 2005), high interstitial pressures due to the leakage of tumor vessels, the absence of a functional lymphatic system in the solid tumor (Boucher et al. 1990), and fibrillar collagen in the extracellular matrix (McKee et al. 2006) further complicate effective and uniform delivery of high-molecular-weight (molecular weight, MW > 2000 Da) drugs. Satisfactory therapeutic outcome requires not only proper drug selection but also effective drug delivery systems.

Various approaches have been explored to maximize drug localization while minimizing systemic toxicity in the region of interest (ROI). These include light, neutron beam, magnetic field, and mechanical energy (Hernot and Klibanov 2008). Ultrasound has also been applied in drug release (Tachibana and Tachibana 2003, Postema and Gilja 2007). In 1976, the cytotoxic effect of nitrogen mustard on mouse leukemia L1210 after sonication was first observed without any mechanical damage to cells, which could not be explained merely by ultrasound-induced hyperthermia (Kremkau et al. 1976). Ultrasound-mediated drug delivery is highly attractive because of the wide acceptance of ultrasound in medicine, easy

and accurate focusing onto deeply seated organs in the body, high efficiency in perturbing cell membranes and increasing their permeability, noninvasiveness, nonviral nature, low cost, absence of ionization for theoretically unlimited treatment, and no effect in the non-sonicated region (ter Haar 2012). This strategy provides an option in treating localized tumors that may be inoperable by surgery either from a physiologic or cosmetic point of view.

In conventional systematic delivery, a drug is initially taken orally or by injecting into a vein, the subcutaneous space, or the muscle. The drug is then absorbed into local blood microcirculation in the intestine or near the injected site where the drug molecules enter the systematic blood circulation and subsequently are transported into different organs or tissues in the body. In local delivery, drugs are often directly injected at the desired site or released from physically implanted drug carriers at the targeted site, such as a tumor. Local delivery can also be achieved via receptor-mediated drug targeting in a vital tumor (i.e., the proliferating vasculature) even if the drugs are injected systematically. However, minimal nonspecific binding or uptake can occur outside the intended targets. Physical and chemical targeting is not always possible in many cases. Whether or not systematic or local administration is used, the molecules, particles, or other formulations in the form of drugs or genes will need to overcome different levels of barriers including factors in systematic circulation, transvascular transport (i.e., extravasation), interstitial transport, and finally intracellular uptake, where the drugs must cross the cell membrane to reach intracellular targets.

Achieving localized high concentrations is often desirable and necessary in most drug delivery applications in order to improve treatment outcome and reduce side effects. Although chemotherapy is one of the most effective cancer treatments, the high doses required for successful tumor elimination also adversely affect healthy tissues in the host, resulting in undesirable side effects and ultimately limiting the overall therapeutic outcome due to dose-limiting toxicity. Furthermore, the lack of effective, targeted delivery of these agents limits their therapeutic utility and potential. The difficulty comes from a series of physical (e.g., high interstitial pressure and factors in the extracellular matrix, such as fibrillar collagen, in solid tumors) and biological (e.g., cell and nuclear membranes) barriers that prevent the effective delivery of the therapeutic agents to their cellular and molecular targets *in vivo* at sufficient and uniform concentrations.

14.2 DRUG/GENE DELIVERY APPROACH

The mechanisms for the ultrasound-mediated drug/gene delivery utilize both thermal and mechanical effects (ter Haar 1999, Dijkmans et al. 2004) (Figure 14.1). First, acoustic radiation force can help agents to penetrate through the vessel wall to the tissue. Second, ultrasound may affect the morphology and properties of the cell membrane (e.g., permeability) due to bubble cavitation, termed as sonoporation. Third, ultrasound-induced hyperthermia has a significant biological effect on cell activities and drug uptake. Last, ultrasound can alter the performance of the drug agent, for example, activating light-sensitive materials of hematoporphyrins to kill cancer cells and inhibit restenosis (Umemura et al. 1996). The increase of the acoustic power used in the application could enhance the efficiency of drug delivery, the propensity of cell lysis, cell membrane permeability, and drug cytotoxicities through elevated temperatures (Ng and Liu 2002).

14.2.1 ACOUSTIC RADIATION FORCE

Acoustic radiation force can push drug-carrying nanoparticles toward a vessel's wall prior to fragmentation, and facilitate their adhesion with receptor ligands (Miri and Mitri 2011). Furthermore, the nonuniform displacement of tissue in the focal zone and transverse waves generated at the interface of tissue and fluid create a steep gradient in shear forces, where the resulting strain works on the cell-to-cell junctions and cellular interfaces. The penetration of various masses (e.g., small molecules, DNA, and nanoparticles) from the fluid into the adjacent epithelium can be increased both systemically and locally under sonication, as well as their rate of effective diffusion through the tissues. The increase in the intercellular gaps for enhanced local interstitial transport and reduction in interstitial fluid pressure in the core of tumors lead to improved extravasation of large molecules and consequent antitumor effects, which are correlated with more widespread induction of necrosis and apoptosis (Ferrara et al. 2007, Rychak et al. 2007, Frenkel 2008) (Figure 14.2).

14.2.2 HYPERTHERMIA

Hyperthermia, the procedure of raising the temperature of tumor-loaded tissue to 40°C–43°C by a variety of approaches including ultrasound, has a remarkable biological effect at the subcellular and cellular levels: decreased DNA synthesis, altered protein synthesis (e.g., heat shock proteins, HSPs), HSP-70 membrane expression, disruption of the microtubule organizing center, varied expression of receptors and binding of growth factors, changes in cell morphology and attachment (Billard et al. 1990, Partanen et al. 2012), apoptosis, and necrosis. Importantly, hyperthermia can also increase tumor blood flow (Karino et al. 1988), vascular permeability to antibodies (Cope et al. 1990, Hosono et al. 1994, Schuster et al. 1995, Hauck et al. 1997), ferritin (Fujiwara and Watanabe 1990) and Evans blue dye (Lefor et al. 1985), and tumor oxygenation (Horsman and Overgaard 1997), which leads to its implementation for solid tumor therapy in combination with thermosensitive drugs. Even non-thermosensitive polymeric

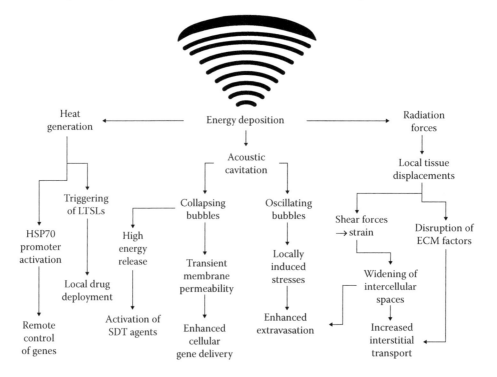

FIGURE 14.1 Schematic diagram representing the effective mechanisms of ultrasound exposures in enhanced drug delivery or gene transfection. HSP, heat shock protein; LTSLs, low-temperature-sensitive liposomes; SDT, sonodynamic therapy; ECM, extracellular matrix. (Reprinted from *Adv. Drug Deliv. Rev.*, 60, Frenkel, V., Ultrasound mediated delivery of drugs and genes to solid tumors, 1193–1208, Copyright 2008, with permission from Elsevier.)

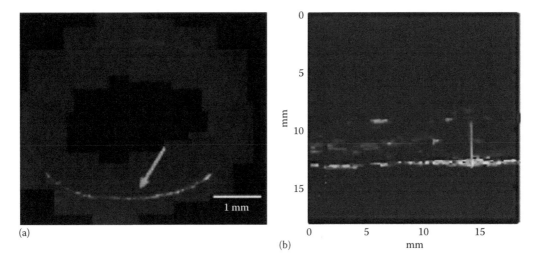

(a)

(b)

FIGURE 14.2 (a) DiI signal (red) on the endothelial cell wall of the swine carotid artery in the fluorescence image and (b) the corresponding ultrasound image at the end of the applied sequence. (From Patil, A.V. et al., *Mol. Imaging*, 10, 238, 2011. With permission.)

carriers and drugs exhibit increased localization in heated tumors because of these physiological effects. The hallmarks of hyperthermia and its pleotropic effects are in favor of its combined use with chemotherapy. Thermal enhancement of drug cytotoxicity (thermosensitization) is based on protein denaturation and aggregation, which lead to the inhibition of the DNA repair machinery and proteins involved in cell cycle regulation without increasing its oncogenic potential (Issels 2008). Sublethal temperatures induce overexpression of inducible HSPs (e.g., HSP-70) and turn tumor cells transiently thermoresistant. Hyperthermia or hyperthermic perfusion produces more tumor cytotoxicity, local control (e.g., recurrent breast cancer and malignant melanoma), and survival (e.g., head and neck lymph-node metastases, glioblastoma, cervical carcinoma) without affecting systemic toxicity, as seen in several Phase III trials. Sophisticated control of temperature in deep body regions spatially as well as temporally will further improve its clinical application (Wust et al. 2002).

The addition of local hyperthermia to radiation therapy has also significantly improved the ability to control superficial malignancies. Properly selected and scheduled anticancer drugs will add substantially to the efficacy of local hyperthermia and radiation for trimodality therapy (Herman et al. 1988) (Figure 14.3).

14.2.3 Bubble Cavitation

At the very low acoustic pressure (mechanical index [MI] of <~0.1), the microbubble oscillates symmetrically and linearly. At an MI of 0.1–0.3, the microbubble has more expansion than compression (nonlinear oscillations) but still keeps the integrity of the shell, which is known as stable cavitation (SC). Higher acoustic pressure (MI = 0.3–0.6) forces microbubble oscillation, with eventual destruction by either gas diffusion during compression or via large shell defects in the inertial cavitation (IC) (see Figure 14.4) (Chomas et al. 2000,

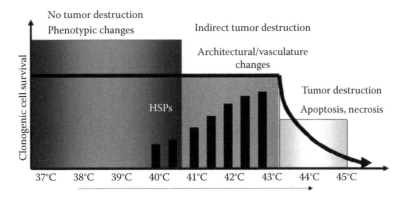

FIGURE 14.3 Effect of deep regional hyperthermia on tumor destruction: no direct cytotoxicity (sublethal temperature) at thermal doses below the threshold (i.e., 43°C for 60 min), architectural and vascular changes at 40°C–43°C leading to an increased blood flow, enhanced drug delivery and drug extravasation (chemosensitization) and higher oxygenation of the tissue (radiosensitization), direct cytotoxicity, and exponential destruction of tumor cells by induction of apoptosis and necrosis with thermal doses above the breakpoint. (Reprinted from *Eur. J. Cancer*, 44, Issels, R.D., Hyperthermia adds to chemotherapy, 2546–2554, Copyright 2008, with permission from Elsevier.)

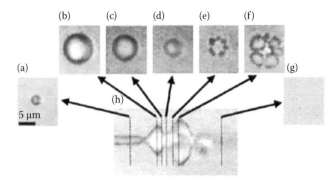

FIGURE 14.4 Optical frame images (a–g) and streak image (h, diameter of the bubble as a function of time) illustrating the oscillation and fragmentation of a microbubble. The bubble has an initial diameter of 3 μm. (Reprinted with permission from Chomas, J.E., Dayton, P.A., May, D., Allen, J., Klibanov, K., and Ferrara, K.W., Optical observation of contrast agent destruction, *Appl. Phys. Lett.*, 77, 1056–1058, Copyright 2000, American Institute of Physics.)

Marmottant et al. 2005, Hernot and Klibanov 2008). However, the specific thresholds of microbubble oscillations depend on the bubble structure (i.e., initial size, wall thickness, and composition) and environment properties (i.e., ambient pressure and temperature) and vary a lot among those used in the lab and clinics. Addition of nano or microparticles (e.g., polymer spheres) in the tumor can lower the IC threshold significantly. Smaller particles have more effects on such a reduction in the cavitation threshold at the same concentration (Larina et al. 2002, Lentacker et al. 2009). Blood has higher viscosity than water and, consequently, a higher cavitation threshold (i.e., about 1.5–2 times as that in water at the frequency of 2.5 MHz). However, polymeric microbubbles show different cavitation characteristics. At the low acoustic pressure, the polymeric microbubble will not oscillate actively because of its stiff shell. If the acoustic pressure exceeds a critical value, defects or cracks will be formed in the shell for gas diffusion and microbubble fragmentation, although complete destruction is rare (Brennen 1995, Sboros 2008). Ambient temperature and pressure, composition of the dissolved gas, frequency and intensity of the acoustic wave, and solution viscosity effect significant changes in acoustic cavitation. The radial motion of bubbles and their ability to damage cells decline as the viscosity of the surrounding fluid increases. The addition of trolox, a water-soluble derivative of vitamin E, renders

human erythrocytes more fragile to mechanical stress due to the enhanced cytolysis with the inclusion of it in the cell membrane (Miller et al. 2003).

Strong physical, chemical, and biological effects are induced by both types of cavitation as the mechanisms of drug delivery (see Figure 14.5) (Miller et al. 1996, Kimmel 2006, Ferrara 2008). Although shock waves, microstreaming, high wall speed of several 100–1000 m/s in the collapse, and microjet produced during IC are short in terms of time, the localized large pressure (e.g., 1000 atm) can generate transient and nonlethal micropores in the endothelial cell, plasma, and nuclear membrane, making them more permeable for drugs (Morgan et al. 2000). During the bubble collapse, the local temperature can reach as high as 5000 K (similar to the surface temperature of the sun) and induce chemical changes in the medium (sonochemistry) (Suslick 1990). The most significant is the generation of highly reactive oxygen species, such as free radicals, that can induce chemical transformations to permeabilize the cell membrane without producing pores on the membrane and affecting the cell viability. Meanwhile, endocytosis and phagocytosis in the uptake and fusions of lipid-shelled microbubbles are important in the intracellular drug absorption (Hernot and Klibanov 2008).

14.3 ACOUSTICALLY ACTIVE DRUG VEHICLES

The most popular vehicles activated by sonication include microbubbles, nanoemulsions, liposomes, and polymeric micelles, which show promising enhancement of localized delivery in solid tumors up to 5–10-folds over traditional methods of delivery (Shortencarier et al. 2004).

14.3.1 MICROBUBBLES

Ultrasound contrast agents (UCAs) in the form of biocompatible microbubbles less than 10 μm in size in order to exit the heart through the pulmonary capillaries are popular in perfusion monitoring to measure vascular density and microvascular flow rate (Cosgrove 2006). A typical dose of UCAs for an echocardiographic evaluation is 10^9–10^{10} microbubbles in a 1–2 mL bolus intravenous injection (Dolan et al. 2009). Microbubbles have a gas core (e.g., perfluoropentane, sulfur hexafluoride, nitrogen) and a highly cohesive and insoluble

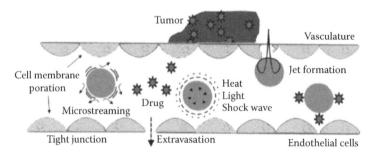

FIGURE 14.5 Schematic diagram of the biological effects and drug release induced during bubble cavitation through the defective tumor microvasculature. Ultrasound can trigger drug release/activation locally by heat, light, shock wave, microstreaming, and microjet.

shell (e.g., protein, phospholipid, polymer), permitting prolonged circulation and preventing nonspecific removal from the circulation by the reticuloendothelial system (RES). The gas core not only has a significant acoustic impedance mismatch for strong echogenicity but is also compressible for bubble cavitation (both SC and IC, depending on whether the acceleration terms are dominated by the pressure of the gas or the inertia of the inrushing liquid). In SC, a bubble oscillates nonlinearly around its equilibrium radius. The stable pulsation of fluid produces microstreaming and shear stress, which are shown to induce transient compromise of cell membranes. However, in IC a bubble is reduced to a minute fraction of its original size, and the gas within dissipates into the surrounding liquid via a violent mechanism. UCAs, whether or not co-administered or encapsulated with pharmaceutical agents, can be intentionally ruptured by ultrasonic waves at a moderately high acoustic pressure at the target sites (Klibanov 2006, Ferrara et al. 2007).

14.3.1.1 Drug-Loaded Microbubbles

There are various ways of entrapping drugs within a microbubble (see Figure 14.6) (Unger et al. 1998a, 2001). Drugs may be incorporated into the membrane or in a shell of microbubbles. Stable and strong deposition of the charged drugs in or onto the microbubbles' shell is realized by electrostatic interactions. However, Kupffer cells, leukocytes, and macrophages have a tendency to capture charged microbubbles, which could substantially decrease their half-life. In addition, drugs can also be embedded into microbubbles. The advantages of drug-loaded microbubbles are that the loaded drugs can be released at the occurrence of IC into the tissue due to the induced shock wave, microstreaming, and microjet. Meanwhile, the process can be tracked with sonography as

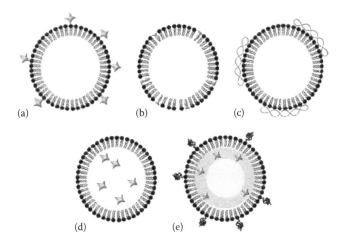

(a)   (b)   (c)

(d)   (e)

FIGURE 14.6 Different approaches of loading drug/DNA into microbubbles: (a) attaching to the membrane, (b) embedding within the membrane, (c) bounding noncovalently to the surface, (d) enclosing inside, and (e) incorporating into an oily film surrounded by a stabilizing layer with a ligand for targeting. (Reprinted from *Eur. J. Radiol.*, 42, Unger, E.C., Matsunaga, T.O., McCreery, T., Schumann, P., Sweitzer, R., and Quigley, R., Therapeutic applications of microbubbles, 160–168, Copyright 2002, with permission from Elsevier.)

the drug carriers are essentially UCAs. However, the monolayer lipid shell (a few nanometers) limits loading the hydrophobic pharmaceuticals and may have a premature release. Although a thicker triglyceride lipid shell and a dissolved drug is a solution, it is only available for hydrophobic drugs (e.g., paclitaxel). Biocompatible and biodegradable polymeric (e.g., pLGA) microbubbles possess a thick and rigid shell for a prolonged circulation half-life and permit a much higher loading capacity of both hydrophobic and hydrophilic drugs whose release rate depends on the drug's properties (e.g., lipophilicity and water solubility).

The aforementioned methods are mostly applicable to highly active drugs, such as gene-based drugs, ensuring the protection of the drug and lowering the propensity of premature release. Therapeutic genes with several thousand pairs and MW over 1 million Da cannot cross the capillary fenestrations of blood vessels (Unger et al. 2001). In addition, intravenous administration is not suitable for genetic materials because of the rapid metabolization by serum esterases. After reaching the tissue targets, genes must penetrate cell membranes, pass cytoplasm, enter the cells' nuclei, and then be digested by lysosomes within the cells. Drawbacks of incorporated naked plasmid DNA (pDNA) and pDNA–polymer complexes released from microbubbles are (a) large microbubble size (3–7 μm for consequent short circulation time and ineffective extravasation into the tumor), (b) the necessity to complex the pDNA with cationic polymers to prevent degradation during fabrication, (c) low loading efficiency of pDNA (~6,700 molecules/bubble) due to the limited number of cationic lipids, and (d) premature release of more than 20% of the encapsulated pDNA (Lentacker et al. 2009).

14.3.1.2 Targeted Microbubbles

Although acoustic radiation force could promote the attachment of microbubbles to endothelial cells in the wave propagation path, it is more attractive to selectively adhere microbubbles to cellular epitopes and receptors of cancer or solid tumors by one or several specific ligands, such as antibodies, carbohydrates, and/or peptides (Figure 14.7) (Hernot and Klibanov 2008). Monoclonal antibodies have a very high specificity and selectivity to a large range of epitopes. In contrast, peptides are low cost and less immunogenic because of their much smaller size (5–15 amino acids). Significant enhancements of microbubble adhesion to activated endothelium (Villanueva et al. 1998), rejecting tissues (Weller et al. 2003), neovasculature endothelium (Weller et al. 2005a), lymph node–related vasculature (Hauff et al. 2004), or activated platelets (Schumann et al. 2002) have been reported. Simultaneous targeting to multiple ligands could synergistically increase the adhesion strength (Weller et al. 2005). There are two ways of coupling ligands to the microbubble shell: covalent (attached to the head of phospholipids directly or via an extended polymer spacer arm), and noncovalent by avidin–biotin bridging due to their wide availability and excellent affinity. However, since avidin carries a strong positive charge in the glycosylate layer, the biodistribution of microbubbles may be altered, resulting in nonspecific adhesion and

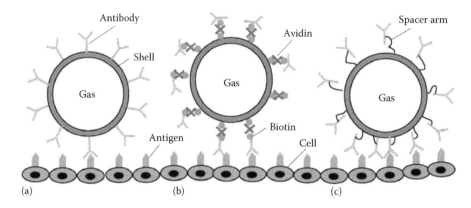

FIGURE 14.7 Targeting microbubble to cancer cells by connecting the receptors on the surface with (a) an antibody, (b) an avidin bridge, and (c) a flexible spacer arm. (Reprinted from *Adv. Drug Deliv. Rev.*, 37, Klibanov, A.L., Targeted delivery of gas-filled microspheres, contrast agents for ultrasound imaging, 139–157, Copyright 1999, with permission from Elsevier.)

introduction of undesired immune response. In comparison, streptavidin may be a better bridging molecule. It is found that connecting an intermediary spacer arm with the ligand at the end as a polyethylene glycol (PEG) molecular tether to the lipid shell indirectly is feasible and has high specificity and targeting. Meanwhile, the tethered ligand could also be buried in a polymeric "overbrush" (Borden et al. 2006). After sonication, the ligand is instantly revealed for targeting to avoid the undesired nonspecific binding and uptake by RES.

Altogether, microbubbles can enhance transport of drugs/genes across vessel walls and cell membranes by the localized bubble cavitation. However, their disadvantages are the relatively low payload and short circulation half-life (minutes for lipid-shelled microbubbles but longer for the polymer types).

14.3.2 NANOEMULSIONS

Although the use of microbubbles is very attractive in drug delivery, especially for its uniqueness of combined diagnosis and targeted therapy with a high effectiveness-to-cost ratio, it lacks an essential prerequisite for effective extravasation into tumor and subsequent drug targeting with a sufficient lifetime in the circulation, because the pore size of most tumors

is usually small (380–780 nm) (Campbell 2006). An alternative solution to the aforementioned problems is developing drug-loaded nanoparticles that can accumulate in tumor and then expand to microbubbles *in situ* under sonication, such as copolymer-stablilized echogenic perfluoropentane (PFP) nanoemulsions. The nanoemulsions are produced from drug-loaded poly(ethylene oxide)-*co*-poly(L-lactide) (PEG-PLLA) or poly(ethylene oxide)-*co*-polycaprolactone (PEG-PCL) micelles. Dodecafluoropentane (DDFP) droplet remains a liquid at the body temperature, although it has a boiling temperature of 29°C at atmospheric pressure, but vaporizes under ultrasound exposure. Acoustic droplet vaporization (ADV), that is, transition of droplet to bubble, is determined by a certain pressure threshold (Zhang and Porter 2010, Reznik et al. 2011). The ADV threshold for PEG-PCL stabilized droplets is lower than that required by PEG-PLLA droplets. The higher echogenicity of microbubbles enables its discrimination from droplet's status in sonography. To prevent excessive phase transition in the vasculature, infusion or injection through a low-gauge needle should be carried out in the application of PFP nanoemulsions. A significant volume expansion in the complete vaporization of the droplet and consequent shrinkage of the microbubble shell may rip off a drug to the local cells (Figure 14.8). All aspects considered, this method

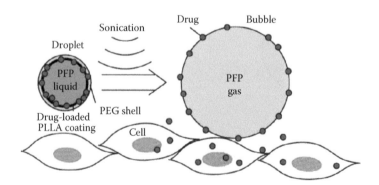

FIGURE 14.8 Schematic diagram of phase transition from nanodroplets to microbubbles and consequent drug release into the neighboring cells under sonication. (Reprinted from *J. Control. Rel.*, 138, Rapoport, N.Y., Kennedy, A.M., Shea, J.E., Scaife, C.L., and Nam, K.-H., Controlled and targeted tumor chemotherapy by ultrasound-activated nanoemulsions/microbubbles, 268–276, Copyright 2009, with permission from Elsevier.)

allows the enhancement of intracellular drug delivery with high ultrasound contrast and drug loading (Rapoport et al. 2009, Zhang and Porter 2010, Reznik et al. 2011).

14.3.3 Liposomes

Liposomes with a typical diameter of 65–120 nm are nontoxic, biodegradable, and non-immunogenic drug delivery vehicles for both hydrophilic and lipophilic drugs, such as doxorubicin (DOX) and vincristine (Huang 2008, Schroeder et al. 2009). Local heating causes the liposome to change from a well-ordered gel to a less-structured liquid crystalline state by inducing thermotropic phase transitions on phospholipids. As a result, the cargo can be released with significant reduction in systemic toxicity and premature degradation or inactivation (Allen 1997). The *in vivo* stability and accumulation at the tumor site can be enhanced to 50–100-fold compared to the free drug (Allen 1997), and small polymeric carriers can respond to both mechanical (i.e., fracture) and thermal (i.e., temperature elevation) activation. PEG liposomes containing DOX have been used in the treatment of Kaposi's sarcoma, refractory ovarian cancer, and breast cancer (Ranson et al. 1997). Furthermore, liposomes are also efficient as nonviral gene carriers (Lentacker et al. 2009).

Different strategies are applied for both drug loading and release from microbubbles and liposomes. Although the gas core occupies a substantial volume of a microbubble, the drug loading capability can be enhanced by shell construction layer by layer or conjugation of drug-entrapped particles. In comparison, drug-specific loading techniques are applied to construct the core of a liposome. Drug release from microbubbles is mainly due to the ultrasound-induced IC, whereas that from a liposome is the consequence of the heat produced by the absorption of acoustic energy. Meanwhile, the gas nuclei will expand and dilate the monolayer boundary or the adjacent bilayer in the strong acoustic field. If the stress induced by the bubble expansion is beyond the elastic threshold of the bounding membrane, the liposome will disrupt and the incorporated contents will be released.

14.3.3.1 Thermosensitive Liposomes

Liposomes are potent nanocarriers to deliver chemotherapeutic drugs to tumors. However, inefficient drug release hinders their application. Thermosensitive liposomes (TSLs) can release drugs upon heat. The thermally responsive macromolecular carrier poly(N-isopropylacrylamide (NIPAAm) can be cheaply and conveniently synthesized through free-radical polymerization (Gasselhuber et al. 2012). However, precisely controlled release of chemo and radiotherapeutic agents conjugated to its structure is difficult. Artificial elastin-like polypeptides (ELPs) are genetically encodable, and polypeptide-based polymers that are recombinantly synthesized in *Escherichia coli* by overexpression of a synthetic gene allow control over the sequence, chain length, and the number and location of reactive side chains (e.g., lysine, cysteine) on the polypeptide (Chilkoti et al. 2002). ELPs exhibit lower critical solution temperature (LCST) than poly(NIPAAm). Traditional TSLs are triggered at 42°C–45°C for drug release over ~30 min. In comparison, temperature ranges of 39°C–40°C can release the payload from low-temperature-sensitive liposomes (LTSLs), as shown in Figure 14.9.

1,2-Distearoyl-*sn*-glycero-3-phosphoethanolamine-*N*-PEG$_{2000}$ (DSPE-PEG$_{2000}$) of 6 mol% and higher caused carboxyfluorescein (CF) release at physiological temperature in dorsal skin flap window chamber models implanted with human BLM melanoma, and the CF leakage increased with DSPE-PEG$_{2000}$ density. TSLs in the size of 80 nm with 5 mol% DSPE-PEG$_{2000}$ were stable at 37°C and released 60% CF in 1 min and almost 100% CF in 1 h at 42°C (Li et al. 2010) (Figure 14.10).

14.3.3.2 Nonthermal Liposomes

Acoustically active lipospheres (AALSs) do not rupture as easily as UCAs because the thicker shell retards the wall velocity in the bubble cavitation. It is necessary to attach a targeting ligand to the liposomes to prevent them from being flushed away by the circulating blood. However, a lower paclitaxel loading efficiency (4 mg/mL of microbubbles) may be an unfavorable outcome of the ligand attachment. Loading and

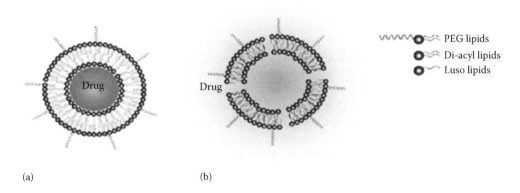

(a) (b)

PEG lipids
Di-acyl lipids
Luso lipids

FIGURE 14.9 Temperature-sensitive liposomes (a) below and (b) above their phase transition temperature, which is typically chosen near 40°C for temperature-sensitive vehicles. Typical liposome diameter is 65–120 nm. (Reprinted with permission from Ferrara, K.W., Borden, M.A., and Zhang, H., Lipid-shelled vehicles: Engineering for ultrasound molecular imaging and drug delivery, *Acc. Chem. Res.*, 42, 881–892, Copyright 2009, American Chemical Society.)

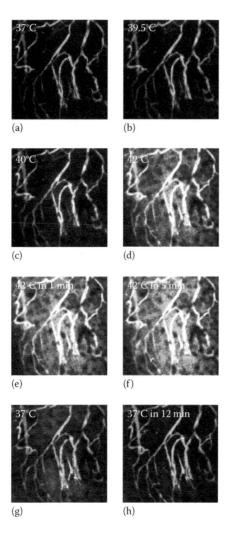

FIGURE 14.10 *In vivo* green carboxyfluorescein (CF) in aqueous phase released from thermosensitive liposomes and red Rho-PE fluorescence–labeled lipid membrane at (a) T_{tissue} = 37°C post injection, (b) T_{tissue} = 39.5°C, (c) 40°C, (d) 42°C, (e) 42°C in 1 min, (f) T_{tissue} = 42°C in 5 min, (g) T_{tissue} = 37°C, and (h) T_{tissue} = 37°C in 12 min. (Reprinted from *J. Control. Rel.*, 143, Li, L., ten Hagen, T.L., Schipper, D., Wijnberg, T.M., van Rhoon, G.C., Eggermont, A.M., Lindner, L.H., and Koning, G.A., Triggered content release from optimized stealth thermosensitive liposomes using mild hyperthermia, 274–279, Copyright 2010, with permission from Elsevier.)

release of hydrophobic drugs in AALSs could be improved by varying the amount of oil and perfluorocarbon gas. In summary, it remains rather difficult to obtain an efficient drug release from AALSs and to also incorporate hydrophilic drugs (Huang and MacDonald 2004). Meanwhile, air-filled echogenic liposomes (ELIPs) with an average size of ~800 nm have been developed to incorporate hydrophilic compounds inside liposomes (Figure 14.11). Drug release from ELIPs occurs relatively easily. When the expansion of ELIPs becomes so high that the elastic limit of the lipid shell is exceeded, drugs can leak out. The incorporation of short-chain lipids onto the ELIPs shell results in a prolonged shell opening and better drug release (Huang and MacDonald 2004). Because of the

flexibility of the lipid monolayer, the ELIP shell can reseal. Therefore, successive expansion cycles can be applied to improve drug release from ELIPs. AALSs are conceptually similar to microbubbles, except for the smaller size for easy extravasation, relatively low encapsulation efficiency, and limited drug release from the shell collapse because a thicker shell retards the wall velocity associated with IC.

14.3.3.3 Liposome-Loaded Microbubble

A more advanced drug-carrying microbubble was achieved by attaching both hydrophilic and hydrophobic drug-loaded nanoparticles to the structure via avidin–biotin interactions (Figure 14.12) (Kheirolomoom et al. 2007, Endo-Takahashi et al. 2012, Negishi et al. 2012). As ~10^5 liposomes are bound to each microbubble, the drug loading capacity of microbubbles can be significantly enhanced. Attachment of polystyrene nanoparticles/liposomes onto the microbubble shell did not hinder IC. Cells close to liposome-loaded microbubbles contained liposome fragments after sonication. This concept can also be used to load small interfering RNA (siRNA) and gene-containing liposomes (lipoplexes) to improve their transfection efficiency. These vehicles combine the advantages of liposomes and microbubbles and overcome the respective individual limitations.

14.3.4 MICELLES

A micelle is an aggregate of surfactant molecules dispersed in a liquid colloid with the hydrophilic "head" regions in contact with the surrounding aqueous solution and the hydrophobic single-tail regions in the micelle center. They are small enough to avoid renal excretion but permit extravasation at the tumor site via the enhanced penetration and retention (EPR) effect (Rapoport 2004). The structure of hydrophobic cores as drug reservoirs and hydrophilic shells ensure micelle solubility in the aqueous medium but prevent micelle aggregation. PEG blocks with a typical length of 1–15 kDa prevent micelle opsonization, thereby avoiding micelle recognition by RES. The thermodynamic and kinetic stability of a micelle is determined by molecular interactions and the length of hydrophobic blocks and the micelle core type, respectively. Soluble polymeric carriers that undergo a LCST phase transition would enable targeting heated tumors and hydrophobically collapse and aggregate at temperatures greater than their LCST (Chilkoti et al. 2002). The self-assembly of amphiphilic block copolymers is activated thermodynamically and is reversible when the concentration reaches the critical micelle concentration (CMC). However, copolymer molecules are individual (unimers) at low concentrations. The encapsulated drug will be released from the prematurely destroyed micelles. The CMC is primarily controlled by the length of a hydrophobic block (the longer the length, the lower the CMC) but less sensitive to the length of a hydrophilic block. However, too high a concentration of the copolymer will initiate micelle aggregation and precipitation (Rapoport 2007). Micelles whose glass transition temperature is higher than the physiological temperature (i.e., 37°C)

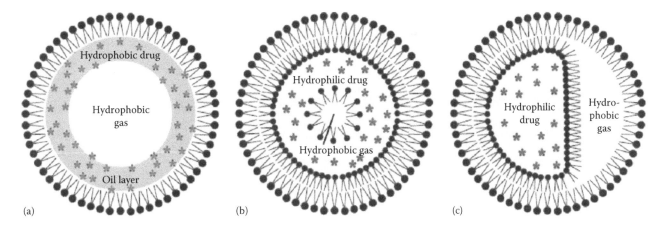

FIGURE 14.11 Schematics of (a) an acoustically active liposphere (AALS) encapsulating hydrophobic drugs and hydrophobic gas. (b and c) Echogenic liposomes (ELIPs) encapsulating hydrophilic drugs and hydrophobic gas. (From Lentacker, I. et al., *Soft Matter*, 5, 2161, 2009.)

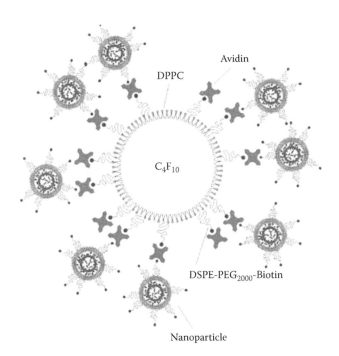

FIGURE 14.12 Schematic diagram of PEGylated nanoparticles being attached to the microbubble with an avidin–biotin binding. Ultrasound irradiation results in the release of intact nanoparticles. (From Lum, A.F.H. et al., *J. Control. Rel.*, 111, 128, 2006.)

may survive for many hours or even days at concentrations below the CMC. In contrast, it takes only a few minutes to dissociate micelles with a "soft" core (e.g., Pluronics). The CMC of this type of micelles is usually low because of direct mixing of the hydrophobic drugs with the soft core. The most outstanding attractions of micelles are the self-assembly of amphiphilic block copolymer molecules and drug encapsulation by simple mixing rather than conjugation. Bubble cavitation can occur, but inside the micelle cores with an increased threshold (Marin et al. 2002). In contrast, drug loading into micelles with solid cores requires a number of more complex techniques (Rapoport 2012, Wan et al. 2012).

14.4 SONOPORATION

14.4.1 INTRACELLULAR DRUG TRANSPORT

Transport of drugs and genes must overcome a series of physical and biological barriers, including the cell plasma membrane, in order to deliver the therapeutic agents to the intracellular and molecular targets. Effective drug therapy extends beyond the first step of increased local drug concentration. Many pharmaceutical agents, including various large molecules (proteins, enzymes, antibodies) and drug-loaded pharmaceutical nanocarriers, are required to exert their therapeutic action inside the cytoplasm or the nucleus or in other specific organelles, such as lysosomes, mitochondria, or the endoplasmic reticulum. However, the lipophilic nature of biological membranes restricts the direct intracellular transport of many of these agents. The cell membrane prevents large molecules, such as peptides, proteins, and DNA, from spontaneously entering cells unless there are active transport mechanisms available, creating a major obstacle to the intracellular delivery of these macromolecular and ionic agents (Figure 14.13).

Viral vectors have high ability of entering the cells and directly delivering their genetic information into the nucleus. Adenoviruses, adeno-associated viruses, lentiviruses, and retroviruses have been used with their own characteristic advantages and disadvantages, depending on the specific application. Adenoviruses induce short-term expression of the delivered nucleotides, which might be due to the adenoviral delivery system itself. In contrast, retroviruses are effective means of stably delivering therapeutic nucleotides to cells, but with the inability to infect nonproliferating cells, random incorporation of the viral DNA into the genomic DNA of the cell which can cause insertional mutagenesis and malignancy, and a potential for replication competent viral particles to emerge, causing cell death. The main limitations of viral vectors are their lack of specificity and the potential for immunogenic and cytotoxic effects. If viral vectors are not administered in anatomically isolated regions, they will disseminate throughout the body, transferring to nontarget

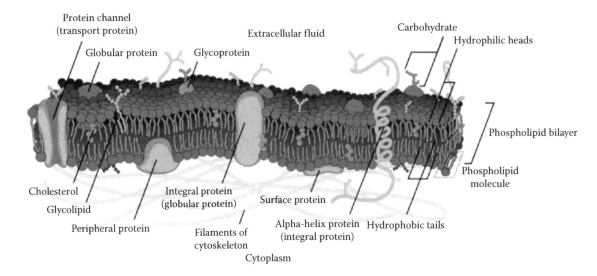

FIGURE 14.13 Anatomical structure of the cell membrane.

cells, wasting large quantities of particles and increasing the possibility of deleterious side effects. Therefore, their clinical use in humans for common and nonanatomically isolated disorders is restricted.

Pharmaceutical and biological approaches have been investigated to efficiently target intracellular sites by using cell-penetrating proteins, peptides, polymers, polymeric nanoparticles, and liposomes (e.g., immunoliposomes, pH-sensitive liposomes, and cationic lipids). The efficacy and specificity toward particular cells is determined by the size, surface charge, composition, and the presence of cell-specific ligands on the carrier. The lipid can fuse with the membrane of the cell and deliver the foreign DNA. Subsequently, transfection rates are high with very little cell mortality. Although lipofection has the potential for extensive *in vivo* use, it also suffers from the lack of site specificity.

External physical energy sources can also be used to enhance intracellular delivery effectively. Electroporation utilizes high-intensity electric fields to create small pores in the cell membrane, thereby allowing the diffusion of DNA, other agents, or their carriers into the cell cytoplasm. Particle bombardment refers to the use of high-speed projectiles (e.g., micrometer-sized gold particles) coated with DNA to ballistically introduce the coated DNA into cells. These physical methods are relatively simple, safe, and can be used in conjunction with other techniques for synergistic benefit, but have a lower transfection efficiency compared to viral methods, and some are intrinsically unsuitable for *in vivo* applications.

Sonoporation employs ultrasound energy to mechanically create nonspecific, transient, and reversible pores to increase the permeability of a cell membrane to enable transient entry of otherwise impermeable compounds into the cytoplasm of surviving cells. Stably enhanced intracellular uptake of agents and the expression of exogenous genes in viable cells after ultrasound exposure both *in vitro* and *in vivo* have been observed. Early investigations utilized 20 kHz ultrasound to show the delivery of fluorescein-labeled dextran molecules and pDNA in cultured cells. Now, megahertz frequency ultrasound, which

has better spatial control and is more appropriate for preclinical and potential clinical applications, has become more popular in the research of sonoporation. In light of the apparent association of cavitation and sonoporation, microbubble UCAs have been used to provide cavitation nuclei at the desired locations and facilitate intracellular delivery with the use of lower acoustic pressures and higher frequencies, which makes sonoporation more suitable for clinical applications. Furthermore, sonoporation can be used in conjunction with viral vectors for targeted transfection if the vectors can be protected via encapsulation as inert carriers until local release and cellular uptake is achieved by sonoporation. Sonoporation refers to transient permeabilization, in which large molecules can be trapped inside surviving cells, in contrast to the permanent permeabilization of nonviable cells indicated by vital stains (e.g., trypan blue dye and porpidium iodide).

Sonoporation can be realized by a variety of ultrasound sources, such as sonicator, shock wave lithotripters, low-intensity pulsed ultrasound (LIPUS), diagnostic ultrasound, and high-intensity focused ultrasound (HIFU), as listed in Table 14.1. The sonicator is used mainly *in vitro* with cell suspensions in a culture dish, chamber, or rotating tube, while the diagnostic scanner, LIPUS, HIFU, and LSW devices can be used *in vivo* for clinical applications. Although there are few direct comparisons of the different sonoporation approaches, the focused ultrasound was found to be significantly better in terms of reduced cell killing and higher transfection percentages. Diagnostic ultrasound images and reception of cavitation-produced acoustic emissions can also provide feedback of treatment in the region of interest.

14.4.2 Mechanism

Most evidence to date illustrate that the mechanisms of sonoporation are associated with passive diffusion-associated transport through the relatively nonspecific large pores in the size of up to several micrometers, the long time required for full recovery of the disrupted membrane (i.e., tens of seconds

TABLE 14.1

Ultrasound Exposure Systems for Sonoporation

Device	Frequency MHz	Intensity or Pressure	*In Vitro*	*In Vivo*
Sonicator	0.02–0.04		Suspension	NA
Lithotripter	0.1–0.3	~10 MPa	Fixed chamber suspension	Malignant tumor
LIPUS	0.5–3.0	0.5–5.0 W/cm²	Culture dish monolayer	Skeletal muscle
		0.8 MPa	Rotating tube	
Pulsed HIFU	0.75–3	~15 MPa	Fixed chamber suspension	Carotid artery
Diagnostic imager	1.3–13	MI = 1.9	Chamber monolayer	Myocardium

to minutes), and the necessity of sufficient extracellular Ca^{2+} and energy to achieve full recovery.

Experimental studies have demonstrated that cyclic acoustic pressure variations generate bubble activity (i.e., bubble oscillation, movement, and collapse). This mechanical process appears to stress the membranes, opening the pores and allowing the transfer of materials into and out of the cell. Reduced cavitation activity results in lower or complete absence of delivery. However, the details of this dynamic and complex process, particularly the interaction of bubbles with cells, are incompletely understood, and precise control of bubble cavitation is challenging. The impact of bubble collapse, fluid microjet formation from asymmetric bubble collapse near a physical boundary, rapid impact and penetration by ballistic microbubble-induced streaming, collisions or coalescence of translating microbubbles with cells, and cavitation-induced sonochemical effects are all associated with sonoporation. Ultrahigh-speed photography with frame rates beyond 1 million frames per second and appropriate magnification allows the observation of details in bubble cavitation (Figure 14.14). An Optison bubble driven by a 20 µs burst of 1 MHz ultrasound at a peak negative pressure of 1.39 MPa was found to grow rapidly in size and collapse to form a microjet penetrating the membrane of a nearby cell (Prentice et al. 2005, Prentice and Campbell 2008). A correlation between endothelial cell deformation, subsequent cell membrane permeability, and transport of propidium iodide (PI) molecules (a membrane integrity probe) into these cells was shown (Van Wamel et al. 2006).

Conventional assays for delivery efficiency and associated side effects, such as loss of cell viability and enhanced apoptosis, have largely relied on post-treatment statistical-based analysis, but they are inadequate in accounting for the impact of the actual transient process of cell poration. The subsequent cellular or functional effects after bubble cavitation monitored in real time can provide important knowledge of this phenomenon. Electrophysiology techniques, such as voltage

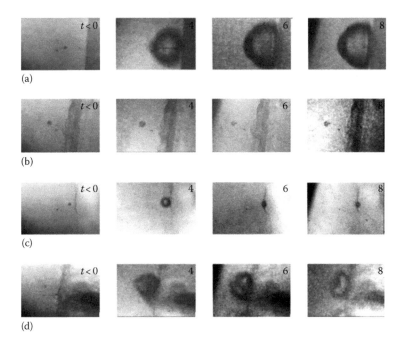

FIGURE 14.14 Ultrahigh-speed sequences showing (a) needle microjet formation (white arrows) from naked microbubbles (black arrows) via a 3.9 MPa (peak negative pressure) pulse, (b) remote jetting from a 5.5 µm microbubble trapped to 36 ± 4 µm from a cell monolayer (γ ≅ 3), (c) contact microjet close to the cell (γ ≅ 1), and (d) contact microjet during the inflation and collapse cycle of a 1.39 MPa pulse at different timings (in microseconds) relative to the instant of cavitation inception; image size 163 µm × 110 µm. (From Prentice, P. and Campbell, P., Application of optical trapping for cavitation studies, NanoScience + Engineering: International Society for Optics and Photonics, 2008, pp. 70381N-70381N-70388. With permission.)

clamp, can measure the ion currents across a cell membrane while maintaining the membrane potential at a set level. As a result, the dynamic biophysical process of sonoporation at the level of a single cell is available with high temporal resolution and sensitivity. As the intact, whole cell membrane has a constant resistance, the transmembrane current is consistent at a certain membrane holding voltage in the absence of activation of endogenous ion channels. During sonoporation, the ultrasound-generated pores in the membrane allow ion flows, reduce the membrane resistance, and change the transmembrane current, which depend on the pore size and the ion concentration gradient across the cell membrane. Xenopus oocytes were exposed to tone-burst ultrasound at 0.96 MHz in the presence of Optison™. The inward transmembrane current measured during the sonoporation process increased at the onset of the sonication, reached its maximum level at the end of the tone burst, and then decayed back to its initial value in 4–10 s (comparable to that of electroporation), indicating the recovery of the cell membrane (Deng et al. 2004). Furthermore, the increase in the acoustic emissions generated by the collapsing bubbles in the focal zone using passive cavitation detection correlated well with the increased amplitude of the inward transmembrane current (Zhou et al. 2008a) (Figure 14.15).

Complete current recovery correlates well with cell survival, while a prolonged duration or nonrecovery of the current always leads to cell death, which is determined by visible loss of intracellular contents and degeneration of cellular integrity. The ultrasound-induced pores in the plasma membrane must reseal to prevent the loss of intracellular contents in order to ensure cell survival, which limits the temporal window for the inward transmembrane passage of extracellular agents to be delivered. Intracellular overload of ions, which is toxic to the cell and can trigger other irreversible and reversible cellular processes, such as apoptosis and calcium oscillation, should be avoided. Therefore, the rate of resealing is one of the key factors determining the uptake efficiency and fate of the cells after sonoporation. Reversible sonoporation occurred at low

to moderate acoustic pressures (<0.5 MPa). High pressure amplitudes and longer exposure times (e.g., 0.7 MPa for 0.5 s) would generate large unrecovered current and subsequently spontaneous cell death (Deng et al. 2004). Compared with tone-burst ultrasound exposures, pulsed exposures can extend the duration of enhanced cell porosity without inducing irreversible membrane disruption. Since the pores will not reseal completely during the interval between individual pulses, the membrane can remain porous during the entire ultrasound exposure, creating a prolonged window of time of enhanced cell porosity to improve delivery while minimizing cell death.

In vitro studies have shown that varying the calcium concentration ($[Ca^{2+}]$), in the extracellular solution affects sonoporation resealing. At 1.8 or 3.0 mM $[Ca^{2+}]$, the transmembrane current returned almost to its initial level by 15 s. Lower than normal $[Ca^{2+}]$ resulted in slower and weaker resealing and often only partial recovery of current. For cells in calcium-free solution, to which 1 mM EGTA was added as a Ca^{2+} chelator, the transmembrane current exhibited no trend of sustained recovery after sonoporation. A threshold of 0.54 mM $[Ca^{2+}]$ was required for complete recovery (Zhou et al. 2008).

Immediate or acute cell death occurs from the mechanical fracture of cells or severe damage of cell membrane integrity (cell lysis). In contrast, subsequent loss of cell viability during sonoporation can occur from membrane disruption when the mechanical fracture of cell does not immediately destroy it, which can result in excessive ion exchange between the intracellular to extracellular space, and the initiation of apoptosis or other injury associated with the ultrasound exposure. The extracellular calcium ion concentration ($[Ca^{2+}]_i$) and dynamics are critically important for post-sonoporation pore resealing, similar to cell repair of membrane wounding. Real-time ratiometric fluorescence imaging of the intracellular $[Ca^{2+}]_i$ with the Ca^{2+}-sensitive indicator dye fura-2 AM, which shifts its absorption spectrum upon binding to Ca^{2+}, could demonstrate the its dynamics. No changes in the emitted fluorescence intensities of the dye were observed in Chinese

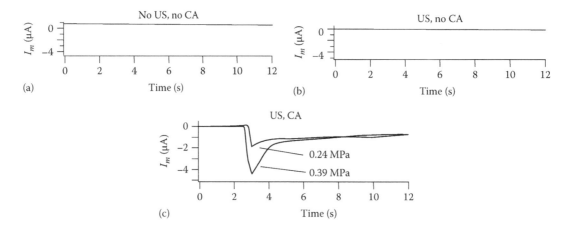

FIGURE 14.15 Ionic currents across the membrane of Xenopus oocytes with a clamped voltage of −50 mV with respect to the ground using (a) no tone-burst ultrasound (US, 0.96 MHz) pulse and no contrast agent (CA) Optison™, (b) US (<0.8 MPa), no CA, and (c) US and CA. (Reprinted from *Ultrasound Med. Biol.*, 30, Deng, C.X., Sieling, F., Pan, H., and Cui, J., Ultrasound-induced cell membrane porosity, 519–526, Copyright 2004, with permission from Elsevier.)

hamster ovary (CHO) cells without ultrasound (US) or with US (1 MHz, 0.2 s, 0.45 MPa) in the absence of Optison™ (5%). In the presence of the UCA, the spatial mean fluorescent intensities within a cell could be tracked as a function of ultrasound exposure (30–120 s) (Kumon et al. 2009). The delays exhibited in the $[Ca^{2+}]_i$ transients particularly demonstrate the generation and propagation of a calcium wave. These observations are most likely the result of Ca^{2+} transport across the cell membrane with nonspecific pores during sonoporation, which is demonstrated by the occurrence of a Ca^{2+} flux even when voltage-dependent L-type Ca^{2+} channels are blocked using verapamil. Thus, the initial change in $[Ca^{2+}]_i$ may not be due to a release of Ca^{2+} from internal stores or buffers (i.e., endoplasmic reticulum, mitochondria, buffering proteins) by the ultrasound exposures. Further studies are required to understand the mechanisms and role of Ca^{2+} in sonoporation.

Sonoporated cells may respond to the disrupted intracellular equilibrium by upregulating cytoplasmic signals related to apoptosis and cell cycle arrest. Ultrasound can also generate calcium transients (i.e., changes of intracellular calcium concentration, oscillations, and spatial waves of $[Ca^{2+}]_i$). Ca^{2+} is an important messenger involved in regulating many cellular processes and functions, such as gene transcription, cell proliferation, fertilization, metabolism, cell migration, wound response, and phagocytosis (Petty 2006). Bubble cavitation (SC and IC) produces an intracellular increase of $[Ca^{2+}]_i$ hundreds of micrometers away in a few seconds. $[Ca^{2+}]_i$ activities affect cell–cell contact, electrical resistance, ZO-1 tight junction protein migration from intracellular sites to the plasma membrane, and tight junction assembly in epithelial and endothelial cells (Park et al. 2010). Meanwhile, mitosis of the cell cycle will be halted through downregulation of the Cdc-2 protein. Apoptosis may be triggered through mitochondria and promote cytochrome-c into the cytoplasm via poly-ADP-ribose polymerase (PARP) cleavage by the activation of the caspase-3 protein. Resealing of the cell membrane is not a simple self-reunion procedure, since the scattered patches on the same size order as the membrane pores were observed on the resealed cells (Zhong et al. 2011). It is hypothesized that an exocytotic patching mechanism may be involved so that the cytoplasmic vesicle will first be delivered to the sonoporation site and then fused together with the membrane, which is similar to the $[Ca^{2+}]_i$ influx into porous cells.

Another barrier of interest is DNA trafficking to the nucleus. Ultrasound could deliver macromolecules into the cytosol but did not promote transport into the nucleus (Guzman et al. 2002). Microscopy illustrated that intracellular DNA was excluded from the nucleus, which is consistent in the other gene therapy methods. Therefore, understanding the intracellular pDNA trafficking would optimize the ultrasound-mediated gene transfection.

14.4.3 IN VITRO STUDIES

In vitro sonoporation studies have been carried out by employing a variety of cell models, fluorescent markers, transgenes, and ultrasound systems, which established the foundation for mechanism investigation, technology development, and translation into preclinical studies using *in vivo* models. The strategy of *in vitro* studies is to provide a simple, yet valuable, and efficient means for investigating the effects of ultrasound parameters on sonoporation outcome (assessed by drug delivery or gene transfection efficiency and cell survival rate) in a well-controlled experimental setup for high-throughput testing without the complexity, difficulty, and expense involved in *in vivo* studies. A typical *in vitro* experimental setup includes detached cells in suspension, a monolayer of cells attached to the bottom of a culture dish, or a custom-made chamber to accommodate a specific experimental configuration. The cells are mixed with the target agents and are treated with ultrasound, may be in the presence of microbubbles (concentration in the range of 1×10^6 to 10×10^6 bubbles/mL). Ultrasound usually propagates through the bottom of the dish or chamber. However, in some situations, the exposure parameters are difficult to determine accurately. For example, the exact rarefactional pressure amplitude (RPA) at the cell monolayer depends on the precise distance between the ultrasound source and the dish, the thickness of the dish bottom, the thickness of the liquid layer in the dish, and the acoustic reflection from the medium's top. In addition, standing waves may be set up in the culture dish.

Sonoporation outcome can be assessed using many established techniques in cell biology or molecular cell biology. Bright-field or fluorescent microscopy is generally used to assess the effects on individual cells, such as changes in morphology, viability, delivery rates of fluorescent markers, or expression of fluorescent reporter genes. Flow cytometry is often used to obtain quantitative results of fluorescent uptake and cell viability in a large population of cells. Confocal microscopy, scanning electron microscopy (SEM), transmission electron microscopy (TEM), and atomic force microscopy (ATM) have been used for high-resolution assessment of the effects on cell membranes due to ultrasound exposures. Molecular cell biology techniques, such as western blot, may also be used to ascertain the protein concentration and the success of therapeutic gene transfection in cells.

It is found that increasing acoustic energy, peak pressure amplitude, or exposure time increases the overall transfection rates at the cost of decreased cell viability. Increasing the number of ultrasound pulses increases transfection to a certain threshold. Increasing the microbubble concentration produces a similar trend, with some investigators suggesting as much as a 3:1 microbubble-to-cell ratio for *in vitro* studies.

However, it is difficult to compare all *in vitro* sonoporation studies due to the wide variety of cell models, ultrasound exposure systems, and the actual sonication conditions with lack of rigorous characterization (e.g., the spatial acoustic field distribution and the peak negative acoustic pressure). Another complexity or difficulty is the control and monitoring of the activity of individual bubbles and their interaction with cells in contrast to the homogeneous distributions of reagent concentration often predictable of dissolved chemical compounds. The random nature of bubble cavitation leads to large statistical variations in experimental results. Furthermore, bubbles tend to move in response to the induced flow in a fluid medium by ultrasound exposures (i.e., acoustic streaming), which further increases the variation in sonoporation study results.

Efficient and well-tolerated internalization of DNA and macromolecules with a diameter no more than 37 nm is allowed by the induction of transient and reversible holes in the cell membrane (sonoporation). The cell structure affects the sonoporation effect and the cell survivability, although the maximum size of molecules for entry as well as the opening period of the pores is not well known. Immediately after sonication, the MAT B III cells showed morphological changes (i.e., smoother surface and smaller size) because of the removal of macromolecules (e.g., glycoproteins or cell-surface receptor CD19) on the surface (Mehier-Humbert et al. 2005). For the cells fixed during sonication, some large pores (on the order of 100 nm) but not small ones (1–10 nm) could be produced, which might be due to the abundant microvilli on the cell surface and the short duration of pore opening (Figure 14.16a). In comparison, fresh red blood cells showed clear pores on their smooth surface after sonication (Figure 14.16b). Therefore, the pore size in sonoporation depends on both the ultrasound parameters and cell types. Small pores are more abundant on the cell membrane, facilitating the internalization of molecules whose amount is inversely proportional to their size, but reseal more quickly after sonication. Large pores on the cell surface may not reseal, and ~15 s disruption on a plasma membrane in smaller cells (e.g., fibroblasts and endothelial cells) is generally lethal. However, the nuclear membrane seems to be unaffected by sonoporation, but nuclear complexes only allow the passive diffusion of small particles up to 9 nm (Miller and Quddus 2000, Miller et al. 2002, Ohl et al. 2006). Unfortunately, the leakage of cytosolic fluid can cause partial escape of internalized molecules. IC dominates

in the ultrasound-mediated DNA transfection through sonoporation, so measuring the IC dose may be helpful in monitoring the delivery effect (Ward et al. 2000, Qiu et al. 2010).

It was observed that macromolecules with different MWs penetrated similarly through the plasma membrane (Larina et al. 2005a). In MCF7 cells, the overall difference between the delivery rates of 10 and 2000 kDa dextrans was about 20%. However, confocal microscopy revealed that intracellular distribution of the delivered macromolecules differed based upon their size. Macromolecules of 2000 kDa did not penetrate through the nuclear membrane, while 10 kDa macromolecules were uniformly distributed within the cells. Macromolecules of MW 70 kDa could penetrate through the nuclear membrane; however, nuclear fluorescence of FITC-dextran was weaker than cytoplasmic fluorescence, suggesting that 70 kDa is close to the threshold of nucleus penetration. At the same sonication conditions, SK-BR-3 human breast carcinoma cells demonstrated only 3.5% death rate, while the value for HeLa human cervix epithelial adenocarcinoma cells was ~40%. Moreover, KM20 human colorectal carcinoma cells demonstrated >70% loading with macromolecular drugs, and only 5% of dead cells, but 10% cells expressing EGFP 24 h after sonication (Larina et al. 2005). Shock waves have been shown to increase the cellular uptake of both small (i.e., adriamycin and fluorescein) and macromolecules (i.e., fluorescein-labeled dextrans, ribosome-inactivating proteins gelonin, and saporin). Utilizing 1-MHz ultrasound at a spatial average peak positive pressure of 0.41 MPa (~10 W/cm²), a better transfection of 50% in the living cells could be achieved, which is similar to the outcome of lipofection. Therefore, ultrasound-induced gene delivery may be more or less suitable to different cell types due to the natural variation, and further optimization of exposure conditions may improve the transfection.

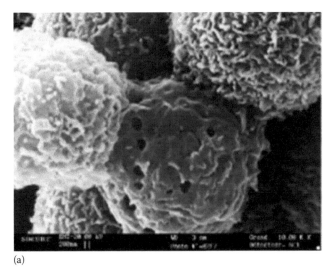

(a)

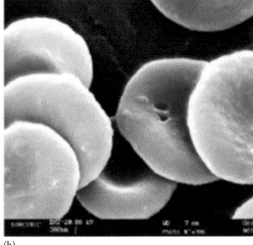

(b)

FIGURE 14.16 Representative pores at (a) MAT B III and (b) red blood cells after sonication ($f = 2.25$ MHz, $p^- = 570$ kPa) in the presence of ultrasound contrast agent (25 particles/cell for MAT B III and 1.2 particles/cell for red blood cells) by scanning electron microscopy at a magnification of 10,000. (Reprinted from *J. Control. Rel.*, 104, Mehier-Humbert, S., Bettinger, T., Yan, F., and Guy, R.H., Plasma membrane poration induced by ultrasound exposure: Implication for drug delivery, 213–222, Copyright 2005, with permission from Elsevier.)

14.4.4 In Vivo Studies

Various small animal studies (i.e., rats and mice) *in vivo* with a few studies involving larger animal models such as dogs have demonstrated the feasibility, versatility, and advantages of sonoporation by combining appropriate vectors (i.e., pDNA, viral vectors, RNA interference, and recombinant proteins) with ultrasound microbubble destruction (1–2 MHz, in pulsed mode, with a duty cycle of 10%–50%, and MI of 1.2–2) for enhancing gene delivery (Mayer and Bekeredjian 2008). Higher MIs or acoustic pressures or low frequency have been shown to correlate with higher levels of acoustic cavitation or bubble destruction and consequently higher transfection rates. Loading of the pDNA on microbubbles can usually be achieved by incubating the DNA with the microbubbles or adding the DNA during their assembly. Because of the confinement and thus the action range of the microbubbles in the vascular space, as well as the lack of active pathways for the pDNA to enter cells beyond the endothelial layer, ultrasound-mediated pDNA delivery is limited. Cardiovascular applications are generally popular in *in vivo* ultrasound gene delivery. Homogeneous transfection using an adenovirus expressing β-galactosidase attached to albumin microbubbles was found in complete rat hearts, which was five times higher than the virus being administered after bubble destruction, which demonstrates the transient nature of the sonoporation-induced permeability (Tsunoda et al. 2005). Co-administration of Optison™ and pDNA for the anticancer gene p53 also provided superior expression in a rat carotid model (Bekeredjian et al. 2005). Similar beneficial results were found using TNF-α siRNA together with albumin microbubbles in a rat myocardial infarction model (Erikson et al. 2003). However, challenges have also been identified from these animal studies. Even though local cardiac transfection can be increased by targeted sonoporation technology, many viral vectors may be released into the systematic circulation and co-transfect the liver or other areas of the body. This could be potentially overcome by using cardiac-specific promoters and selecting cardiotropic surface domains in the design of vectors. Another potential side effect is hemorrhage during ultrasonic irradiation (Figure 14.17).

However, sonoporation has not yet been used in the clinic despite considerable momentum and promising progress both *in vitro* and *in vivo* demonstrating its potential for a wide variety of medical applications. A number of problems need to be solved for further development, including optimizing the sonoporation parameters and biochemical conditions for high delivery efficiency and consistent outcome, correlating *in situ* results with those *in vivo*, and understanding and predicting the causes of any downstream cellular and organ-level bioeffects. The relatively large size of UCAs (micrometer or submicrometer, optimal for MHz ultrasound activation) restricts their application in the vascular space to the endothelial cells, rather extravasating beyond the capillaries. In addition, the microbubble concentration depends on a number of parameters in a physiological environment, such as the local blood

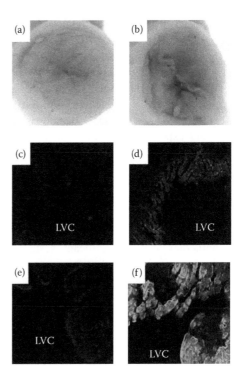

FIGURE 14.17 Expression of transgene product in cardiomyocytes at the subendocardial layer and anterior-septal wall of mice given an intra-LV injection with pGEG.b/BR14 in (a, c, e) control and (b, d, f) sonicated region at 2.0 W, 50% duty cycle 4 days after the transfection shown by stereomicroscopic images of X-gal-stained cross sections (a,b: 4×) and fluorescence microscopic images of cryosections in left ventricular cavity (LVC) (c,d: 40×, e,f: 100×). (Reprinted from *Biochem. Biophys. Res. Commun.*, 336, Tsunoda, S., Mazda, O., Oda, Y., Iida, Y., Akabame, S., Kishida, T., Shin-Ya, M., Asada, H., Gojo, S., and Imanishi, J., Sonoporation using microbubble BR14 promotes pDNA/siRNA transduction to murine heart, 118–127, Copyright 2005, with permission from Elsevier.)

flow and stability of the lipid shell. The low concentration of microbubbles, the tradeoff between cell death and delivery, and low transfection efficiency could limit the use of sonoporation, especially if a high percentage of affected cells is necessary for a specific application, such as cardiovascular diseases. Sonoporation also has some side effects associated with bubble cavitation. Ultrasound exposure *in vivo* induces cell death, local hemorrhages, and possibly inflammation, resulting in an adverse effect on capillary integrity.

To date, the majority of studies on sonoporation have utilized systemic co-injection of microbubbles and agents (pDNA, drug carriers, or imaging markers) followed by targeted ultrasound exposures. Multifunctional vehicles that can carry a drug payload, be activated by ultrasound to locally release a drug, and interact with ultrasound effectively to increase cell membrane permeability have been tested recently. The carriers can also be fabricated to serve as imaging agents for ultrasound, positron emission tomography (PET), magnetic resonance imaging (MRI), or optical imaging. In addition, they could also possess ligands to bind specific receptors on cell surfaces to achieve more desirable biodistribution at the targeted tissue site or cells. Such a development provides both a challenge and an opportunity for interdisciplinary efforts and innovation in sonoporation.

Modifying cellular structure through chemical means before sonication may alter cellular integrity against the mechanical effects of ultrasound. Lidocaine, which can increase cell membrane fluidity, was shown to alter the bioeffects of ultrasound, such as enhancing sonotransfection. Ultrasound exposures may alter the cellular integrity of cells, and the ultrastructural effects observed with TEM include mitochondrial swelling, chromatin condensation, and destruction of the plasma membrane.

14.5 TARGETED DRUG DELIVERY FOR CANCER

Since chemotherapy causes significant systemic toxicity, improvement of targeted drug delivery has a great impact in clinics. Low-intensity ultrasound was found to alter the cell membrane as minor disruption of cell surface and disappearance of microvilli with SEM, thus resulting in anticancer drug uptake into carcinoma cells, enhanced intracellular delivery after intraperitoneal or intratumoral administration, and potentiated cytotoxicity (Kennedy 2005). Cell death after treatment was also shown to associate with apoptosis (Yu et al. 2001, Larkin et al. 2008). In Yoshida sarcoma-bearing Donryu rats, combining ultrasound exposures (1.92 MHz) with daunomycin (1 mg/kg) or adriamycin (10 mg/kg) resulted in enhanced tumor growth delay, which prolonged the lives of animals (Yumita et al. 1987). Daunomycine alone had no significant effect, while adriamycin had only a minimal effect that was less than that of ultrasound alone. Ultrasound (1.733 MHz, 2 W/cm^2) in combination with adriamycin (20 mg/kg) resulted in a significant decrease in the doubling time of both fibrosarcoma RIF-1 and melanoma B-16 tumor cells when compared to the untreated controls, adriamycin alone, ultrasound alone, and adriamycin combined with hyperthermia (41°C) (Saad and Hahn 1992). Nonthermal effects of ultrasound may even be combined with a hyperthermia source to enhance antitumor effects, such as thiotepa (2 mg/kg) or cyclophosphamide (50 mg/kg) in transitional cell TCT-4909 tumors in Fischer 344 rats (Longo et al. 1983); cyclophosphamide (50 mg/kg), melphalan (4–6 mg/kg), or procarbazine (100 mg/kg) in mammary duct cell carcinoma and lung carcinoma in mice; methotrexate in murine transitional cell carcinomas; and cisplatin (8 mg/kg) in tumor-bearing C3H mice (Yuh et al. 2005). Heat-sensitive liposomes with encapsulated chemotherapeutic drugs with prolonged intravascular circulation time have already been developed and investigated. Because the permeability of malignant tumor vessels are high (neo-angiogenesis), liposomes will accumulate in the interstitial space in the tumor (Allen 1997). A reduction in the size of Morris hepatoma 3924A tumors up to 75% and improved survival in Harlan Sprague–Dawley rats were found by combining adriamycin with HIFU ablation (Yang et al. 1991), where the concentration of energy allowed temperatures to be reached in order to cause the destruction of tissue by the process of coagulative necrosis. Meanwhile, HIFU-induced hyperthermia and bubble cavitation will enhance both extravasation and drug release (Kong et al. 2001, Dittmar et al. 2005). Mean DOX concentration in tumors treated with HIFU pulses was significantly higher (124%) than that in control group. Extravasation of dextran-fluorescein isothiocyanate was observed in the vasculature of HIFU-treated tumors but not in that of untreated tumors (Yuh et al. 2005). However, a recent study of delivery of liposome-encapsulated DOX by HIFU in a mouse breast cancer model did not achieve the expected results (Frenkel et al. 2006).

14.5.1 USE OF MICROBUBBLE

Addition of microbubbles in the ultrasound-mediated drug delivery could decrease the threshold of bubble cavitation and transiently increase capillary permeability in tumor cells (Bekeredjian et al. 2007). Early studies of drug delivery involved the co-administration of a drug and microbubbles followed by sonication. Then drug-loaded vehicles were investigated for local release and cell uptake simultaneously only in the sonication area. Studies demonstrated that 20 kHz ultrasound irradiation in combination with the nanoparticle and chemotherapeutic agent 5-fluorouracil (FU) injections significantly suppressed tumor growth, and even completely eliminated the tumor 2 months later, while the control volume increased and the tumor regrew from the peripheral points of the irradiated tumors ($p < 0.05$). Histological studies of H&E stained sections of control tumor revealed viable tumor cells with well-defined nuclei. Treatment with 5-FU alone produced small necrotic regions in the tumor, most commonly near blood vessels. Ultrasound irradiation, in combination with 5-FU and polystyrene nanoparticle injection (100–280 nm in diameter and concentration up to 0.2% w/w), resulted in dramatic tissue necrosis that was noted throughout the tumor (Larina et al. 2005).

Optimization of ultrasound frequency, intensity, duty cycle, time of irradiation, and concentration of Optison® could lead to 73.5% ± 3.3%, 72.7% ± 0.9%, and 62.7% ± 2.1% delivery of 10, 70, and 2000 kDa macromolecules in the MCF7 cell line, respectively, and 36.7% ± 4.9% of cell transfection, while dead cell count was only 13.5% ± 1.6% (Larina et al. 2005). These results suggest that optimized treatment parameters provide efficient drug and gene delivery to cancer cells and can be utilized in further *in vivo* experiments (see Figure 14.18).

Bubble cavitation-induced convection can transport drugs or particles over several tens of micrometers from the vessel wall. The change in capillary permeation is strongly dependent on the cavitation dose. An acoustic pressure of 0.75 MPa at 1 MHz can produce capillary rupture in the intact rat muscle microcirculation with high microbubble concentration, and the pressure required increases proportionally to the driving frequency. Combining radiation force with destructive ultrasound pulses has increased the deposition of oil on cell membranes *in vitro* by 10-fold in comparison to ultrasound alone (Ferrara 2008).

Hemolysis, microvascular leakages, capillary ruptures, petechial hemorrhages, mild elevations of troponin-T in blood, cardiomyocyte injury, inflammatory cell infiltrations, and

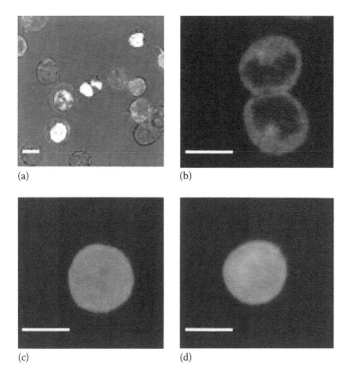

(a)   (b)

(c)   (d)

FIGURE 14.18 Confocal microscopy of MCF7 cells loaded with FITC-dextran by sonication (3 MHz, 20% duty cycle, 3 W/cm², 1 min with 200 μL of Optison per 1 mL sample). (a) Red staining (propidium iodide) marks nuclei of dead or dying cells, while green indicates distribution of 70 kDa FITC-dextran. Single cells loaded with (b) 2000 kDa, (c) 70 kDa, and (d) 10 kDa FITC dextran. Bars indicate 10 μm. (From Larina, I.V. et al., *Anticancer. Res.*, 25, 149, 2005.)

premature ventricular contractions were also observed in both *in vitro* and *in vivo* experiments after the sonication with UCAs (Ng and Liu 2002). The extent of these bioeffects is influenced by several factors, such as the concentration of UCAs, the drug-delivery method (intra-arterial vs. intravenous), the characteristics of acoustic field (i.e., frequency, pressure, beam size, duty cycle, and exposure time), properties of the targeted tissue, and the ultrasound imaging mode (intermittent).

14.5.2 Use of Nanoemulsion

An ovarian carcinoma tumor disappeared completely after four sonications with systemic injections of the nanodroplet-encapsulated PTX, nbGEN (20 mg/kg as PTX), in 2 weeks. PTX alone was tightly retained by the nanodroplet carrier *in vivo*, which avoided its unintentional release to normal tissues. Thus, the unsonicated tumor grew at the same rate as control. A similar phenomenon was also found in the pancreatic cancer model. It is important to note that the number of metastatic sites was substantially lower in the treated groups, indicating the potential of ultrasound-induced immune response, whose mechanism is unclear and under investigation (Rapoport et al. 2009).

14.5.3 Use of Liposome

Liposomes have a long half-life time in the circulation, resulting in a high-resolution PET image of the vascular structure (Figure 14.19). The accumulation of liposomes with a vascular

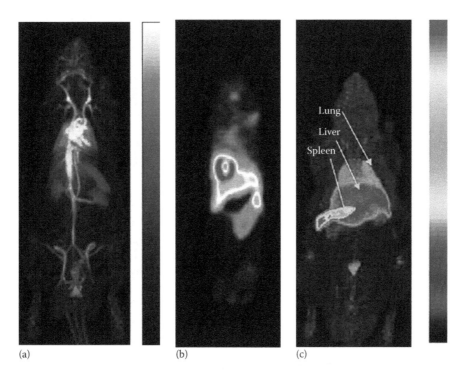

(a)   (b)   (c)

FIGURE 14.19 Positron emission tomography (PET) 90 min maximum intensity projection images of (a) long-circulating liposomes, (b) short heart-targeted peptide (Cys-Arg-Pro-Pro-Arg)–coated liposomes, and (c) microbubbles in a rat model. Particles were radiolabeled by incorporating [¹⁸F]FDP into the particle shell. (Reprinted with permission from Ferrara, K.W., Borden, M.A., and Zhang, H., Lipid-shelled vehicles: Engineering for ultrasound molecular imaging and drug delivery, *Acc. Chem. Res.*, 42, 881–892, Copyright 2009, American Chemical Society.)

targeting ligand on the shell at the heart endothelium can occur quickly (i.e., within 100 s in the mouse heart). In comparison, lipid-shelled microbubbles can circulate for a few minutes, with the shell accumulating in the liver and spleen and the gas core exhaled through the lungs (Tartis et al. 2008).

Acoustically active liposomes can have drug encapsulation efficiency as high as 15%, and drug is released by 1 MHz ultrasound at 2 W/cm^2 for 10 s. The sensitivity of liposomes to ultrasound stimulation could be increased by the inclusion of 4% diheptanolyphosphatidylcholine (DHPC), which has a high correlation with the loss of the gaseous core in the acoustic field. Overall, the current encapsulation and triggered release techniques in liposomes have a high efficiency for great potential in drug delivery.

Hyperthermia is well known for producing antitumor effects on its own and in combination with other modalities. For example, greater antitumor effects in Lewis lung carcinoma tumor-bearing mice were observed when interleukin-2 therapy was administered with local hyperthermia. Ultrasound can be used to provide hyperthermia and to enhance DOX delivery and consequent antitumor effects when combining heat-sensitive liposomes and nondestructive pulsed HIFU exposures in a murine adenocarcinoma model (Dromi et al. 2007). Initially, there was no release of DOX from either LTSLs or non-thermal sensitive liposomes (NTSLs). After 2 min of incubation, the LTSLs began releasing DOX at a temperature of 39°C (~35% of payload), more at 42°C (~50%), and approached 100% after 12 min. In contrast, the NTSLs did not release any detectable levels of encapsulated DOX at 42°C after 12 min. When LTSLs were exposed to pulsed HIFU, a three- to four-fold increase in concentration of DOX in a more rapid manner was found *in vivo* (Dromi et al. 2007).

14.5.4　Use of Micelles

In the first clinical trial of a micellar drug in Japan, DOX was chemically conjugated to PEG–P(Asp) chains and protected from being digested by erythrocytes. This micelle manifested a higher DOX concentration in the solid tumor and exhibited a significant decrease in cardiotoxicity at the DOX dose of 100 mg/kg (Kataoka et al. 2000, Rapoport 2007). In rats receiving DOX-encapsulated micelles, the sonicated tumor grew less than the bilateral tumors, despite large variation among each group and incomplete tumor regression during the time investigated. In comparison, the tumor volume increased exponentially over time in the negative control group. The slowed tumor growth was due to either increased DOX concentration in the vicinity of the tumor or to the interaction between the ultrasound, the tumor, and DOX leached from the micelle. Ultrasound promotes the extravasation into the tumor capillaries of any Plurogel drug carriers. The increased permeability of angiogenic vessels and the quantity of DOX from stable micelles improved hepatic colorectal metastases in a mouse model. If defects occurred in the cell membrane, ultrasound could enhance the drug uptake to the tumor, irrespective of whether being released from a nearby Plurogel by sonication or having previously diffused out of the Plurogel.

Because extensive cell lysis can cause a massive release of lysosomal enzymes and acute inflammation, there should be a compromise between the enhanced drug uptake and cell lysis for the optimal outcome.

Drug release from micelles could be due to either the diffusion out of micelles or micelle perturbation/degradation by ultrasound. The released drug was quickly re-encapsulated in the interval between sonication pulses so as to minimize the side effects to normal tissues (Figure 14.20). At pulse durations longer than the threshold, up to 10% release of DOX is close to that in the continuous wave (CW) mode (Husseini et al. 2000). Pulsed wave (PW) mode is preferable to CW mode by virtue of minimizing potential overheating in the clinical practice. Longer pulses and shorter intervals can keep a high concentration of releasing drug consistently throughout the sonication. The drug release from Pluronic micelles, which have comparable characteristic times of the intracellular drug uptake, increases with the longer pulse duration, but not the pulse interval. The drug re-encapsulation rate does not depend on the pulse duration. If the intervals are long enough, complete re-encapsulation of the released drug occurs.

The optimal stability, low toxicity, and long half-life time of encapsulated DOX were achieved by mixing Pluronic P-105 and polyethylene oxide (PEO)-diacylphospholipid. In addition to the increased intracellular uptake, a uniform distribution of the micelles and drug in the tumor volume was found after the sonication via the EPR effect, without molecular targeting (see Figure 14.21). For the stabilized Pluronic micelles, the optimal sonication was 4–8 h after the injection (Munshi et al. 1997). Despite excellent performance *in vitro*, the Plurogels induced significant cell dehydration *in vivo*.

Multidrug resistance (MDR) impedes successful chemotherapy of metastatic diseases. One of the well-understood mechanisms of MDR is overexpression of P-glycoprotein (P-gp), which confers resistance to tumor cells by extruding many structurally and functionally unrelated hydrophobic anticancer drugs using the energy of ATP hydrolysis. Most of the P-gp modulating agents, such as verapamil, are cytotoxic and lack specificity due to the abundance of P-gp. The enhancement of intracellular uptake and cytotoxicity of DOX, as well as inhibition of drug efflux produced by ultrasound hyperthermia, were far better than that produced by the P-gp modulator. Simultaneously, Pluronic micelles also shielded the drug uptake. These two processes compete with each other, depending on the micelle concentration and the micelle–drug interaction. Use of Pluronic F-128 at concentration up to 1 wt.% has obtained clinical approval from the Federal Drug Administration (FDA). Comparative studies of chemotherapy with and without the use of ultrasound (20 and 70 kHz) have been carried out in mice bearing highly MDR human colon KM20 tumors (Larina et al. 2005). Sonication alone did not produce tumor regression or affect the tumor growth rate significantly. Without the anticancer drug, sonication in combination with the nanoparticles resulted in a temporal decrease of tumor volume. However, the tumor regression occurred several days later, suggesting that it was due to mechanical damage to tumor blood vessels by bubble

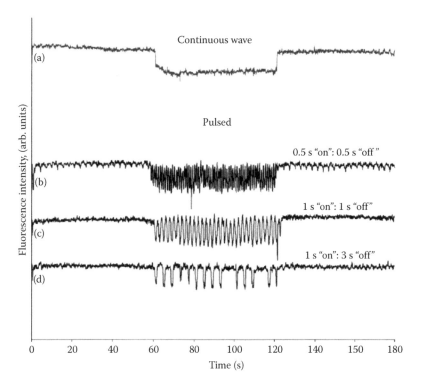

FIGURE 14.20 Profile of doxorubicin (DOX) release profiles from 10% Pluronic micelles under continuous wave (CW) or pulsed 20 kHz ultrasound at an acoustic intensity of 0.058 W/cm^2. The decrease of DOX fluorescence intensity was used to indicate the concentration of DOX in the aqueous medium. (Reprinted from *J. Control. Rel.*, 69, Husseini, G.A., Myrup, G., Pitt, W.G., Christensen, D.A., and Rapoport, N., Factors affecting acoustically triggered release of drugs from polymeric micelles, 43–52, Copyright 2000, with permission from Elsevier.)

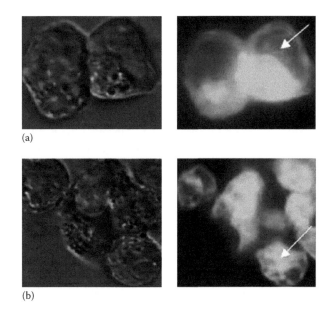

FIGURE 14.21 Subcellular trafficking of the Ruboxil (Rb). (a) No penetration into the cell nuclei (indicated by arrow). (b) Effective accumulation in 1% Pluronic P-105 micelles in the cell nuclei of the multidrug-resistant ovarian carcinoma cells. (Rapoport, N., Marin, A., Luo, Y., Prestwich, G., and Muniruzzaman, M.: Intracellular uptake and trafficking of pluronic micelles in drug-sensitive and MDR cells: Effect on the intracellular drug localization. *J. Pharm. Sci.*, 2002. 91. 157–170. Copyright Wiley-VCH Verlag GmbH & Co. KGaA. Reproduced with permission.)

cavitation (SC and IC). There was a statistical difference between sonicated and control tumors ($p < 0.05$) for up to 2 months in mice. No metastatic tumors were observed in the internal organs.

Drug release from micelles decreases with driving frequency (i.e., from 20 kHz to 3 MHz) and increases linearly with acoustic intensity for the stabilized Pluronic micelles. However, there is no acoustic intensity threshold in drug release from micelles, which suggests that IC is not the sole mechanism involved. Decreased drug efflux and increased nuclear transport resulted in a dramatic sensitization of the MDR cells in the presence of Pluronic micelles, more than that on noncancerous cells. Despite the decreased intracellular drug uptake by Pluronic micelles due to the induced deactivation of drug efflux pumps by the decreased energy in the MDR cells, the high growth inhibition was almost the same for micellar-encapsulated and the free drug, which may be due to the cytostatic action of Pluronic micelles overcoming the cytotoxic action of the drug. Such an inhibition effect on MDR cell growth was higher than that on drug-sensitive cells. The presence of Pluronic unimers rather than that of micelles, as the absence of drugs could enhance both cytostatic and cytotoxic actions. All considered, localized and controlled drug delivery to drug-sensitive and MDR tumors by the synergy of drug, micelle, and ultrasound techniques that can effectively deliver drugs is feasible and will be further evaluated in the coming *in vivo* experiments.

14.6 GENE THERAPY

Genes are specific sequences of bases that contain the heredity information of every living cell and instructions for producing specific proteins. In certain pathological conditions (e.g., genetic disorders), a particular gene may be altered or absent. Gene therapy is a treatment method that intends to replace a malfunctioning gene or provide a missing gene by delivering exogenous genes to the treated cells for both inherited and acquired diseases, such as severe combined immunodeficiency disease (SCID), cystic fibrosis, and some forms of cancer. Alternatively, the delivered genes may be connected with some targeting molecules in order to interfere with the cell function, such as production of angiogenesis inhibitors in cancer treatment to obstruct, degrade, and finally destroy the tumor. This concept was introduced to modern medicine a few decades ago. Many diseases have a genetic basis to be treated by corrective genes, and many medicinal foreign proteins could be introduced by reprogramming the cells to produce curative drugs *in situ*. The genetic materials could be dissected, recombined, and packaged into artificial vectors in several different ways. The major technical problem with gene therapy is to provide a method to transfer the large charged molecules into the nucleus of the treated cells without damaging the cells. The simplest form may be the bacteria-derived plasmid, which is an extrachromosomal DNA molecule and not connected to the cell's DNA for autonomous replication. Plasmids are usually adjusted in size and in the form of a circle, coil, or as a package. Packaging can be accomplished by the condensation of the pDNA by various molecules such as lipids or polylysine. Large molecules of DNA contain a specific gene or genes, plus promoter sequences to yield gene expression by the production of the coded protein. The functioning DNA protein can be expressed for some time after reaching the interior of the cell. Some cationic liposomes can be used to deliver DNA into the cells by facilitating passage through the cell membrane. For stable transfection, the gene is actually incorporated into the genome, and thus becomes a permanent feature of the genetic makeup of the cells, which is ideal for inserting missing genes but difficult to achieve safely. In contrast, transient transfection is a simpler and safer means of gene therapy, for which the therapeutic gene is not integrated into the chromosomes, and expression declines over days or weeks (Figure 14.22).

One popular method is to use a virus as a vector to penetrate the cell membrane and insert the gene or plasmid into the cells. Viruses are modified to program replication, include the therapeutic genetic material, and moderate their toxicity. Viral-mediated transfection allows high transfection rates in comparison to DNA alone. Adenoviruses and retroviruses are used in gene therapy of cardiovascular disease and cancer therapy, respectively. Although numerous animal studies and even some human studies have been carried out with some success, severe adverse effects associated with this technique were also reported, such as viral mutations, the nonspecific action of some vectors, and severe immunological reactions (i.e., inflammatory response), particularly for adenoviruses.

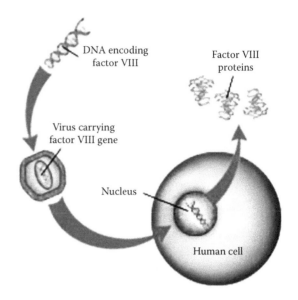

FIGURE 14.22 Schematic diagram of gene therapy.

The stable transfection strategies also have the potential for induction of new mutations for a new disease when the genetic material is randomly inserted into the genome. Therefore, the FDA has not approved its use in clinics.

There are several potential nonviral alternatives of gene therapy, such as lipofection, electroporation, particle bombardment (i.e., gene gun), and sonoporation. In lipofection, which is one of the most widely used nonviral approaches, cationic lipids encapsulate the negatively charged DNA and allow DNA transfer through the cell membrane with high transfection rates but minimal cellular toxicity, particularly *in vitro*. However, lipofection has similar problems as the viral vectors, such as poor control of spatial delivery to the desired tissue. Electroporation utilizes high-voltage electric fields to open pores in cell membranes, allowing the transfer of DNA to the interior. Although it affords some degree of spatial targeting, placing electrodes *in vivo* is challenging and invasive. A straightforward modality involves the bombardment of tissue by particles to inject foreign DNA into cells, which provides highly accurate delivery to targeted tissue but is limited to the applications on superficial targets, such as skin cancer (Figure 14.23).

One safer alternative to viral transfection is ultrasonic radiation because of the nonthermal bioeffects in sonoporation. Damage to the living cell membrane through certain amount of sonication would result in high permeability and possibility of large molecular transportation to the cytoplasm and even the nucleus. If the acoustic intensity is not too high, the cells may reverse the damage and resume their normal activity within 24 h after sonication. If the molecule is an active DNA vector, then this phenomenon can subsequently modify the gene expression enhancement and the DNA vectors would have created a new therapeutic regimen of potentially wide applicability. *In vitro* results for ultrasound-mediated gene transfer appear very promising but have proven difficult to translate *in vivo*. Minimal cavitation activity *in vivo* may be due to a dearth of cavitation nucleation sites. However, sonoporation

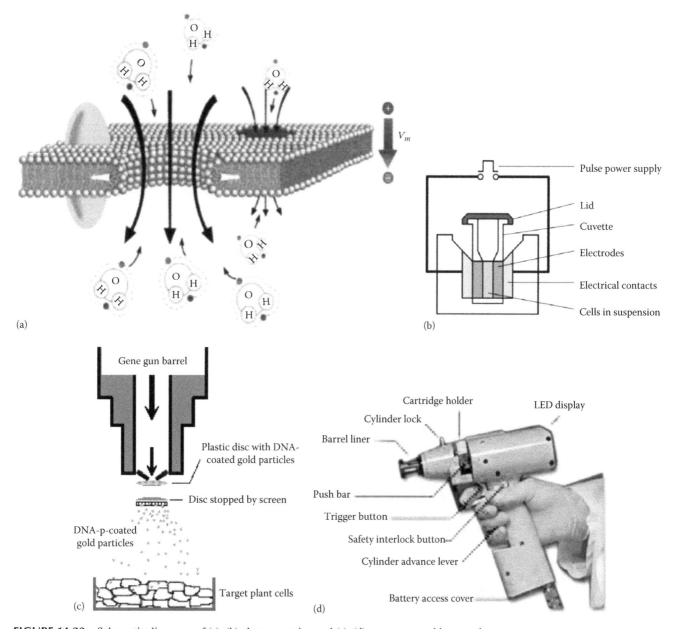

FIGURE 14.23 Schematic diagrams of (a), (b) electroporation and (c), (d) gene gun used in gene therapy.

and DNA delivery have been demonstrated in several different tissues by augmenting DNA-laden microbubble contrast agents. In order to verify the success of gene transfection, fluorescent molecules, which produce light of a different color under special illumination, are usually attached to the therapeutic genes. By using proper filtering, this fluorescence in the cells can be observed through a the microscope. The intensity and duration of the sonication affect the percentage of the successful penetration. Furthermore, it was also shown that the use of contrast materials augments the transfection efficiency (Lawrie et al. 2000). The potential of ultrasound-mediated gene therapy for cancer (Li et al. 2009, Suzuki et al. 2010), myocardial infarction (Khan et al. 2003, Fujii et al. 2009), restenosis (Lawrie et al. 2000), transplant rejection (Azuma et al. 2003), renal fibrosis (Lan et al. 2003), and stem cell stimulation for dental wound repair (Branski et al. 2009)

have been investigated. Significant research has been made on the outcomes of this method, including the most appropriate ultrasound modes, nucleation agents, gene vectors, and disease targets, and prospective results point to the clinical application of this emerging therapeutic ultrasound technology (Newman et al. 2001) (Figure 14.24).

Microbubbles (e.g., Levovist®, Albunex®, and Optison®) with ultrasound can give up to sixfold increase in the beta-Gal gene transfection to vascular smooth muscle cells (VSMCs). If adenovirus genes are loaded on the lipid shell of microbubbles, a 10-fold increase could be achieved in a mouse heart model. Delivery to a specific site can be aided using the targeted microbubbles. Bubble cavitation (both SC and IC) leads to a fusion between viral the vectors and cell membranes rather than a forced deposition of intact virus particles inside the cytoplasm. In this way, the virus can unpack

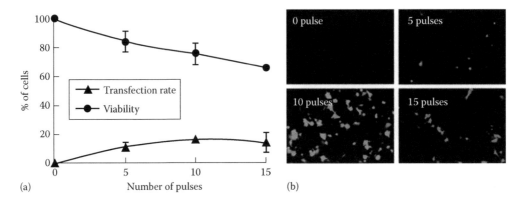

FIGURE 14.24 (a) Cell viability, transfection rate, and (b) green fluorescent protein (GFP) in HeLa cells using 1 MHz ultrasound, 0.5 Hz pulse repetition frequency, 50% duty factor, and at different number of pulses. (Feril, L.B., Ogawa, R., Tachibana, K., and Kondo, T.: Optimized ultrasound-mediated gene transfection in cancer cells. *Cancer Sci.*, 2006. 97. 1111–1114. Copyright Wiley-VCH Verlag GmbH & Co. KGaA. Reproduced with permission.)

and easily proceed to the nucleus. Also remarkable was the rather uniform gene transfer in the cardiac tissue, probably due to the ability of adeno-associated virus (AAV) serotypes to cross blood vessel barriers and spread through the extracellular matrix (Lentacker et al. 2009). Moreover, adverse immune responses can be drastically reduced as the virus is shielded from the immune system. Fifty percent transfection efficiency was achieved, which was comparable to that achieved using lipofection. As a physical method, ultrasound does not have the limitations in pDNA uptake into cells by lipofection, such as the net charge of the cationic lipids, DNA complexes and concentrations, pH, and concentration of electrolytes. The optimization of different cationic lipids used for each lipofection experiment has led to inconsistent results (Ng and Liu 2002). Although relatively large amounts of pDNA are needed for competitive transfection rate in sonication, the *in vivo* site specificity generally outweighs this small disadvantage (Greenleaf et al. 1998).

In addition, the use of magnetic resonance–guided HIFU (MRgHIFU)–induced local hyperthermia in gene therapy also holds much promise, similar to its application for heat-activated chemotherapy. Temperature-mediated control of gene expression has shown its feasibility by combining MRgHIFU with thermosensitive promoters to regulate local gene transcription and subsequent expression in preclinical studies (Moonen 2007). Heat shock protein promoters (e.g., HSP70) have been widely employed in gene therapy strategies because they are both heat inducible and efficient (i.e., several thousand fold in response to hyperthermia) in initiating and regulating transcription of therapeutic genes both *in vitro* in C6 cells and *in vivo* in muscle. Spatial and temporal control of therapeutic transgene expression in the targeted tissue is among the key challenges in such gene therapy technique since the promoter activity is highly temperature dependent in the range of 40°C–45°C. Mesenchymal stem cells obtained from the bone marrow of rats were transfected with hsp-luc to express the bioluminescent luciferase gene under the control of an HSP70 promoter. The transformed mesenchymal stem cells were injected in the left renal artery, and transgenic expression

was induced by MRI-controlled HIFU hyperthermia (Braiden et al. 2000). Local induction of the luciferase gene as *in vivo* bioluminescence by MR-guided HIFU in a transgenic mouse (Deckers et al. 2009). Transgene expression controlled by ultrasonic heating using a minimally invasive approach was investigated in the prostate. Ultrasound imaging was used to guide the injection of an adenoviral vector, which contained a reporter gene for luciferase under the control of the HSP70B promoter, into both lobes of the prostate gland in three beagle dogs. Afterwards, the prostate lobes were heated using an ultrasound transducer under the guidance of an MRI system. High levels of luciferase expression were observed only in the sonicated areas (Silcox et al. 2003) (Figure 14.25).

14.7 SONOPHORESIS

14.7.1 Transdermal Drug Delivery

Oral administration (i.e., by swallowing pills) and percutaneous hypodermic injection are routine means of drug administration to the circulation system. However, they are associated with some shortcomings, such as premature drug metabolization by the first-pass effect of the liver and gut wall, the gastric degradation of the drugs, unstable plasma level, difficulty in immediate termination, painfulness, dangerous medical waste, the risk of disease transmission by needle reuse, and associated pain (Prausnitz and Langer 2008). As soon as the drug reaches the blood, its concentration peaks rapidly and then starts to decay exponentially as a result of degradation or physiological process. Furthermore, without targeting, the drug will be carried by the cardiovascular system to almost every organ in the body mostly to irrelevant regions. Hence, drug delivery efficiencies in low and high concentrations are needed. Human skin has a large surface area for absorption (approximately 2 m²), relatively low proteolytic activity, and is a most readily accessible site for drug delivery across the skin and into systemic circulation. Transdermal drug delivery (TDD), that is, transporting therapeutic agents through the epidermis, is noninvasive, and the systems can

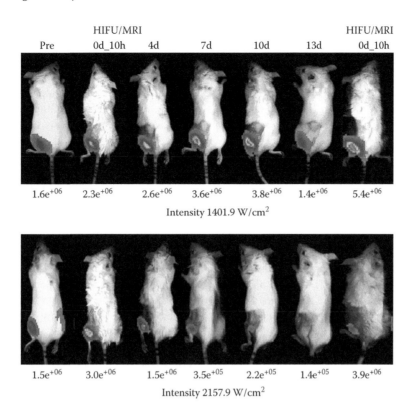

HIFU/MRI   HIFU/MRI

Pre   0d_10h   4d   7d   10d   13d   0d_10h

$1.6e^{+06}$ $2.3e^{+06}$ $2.6e^{+06}$ $3.6e^{+06}$ $3.8e^{+06}$ $1.4e^{+06}$ $5.4e^{+06}$

Intensity 1401.9 W/cm^2

$1.5e^{+06}$ $3.0e^{+06}$ $1.5e^{+06}$ $3.5e^{+05}$ $2.2e^{+05}$ $1.4e^{+05}$ $3.9e^{+06}$

Intensity 2157.9 W/cm^2

FIGURE 14.25 Luciferase activity after the 20 s HIFU exposure at the intensity of application of 1401.9 and 2157.9 W/cm^2 with follow-up to 14 days and the second HIFU application. (Reprinted from *Ultrasonics*, 49, Hundt, W., Steinbach, S., Mayer, D., and Bednarski, M.D., Modulation of luciferase activity using high intensity focused ultrasound in combination with bioluminescence imaging, magnetic resonance imaging and histological analysis in muscle tissue, 549–557, Copyright 2009, with permission from Elsevier.)

be self-administered. The other advantages are strict control of transdermal penetration rate, long release, rapid termination, low risk of infection, less anxiety and pain, higher patient compliance, low cost, and not immunologically sensitizing, despite potential disadvantages of minor tingling, irritation, and burning. The steady permeation of TDD allows maintaining long-lasting and more consistent serum drug levels, often a goal of therapy. However, low permeability of human skin due to the stratum corneum (the uppermost layer of the skin) is the major reason for the poor molecular diffusion across mammalian epidermis. The molecular size of protein and peptide drugs precludes their passive TDD in concentrations required to elicit desirable pharmacological effects (Figure 14.26).

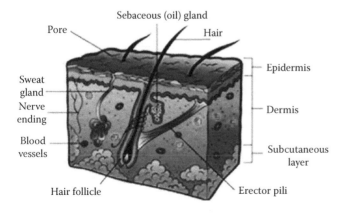

FIGURE 14.26 Anatomy of the human skin.

Thus, a variety of chemical and physical approaches have been investigated to reduce the barrier properties of the stratum corneum and to enhance transdermal transport. Three generations of development have taken place in TDD. Chemical enhancers, such as dimethyl sulfoxide (DMSO), 1-dodecylazacycloheptan-2-one, azone, surfactants, solvents and binary polar and apolar systems, can temporarily alter the barrier properties of the stratum corneum (intercellular lipids and corneocytes) (Williams and Barry 2004). However, they have not been accepted widely either because of suspected pharmacologic activity or unresolved safety questions (Barry 1987). Micro-needle arrays were designed and applied to create a physical pathway through the upper epidermis (which contains no nerves), but not deep enough to cause any pain to the receptors in the dermis (Sivamani et al. 2007). These silicon needles may have a dimension of about 150 μm in length and 80 μm in diameter. Removal of the stratum corneum barrier by controlled ablation, such as with erbium:yttrium–aluminum garnet (YAG) laser and radio frequency, has also been investigated to increase the skin permeability. The molecular size, lipophilicity, and sequence of the peptides to be delivered all played important roles in modulating the enhancement of delivery and consequent activity. There was a threefold increase of antibodies in the serum found in mouse skin using laser ablation followed by skin vaccination with a lysozyme antigen (Lee et al. 2008b). Uniform microchannels (MCs) were created by placing an alternating electrical current transferred through microelectrodes, thus ablating the cell underneath each electrode in rat and guinea pig (GP) skin and consequently forming microscopic passages in the stratum corneum and outer dermis. Creation of MCs in the outer layers of the skin enabled efficient delivery of human growth hormone, with a bioavailability of 75% (rats) or 33% (GPs) relative to subcutaneous injection (Levin et al. 2005). Iontophoresis involves the application of a small electric current (usually 0.5 mA/cm^2) to a drug reservoir placed directly on the surface of the skin. The same charge of the electrode as the solute results in an effective repulsion of solute molecules away from the electrode into and through the skin (Prausnitz et al. 1993). Electroporation involves the application of large electrical pulses (>50 V, 10 µs to 100 ms) to the skin in order to form transient pores in the stratum corneum and increase the skin permeability by several orders, which is mainly attributed to the electrophoretic movement and diffusion through the newly created aqueous pathways (Denet et al. 2004). These techniques are mostly reversible a few hours after the termination without permanent damage, except microneedles. Estradiol, fentanyl, lidocaine, and testosterone are amenable to administration by the transdermal route for the treatment of local allergies or infections.

14.7.2 Ultrasound Effects on the Skin for Facilitating Transdermal Drug Delivery

Various investigators have reported histological studies of animal skin exposed to ultrasound under various conditions to assess the effect of ultrasound on living skin cells. Histological studies of hairless rat or rabbit skin exposed to low-intensity ultrasound, such as 1 MHz and 2 W/cm^2 (Levy et al. 1989), 105 kHz and 5 kPa (Tachibana 1992), 20 kHz and 12.5–225 mW/cm^2 (Mitragotri et al. 1995), 48 kHz and 0.5 W/cm^2 (Yamashita et al. 1997), did not show damage to the epidermis and the underlying living tissues. Using SEM, it was found that the ultrasound-induced effect on murine skin was much more significant than on human skin (Yamashita et al. 1997). The outer layer in the murine stratum corneum had completely been removed, and pores were generated. In contrast, in the human skin, the absence of some of the keratinocytes around hair follicles was observed. Similar conclusions were made in another study using low-frequency and low-intensity ultrasound (20 kHz and <2.5 W/cm^2) on hairless rat and human skin (Boucaud et al. 2001). Slight and transient erythema was initially observed in hairless rats, whereas deep lesions (i.e., dermal and muscle necrosis) were observed 24 h later. In contrast, lesions could not be seen when a heat source was applied to the skin, which indicates that these delayed and deep lesions are not only attributable to the increase in temperature at the skin surface during sonication. However, no histological changes were observed in the human samples using TEM. Ultrasound can disrupt the lipid structure in the stratum corneum and thereby increase permeability utilizing either noncavitation or cavitation effects. Numerous studies have illustrated that sonophoresis is generally safe with no negative short- or long-term side effects (Figure 14.27).

The first success on sonophoresis or phonophoresis—which is defined as the movement of drugs through intact skin and into soft tissue under or following the influence of an ultrasonic perturbation—was the treatment of polyarthritis of the hand's digital joints using hydrocortisone ointment with sonication in the 1950s (Fellinger 1954). Since then, sonophoresis has been used in sports medicine as a physical therapy (Byl 1995, Mitragotri et al. 1995). In the past two decades, with the continuous development and research in sonophoresis, its clinical potential at various frequencies for different molecules has been evaluated. Typical enhancement of sonophoresis is about 10-fold, which might be sufficient for local delivery of certain drugs, such as hydrocortisone, but not for the systemic delivery of most drugs. Ultrasound pretreatment using SonoPrep® (Echo therapeutics, Inc., Franklin, MA) for 15 s followed by 5 min of 4% liposomal lidocaine cream significantly reduced patients' perception of pain from the intravenous cannulation with no adverse side effects in any of 87 participants during the follow-up period (Becker et al. 2005). Therefore, in August 2004, Sontra medical received 510(k) clearance from the FDA to market the SonoPrep ultrasonic skin permeation device and procedure tray for use with topical lidocaine. It is to be noted that extensive studies have been carried out for insulin, heparin, and tetanus toxoid vaccine in the frequency range of 0.75–3.0 MHz and an intensity range of 0–2.4 W/cm^2 (Prausnitz and Langer 2008, Polat et al. 2011).

The use of high-frequency sonophoresis (HFS, ≥0.7 MHz) was tried as early as the 1950s, while low-frequency sonophoresis (LFS, 20–100 kHz) has been investigated during the past two decades. Both HFS and LFS improved current methods of local, regional, and systemic drug delivery, or even may be suitable for vaccine delivery in the future (Langer 2000, Polat et al. 2011). The exact mechanisms are still not well understood but may include thermal, cavitation, and mechanical effects (the induction of active transport, force convection). Bubble cavitation and mechanical effect may be the underlying dominant mechanisms for LFS and HFS, respectively. Cavitation causes disorder of the intercellular lipids in the stratum corneum in LFS, resulting in significant water penetration into the disordered lipid region and the formation of aqueous channels across the keratinocytes rather than the hair follicles. Since the effects of ultrasound-induced cavitation on the skin are highly heterogeneous, regions of high permeability are localized and 80-fold more permeable and 5000-fold less electrical resistive than the surrounding skin (Kushner et al. 2004). Fluorescent-labeled hydrophilic dextran molecules (40 kDa) in hairless rat skin after LFS (41 kHz, 120 mW/cm^2, 5 min) had a penetration depth of 20 µm, as seen by confocal microscopy (Morimoto et al. 2005). Several crack-like structures were observed in the sonicated skin, which suggested a degree of structural alteration followed by convective solvent flow probably via both corneocytes and lipids of the stratum corneum. The permeable skin areas by LFS (20 kHz) were discrete (Alvarez-Román et al. 2003). In some regions, substantial delivery enhancement of the fluorophore was seen below the stratum corneum. Such localized and permeable regions become

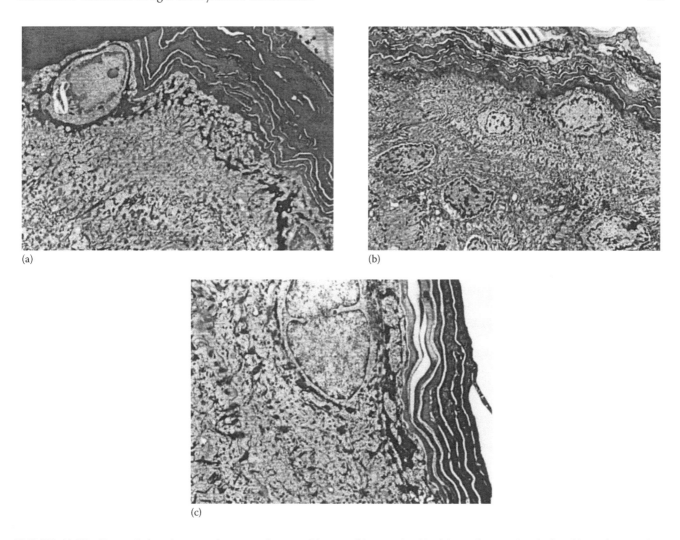

(a)

(b)

(c)

FIGURE 14.27 Transmission electron microscopy images of human skin samples (a) without ultrasound and after (b) continuous ultrasound (2.5 W/cm², 10 min) and (c) pulsed ultrasound (2.5 W/cm², 1 h, 10% duty cycle). (Reprinted from *Int. J. Pharm.*, 228, Boucaud, A., Machet, L., Arbeille, B., Machet, M., Sournac, M., Mavon, A., Patat, F., and Vaillant, L., *In vitro* study of low-frequency ultrasound-enhanced transdermal transport of fentanyl and caffeine across human and hairless rat skin, 69–77, Copyright 2001, with permission from Elsevier.)

more homogeneously distributed with increase in driving frequency of the ultrasound, whose optimum was found around 60 kHz. The heterogeneity of TDT during LFS (20 kHz, 2.4 W/cm²) was also confirmed using 20 nm quantum dots (QDs) (Paliwal et al. 2006). QDs distributed in several discrete pockets spanning about 40–80 µm in width and up to 60 µm in depth in fluorescence microscopy, while a high heterogeneity in QD distribution was shown with electron microscopy. The presence of QD-localized pockets (up to 50 nm wide and 300 nm long) was observed within the intercellular lipids and corneodesmosome junctions of the stratum corneum, and occasionally in corneocytes. Electron micrographs of untreated skin showed that sonication significantly increased the frequency of occurrence as well as the size of the scattered and nonconnected defects in the bilayers of the stratum corneum. Furthermore, sonication in the presence of sodium lauryl sulfate (SLS) induced similar but more pronounced dilatory defects in the ultrastructure of the stratum corneum. Although the areal density of defects significantly

increased in the stratum corneum after LFS, their number density did not change. Noncavitation ultrasound is found to generally enhance small lipophilic compounds in sonophoresis, while localized shock waves and high-speed liquid microjets due to inertial bubble cavitation at the stratum corneum could increases skin permeability for up to many hours without damaging deeper tissues. In contrast, tissue heating is not a big concern at the energy level currently used. The effect of low-frequency ultrasound (<100 kHz) on transdermal transport is more significant than that of high frequencies (i.e., 10 MHz) (Merino et al. 2003). LFS has been used to enhance the transport of various low-MW drugs as well as high-MW proteins (e.g., insulin, γ-interferon, and erythropoietin) across human cadaver skin *in vitro* with the enhancement of up to 1000-fold (Mitragotri et al. 1996). Pulsed LFS (20 kHz, 225 mW/cm², 100 ms pulses applied every second) delivered insulin solution (100 U/mL) to diabetic hairless rats and reduced the blood glucose level from ~400 to 200 mg/dL in 30 min (Mitragotri et al. 1995). Similar impressive effects, namely a 75% reduction in glucose levels, were found at an energy dose of 900 J/cm² (Boucaud et al. 2001).

To obtain the maximum transportation efficiency, several new drug carriers have been investigated. Hydrophilic and lipophilic drugs can be encapsulated to the lipid walls or the aqueous interior of the liposomes in diameters of about 25–10,000 nm, respectively, which is an interdisciplinary topic of great interest (Touitou et al. 1994). Ethosomes are noninvasive delivery carriers that enable drugs to reach the deep skin layers and/or the systemic circulation (Godin and Touitou 2003). If methotrexate (MTX), which is an anti-psoriatic, anti-neoplastic, and highly hydrosoluble agent with limited transdermal permeation, is loaded with ethosomal carriers in the optimal size of 143 ± 16 nm, transdermal flux can be enhanced to 57.2 ± 4.34 μg/cm²/h, while decreasing the lag time of 0.9 h across human cadaver skin to 0.9 h (Dubey et al. 2007). Microparticles and nanoparticles consisting of proteins or biodegradable and biocompatible polymers, such as polylactide and poly(lactic-*co*-glycolic) acid (PLGA), can incorporate drugs or antigens in the form of solid solutions or dispersions and have been shown to enhance the delivery of certain drugs. Biodegradable nanoparticles (150–250 nm) have been used for transdermal DNA delivery using a low-pressure gene gun into the mouse skin for enhanced gene expression (Lee et al. 2008b).

Ultrasound has been used as a brief pretreatment procedure followed by passive diffusion (Mitragotri and Kost 2004). One of the challenges is that the degree of skin permeabilization needs to be determined prior to drug placement, which can be quantified by electrical conductance measurements on the skin. Formal relationships between skin conductance (or resistance) and skin permeability have been developed *in vitro*, *in vivo*, or clinical experiments to determine the threshold. Pretreatment of the skin in rats by low-frequency ultrasound (20 kHz, 7 W/cm²) followed by the placement of insulin (500 U/mL) on the treated skin was found to increase the conductivity of the skin by about 60-fold. As a result, glucose level in rats pretreated with ultrasound decreased by about 80% within 2 h. However, no change in blood glucose levels was found when insulin was placed on untreated skin. Such an effect was also found in macromolecule transport of low-MW heparin (LMWH) in rats, as measured by monitoring the anti-Xa (aXa) activity in blood. In the ultrasound-treated animals, serum activity of aXa increased slowly for about 2 h, after which it increased rapidly before achieving a steady-state level after 4 h of about 2 U/mL in contrast to intravenous or subcutaneous administrations of the agent, which resulted in more transient changes in aXa activity. Pretreatment with low-frequency ultrasound (55 kHz, 4–14 s) could shorten the lag time for the topical analgesic cream EMLA™ (a mixture of lignocaine and prilocaine, AstraZeneca International, Wilmslow, U.K.) to induce local analgesia on 42 human subjects in about 60 min (Katz et al. 2004).

In order to facilitate sonophoresis, wearable ultrasound devices have been developed. A low-profile, light, cymbal-array flextensional transducer, which integrates two metal caps exposed to a lead–zirconate–titanate (PZT) ceramic, was

FIGURE 14.28 A two-by-two array of cymbal transducers connected in parallel and encased in URALITE® polymer. (Reprinted from *Ultrasound Med. Biol.*, 29, Smith, N.B., Lee, S., and Shung, K.K., Ultrasound-mediated transdermal *in vivo* transport of insulin with low-profile cymbal arrays, 1205–1210, Copyright 2003, with permission from Elsevier.)

proposed (Smith et al. 2003). The fundamental mode of vibration is the flexing of the end cap caused by the radial motion of the ceramic and the axial motion of the piezoelectric disk. In order to improve delivery efficiency, a rectangular cymbal design was developed to achieve a broad spatial intensity field without increasing the size of the device for the spatial-peak temporal-peak intensity (I_{SPTP}). The glucose level decreased faster on using the cymbal array in hyperglycemic rabbits. The cymbal array (2 × 2 and 3 × 3 transducers) ultrasound (37 × 37 × 7 mm³, 22 g, 20 kHz, and 100 mW/cm²) was used to deliver insulin to hyperglycemic rats, rabbits (Lee et al. 2004), and pigs (Park et al. 2007). Preliminary *in vivo* studies showed the feasibility and potential of the light-weight cymbal array transducer in human applications (Figure 14.28).

14.7.3 TRANSDERMAL IMMUNIZATION AND GENE THERAPY

Transcutaneous immunization is a technique in which vaccine antigens in solution are applied on the skin to induce an antibody response without systemic or local toxicity. The primary advantage of transcutaneous immunization is the presentation of immunogens to antigen presenting cells (APCs) within the skin, specifically Langerhans cells that are highly potent immune cells populating within the epidermis. Langerhans cells are in close proximity to the outermost layer of the skin (the stratum corneum) and represent a network of immune cells that occupies about 20% of the skin's total surface area despite composing only 1% of the epidermis' cell population. Langerhans cells initiate immune responses by acting as professional APCs, taking up and processing antigens and subsequently presenting antigenic peptides to naive T cells in the lymph nodes. Transcutaneous immunization has been shown to generate both systemic (IgG/IgM response) and mucosal immunity (IgA response), whereas conventional needle-based injections often generates only systemic immunity. It showed that ultrasound pretreatment (20 kHz, 100 J/cm^2) could induce a significant increase in IgG response and activation of Langerhans cells in the epidermis using tetanus toxoid in BALB/c mice (Tezel et al. 2005). LFS as a transcutaneous immunization adjuvant has higher delivery efficiency in comparison to subcutaneous injection (1.3 μg vs. 10 μg) and eliminates the requirement of toxin to elicit an immune response.

Gene therapy is a technique for correcting disease-causing genes in the target cell by the normal genes in a carrier molecule (vector). Topical delivery of the vector–gene complex can be used to target cells within the skin, as well as for reaching systemic circulation. There are almost 100 diseases within the skin associated with defective genes that need cutaneous gene therapy, and the most obvious candidates are the severe forms of particular genodermatoses (monogenic skin disorders), such as epidermolysis bullosa and ichthyosis. Cutaneous wounds, such as severe burns and skin wounds of diabetic origin, are also candidates, requiring the penetration of large complexes into or through the skin.

14.7.4 Analytes Monitoring

Similar to drug delivery to the skin, considerable efforts have been made in extracting substances across the skin without using a needle. Frequent monitoring and tight metabolic control of blood glucose levels, particularly glucose, can reduce microvascular complications and subsequent comorbidities in diabetic patients. Self-monitoring with finger sticks provides intermittent data at best but results in poor compliance. Implantable sensors, minimally invasive skin microporation, laser or miniaturized lancets, near-infrared spectroscopy, transdermal permeation enhancers, and reverse iontophoresis have been developed as painless and convenient methods. However, one of the challenges in noninvasive transdermal diagnostics is obtaining sufficient quantities of the analyte.

A single, short sonication (less than 2 min) was sufficient to extract glucose noninvasively across the skin of type 1 diabetic patients. It is found that the skin permeability remained high for about 15 h after sonophoresis and decreased to its normal value in 24 h. After skin permeability enhancement with a cymbal array (20 kHz, I_{SPTP} = 100 mW/cm^2, 20% duty cycle) on anesthetized Sprague–Dawley rats, an electrochemical glucose sensor was placed on the abdomen to measure the glucose concentrations through the skin (Lee et al. 2004). The good correlation of the result with that collected from the jugular vein showed the feasibility of quantifying glucose concentrations. The site-to-site variability of skin permeability after sonication in the same patient was about the same as patient-to-patient variability, which indicates the necessity of calibration between transdermal glucose flux and

blood glucose values for accurate prediction. Based on such a calibration, the mean relative error between transdermal glucose flux and blood glucose values was 17% (Lavon and Kost 2004). Blood glucose flux readings after sonophoresis (lasting about 10 s) were measured every 20 min over an 8 h period in 10 diabetic patients. Completed datasets had a correlation coefficient of 0.84 (n = 241). No patients complained of pain or irritation at any time during the study. This preliminary result demonstrated that continuous monitoring of glucose flux through ultrasonically permeable skin was accurate, safe, and feasible after single or multiple calibrations in diverse clinical settings (Chuang et al. 2004).

14.7.5 Synergy with Other Modalities

Sonophoresis could potentially enhance transport synergistically with various other means of transport enhancement, such as with chemical enhancers, iontophoresis, and electroporation (Mitragotri 2000). The synergistic effect of LFS with 1% sodium lauryl sulfate (SLS), a surfactant, had shown a 200-fold increase in skin permeability to mannitol, while the corresponding increase for ultrasound alone or SLS alone was eight- and threefold, respectively. The reason may be the alteration of the pH profile, which resulted in improved SLS lipophylic solubility, penetration, and dispersion (Mitragotri et al. 2000). HFS (1 MHz) could not only reduce the threshold voltage for effective electroporation but also increase TDT at a given electroporation voltage. The synergistic effect may be due to the enhanced penetration of the molecules by ultrasound through pores in the skin created by electroporation (Mitragotri 2000). Combining sonication and iontophoresis induces higher delivery enhancement than the contribution of each of them. In such a synergy, ultrasonic pretreatment reduces skin resistance so that a lower voltage can be used effectively, which results in lower power requirements and less skin damage.

14.8 SONOTHROMBOLYSIS

14.8.1 Dissolution of Blood Clot

Clot formation is a vital mechanism that prevents excessive blood loss from the body. However, unwanted clotting contributes significantly to the lethality of cardiovascular disease, which is the primary cause of death and disability globally. A clot within a vessel in the brain causes ischemic stroke, within the coronary artery causes myocardial infarction, and within the peripheral veins causes deep vein thrombosis (DVT). Acute ischemic stroke occurs frequently worldwide, and there are about 780,000 new stroke cases per year in the United States. Mortality due to stroke is the third leading cause of death. Approximately 70% of strokes are ischemic, and the remaining 30% are hemorrhagic. Risk factors for ischemic stroke include age, race, sex and family history, hypertension, diabetes, smoking, and atrial fibrillation. Even mild hypertension (systolic blood pressure >115 mmHg) is associated with increased risk. Diabetes increases the individual's risk for stroke by a factor of 2–6, and nearly 40% of ischemic strokes can be at least partially attributed to the development of diabetes. Smoking is a well-known modifiable risk factor for cardiovascular disease in general, with the overall relative risk of 1.9 for ischemic stroke (a larger attributable risk for younger smokers). In addition, approximately 1 million Americans experience a myocardial infarction every year. Modifiable risk factors for this disease include hypertension, smoking, high cholesterol, and diabetes. Substantial reductions in mortality of up to 71% have been achieved in the acute treatment of this disease. DVT is a common disorder in which the patient suffers from blood clots in the larger veins, typically in the lower extremities, and could be fatal if the resulting embolization of the clot flows to the lungs. In addition, the long-term presence of DVT can damage the venous structures, resulting in venous insufficiency and chronic lower extremity edema. The overall incidence of DVT is 79 per 100,000 per year. Risk factors include cancer, recent surgery or immobilization, pregnancy, the presence of clotting disorders, and the use of hormone replacement therapy or oral contraceptives.

Thrombolysis, that is, the dissolution and breaking up of a clot using thrombolytics in medicine, could limit the damage caused by the blockage of the blood vessel. Tissue plasminogen activator (tPA) is a protein manufactured by vascular endothelial cells that regulates clot breakdown in the body (as does urokinase) using recombinant biotechnology techniques. tPA is a serine protease consisting of 527–530 amino acid residues and acts by hydrolyzing the peptidic bond between residues R561 and V562 site on plasminogen, thus converting this zymogen into its active form plasmin. Although tPA exhibits some activity for plasminogen in plasma, the presence of clot and fibrin greatly increases its efficacy. It is inhibited by plasminogen activator inhibitor (PAI-1) and has a half-life *in vivo* of approximately 4 min. Recombinant tissue plasminogen activator (rt-PA) is the recombinant manufactured form of tPA, which is referred to as the endogenous form of this protein, and is the prototypical drug for plasminogen activators, converting plasminogen to plasmin *in vivo*. The plasmin cleaves the fibrin mesh of a thrombus into fibrin degradation products. Streptokinase is a 47 kD plasminogen activator consisting of 414 residues, and first forms a complex with freely circulating plasminogen in the plasma for its fibrinolytic activity with an elimination half-life of about 90 min. It is commonly used in Europe and Canada for thrombolysis in myocardial infarction. Urokinase is a plasminogen activator obtained from cultured human renal cells with a half-life of about 15 min, and was the major thrombolytic in the United States until the advent of streptokinase and rt-PA. It currently only carries the indication of use in pulmonary embolus. Reteplase (rPA) is a mutant form of tPA and represents one of the first attempts to extend the pharmacologic half-life of the drug. In clinics, rPA exhibits greater vessel patency rates in acute myocardial infarction treatment compared with rt-PA. Tenecteplase (TNK-tPA) is also a tPA mutant with increased plasma half-life, improved fibrin specificity, and reduced degree of inhibition by PAI-1, whose characteristics allow single-bolus administration for the treatment of myocardial infarction.

Currently, the only FDA-approved therapy for acute ischemic stroke is intravenous (IV) administration of rt-PA within 3 h of stroke onset. The thrombolytic efficacy of rt-PA increased monotonically as a function of rt-PA concentration up to a threshold, and higher increases in rt-PA concentration did not increase thrombolytic efficacy, which may be due to plasminogen depletion. Overall, this therapy is moderately efficacious, recanalizing only 30%–40% of occluded major intracranial arteries within the first 1–2 h after initiation of treatment, and results in a 30% greater chance of little or no disability as compared to control at 3 months follow-up. However, there is a 6.4% incidence of intracerebral hemorrhagic complications in patients receiving this rt-PA treatment, which may be due to a higher rt-PA dose in clinical thrombolysis. Therapy for myocardial infarction includes medications such as aspirin, oxygen, nitrates, beta blockers, and angiotensin-converting enzyme (ACE) inhibitors. The primary therapy is early reperfusion either using thrombolytic medications such as rt-PA or the use of percutaneous cardiac intervention (PCI). Thrombolytic treatment of acute myocardial infarction is common, but with a substantial rate of failure of up to 40%, while PCI is preferred for acute cardiogenic shock. Current treatment for DVT is primarily the prevention of pulmonary embolism of the culprit clot. Anticoagulation medications (e.g., coumadin) are administered for months, and the prothrombin time is monitored to determine clotting tendency of blood, with longer times indicating a predisposition of the patient to hemorrhage.

Because of the need for improved clinical treatments of thrombo-occlusive disease, ultrasound has been investigated either alone or in combination with thrombolytic drugs to enhance recanalization. The first method involves the use of intravascular ultrasonically driven wire tips both for the acceleration of dissolution of blood clots and for the acceleration of dissolution of clots by thrombolytic drugs (Tachibana and Tachibana 2001, Siegel and Luo 2008). Specialized tips have been designed for use in dentistry (Walmsley et al. 1988). Catheter-based sonothrombolysis can be used within a few days of thrombus formation using rt-PA. Success rates were 80% in patients with iliofemoral occlusive disease, but 5%–10% of symptomatic hemorrhage. Catheter-administered rt-PA resulted in complete clot lysis (≥90%) in 37 of 53 cases with acute DVT and an overall lysis of 91% (Parikh et al. 2008) as well as improved 6-month iliofemoral patency (65.9% vs. 47.4% using standard treatment) in 189 patients (Verhaeghe et al. 1997, Enden et al. 2012). Because standard interventional cardiology techniques using balloon angioplasty and coronary arterial stents have excellent procedural success rates in the treatment of thrombotic coronary artery, there appears to be little current clinical need for a catheter-based ultrasound thrombolysis system to treat acute myocardial infarction. However, the Interventional Management of Stroke (IMS) prospective clinical study has indicated encouraging results for the EKOS micro-infusion catheter (EKOS Corp., Bothell, WA), which provided pulsed ultrasound in combination with rt-PA infusions for the treatment of ischemic stroke. Recently, the efficacy of 2 MHz pulsed transcranial

Doppler ultrasound-enhanced thrombolysis using IV rt-PA in acute ischemic stroke patients (sonothrombolysis) exhibited greater early recanalization of the affected cerebral vessel and a trend toward improved neurologic outcome at 3 months. No clear dependence of the observed thrombolytic enhancement on ultrasound duty cycle was evident. Ultrasound without rt-PA did not significantly enhance thrombolysis (Holland et al. 2002). The combination of a galactose-based UCA with 2 MHz ultrasound and rt-PA in acute ischemic stroke patients had significantly greater 2 h recanalization rate (54%) in comparison to those with rt-PA and ultrasound (41%) and rt-PA alone (24%). There seemed an improved neurologic outcome at 24 h in the rt-PA, ultrasound, and UCA groups, but it was not significant (Molina et al. 2006). However, in the Transcranial Low-Frequency Ultrasound Mediated Thrombolysis in Brain Ischemia (TRUMBI) Study Phase II trial using 300 kHz ultrasound in acute ischemic stroke patients, sonothrombolysis showed no improvement but a greater rate of symptomatic intracerebral hemorrhage (ICH: ~35%) compared with those treated with standard IV rt-PA alone (Daffertshofer et al. 2005). In an *in vivo* study of canine myocardial infarction, rt-PA was administered with 27-kHz transcutaneous ultrasound exposure after an occlusion of the proximal left-anterior descending (LAD) artery was induced by electrocoagulation (Siegel et al. 2000). The mean thrombolysis in myocardial infarction (TIMI) flow grade in the t-PA and ultrasound group was high in comparison to the t-PA alone group (2.42 ± 1.9 vs.

0.92 ± 1.4 at 90 min and 2.58 ± 0.9 vs. 0.75 ± 1.4 at 180 min, respectively). Pathological examination confirmed the angiographic patency rate and did not reveal injury secondary to ultrasound in the skin, soft tissues, heart, or lungs. This conclusion was also confirmed in preclinical experiments among a total of 25 patients with myocardial infarction. No unanticipated major adverse events were observed after ultrasound exposure (Cohen et al. 2003) (Figure 14.29).

14.8.2 Mechanisms of Sonothrombolysis

Theoretical simulation has demonstrated that the process of thrombolysis is limited by diffusion of fibrinolytic enzymes into the clot due to the low porosity of its surface. Thus, it is hypothesized that mechanisms of enhanced thrombolysis during the ultrasonic exposure is the increased transport and penetration of fibrinolytic enzymes into the clot by acoustic streaming (Lauer et al. 1992) and bubble cavitation (Everbach and Francis 2000, Suchkova et al. 2002, Pieters et al. 2004, Chuang et al. 2013). In addition, the 6°C temperature rise caused by ultrasound may contribute to an increase in enzymatic action and consequent increase in lytic rate (Sakharov et al. 2000). However, in another study, the measured temperature elevations were well below the threshold for enhancing thrombolysis (Nahirnyak et al. 2007). Temperature elevation was also assessed experimentally in human and porcine clots exposed to 0.12–3.5 MHz pulsed ultrasound *in vitro* with a

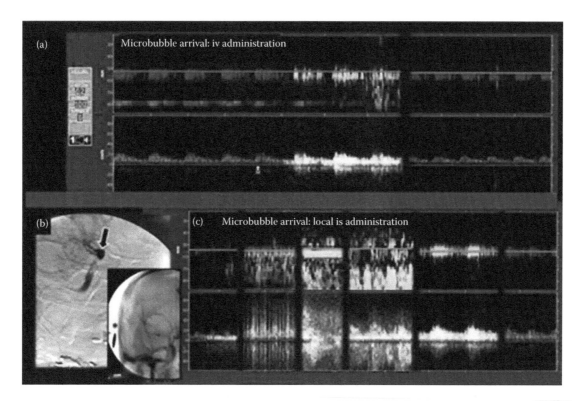

FIGURE 14.29 (a) Moderate transcranial Doppler (TCD) signal enhancement after systemic intravenous microbubble (MB) administration to occluded artery. (b) Angiogram showing intraclot site (arrow) through microcatheter. (c) Massive signal enhancement during MB infusion. (Ribo, M., Molina, C.A., Alvarez, B., Rubiera, M., Alvarez-Sabin, J., and Matas, M.: Intra-arterial administration of microbubbles and continuous 2-MHz ultrasound insonation to enhance intra-arterial thrombolysis. *J. Neuroimag.*, 2010. 20. 224–227. Copyright Wiley-VCH Verlag GmbH & Co. KGaA. Reproduced with permission.)

peak-to-peak pressure of 0.25 MPa and 80% duty cycle, which was 0.25°C and 0.33°C ± 0.04°C, respectively. The role of stable and inertial cavitation as possible mechanism for ultrasound (120 kHz)-enhanced rt-PA thrombolysis was investigated *in vitro* (Datta et al. 2008). The mass loss was 26% in the stable cavitation regime, 20.7% in the combined stable and inertial cavitation regime, 13.7% in the no cavitation regime, and 13.7% of rt-PA alone. These results indicate a key role of stable bubble activity in the enhancement of thrombolysis. In comparison, 27 kHz ultrasound has been found to disrupt peripheral arterial and venous thrombi in several animal models (i.e., coronary occlusion) even in the absence of rt-PA (Sobbe et al. 1974).

Pulsed HIFU exposures have demonstrated the ability to enhance thrombolysis noninvasively while using rt-PA. Increased gaps between cells and more exposed fibrin on the surface of clots were shown *in vitro* using SEM, and improved binding and penetration of the rt-PA was observed with pretreatment of pulsed-HIFU. Such significantly improved thrombolytic rates could also be reproduced in the marginal ear vein of rabbits. Preliminary histological analysis showed the exposures to be safe, without deleterious effects in the endothelium of the treated vessels or in the surrounding tissue. In this situation, acoustic radiation forces and the repetitive displacements are a novel mechanism for enhancing the permeability of soft tissues, which enable better delivery of both systemically and locally administered agents. Furthermore, after 2 min of HIFU treatment (500 kHz, 25% duty cycle, pulse duration of 200 μs, spatial peak temporal average intensity of 35 W/cm^2), no residual intraluminal clot was evident sonographically in an *ex vivo* porcine artery, nor was any histological damage to the arterial wall evident (Rosenschein et al. 2000). More recently, histotripsy (1 MHz, 5-cycle pulses at a pulse repetition rate of 1 kHz, peak rarefactional pressure greater than 6 MPa), that is, the fractionation of soft tissue through controlled cavitation using short HIFU pulses, was also confirmed for its ability of thrombolysis with ultrasound energy alone (Maxwell et al. 2009). Canine blood clots were lysed effectively *in vitro* in 1.5–5 min such that 96% of the debris was smaller than 5 μm in diameter (Figure 14.30).

14.8.3 NEW TREATMENT STRATEGIES

The introduction of lipid-encapsulated microbubbles in sonothrombolysis could augment the dissolution in an *in vitro* human blood clot model (Tachibana and Tachibana 1999) and the recanalization of arteriovenous graft thrombi in a canine model (Xie et al. 2005). A significantly greater clearing of thrombus using higher PW ultrasound intensities (10 W/cm^2) compared to lower CW ultrasound intensities (0.4–0.6 W/cm^2) was found. Both of them were attributed to the cavitation effects. Stable cavitation was sustained using an intermittent infusion of a commercial contrast agent and correlated with enhanced thrombolysis; meanwhile, inertial cavitation was found to be counterproductive for enhanced thrombolysis, which might be due to the lack of sustained bubble activity throughout the entire exposure (Datta et al. 2008).

Because of the diffusion and infusion of microbubbles in the circulating system, their concentration in the region of interest may not be high enough for significant cavitation effects. Targeted echogenic immunoliposomes (ELIPs) are novel agents that may permit the evaluation of vasoactive and pathologic endothelium and can highlight thrombus or plaque rupture in a B-mode ultrasound image if coupled with antifebrin or D-Dimer antibody (Bangham 1993). In addition, echogenic liposomes have the potential to encapsulate a drug for targeted therapeutic delivery efficiently and safely. A multistep, acoustic biotinylated, lipid-coated, perfluorocarbon nanoemulsion could be successfully targeted to thrombi *in vitro* and infiltrate arterial walls and localize tissue factor expression *in vivo* while maintaining ultrasound contrast (Lanza et al. 2000). Targeted microbubbles containing perfluorobutane (aserosomes) with a diameter of 2 μm could bind to thrombus *in vitro* (Unger et al. 1998).

The delivery of a thrombolytic drug to target thrombi in an efficient and safe way is the principal goal of a clinically useful strategy. Localized pharmacotherapeutic delivery has distinct advantages of increased concentrations at the specific site, less dose required, and decreased toxicities. Ultrasound can be used to fragment targeted drug carriers (e.g., liposomes) to release the drug to produce a large temporal peak for improved therapeutic efficacy. Pulsed-HIFU exposures were used to deploy DOX from LTSLs in preclinical murine tumors for enhanced antitumor effect (Dromi et al. 2007). ELIPs could load rt-PA and generate echoes by binding to specific fibrin of an *in vitro* porcine clot using a clinical diagnostic scanner at a lower exposure output level (MI = 0.04), while a therapeutic concentration of rt-PA could be released with pulsed 6.0 MHz color Doppler ultrasound at an MI of >0.43 (Tiukinhoy-Laing et al. 2007, Klegerman et al. 2008, Smith et al. 2010). The lytic efficacy of rt-PA-loaded ELIP is comparable to that of rt-PA alone *in vitro* (Shaw et al. 2009).

Because of the wide range of ultrasound devices and parameters being employed, the optimal values, such as frequency (20 kHz–2 MHz), wave mode (pulsed or continuous), and the intensity, are unknown. More clinical trials of sonothrombolysis in different countries are required to define clinical feasibility and identify potential risks. Elucidation of the mechanism and potential risks will provide the foundation to develop new treatment strategies. Overall, sonothrombolysis has great promise in the treatment of acute and chronic thrombotic diseases.

14.9 DRUG DELIVERY ACROSS THE BLOOD–BRAIN BARRIER (BBB)

14.9.1 BLOOD–BRAIN BARRIER

The blood–brain barrier (BBB) is an endothelial barrier present in the brain microvasculature and consists of tight junctions around all capillaries in the central nervous system (CNS), which separates circulating blood from the brain's extracellular fluid and therefore prevents various substances (e.g., polar molecular) from leaking into the brain in a chemically stable

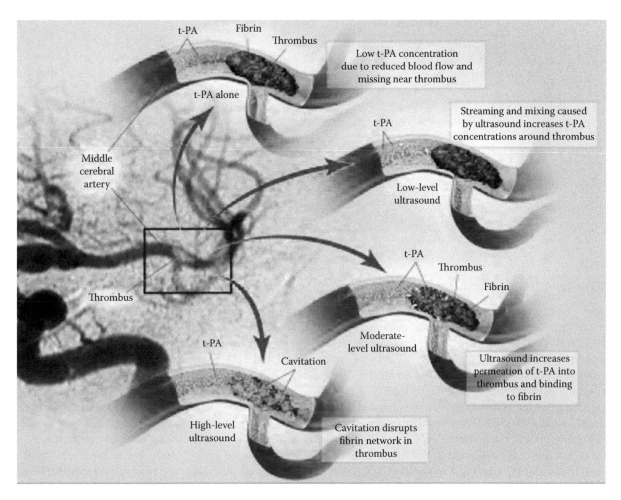

FIGURE 14.30   Schematic diagram of sonothrombolysis for blood clot in the cerebral vessel.

environment (Figure 14.31). But for the BBB, the microvasculature in the brain would enable superior drug delivery than normally occurs, owing to its well-branched network of vessels that meander for more than 600 km through this organ, providing oxygen and nutrients practically at the doorstep of every cell. Electron microscopy studies have illustrated the different structure of the endothelial cells (ECs) in the brain from those in the peripheral tissue: very few pinocytotic vesicles for macromolecule transport, thus limiting the amount of transcellular transport, and lack of fenestrations, thus effectively blocking the free diffusion of water-soluble solutes from the blood and severely restricting the amount of paracellular transport. In addition, the "sink effect" of cerebrospinal fluid (CSF) allows the passage of substances from the interstitial fluid to the CSF, with the CSF constantly circulating and carrying substances away. The brain can remove substances from the CNS. Efflux transporter proteins, such as P-glycoprotein (Pgp), the breast cancer–related protein (BCRP), and multidrug resistance proteins (MRP), are active in the BBB for the removal of compounds. An additional barrier that a systemically administered drug encounters before entering the CNS is known as the blood–cerebrospinal fluid barrier (BCB). The BCB is located in the epithelium of the choroids plexus, where blood-borne solutes are able to move out into the extracellular space through the paracellular pathway but are prevented from entering the CSF by the apical bands of tight junctions between adjacent choroidal epithelial cells. Although this barrier system is important for the maintenance of the homeostasis of the brain, it prevents the delivery of therapeutic agents into the brain and complicates the treatment of CNS disorders (Kinoshita 2006). Thus, drug delivery across the BBB is one of the most important key factors in the treatment of diseases of the CNS. The intracellular space in the BBB is so tight that it restricts the entry of large substances for maintaining the internal milieu of the brain, such as neuropharmaceutical agents for CNS disorders (e.g., brain cancer, brain trauma, viral infections of the brain, stroke, Alzheimer's disease, dementia, Parkinson's disease, Huntington's disease, mood disorder, AIDS, viral and bacterial meningitis, amyotrophic lateral sclerosis, multiple sclerosis, and spinal cord injury), although a selective entry of small-sized nutrients and minerals across BBB is possible with the presence of multiple endogenous transporters. Although there are more than 7000 drugs in the Comprehensive Medicinal Chemistry (CMC) database, only 5% of those drugs are active and effective in three CNS diseases: affective disorders, chronic pain, and epilepsy.

In order to overcome BBB, several novel approaches, such as nanoparticles, liposomes, antibody-mediated delivery approaches, application of genomics, modifying drug formats

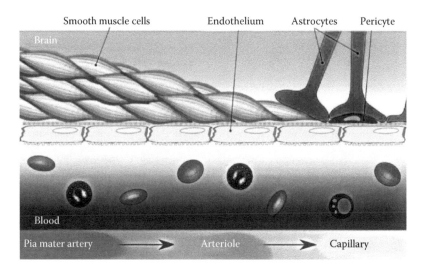

FIGURE 14.31 Schematic diagram of the blood–brain barrier.

or attaching with a carrier, or focusing on the delivery methods such as intracarotid injection and direct catheter insertion into the brain (Alam et al. 2010), are in development to target the brain efficaciously and effectively. All of them can be classified into three strategies: modulating the BBB by drugs, circumventing the BBB, and disrupting the BBB.

Permeability of the cerebral capillaries is favored by low molecular weight (150–500 Da), lipophilicity (log octanol/water partition coefficient between −0.5 and 6.0), and the absence of ionization at physiological pH (7.4). The key factor affecting the ability of a drug to enter the CNS via transcellular passive diffusion seems to be lipid solubility. The polar molecule renders the drug water soluble and, therefore, decreases its ability to cross the EC membrane and the BBB. Lipid solubility can be transiently increased by masking some of the polar functional groups with nonpolar groups to form a prodrug. Removal of the masking groups by specific enzymes reverses the drug back to its active form. However, increasing lipid solubility of drugs has the great disadvantage of increasing plasma protein binding and uptake by other tissues and hence can increase systemic toxicity. An alternative strategy for increasing drug lipophilicity is to render the drug acceptable to the BBB transport systems by designating it as a pseudo-substrate for these transporters. The large neutral amino acid (LNAA) carrier system is capable of transporting endogenous as well as exogenous LNAAs with great structural diversity, such as the L-dopa neutral amino acid drug in treating Parkinson's disease. Macromolecular (i.e., large peptides and proteins) delivery across the BBB can be effected through transcytosis (endocytosis) using a modified protein or receptor-specific monoclonal antibody (i.e., mAb, OX26), called chimeric peptide technology. The delivery capability of this system is generally low, as the use of a chimeric peptide results in only one peptide or protein molecule being delivered per antibody, and the sophisticated chemistry involved has a high cost. Liposomes and nanoparticles can hold large quantities of high-MW drugs, peptides, and genes. Liposomes up to 500 nm in diameter can carry up to 10,000 drug molecules, or more if conjugated to a BBB delivery vector using

chimeric technology. However, one drawback is that liposomes are rapidly removed from the bloodstream as a result of the uptake by the RES, which can be mitigated by attaching PEG to the outer surface of the liposomes as a stealth coating. If a brain-specific promoter is utilized, the enzyme expression becomes restricted to the brain. Poly(butyl)cyanoacrylate (PBCA) nanoparticles (typically 250 nm in diameter) coated with Tween 80 (polysorbate 80) have been used to deliver several drugs to the CNS with a good degree of success *in vivo*. When injected intravenously, the nanoparticles become further coated with absorbed plasma proteins, especially apolipoprotein E (Apo-E), which is mistaken for low-density lipoprotein (LDL) particles by the cerebral endothelium and is internalized by the LDL uptake system. The major disadvantages of drug modification are its lack of site specificity and the fact that not all drugs can be modified.

The most direct way of bypassing the BBB is to deliver the therapeutic agents directly either into the brain's interstitium or into the CSF, such as direct injection, intracranial implants, and slow-release devices, which can theoretically yield high CNS drug concentrations with minimal systemic exposure and toxicity. Drawbacks for this delivery strategy are invasiveness, potential damage to brain tissue both around the implant and along the track of an introductory catheter, and a rapid volume injection directly into brain parenchyma (Kinoshita 2006). Furthermore, a slow rate of drug distribution within the CSF and interstitial fluid is observed, especially for macromolecules.

Osmotic BBB disruption is one of the earliest techniques developed to disrupt the tight junctions and the first to be used in humans more than 20 years ago. The intracarotid infusion of a hypertonic solution such as mannitol induces cell shrinkage by osmotically drawing water out of the ECs, which results in the disengagement of the proteins in the tight junctions for 20–30 min and increasing drug delivery by 10- to 100-fold. The main disadvantages of this procedure are that not only the transport of the drug but also of all the other solutes in the blood is increased, and possible toxicity even if the disruption is very transient. Chemical modulation, such as with the peptide bradykinin and the synthetic bradykinin analog RMP-7, can modulate the tight junctions by elevating intracellular free calcium levels. This method has the nonselective BBB disruption as osmotic technology.

14.9.2 BBB DISRUPTION BY ULTRASOUND

It was found that ultrasound can temporarily modulate the BBB selectively at targeted locations with some precision in a number of animal studies, but not globally (Hynynen et al. 2003, Meairs and Alonso 2007, Hynynen 2008, Vykhodtseva et al. 2008). Initial investigation of ultrasound-induced lesions in the mammalian brain found BBB disruption (Bakay et al. 1956, Vykhodtseva et al. 1995), which was illustrated as brain parenchyma heavily stained with Trypan blue at the periphery of ultrasound-induced thermal lesions (Patrick et al. 1990) or even with no evidence of a discrete lesion (Ballantine et al. 1956, Vykhodtseva et al. 1994). So, transcranial ultrasound exposure is suggested to deliver chemotherapy agents to brain tumors in combination with thermal ablation therapy, but the difficulty is the production of a predictable BBB disruption without brain damage. However, reproducible and reliable opening of the BBB without or only with an acceptable magnitude of tissue damage is challenging (Patrick et al. 1990).

Recent advances were able to induce reproducible, consistent, and localized BBB disruption in rabbit through a bone window without permanent, obvious damage to the brain parenchyma by a low continuous acoustic exposure (10 s, 1.63 MHz, 1.1 W) or pulsed sonication (0.2–11.5 W, temporal and spatial peak intensity range of 16–690 W/cm^2, burst length of 10–100 ms, pulse repetition frequency of 1 Hz, 20 s) under the guidance of T2-weighted MRI and control program with a bolus of a UCA injected intravenously via the marginal ear vein (0.05 mL/kg) 10 s before sonication (Hynynen et al. 2001). BBB disruption was confirmed through the leakage of the MR contrast agent into the brain parenchyma, as shown by in the signal intensity on the T1-weighted images after sonication, which increased with the applied power, continuous but declining opening of the BBB, and total reverse on the next day. Postmortem light microscopy showed damage to the brain in about 70% of the sonicated locations at the highest output. Since the highest temperature increase was only about 5°C, the major reason of BBB disruption may be due to acoustic cavitation rather than thermal effects (Hynynen et al. 2001). There was no neural damage, but in many cases, there was red blood cell extravasation, indicating some vascular damage, at the lowest pressure. With the presence of microbubbles, the energy required to disrupt BBB is roughly two orders of magnitude smaller than with ultrasound alone (Hynynen et al. 2001, McDannold et al. 2008, O'Reilly and Hynynen 2012). The disrupted BBB of the sonicated rabbit brain repaired itself within approximately 6 h, and long-term follow-up showed nearly no damage (McDannold et al. 2005). So, the ultrasound-induced BBB disruption is a transient and reversible event (Hynynen et al. 2005). The reliability and reproducibility of BBB opening were much higher than using

ultrasound alone. Furthermore, the pressure threshold for BBB opening becomes lower for decreasing driving frequencies (i.e., 0.25 MPa at 260 kHz), although at the expense of a relatively large focus (Vykhodtseva et al. 2008).

Electron microscopy revealed enhanced transcellular pathway with ultrasound exposure. Drugs were taken up by the endothelial cells by means of vesicle transportation through the endothelial cells and on the loosening or destruction of the tight junctions (Hynynen et al. 2005). The delivery of 1,3-bis(2-chloroethyl)-1-nitrosourea (BCNU) to rodent gliomas for suppressing tumor progression was significantly improved without damage to normal tissue as indicated in MRI and histology. This novel technique promises a more effective and tolerable means of tumor therapy in a localized manner with lower therapeutic doses and concurrent clinical monitoring, and the delivered antibodies do not lose their innate function (Chen et al. 2010, Liu et al. 2010). Herceptin (Kinoshita et al. 2006a), D4-receptor antibodies (Kinoshita et al. 2006b), DOX (Treat et al. 2007), and methotrexate (Mei et al. 2009) for cancer treatment, antiamyloid-beta antibodies for Alzheimer's (Jordão et al. 2010), and stem cells for neuronal regeneration (Burgess et al. 2011) have been delivered through the BBB using focused ultrasound.

However, one major drawback is that the ultrasound energy is highly absorbed by the skull, resulting in thermal damage of the tissue adjacent to it and substantial attenuation of the ultrasound beam, and distortion and shifts of the propagating waves, making acoustic energy focusing at a precise location in the brain extremely challenging. The cranium of humans is much thicker than that of mice or rats, which means that a more careful and delicate control of ultrasound is necessary to form a target inside the brain. Recent advances in ultrasound technology combined with the use of magnetic resonance imaging for targeting and monitoring have made the propagation of focused ultrasound through the intact skull possible (Clement and Hynynen 2002, Clement et al. 2005, Marquet et al. 2009). Computed tomography–derived preliminary information could be registered with MR images to correct for wave distortion in bone, in particular to determine the driving signal (amplitude and phase) for each transducer element to achieve a well-focused beam through the skull. However, until now no clinical trials have been performed using ultrasound-induced BBB disruption techniques.

14.9.3 MECHANISMS OF BBB DISRUPTION BY ULTRASOUND

The biophysiological mechanisms of BBB disruption are currently unknown, so more studies are needed in order to better understand the processes involved. A large number of scenarios may contribute in full or in part to BBB opening. Commercially available UCAs have a wide range of size distribution. For example, although the mean diameter of Optison is 2–4.5 μm, the maximum value could be as large as 32 μm. These large bubbles may be able to produce temporary blood vessel occlusions (microembolization), and this transient ischemia is the possible reason for the BBB opening in capillaries of rats when Optison was introduced directly into the cerebral arterial circulation without ultrasound exposure (George Mychaskiw et al. 2000). Increased BBB permeability was confirmed through the distribution of Evan blue–labeled albumin (67 kDa and radius 3.5 nm) by transcytosis. However, the large bubbles are usually filtered away from the lung after being introduced intravenously. In the acoustic field, microbubbles may grow several times their original diameter before collapsing and fill the vessel cross section. Stable cavitation can generate cell membrane deformation and produce red blood cell (RBC) extravasation by exerting sufficiently large stresses on the vessel wall to rupture the tight junctions without affecting cell viability to enhance both the intracellular and paracellular passages across the BBB (McDannold et al. 2006). Bubble oscillations in the microvasculature introduce fluctuations in the systemic pressure and blood flow to which the brain may respond by passive adjustment of the vessel wall diameter. Arteriole vasoconstriction leads to reduce blood flow and alter BBB permeability (Brightman et al. 1970). In addition, hemorrhages are usually attributed to bubbles undergoing inertial collapse in the intensified field against the blood vessel wall, which probably occurs predominantly in the largest vessels and is suppressed significantly in the smaller vessels (Liu et al. 2008). At a lower driving frequency, more bubbles become available for inertial cavitation. Overall, the BBB disruption could occur only with stable cavitation and without the presence of inertial cavitation (McDannold et al. 2006).

14.9.4 Animal Studies

Extensive investigations have been carried out on optimizing the exposure parameters (i.e., pressure threshold, driving frequency, and concentration of UCA) for ultrasound-mediated BBB opening under MR guidance in the normal (i.e., nonpathological) animal brain using comprehensive histology. The possibility of BBB opening in rabbits exposed by ultrasound (690 kHz, pulse duration of 10 ms, pulse repetition frequency of 1 Hz, and sonication duration of 20 s) at pressure amplitudes of 0.4, 0.8, and 1.4 MPa was 50%, 90%, and 100%, respectively (Hynynen et al. 2005). Histological findings revealed areas of RBC extravasation in 42 out of

48 locations analyzed, ranging from a few scattered erythrocytes (i.e., 0.5 MPa) to extensive extravasation and mild damage to the brain parenchyma (i.e., 1.4 MPa). The threshold for BBB disruption, described as with 50% probability, occurred at the pressure amplitudes of 0.25, 0.60, 0.61, and 0.69 MPa at a driving frequency of 260 kHz, 1.5 MHz, 1.63 MHz, and 2.1 MHz, respectively, while the other parameters were the same (Hynynen et al. 2001, 2006, McDannold et al. 2005, 2007, McKee et al. 2006, Treat et al. 2007). Thus, it is estimated that an MI of 0.47 is required for BBB disruption in the normal rabbit brain as monitored by MR imaging and histology observations. The size and the amount of Evans blue extravasation in rat brain increased concomitantly with UCA dose (0, 30, 60, 90 µL/kg) and applied pressure (0.9 and 1.2 MPa) at 1 MHz with a pulse length of 10 ms, a pulse repetition frequency of 1 Hz, and a duration of 30 s (Yang et al. 2007), and their light microscopy images did not reveal any cell damage at the pressure amplitude of 0.9 MPa and the UCA dose of 30 µL/kg, but did reveal apoptotic cells at the higher dose of 60 µL/kg. BBB disruption was accompanied by extravasation of some RBCs at the pressure amplitude of 1.2 MPa, and the number of extravasated RBCs increased with UCA doses. The apoptotic cells were mainly localized to the blood vessels, with only a few appearing outside the sonicated areas. It is suggested that magnetic resonance susceptibility–weighted imaging (MR-SWI) is more sensitive than standard T2-weighted and contrast-enhanced T1-weighted MR imaging techniques in detecting hemorrhage after brain sonication in the presence of microbubbles (Figure 14.32).

Using sonication at a frequency of 2.0 MHz, an intensity of 485 W/cm², and a duration of 0.2 s with no presence of UCA, reversible BBB disruption could be achieved in 29 of 55 applications, tissue damage in 11 of 29 applications, and RBC extravasation in 3 of the 11 damage cases (Mesiwala et al. 2002). Therefore, in comparison to UCA-aided BBB disruption, fewer cases of RBC extravasation were detected for a comparable range of exposure pressures and frequencies. BBB disruption in this study persisted for ~72 h. Furthermore, the thermal damage threshold of normal tissue in BBB disruption was estimated to be between 48.0°C and 50.8°C, as monitored by MR thermometry at the ultrasound exposure (frequency of 1.63 MHz, acoustic power in the range of 1.4–21.7 W, and sonication duration of 30 s). The temperature information was

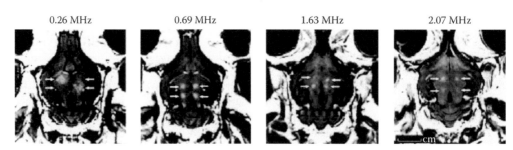

| 0.26 MHz | 0.69 MHz | 1.63 MHz | 2.07 MHz |

FIGURE 14.32 Focal contrast enhancement (arrows) in T1-weighted MR images perpendicular to the direction of the ultrasound beam, demonstrating local blood–brain barrier disruption in the rabbit brain produced by focused ultrasound (FUS) at the frequency of 2.1, 1.63, 0.69, and 0.26 MHz, respectively. (Reprinted from *Ultrasonics*, 48, Vykhodtseva, N., McDannold, N., and Hynynen, K., Progress and problems in the application of focused ultrasound for blood-brain barrier disruption, 279–296, Copyright 2008, with permission from Elsevier.)

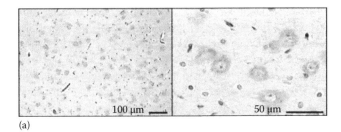

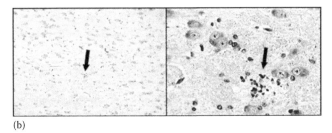

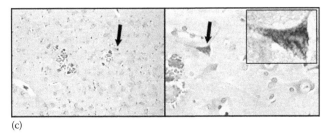

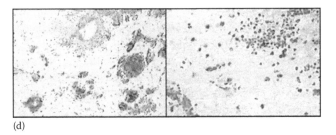

(a)

(b)

(c)

(d)

FIGURE 14.33 (a) Grade 0: no damage; (b) Grade 1: a few extravasated erythrocytes; (c) Grade 2: microscopic areas of perivascular extravasations, some of them associated with evident damage to the brain parenchyma (insert: magnification of an ischemic cell); (d) Grade 3: extensive extravasation; hemorrhagic lesions (infarct) in the histology of VAF–toluidine sections after ultrasound exposure. (Reprinted from *Neuroimage*, 24, Hynynen, K., McDannold, N., Sheikov, N.A., Jolesz, F.A., and Vykhodtseva, N., Local and reversible blood-brain barrier disruption by noninvasive focused ultrasound at frequencies suitable for trans-skull sonications, 12–20, Copyright 2005, with permission from Elsevier.)

found to be better than the ultrasound parameters at predicting damage (McDannold et al. 2004) (Figure 14.33).

An intravenously administered antibody that targeted the dopamine D4 receptor in the brain could cross the BBB in a mouse model through the intact skull and recognize its antigens using MR-guided focused ultrasound (FUS) (driving frequency of 690 kHz, −6 dB beam size of 2.3 mm × 14 mm, burst length of 10 ms, repetition frequency of 1 Hz, sonication duration of 40 s, and varied acoustic pressure of 0.6–1.1 MPa) and UCA injection (Kinoshita et al. 2006a). BBB disruption was confirmed by contrast-enhanced T1-weighted MR

and postmortem by Trypan blue staining. Little or no hemorrhage in the ultrasound focal region was observed at a pressure amplitude of <0.8 MPa. All mice showed major tissue damage when the pressure exceeded 0.8 MPa. Using the same protocol, the feasibility of delivering the chemotherapeutic agent Herceptin (150 kDa), which is effective in the treatment of breast cancer metastasis, to the normal murine brain was also investigated (Kinoshita et al. 2006b). The amount of Herceptin delivered to the brain parenchyma was 2504 and 3257 ng at the peak negative pressures of 0.8 and 0.6 MPa, respectively. There was only a few scattered extravasated RBCs and apoptotic cells under an exposure of 0.6 MPa, and the number increased after the 0.8 MPa sonication at sites of the most severe extravasation. VAF (vanadium acid fuchsin) staining showed no major ischemic damage. In addition, the delivery of the chemotherapeutic agent DOX, encapsulated in long-circulating pegylated liposomes, was also examined in the normal rat brain (Treat et al. 2007). A pressure of 1.1 MPa was required to induce BBB disruption at a driving frequency of 1.5 or 1.7 MHz, a pulse length of 10 ms, a pulse repetition frequency of 1 Hz (duty cycle 1%), an insonation time of 30 s, and 0.1 mL/kg Optison injections. The DOX concentration was 886 ± 327 ng/g tissue in the brain with minimal adverse effects and 251 ± 119 ng/g in the control group. With higher UCA doses, DOX concentration was up to 5366 ± 659 ng/g tissue, but with more significant tissue damage (Figure 14.34).

TEM was employed for analyzing the transport of the endogenous immunoglobulin G (IgG) and horseradish peroxidase (HRP, 40 kDa, 5 nm) across the BBB in the rabbit brain 1–2 h after sonication. Several routes of IgG passage have been revealed: transcytosis, transendothelial opening (fenestration and channel formation), widening of interendothelial clefts and opening of part of tight junctions, and free passage through the injured endothelial lining. As for HRP, the BBB disruption occurred in capillaries, arterioles, and venules, demonstrating similar routes of transport. The mean endothelial pinocytotic densities (the number of HRP-containing vesicles per μm^2 of the cell cytoplasm) were 0.9 and 1.05 vesicles/μm^2 1 h after ultrasound exposure at frequencies of 690 and 260 kHz in the capillaries, 1.63 and 2.43 vesicles/μm^2 in the arterioles, 0.7 and 0.14 vesicles/μm^2 in the control, respectively (Sheikov et al. 2006). Although fewer HRP-positive vesicles were observed in the venules, transport of HRP through interendothelial clefts was often seen in venules. Disassembling of the tight junctions' molecular structure was observed up to 4 h after insonation and permitted the paracellular passage of agents with molecular weights up to at least 40 kDa.

Multiphoton confocal microscopy could demonstrate the vascular effects of ultrasound-enhanced BBB disruption in the normal murine brain in real time, including vasoconstriction, blood flow disruption, and various kinds of leakage from the capillaries, which occur in a relatively short time scale (less than 1 min) and cannot be observed by MR imaging or histology. Vasoconstriction was observed in 14 of the 16 cases at the onset of the ultrasound exposure (frequency of 1.029 MHz, pulse duration of 10 ms, pulse repetition frequency of 1 Hz, insonation duration of 45–60 s, acoustic

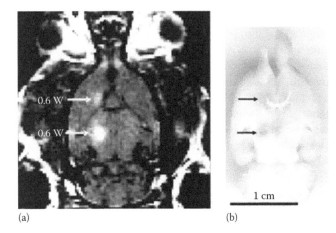

(a)   (b)

FIGURE 14.34 (a) Contrast-enhanced transversal (axial) T1-weighted MR fast spin-echo image after sonication at acoustic power of 0.6 W. (b) Leakage of Trypan blue (arrows) from the vasculature into the brain parenchyma fixed in 10% buffered formalin phosphate 4 h after sonication. (Treat, L.H., McDannold, N., Vykhodtseva, N., Zhang, Y., Tam, K., and Hynynen, K.: Targeted delivery of doxorubicin to the rat brain at therapeutic levels using MRI-guided focused ultrasound. *Int. J. Cancer.* 2007. 121. 901–907. Copyright Wiley-VCH Verlag GmbH & Co. KGaA. Reproduced with permission.)

power of 0.2 W, peak focal pressure of 0.2 MPa, and 0.01 mL of Optison). Many vessels, including capillaries, did not constrict despite being affected along their entire length. The vasomotor response included an initial constriction, reducing the diameter on average by 56% ± 12%, followed by a recovery time lasting 21–615 s. Longer vasospasms were observed after the administration of UCA, which suggested the longer distribution of microbubbles throughout the brain's microvasculature. Blood flow disruption (i.e., 4 min to completely perfuse Evan blue vs. 9 s in nonsonicated region) was seen in 11 of the 16 locations, and was always associated with vasospasm. Extravasation of two dextran-conjugated fluorophores, namely Alexa Fluor 488 green (10 kDa) and Texas Red (70 kDa), occurred via hemorrhage, focal compromise of the vessel wall (micro-disruption), and slow leakage from patent vessels without RBC extravasation (Raymond et al. 2006). During microdisruption, the fluorophores spread rapidly (3–9 s) from the vessel wall and diffused radially outward even without vasoconstriction and/or blood flow disruption. Slow disruption was seen as a gradual rise in extravascular signal that started at the onset of ultrasound exposure without visible disruption of the endothelium.

14.10 SONODYNAMIC THERAPY

14.10.1 FROM PHOTODYNAMICS TO SONODYNAMICS

The use of light for therapy began in 1900, combining acridine orange and light to destroy paramecium (Raab 1900). In the twentieth century, photosensitizing agents, mainly nonporphyrin photosensitizers, were initially used to detect tumor tissue by observing the intratumoral fluorescence. Until the 1960s, the field of photodynamic therapy (PDT) lay fallow,

when the hematoporphyrin derivative (HpD) was found to destroy tumor tissue (Lipson et al. 1961). The cytotoxic product of the photochemical reaction was identified to be singlet oxygen (Weishaupt et al. 1976). Porphyrin photosensitizers were then examined as photosensitizers because they are efficient singlet oxygen generators and have absorption maxima in the red portion of the electromagnetic spectrum. Photofrin1, a purified version of HpD, was approved for use in the United States against early and late-stage lung cancers and esophageal cancers and dysplasias (Dougherty 1996). Nearly 10,000 patients in the United States, Canada, the Netherlands, Japan, France, and Italy have undergone PDT against bladder cancers, brain cancers, breast metastases, skin cancers, gynecological malignancies, colorectal cancers, thoracic malignancies, as well as oral, head, and neck cancers (Dougherty et al. 1998, Pandey and Zheng 2000). However, because of the limited penetration depth of light, the tumor should be close to light source in PDT.

The terminology "sonodynamic therapy" (SDT) appears contextually aligned with the PDT. SDT combines ultrasound with a sonosensitive agent derived from chlorophyll that was originally used as light-activated chemicals in the treatment of solid tumors and cancers (Yumita et al. 1994, Kuroki et al. 2007). However, the use of ultrasound is more complicated than using light because it can potentially alter cell membrane permeability (sonoporation) and generate acoustic cavitation which can produce free radicals and light (sonoluminescence). If the acoustic intensity is high enough, the induced temperature elevation could become a predominant player in SDT. The agents themselves have no antitumor ability but exhibit it only by the sonochemistry. Therefore, much less risk of adverse effects is expected on normal tissues. In comparison to PDT, the prime advantage is its ability to penetrate into deeper tissue targets in clinics.

14.10.2 MECHANISM

Despite a large number of both *in vitro* and *in vivo* reports on SDT, the mechanism of how ultrasound augments the activity of anticancer drugs and other agents remains poorly understood. Ultrasound itself can cause cellular damage, either lethal or sublethal. Sublethally damaged cells are more susceptible to anticancer agents, even administered at a normally sublethal dose. An increase in lipid peroxidation products induced by ultrasound was found to be associated with suppression of tumor cell growth, and there was marked, but transient depletion of intracellular thiols (expressed as glutathione equivalents) in the surviving cells. Modulation of glutathione (sulfhydrl-containing tripeptide, known to detoxify free radicals) by ultrasound exposures affects the potentiation and protection of adriamycin. Therefore, depletion in glutathione renders the cells more sensitive to the agent. Sonosensitive drugs (e.g., protoporphyrin IX) create a short-lived cytotoxic reactive oxygen species (ROS) (i.e., superoxide radicals and singlet oxygen) to induce peroxidation of membrane lipids and irreversible alteration of the target tissue. Acoustic cavitation potentiates redox (oxidation–reduction) cycling drugs by chemical reduction and, thereafter, by the production of OH^- via Fenton's pathway, which may potentiate drugs containing quinine (e.g., adriamycin, daunomycin, and mitomycin C). Production of free radicals in the extracellular medium during bubble collapse of IC with local large temperature elevations is proposed as the most likely mechanism of the chemical activation of the sonosensitizer either by direct pyrolysis or due to the reaction with H and OH radicals, formed by the pyrolysis of water (Mišík and Riesz 2000). These free radicals (mostly carbon centered) react with oxygen to form peroxyl and alkoxyl radicals, which have low reactivity with organic components dissolved in biological media and therefore have a higher probability of reaching critical cellular targets, unlike OH radicals and H atoms formed by pyrolysis. Ultrasound exposure may also introduce electronic excitation of porphyrins and a photochemical process in the cavitation process. Pyrolysis of water vapor generates the very reactive hydroxyl radical and hydrogen atom, which may recombine or react with volatile solute molecules to generate new free radicals and react with oxygen to produce the hydroperoxide radical. The primary radicals can also react with nonvolatile solutes. Longer-lived residual chemical compounds may also be produced during ultrasound exposure to induce bioeffects in cells. Single-strand breaks could be induced in CHO cells not only during sonication (Miller et al. 1995) but also after sonolysis presumably through the action of residual H_2O_2 (Miller et al. 1991). Phosphate-buffered saline has significantly more hydroxyl radicals and H_2O_2, primarily due to the scavenging effect of glucose and hydrophobic amino acids (Trp, Phe, Tyr, Leu, Val, Met) (Mišík and Riesz 1999). The genotoxic free radicals in vivo are related to the transition-metal-ion-assisted conversion of superoxide radical anion and H_2O_2 to hydroxyl radicals, which in turn almost indiscriminately react with the nuclear material. Sonication significantly enhances the hydroxyl radical production from redox-cycling drugs (e.g., adriamycin or DOX and mitomycin C) in comparison to those on non-redox-cycling ones (e.g., 5-fluorouracil and methotrexate) via Fenton's reaction. Superoxide radical ions could increase the release of iron from ferritin and provide a pool of active Fe^{2+} to catalyze the Fenton reaction [88]. Unlike ionizing radiation or photodynamic exposure where free radicals and singlet oxygen, respectively, can be produced intracellularly, exposure of cells to ultrasound results in extracellular production of free radicals. Hydroxyl radicals either formed primarily during collapse of cavitation bubbles in aqueous media or secondarily by Fenton conversion of H_2O_2 have limited migration range of only 1.5–9 nm because of their high reactivity and very short lifetime. In contrast, alkoxyl and peroxyl radicals could be cytotoxic factors and migrate over significant distances through the biological milieu by virtue of their longer lifetimes and higher selectivity (Mišík and Riesz 2000). The alkylperoxyl radicals may abstract bisallylic hydrogen atoms from polyunsaturated fatty acid chains of the membrane lipids rapidly by the initiation of lipid peroxidation. Furthermore, sonoluminescent light can introduce electronic excitation of the sensitizer and energy transfer to oxygen to generate the highly reactive singlet

molecular oxygen. However, during a hematoporphyrin study, singlet oxygen fluorescence was not easily detected during ultrasound exposure in comparison to laser excitation.

14.10.3 IN VITRO STUDIES

Sonosensitizers are usually added to the cell suspension and incubated for a given period prior to sonication in vitro. In order to determine the correlation between the acoustic pressure (or intensity) distribution and the therapeutic outcome, standing waves should be minimized by utilizing ultrasound absorbers in the experimental setup. The addition of bubble nuclei by saturating carefully prepared cell samples with air or by adding UCAs can facilitate the acoustic activation in the cell suspension investigation. Bioeffects such as cell viability can be assayed immediately after sonication, while the others may require incubation before becoming detectable, such as 6–12 h for assessing apoptosis.

A subtherapeutic dose of adriamycin (40–160 μM) did not cause cell damage to sarcoma 180 cells, but the cell death was three times greater after ultrasound exposure (1.93 MHz, 6 W/cm^2) than obtained with ultrasound alone, which was highly correlated with ultrasound-generated nitroxide. The synergistic loss of cell viability was prevented in the presence of 10 mM histidine (a $^1O_2$ and OH scavenger), but not in the presence of 100 mM mannitol (Umemura et al. 1997). A similar observation was found in sarcoma 180 cells using 4'-O-tetrahydropyranyladriamycin (THP-adriamycin), a less cardiotoxic adriamycin derivative, at doses of 16, 40, or 80 μM (Yumita et al. 1999) and photofrin II at a dose of 80 μg/mL (Umemura et al. 1999, Yumita et al. 2000). ATX-70 (a gallium-porphyrin analog) and hematoporphyrin (Hp) at 50 μg/mL, which do not exhibit any known anticancer effects or cytotoxicity, were more effective against sarcoma 180 cells in combination with ultrasound than with ultrasound (1.93 MHz, 4.5 W/cm^2) alone, which may be due to the enhanced apoptosis and caspase 8 activation (Umemura et al. 1990, 1993). Such a synergistic effect increased with the presence of D_2O, which prolongs the lifespan of singlet oxygen. However, combining 50 kHz ultrasound with ATX-70 in human leukemia HL-525 cells enhanced the toxicity to the cells, but not radical intermediates, as observed using EPR spin trapping. The addition of POBN (a free-radical scavenger) did not reduce the enhanced toxicity of the treatments (Miyoshi et al. 1997). However, intracellular protoporphyrin IX (another porphyrin derivative) did not increase by microbubble-enhanced ultrasound as extracellular intake, but it did with 5-aminolevulinic acid or ultrasound-induced hyperthermia in a moderately acidic environment (pH = 6.6) (Wang et al. 2008).

Afterward, the dose-dependent (ultrasound and agent) relationship of sonosensitive activation and cell survival was investigated. Hematoporphyrin (the most common photodynamic sensitizer) with 1.92 MHz ultrasound at intensities of 1.27 and 3.18 W/cm^2 enhanced the death rate of mouse sarcoma and rat ascites 130 tumor cells from 30% to 99% and from 50% to 95%, respectively (Yumita et al. 1989). The inhibitory effects of hematoporphyrin derivatives on the tumor cell growth were confirmed by the cell's morphological changes, cytochrome *c* oxidase activity, and degradation of DNA. No cytotoxicity or structural modification of hematoporphyrin derivative in sonication was observed on *in vitro* human colorectal adenocarcinoma cells (HT-29) and CHO cells. Low frequency-ultrasound (270 kHz) exposure for 60 s at intensities of 0.15, 0.3, and 0.45 W/cm^2 and Photofrin II decreased the survival of HL-60 cells by 92.9% ± 1.5% vs. 49.6% ± 5.1% (without vs. with Photofrin), 82.3% ± 2.2% vs. 34.5% ± 3.1%, and 77.0% ± 7.2% vs. 27.4% ± 3.0%, respectively (Tachibana et al. 1993). The survival rate of MT-2 cells under ultrasound exposure (450 kHz at an intensity of 0.3–0.5 W/cm^2) was found to be inversely proportional to the amount of Photofrin (Uchida et al. 1997) (Figure 14.35).

14.10.4 *In Vivo* Studies

Studies on SDT have also been carried out *in vivo*, where the environment of interaction of agents, cells, and ultrasound energy is more complicated, to prove it as a clinically viable procedure. ATX-70 demonstrated synergistic effects with ultrasound in colon-26 carcinomas in mice, and also improved the survival of mice bearing squamous cell carcinomas (Sasaki et al. 1998). Consistent *in vitro* studies showed augmented antitumor effects using ultrasound, Photofrin II, and hematoporphyrin against solid colon-26 sarcoma in 180 mice (Yumita et al. 2000). Protoporphyrin IX was found to accumulate in murine tumors several hours after administration with the tumor-to-tissue and tumor-to-plasma concentration

ratio peaking at about 24 h, and administration of 200 mg/kg 5-aminolaevulinic acid (ALA) could increase the concentration 1–6 h after administration (van den Boogert et al. 1998). These pharmacokinetic studies were important in optimizing the effects of ultrasound in sonodynamic therapy. Significant tumor growth inhibition and increased survival of the sonodynamically treated tumor-bearing mice were found compared to those with ultrasound exposure alone. The peripheral mononuclear cells in the blood of acute-type adult T-cell leukemia patients after ultrasound exposure (0.3 W/cm^2, 60 s) alone and together with 100 μg/mL of Photofrin were 69.4% ± 22.5% and 30.0% ± 23.0%, respectively (Uchida et al. 1997). In contrast, there was no significant difference of survival rates of normal human peripheral mononuclear cells between ultrasound-treated groups with and without Photofrin. Therefore, SDT may be an effective and extracorporeal blood treatment modality for acute-type adult T-cell leukemia patients (Figure 14.36).

14.11 DISCUSSION AND FUTURE DIRECTIONS

Ultrasound-mediated drug delivery may use microbubbles or liposomes for both diagnosis and therapy at the tumor-specific region. In this modality, drug vehicles target specific tumors precisely, and then ultrasound triggers the drug release in a temporally controlled manner, enhancing drug diffusion through the vessel wall for a more uniform distribution in the tumor and perturbing cell membranes for intracellular drug uptake. A variety of solid tumors, such as the

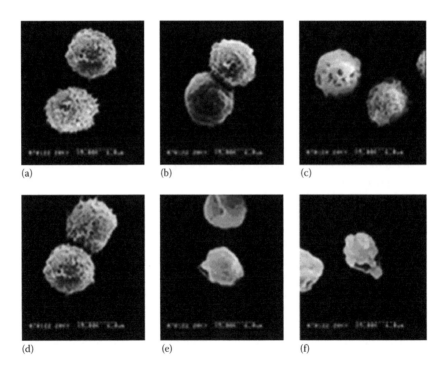

(a)   (b)   (c)

(d)   (e)   (f)

FIGURE 14.35 Surface electron microscopy images of (a) control Ehrlich ascites carcinoma cells, (b,c) cells treated with ultrasound (1.34 MHz, continuous mode, 60 s) 2 and 4 h later, respectively, (d) cells 4 h after treated with 20 μM protoporphyrin IX (PPIX), and (e,f) cells irradiated with ultrasound plus 20 μM PPIX after 2 and 4 h of culture, respectively. (Reprinted from *Ultrason. Sonochem.*, 16, Zhao, X., Liu, Q., Tang, W., Wang, X., Wang, P., Gong, L., and Wang, Y., Damage effects of protoporphyrin IX–sonodynamic therapy on the cytoskeletal F-actin of Ehrlich ascites carcinoma cells, 50–56, Copyright 2009, with permission from Elsevier.)

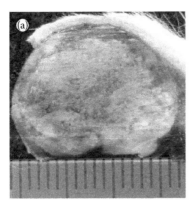
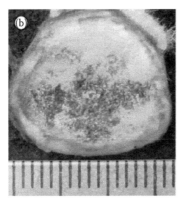
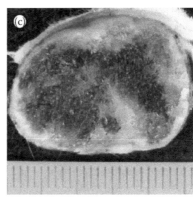

FIGURE 14.36 Histotopographic sections of glioma C6 rat tumor after sonodynamic therapy with Photolon at acoustic density of (a) 0.4, (b) 0.7, and (c) 1.0 W/cm², the corresponding necrosis area is 61.04% ± 4.77%, 82.65% ± 2.41%, 79.71% ± 4.66% (*n* = 11–13), respectively. (From Tserkovsky, D.A. et al., *Exp. Oncol.*, 34, 332, 2012.)

breast, colon, ovary, uterus, and larynx, have been treated with great success (Ferrara 2008). In comparison with conventional drug delivery approach, this modality has several important and attractive aspects. Microbubbles or echogenic agents have already been used widely in clinical sonography. It is possible to selectively target and deliver drug to the sonicated area at the deeply seated tumor using image guidance (e.g., ultrasound, CT, MRI, or PET) with a high positioning accuracy, which is beneficial in noninvasive treatment of a localized tumor. Ultrasound exposure is a noninvasive and extracorporeal method in comparison to RF or microwave ablation, which require the insertion of an interstitial needle or antenna. Delivery of various agents (e.g., small molecules, DNA, and nanoparticles) to different tissues with satisfactory therapeutic efficacy by either mechanical, thermal, or their combined effects is possible. The drug release was found to be rapid at the sonicated region only, and such target specificity is much better than with current chemotherapy. It could also be combined with other adjuvant therapies to avoid or reduce metastasis. However, the drawback is the relatively small volumes (a few millimeters in the lateral direction and about 1 cm in the axial direction) that can be treated. With the availability of image-guided HIFU system in clinics, ultrasound-mediated drug delivery may be rapidly applied to patients with only changes of the operation parameters on existing HIFU devices. However, this technology is still in its infancy, and the road ahead presents a number of exciting opportunities, but also pitfalls that will need to be circumvented.

A major challenge is to understand the mechanisms involved in this technology in order to optimize the ultrasound parameters. For example, for effective molecular internalization facilitated by the increased cell permeability, the plasmid should be close to the target because the cells return almost immediately to their initially impermeable state after sonication. Therefore, the transient pore opening limits the efficiency and quantity of delivery of the drug. If the IC-induced pores in the cell membrane are too large to be resealed quickly, the cell may lose vital cytoplasmic compound. The primary cause of cell death of sonoporation is likely to be a mechanical effect, resulting from shear stresses to the cell membranes by violent IC or microstreaming, which was supported by the observation that a large number of cells disappeared shortly after exposure (approximately 10 min), but before any slow biochemical processes (e.g., oncosis and apoptosis). Therefore, there should be a tradeoff between these two phenomena for high drug penetration through the membrane into cell cytoplasm and cell nucleus, as well as sufficient cell viability. The dominant effects should be monitored throughout the whole process of drug delivery, such as bubble cavitation (SC and IC) activities as the dose of the mechanical effect, and then used as a feedback in a closed-loop control system for effective and consistent outcome. Similarly, the temperature elevation would also be measured to present the thermal effect when temperature-sensitive drug vehicles are used.

In order to improve the performance of ultrasound-mediated drug/gene delivery, efforts are focused to enhance the stability of the vesicle, increase drug/gene loading, and improve targeting (Huang and MacDonald 2004, Ferrara et al. 2009). Air sensitivity in liposome can be increased by adding short-chain lipids (i.e., DHPCs) without entrapment of air or a calcein marker. Although DHPCs can destabilize bilayers by lowering their lysis tension, such a phenomenon is not the critical factor for lytic events. A general drawback of microbubbles as drug carriers is the rather small space available for loading the agents. Targeting ligands may prevent an efficient loading on the shell with agents, especially via electrostatic interactions between the drug and the shell. Drug release is restricted to the sonicated areas with low acoustic intensity, and therefore allows a low drug dose. Their relatively short *in vivo* half-life limits therapeutic irradiation. Because microbubbles have a broad size distribution, they will not respond evenly to a given ultrasound frequency. Small submicrometer-sized bubbles might be able to extravasate more easily, especially in the tumor vasculature (380–780 nm of pore size), but they can be rapidly taken up by RES (Xing et al. 2010, Lapotko 2011, Cavalli et al. 2012). In contrast, larger microbubbles can carry a higher drug payload and are more easily destroyed at relatively low acoustic intensities (Lentacker et al. 2009).

An important advantage of ELPs over other thermosensitive carriers is that drug accumulates because of the phase transition of the ELPs rather than by triggered drug release (Chilkoti et al. 2002). Therefore, a concentration gradient is not required to drive accumulation, and the ELP-drug conjugate injected at a low concentration aggregates in the heated tumor. Second, in comparison to antibody and other affinity-targeting approaches, the aggregation of ELP-radionuclide conjugates can directly target the heated tumor microvasculature, and circumvent the barriers associated with extravasation. Third, the clinical implementation does not require any concomitant development in hyperthermia. Finally, the aggregated ELP will resolubilize because of reversion of the phase transition after cessation of hyperthermia.

Electroporation, a popular experimental procedure in gene transfection, requires manipulating electric fields by setting and positioning two electrodes to the cells. In comparison, sonoporation requires a much simpler setting, especially *in vivo* (Mehier-Humbert et al. 2005). Bubble cavitation (SC and IC) may damage endothelial cells in both arteries and capillaries, through which pDNA has been transfected effectively into cardiomyocytes. Because of its small size, siRNA could pass through the coronary arteries and arterioles, resulting in efficient transduction from the bloodstream into arterial/arteriolar walls including the smooth muscle cells. Although pDNA permeates the endothelium more slowly, its transduction is predominant at the capillary bed with intense IC. The efficacy of gene transduction depends on the types and properties of the cells because neuroectodermal and limb ectodermal cells tend to express exogenous genes after a relatively short period of sonication. As part of such differences of gene transduction efficacy, it will be important to regulate the state of diffusion or the retention of an injected solution to the target. A higher concentration of injected DNA may increase the viscosity of the DNA/microbubble mixture, and lead to the failure of increment of gene transduction efficiency in neural tubes.

The poor water solubility of some sonosensitizers, which is mainly due to the physicochemical properties of hypocrellin, leads to their easy aggregation in aqueous media and limits their clinical applications. As a result, their concentration in a specific target may not be sufficient for the therapeutic requirement. One of the solutions is nanotechnology in medicine, which is able to manipulate molecules and supramolecular structures to produce programmed functions. Encapsulated sonosensitiser by hydrophobic nanoparticles has been shown to enhance the circulation time and prevent uptake by RES. The enormous surface area can be modified for an array of functionalities with diverse chemical or biochemical properties. In addition, the presence of nanoparticles in the appropriate size and amount in the liquid could provide nucleation sites, decrease the cavitation threshold, and increase the liquid temperature during sonication because of their surface roughness. Gold nanoparticles are highly attractive for their low toxicity, good uptake by mammalian cells, and antiangiogenetic properties. In the presence of gold nanoparticles, the relaxation time of protoporphyrin IX becomes longer, which favors an efficient production of ROS (Perez et al. 2008). Inorganic nanoparticles, such as titanium dioxide (TiO_2), strongly interact with light for SDT. Although TiO_2 cannot be used as an SDT drug alone because of its insufficient selectivity and low efficiency in cancer cells, it can work as an effective sonocatalyzer in the treatment of bladder cancer and in glioma cell lines (Yamaguchi et al. 2011). Since the long-term side effects of inorganic nanoparticle accumulation are still unknown, luminescent silica nanoparticles that decompose in the aqueous medium in hours may also work as sonosensitizer.

Overall, the research and application of ultrasound-mediated drug/gene delivery has developed tremendously during the past decade, and this advance will continue in research labs, academia, and industry. The future technological growth may be concentrated on, but not limited to, the following aspects: the understanding of the interaction of ultrasound with acoustically activated drug vehicles and the subsequent bioeffects; the development of drug vehicles for effective penetration through endothelial junctions that are typically damaged in systemic inflammation or tumor vasculature to target specific receptors rather than the vascular endothelium (high specificity), more effective drug/gene loading, and sufficient circulating lifespan; response of cancer/tumor cells to ultrasound (sonoporation) and drug (apoptosis); the optimization of ultrasound parameters for drug diagnosis and release in different organs with tradeoff between cell viability and drug/gene transduction; *in situ* real-time control by detecting the IC, temperature, or drug concentration; and large animal experiments and translation into the clinic.

The most common cancer treatment, chemotherapy, is often limited by cytotoxic effects on normal tissues, often associated with high doses. Therefore, it is highly desirable to reduce the dosage or frequency of administration by enhancing the effectiveness of drugs to the specific target. The application of ultrasound in the delivery of several therapeutic classes (e.g., chemotherapeutic, thrombolytic, and DNA-based drugs) acting via bubble cavitation (SC and IC), radiation forces, and/or heat has recently gained impetus. The advantages of this novel technology include noninvasiveness, low cost, easy operation, good focusing and penetration inside the body, and no radiation. Imaging methods (e.g., sonography and MRI) may be particularly helpful in defining the target, for determining local drug concentration, and for evaluation of efficacy and temperature. Recent successes illustrate ultrasound as a valuable future therapeutic tool for drug delivery by lowering the administration dosage of anticancer drugs. However, this technology is still in its infancy. Progression to clinical use will depend on several requirements. The core requirement is the appropriate design of drug vehicles and assemblies, such as microbubbles, liposomes, and micelles that are capable of carrying a sufficient payload, yet maintaining specific response to acoustic energy. Conjugating tumor- and disease-specific antigens or binding ligands to the surface of drug vehicles can improve their specificity for targets without compromising either echogenicity or payload capacity. Prolonged circulation and minimal removal by RES are also required. Only by overcoming these obstacles can ultrasound translate from the bench to the bedside. An increasing body of knowledge has been acquired on the interaction of ultrasound energy with tissues, therapeutic agents, and drug carriers for enhanced therapeutic efficacy and efficiency. With advances in the transducer and sonography technologies, a variety of ultrasound-based therapeutic applications have been developed and continue to be proposed. For technical improvement and translation from one tissue type to another, an in-depth understanding of the ultrasound mechanisms and physical characteristics of the tissues will be required, as well as collecting both *in vitro* and *in vivo* data and developing mathematical models to optimize the treatment protocols. It must be noted that this technique has mainly been applied *in vitro* or in small animal studies to date. There is a great need in the field for large animal and human studies. Moreover, ultrasound-mediated gene delivery seems to outweigh drug delivery in certain applications, such as cardiovascular diseases. Altogether, ultrasound-mediated drug/gene delivery is a promising technology that attracts the interests of both scientific research and medical applications.

REFERENCES

Alam MI, Beg S, Samad A, Baboota S, Kohli K, Ali J, Ahuja A, Akbar M. Strategy for effective brain drug delivery. *European Journal of Pharmaceutical Sciences* 2010;40:385–403.

Allen TM. Liposomes: Opportunities in drug delivery. *Drugs* 1997;54:8–14.

Alvarez-Román R, Merino G, Kalia Y, Naik A, Guy RH. Skin permeability enhancement by low frequency sonophoresis: Lipid extraction and transport pathways. *Journal of Pharmaceutical Sciences* 2003;92:1138–1146.

Awada A, Piccart M. Strategies offering protection from the toxic effects of anticancer treatments with a focus on chemoprotective agents. *Current Opinion in Oncology* 2000;12:289–296.

Azuma H, Tomita N, Kaneda Y, Koike H, Ogihara T, Katsuoka Y, Morishita R. Transfection of NFκB-decoy oligodeoxynucleotides using efficient ultrasound-mediated gene transfer into donor kidneys prolonged survival of rat renal allografts. *Gene Therapy* 2003;10:415–425.

Bakay L, Hueter T, Ballantine H, Sosa D. Ultrasonically produced changes in the blood-brain barrier. *AMA Archives of Neurology and Psychiatry* 1956;76:457–467.

Ballantine H, Hueter T, Nauta W, Sosa D. Focal destruction of nervous tissue by focused ultrasound: Biophysical factors influencing its application. *Journal of Experimental Medicine* 1956;104:337–360.

Bangham A. Liposomes: The Babraham connection. *Chemistry and Physics of Lipids* 1993;64:275–285.

Barry BW. Mode of action of penetration enhancers in human skin. *Journal of Controlled Release* 1987;6:85–97.

Becker BM, Helfrich S, Baker E, Lovgren K, Minugh PA, Machan JT. Ultrasound with topical anesthetic rapidly decreases pain of intravenous cannulation. *Academic Emergency Medicine* 2005;12:289–295.

Bekeredjian R, Grayburn PA, Shohet RV. Use of ultrasound contrast agents for gene or drug delivery in cardiovascular medicine. *Journal of the American College of Cardiology* 2005;45:329.

Bekeredjian R, Kroll RD, Fein E, Tinkov S, Coester C, Winter G, Katus HA, Kulaksiz H. Ultrasound targeted microbubble destruction increases capillary permeability in hepatomas. *Ultrasound in Medicine and Biology* 2007;33:1592–1598.

Billard B, Hynynen K, Roemer R. Effects of physical parameters on high temperature ultrasound hyperthermia. *Ultrasound in Medicine and Biology* 1990;16:409–420.

Borden MA, Sarantos MR, Stieger SM, Simon SI, Ferrara KW, Dayton PA. Ultrasound radiation force modulates ligand availability on targeted contrast agents. *Molecular Imaging* 2006;5:139–147.

Boucaud A, Machet L, Arbeille B, Machet M, Sournac M, Mavon A, Patat F, Vaillant L. In vitro study of low-frequency ultrasound-enhanced transdermal transport of fentanyl and caffeine across human and hairless rat skin. *International Journal of Pharmaceutics* 2001;228:69–77.

Boucher Y, Baxter LT, Jain RK. Interstitial pressure gradients in tissue-isolated and subcutaneous tumors: Implications for therapy. *Cancer Research* 1990;50:4478–4484.

Braiden V, Ohtsuru A, Kawashita Y, Miki F, Sawada T, Ito M, Cao Y, Kaneda Y, Koji T, Yamashita S. Eradication of breast cancer xenografts by hyperthermic suicide gene therapy under the control of the heat shock protein promoter. *Human Gene Therapy* 2000;11:2453–2463.

Branski LK, Gauglitz GG, Herndon DN, Jeschke MG. A review of gene and stem cell therapy in cutaneous wound healing. *Burns* 2009;35:171–180.

Brennen CE. *Cavitation and Bubble Dynamics*. New York: Oxford University Press, 1995.

Brightman M, Klatzo I, Olsson Y, Reese T. The blood-brain barrier to proteins under normal and pathological conditions. *Journal of the Neurological Sciences* 1970;10:215–239.

Burgess A, Ayala-Grosso CA, Ganguly M, Jordão JF, Aubert I, Hynynen K. Targeted delivery of neural stem cells to the brain using MRI-guided focused ultrasound to disrupt the blood-brain barrier. *PLoS One* 2011;6:e27877.

Byl NN. The use of ultrasound as an enhancer for transcutaneous drug delivery: Phonophoresis. *Physical Therapy* 1995;75:539–553.

Campbell R. Tumor physiology and delivery of nanopharmaceuticals. *Anti-Cancer Agents in Medicinal Chemistry* 2006;6:503–512.

Cavalli R, Bisazza A, Trotta M, Argenziano M, Civra A, Donalisio M, Lembo D. New chitosan nanobubbles for ultrasound-mediated gene delivery: Preparation and in vitro characterization. *International Journal of Nanomedicine* 2012;7:3309–3318.

Chen P-Y, Liu H-L, Hua M-Y, Yang H-W, Huang C-Y, Chu P-C, Lvu L-A et al. Novel magnetic/ultrasound focusing system enhances nanoparticle drug delivery for glioma treatment. *Neuro-Oncology* 2010;12:1050–1060.

Chilkoti A, Dreher MR, Meyer DE, Raucher D. Targeted drug delivery by thermally responsive polymers. *Advanced Drug Delivery Reviews* 2002;54:613–630.

Chomas JE, Dayton PA, May D, Allen J, Klibanov K, Ferrara KW. Optical observation of contrast agent destruction. *Applied Physics Letters* 2000;77:1056–1058.

Chuang H, Taylor E, Davison TW. Clinical evaluation of a continuous minimally invasive glucose flux sensor placed over ultrasonically permeated skin. *Diabetes Technology and Therapeutics* 2004;6:21–30.

Chuang Y-H, Cheng P-W, Li P-C. Combining radiation force with cavitation for enhanced sonothrombolysis. *IEEE Transactions on Ultrasonics, Ferroelectrics and Frequency Control* 2013;60:97–104.

Clement G, Hynynen K. A non-invasive method for focusing ultrasound through the human skull. *Physics in Medicine and Biology* 2002;47:1219.

Clement GT, White PJ, King RL, McDannold N, Hynynen K. A magnetic resonance imaging–compatible, large-scale array for trans-skull ultrasound surgery and therapy. *Journal of Ultrasound in Medicine* 2005;24:1117–1125.

Cohen MG, Tuero E, Bluguermann J, Kevorkian R, Berrocal DH, Carlevaro O, Picabea E, Hudson MP, Siegel RJ, Douthat L. Transcutaneous ultrasound-facilitated coronary thrombolysis during acute myocardial infarction. *American Journal of Cardiology* 2003;92:454–457.

Cope DA, Dewhirst MW, Friedman HS, Bigner DD, Zalutsky MR. Enhanced delivery of a monoclonal antibody F (ab′) 2 fragment to subcutaneous human glioma xenografts using local hyperthermia. *Cancer Research* 1990;50:1803–1809.

Cosgrove D. Ultrasound contrast agents: An overview. *European Journal of Radiology* 2006;60:324–330.

Curti BD. Physical barriers to drug delivery in tumors. In: Chabner BA, ed., *Cancer Chemotherapy and Biotherapy: Principles and Practice*, Philadelphia, PA: Lippincott-Raven, 1996, pp. 709–719.

Daffertshofer M, Gass A, Ringleb P, Sitzer M, Sliwka U, Els T, Sedlaczek O, Koroshetz WJ, Hennerici MG. Transcranial low-frequency ultrasound-mediated thrombolysis in brain ischemia increased risk of hemorrhage with combined ultrasound and tissue plasminogen activator: Results of a phase II clinical trial. *Stroke* 2005;36:1441–1446.

Datta S, Coussios C-C, Ammi AY, Mast TD, de Courten-Myers GM, Holland CK. Ultrasound-enhanced thrombolysis using Definity® as a cavitation nucleation agent. *Ultrasound in Medicine and Biology* 2008;34:1421–1433.

Deckers R, Quesson B, Arsaut J, Eimer S, Couillaud F, Moonen CT. Image-guided, noninvasive, spatiotemporal control of gene expression. *Proceedings of the National Academy of Sciences* 2009;106:1175–1180.

Denet AR, Vanbever R, Preat V. Skin electroporation for transdermal and topical delivery. *Advanced Drug Delivery Review* 2004;56:659–674.

Deng CX, Sieling F, Pan H, Cui J. Ultrasound-induced cell membrane porosity. *Ultrasound in Medicine and Biology* 2004;30:519–526.

Dijkmans PA, Juffermans LJ, Musters RJ, van Wamel A, ten Cate FJ, van Gilst W, Visser CA, de Jong N, Kamp O. Microbubbles and ultrasound: From diagnosis to therapy. *European Journal of Echocardiography* 2004;5:245–256.

Dittmar KM, Xie J, Hunter F, Trimble C, Bur M, Frenkel V, Li KC. Pulsed high-intensity focused ultrasound enhances systemic administration of naked DNA in squamous cell carcinoma model: Initial experience. *Radiology* 2005;235:541–546.

Dolan MS, Gala SS, Dodla S, Abdelmoneim SS, Xie F, Cloutier D, Bierig M, Mulvagh SL, Porter TR, Labovitz AJ. Safety and efficacy of commercially available ultrasound contrast agents for rest and stress echocardiography: A multicenter experience. *Journal of the American College of Cardiology* 2009;53:32–38.

Dougherty TJ. A brief history of clinical photodynamic therapy at Roswell Park Cancer Institute. *Journal of Clinical Laser Medicine and Surgery* 1996;14:219–221.

Dougherty TJ, Gomer CJ, Henderson BW, Jori G, Kessel D, Korbelik M, Moan J, Peng Q. Photodynamic therapy. *Journal of the National Cancer Institute* 1998;90:889–905.

Dromi S, Frenkel V, Luk A, Traughber B, Angstadt M, Bur M, Poff J, Xie JW, Libutti SK, Li KCP, Wood BJ. Pulsed-high intensity focused ultrasound and low temperature-sensitive liposomes for enhanced targeted drug delivery and antitumor effect. *Clinical Cancer Research* 2007;13:2722–2727.

Dubey V, Mishra D, Dutta T, Nahar M, Saraf DK, Jain NK. Dermal and transdermal delivery of an anti-psoriatic agent via ethanolic liposomes. *Journal of Controlled Release* 2007;123:148–154.

Enden T, Haig Y, Kløw N-E, Slagsvold C-E, Sandvik L, Ghanima W, Hafsahl G, Holme PA, Holmen LO, Njaastad AM. Long-term outcome after additional catheter-directed thrombolysis versus standard treatment for acute iliofemoral deep vein thrombosis (the CaVenT study): A randomised controlled trial. *Lancet* 2012;379:31–38.

Endo-Takahashi Y, Negishi Y, Kato Y, Suzuki R, Maruyama K, Aramaki Y. Efficient siRNA delivery using novel siRNA-loaded bubble liposomes and ultrasound. *International Journal of Pharmaceutics* 2012;422:504–509.

Erikson JM, Freeman GL, Chandrasekar B. Ultrasound-targeted antisense oligonucleotide attenuates ischemia/reperfusion-induced myocardial tumor necrosis factor-alpha. *Journal of Molecular and Cellular Cardiology* 2003;35:119–130.

Everbach EC, Francis CW. Cavitational mechanisms in ultrasound-accelerated thrombolysis at 1 MHz. *Ultrasound in Medicine and Biology* 2000;26:1153–1160.

Fellinger KK. *und Therapie des Chronischen Gelenkhuematismum.* Maudrich, Austria, 1954, pp. 549–552.

Feril LB, Ogawa R, Tachibana K, Kondo T. Optimized ultrasound-mediated gene transfection in cancer cells. *Cancer Science* 2006;97:1111–1114.

Ferrara K, Pollard R, Borden M. Ultrasound microbubble contrast agents: Fundamentals and applications to gene and drug delivery. *Annual Review of Biomedical Engineering* 2007;9:415–447.

Ferrara KW. Driving delivery vehicles with ultrasound. *Advanced Drug Delivery Reviews* 2008;60:1097–1102.

Ferrara KW, Borden MA, Zhang H. Lipid-shelled vehicles: Engineering for ultrasound molecular imaging and drug delivery. *Accounts of Chemical Research* 2009;42:881–892.

Floyd JD, Nguyen DT, Lobins RL, Bashir Q, Doll DC, Perry MC. Cardiotoxicity of cancer therapy. *Journal of Clinical Oncology* 2005;23:7685–7696.

Frenkel V. Ultrasound mediated delivery of drugs and genes to solid tumors. *Advanced Drug Delivery Reviews* 2008;60:1193–1208.

Frenkel V, Etherington A, Greene M, Quijano J, Xie J, Hunter F, Dromi S, Li KC. Delivery of liposomal doxorubicin (Doxil) in a breast cancer tumour model: Investigation of potential enhancement by pulsed-high intensity focused ultrasound exposure. *Academic Radiology* 2006;13:469–479.

Fujii H, Sun Z, Li S-H, Wu J, Fazel S, Weisel RD, Rakowski H, Lindner J, Li R-K. Ultrasound-targeted gene delivery induces angiogenesis after a myocardial infarction in mice. *JACC: Cardiovascular Imaging* 2009;2:869–879.

Fujiwara K, Watanabe T. Effects of hyperthermia, radiotherapy and thermoradiotherapy on tumor microvascular permeability. *Pathology International* 1990;40:79–84.

Gasselhuber A, Dreher MR, Partanen A, Yarmolenko PS, Woods D, Wood BJ, Haemmerich D. Targeted drug delivery by high intensity focused ultrasound mediated hyperthermia combined with temperature-sensitive liposomes: Computational modelling and preliminary in vivo validation. *International Journal of Hyperthermia* 2012;28:337–348.

George Mychaskiw I, Badr AE, Tibbs R, Clower BR, Zhang JH. Optison (FS069) disrupts the blood-brain barrier in rats. *Anesthesia and Analgesia* 2000;91:798–803.

Godin B, Touitou E. Ethosomes: New prospects in transdermal delivery. *Critical Reviews in Therapeutic Drug Carrier Systems* 2003;20:63–102.

Greenleaf WJ, Bolander ME, Sarkar G, Goldring MB, Greenleaf JF. Artificial cavitation nuclei significantly enhance acoustically induced cell transfection. *Ultrasound in Medicine and Biology* 1998;24:587–595.

Guzman HR, Nguyen DX, McNamara AJ, Prausnitz MR. Equilibrium loading of cells with macromolecules by ultrasound: Effects of molecular size and acoustic energy. *Journal of Pharmaceutical Sciences* 2002;91:1693–1701.

Hauck M, Coffin D, Dodge R, Dewhirst M, Mitchell J, Zalutsky M. A local hyperthermia treatment which enhances antibody uptake in a glioma xenograft model does not affect tumour interstitial fluid pressure. *International Journal of Hyperthermia* 1997;13:307–316.

Hauff P, Reinhardt M, Briel A, Debus N, Schirner M. Molecular targeting of lymph nodes with L-selectin ligand-specific us contrast agent: A feasibility study in mice and dogs. *Radiology* 2004;231:667–673.

Herman TS, Teicher BA, Jochelson M, Clark J, Svensson G, Coleman CN. Rationale for use of local hyperthermia with radiation therapy and selected anticancer drugs in locally advanced human malignancies. *International Journal of Hyperthermia* 1988;4:143–158.

Hernot S, Klibanov AL. Microbubbles in ultrasound-triggered drug and gene delivery. *Advanced Drug Delivery Reviews* 2008;60:1153–1166.

Holland CK, Vaidya SS, Coussios CC, Shaw GJ. Thrombolytic effects of 120-kHz and 1-MHz ultrasound and tissue plasminogen activator on porcine whole blood clots. *Journal of the Acoustical Society of America* 2002;112:2370.

Horsman M, Overgaard J. Can mild hyperthermia improve tumour oxygenation? *International Journal of Hyperthermia* 1997;13:141–147.

Hosono MN, Hosono M, Endo K, Ueda R, Onoyama Y. Effect of hyperthermia on tumor uptake of radiolabeled anti-neural cell adhesion molecule antibody in small-cell lung cancer xenografts. *Journal of Nuclear Medicine* 1994;35:504–509.

Howlader N, Noone AM, Krapcho M, Neyman N, Aminou R, Waldron W, Altekruse SF, Kosary CL, Ruhl J, Tatalovichz. *SEER Cancer Statistics Review*, National Cancer Institute, 2011, pp. 1975–2008.

Huang SL. Liposomes in ultrasonic drug and gene delivery. *Advanced Drug Delivery Reviews* 2008;60:1167–1176.

Huang SL, MacDonald RC. Acoustically active liposomes for drug encapsulation and ultrasound-triggered release. *Biochimica et Biophysica Acta* 2004;1665:134–141.

Hundt W, Steinbach S, Mayer D, Bednarski MD. Modulation of luciferase activity using high intensity focused ultrasound in combination with bioluminescence imaging, magnetic resonance imaging and histological analysis in muscle tissue. *Ultrasonics* 2009;49:549–557.

Husseini GA, Myrup G, Pitt WG, Christensen DA, Rapoport N. Factors affecting acoustically triggered release of drugs from polymeric micelles. *Journal of Controlled Release* 2000;69:43–52.

Hynynen K. Ultrasound for drug and gene delivery to the brain. *Advanced Drug Delivery Reviews* 2008;60:1209–1217.

Hynynen K, McDannold N, Sheikov NA, Jolesz FA, Vykhodtseva N. Local and reversible blood-brain barrier disruption by noninvasive focused ultrasound at frequencies suitable for trans-skull sonications. *Neuroimage* 2005;24:12–20.

Hynynen K, McDannold N, Vykhodtseva N, Jolesz F. Non-invasive opening of BBB by focused ultrasound. *Acta Neurochirurgica. Supplement* 2003;86:555–558.

Hynynen K, McDannold N, Vykhodtseva N, Jolesz FA. Noninvasive MR imaging-guided focal opening of the blood-brain barrier in rabbits. *Radiology* 2001;220:640–646.

Hynynen K, McDannold N, Vykhodtseva N, Raymond S, Weissleder R, Jolesz FA, Sheikov N. Focal disruption of the blood-brain barrier due to 260-kHz ultrasound bursts: A method for molecular imaging and targeted drug delivery. *Journal of Neurosurgery* 2006;105:445–454.

Issels RD. Hyperthermia adds to chemotherapy. *European Journal of Cancer* 2008;44:2546–2554.

Jain RK. Normalization of tumor vasculature: An emerging concept in antiangiogenic therapy. *Science* 2005;307:58–62.

Jordão JF, Ayala-Grosso CA, Markham K, Huang Y, Chopra R, McLaurin J, Hynynen K, Aubert I. Antibodies targeted to the brain with image-guided focused ultrasound reduces amyloid-β plaque load in the TgCRND8 mouse model of Alzheimer's disease. *PLoS One* 2010;5:e10549.

Karino T, Koga S, Maeta M. Experimental studies of the effects of local hyperthermia on blood flow, oxygen pressure and pH in tumors. *Japanese Journal of Surgery* 1988;18:276–283.

Kataoka K, Matsumoto T, Yokoyama M, Okano T, Sakurai Y, Fukushima S. Doxorubicin-loaded poly(ethylene glycol)-poly(beta-benzyl-L-aspartate) copolymer micelles: Their pharmaceutical characteristics and biological significance. *Journal of Controlled Release* 2000;64:143–153.

Katz NP, Shapiro DE, Herrmann TE, Kost J, Custer LM. Rapid onset of cutaneous anesthesia with EMLA cream after pretreatment with a new ultrasound-emitting device. *Anesthesia and Analgesia* 2004;98:371–376.

Kennedy JE. High-intensity focused ultrasound in the treatment of solid tumors. *Nature Reviews. Cancer* 2005;5:321–327.

Khan T, Sellke F, Laham R. Gene therapy progress and prospects: Therapeutic angiogenesis for limb and myocardial ischemia. *Gene Therapy* 2003;10:285–291.

Kheirolomoom A, Dayton PA, Lum AFH, Little E, Paoli EE, Zheng HR, Ferrara KW. Acoustically-active microbubbles conjugated to liposomes: Characterization of a proposed drug delivery vehicle. *Journal of Controlled Release* 2007;118:275–284.

Kimmel E. Cavitation bioeffects. *Critical Reviews in Biomedical Engineering* 2006;34:105.

Kinoshita M. Targeted drug delivery to the brain using focused ultrasound. *Top Magnetic Resonance Imaging* 2006;17:209–215.

Kinoshita M, McDannold N, Jolesz FA, Hynynen K. Noninvasive localized delivery of Herceptin to the mouse brain by MRI-guided focused ultrasound-induced blood–brain barrier disruption. *Proceedings of the National Academy of Sciences* 2006a;103:11719–11723.

Kinoshita M, McDannold N, Jolesz FA, Hynynen K. Targeted delivery of antibodies through the blood–brain barrier by MRI-guided focused ultrasound. *Biochemical and Biophysical Research Communications* 2006b;340:1085–1090.

Klegerman ME, Zou Y, Mcpherson DD. Fibrin targeting of echogenic liposomes with inactivated tissue plasminogen activator. *Journal of Liposome Research* 2008;18:95–112.

Klibanov AL. Targeted delivery of gas-filled microspheres, contrast agents for ultrasound imaging. *Advanced Drug Delivery Reviews* 1999;37:139–157.

Klibanov AL. Microbubble contrast agents: Targeted ultrasound imaging and ultrasound-assisted drug-delivery applications. *Investigative Radiology* 2006;41:354–362.

Kong G, Braun RD, Dewhirst MW. Characterization of the effect of hyperthermia on nanoparticle extravasation from tumour vasculature. *Cancer Research* 2001;61:3021–3032.

Kremkau FW, Kaufmann JS, Walke MM, Burch PG, Spurr CL. Ultrasonic enhancement of nitrogen mustard cytotoxicity in mouse leukemia. *Cancer* 1976;37:1643–1647.

Kumon RE, Aehle M, Sabens D, Parikh P, Han Y, Kourennyi D, Deng C. Spatiotemporal effects of sonoporation measured by real-time calcium imaging. *Ultrasound in Medicine and Biology* 2009;35:494–506.

Kuroki M, Hachimine K, Abe H, Shibaguchi H, Maekawa SI, Yanagisawa J, Kinugasa T, Tanaka T, Yamashita Y. Sonodynamic therapy of cancer using novel sonosensitizers. *Anticancer Research* 2007;27:3673–3677.

Kushner J, Blankschtein D, Langer R. Experimental demonstration of the existence of highly permeable localized transport regions in low-frequency sonophoresis. *Journal of Pharmaceutical Sciences* 2004;93:2733–2745.

Lan HY, Mu W, Tomita N, Huang XR, Li JH, Zhu H-J, Morishita R, Johnson RJ. Inhibition of renal fibrosis by gene transfer of inducible Smad7 using ultrasound-microbubble system in rat UUO model. *Journal of the American Society of Nephrology* 2003;14:1535–1548.

Langer R. Biomaterials in drug delivery and tissue engineering: One laboratory's experience. *Accounts of Chemical Research* 2000;33:94–101.

Lanza GM, Abendschein DR, Hall CS, Scott MJ, Scherrer DE, Houseman A, Miller JG, Wickline SA. In vivo molecular imaging of stretch-induced tissue factor in carotid arteries with ligand-targeted nanoparticles. *Journal of the American Society of Echocardiography* 2000;13:608–614.

Lapotko D. Plasmonic nanobubbles as tunable cellular probes for cancer theranostics. *Cancers* 2011;3:802–840.

Larina I, Evers B, Bartels C, Ashitkov T, Larin K, Esenaliev R. Ultrasound-enhanced drug delivery for efficient cancer therapy. 24th Annual Conference and the Annual Fall Meeting of the Biomedical Engineering Society EMBS/BMES, Houston, TX, Oct. 23–26, 2002, pp. 492–493.

Larina IV, Evers BM, Ashitkov TV, Bartels C, VLarin KV, Esenaliev RO. Enhancement of drug delivery in tumors by using interaction of nanoparticles with ultrasound radiation. *Technology in Cancer Research and Treatment* 2005a;4:217–226.

Larina IV, Evers BM, Esenaliev RO. Optimal drug and gene delivery in cancer cells by ultrasound-induced cavitation. *Anticancer Research* 2005b;25:149–156.

Larkin JO, Casey GD, Tangney M, Cashman J, Collins CG, Soden DM, O'Sullivan GC. Effective tumor treatment using optimized ultrasound-mediated delivery of bleomycin. *Ultrasound in Medicine and Biology* 2008;34:406–413.

Lauer CG, Burge R, Tang DB, Bass BG, Gomez ER, Alving BM. Effect of ultrasound on tissue-type plasminogen activator-induced thrombolysis. *Circulation* 1992;86:1257–1264.

Lavon I, Kost J. Ultrasound and transdermal drug delivery. *Drug Discovery Today* 2004;9:670–676.

Lawrie A, Brisken A, Francis S, Cumberland D, Crossman D, Newman C. Microbubble-enhanced ultrasound for vascular gene delivery. *Gene Therapy* 2000;7:2023–2027.

Lee P-W, Peng S-F, Su C-J, Mi F-L, Chen H-L, Wei M-C, Lin H-J, Sung H-W. The use of biodegradable polymeric nanoparticles in combination with a low-pressure gene gun for transdermal DNA delivery. *Biomaterials* 2008a;29:742–751.

Lee S, Newnham RE, Smith NB. Short ultrasound exposure times for noninvasive insulin delivery in rats using the lightweight cymbal array. *IEEE Transactions on Ultrasonics, Ferroelectrics and Frequency Control* 2004;51:176–180.

Lee W-R, Pan T-L, Wang P-W, Zhuo R-Z, Huang C-M, Fang J-Y. Erbium: YAG laser enhances transdermal peptide delivery and skin vaccination. *Journal of Controlled Release* 2008b;128:200–208.

Lefor AT, Makohon S, Ackerman NB. The effects of hyperthermia on vascular permeability in experimental liver metastasis. *Journal of Surgical Oncology* 1985;28:297–300.

Lentacker I, De Smedt SC, Sanders NN. Drug loaded microbubble design for ultrasound triggered delivery. *Soft Matter* 2009;5:2161–2170.

Levin G, Gershonowitz A, Sacks H, Stern M, Sherman A, Rudaev S, Zivin I, Phillip M. Transdermal delivery of human growth hormone through RF-microchannels. *Pharmaceutical Research* 2005;22:550–555.

Levy D, Kost J, Meshulam Y, Langer R. Effect of ultrasound on transdermal drug delivery to rats and guinea pigs. *Journal of Clinical Investigation* 1989;83:2074.

Li L, ten Hagen TL, Schipper D, Wijnberg TM, van Rhoon GC, Eggermont AM, Lindner LH, Koning GA. Triggered content release from optimized stealth thermosensitive liposomes using mild hyperthermia. *Journal of Controlled Release* 2010;143:274–279.

Li YS, Davidson E, Reid CN, McHale AP. Optimising ultrasound-mediated gene transfer (sonoporation) *in vitro* and prolonged expression of a transgene *in vivo*: Potential applications for gene therapy of cancer. *Cancer Letters* 2009;273:62–69.

Lipson RL, Baldes EJ, Olsen AM. The use of a derivative of hematoporphyrin in tumor detection. *Journal of the National Cancer Institute* 1961;26:1–11.

Liu H-L, Hua M-Y, Chen P-Y, Chu P-C, Pan C-H, Yang H-W, Huang C-Y, Wang J-J, Yen T-C, Wei K-C. Blood-brain barrier disruption with focused ultrasound enhances delivery of chemotherapeutic drugs for glioblastoma treatment. *Radiology* 2010;255:415–425.

Liu H-L, Wai Y-Y, Chen W-S, Chen J-C, Hsu P-H, Wu X-Y, Huang W-C, Yen T-C, Wang J-J. Hemorrhage detection during focused-ultrasound induced blood-brain-barrier opening by using susceptibility-weighted magnetic resonance imaging. *Ultrasound in Medicine and Biology* 2008;34:598–606.

Longo FW, Tomashefsky P, Rivin BD, Tannenbaum M. Interaction of ultrasonic hyperthermia with two alkylating agents in a murine bladder tumor. *Cancer Research* 1983;43:3231–3235.

Lum AFH, Borden MA, Dayton PA, Kruse DE, Simon SI, Ferrara KW. Ultrasound radiation force enables targeted deposition of model drug carriers loaded on microbubbles. *Journal of Controlled Release* 2006;111:128–134.

Marin A, Sun H, Husseini GA, Pitt WG, Christensen DA, Rapoport NY. Drug delivery in pluronic micelles: Effect of high-frequency ultrasound on drug release from micelles and intracellular uptake. *Journal of Controlled Release* 2002;84:39–47.

Marmottant P, Meer Svd, Emmer M, Versluis M, Jong Nd, Hilgenfeldt S, Lohse D. A model for large amplitude oscillations of coated bubbles accounting for buckling and rupture. *Journal of the Acoustical Society of America* 2005;118:3499–3505.

Marquet F, Pernot M, Aubry J, Montaldo G, Marsac L, Tanter M, Fink M. Non-invasive transcranial ultrasound therapy based on a 3D CT scan: Protocol validation and in vitro results. *Physics in Medicine and Biology* 2009;54:2597.

Maxwell AD, Cain CA, Duryea AP, Yuan L, Gurm HS, Xu Z. Noninvasive thrombolysis using pulsed ultrasound cavitation therapy–histotripsy. *Ultrasound in Medicine and Biology* 2009;35:1982–1994.

Mayer CR, Bekeredjian R. Ultrasonic gene and drug delivery to the cardiovascular system. *Advanced Drug Delivery Reviews* 2008;60:1177–1192.

McDannold N, Vykhodtseva N, Hynynen K. Targeted disruption of the blood-brain barrier with focused ultrasound: Association with cavitation activity. *Physics in Medicine and Biology* 2006;51:793–801.

McDannold N, Vykhodtseva N, Hynynen K. Use of ultrasound pulses combined with definity for targeted blood-brain barrier disruption: A feasibility study. *Ultrasound in Medicine and Biology* 2007;33:584–590.

McDannold N, Vykhodtseva N, Hynynen K. Effects of acoustic parameters and ultrasound contrast agent dose on focused-ultrasound induced blood-brain barrier disruption. *Ultrasound in Medicine and Biology* 2008;34:930–937.

McDannold N, Vykhodtseva N, Jolesz FA, Hynynen K. MRI investigation of the threshold for thermally induced blood–brain barrier disruption and brain tissue damage in the rabbit brain. *Magnetic Resonance in Medicine* 2004;51:913–923.

McDannold N, Vykhodtseva N, Raymond S, Jolesz FA, Hynynen K. MRI-guided targeted blood-brain barrier disruption with focused ultrasound: Histological findings in rabbits. *Ultrasound in Medicine and Biology* 2005;31:1527–1537.

McKee TD, Grandi P, Mok W, Alexandrakis G, Insin N, Zimmer JP, Bawendi M, Boucher Y, Breakefield XO, Jain RK. Degradation of fibrillar collagen in a human melanoma xenograft improves the efficacy of an oncolytic herpes simplex virus vector. *Cancer Research* 2006;6:2509–2513.

Meairs S, Alonso A. Ultrasound, microbubbles and the blood–brain barrier. *Progress in Biophysics and Molecular Biology* 2007;93:354–362.

Mehier-Humbert S, Bettinger T, Yan F, Guy RH. Plasma membrane poration induced by ultrasound exposure: Implication for drug delivery. *Journal of Controlled Release* 2005;104:213–222.

Mei J, Cheng Y, Song Y, Yang Y, Wang F, Liu Y, Wang Z. Experimental study on targeted methotrexate delivery to the rabbit brain via magnetic resonance imaging–guided focused ultrasound. *Journal of Ultrasound in Medicine* 2009;28:871–880.

Merino G, Kalia YN, Delgado-Charro MB, Potts RO, Guy RH. Frequency and thermal effects on the enhancement of transdermal transport by sonophoresis. *Journal of Controlled Release* 2003;88:85–94.

Mesiwala AH, Farrell L, Wenzel HJ, Silbergeld DL, Crum LA, Winn HR, Mourad PD. High-intensity focused ultrasound selectively disrupts the blood-brain barrier *in vivo*. *Ultrasound in Medicine and Biology* 2002;28:389–400.

Miller D, Thomas R, Buschbom R. Comet assay reveals DNA strand breaks induced by ultrasonic cavitation *in vitro*. *Ultrasound in Medicine and Biology* 1995;21:841–848.

Miller D, Thomas R, Frazier M. Ultrasonic cavitation indirectly induces single strand breaks in DNA of viable cells *in vitro* by the action of residual hydrogen peroxide. *Ultrasound in Medicine and Biology* 1991;17:729–735.

Miller DL, Pislaru SV, Greenleaf JF. Sonoporation: Mechanical DNA delivery by ultrasonic cavitation. *Somatic Cell and Molecular Genetics* 2002;27:115–134.

Miller DL, Quddus J. Sonoporation of monolayer cells by diagnostic ultrasound activation of contrast-agent gas bodies. *Ultrasound in Medicine and Biology* 2000;26:661–667.

Miller MW, Miller DL, Brayman AA. A review of *in vitro* bioeffects of inertial ultrasonic cavitation from a mechanistic perspective. *Ultrasound in Medicine and Biology* 1996;22:1131–1154.

Miller MW, Miller WM, Battaglia LF. Biological and environmental factors affecting ultrasound-induced hemolysis *in vitro*: 3. antioxidant (Trolox®) inclusion. *Ultrasound in Medicine and Biology* 2003;29:103–112.

Miri AK, Mitri FG. Acoustic radiation force on a spherical contrast agent shell near a vessel porous wall – theory. *Ultrasound in Medicine and Biology* 2011;37:301–311.

Mišík V, Riesz P. Free radical intermediates in sonodynamic therapy. *Annals of the New York Academy of Sciences* 2000;899:335–348.

Mišík Vr, Riesz P. EPR characterization of free radical intermediates formed during ultrasound exposure of cell culture media. *Free Radical Biology and Medicine* 1999;26:936–943.

Mitragotri S. Synergistic effect of enhancers for transdermal drug delivery. *Pharmaceutical Research* 2000;17:1354–1359.

Mitragotri S, Blankschtein D, Langer R. Ultrasound-mediated transdermal protein delivery. *Science* 1995;269:850–853.

Mitragotri S, Blankschtein D, Langer R. Transdermal drug delivery using low-frequency sonophoresis. *Pharmaceutical Research* 1996;13:411–420.

Mitragotri S, Kost J. Low-frequency sonophoresis: A review. *Advanced Drug Delivery Reviews* 2004;56:589–601.

Mitragotri S, Ray D, Farrell J, Tang H, Yu B, Kost J, Blankschtein D, Langer R. Synergistic effect of low-frequency ultrasound and sodium lauryl sulfate on transdermal transport. *Journal of Pharmaceutical Sciences* 2000;89:892–900.

Miyoshi N, Mišík V, Riesz P. Sonodynamic toxicity of gallium-porphyrin analogue ATX-70 in human leukemia cells. *Radiation Research* 1997;148:43–47.

Molina CA, Ribo M, Rubiera M, Montaner J, Santamarina E, Delgado-Mederos R, Arenillas JF, Huertas R, Purroy F, Delgado P. Microbubble administration accelerates clot lysis during continuous 2-MHz ultrasound monitoring in stroke patients treated with intravenous tissue plasminogen activator. *Stroke* 2006;37:425–429.

Moonen CT. Spatio-temporal control of gene expression and cancer treatment using magnetic resonance imaging-guided focused ultrasound. *Clinical Cancer Research* 2007;13:3482–3489.

Morgan KE, Allen JE, Dayton PA, Chomas JE, Klibaov AL, Ferrara KW. Experimental and theoretical evaluation of microbubble behavior: Effect of transmitted phase and bubble size. *IEEE Transactions on Ultrasonics, Ferroelectrics and Frequency Control* 2000;47:1494–1509.

Morimoto Y, Mutoh M, Ueda H, Fang L, Hirayama K, Atobe M, Kobayashi D. Elucidation of the transport pathway in hairless rat skin enhanced by low-frequency sonophoresis based on the solute–water transport relationship and confocal microscopy. *Journal of Controlled Release* 2005;103:587–597.

Munshi N, Rapoport N, Pitt WG. Ultrasonic activated drug delivery from Pluronic P-105 micelles. *Cancer Letters* 1997;118:13–19.

Nahirnyak V, Mast TD, Holland CK. Ultrasound-induced thermal elevation in clotted blood and cranial bone. *Ultrasound in Medicine and Biology* 2007;33:1285–1295.

Negishi Y, Endo-Takahashi Y, Matsuki Y, Kato Y, Takagi N, Suzuki R, Maruyama K, Aramaki Y. Systemic delivery systems of angiogenic gene by novel bubble liposomes containing cationic lipid and ultrasound exposure. *Molecular Pharmaceutics* 2012;9:1834–1840.

Newman CM, Lawrie A, Brisken AF, Cumberland DC. Ultrasound gene therapy: On the road from concept to reality. *Echocardiography* 2001;18:339–347.

Ng KY, Liu Y. Therapeutic ultrasound: Its application in drug delivery. *Medicinal Research Reviews* 2002;22:204–223.

O'Reilly MA, Hynynen K. Ultrasound enhanced drug delivery to the brain and central nervous system. *International Journal of Hyperthermia* 2012;28:386–396.

Ohl CD, Arora M, Ikink R, De Jong N, Versluis M, Delius M, Lohse D. Sonoporation from jetting cavitation bubbles. *Biophysical Journal* 2006;91:4285–4295.

Paliwal S, Menon GK, Mitragotri S. Low-frequency sonophoresis: Ultrastructural basis for stratum corneum permeability assessed using quantum dots. *Journal of Investigative Dermatology* 2006;126:1095–1101.

Pandey RK, Zheng G. *Porphyrins as Photosensitizers in Photodynamic Therapy*. San Diego, CA: Academic Press, 2000.

Parikh S, Motarjeme A, McNamara T, Raabe R, Hagspiel K, Benenati JF, Sterling K, Comerota A. Ultrasound-accelerated thrombolysis for the treatment of deep vein thrombosis: Initial clinical experience. *Journal of Vascular and Interventional Radiology* 2008;19:521–528.

Park E, Werner J, Smith NB. Ultrasound mediated transdermal insulin delivery in pigs using a lightweight transducer. *Pharmaceutical Research* 2007;24:1396–1401.

Park J, Fan Z, Kumon RE, El-Sayed MEH, Deng CX. Modulation of intracellular CA^{2+} concentration in brain microvascular endothelial cells in vitro by acoustic cavitation. *Ultrasound in Medicine and Biology* 2010;36:1176–1187.

Partanen A, Yarmolenko PS, Viitala A, Appanaboyina S, Haemmerich D, Ranjan A, Jacobs G, Woods D, Enholm J, Wood BJ. Mild hyperthermia with magnetic resonance-guided high-intensity focused ultrasound for applications in drug delivery. *International Journal of Hyperthermia* 2012;28:320–336.

Patil AV, Rychak JJ, Klibanov AL, Hossack JA. A real-time technique for improving molecular imaging and guiding drug delivery in large blood vessels: In vitro and ex vivo results. *Molecular Imaging* 2011;10:238.

Patrick JT, Nolting MN, Goss SA, Dines KA, Clendenon JL, Rea MA, Heimburger RF. Ultrasound and the blood-brain barrier. In: Bicher HI, McLaren JR, Pigliucci GM, eds., *Consensus on Hyperthermia for the 1990s*. Advances in Experimental Medicine and Biology, Vol. 267. New York: Springer-Verlag, 1990, pp. 369–381.

Perez JLJ, Orea AC, Gallegos ER, Fuentes RG. Photoacoustic spectroscopy to determine in vitro the nonradiative relaxation time of protoporphyrin IX solution containing gold metallic nanoparticles. *European Physical Journal Special Topics* 2008;152:353–356.

Petty HR. Spatiotemporal chemical dynamics in living cells: From information trafficking to cell physiology. *Biosystems* 2006;83:217–224.

Pieters M, Hekkenberg RT, Barrett-Bergshoeff M, Rijken DC. The effect of 40 kHz ultrasound on tissue plasminogen activator-induced clot lysis in three *in vitro* models. *Ultrasound in Medicine and Biology* 2004;30:1545–1552.

Polat BE, Hart D, Langer R, Blankschtein D. Ultrasound-mediated transdermal drug delivery: Mechanism, scope, and emerging trends. *Journal of Controlled Release* 2011;152:330–348.

Postema M, Gilja OH. Ultrasound-directed drug delivery. *Current Pharmaceutical Biotechnology* 2007;8:355–361.

Prausnitz MR, Bose VG, Langer R, Weaver JC. Electroporation of mammalian skin: A mechanism to enhance transdermal drug delivery. *Proceedings of the National Academy of Sciences of the United States of America* 1993;90:10504–10508.

Prausnitz MR, Langer R. Transdermal drug delivery. *Nature Biotechnology* 2008;26:1261–1268.

Prentice P, Campbell P. Application of optical trapping for cavitation studies. Proc. SPIE, 7038, San Diego, CA, August 10, 2008, *Optical Trapping and Optical Micromanipulation V*, 2008. pp. 70381N-70388.

Prentice P, Cuschieri A, Dholakia K, Prausnitz M, Campbell P. Membrane disruption by optically controlled microbubble cavitation. *Nature Physics* 2005;1:107–110.

Qiu Y, Luo Y, Zhang Y, Cui W, Zhang D, Wu J, Zhang J, Tu J. The correlation between acoustic cavitation and sonoporation involved in ultrasound-mediated DNA transfection with polyethylenimine (PEI) in vitro. *Journal of Controlled Release* 2010;145:40–48.

Raab O. Ueber die wirkung fluoreszierender stoffe auf infusorien. *Zeitschrift fur Biologie* 1900;39:524–546.

Ranson MR, Carmichael J, O'Byrne K, Stewart S, Smith D, Hwell A. Treatment of advanced breast cancer with sterically stabilized liposomal doxorubicin: Results of a multicenter phase II trial. *Journal of Clinical Oncology* 1997;15:3185–3191.

Rapoport N. Combined cancer therapy by micellar-encapsulated drug and ultrasound. *International Journal of Pharmaceutics* 2004;277:155–162.

Rapoport N. Physical stimuli-responsive polymeric micelles for anti-cancer drug delivery. *Progress in Polymer Science* 2007;32:962–990.

Rapoport N. Ultrasound-mediated micellar drug delivery. *International Journal of Hyperthermia* 2012;28:374–385.

Rapoport N, Marin A, Luo Y, Prestwich G, Muniruzzaman M. Intracellular uptake and trafficking of pluronic micelles in drug-sensitive and MDR cells: Effect on the intracellular drug localization. *Journal of Pharmaceutical Sciences* 2002;91:157–170.

Rapoport NY, Kennedy AM, Shea JE, Scaife CL, Nam K-H. Controlled and targeted tumor chemotherapy by ultrasound-activated nanoemulsions/microbubbles. *Journal of Controlled Release* 2009;138:268–276.

Raymond SB, Skoch J, Hynynen K, Bacskai BJ. Multiphoton imaging of ultrasound/Optison mediated cerebrovascular effects in vivo. *Journal of Cerebral Blood Flow and Metabolism* 2006;27:393–403.

Reznik N, Williams R, Burns PN. Investigation of vaporized submicron perfluorocarbon droplets as an ultrasound contrast agent. *Ultrasound in Medicine and Biology* 2011;37:1271–1279.

Ribo M, Molina CA, Alvarez B, Rubiera M, Alvarez-Sabin J, Matas M. Intra-arterial administration of microbubbles and continuous 2-MHz ultrasound insonation to enhance intra-arterial thrombolysis. *Journal of Neuroimaging* 2010;20:224–227.

Rosenschein U, Furman V, Kerner E, Fabian I, Bernheim J, Eshel Y. Ultrasound imaging–guided noninvasive ultrasound thrombolysis preclinical results. *Circulation* 2000;102:238–245.

Rychak JJ, Klibanov AL, Ley KF, Hossack JA. Enhanced targeting of ultrasound contrast agents using acoustic radiation force. *Ultrasound in Medicine and Biology* 2007;33:1132–1139.

Saad AH, Hahn GM. Ultrasound-enhanced effects of adriamycin against murine tumors. *Ultrasound in Medicine and Biology* 1992;18:715–723.

Sakharov DV, Hekkenberg RT, Rijken DC. Acceleration of fibrinolysis by high-frequency ultrasound: The contribution of acoustic streaming and temperature rise. *Thrombosis Research* 2000;100:333–340.

Sasaki K, Yumita N, Nishigaki R, Umemura S-i. Antitumor effect sonodynamically induced by focused ultrasound in combination with Ga-porphyrin complex. *Cancer Science* 1998;89:452–456.

Sboros V. Response of contrast agents to ultrasound. *Advanced Drug Delivery Reviews* 2008;60:1117–1136.

Schroeder A, Kost J, Barenholz y. Ultrasound, liposomes, and drug delivery: Principles for using ultrasound to control the release of drugs from liposomes. *Chemistry and Physics of Lipids* 2009;162:1–16.

Schumann PA, Christiansen JP, Quigley RM, McCreery TP, Sweitzer RH, Unger EC, Lindner JR, Matsunaga TO. Targeted-microbubble binding selectively to GPIIb IIIa receptors of platelet thrombi. *Investigative Radiology* 2002;37:587–593.

Schuster J, Zalutsky M, Noska M, Dodge R, Friedman H, Bigner D, Dewhirst M. Hyperthermic modulation of radiolabelled antibody uptake in a human glioma xenograft and normal tissues. *International Journal of Hyperthermia* 1995;11:59–72.

Shaw GJ, Meunier JM, Huang S-L, Lindsell CJ, McPherson DD, Holland CK. Ultrasound-enhanced thrombolysis with tPA-loaded echogenic liposomes. *Thrombosis Research* 2009;124:306–310.

Sheikov N, McDannold N, Jolesz F, Zhang Y-Z, Tam K, Hynynen K. Brain arterioles show more active vesicular transport of blood-borne tracer molecules than capillaries and venules after focused ultrasound-evoked opening of the blood-brain barrier. *Ultrasound in Medicine and Biology* 2006;32:1399–1409.

Shortencarier MJ, Dayton PA, Bloch SH, Schumann PA, Matsunaga TO, Ferrara KW. A method for radiation-force localized drug delivery using gas-filled liposomes. *IEEE Transactions on Ultrasonics, Ferroelectrics and Frequency Control* 2004;51:822–831.

Siegel RJ, Atar S, Fishbein MC, Brasch AV, Peterson TM, Nagai T, Pal D, Nishioka T, Chae J-S, Birnbaum Y. Noninvasive, transthoracic, low-frequency ultrasound augments thrombolysis in a canine model of acute myocardial infarction. *Circulation* 2000;101:2026–2029.

Siegel RJ, Luo H. Ultrasound thrombolysis. *Ultrasonics* 2008;48:312–320.

Silcox C, Smith R, King R, McDannold N, Bromley P, Walsh K, Hynynen K. MRI-monitored ultrasonic heating allows for the spatially controlled in vivo expression of the transgene luciferase in canine prostate. *2003 IEEE Symposium on Ultrasonics*, IEEE, 2003, pp. 1002–1005, Oct. 5–8, Honolulu, HI.

Sivamani RK, Liepmann D, Maibach HI. Microneedles and transdermal applications. *Expert Opinion in Drug Delivery* 2007;4:19–25.

Smith DA, Vaidya SS, Kopechek JA, Huang S-L, Klegerman ME, McPherson DD, Holland CK. Ultrasound-triggered release of recombinant tissue-type plasminogen activator from echogenic liposomes. *Ultrasound in Medicine and Biology* 2010;36:145–157.

Smith NB, Lee S, Shung KK. Ultrasound-mediated transdermal *in vivo* transport of insulin with low-profile cymbal arrays. *Ultrasound in Medicine and Biology* 2003;29:1205–1210.

Sobbe A, Stumpff U, Trübestein G, Figge H, Kozuschek W. Die Ultraschall-Auflösung von Thromben. *Klinische Wochenschrift* 1974;52:1117–1121.

Suchkova V, Carstensen EL, Francis CW. Ultrasound enhancement of fibrinolysis at frequencies of 27 to 100 kHz. *Ultrasound in Medicine and Biology* 2002;28:377–382.

Suslick KS. Sonochemistry. *Science* 1990;247:1439–1445.

Suzuki R, Namai E, Oda Y, Nishiie N, Otake S, Koshima R, Hirata K, Taira Y, Utoguchi N, Negishi Y. Cancer gene therapy by IL-12 gene delivery using liposomal bubbles and tumoral ultrasound exposure. *Journal of Controlled Release* 2010;142:245–250.

Tachibana K. Transdermal delivery of insulin to alloxan-diabetic rabbits by ultrasound exposure. *Pharmaceutical Research* 1992;9:952–954.

Tachibana K, Kimura N, Okumura M, Eguchi H, Tachibana S. Enhancement of cell killing of HL-60 cells by ultrasound in the presence of the photosensitizing drug Photofrin II. *Cancer Letters* 1993;72:195–199.

Tachibana K, Tachibana S. Application of ultrasound energy as a new drug delivery system. *Japanese Journal of Applied Physics* 1999;38:3014.

Tachibana K, Tachibana S. The use of ultrasound for drug delivery. *Echocardiography* 2001;18:323–328.

Tartis MS, Kruse DE, Zheng HR, Zhang H, Kheirolomoom A, Marik J, Ferrara KW. Dynamic microPET imaging of ultrasound contrast agents and lipid delivery. *Journal of Controlled Release* 2008;131:160–166.

ter Haar G. Therapeutic ultrasound. *European Journal of Ultrasound* 1999;9:3–9.

ter Haar G. Ultrasound mediated drug delivery: A 21st century phoenix? *International Journal of Hyperthermia* 2012;28:279.

Tezel A, Paliwal S, Shen Z, Mitragotri S. Low-frequency ultrasound as a transcutaneous immunization adjuvant. *Vaccine* 2005;23:3800–3807.

Tiukinhoy-Laing SD, Buchanan K, Parikh D, Huang S, Macdonald RC, Mcpherson DD, Klegerman ME. Fibrin targeting of tissue plasminogen activator-loaded echogenic liposomes. *Journal of Drug Targeting* 2007;15:109–114.

Touitou E, Junginger HE, Weiner ND, Nagai T, Mezei M. Liposomes as carriers for topical and transdermal delivery. *Journal of Pharmaceutical Science* 1994;83:1189–1203.

Treat LH, McDannold N, Vykhodtseva N, Zhang Y, Tam K, Hynynen K. Targeted delivery of doxorubicin to the rat brain at therapeutic levels using MRI-guided focused ultrasound. *International Journal of Cancer* 2007;121:901–907.

Tserkovsky DA, Alexandrova EN, Chalau VN, Istomin YP. Effects of combined sonodynamic and photodynamic therapies with photolon on a glioma C6 tumor model. *Experimental Oncology* 2012;34:332–335.

Tsunoda S, Mazda O, Oda Y, Iida Y, Akabame S, Kishida T, Shin-Ya M, Asada H, Gojo S, Imanishi J. Sonoporation using microbubble BR14 promotes pDNA/siRNA transduction to murine heart. *Biochemical and Biophysical Research Communications* 2005;336:118–127.

Uchida T, Tachibana K, Hisano S, Morioka E. Elimination of adult T cell leukemia cells by ultrasound in the presence of porfimer sodium. *Anti-Cancer Drugs* 1997;8:329–335.

Umemura S, Kawabata K, Sasaki K, Yumita N, Umemura K, Nishigaki R. Recent advances in sonodynamic approach to cancer therapy. *Ultrasonics Sonochemistry* 1996;3: S187–S191.

Umemura S-i, Yumita N, Nishigaki R. Enhancement of ultrasonically induced cell damage by a gallium-porphyrin complex, ATX-70. *Cancer Science* 1993;84:582–588.

Umemura S-i, Yumita N, Nishigaki R, Umemura K. Mechanism of cell damage by ultrasound in combination with hematoporphyrin. *Cancer Science* 1990;81:962–966.

Umemura S-i, Yumita N, Okano Y, Kaneuchi M, Magario N, Ishizaki M, Shimizu K, Sano Y, Umemura K, Nishigaki R. Sonodynamically-induced in vitro cell damage enhanced by adriamycin. *Cancer Letters* 1997;121:195–201.

Umemura S-i, Yumita N, Umemura K, Nishigaki R. Sonodynamically induced effect of rose bengal on isolated sarcoma 180 cells. *Cancer Chemotherapy and Pharmacology* 1999;43:389–393.

Unger EC, Hersh E, Vannan M, Matsunaga TO, McCreery T. Local drug and gene delivery through microbubbles. *Progress in Cardiovascular Diseases* 2001;44:45–54.

Unger EC, Matsunaga TO, McCreery T, Schumann P, Sweitzer R, Quigley R. Therapeutic applications of microbubbles. *European Journal of Radiology* 2002;42:160–168.

Unger EC, McCreery TP, Sweitzer RH, Caldwell VE, Wu Y. Acoustically active liposhperes containing paclitaxel: A new therapeutic ultrasound contrast agent. *Investigative Radiology* 1998a;33:886.

Unger EC, McCreery TP, Sweitzer RH, Shen D, Wu G. In vitro studies of a new thrombus-specific ultrasound contrast agent. *American Journal of Cardiology* 1998b;81:58G–61G.

van den Boogert J, Hillegersberg Rv, de Rooij FW, de Bruin RW, Edixhoven-Bosdijk A, Houtsmuller AB, Siersema PD, Paul Wilson J, Tilanus HW. 5-Aminolaevulinic acid-induced protoporphyrin IX accumulation in tissues: Pharmacokinetics after oral or intravenous administration. *Journal of Photochemistry and Photobiology B: Biology* 1998;44:29–38.

Van Wamel A, Kooiman K, Harteveld M, Emmer M, Ten Cate FJ, Versluis M, De Jong N. Vibrating microbubbles poking individual cells: Drug transfer into cells via sonoporation. *Journal of Controlled Release* 2006;112:149–155.

Verhaeghe R, Stockx L, Lacroix H, Vermylen J, Baert A. Catheter-directed lysis of iliofemoral vein thrombosis with use of rt-PA. *European Radiology* 1997;7:996–1001.

Villanueva FS, Jankowski RJ, Klibanov S, Pina ML, Alber SM, Watkins SC, Brandenburger GH, Wagner WR. Microbubbles targeted to intercellular adhesion molecule-1 bind to activated coronary artery endothelial cells. *Circulation* 1998;98:1–5.

Vykhodtseva N, Hynynen K, Damianou C. Pulse duration and peak intensity during focused ultrasound surgery: Theoretical and experimental effects in rabbit brain *in vivo*. *Ultrasound in Medicine and Biology* 1994;20:987–1000.

Vykhodtseva N, Hynynen K, Damianou C. Histologic effects of high intensity pulsed ultrasound exposure with subharmonic emission in rabbit brain *in vivo*. *Ultrasound in Medicine and Biology* 1995;21:969–979.

Vykhodtseva N, McDannold N, Hynynen K. Progress and problems in the application of focused ultrasound for blood-brain barrier disruption. *Ultrasonics* 2008;48:279–296.

Walmsley A, Laird W, Williams A. Dental plaque removal by cavitational activity during ultrasonic scaling. *Journal of Clinical Periodontology* 1988;15:539–543.

Wan CPL, Jackson JK, Pirmoradi FN, Chiao M, Burt HM. Increased accumulation and retention of micellar paclitaxel in drug-sensitive and P-Glycoprotein–expressing cell lines following ultrasound exposure. *Ultrasound in Medicine and Biology* 2012;38:736–744.

Wang XB, Liu QH, Wang P, Tang W, Hao Q. Study of cell killing effect on S180 by ultrasound activating protoporphyrin IX. *Ultrasonics* 2008;48:135–140.

Ward M, Wu J, Chiu JF. Experimental study of the effects of optison® concentration on sonoporation *in vitro*. *Ultrasound in Medicine and Biology* 2000;26:1169–1175.

Weishaupt KR, Gomer CJ, Dougherty TJ. Identification of singlet oxygen as the cytotoxic agent in photo-inactivation of a murine tumor. *Cancer Research* 1976;36:2326–2329.

Weller GE, Lu E, Csikari MM, Klibanov AL, Fischer D, Wagner WR, Villanueva FS. Ultrasound imaging of acute cardiac transplant rejection with microbubbles targeted to intercellular adhesion molecule-1. *Circulation* 2003;108:218–224.

Weller GE, Villanueva FS, Tom EM, Wagner WR. Targeted ultrasound contrast agents: In vitro assessment of endothelial dysfunction and multi-targeting to ICAM-1 and sialyl Lewis. *Biotechnology and Bioengineering* 2005a;92:780–788.

Weller GE, Wong MK, Modzelewski RA, Lu E, Klibanov AL, Wagner WR, Villanueva FS. Ultrasonic imaging of tumor angiogenesis using contrast microbubbles targeted via the tumor-binding peptide arginine-arginine-leucine. *Cancer Research* 2005b;65:533–539.

Williams AC, Barry BW. Penetration enhancers. *Advanced Drug Delivery Review* 2004;56:603–618.

Wust P, Hildebrandt B, Sreenivasa G, Rau B, Gellermann J, Riess H, Felix R, Schlag PM. Hyperthermia in combined treatment of cancer. *Lancet Oncology* 2002;3:487–497.

Xie F, Tsutsui JM, Lof J, Unger EC, Johanning J, Culp WC, Matsunaga T, Porter TR. Effectiveness of lipid microbubbles and ultrasound in declotting thrombosis. *Ultrasound in Medicine and Biology* 2005;31:979–985.

Xing Z, Wang J, Ke H, Zhao B, Yue X, Dai Z, Liu J. The fabrication of novel nanobubble ultrasound contrast agent for potential tumor imaging. *Nanotechnology* 2010;21:145607.

Yamaguchi S, Kobayashi H, Narita T, Kanehira K, Sonezaki S, Kudo N, Kubota Y, Terasaka S, Houkin K. Sonodynamic therapy using water-dispersed TiO_2-polyethylene glycol compound on glioma cells: Comparison of cytotoxic mechanism with photodynamic therapy. *Ultrasonics Sonochemistry* 2011;18:1197–1204.

Yamashita N, Tachibana K, Ogawa K, Tsujita N, Tomita A. Scanning electron microscopic evaluation of the skin surface after ultrasound exposure. *Anatomical Record* 1997;247:455–461.

Yang F-Y, Fu W-M, Yang R-S, Liou H-C, Kang K-H, Lin W-L. Quantitative evaluation of focused ultrasound with a contrast agent on blood-brain barrier disruption. *Ultrasound in Medicine and Biology* 2007;33:1421–1427.

Yang R, Reilly CR, Rescorla FJ, Faught PR, Sanghvi NT, Fry FJ, Franklin Jr TD, Lumeng L, Grosfeld JL. High-intensity focused ultrasound in the treatment of experimental liver cancer. *Archives of Surgery* 1991;126:1002.

Yu T, Wang Z, Jiang S. Potentiation of cytotoxicity of adriamycin on huaman ovarian carcinoma cell line 3AO by low-level ultrasound. *Ultrasonics* 2001;39:307–309.

Yuh EL, Shulman SG, Mehta SA, Xie J, Chen L, Frenkel V, Bednarski MD, Li KC. Delivery of systemic chemotherapeutic agent to tumors by using focused ultrasound: Study in a murine model. *Radiology* 2005;234:431–437.

Yumita N, Kaneuchi M, Okano Y, Nishigaki R, Umemura K, Umemura S. Sonodynamically induced cell damage with 4′-*O*-tetrahydropyrangladriamycin, THP. *Anticancer Research* 1999;19:281–284.

Yumita N, Nishigaki R, Umemura K, Morse PD, Swartz HM, Cain CA, Umemura S. Sonochemical activation of hematoporphyrin: An ESR study. *Radiation Research* 1994;138:171–176.

Yumita N, Nishigaki R, Umemura K, Umemura S-i. Hematoporphyrin as a sensitizer of cell-damaging effect of ultrasound. *Cancer Science* 1989;80:219–222.

Yumita N, Nishigaki R, Umemura S-i. Sonodynamically induced antitumor effect of Photofrin II on colon 26 carcinoma. *Journal of Cancer Research and Clinical Oncology* 2000;126:601–606.

Yumita N, Okumura A, Nishigaki R, Umemura K, Umemura S-i. The combination treatment of ultrasound and antitumor drugs on yoshida sarcoma. *Thermal Medicine* 1987;3:175–182.

Zhang P, Porter T. An in vitro of a phase-shift nanoemulsion: A potential nucleation agent for bubble-enhanced HIFU tumor ablation. *Ultrasound in Medicine and Biology* 2010;36:1856–1866.

Zhao X, Liu Q, Tang W, Wang X, Wang P, Gong L, Wang Y. Damage effects of protoporphyrin IX–sonodynamic therapy on the cytoskeletal F-actin of Ehrlich ascites carcinoma cells. *Ultrasonics Sonochemistry* 2009;16:50–56.

Zhong W, Sit WH, Wan JMF, Yu ACH. Sonoporation induces apoptosis and cell cycle arrest in human promyelocytic leukemia cells. *Ultrasound in Medicine and Biology* 2011;37:2149–2159.

Zhou Y, Cui J, Deng CX. Dynamics of sonoporation correlated with acoustic cavitation activities. *Biophysical Journal* 2008a;94:L51–L53.

Zhou Y, Shi J, Cui J, Deng CX. Effects of extracellular calcium on cell membrane resealing in sonoporation. *Journal of Controlled Release* 2008b;126:34–43.

15 Future of Therapeutic Ultrasound

Since its introduction, therapeutic ultrasound has made significant progress in its application to medicine. This emerging modality has the primary advantage, with respect to others, of being able to penetrate deep into the body and deliver acoustic energy to a specific region with millimeter accuracy and without damaging the surrounding or intervening tissue. Through the specific selections of frequency, amplitude, waveform, and treatment protocols, ultrasound can induce thermal and mechanical effects or mediate a range of biological effects that vary from localized, temporary changes in membrane permeability to more permanent damages to cells and tissue ablation. In addition, ultrasound imaging can simultaneously provide real-time information for diagnosis, targeting, and monitoring of therapy. Therefore, its versatility and noninvasive nature make therapeutic ultrasound an appealing method for a variety of therapies, which are greatly contributing to improving its broad clinical acceptance. Several technologies have been tried in clinics with great success, such as shock wave lithotripsy (SWL), sonophoresis, sonothrombolysis, and high-intensity focused ultrasound (HIFU). They were found to be more effective than the other existing modalities, with less morbidity and lower cost. For example, HIFU for thermal ablation of tumors has shown great promise in the treatment of prostate cancer, and it has also shown positive results in the treatment of various solid tumors and cancers as well as brain disorders. However, the ability to thermally ablate tissue by HIFU is limited to localized lesions or tumors. Systemic administration of drugs is a more generalized treatment. Synergistic effects of ultrasound, either thermal or mechanical, in combination with sonosensitizers, anticancer drugs, chemotherapeutic agents, antibiotics, thrombolytics, DNA-based drugs, and many other agents provide novel horizons for advanced clinical applications and revolutionize the treatment of cancers and tumors or other diseases because of the various advances in this field. A variety of drug vehicles, such as ultrasound contrast agents (microbubbles), liposomes, and proteinaceous microspheres, have also been developed to significantly enhance the delivery outcome, for example, by lowering the cavitation threshold, and hold great promise in future applications. Consequently, the anticancer agent can be used at a safe dose, and an augmented effect can be attained with fewer side effects at a targeted site (e.g., a tumor). Clinical trials have already been conducted on patients with head and neck tumors and breast adenocarcinoma, resulting in complete response in some patients and partial response in others. Ultrasound-enhanced gene delivery can make monumental contributions to the treatment of cardiovascular disease, cancer, autoimmune diseases, and many more. Formulations of therapeutic agents with a strong affinity to cancer cells over normal cells would provide greater promise for a more effective treatment using an ultrasound-mediated delivery system.

Systemic immunomodulatory effects in patients (e.g., those with posterior choroidal melanoma) can also be stimulated after certain ultrasound exposures, such as ultrasound-induced hyperthermia and HIFU ablation. Not all of these technologies have yet reached a routine level of application, but they all show great promise.

There seems to be three generations in the progressive research work of therapeutic ultrasound. The first generation emphasized the bioeffect of ultrasound, although the initial aim was to prove that ultrasound imaging, especially in obstetrics, could provide diagnostic information without inducing significant cellular effects. Ultrasound applications in synergy with the available thermal and mechanical effects would generate beneficial outcomes. A spectrum of biological changes is achieved depending on the acoustic intensity, pulse length, driving frequency, exposure duration, concentration of microbubbles, and may be other physical or environment parameters. The induced biological effect is mostly reversible and may be beneficial at low intensities (<100 mW/cm^2). In comparison, very high intensities ($>1,000$ W/cm^2) can produce irreversible mechanical damage and instantaneous tissue necrosis. The second generation deliberately and innovatively exploited more aggressive ultrasound applications for direct intervention, including SWL, HIFU, thrombolysis, tissue healing and rejuvenation, cancer treatment, immune adjuvancy, arteriogenesis, and hemostatis, after extensive investigations and identification of the induced bioeffects that are most desirable and that must be avoided. As a result, a number of ultrasound systems have been developed, tested, and applied in clinics. A third emerging generation involves indirect ultrasound application to stimulate tissues actively or enhance the efficacy of bioacoustic therapy, which needs more understanding of cellular or tissue response in biology, such as ultrasound-stimulated immune response to cancer invasion.

In the early days of therapeutic ultrasound, the equipment used was rather primitive, but significant and rapid developments have been made in recent years. There are several technical challenges identified in order to fulfill the potential of therapeutic ultrasound, such as acoustic access, real-time imaging, treatment planning, dosimetry, and marketability. Transducer development, such as phased arrays, makes it possible to obtain high focal gains from existing acoustic windows. Transducers are becoming more powerful, smaller, and more efficient, and they may also need to be designed to control and optimize specific acoustic phenomena and other activities within the body. Intraluminal compact and integrated systems in a catheter format for diagnostic and image-guided ultrasound therapy can be envisioned soon. Novel methods of real-time imaging and optimization of HIFU treatment are being developed. Elastography with the use of color Doppler and MRI can monitor the subsequent biological changes and increase the effectiveness and spatial accuracy of HIFU ablation. Numerical modeling is helpful in treatment planning. Research is still needed in the optimal strategy of energy delivery, such as the technique of gradual pulsed exposures to achieve a cumulative therapy outcome.

The output and performance of an ultrasound system should be measured accurately and regularly. In order to make comparisons between different models, more details of system construction and treatment protocols must be described clearly. Inconsistencies in their publication may be due to the use of various ultrasound parameters to activate different biochemical pathways (such as apoptosis vs. necrosis). The optimal treatment protocol, including frequency and dosage, for each device and target should be found. Otherwise, widespread use of therapeutic ultrasound will be restricted. Adequate randomized, double-blinded, placebo-controlled clinical studies are necessary to evaluate the performance of ultrasound therapy and its potential for clinical translation. It is critical to consider not only the statistical significance of all trials but also the magnitude of the treatment effect. However, determination of the magnitude of a clinically important difference is difficult because many factors, including the natural history of the disease, the reference therapy, potential adverse reactions and inconvenience of treatment, personal preferences and costs, equipment and time spent, may all have contributions. In addition, the absolute difference in success rates between intervention groups may depend on the population baseline. Insufficient details of clinical trials, including aspects of trial validity, inclusion and exclusion criteria, treatment parameters, and outcome measures, often hamper the quality assessment. A more complete and informative report may not only result in higher validity scores but also reveal additional flaws in the design or conduct of trials. Therefore, guidelines for the reporting of trials are preferred and would ensure adequate data presentation and analysis.

Understanding physical mechanisms underlying therapeutic ultrasound with tissue will lead to a resurgence of interest in this technology in order to fully realize its potential and is integral to clinical success for the generation of specific beneficial effects. Effort of specific targeting and localized release requires an understanding of the strengths and weaknesses of ultrasonic bioeffects, which is still at a mechanistic level. The poor understanding of the underlying mechanism hinders foresight into the future of this technology. For example, after the initial success of SWL in fragmentating kidney stones in an extracorporeal way in the mid-1980s, the design of the second-generation lithotripter was mostly based on the assumption that increasing the amplitude of the focus of devices and reducing the focal size could reduce the required number of shocks and avoid renal damage by restricting high acoustic pressures to the stone. Although it seems reasonable, the mechanism of cavitation in SWL is not known, and the efficacy and safety of new lithotripter are not as expected by many clinical measures. In comparison, the role of cavitation in HIFU was realized after extensive investigation. Efforts have been made to monitor its formation in B-mode ultrasound image,

to control HIFU treatment, to utilize it for effective generation of tissue necrosis, and to avoid it in unintended regions. Plenty of research activities have aimed at both measuring and controlling ultrasound-induced bioeffects due to the primary effects of ultrasound itself or secondary effects, such as shock waves, fluid shearing, and sonochemistry. The relative importance of each in different applications also varies significantly. In addition, ultrasonic effects demonstrated *in vitro* (e.g., isolated cell and single bubble) may not be reproduced *in vivo* (e.g., multicellular tissue exposure to bubble clouds) because perfusion of cavitation nuclei (i.e., microbubbles) and stimulation of bubble activity inside a solid tissue are difficult. Several cellular responses, such as upregulation of key proteins (e.g., release of heat shock protein), initiation of apoptosis, and increase in intracellular calcium levels under ultrasound exposure, have been implicated. The response of cells and their membranes to ultrasound may aid the investigation and application of drug release without collateral damage to the target or adjacent tissues. Mechanistically, the exact biochemical pathways behind ultrasound therapies are not completely known, and many studies show conflicting results. At the tissue level, most of ultrasound's beneficial effects are mediated by an inflammatory response resulting in a coordinated recruitment of cells for accelerated tissue regeneration, activation of immune cells in vaccination, and the stimulation of vasculature growth. Thus, it is difficult to transform *in vitro* results to animal studies directly due to significant differences in ultrasound's bioeffects at the cellular and tissue levels and *in vivo*. Most investigations are still highly experimental and far from being applicable in the clinical situation. There are many gaps to be filled with extensive clinical research rather than laboratory work.

More modeling and experimental investigation of cavitation physics, such as the effect of microbubble size, internal gas composition, membrane or wall thickness and mechanical properties (modulus, shear strength, viscosity, etc.), acoustic frequency, and pressure amplitude on the expansion and collapse of microbubble, are required in order to optimize cavitation or drug delivery. Although ultrasound contrast agents are available in the market, their different compositions and properties, as well as a large range of acoustic parameters reported, make the direct conclusion of published results difficult. Such an investigation allows the optimization of ultrasound design and treatment protocol. In addition, simulation of nonlinear acoustic wave propagation needs more work. The most popular model used in HIFU work is the Khokhlov–Zabozotskaya–Kuznetsov (KZK) equation. However, this equation is meant only for axial symmetric source with limited converging angle. The use of the phased-array transducer, the presence of complicated tissue structure in the propagation pathway (e.g., the rib cage), the heterogeneity of tissue, and multiple interfaces as shown in the real clinical situation should be considered to evaluate the degree of focus shifting, waveform distortion, influence to the temperature elevation and bubble cavitation, and so on.

Although both biotin and streptavidin have been suggested for bubble targeting, they may induce an antigenic response. So, a self-assembled biological system and a chemical system such as maleimide–thiol bond formation are under investigation to avoid this problem. Decorating DNA on the exterior shell of a microbubble stabilized by a surfactant with the capability of drug load inside, either in an oil or aqueous phase, is a hot topic. Gas liposomes or microbubbles composed of native proteins (such as albumin) or poly(ethylene oxide) chains on the surface may be technically advantageous since they may not be recognized and cleared from the circulatory system. Because the targeting molecules used are different for each application due to the uniqueness of individualized antibodies, the attachment of a generic binder to the microbubbles may be a solution. Meanwhile, the target-specific antibodies could act as a complementary binder and be mixed with the generic microbubbles.

Finally, strategies, infrastructural resources, and collaboration are required at local, national, and international levels to fulfill the ultimate goals of ultrasound therapy. It is important to foster this research environment because there are many questions to be answered and highlighted for translation into clinics and wide application in health care.

Index

T - #0562 - 071024 - C21 - 279/216/17 - PB - 9780367658663 - Gloss Lamination